# Growth and Development of Short Rotation Woody Crops for Rural and Urban Applications

# Growth and Development of Short Rotation Woody Crops for Rural and Urban Applications

Editors

**Ronald S. Zalesny, Jr.**
**Andrej Pilipović**

MDPI • Basel • Beijing • Wuhan • Barcelona • Belgrade • Manchester • Tokyo • Cluj • Tianjin

*Editors*

Ronald S. Zalesny, Jr.
USDA Forest Service
Northern Research Station
Rhinelander
United States

Andrej Pilipović
University of Novi Sad
Institute of Lowland Forestry
and Environment
Novi Sad
Serbia


*Editorial Office*
MDPI
St. Alban-Anlage 66
4052 Basel, Switzerland


This is a reprint of articles from the Special Issue published online in the open access journal *Forests* (ISSN 1999-4907) (available at: www.mdpi.com/journal/forests/special_issues/growth_development_woody_crops).

For citation purposes, cite each article independently as indicated on the article page online and as indicated below:

LastName, A.A.; LastName, B.B.; LastName, C.C. Article Title. *Journal Name* **Year**, *Volume Number*, Page Range.

**ISBN 978-3-0365-4502-8 (Hbk)**
**ISBN 978-3-0365-4501-1 (PDF)**


Cover image courtesy of Ronald S. Zalesny

# Contents

# About the Editors

**Ronald S. Zalesny, Jr.**

Ronald S. Zalesny Jr. is a Supervisory Research Plant Geneticist at the USDA Forest Service, Northern Research Station, Rhinelander, Wisconsin, USA. He specializes in genetic and physiological mechanisms of poplars, willows, and other short rotation woody crops grown to maximize ecosystem services for phytoremediation, phyto-recurrent selection, and associated phytotechnologies.

**Andrej Pilipović**

Andrej Pilipović is a Senior Research Fellow at the University of Novi Sad, Institute of Lowland Forestry and Environment, Novi Sad, Serbia. He specializes in the use of poplar, willow, and other forest tree species for growth in altered and unfavorable environments for various uses such as phytoremediation, biomass production, and carbon sequestration.

# Preface to "Growth and Development of Short Rotation Woody Crops for Rural and Urban Applications"

Woody biomass from short rotation woody crops (SRWCs) plays a substantial role in feedstock production for alternative energy sources throughout the world, thus helping to mitigate climate change driven by excessive use of fossil fuels. Establishment of these biomass production systems presents the basis for more efficient development of renewable energy sources while avoiding impacts to essential ecosystem services (e.g., additional emissions of carbon dioxide ($CO_2$) into the atmosphere). In addition to these bioenergy-related uses, the increase of degraded land such as industrial brownfields and municipal landfills has prompted the integration of biomass production with phytotechnologies to produce income, sequester carbon, and clean the environment. Recognizing the need for information linking the silviculture of intensive forestry with the provision of ecosystem services, this special issue focuses on the growth and development of SRWCs grown for all types of applications along the rural to urban continuum (i.e., forest buffers, forest health screening, phytoremediation, short rotation coppice, volume production, wastewater reuse).

**Ronald S. Zalesny, Jr. and Andrej Pilipović**
*Editors*

*Editorial*

# Growth and Development of Short-Rotation Woody Crops for Rural and Urban Applications

Ronald S. Zalesny, Jr. [1,*] and Andrej Pilipović [2]

1   USDA Forest Service, Northern Research Station, Institute for Applied Ecosystem Studies, Rhinelander, WI 54501, USA
2   Institute of Lowland Forestry and Environment, University of Novi Sad, 21000 Novi Sad, Serbia; andrejp@uns.ac.rs
*   Correspondence: ronald.zalesny@usda.gov

**Citation:** Zalesny, R.S., Jr.; Pilipović, A. Growth and Development of Short-Rotation Woody Crops for Rural and Urban Applications. *Forests* **2022**, *13*, 867. https://doi.org/10.3390/f13060867

Received: 26 May 2022
Accepted: 28 May 2022
Published: 31 May 2022

**Publisher's Note:** MDPI stays neutral with regard to jurisdictional claims in published maps and institutional affiliations.

## 1. Introduction

Woody biomass from short-rotation woody crops (SRWCs) plays a substantial role in feedstock production for alternative energy sources throughout the world, thus helping to mitigate climate change driven by excessive use of fossil fuels. The establishment of these biomass production systems presents the basis for more efficient development of renewable energy sources while avoiding impacts (e.g., additional emissions of carbon dioxide ($CO_2$) into the atmosphere) on essential ecosystem services such as clean water and healthy soils. In addition to these bioenergy-related uses, the increase of degraded land such as industrial brownfields and municipal landfills has prompted the integration of biomass production with phytotechnologies to produce income, sequester carbon, and clean the environment. Recognizing the need for information linking the silviculture of intensive forestry with the provision of ecosystem services, this Special Issue focused on the growth and development of SRWCs grown for numerous applications in rural and urban areas.

There are a total of 20 papers in the Special Issue representing 13 countries and four genera (*Phalaris* L., *Populus* L., *Robinia* L., *Salix* L.) (Figure 1; Table 1). In addition to the development and management of a *Salix* cultivar database [1], rural and urban applications represented in the Special Issue include: (a) *forest buffers* [2], (b) *forest health screening* [3,4], (c) *phytoremediation* [5–7], (d) *short rotation coppice* [8–15], (e) *volume production* [16–18], and (f) *wastewater reuse* [19,20] (Table 1). There were >130 genotypes from 27 genomic groups tested across all studies (Table 2), representing the importance of phyto-recurrent selection and other methods to choose clones for local and regional biomass production systems whose methodologies and approaches are relevant worldwide. Our objective in this editorial was to summarize each of the studies included in the Special Issue, which is included in the following section.

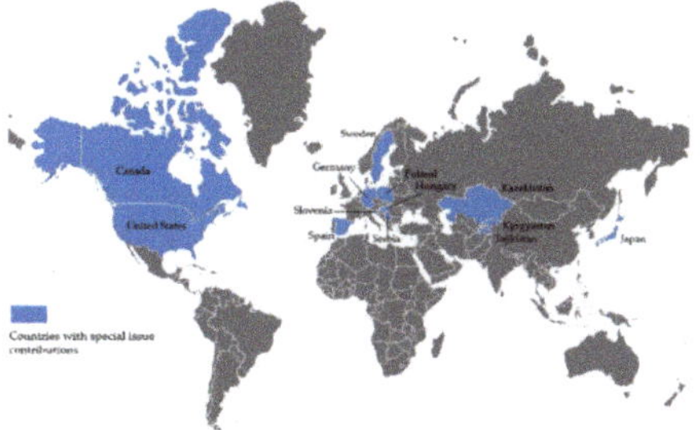


**Figure 1.** Countries with manuscript contributions in the Special Issue on the Growth and Development of Short-Rotation Woody Crops for Rural and Urban Applications (https://www.mdpi.com/journal/forests/special_issues/growth_development_woody_crops; accessed on 25 May 2022).

**Table 1.** Applications of short-rotation woody crops tested worldwide and described in the contributions of the Special Issue on the Growth and Development of Short-Rotation Woody Crops for Rural and Urban Applications (https://www.mdpi.com/journal/forests/special_issues/growth_development_woody_crops; accessed on 25 May 2022).

| Application | Genus | Location | Contribution | DOI |
|---|---|---|---|---|
| Cultivar Database [1] | *Salix* | Global | McGovern et al. [1] | https://doi.org/10.3390/f12050631 |
| Forest Buffers | *Populus* | Canada | Fortier et al. [2] | https://doi.org/10.3390/f12020122 |
| Forest Health Screening | *Populus* | Serbia | Zlatković et al. [3] | https://doi.org/10.3390/f11101080 |
| | *Populus* | Serbia | Galović et al. [4] | https://doi.org/10.3390/f12050636 |
| Phytoremediation | *Populus* | United States | Zalesny et al. [5] | https://doi.org/10.3390/f12040430 |
| | *Populus* | United States | Pilipović et al. [6] | https://doi.org/10.3390/f12040474 |
| | *Populus* | Canada | Hu et al. [7] | https://doi.org/10.3390/f12050572 |
| Short Rotation Coppice | *Salix* | Japan | Han et al. [8] | https://doi.org/10.3390/f11050505 |
| | *Salix* | Japan | Harayama et al. [9] | https://doi.org/10.3390/f11080809 |
| | *Populus* | Canada | Thiffault et al. [10] | https://doi.org/10.3390/f11070785 |
| | *Populus* | Spain | González et al. [11] | https://doi.org/10.3390/f11111133 |
| | *Robinia* | Spain | González et al. [11] | https://doi.org/10.3390/f11111133 |
| | *Populus* | Spain | Oliveira et al. [12] | https://doi.org/10.3390/f11121352 |
| | *Populus* | Germany | Landgraf et al. [13] | https://doi.org/10.3390/f11101048 |
| | *Populus* | Kazakhstan | Thevs et al. [14] | https://doi.org/10.3390/f12030373 |
| | *Populus* | Kyrgyzstan | Thevs et al. [14] | https://doi.org/10.3390/f12030373 |
| | *Populus* | Tajikistan | Thevs et al. [14] | https://doi.org/10.3390/f12030373 |
| | *Populus* | Hungary | Schiberna et al. [15] | https://doi.org/10.3390/f12050623 |
| Volume Production | *Robinia* | Poland | Kraszkiewicz [16] | https://doi.org/10.3390/f12040470 |
| | *Populus* | United States | Ghezehei et al. [17] | https://doi.org/10.3390/f12070869 |
| | *Phalaris* | Sweden | Mola-Yudego et al. [18] | https://doi.org/10.3390/f12070897 |
| Wastewater Reuse | *Salix* | Hungary | Kolozsvári et al. [19] | https://doi.org/10.3390/f12040457 |
| | *Salix* | Slovenia | Istenič and Božič [20] | https://doi.org/10.3390/f12050554 |

[1] McGovern et al. [1] did not describe the growth and development of short-rotation woody crops for rural and urban applications but rather a database of *Salix* cultivars that can be used globally for genotype management and selection.

**Table 2.** Genomic groups, taxonomic sections, and genotypes of *Populus*, *Robinia*, and *Salix* tested worldwide and described in the contributions of the Special Issue on the Growth and Development of Short-Rotation Woody Crops for Rural and Urban Applications (https://www.mdpi.com/journal/forests/special_issues/ growth_development_woody_crops, accessed on 25 May 2022).

| Genomic Group [1,2] | Section [3] | Genotype [4] | Contribution |
| --- | --- | --- | --- |
| *P. alba* 'A' | *Populus* | '111PK', 'Ozolin' | [11,14] |
| *P. balsamifera* 'B' | *Tacamahaca* | See [7] | [7] |
| *P. deltoides* 'D' | *Aigeiros* | '7300502', '89M060', 'Antonije', 'Bora', 'Lux', 'PE19/66', 'Samsun', 'Viriato' | [4–6,12,14] |
| *P. nigra* 'N' | *Aigeiros* | 'Tr 56/75', 'Bordils', 'Lombardo Ieones', 'Mirza Terek', 'Pyramidalis' | [12,14] |
| *P. pamirica* 'P' | *Tacamahaca* | Not specified | [14] |
| *P. simonii* 'S' | *Tacamahaca* | Not specified | [14] |
| *P. trichocarpa* 'T' | *Tacamahaca* | 'Fritzi Pauley', 'Muhle Larsen', 'Trichobel', 'Weser 6' | [13,14] |
| *P. alba* × *P. tremula* 'AT$_{tremula}$' | *Populus* × *Populus* | '4×Göttingen', 'P1' | [13] |
| *P. deltoides* × *P. deltoides* 'DD' | *Aigeiros* × *Aigeiros* | '140', '356' | [17] |
| *P. deltoides* × *P. maximowiczii* 'DM' | *Aigeiros* × *Tacamahaca* | '230', 'DM114', 'NC14106' | [5,6,17] |
| *P. deltoides* × *P. nigra* 'DN' | *Aigeiros* × *Aigeiros* | '9732-11', '9732-24', '9732-31', '9732-36', '99038022', '99059016', '2000 Verde', 'AF2', 'AF13', 'AF15', 'AF16', 'AF17', 'AF18', 'AF19', 'AF20', 'AF24', 'AF24', 'AF28', 'Agathe-F', 'BL Constanzo', 'Bellini', 'Blanc du Poitou', 'Branagesi', 'B-1M', 'Campeador', 'Canadá Blanco', 'DN2', 'DN5', 'DN34', 'DN177', 'Dorskamp', 'E-298', 'Flevo', 'Guardi', 'H-8', 'H-11', 'H-17', 'H-33', 'H-328', 'Harff', 'Heidemij', 'I-214', 'I-45/51', 'I-454/40', 'Isières', 'Jacometti 78 B', 'Koltay', 'Kopecky', 'Kornik-21', 'Luisa Avanzo', 'MC', 'Oudenberg', 'Orion', 'Pannonia', 'Robusta', 'Tiepolo', 'Triplo', 'Veronese', 'Vesten' | [3–6,12–15] |
| *P. maximowiczii* × *P. nigra* 'MN' | *Tacamahaca* × *Aigeiros* | 'Rochester' | [13] |
| *P. maximowiczii* × *P. trichocarpa* 'MT' | *Tacamahaca* × *Tacamahaca* | 'Androscoggin', 'Fastwood-1', 'Fastwood-2', 'Matrix-11', 'Matrix-24', 'Matrix-49', 'NE42' | [13,14] |

3

**Table 2.** *Cont.*

| Genomic Group [1,2] | Section [3] | Genotype [4] | Contribution |
|---|---|---|---|
| *P. nigra* × *P. maximowiczii* 'NM' | *Aigeiros* × *Tacamahaca* | 'Max-1', 'Max-3', 'Max-4', 'NM2', 'NM5', 'NM6' | [5,6,13,14] |
| *P. trichocarpa* × *P. deltoides* 'TD' | *Tacamahaca* × *Aigeiros* | 'Beaupre', 'Boelare', 'Raspalje', 'Unal', '185', '49-177' | [12,17] |
| *P. tremula* × *P. tremuloides* 'T$_{tremula}$T$_{tremuloides}$' | *Populus* × *Populus* | 'Esch5' | [13] |
| (*P. deltoides* × *P. nigra*) × *P. maximowiczii* 'DN×M' | (*Aigeiros* × *Aigeiros*) × *Tacamahaca* | 'DN × M-915508' | [2] |
| *P. laurifolia* × (*P. deltoides* × *P. nigra*) 'L×DN' | *Tacamahaca* × (*Aigeiros* × *Aigeiros*) | 'Kazakhstani', 'Kyzyl-Tan' | [14] |
| (*P. maximowiczii* × *P. deltoides*) × *P. trichocarpa* 'MD×T' | (*Tacamahaca* × *Aigeiros*) × *Tacamahaca* | Not specified | [10] |
| *P. trichocarpa* × (*P. trichocarpa* × *P. deltoides*) 'T×TD' | *Tacamahaca* × (*Tacamahaca* × *Aigeiros*) | 'AF8' | [13] |
| (*P. trichocarpa* × *P. deltoides*) × *P. alba* 'TD×A' | (*Tacamahaca* × *Aigeiros*) × *Populus* | 'I-114/69' | [12] |
| (*P. trichocarpa* × *P. deltoides*) × *P. nigra* 'TD×N' | (*Tacamahaca* × *Aigeiros*) × *Aigeiros* | 'AF6', 'Monviso' | [13] |
| *R. pseudoacacia* 'P$_{robinia}$' | na [5] | 'Nyirsegy' | [11] |
| *S. alba* 'A$_{salix}$' | na | 'Naperti', 'V-160' | [19,20] |
| *S. pet-susu* 'P$_{salix}$' | na | 'P.C.51', 'P.G.12', 'P.G.D', 'P.I.62', 'P.I.81', 'P.I.82', 'P.T.59' | [8,9] |
| *S. sachalinensis* 'S$_{salix}$' | na | 'S.I.27', 'S.I.44', 'S.I.67', 'S.S.3', 'S.T.27' | [8,9] |
| *S. alba* × *S. alba* 'A$_{salix}$A$_{salix}$' | na | 'V-052', 'V-093' | [20] |

[1] Species authorities (*Populus, Robinia, Salix*): *P. alba* L.; *P. balsamifera* L.; *P. deltoides* Bartr. Ex Marsh; *P. laurifolia* Ledeb.; *P. maximowiczii* A. Henry; *P. nigra* L.; *P. pamirica* Komarov; *P. simonii* Carrière; *P. tremula* L.; *P. tremuloides* Michx.; *P. trichocarpa* Torr. et. Gray; *R. pseudoacacia* L.; *S. alba* L.; *S. pet-susu* Kimura; *S. sachalinensis* F. Schmidt. [2] Genomic group synonyms (*Populus*): *P. alba* × *P. tremula* = *P.* × *canescens* (Aiton) Sm.; *P. deltoides* × *P. nigra* = *P.* × *canadensis* Mönch = *P.* × *euramericana* (Dode) Guinier; *P. trichocarpa* × *P. deltoides* = *P.* × *generosa* Henry. [3] Section authorities (*Populus*): Aigeiros Duby; *Populus* L.; *Tacamahaca* Spach. [4] Genotype synonyms: *Populus*: 'Agathe F' = 'E-298'; 'Antonije' = '182/81'; 'Bora' = 'B229'; 'DN34' = 'Eugenei'; 'I-214' = 'Campeador', 'NE42' = 'Hybride 275' = 'H-275'; 'Pannonia' = 'M1', 'Pyramidalis' may be 'Mirza Terek'. *Salix*: 'P.I.81' = 'P81'; 'P.I.82' = 'P82'; 'S.I.27' = 'S27'; 'S.I.67' = 'S67'. [5] na = not applicable.

## 2. Applications from Around the Globe

McGovern et al. [1] described a proof-of-concept of an SQL database to store existing information on *Salix* cultivars and to allow users to compare and submit new *Salix* cultivar entries. The development and management of this *cultivar database* have the potential to enhance an existing checklist for *Salix* cultivars that includes 968 epithet records in a Microsoft Excel spreadsheet format. This existing checklist has been maintained since 2015 by the International Commission on Poplars and Other Fast-Growing Trees Sustaining People and the Environment of the Food and Agriculture Organization of the United Nations (UN FAO) (https://www.fao.org/ipc/en/; accessed on 25 May 2022), highlighting the global reach of their work.

Fortier et al. [2] conducted a study in Canada on the use of hybrid poplars in *forest buffers* to reduce firewood harvest pressure in woodlots while improving ecosystem services related to soils, water, and carbon. They evaluated the natural drying and chemical characteristics of hybrid poplar firewood produced from bioenergy buffers and then compared those results to *Populus tremuloides* Michx., *Acer rubrum* L., and *Fraxinus americana* L. from adjacent woodlots. They determined that hybrid poplar buffers could be used as firewood feedstock in the fall and spring when heat demand is less intense than in the colder winter months.

Zlatković et al. [3] used a *forest health screening* approach to identify a bacterial pathogen (*Lonsdalea populi*) causing cankering of two-year-old hybrid poplar in the Vojvodina province in Serbia. This was the first report of *L. populi* causing bacterial canker disease in the country as well as throughout southeastern Europe. The cankering was observed on stems and branches and consisted of a soft, watery, colorless fluid that smelled rotten and flowed from bark fissures. Two weeks after being observed, the cankers caused crown dieback. These results are important for the region and Serbia, given the implications for the potential need to screen for *L. populi* in poplar breeding and testing programs.

Galović et al. [4] used a *forest health screening* approach to test the variability among three hybrid poplar genotypes in their ability to tolerate salts in halomorphic soils such as those in the Vojvodina province in Serbia. The clones were hydroponically subjected to NaCl concentrations ranging from 150 to 450 mM, and biochemical responses were quantified in the leaves via estimation of radical scavenging capacities and accumulation of total phenolic content and flavonoids. Using molecular genetic approaches, they reported that two of the three clones were highly salt-tolerant and exhibited potential for phytoremediation of halomorphic soils and other saline environments.

Zalesny et al. [5] evaluated the genotype × environment interactions of hybrid poplars growing at sixteen *phytoremediation* buffer systems (i.e., phyto buffers) in the Great Lakes Basin in the United States (Figure 2). They tested health, growth, and volume during establishment (i.e., ages one to four years) and identified generalist clones exhibiting superior performance across a broad range of phyto buffers as well as specialist genotypes that were adapted to local soil and climate conditions. They concluded that a combination of these response groups would enhance the potential for phytoremediation best management practices that are regionally developed and yet globally relevant.

Pilipović et al. [6] studied hybrid poplars at the *phytoremediation* buffer systems (i.e., phyto buffers) described by Zalesny et al. [5]. They compared the establishment potential of promising hybrid poplar clones developed at the University of Minnesota Duluth's Natural Resources Research Institute (NRRI) with experimental genotypes with a rich history of testing and common genotypes used for commercial and/or research purposes in the midwestern United States. Overall, certain NRRI clones had exceptional survival and growth relative to experimental or common clones across at least ten phyto buffers, indicating their potential for use in geographically robust phytotechnologies.

**Figure 2.** Poplar phytoremediation buffer system in the Great Lakes Basin, United States. From Zalesny et al. [5]. Photo courtesy of Paul Manley, Missouri University of Science and Technology.

Hu et al. [7] described field testing of salt-tolerant balsam poplar (*Populus balsamifera* L.) clones used for reclamation around end-pit lakes associated with bitumen extraction in northern Alberta, Canada. They used ***phytoremediation*** approaches to select genetically suitable native balsam poplar clones screened in the greenhouse and at field sites with a tolerance to salty process-affected water resulting from the hot-water bitumen extraction process at oil sands mine sites. Overall, their work elucidated an integrated system for choosing balsam poplar for oil sand reclamation, providing information showing the advantage of deploying selected native material versus unselected genotypes.

Han et al. [8] tested the influence of mulching and cutback (i.e., coppicing) on the suppression of weed competition and their interactive effects on biomass productivity of ***short-rotation coppice (SRC)*** willow in northern boreal Hokkaido, Japan. Trees were harvested after three years of growth following cutback, and those grown with mulch exhibited 1130% greater biomass production than those exposed to weed competition. In these non-mulched plots, weed biomass was 800% greater than willows. Overall, their results showed that SRC willow is a biomass feedstock alternative in the region if used with mulching to sustain complete weed control.

Harayama et al. [9] estimated the yield loss of ***short-rotation coppice (SRC)*** willow from deer browse in northern boreal Hokkaido, Japan. They allowed deer browsing to occur after the first summer of the second coppice cycle and subsequently recorded the number of sprouting stems and the number of deer-browsed stems. Then, after three years, they quantified yield losses and reported 80% reductions in yield after browsing of only a single stem per parent root system. At the stand scale, these yield losses were as high as 6 tons ha$^{-1}$ year$^{-1}$ (dry biomass), suggesting the need for silvicultural prescriptions that include control of deer browsing.

Thiffault et al. [10] tested two intensive mechanical site preparation treatments versus a control with no site preparation to assess the survival, growth, and nutritional status of ***short-rotation coppice (SRC)*** poplars in Québec, Canada. They also assessed differences among treatments for inorganic soil N. After four growing seasons, survival was nearly twice as high for both treatments (mounding = 99%; V-blade = 91%) relative to the control (48%), and trees exhibited 155% (mounding) and 91% (V-blade) greater diameter than control trees. Overall, they reported mounding as being the best treatment given higher survival and growth along with the lowest erosion potential.

González et al. [11] quantified mid-rotation nutrient contributions from leaf litter of ***short-rotation coppice (SRC)*** of white poplar, black locust, and an even mix of both species on the Iberian Peninsula of Spain. They reported white poplar exhibited 32% and 20%

greater leaf biomass than black locust and the species mix, respectively. White poplar had 15% more leaf carbon than the other two treatments, which did not differ from one another. Contributions of individual macronutrients were highly variable across species and the mix, leading to their results recommending deploying mixtures of species to achieve a potential reduction in the amount of mineral fertilization required at the stand level.

Oliveira et al. [12] reviewed the potential of *short-rotation coppice (SRC)* poplars in Mediterranean conditions and in Spain as sustainable biomass feedstock production systems for a circular bioeconomy that is robust to global change. They reviewed these SRCs for their abilities to provide quality biomass with predictable yield and periodicity across the landscape. In their analysis, they considered: genetic plant material, planting designs, site maintenance activities, yield prediction, biomass characterization, and ecosystem services. Despite recent advances, they concluded more work on these components is necessary to develop a circular bioeconomy at regional and national levels.

Landgraf et al. [13] tested the survival, growth, and biomass production of 37 poplar genotypes grown as *short-rotation coppice (SRC)* in northeastern Germany. In addition to first-year survival, they reported results after the first and second coppice cycles, with three years for each cycle. Overall, their varieties exhibited broad variation in all traits, with the top seven clones having at least 11 Mg ha$^{-1}$ year$^{-1}$ of aboveground dry biomass after the second coppice cycle being recommended for commercial use. Six varieties had less than 4 Mg ha$^{-1}$ year$^{-1}$. In general, biomass yield increased from the first to the second harvest, although some varieties produced less biomass in subsequent years.

Thevs et al. [14] estimated the growth rates and biomass production of 30 poplar genotypes grown as *short-rotation coppice (SRC)* across nine sites in Kyrgyzstan, Kazakhstan, and Tajikistan in central Asia. There was a difference in genomic group performance based on elevation, with *P. deltoides* × *P. nigra* and *P. nigra* × *P. maximowiczii* clones exhibiting the greatest stem volumes and biomass yields at lower elevations, and *P. maximowiczii* × *P. trichocarpa* and pure *P. trichocarpa* genotypes performing the best at higher elevations. They concluded that many of the cultivars tested could be incorporated into SRC and agroforestry applications.

Schiberna et al. [15] reviewed the biomass production potential of *short-rotation coppice (SRC)* poplars in Hungary. Based on the literature-derived values for site characteristics, yield, and costs, they developed an economic model to predict the financial performance of these biomass feedstock production systems. They reported break-even yields ranging between 6 and 8 Mg ha$^{-1}$ year$^{-1}$ of aboveground dry biomass on shorter rotations with an evenly distributed cash flow. In addition to SRC applications, they also discussed the potential of extending industrial rotations to range from 20 to 25 years to produce high-quality veneer logs, which are currently limited to rotations of up to 15 years.

Kraszkiewicz [16] quantified the growth and *volume production* of 14 black locust stands varying in soil and climate conditions in Małopolska Kraina, southeastern Poland. The biomass volume of the stands was similar to that of natural forests. In addition, four of these stands were 4 to 8 years old and exhibited a stand height (2 to 8 m) and diameter (4.5 to 12.0 cm) consistent with short-rotation poplar and willow systems in the region. Based on his results, he concluded that black locusts can be complementary to poplars and willows as bioenergy feedstocks to produce medium-sized timber on marginal lands not suitable for most tree species.

Ghezehei et al. [17] estimated the *volume production* and profitability of poplars grown with different planting densities and fertilization treatments across three sandy coastal sites in North Carolina in the United States. Overall, survival ranged from 62 to 93%, and the mean annual increment of green stem biomass of six-year-old trees ranged from 9 to 25 Mg ha$^{-1}$ year$^{-1}$ across densities. Fertilization increased volume production on fertile soils but not at marginal sites. Given economic barriers of establishment costs and weed control with higher planting densities, their calculated break-even price was 27 USD Mg$^{-1}$ (delivered). Weed control was more important than fertilizer for determining this threshold.

Mola-Yudego et al. [18] compared the *volume production*, land-use patterns, and climatic profiles of reed canary grass versus traditional energy crops (i.e., poplars and willows) in Sweden. Reed canary grass is grown in colder climates in areas that have lower agricultural productivity than poplars and willows, yet they found its mean yields of 6 Mg ha$^{-1}$ year$^{-1}$ (experimental) and 3.5 Mg ha$^{-1}$ year$^{-1}$ (commercial) to be similar. Nevertheless, they concluded that broad-scale application of reed canary grass may be hindered as its land area for production is more sensitive to policy incentives than short rotation woody crops (i.e., due to insufficient markets and lack of compensation for ecosystem services).

Kolozsvári et al. [19] conducted a *wastewater reuse* project utilizing fish farm effluent as irrigation and fertilization for the production of short-rotation energy willow in Hungary. Comparing two fertigation sources (i.e., effluent water and freshwater), they reported that effluent water increased willow yield. The phytoextraction of nutrients was tissue-specific, with nitrogen and sodium being taken up into leaves and phosphorus accumulating in the stems. There was an inverse relationship between phosphorus uptake and irrigation volume. Trees irrigated with effluent water were healthier than those with freshwater, indicating the potential for wastewater reuse to increase willow production.

Istenič and Božič [20] tested the potential for *wastewater reuse* in an evaporative willow system (EWS) accepting primary treated municipal wastewater in a sub-Mediterranean climate in Slovenia. Willows receiving wastewater exhibited greater growth and biomass than untreated controls. The nutrient recovery potential of the EWS was high, with the uptake of nitrogen (48%) and phosphorus (45%) being greater in willows than in other plants used for wastewater treatment. Trees from one genotype had the least biomass and the greatest nutrient uptake, leading to the need for clonal selection to maximize the biomass production of EWS while mitigating discharge to surface and groundwater.

### 3. Concluding Remarks

As highlighted above, there is great potential for SRWCs to be included in biomass feedstock portfolios and environmental applications in rural and urban areas. Coupled with engineering approaches, the green solutions presented in this Special Issue offer an opportunity to sustainably produce biomass for bioenergy, biofuels, and bioproducts while reducing impacts from anthropogenic activities on local- and landscape-level ecosystem services. In addition, integrating biomass production with phytotechnologies offers potential pollution solutions for increasing community health and livelihoods.

**Author Contributions:** Conceptualization, A.P. and R.S.Z.J.; writing—original draft preparation, R.S.Z.J.; writing—review and editing, A.P. and R.S.Z.J.; visualization, R.S.Z.J.; funding acquisition, R.S.Z.J. All authors have read and agreed to the published version of the manuscript.

**Funding:** This work was funded by the Great Lakes Restoration Initiative (GLRI; Template #738 Landfill Runoff Reduction).

**Acknowledgments:** The findings and conclusions in this publication are those of the authors and should not be construed to represent any official USDA or United States Government determination or policy. We are grateful to all of the authors and reviewers of the individual manuscripts; without you this Special Issue would not be possible. We thank E.R. Rogers and R.A. Vinhal for reviewing earlier versions of this manuscript.

**Conflicts of Interest:** The authors declare no conflict of interest.

## References

1. McGovern, P.; Kuzovkina, Y.; Soolanayakanahally, R. Short communication: IPC *Salix* cultivar database proof-of-concept. *Forests* **2021**, *12*, 631. [CrossRef]
2. Fortier, J.; Truax, B.; Gagnon, D.; Lambert, F. Natural drying and chemical characteristics of hybrid poplar firewood produced from agricultural bioenergy buffers in southern Québec, Canada. *Forests* **2021**, *12*, 122. [CrossRef]
3. Zlatković, M.; Tenorio-Baigorria, I.; Lakatos, T.; Tóth, T.; Koltay, A.; Pap, P.; Marković, M.; Orlović, S. Bacterial canker disease on *Populus* × *euramericana* caused by *Lonsdalea populi* in Serbia. *Forests* **2020**, *11*, 1080. [CrossRef]

4. Galović, V.; Kebert, M.; Popović, B.; Kovačević, B.; Vasić, V.; Joseph, M.; Orlović, S.; Szabados, L. Biochemical and gene expression analyses in different poplar clones: The selection tools for afforestation of halomorphic environments. *Forests* **2021**, *12*, 636. [CrossRef]
5. Zalesny, R.; Pilipović, A.; Rogers, E.; Burken, J.; Hallett, R.; Lin, C.; McMahon, B.; Nelson, N.; Wiese, A.; Bauer, E.; et al. Establishment of regional phytoremediation buffer systems for ecological restoration in the Great Lakes Basin, USA. I. Genotype × environment interactions. *Forests* **2021**, *12*, 430. [CrossRef]
6. Pilipović, A.; Zalesny, R.; Rogers, E.; McMahon, B.; Nelson, N.; Burken, J.; Hallett, R.; Lin, C. Establishment of regional phytoremediation buffer systems for ecological restoration in the Great Lakes Basin, USA. II. New clones show exceptional promise. *Forests* **2021**, *12*, 474. [CrossRef]
7. Hu, Y.; Kamelchuk, D.; Krygier, R.; Thomas, B. Field testing of selected salt-tolerant screened balsam poplar (*Populus balsamifera* L.) clones for use in reclamation around end-pit lakes associated with bitumen extraction in northern Alberta. *Forests* **2021**, *12*, 572. [CrossRef]
8. Han, Q.; Harayama, H.; Uemura, A.; Ito, E.; Utsugi, H.; Kitao, M.; Maruyama, Y. High biomass productivity of short-rotation willow plantation in boreal Hokkaido achieved by mulching and cutback. *Forests* **2020**, *11*, 505. [CrossRef]
9. Harayama, H.; Han, Q.; Ishihara, M.; Kitao, M.; Uemura, A.; Sasaki, S.; Yamada, T.; Utsugi, H.; Maruyama, Y. Estimation of yield loss due to deer browsing in a short rotation coppice willow plantation in northern Japan. *Forests* **2020**, *11*, 809. [CrossRef]
10. Thiffault, N.; Elferjani, R.; Hébert, F.; Paré, D.; Gagné, P. Intensive mechanical site preparation to establish short rotation hybrid poplar plantations—A case-study in Québec, Canada. *Forests* **2020**, *11*, 785. [CrossRef]
11. González, I.; Sixto, H.; Rodríguez-Soalleiro, R.; Oliveira, N. Nutrient contribution of litterfall in a short rotation plantation of pure or mixed plots of *Populus alba* L. and *Robinia pseudoacacia* L. *Forests* **2020**, *11*, 1133. [CrossRef]
12. Oliveira, N.; Pérez-Cruzado, C.; Cañellas, I.; Rodríguez-Soalleiro, R.; Sixto, H. Poplar short rotation coppice plantations under Mediterranean conditions: The case of Spain. *Forests* **2020**, *11*, 1352. [CrossRef]
13. Landgraf, D.; Carl, C.; Neupert, M. Biomass yield of 37 different SRC poplar varieties grown on a typical site in northeastern Germany. *Forests* **2020**, *11*, 1048. [CrossRef]
14. Thevs, N.; Fehrenz, S.; Aliev, K.; Emileva, B.; Fazylbekov, R.; Kentbaev, Y.; Qonunov, Y.; Qurbonbekova, Y.; Raissova, N.; Razhapbaev, M.; et al. Growth rates of poplar cultivars across central Asia. *Forests* **2021**, *12*, 373. [CrossRef]
15. Schiberna, E.; Borovics, A.; Benke, A. Economic modelling of poplar short rotation coppice plantations in Hungary. *Forests* **2021**, *12*, 623. [CrossRef]
16. Kraszkiewicz, A. Productivity of black locust (*Robinia pseudoacacia* L.) grown on varying habitats in southeastern Poland. *Forests* **2021**, *12*, 470. [CrossRef]
17. Ghezehei, S.; Ewald, A.; Hazel, D.; Zalesny, R.; Nichols, E. Productivity and profitability of poplars on fertile and marginal sandy soils under different density and fertilization treatments. *Forests* **2021**, *12*, 869. [CrossRef]
18. Mola-Yudego, B.; Xu, X.; Englund, O.; Dimitriou, I. Reed canary grass for energy in Sweden: Yields, land-use patterns, and climatic profile. *Forests* **2021**, *12*, 897. [CrossRef]
19. Kolozsvári, I.; Kun, Á.; Jancsó, M.; Bakti, B.; Bozán, C.; Gyuricza, C. Utilization of fish farm effluent for irrigation short rotation willow (*Salix alba* L.) under lysimeter conditions. *Forests* **2021**, *12*, 457. [CrossRef]
20. Istenič, D.; Božič, G. Short-rotation willows as a wastewater treatment plant: Biomass production and the fate of macronutrients and metals. *Forests* **2021**, *12*, 554. [CrossRef]

*forests*

*Communication*

# Short Communication: IPC *Salix* Cultivar Database Proof-of-Concept

Patrick N. McGovern [1,*], Yulia A. Kuzovkina [2] and Raju Y. Soolanayakanahally [3]

1   Private Nurseryman and Middleware Administrator, Grand Rapids, MI 49525, USA
2   Department of Plant Science, University of Connecticut, Storrs, CT 06269, USA; jkuzovkina@uconn.edu
3   Indian Head Research Farm, Agriculture and Agri-Food Canada, Indian Head, SK S0G 2K0, Canada; raju.soolanayakanahally@canada.ca
*   Correspondence: pmcgover@gmail.com

**Citation:** McGovern, P.N.; Kuzovkina, Y.A.; Soolanayakanahally, R.Y. Short Communication: IPC *Salix* Cultivar Database Proof-of-Concept. *Forests* **2021**, *12*, 631. https://doi.org/10.3390/f12050631

Academic Editor: Timothy A. Martin

Received: 13 February 2021
Accepted: 12 May 2021
Published: 17 May 2021

**Abstract:** A variety of *Salix* L. (Willow) tree and shrub cultivars provide resources for significant commercial markets such as bioenergy, environmental applications, basket manufacturing, and ornamental selections. The International Poplar Commission of the Food and Agriculture Organization (IPC FAO) has maintained the Checklist for Cultivars of *Salix* L. (Willow) since 2015 and now lists 968 epithet records in a Microsoft Excel spreadsheet format. This Proof-of-Concept (POC) investigates using an SQL database to store existing IPC *Salix* cultivar information and provide users with a format to compare and submit new *Salix* cultivar entries. The original IPC data were divided into three separate tables: Epithet, Species, and Family. Then, the data were viewed from three different model perspectives: the original *Salix* IPC spreadsheet data, the Canadian (PWCC), and the Open4st database. Requirements for this process need to balance database integrity rules with the ease of adding new *Salix* cultivar entries. An integrated approach from all three models proposed three tables: Epithet, Family, and Pedigree. The Epithet and Family tables also included Species data with a reference to a website link for accepted species names and details. The integrated process provides a more robust method to store and report data, but would require dedicated IT personnel to implement and maintain long-term. A potential use case scenario could involve users submitting their Checklist entries to the Salix administrator for review; the entries are then entered into a test environment by IT resources for final review and promotion to a production online environment. Perhaps the most beneficial outcome of this study is the investigation of various strategies and standards for Epithet and Family recording processes, which may benefit the entire *Populus* and *Salix* communities.

**Keywords:** proof-of-concept: use case; spreadsheet; CSV file; SQL; database; data integrity; GitHub; Linux

## 1. Introduction

Tree breeding projects are expensive, span multiple years, and may require data from historical tree generations. These are worthy justifications to use a robust cultivar process to track epithets, families, and related details. Accurate cultivar details could provide supportive data for plant patent applications. Pedigree data from multi-generational breeding may avoid inbreeding mistakes and save time in future years. Epithet and family relationships can also be associated with nursery, field trials, and statistical views.

An online Google search for the terms "tree cultivar database" returned over one million results spanning a variety of horticultural species and topics such as urban trees, an avocado variety database, and arboretum collections. Another online Google search for "botanical database schema" returned over 500,000 results also spanning diverse subjects that addressed specific organizational needs and unrelated topics. These wide-ranging results may help explain why botanical organizations design and create their own plant databases for their specific user and technical requirements.

The venerable spreadsheet computer application is often the tool of choice for quickly storing and analyzing a variety of data. College courses have long included spreadsheet training, allowing for the widespread acceptance of spreadsheets for home, business, and scientific data. There are a variety of spreadsheet applications with proprietary file formats. However, the comma-separated (CSV) file format can be exported from most spreadsheets using a comma or other character as a field separator. These CSV text files provide a somewhat universal format to exchange data files without proprietary dependencies. Nevertheless, the ubiquitous nature of spreadsheets presents risks in terms of data integrity by not enforcing data accuracy, and in terms of the challenge of managing many different files with related data over time.

### 1.1. Characteristics of the Original IPC Salix Spreadsheet Data

In 2013, the International Poplar Commission of the Food and Agriculture Organization (IPC FAO) was appointed as the International Cultivar Registration Authority (ICRA) for willows. The first edition of the Checklist for Cultivars of *Salix* L. (Willow) [1,2] was compiled in 2015 in Microsoft Word format to promote a standardized registration process for new cultivar epithets. Eight hundred and fifty-four cultivar epithets with accompanying information were included in the first edition of the Checklist. Since then, more epithets have been added to the Checklist, which had grown to 968 records by 2020 when it was converted to a spreadsheet in Microsoft Excel format. Some duplications of epithets were revealed, making it difficult to discern unique epithet, species, and family names. There are 27 epithet names that are duplicated multiple times, including 'Pendula' (11 times) and 'Pyramidalis' (3 times). The original *Salix* spreadsheet contains 39 columns providing a flat one-line view of each epithet that does not enforce relationships between parents and other related epithets. Epithet records are encapsulated with opening and closing single quote characters [3] (e.g., 'Abbey's Harrison'). There are seven trademark names in the epithet column suffixed with the trademark sign (™) and one with the registered trademark symbol (®).

The original species field data is a mix of *Salix* species and hybrid names having 91 unique name combinations. There were 21 hybrid names that included the Punycode "×" multiplication character denoting a hybrid name. A number of the species name records were suffixed with spaces and single quote characters. There were 25 epithets with null (empty) species values and 236 with "S." representing an unknown *Salix* species name. The original *Salix* spreadsheet lists 122 records with family associations by listing the parents in the "mother" or "father" columns. The parents were a combination of species, hybrid names, and cultivars marked with single quote characters.

It should be noted that the compilation of the Checklist for Cultivars of *Salix* L. (Willow) was the first attempt to assemble scattered records from existing references, and not through direct communication with the cultivar developer and breeders. Therefore, the Checklist records lack the standard details present in the other databases investigated in this study. For example, there were limited data on pedigree and parents of most hybrid cultivars that were not identified at the clonal level. Also, there were no seedlot records or experimental trials.

### 1.2. Characteristics of the Canadian Database Model

The Canadian *Populus* and *Salix* Clone Directory [4] was produced by the Poplar and Willow Council of Canada (PWCC). The PWCC is a non-profit organization established in 1977 for wise use, conservation, and sustainable production of poplar and willow genetics resources. It stores Canadian germplasm data for clones, pollen, seedlot, and progeny. The original database was in a hard copy state since 1986, then transferred to electronic format with approximately 1000 new entries contributed by forest companies, governments, and private industry, for a total of over 6600 entries. Between 2015 and 2018, the database was converted to an online Microsoft Access database format and now contains over 26,419 records searchable by 25 column headings.

The established nature, size, and evolution of the Canadian database could provide a basis for standards for similar databases. Some Canadian database columns with a summary of their descriptions [5] and noteworthy database observations are:

- **ID:** A unique number for each entry. This column may allow duplicate "Name" column entries without conflict.
- **Name:** The family population name or a clone number or name.
- **Family:** Population entries display seedlot number. Clone entries display family numbers if known. This column data may associate with other family columns such as "Family comments", Male and Female parent columns, Male and Female Clone, and "Year Bred" columns.
- **Sex:** Four single-character designations: M (Male), F (Female), U (Unknown), or B (Both).
- **Genus/Material Type:** Five two-character designations for the genus or material type.
- **Source Type:** Multi-character categories describing source types including sib types, cuttings, wild, or NA (Not Available/Not Applicable).
- **Category:** Multi-character classifications of *Populus* or *Salix* germplasm including sib types, cuttings, wild, native, or NA (Not Available/Not Applicable).
- **Female and Male Parent Columns:** The female or male species name with parentheses used to distinguish between parents in three-species clones.
- **Female and Male Parent Clone Columns:** The female or male clone name or number used to produce the clone. Parentheses are used to distinguish between parents in three-species clones.
- **Year Bred:** The four-digit year when the breeding or collection took place or NA (Not Available/Not Applicable).
- **Year Selected:** The four-digit year when the material was selected or NA (Not Available/Not Applicable).
- **Year Released:** The four-digit year when the material was released for commercial use or NA (Not Available/Not Applicable).
- **Hybrid Designation:** Hybrid names are preceded with the Punycode "×" multiplication character to improve database searching. However, caution is urged since the exact hybrid lineage has not been scientifically verified for some of the earlier data. This column lists an applicable hybrid name that may associate with the listed parental species name entries.
- Many column entries only allow specific character set entries using SQL check constraints to help maintain data entry integrity. Examples include the "Sex" (single character), "Category" (variable characters), and "Year selected" (4 digits). Some fields also allow "NA" values that designate "Not Available/Not Applicable" data.
- The Canadian database records have a flat single table appearance similar to a database view making it possible to display them in a single spreadsheet worksheet. This single table view process makes it easier for users to understand and view the data simply using the various search fields to access the entire database. Searching the database is case insensitive and does not require the exact case of the intended values. Below are sample searches that help describe the data:
- Searching the Clone Directory Database [6] for the clone "Name", "a69" can be entered as, "A69" or "a69" both returning 65 results.
- The aforementioned search returns "AK50" as a "Name" column value without a space, while other "AK" entries have spaces.
- Searching the "Name" fields for "ak" or "AK" both return 39 results including "AK50".
- Searching the "Name" fields for "ak 30" and the "Female Parent Clone" column for "a69" returns two "AK 30" "Name" records. This is possible because the "ID" fields are different. Also, note that the "Current Status" and "Data Source" values are different and may explain the discrepancy.
- Searching the "Female Parent Clone" column for "473-5070" returns ten records with the "Female Parent Clone" value of: "473-5070; FR17; Fraser River South-BC, Canada".

However, searching the "Name" column for "473-5070" or "FR17" returns no records. The database search tool may not be able to identify parents also included in the "Name" clone column.

*1.3. Characteristics of the Open4st Database Model*

The Open4st database [7] was developed in March 2011 to provide online access to the Open4st project clones, families, and related experimental data [8]. It uses the open-source PostgreSQL relational database [9] and the online DBKiss database application [10] to allow read-only access to the database tables and views. It contains an SQL editor that allows custom SQL queries for more in-depth custom reports and can save previous queries for later retrieval. This database process is designed to be used as a central repository by importing data via CSV files, creating SQL queries for specific reports and views for "big picture" cumulative annual summaries that can be exported back to CSV files for R Programming and further user analysis.

An open-source development copy of the Open4st database [11] (aka "r4st") is available via the pmcgover/24dev-demo GitHub public repository. It is a prototyping add-on process for the OSGeo Live DVD [12] that allows users to review, modify, and execute the open-source code using the MIT license [13]. To access this material, users can review the GitHub 24dev-demo documentation, download the latest 24dev-demo release, and install it on an OSGeoLive system [12].

The r4st database documentation provides a high-level description of the 24dev-demo process. The r4st/csv folder contains the CSV files that are loaded into the database with scripts from the r4st/bin folder to create the tables and views. This is essentially a build process that drops and recreates all of the existing data each time the scripts are activated. This allows for easy modifications but could be deactivated and used long term without the build process. The final build step creates a single database dump file of the entire database that can be copied to different server environments and optionally configured from the default write access to read-only access. Key Open4st tables include the Plant (epithet details), Family (parent details), Pedigree (parent/child lineage details), test_detail (annual nursery details), taxa (clone or family species details), and field_trial (field trial details) tables. Related data are associated between tables using foreign key relationships.

Figure 1 shows the current online Open4st DBKiss database application that contains 11 tables and 40 views. Below are key usage notes for the DBKiss database application [10] used throughout this POC.

- Database Tables are displayed on the left side and views on the right. Clicking on a table or a view displays its contents similar to a spreadsheet. Database tables in this context are based on actual spreadsheet data that are loaded into the database.
- Database views can display data from different tables in a variety of reports and summaries with different levels of aggregation.
- Clicking on the view "Count" header displays the count of each view.
- Below are navigation options for the tables or views (Figure 2: Open4st Plant Table Listing):
  - Selecting the "full content" box and clicking the "Search" button expands rows to include all data.
  - Export any given page by clicking one of the "Export to CSV:" options.
  - Click on any column header to reverse the entire column sort order.
  - The user can search any column by entering text in the search box below the "All tables" link, then select the related column from the next drop-down list. Searches are case insensitive.
- The Execute SQL pages allow the user to enter SQL Queries to further refine the search. This is an advanced feature.

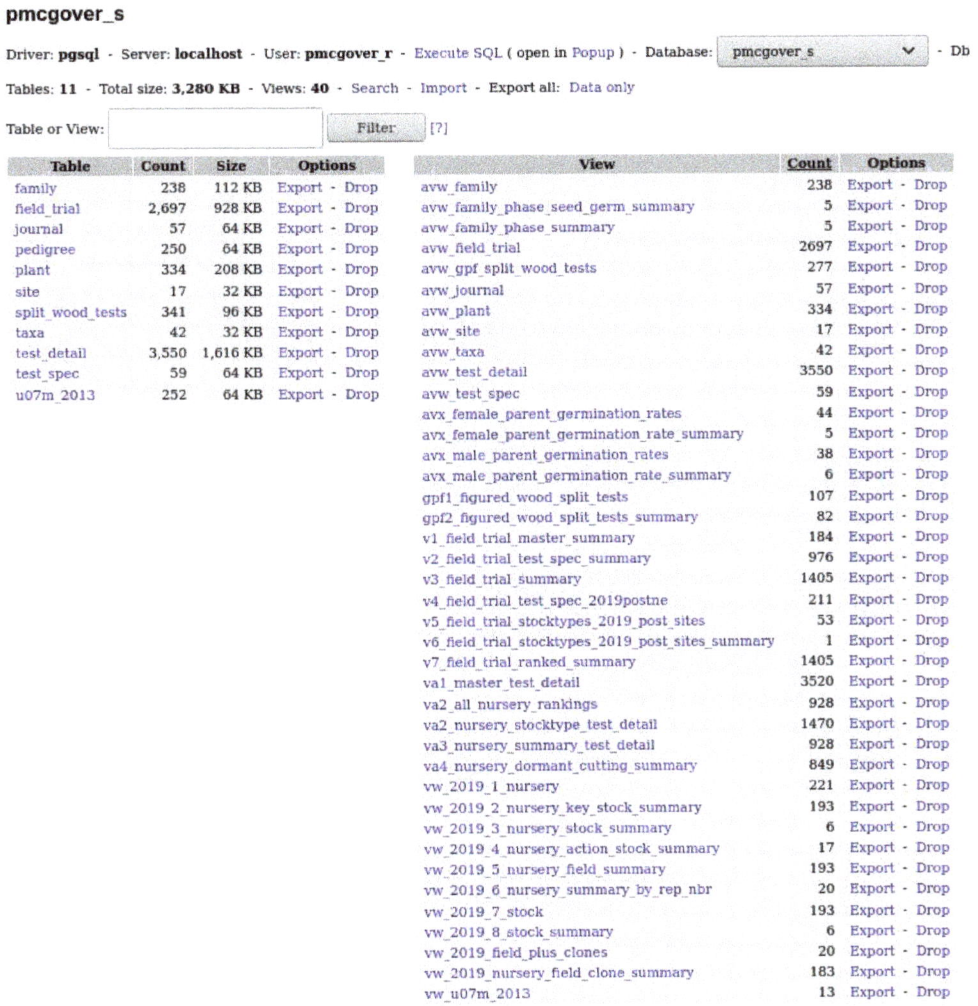

## pmcgover_s

Driver: **pgsql** - Server: **localhost** - User: **pmcgover_r** - Execute SQL ( open in Popup ) - Database: pmcgover_s - Db

Tables: **11** - Total size: **3,280 KB** - Views: **40** - Search - Import - Export all: Data only

Table or View: [ Filter ] [?]

| Table | Count | Size | Options |
|---|---|---|---|
| family | 238 | 112 KB | Export - Drop |
| field_trial | 2,697 | 928 KB | Export - Drop |
| journal | 57 | 64 KB | Export - Drop |
| pedigree | 250 | 64 KB | Export - Drop |
| plant | 334 | 208 KB | Export - Drop |
| site | 17 | 32 KB | Export - Drop |
| split_wood_tests | 341 | 96 KB | Export - Drop |
| taxa | 42 | 32 KB | Export - Drop |
| test_detail | 3,550 | 1,616 KB | Export - Drop |
| test_spec | 59 | 64 KB | Export - Drop |
| u07m_2013 | 252 | 64 KB | Export - Drop |

| View | Count | Options |
|---|---|---|
| avw_family | 238 | Export - Drop |
| avw_family_phase_seed_germ_summary | 5 | Export - Drop |
| avw_family_phase_summary | 5 | Export - Drop |
| avw_field_trial | 2697 | Export - Drop |
| avw_gpf_split_wood_tests | 277 | Export - Drop |
| avw_journal | 57 | Export - Drop |
| avw_plant | 334 | Export - Drop |
| avw_site | 17 | Export - Drop |
| avw_taxa | 42 | Export - Drop |
| avw_test_detail | 3550 | Export - Drop |
| avw_test_spec | 59 | Export - Drop |
| avx_female_parent_germination_rates | 44 | Export - Drop |
| avx_female_parent_germination_rate_summary | 5 | Export - Drop |
| avx_male_parent_germination_rates | 38 | Export - Drop |
| avx_male_parent_germination_rate_summary | 6 | Export - Drop |
| gpf1_figured_wood_split_tests | 107 | Export - Drop |
| gpf2_figured_wood_split_tests_summary | 82 | Export - Drop |
| v1_field_trial_master_summary | 184 | Export - Drop |
| v2_field_trial_test_spec_summary | 976 | Export - Drop |
| v3_field_trial_summary | 1405 | Export - Drop |
| v4_field_trial_test_spec_2019postne | 211 | Export - Drop |
| v5_field_trial_stocktypes_2019_post_sites | 53 | Export - Drop |
| v6_field_trial_stocktypes_2019_post_sites_summary | 1 | Export - Drop |
| v7_field_trial_ranked_summary | 1405 | Export - Drop |
| va1_master_test_detail | 3520 | Export - Drop |
| va2_all_nursery_rankings | 928 | Export - Drop |
| va2_nursery_stocktype_test_detail | 1470 | Export - Drop |
| va3_nursery_summary_test_detail | 928 | Export - Drop |
| va4_nursery_dormant_cutting_summary | 849 | Export - Drop |
| vw_2019_1_nursery | 221 | Export - Drop |
| vw_2019_2_nursery_key_stock_summary | 193 | Export - Drop |
| vw_2019_3_nursery_stock_summary | 6 | Export - Drop |
| vw_2019_4_nursery_action_stock_summary | 17 | Export - Drop |
| vw_2019_5_nursery_field_summary | 193 | Export - Drop |
| vw_2019_6_nursery_summary_by_rep_nbr | 20 | Export - Drop |
| vw_2019_7_stock | 193 | Export - Drop |
| vw_2019_8_stock_summary | 6 | Export - Drop |
| vw_2019_field_plus_clones | 20 | Export - Drop |
| vw_2019_nursery_field_clone_summary | 183 | Export - Drop |
| vw_u07m_2013 | 13 | Export - Drop |

**Figure 1.** The Open4st Online Database: Table and View Listing.

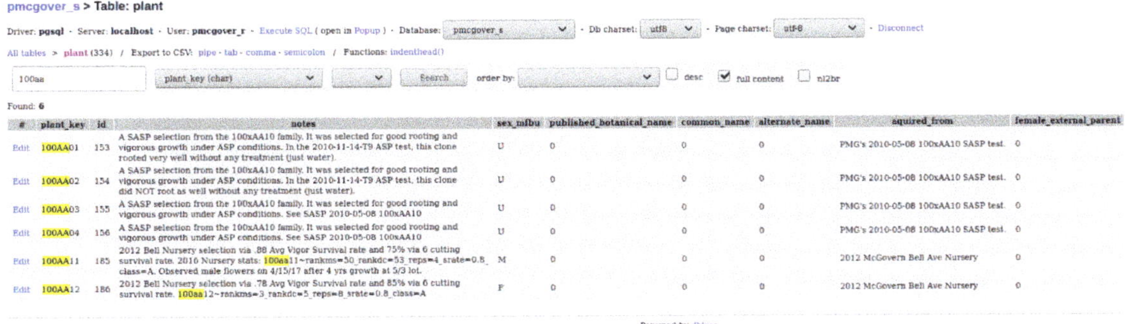

pmcgover_s > Table: plant

Driver: **pgsql** - Server: **localhost** - User: **pmcgover_r** - Execute SQL ( open in Popup ) - Database: pmcgover_s - Db charset: utf8 - Page charset: utf-8 - Disconnect

All tables > plant (334) / Export to CSV: pipe - tab - comma - semicolon / Functions: indenthead()

Found: 6

| # | plant_key | id | notes | sex_mfbu | published_botanical_name | common_name | alternate_name | aquired_from | female_external_parent |
|---|---|---|---|---|---|---|---|---|---|
| Edit | 100AA01 | 153 | A SASP selection from the 100xAA10 family. It was selected for good rooting and vigorous growth under ASP conditions. In the 2010-11-14-T9 ASP test, this clone rooted very well without any treatment (just water). | U | 0 | 0 | 0 | PMG's 2010-05-08 100xAA10 SASP test. | 0 |
| Edit | 100AA02 | 154 | A SASP selection from the 100xAA10 family. It was selected for good rooting and vigorous growth under ASP conditions. In the 2010-11-14-T9 ASP test, this clone did NOT root as well without any treatment (just water). | U | 0 | 0 | 0 | PMG's 2010-05-08 100xAA10 SASP test. | 0 |
| Edit | 100AA03 | 155 | A SASP selection from the 100xAA10 family. It was selected for good rooting and vigorous growth under ASP conditions. See SASP 2010-05-08 100xAA10 | U | 0 | 0 | 0 | PMG's 2010-05-08 100xAA10 SASP test. | 0 |
| Edit | 100AA04 | 156 | A SASP selection from the 100xAA10 family. It was selected for good rooting and vigorous growth under ASP conditions. See SASP 2010-05-08 100xAA10 | U | 0 | 0 | 0 | PMG's 2010-05-08 100xAA10 SASP test. | 0 |
| Edit | 100AA11 | 185 | 2012 Bell Nursery selection via .88 Avg Vigor Survival rate and 75% via 6 cutting survival rate. 2016 Nursery stats: 100aa11~rankms=50_rankdc=53_reps=4_srate=0.8_class=A. Observed male flowers on 4/15/17 after 4 yrs growth at 5/3 lot. | M | 0 | 0 | 0 | 2012 McGovern Bell Ave Nursery | 0 |
| Edit | 100AA12 | 186 | 2012 Bell Nursery selection via .78 Avg Vigor Survival rate and 85% via 6 cutting survival rate. 100aa12~rankms=3_rankdc=5_reps=8_srate=0.8_class=A | F | 0 | 0 | 0 | 2012 McGovern Bell Ave Nursery | 0 |

Powered by dbkiss

**Figure 2.** The Open4st Online Database: Plant Table Listing.

The Open4st database was developed for a restricted set of users. It does not have time-tested exposure or follow column naming standards similar to the IPC *Salix* or Canadian databases. The species data is derived from a "Taxa" table, which does not follow consistent binomial or species naming conventions.

*1.4. Characteristics of the Integrated IPC Salix Cultivar Database Proof-of-Concept*

Each database model has pros and cons for its respective processes. Given these contrasting models, the intent of this POC is to demonstrate an SQL database to store existing IPC *Salix* cultivar information and provide users with a format to compare and submit new *Salix* cultivar entries. An integrated approach from all three models is proposed using three tables: Epithet, Family, and Pedigree.

The database tables can be displayed online or generated from a desktop application with desired views copied to an online spreadsheet. Users can view the flat inline Epithet and Family records to understand relationships and allow them to submit their new *Salix* cultivar epithet and family entries. A potential scenario could involve users submitting their Checklist entries to the *Salix* administrator for review; the entries are then entered into a test environment by IT resources for final review and promotion to a production online environment.

*1.5. Integrated IPC Salix Cultivar Database Use Case*

The following use case scenario describes the interactions, events, and flow steps between the actors (participants) and the various systems (Figure 3).

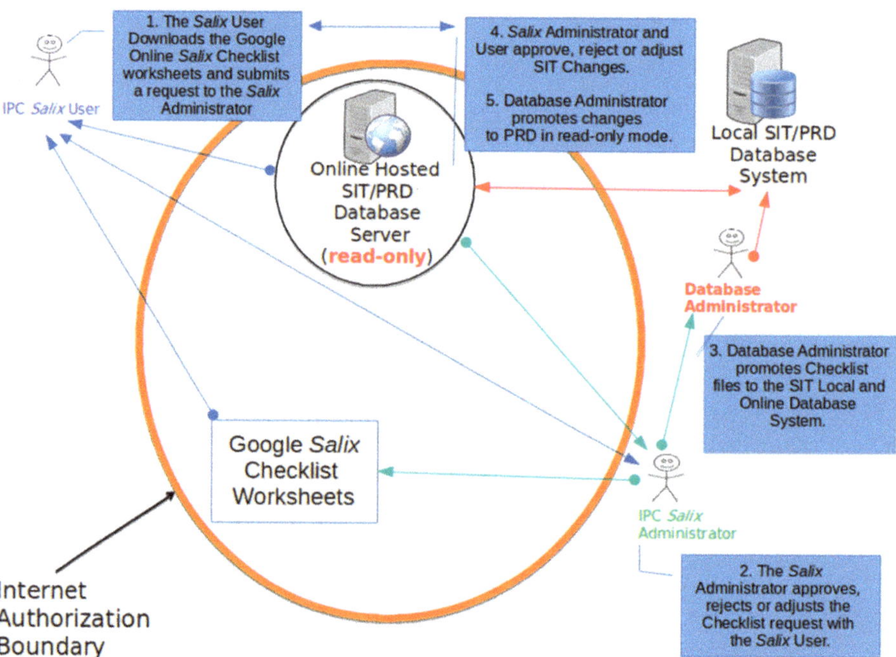


**Figure 3.** *Salix* Checklist use case flow diagram.

1.5.1. Brief Description

This use case describes a local and online database system to store IPC *Salix* Cultivar information populated by a user Checklist registration process.

### 1.5.2. Actors

1. IPC *Salix* Administrator: (e.g., Yulia Kuzovkina).
2. IPC *Salix* User: A user submitting an IPC *Salix* Checklist request that should be approved by Kuzovkina before it is uploaded to the database.
3. Database Administrator (DBA): (e.g., Patrick McGovern).
4. Local Database System: Located on the DBA's local desktop computer with production (PRD) and pre-production (SIT or Staging) database environments.
5. Online Database System: Located on an online hosted database server with SIT and PRD database environments.
6. Online *Salix* Checklist Views: The checklist views vw3_checklist_root_level_epithet [14] and vw4_checklist_epithet_family [15], stored on the Online and Local Database System.
7. Online *Salix* Checklist Worksheets: The user-accessible Google worksheets with the Online *Salix* Checklist View data that is documented and formatted to allow users to copy and submit their own root-level epithet and epithet/family level records. See Google worksheets: vw3_checklist_root_level_epithet [16] vw4_checklist_epithet_family [17] and a CollaboratedChecklist [18] for all parties to collaborate on the proposed Checklist entries.

### 1.5.3. Triggers

The system is triggered when the IPC *Salix* User submits their IPC *Salix* Cultivar files to the IPC *Salix* Administrator for approval.

### 1.5.4. Pre-Conditions

1. The Local and Online Database Systems have pre-populated table and view data (e.g., Epithet, Family and Pedigree tables, and Checklist views).
2. The Online Salix Cultivar Worksheets (vw3_checklist_root_level_epithet [16] and vw4_checklist_epithet_family [17]) have the latest version of the vw3_checklist_root_level_epithet [14] and vw4_checklist_epithet_family [15] data from the Online Database System and are available for IPC Salix user access.

### 1.5.5. Basic Flow of Events

1. The IPC *Salix* Users download the Google Online *Salix* Cultivar Worksheet with the vw3_checklist_root_level_epithet and vw4_checklist_epithet_family view data and submit their new root epithet and/or family epithet *Salix* record request to the IPC *Salix* Administrator.
2. The IPC *Salix* Administrator reviews the IPC *Salix* user Cultivar submission and communicates with the user to approve, reject, or adjust the request.
3. The approved Cultivar files are entered by the DBA into the SIT Local Database System, which is then copied to the SIT read-only Online Database System viewable by all users.
4. The IPC *Salix* Administrator and user approve, reject, or adjust the SIT changes.
5. The approved changes are then promoted by the DBA and copied to the PRD Local and Online read-only Database systems.

### 1.5.6. Special Requirements

1. The complexity of the system may likely require dedicated IT personnel to implement and maintain this process long term.
2. Any other system changes are also tested first in SIT, and then promoted to PRD. Version control for database and/or Checklist file updates are considered.
3. The Local Database System can be tested locally via the 24dev-demo [11] add-on with the OSGeo-Live DVD [12], USB flash drive [19], or the VirtualBox application.

*1.6. Technical Information*

Additional methodologies, technical details, and screencast videos are also available [20].

**Author Contributions:** Conceptualization, methodology, software, formal analysis, and writing—original draft preparation, P.N.M.; validation, writing—review and editing, Y.A.K. and R.Y.S. All authors have read and agreed to the published version of the manuscript.

**Funding:** This research received no external funding.

**Data Availability Statement:** The data presented in this study can be available on request from the corresponding author.

**Acknowledgments:** We thank Ron Zalesny, Jeffery Jackson, Bradford Bender, and Harvey Anderson for their contributions to this project. Special thanks to the Canadian Forest Service for sharing their database, plant materials, and expertise over the years. The authors also thank the Poplar and Willow Council of Canada for providing access to their clone directory database and Samir Patel for his editorial advice and MDPI formatting expertise.

**Conflicts of Interest:** The authors declare no conflict of interest.

# References

1. Kuzovkina, Y.A. Checklist for Cultivars of *Salix* L. (Willow). 2015. Available online: http://www.fao.org/forestry/44983-0370ab0c9786d954da03a15a8dd4721ed.pdf (accessed on 15 January 2021).
2. Kuzovkina, Y.A. Compilation of the Checklist for Cultivars of *Salix* (Willow). *Hortic. Sci.* **2015**, *50*, 1608–1609. [CrossRef]
3. Butterick, M. Straight and Curly Quotes. Available online: https://practicaltypography.com/straight-and-curly-quotes.html (accessed on 15 January 2021).
4. Poplar and Willow Council of Canada. Populus and Salix Clone Directory for Canada. 2018. Available online: http://www.poplar.ca/clone-directory (accessed on 15 January 2021).
5. Poplar and Willow Council of Canada. Headings and Descriptions for the Poplar & Willow Clone Directory Database. 2018. Available online: http://www.poplar.ca/clone-directory-descriptions (accessed on 15 January 2021).
6. Poplar and Willow Council of Canada. Clone Directory Database. 2018. Available online: http://www.poplar.ca/clone-directory-database (accessed on 15 January 2021).
7. McGovern, P.N. Using and Accessing the r4st Database. 2020. Available online: https://sites.google.com/site/open4st/faq/what-is-the-open4st-database (accessed on 15 January 2021).
8. Open4st. Aspen Trees: A Synergistic Resource. 2010. Available online: https://sites.google.com/site/open4st (accessed on 15 January 2021).
9. PostgreSQL. The World's Most Advanced Open Source Relational Database. Available online: https://www.postgresql.org/ (accessed on 15 January 2021).
10. Tomczak, C. DBKiss. 2019. Available online: https://github.com/cztomczak/dbkiss (accessed on 18 January 2021).
11. McGovern, P.N. GitHub 24dev-Demo. 2020. Available online: https://github.com/pmcgover/24dev-demo (accessed on 18 January 2021).
12. OSGEOLive. OSGEOLive. 2011. Available online: https://live.osgeo.org/en/index.html (accessed on 18 January 2021).
13. Open Source Initiative. The MIT License. Available online: https://opensource.org/licenses/MIT (accessed on 18 January 2021).
14. McGovern, P.N. View vw3_Checklist_Root_Level_Epithet. 2020. Available online: http://pmcgovern.us/ipc/ipcro2.php?viewtable=vw3_checklist_root_level_epithet (accessed on 18 January 2021).
15. McGovern, P.N. View Vw4_Checklist_Epithet_Family. 2020. Available online: http://pmcgovern.us/ipc/ipcro2.php?viewtable=vw4_checklist_epithet_family&search=&column=&column_type=&order_by=&full_content=on (accessed on 18 January 2021).
16. McGovern, P.N. Worksheet vw3_Checklist_Root_Level_Epithet. 2020. Available online: https://docs.google.com/spreadsheets/d/1z05wOF6q5OFMboB7BdNih5TaAFRM__K0yhNzuq8Tq4U/edit#gid=2137439597 (accessed on 18 January 2021).
17. McGovern, P.N. Worksheet vw4_Checklist_epithet_Family. 2020. Available online: https://docs.google.com/spreadsheets/d/1z05wOF6q5OFMboB7BdNih5TaAFRM__K0yhNzuq8Tq4U/edit#gid=76850055 (accessed on 18 January 2021).
18. McGovern, P.N. Worksheet CollaboratedChecklist. 2020. Available online: https://docs.google.com/spreadsheets/d/1z05wOF6q5OFMboB7BdNih5TaAFRM__K0yhNzuq8Tq4U/edit#gid=1533711583 (accessed on 18 January 2021).
19. VirtualBox. Welcome to VirtualBox.org! Available online: https://www.virtualbox.org (accessed on 18 January 2021).
20. McGovern, P.N.; Kuzovkina, Y.A.; Soolanayakanahally, R.Y. Technical Information: IPC Salix Cultivar Database Proof-of-Concept. 2021. Available online: https://sites.google.com/site/open4st/faq/technical-information-ipc-salix-cultivar-database-proof-of-concept (accessed on 28 April 2021).

*Article*

# Natural Drying and Chemical Characteristics of Hybrid Poplar Firewood Produced from Agricultural Bioenergy Buffers in Southern Québec, Canada

Julien Fortier [1,*], Benoit Truax [1], Daniel Gagnon [1,2] and France Lambert [1]

1    Fiducie de Recherche sur la Forêt des Cantons-de-l'Est/Eastern Townships Forest Research Trust, 1 rue Principale, Saint-Benoît-du-Lac, QC J0B 2M0, Canada; btruax@frfce.qc.ca (B.T.); daniel.gagnon@uregina.ca (D.G.); france.lambert@frfce.qc.ca (F.L.)
2    Department of Biology, University of Regina, 3737 Wascana Parkway, Regina, SK S4S 0A2, Canada
*    Correspondence: fortier.ju@gmail.com

**Citation:** Fortier, J.; Truax, B.; Gagnon, D.; Lambert, F. Natural Drying and Chemical Characteristics of Hybrid Poplar Firewood Produced from Agricultural Bioenergy Buffers in Southern Québec, Canada. *Forests* **2021**, *12*, 122. https://doi.org/10.3390/f12020122

Academic Editors: Ronald S. Zalesny and Andrej Pilipović
Received: 14 December 2020
Accepted: 20 January 2021
Published: 23 January 2021

**Publisher's Note:** MDPI stays neutral with regard to jurisdictional claims in published maps and institutional affiliations.

**Abstract:** Implementing bioenergy buffers on farmland using fast-growing tree species could reduce firewood harvest pressure in woodlots and increase forest connectivity, while improving carbon sequestration, phytoremediation, stream habitats, soil stabilization and hydrological regulation. The objective of the study was to evaluate the natural drying and chemical characteristics of hybrid poplar firewood produced from bioenergy buffers, and to compare these characteristics with those of native species harvested in adjacent woodlots. In Trial A, 110 cm-long unsplit logs (a feedstock for biomass furnaces) were produced to evaluate the effect of log diameter class on firewood quality. In this trial, hybrid poplar firewood characteristics were also compared with *Populus tremuloides*, *Acer rubrum* and *Fraxinus americana*. In Trial B, the effect of hybrid poplar genotype and cover treatment was evaluated on the moisture content of short split logs (40 cm long). Firewood of satisfactory quality was produced on a yearly cycle for short split logs, and on a biannual cycle for long unsplit logs. Covering short split log cords with metal sheeting lowered the final moisture content (from 20.7% to 17.3%) and reduced its variability, while genotype did not significantly affect final moisture content. In Trial A, larger-diameter logs from hybrid poplar had lower element concentrations, but slightly higher moisture content after two years. A two-fold variation in N concentration was observed between diameter classes, suggesting that burning larger poplar logs would minimize atmospheric N pollution. Heating value, carbon and calcium concentrations increased following the seasoning of hybrid poplar firewood. After the first seasoning year outdoors, hybrid poplar had the highest moisture content (33.1%) compared to native species (24.1–29.5%). However, after the second seasoning year in an unheated warehouse, the opposite was observed (14.3% for hybrid poplar vs. 15.0–21.5% for native species). Heating value, carbon and nitrogen concentrations were similar between tree species, while high phosphorus and base cation concentrations characterized hybrid poplar, suggesting higher ash production. Poplar bioenergy buffers could provide a complementary source of firewood for heating in the fall and in the spring, when the heat demand is lower than during cold winter months.

**Keywords:** fuelwood; seasoning; log diameter; splitting; heating value; moisture content; agroforestry; red maple; white ash; trembling aspen

## 1. Introduction

Compared to wood chip or pellet production, firewood requires little processing and equipment, and is typically seasoned outdoors [1]. It is therefore an attractive solid biofuel for private landowners and farmers, as it can be produced at low cost. On private forestland and farmland of Northeastern America, most bioenergy feedstock comes from firewood that is harvested in woodlots, and very little land area is dedicated to bioenergy plantations of fast-growing woody species [2–4]. Because several hardwood species (birches, maples, ashes, beech, oaks, etc.) are locally abundant, deciduous species with a low wood density such as poplars (*Populus* sp.) are often disregarded as a source of firewood.

In regions where agriculture dominates the landscape, rural communities are more than ever facing major environmental challenges related to water quality decline, stream habitat protection, forest habitat loss and fragmentation, and climate change [5–7]. The large-scale implementation of bioenergy buffers along agricultural riparian zones, and field margins could be a solution to rapidly address these challenges, as it would reduce firewood harvesting pressure in woodlots and increase forest patch connectivity, while creating opportunities for carbon sequestration, agricultural pollutants removal, stream habitat improvement and hydrological regulation [8–13]. In Northeastern America, hybrid poplars (*Populus* × spp.) are especially promising for the design of bioenergy buffers, as they create a forest canopy within a decade, and provide high woody biomass yields even on more marginal land [9,14,15]. In this context, it is important to validate that firewood of satisfactory quality can be produced in hybrid poplar buffers in order to stimulate their adoption by the farming community.

Moisture content of seasoned firewood is negatively linked to the amount of heat produced during combustion [16]. Consequently, when moisture content of firewood is low, less feedstock is needed for heating, which reduces wood burning impacts on greenhouse gases and on the thermal load to the atmosphere [17]. The combustion of inadequately seasoned wood also increases creosote accumulation in chimneys and the release of air pollutants (fine particles from smoke, carbon monoxide, benzene, formaldehyde and polycyclic aromatic hydrocarbons), which creates safety and health issues [18,19]. In North America, targeted moisture content for seasoned firewood is generally 20% or less on a wet weight basis [19]. Short split logs (±30–40 cm in length), used in wood stoves or small wood furnaces, usually dry within a year when seasoned adequately [18]. However, more and more houses and farm buildings are now equipped with large biomass furnaces that burn longer wood logs (i.e., 100–150 cm of length), the seasoning of which generally takes two years under a cold temperate climate (A. Couture, Sequoia Industries, pers. comm.). Very few studies have measured the moisture content of different log sizes from hybrid poplars during seasoning. Studies carried out in Italy have shown little effect of log size on moisture content loss of *P. deltoides* × *P. nigra* [1,20], which contrasted with conclusions reached in Oregon, United States [21]. Generally, the drying rate of fuelwood decreases as log diameter or length increase [18,22]. Covering the top of firewood piles during outdoor seasoning has also been recommended to lower the moisture content of hardwood and softwood species [18,23].

Energy content (i.e., heating value) is another important characteristic of firewood. Minor variations in heating values are generally observed between tree species, but softwood species tend to have slightly higher heating values than hardwoods due to their higher concentration in extractives and lignin [16]. There is also evidence of both positive and negative changes in the heating value of woody biomass during seasoning [16,24]. However, no studies have measured heating value variations of poplar firewood during seasoning, nor the effect of log size on the properties of this fuelwood.

Elemental characterization of biomass can provide information about the potential environmental and operational impact of feedstock. During combustion, biomass nitrogen (N) can be transformed into nitric oxides, and nitrous oxide to a lesser extent [25]. Both of these gases deplete the ozone layer, while nitric oxides and nitrous oxide respectively contribute to acid rain depositions and global warming [26,27]. Sulphur oxide production, which also increases acid rain, is generally limited during woody biomass combustion, as most sulphur is embedded in the ashes [25]. However, alkali metals can react with sulphate or chlorine, and lead to the formation of salts that cause fouling, slagging and corrosion problems in combustion appliances [25]. Moreover, biomass with high nutrient content leads to higher ash production, which increases the frequency of equipment maintenance [25]. Generally, macronutrient concentrations in the stem of hybrid poplars tend to be high compared to other tree species [28]. There is also a general decline in the concentration of most macronutrients as tree stem diameter increases [29], which suggests that producing wood logs of larger diameters would improve the elemental properties of firewood. Finally, significant nutrient leaching from woody biomass piles can occur during outdoor seasoning, but changes in elemental composition of woody biomass during seasoning have rarely been studied [16].

In this study, we measured the natural drying and chemical characteristics of hybrid poplar firewood produced from bioenergy buffers located on farmland in southern Québec, Eastern Canada (Köppen climate zone Dfb, i.e., warm-summer humid continental climate). In a first trial, 110 cm-long unsplit wood logs were produced from hybrid poplar bioenergy buffers (genotype DN × M-915508), but also from native species harvested from adjacent woodlots (trembling aspen, *Populus temuloides* Michx., red maple, *Acer rubrum* L., and white ash, *Fraxinus americana* L.). These firewood logs were seasoned outdoors for a year, and indoors in an unheated warehouse for another year. Their moisture content (wet weight basis) was sampled at the end of each year. Hybrid poplar logs were separated into three diameter classes, and chemical characteristics of hybrid poplar biomass were measured at harvest and one year after outdoor seasoning. At the end of the first year of outdoor drying, the chemical characteristics of native species were also compared to those of hybrid poplars. In a second trial, we measured, after one year of outdoor seasoning, the effects of hybrid poplar genotype and firewood cover treatment on the moisture content of 40 cm-long split logs.

## 2. Materials and Methods

### 2.1. Site Description

This study took place in the municipality of St-Benoît-du-Lac, a 216-ha property owned by a Benedictine monastic community and located in the Estrie region of southern Québec, Eastern Canada (45°10′N; 72°16′ W). In 2011, 15 m-wide bioenergy buffers were planted (1666 trees/ha) with three hybrid polar genotypes (DN × M-915508, D × N-3570, M × B-915311), downslope of 45 ha of hayfields. These multifunctional buffers were established with the objective of reducing non-point source pollution from upslope fields, increasing carbon sequestration and producing firewood for the biomass furnaces that provide heat to the Abbey buildings. After eight years, the firewood volume production capacity of the bioenergy buffers ranged from 20.5–29.3 m$^3$/ha/yr, depending on genotype, planting stock type and deer protection treatments [9].

### 2.2. Firewood Harvesting, Processing and Sampling

#### 2.2.1. Trial A—110 cm Long Unsplit Wood Logs

From 7–14 November 2017, a total of 110 hybrid poplar trees were harvested from 7-year-old bioenergy buffers located along different cultivated fields. Trees from genotype DN × M-915508 were cut into log sections of 110 cm, which is the log dimension required for burning in the Abbey biomass furnaces. Wood logs were separated into three different diameter classes: (1) small end diameter ≥13 cm; (2) small end diameter between 8 cm and 12.9 cm; and (3) small end diameter between 3 cm and 7.9 cm. On 15 November 2017, poplar logs were stacked into metal racks that were designed to be directly inserted into the furnaces (Figure 1). Log piles in metal racks were ±80 cm wide by ±135 cm in height. For diameter classes 1 and 2, five metal racks were prepared, while only two racks were prepared for diameter class 3.

**Figure 1.** (**a**) 7-year-old bioenergy buffers located downslope of hayfields; (**b**) Hybrid poplar firewood log stacking in metal racks according to the different diameter classes. At the end of the study, firewood logs were used to heat the St-Benoît-du-Lac Abbey buildings.

During the same period, native deciduous species growing in woodlots adjacent to the hayfields were also felled and processed into 110 cm wood logs that were stacked in the same metal racks used for hybrid poplars. Three to four trees from trembling aspen (*Populus tremuloides*), red maple (*Acer rubrum*) and white ash (*Fraxinus americana*) were felled and processed, which produced one metal rack of stacked logs from each species. The mean age of the felled trees was determined by a ring count on a representative tree from each species: 36 years for trembling aspen; 34 years for red maple; and 47 years for white ash.

On 16 November 2017, wood logs staked in metal racks were placed along a gravel road bordering an open field, with log ends being positioned parallel to the dominant winds. No cover was put on racks, and wood logs were left outdoors for almost year (until 4 November 2018). Wood log racks were then placed in an unheated warehouse for another year, prior to being used as feedstocks for the Abbey.

On 14 November 2017, six hybrid poplar logs were sampled (wood disc ±2.5 cm thick taken with a chainsaw) from each diameter class to determine initial chemical characteristics. Due to unfavorable weather conditions during tree felling, the initial moisture content of wood logs was not measured, but was estimated from data in Trial B for the case of hybrid poplar and from the literature for the native deciduous species. On 6 November 2018, after approximately one year of seasoning outdoors, hybrid poplar and native deciduous species logs were sampled for moisture content and wood chemistry. Wood discs were collected halfway between the middle and the endpoint of logs. For each log sampled, two discs were collected, one that was immediately weighed (±0.1 g) and another one for chemical analyses. Six logs per diameter classes were sampled for hybrid poplar, and six logs were sampled for the native species (aspen, maple and ash). On 22 October 2019, after approximately two years of seasoning, only moisture content was sampled using the procedure mentioned above. Ten logs per diameter class were sampled for hybrid poplar, and 10 logs were sampled for the native species.

Sampled logs were taken from different positions in metal racks and covered a wide range of diameters. The midpoint diameter of all sampled logs was recorded using perpendicular caliper measurements. Log subsamples were put in paper bags and taken to the lab, where they were air-dried for two months. The subsamples that were fresh-weighed during sampling were then oven-dried (95 °C) until a constant mass was reached (after 24 h), and immediately weighed once out of the oven. The moisture content of logs was calculated on a wet weight basis: Moisture content (%) = (Weight of water/Total weight) × 100.

### 2.2.2. Trial B—40 cm Long Split Wood Log

From 5–9 November 2018, a total of 39 hybrid poplar trees were harvested from 8-year-old bioenergy buffers located along different cultivated fields. Twenty trees from genotype D × N-3570 and 19 trees from genotype M × B-915311 were felled. Depending on tree size, two to four subsamples (stem discs) per tree were taken along different stem sections (see Truax et al. [9] for additional details). Stem wood subsamples were immediately fresh-weighed in the field. The subsamples were put in paper bags and taken to the lab, where they were air-dried for two months. The subsamples were then oven-dried (95 °C) until a constant mass was reached (after 24 h) and immediately weighed. This allows measurements of moisture content of freshly harvested hybrid poplars.

The 39 felled hybrid poplars were cut into 80 cm-long sections. Logs were stacked on wood poles and left at the buffer's margins until spring. On 22 May 2019, hybrid poplar logs were collected and cut into smaller logs (40 cm in length), which are typically used in wood stoves or small biomass furnaces. Logs were then manually split with an axe, except for the small diameter logs. On 23 May 2019, logs were stacked bark side up to form four wood cords ($\pm$240 cm long $\times$ 120 cm of height $\times$ 40 cm wide): two cords per genotype, with one cord per genotype being covered with a roofing metal sheet. Cords were stacked on wood posts in a well-aerated gravel parking lot with full sunlight exposure. On 22 October 2019, 16 split logs of various sizes were collected from each genotype $\times$ cover treatment at different positions in the cords and brought to the lab. Only logs split into two pieces were selected for subsampling. On 24 October 2019, a subsample (i.e., a wood slice $\pm$2.5 cm thick) from each log was taken in the middle of the log and immediately weighed fresh. The subsamples were then oven-dried (95 °C) until a constant mass was reached (after 24 h), and immediately weighed once out of the oven. Moisture content of logs was calculated on a wet weight basis. A summary of key dates related to firewood processing and sampling is presented in Table 1 for both trials.

**Table 1.** A summary of key dates related to firewood processing and sampling for Trials A and B.

| Trial A—Unsplit 110 cm Long Logs<br>(Hybrid Poplar, Trembling Aspen, Red Maple, White Ash) | Dates |
|---|---|
| Tree felling and log processing | 7–14 November 2017 |
| Log stacking in metal racks | 14 November 2017 |
| Outdoor seasoning | 14 November 2017–4 November 2018 |
| Warehouse seasoning (unheated) | 4 November 2018–22 October 2019 |
| Chemistry sampling (fresh, hybrid poplar only) | 7–14 November 2017 |
| Chemistry and moisture sampling (after 1 year, 4 species) | 6 November 2018 |
| Moisture sampling (after 2 years, 4 species) | 22 October 2019 |
| **Trial B—Split 40 cm Long Logs**<br>**(2 Hybrid Poplar Genotypes × 2 Cover Treatments)** | **Dates** |
| Tree felling and moisture sampling | 5–9 November 2018 |
| Log processing and splitting | 22 May 2019 |
| Log stacking | 23 May 2019 |
| Outdoor seasoning | 5–9 November 2018–22 October 2019 |
| Moisture sampling (after 1 year) | 22 October 2019 |

### 2.3. Meteorological Data during the Trials

A continental moderate-subhumid climate characterizes the study site [30], and more generally, the southern Québec region belongs to the Köppen climate zone Dfb (warm-summer humid continental climate). For the duration of the study, average monthly temperatures and total precipitations are presented in Figure 2, along with 30 years climatic normals (1981–2010). Meteorological data were obtained from the nearest meteorological station of Magog [31,32].

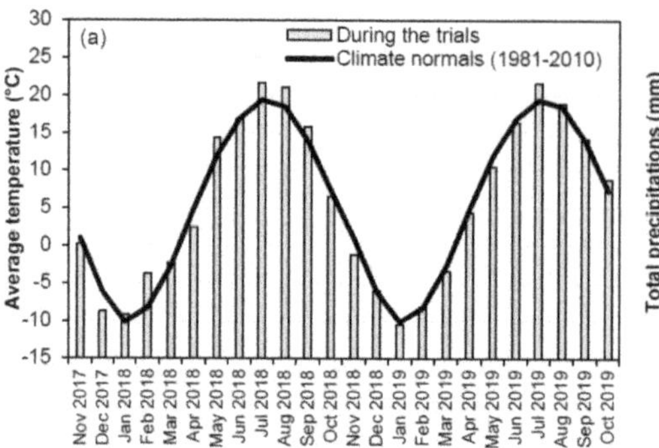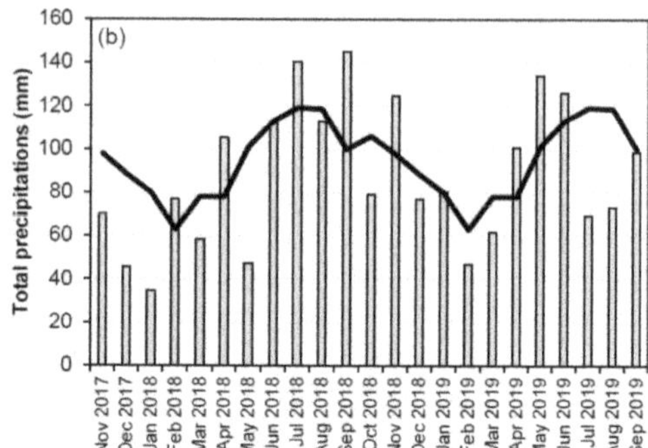



**Figure 2.** (**a**) Average monthly temperatures; and (**b**) total monthly precipitations during the firewood seasoning trials in comparison to 30 years climatic normals. Data were taken from the nearest meteorological station (Magog, QC, Canada). Precipitation data for October 2019 are not shown because an extreme rain event occurred on 31 October (79 mm) after the last moisture content sampling in Trial B (22 October 2019).

### 2.4. Chemical Analyses of Firewood

Firewood subsamples used for chemical analyses were oven-dried at 65 °C and ground in a mill (Pulverisette 15, Fritsch, Idar-Oberstein, Germany) to a particle size of <0.5 mm. C, N and S concentrations were determined by the analytic lab of the Institut des Sciences de la Forêt Tempérée (ISFORT) in Ripon (QC, Canada), with a TruMac CNS analyzer (LECO corporation, MI, USA). P, K, Ca and Mg concentrations and lower heating value (LHV) of woody biomass samples were analyzed by the Centre Technologique des Résidus Industriels (CTRI) in Rouyn-Noranda (QC, Canada). A microwave plasma atomic emission spectrometer (4200 MP-AES, Agilent Technologies, Santa Clara, CA, USA) was used for P, K, Ca and Mg concentration determination, and a bomb calorimeter (6400 Calorimeter, Parr Instrument Company, Moline, IL, USA) was used to determine the LHV of firewood samples on a dry weight basis.

### 2.5. Statistical Analyses

#### 2.5.1. Trial A

The dataset related to hybrid poplar firewood was first analyzed using a one-way analysis of variance (ANOVA) to test the effect of diameter class on firewood characteristics (chemistry at harvest and after one year, and moisture content after one year and two years). $n = 6$ per diameter class for all variables, except for moisture content after two years ($n = 10$ per diameter class). A two-way ANOVA was used to test the effect of diameter class and drying treatment (freshly harvested vs. one year of drying) and the interaction effect on firewood chemistry (6 logs per diameter class × 3 diameter classes × 2 drying treatments). Following all ANOVAs, the normality of residuals distribution was verified using the Shapiro-Wilk W-test ($p < 0.05$), skewness ($<|1|$) and kurtosis ($<|1|$). A few transformations were done to meet this assumption, and ANOVAs were rerun. For the one-way ANOVA, a reciprocal transformation was done on firewood moisture content data after two years, while for the two-way ANOVA, a logarithmic (ln) transformation was done on firewood S and P concentrations [33].

For the dataset related to firewood characteristics of hybrid poplar and the three native species, a one-way ANOVA was used to test the effect of tree species on firewood characteristics after one and two years of drying. Observations from the 3.0–7.9 cm diameter class for hybrid poplar were removed from the data set, as wood logs of such a diameter were almost absent for the other species. In the final dataset analyzed, there was no significant species effect on wood log diameter sampled after one year ($p = 0.96$), and after two years of drying ($p = 0.83$). To meet the assumption of normality in residuals distribution, a reciprocal transformation was done on firewood moisture content data after two years of drying, and a logarithmic (ln) transformation was done on firewood P concentration after one year of drying. Because sample size was unequal between the four tree species ($n$ = 12 and 20 for hybrid poplar and $n$ = 6 and 10 for the other species for firewood characteristics measured after one year and two years of drying respectively), Tukey's HSD test ($\alpha$ = 0.05) was used as a means separation procedure [34].

A correlation analysis, using linear least square regressions, was also done to explore potential relationships between firewood log diameter and moisture content or chemical characteristics. After graphical exploration of the data, non-linear trends were observed for the relationships between wood log diameter and elemental concentrations. Choice of the final relationships presented was made on the basis of highest fit (i.e., $R^2$) and normality in residuals distribution (Shapiro-Wilk W-test).

### 2.5.2. Trial B

A one-way ANOVA was used to test the effect of hybrid poplar genotype on initial moisture content of harvested trees ($n$ = 20 for genotype D × N-3570 and $n$ = 19 for genotype M × B-915311). A two-way ANOVA was used to test the effect of hybrid poplar genotype and firewood cover treatment (uncovered vs. covered), and the interaction effect between those factors on the final moisture content of split logs ($n$ = 64, 16 logs per genotype per treatment × 2 genotypes × 2 treatments). Given that the Shapiro-Wilk W-test is often inappropriate for testing residuals normality for larger sample sizes ($n$ > 50) [35], we used normal quantile-quantile plots (Q-Q plots) as a diagnostic tool for verifying normality of residuals distribution [36]. Two outliers (extreme values) were detected in the uncovered treatment for genotype D × N-3570. Outliers were removed from the data set, and the ANOVA was rerun. Given that the significance of tested effects was the same between the ANOVAs with and without the outliers, results from the dataset containing the outliers are presented. All statistical analyses were done using JMP (version 11) from SAS Institute (Cary, NC, USA).

## 3. Results

### 3.1. Trial A

For hybrid poplar unsplit firewood logs (110 cm long), a significant diameter class effect was observed on moisture content after one year ($p$ < 0.001) and two years ($p$ = 0.004) of drying, with a decreasing moisture content being observed for the smaller diameter logs (Figure 3a). Moisture content differences between logs from different diameter classes were also much larger after one year of drying outdoors (ranging 38.5% to 26.0% between diameter classes) than after the second year, where seasoning took place in an unheated warehouse (ranging 15.4% to 12.7% between diameter classes). Those trends were reflected in the slope of the linear relationships between wood log diameter and moisture content after one and two years of drying (Figure 4a).

Across the four species, hybrid poplar firewood logs had the highest moisture content after one year of drying, followed by trembling aspen, red maple and white ash (Figure 3b). A very different pattern was observed after two years of drying, since hybrid poplar was the species with the lowest moisture content (14.3%), and white ash had the highest moisture content (21.5%). However, the moisture contents of trembling aspen and red maple were not statistically different from the moisture content of hybrid poplar after two years of drying. For trembling aspen and red maple, there were also strong positive linear relationships between wood log diameter and moisture content after one year and two years of drying (Figure 4b,c), while no significant relationship was observed for white ash.

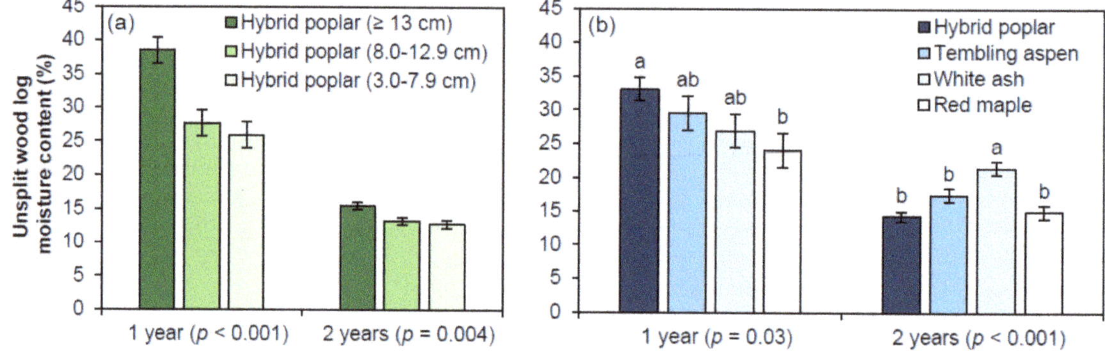

**Figure 3.** (**a**) Diameter class effect on the moisture content of unsplit hybrid poplar firewood logs (110 cm in length) after one year and two years of drying (genotype DN × M-915508); (**b**) Tree species effect on the moisture content of unsplit firewood logs (110 cm in length) after one and two years of drying. In panel (**b**), means with different letters are significantly different ($\alpha = 0.05$; Tukey's HSD test). Vertical bars represent the standard error of the mean. During the first year, wood logs were stored outdoors in an open field with no protection from precipitation. During the second year, wood logs were stored in an unheated warehouse. Moisture content at harvest is in the order of 59% for *P. maximowiczii* hybrids (see Figure 6 legend), 62% for trembling aspen, 41% for red maple and 31% for white ash [16]. To allow proper comparison of hybrid poplar with native species (panel b), the 3.0–7.9 cm diameter class was removed from the hybrid poplar data set (see Section 2.5.1).

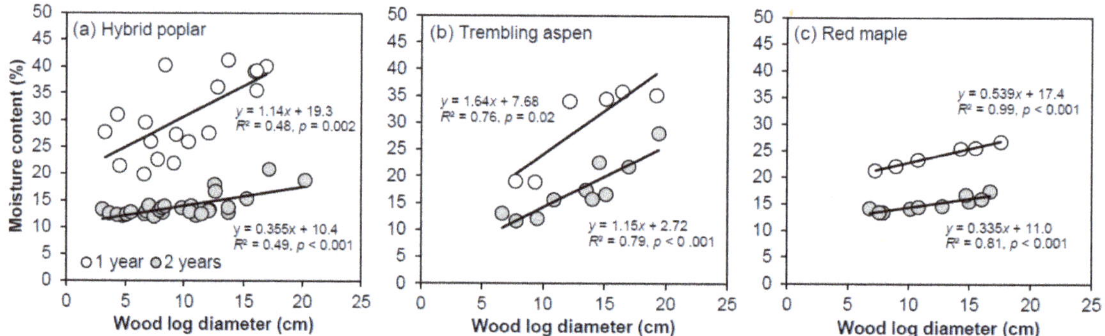

**Figure 4.** Linear relationships between the diameter of unsplit firewood logs (110 cm of length) and their moisture content after one year and two years of drying for: (**a**) hybrid poplar (genotype DN × M-915508); (**b**) trembling aspen; and (**c**) red maple. No significant relationship was found for white ash. For the one- and two-year relationships, respectively, $n = 18$ and $n = 30$ for poplar, and $n = 6$ and $n = 10$ for aspen and maple.

Chemical characteristics of freshly harvested and seasoned hybrid poplar firewood was also significantly affected by the diameter class (Table 2). A significant decline in N, S, P and Mg concentrations was observed, with increasing diameter classes for freshly harvested logs and for the logs seasoned outdoors for one year (Table 2). Furthermore, the regression analyses showed significant non-linear trends (i.e., negative logarithmic relationships) between the above-mentioned variables (Figure 5). LHV showed a slight linear decline with increasing log diameter, but only for freshly harvested logs, as no significant effect of diameter was observed on LHV after one year of seasoning (Table 2, Figure 5a). Ca concentration increased with the diameter of freshly harvested logs, but not for seasoned logs (Table 2, Figure 5d). There was also a significant drying treatment effect for LHV ($p < 0.001$), with seasoned logs having slightly higher heating value than freshly harvested logs (18.61 MJ/kg vs. 19.13 MJ/kg). Such a trend was also observed for the C and Ca concentrations (Table 2).

**Table 2.** Diameter class and drying treatment effects on chemical properties of unsplit hybrid poplar firewood logs (110 cm in length). S.E. = standard error of the mean. *p*-values in bold denote a significant effect.

| | $LHV_{dry}$ (MJ/kg) | | C (g/kg) | | N (g/kg) | | S (g/kg) | | P (g/kg) | | K (g/kg) | | Ca (g/kg) | | Mg (g/kg) | |
|---|---|---|---|---|---|---|---|---|---|---|---|---|---|---|---|---|
| Diameter Class | Fresh | 1 year | Fresh | 1 year | Fresh | 1 year | Fresh | 1 year | Fresh | 1 year | Fresh | 1 year | Fresh | 1 year | Fresh | 1 year |
| 3.0–7.9 cm | 18.75 | 19.19 | 507.1 | 519.3 | 3.50 | 3.25 | 0.274 | 0.258 | 0.544 | 0.544 | 1.80 | 1.85 | 2.48 | 3.20 | 0.547 | 0.610 |
| 8.0–12.9 cm | 18.62 | 19.19 | 512.5 | 521.3 | 2.46 | 2.15 | 0.187 | 0.173 | 0.398 | 0.358 | 1.85 | 1.75 | 2.67 | 3.13 | 0.481 | 0.486 |
| $\geq$13 cm | 18.47 | 19.00 | 514.8 | 515.0 | 1.85 | 1.62 | 0.166 | 0.132 | 0.316 | 0.263 | 2.05 | 1.96 | 3.14 | 3.40 | 0.444 | 0.466 |
| S.E. Class | 0.05 | 0.15 | 3.2 | 1.6 | 0.13 | 0.25 | 0.009 | 0.017 | 0.015 | 0.049 | 0.10 | 0.16 | 0.12 | 0.17 | 0.023 | 0.025 |
| *p*-value Class | **0.002** | 0.60 | 0.24 | **0.04** | **<0.001** | **0.001** | **<0.001** | **<0.001** | **<0.001** | **0.003** | 0.21 | 0.69 | **0.005** | 0.54 | **0.02** | **0.002** |
| Mean | 18.61 | 19.13 | 511.5 | 518.5 | 2.60 | 2.34 | 0.209 | 0.188 | 0.419 | 0.388 | 1.90 | 1.85 | 2.76 | 3.24 | 0.490 | 0.521 |
| S.E. Drying | 0.06 | | 1.4 | | 0.12 | | 0.008 | | 0.021 | | 0.08 | | 0.09 | | 0.014 | |
| *p*-value Drying | **<0.001** | | **0.002** | | 0.12 | | **0.02** | | 0.07 | | 0.68 | | **<0.001** | | 0.13 | |
| *p*-value Class × Drying | 0.82 | | 0.06 | | 0.98 | | 0.46 | | 0.61 | | 0.82 | | 0.32 | | 0.47 | |

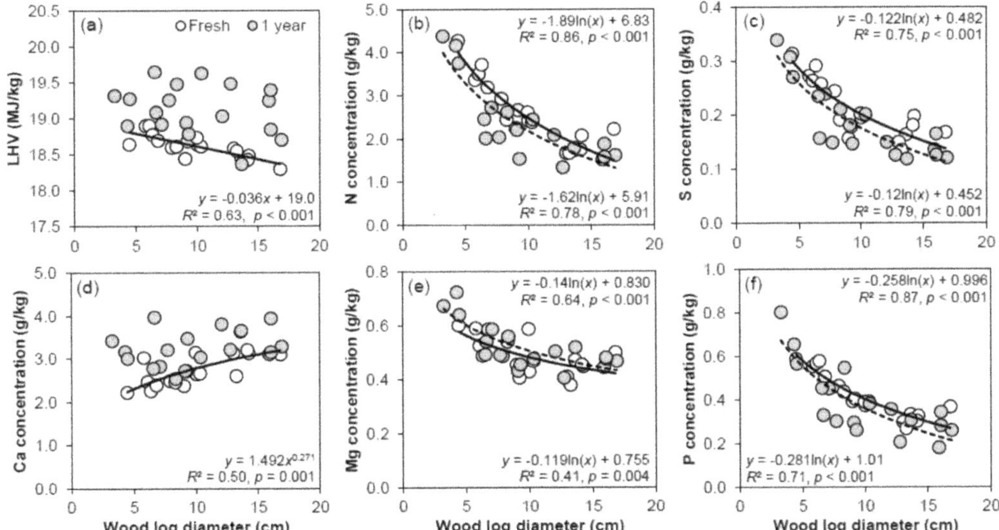



**Figure 5.** Relationships between hybrid poplar firewood log diameter (110 cm in length) and their (**a**) lower heating value ($LHV_{dry}$) or (**b–f**) elemental concentration at harvest (Fresh, solid line) and after one year of drying outdoors (1 year, dashed line). *n* = 18 for each relationship.

Results in Table 3 showed that LHV, C and N concentrations were similar for hybrid poplar, trembling aspen, red maple and white ash firewood, after one year of outdoor seasoning. White ash was the species with the highest S concentration, while P, K, Ca and Mg were more concentrated in hybrid poplar wood. P, K and Mg concentrations of hybrid poplar firewood were almost two times higher than what was observed for trembling aspen (Table 3).

**Table 3.** Diameter and chemical properties ($\pm$S.E. of the mean) of unsplit firewood logs (110 cm in length). Means not connected with the same letter are significantly different ($\alpha$ = 0.05, Tukey's HSD test). *n* = 12 for hybrid poplar and *n* = 6 for the other species. *p*-values in bold denote a significant effect.

| Tree Species | Diameter (cm) | $LHV_{dry}$ (MJ/kg) | C (g/kg) | N (g/kg) | S (g/kg) | P (g/kg) | K (g/kg) | Ca (g/kg) | Mg (g/kg) |
|---|---|---|---|---|---|---|---|---|---|
| H. poplar | 12.3 ± 1.1 | 19.09 ± 0.08 | 518.1 ± 1.2 | 1.89 ± 0.12 | 0.153 ± 0.008 b | 0.310 ± 0.020 a | 1.85 ± 0.08 a | 3.27 ± 0.14 a | 0.476 ± 0.016 a |
| T. aspen | 13.3 ± 1.6 | 18.92 ± 0.12 | 517.5 ± 1.6 | 1.65 ± 0.18 | 0.133 ± 0.012 b | 0.167 ± 0.029 b | 0.94 ± 0.12 b | 3.14 ± 0.19 a | 0.278 ± 0.023 b |
| R. maple | 12.4 ± 1.6 | 18.75 ± 0.12 | 520.3 ± 1.6 | 1.84 ± 0.18 | 0.139 ± 0.012 b | 0.247 ± 0.029 ab | 1.08 ± 0.12 b | 1.82 ± 0.19 b | 0.193 ± 0.023 b |
| W. ash | 12.9 ± 1.6 | 18.88 ± 0.12 | 520.7 ± 1.6 | 1.94 ± 0.18 | 0.199 ± 0.012 a | 0.180 ± 0.029 b | 1.73 ± 0.12 a | 2.63 ± 0.19 a | 0.252 ± 0.023 b |
| *p*-value | 0.96 | 0.12 | 0.40 | 0.64 | **0.002** | **<0.001** | **<0.001** | **<0.001** | **<0.001** |

### 3.2. Trial B

The dry weight of selected logs for moisture content subsampling did not differ significantly between the cover treatments ($p = 0.74$), between the genotypes ($p = 0.37$) and between the genotype/cover treatment combinations ($p = 0.89$). For the short logs (40 cm of length) that were split on 22 May 2019, there was a significant cover treatment effect ($p < 0.001$) on moisture content measured five months later on 22 October 2019 (Figure 6a). Overall, covered logs had lower moisture content (17.3%) than uncovered logs (20.7%). There was a trend towards slightly higher moisture content ($p = 0.11$) for genotype D × N-3570 than for genotype M × B-915311 across the cover treatments. This trend was related to the presence of two extreme values for genotype D × N-3570 in the uncovered treatment (Figure 6b). When those outliers were excluded from the ANOVA, the genotype and the interaction effects were far from significance ($p = 0.51$ and $p = 0.58$, respectively), while the cover treatment effect remained highly significant ($p < 0.001$).

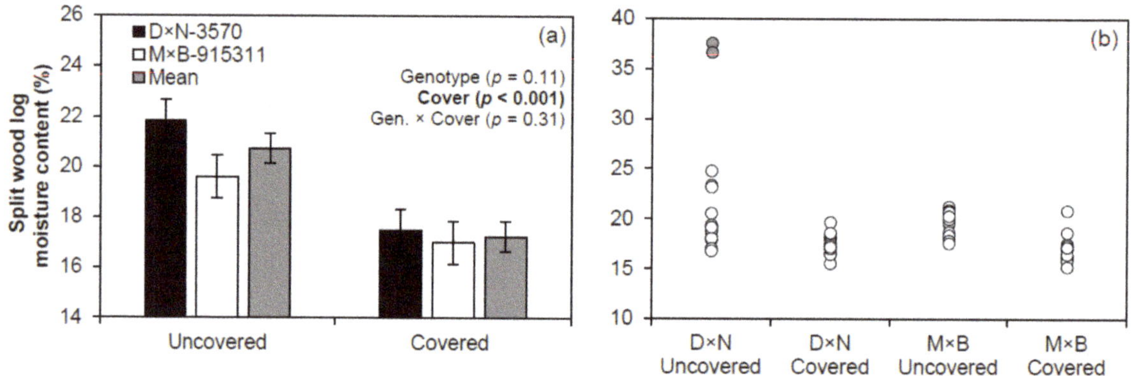


**Figure 6.** (**a**) Genotype and cover treatment effects on the final moisture content of split hybrid poplar firewood logs (40 cm in length). Vertical bars represent the standard error of the mean; (**b**) Scatter plot of observations for moisture content of firewood logs for the different genotype/treatment combinations, with two outliers being represented by grey circles. For each genotype/treatment combination, 16 logs were sampled on 22 October 2019. Logs were split on 22 May 2019. Initial moisture content of harvested trees was statistically different between the genotypes ($p < 0.001$): $57.3 \pm 0.3\%$ for genotype D × N-3570 ($n = 20$) and $58.9 \pm 0.3\%$ for genotype M × B-915311 ($n = 19$).

## 4. Discussion

This study, conducted in a warm-summer humid continental climate, showed that hybrid poplar firewood harvested in the middle of the fall can reach satisfactory moisture content ($\pm 20\%$) for the next heating season, when it is split and processed in small logs (40 cm-long) in the spring. Covering the top of firewood cords with metal sheeting from the moment they were split (late May) significantly decreased the overall moisture content, from 20.7% down to 17.3% (Figure 4). Covering split logs was also important to reach a more homogeneous moisture content for genotype D × N-3570 (Figure 6b). In the uncovered treatment, we observed moisture content of 36.5% and 37.5% for two logs of genotype D × N-3570 that had a thick and furrowed bark. Such bark traits likely contributed to rewetting when logs are stacked bark side up, as we did. Conversely, the maximal moisture content observed for uncovered logs of genotype M × B-915311, which has a smooth bark, was 21.3%. Thus, if covering or roofing hybrid poplar logs just after splitting is not possible, it would be recommended to use smooth-barked genotypes to allow better percolation of rain water through the wood cords. Stacking logs bark side down is also an option that needs to be investigated with uncovered firewood from furrowed-barked genotypes. However, uncovered split logs piled bark side down generally regain more moisture during rainfall events compared to logs stacked bark side up [37]. In all cases, once logs are dry enough, split firewood should be stacked under a well-aerated shelter no later than October to prevent rewetting [18], which is a major issue for split poplar logs [38].

For the long unsplit logs (110 cm in length), a two-year cycle was necessary to produce firewood with satisfactory moisture content before the heating season (Figure 3). In our study, the second seasoning year took place in a well-aerated unheated warehouse (Table 1), which prevented wood rewetting for a whole year. One year of outdoor drying was clearly not enough to reach acceptable moisture content, even for the smaller diameter logs, although moisture content decreased substantially with log diameter (Figures 3 and 4). Such a trend was related to the fact that larger diameter logs dry more slowly [22], and tend to have a higher moisture content at harvest [39]. We also observed that moisture variation across the diameter classes was much larger after the first seasoning year than after the second year, as depicted by the slopes of regressions in Figure 4a. This reflects the convergence of wood drying curves with time, as wood log moisture content approaches the equilibrium point with ambient air [1,20].

In this case study, the abundant precipitation received during the summer of 2018 (Figure 2b) may have slowed log drying, with cumulative precipitations being negatively related to moisture content loss in unsplit hybrid poplar logs [21]. As an indication of the high moisture content of hybrid poplar logs during the first summer of seasoning, we observed abundant sprouts, which remained alive until August on the larger diameter logs (Figure 7). Conversely, two weeks after splitting and stacking the short hybrid poplar logs, no sprouting was observed, despite the fact that emerging sprouts were seen when the wood was split. This suggests high moisture content loss during the first weeks following splitting, as was also observed in other studies [20]. Moreover, important fungal colonization (mycelium and fruiting bodies) was observed on hybrid poplar logs from all diameter classes after outdoor seasoning (Figure 8), with moisture content of 30–50% optimizing the growth of wood-decay fungi [16]. By causing mass loss of up to 6% per year on long unsplit poplar logs, fungal growth can substantially reduce the volumetric energy content of unsheltered poplar fuelwood [38]. Therefore, if long unsplit poplar logs cannot be sheltered after the first seasoning year, this firewood production avenue would not be recommended. Additional studies are also needed to evaluate to which extent splitting and covering or sheltering long logs from hybrid poplar would improve firewood quality and allow feedstock production on an annual cycle, thereby reducing by half the storage space needed. An investment of approximately $ 3000 CAD would be needed to acquire a tractor-powered log splitter, which would be capable of processing logs of up to 120 cm in length [40].

**Figure 7.** (**a**) On 7 June 2018, unsplit hybrid poplar logs had abundant sprouts, which remained alive until August on the larger diameter logs; and (**b**) on 4 June 2019, no sprouting was observed on short split logs that were processed on 22 May 2019.

**Figure 8.** (**a**) Fruiting bodies; and (**b**) mycelium observed following fungal colonization of hybrid poplar logs (110 cm length) after one year of outdoor seasoning.

Among the four species tested, unsplit logs from hybrid poplar had the lowest moisture content after two years, despite the opposite trend being observed after one year (Figure 3b). Furthermore, although moisture content was not significantly different between hybrid poplar, trembling aspen and red maple, all these species had significantly lower moisture content than white ash after two years. Besides, at the end of the trial, log diameter was found to be a strong predictor of moisture content for all species, except for white ash (Figure 4). Thus, if splitting is not feasible with long logs, the use of smaller diameter pieces from species with a lower wood density (i.e., soft maples and poplars) would lead to greater net energy value on a mass basis, if we assume that LHV does not vary with log size following seasoning. Our results also showed that after two years, moisture content variations across the diameter range sampled were narrower for hybrid poplar than for trembling aspen, the larger logs of which had a moisture content of up to 28% (Figure 4b). This would be consistent with the fact that heartwood dries much more slowly than the sapwood in poplars [41], and that the proportion of heartwood in 7-year-old hybrid poplars is smaller than in slower growing forest-grown aspens of similar diameter.

After one year of seasoning, N concentration of hybrid poplar firewood showed a large decrease, with increasing log diameter (Table 2, Figure 5). Therefore, producing larger diameter logs in more widely spaced plantations could be a solution to minimize N oxides emissions by wood burning. On the other hand, the production of smaller diameter logs would provide feedstock with lower moisture content (Figure 3a) and higher net heating value. It would also allow more N and P to be exported from agricultural buffers (Table 2), which is of great importance to maintain the nutrient retention effectiveness of buffer strips in the long term [42]. Sulphur concentration in poplar firewood also largely declined with log diameter (Table 2, Figure 5). However, S concentrations reported in this study are well below the threshold values above which biomass S causes concerns for air quality (>2 g/kg) or appliance corrosion (>1 g/kg) [25]. Additionally, after one year of seasoning, LHV and N concentration of hybrid poplar firewood was in the range of values observed for trembling aspen, red maple and white ash (Table 3). Conversely, P, K, Ca and Mg concentrations were the highest for hybrid poplar firewood, suggesting that its combustion may produce more ashes than the native species studied. Still, these conclusions are based on comparisons with a single hybrid poplar genotype, and do not reflect the wide genotypic variability in stem nutrient concentrations of hybrid poplars [9,28].

The LHV of freshly harvested hybrid poplar wood observed for genotype DN × M-915508 across the three diameter classes (18.61 MJ/kg) was similar to values reported on a whole-stem basis for 8-year-old trees of genotype D × N-3570 (18.65 MJ/kg) and genotype M × B-915311 (18.69 MJ/kg) grown at the same site [9]. This contrasts with the range of higher heating values observed between several genotypes in a short rotation coppice in Chile (17.7–20.8 MJ/kg) [43]. Interestingly, we observed a slight decline in the LHV of freshly harvested hybrid poplar logs with increasing diameter, a trend no longer significant after seasoning (Table 2, Figure 5a). At harvest, hybrid poplar branches also had slightly higher LHV compared to stem wood with bark [9]. More elevated heating values in smaller diameter woody tissues may be related to their higher content in energy-rich extractives, as observed in *Liriodendron tulipifera* L. [44]. Overall, the heating value of hybrid poplar logs also increased by 2.8% following the one-year seasoning period outdoors. However, we did not measure if this increasing trend in LHV occurred after the second seasoning year. Following seasoning, both increases and decreases in energy values have been reported for different woody biomass fuels [16,45]. The volatilization and oxidation of wood extractives reduce the heating value, while the opposite trend occurs following the preferential biological degradation of carbohydrate polymers, as this process increases the lignin concentration of woody biomass [16]. As in other studies [24,46], we observed a significant increase (of 1.4%) in the C concentration of poplar fuelwood following seasoning, which is consistent with the increase in the LHV observed (Table 2).

At harvest, Ca concentration slightly increased with poplar log diameter, which contrasted with the trend observed for other macronutrients (Figure 5). In another study, Ca concentration in young hybrid aspens (*P. tremula* × *P. tremuloides*) was found to be fairly constant along the stem [47]. Surprisingly, the diameter effect on Ca concentration was no longer significant following the seasoning period, and seasoned hybrid poplar firewood had a higher Ca concentration (Table 2). Mass loss during seasoning could have led to such an increase in Ca concentration, although the opposite trend was observed for S concentration. Different and interacting factors likely contribute to the changes observed in element concentration following seasoning, including mass loss, nutrient loss from sprouting, nutrient leaching induced by precipitation, fungal growth and endophyte activity.

Finally, we may have slightly underestimated the moisture content of firewood as moisture content was measured after oven-drying samples at 95 °C, and not at 105 °C, which is preferable [48]. Moreover, we sampled very few individuals for each native woodlot species, as the focus of this study was hybrid poplar. Thus, conclusions regarding hybrid poplar comparisons with other species are limited by this low sampling effort. Some operational aspects could also be improved. In both trials, we harvested wood in early November. However, because little drying occurs during the fall and winter months, harvesting firewood in late winter or early spring is recommended [18]. Softwood species like Norway spruce (*Picea abies* K.) can even be harvested in early June and be ready for the heating season, when small split logs are properly sheltered. Harvesting poplars in late spring with their foliage may also accelerate stem moisture loss through transpirational drying [16], while maximizing nutrient exportation from agricultural buffers [49]. On the other hand, if heavy machinery is used, a frozen-ground harvest would be preferable to maintain the integrity of buffer soils and reduce sediment inputs to streams [50].

## 5. Conclusions

This study, done in a warm-summer humid continental climate, showed that hybrid poplar firewood of satisfactory quality can be produced on a yearly cycle for short split logs and on a biannual cycle for long unsplit logs. Covering split logs was important to lower moisture content and reduce its variability. For the long unsplit poplar logs, moisture content and element concentrations showed opposite trends in relation to log diameter, with larger logs containing fewer mineral elements, but higher moisture content and thus lower net heating value. Based on wood elemental composition, we found little evidence that the use of hybrid poplar firewood would be more problematic than the use of firewood produced with native tree species, except that it may generate more ashes given its high P and base cation content. Given the rapid combustion of the low-density poplar wood, this feedstock would be especially suited for heating in the fall and in the spring, when the heat demand is lower than during cold winter months. The large-scale implementation of high-yielding bioenergy buffers could improve ecosystem services provision on farmland (C sequestration, non-point source pollution control, hydrological regulation, increase forest habitat connectivity), while reducing the firewood harvesting pressure in natural forest habitats. This could create opportunities for forest habitat conservation in agriculture-dominated landscapes.

**Author Contributions:** Conceptualization, J.F., B.T. and D.G.; Methodology, J.F. and B.T.; Formal Analysis, J.F.; Investigation, J.F. and B.T.; Data Curation, J.F.; Writing—Original Draft Preparation, J.F.; Writing—Review & Editing, B.T., D.G. and F.L.; Visualization, J.F.; Project Administration, F.L. and B.T.; Funding Acquisition, B.T., D.G., F.L. and J.F. All authors have read and agreed to the published version of the manuscript.

**Funding:** This research and the APC were funded by Agriculture and Agri-food Canada, Canada (AGGP2-001) (Agricultural Greenhouse Gas Program).

**Data Availability Statement:** The data presented in this study are available on request from the corresponding author.

**Acknowledgments:** We are grateful to the Benedictine Community of St-Benoît-du-Lac for having allowed this project to be carried out on their property. We thank Agriculture and Agri-food Canada, Canada (AGGP2-001) (Agricultural Greenhouse Gas Program), Natural Resources Canada, the Ministère des Forêts, de la Faune et des Parcs du Québec (Chantier sur la Forêt Feuillue program) and CRÉ-Estrie (Volet 2 program), as well as Tree Canada, for the funding received. R. Pouliot and O. Srdjan (ISFORT, UQO, Ripon, QC, Canada) and the staff from the CTRI (Rouyn-Noranda, QC, Canada) are thanked for the chemical analyses of wood samples. Finally, we wish to thank tree planters and field assistants (O. Dubuc, J.D. Careau, L. Godbout, Brother L. Lamontagne, Y. Lauzière, J. Lemelin, M.A. Pétrin), as well as the Berthier Nursery (MFFP) for the planting stocks.

**Conflicts of Interest:** The authors declare no conflict of interest.

## References


1. Manzone, M. Energy and moisture losses during poplar and black locust logwood storage. *Fuel Process. Technol.* **2015**, *138*, 194–201. [CrossRef]
2. Côté, M.-A.; Gilbert, D.; Nadeau, S. Characterizing the profiles, motivations and behaviour of Quebec's forest owners. *For. Policy Econ.* **2015**, *59*, 83–90. [CrossRef]
3. Doornbos, J.; Richardson, J.; van Oosten, C. Activities related to poplar and willow cultivation and utilization in Canada. In *Canadian Report to the 25th Session of the International Poplar Commission*; Poplar and Willow Council of Canada: Edmonton, AB, Canada, 2016.
4. Butler, B.J.; Hewes, J.H.; Dickinson, B.J.; Andrejczyk, K.; Butler, S.M.; Markowski-Lindsay, M. Family forest ownerships of the united states, 2013: Findings from the USDA Forest Service's national woodland owner survey. *J. For.* **2016**, *114*, 638–647. [CrossRef]
5. Jobin, B.; Latendresse, C.; Baril, A.; Maisonneuve, C.; Boutin, C.; Côté, D. A half-century analysis of landscape dynamics in southern Québec, Canada. *Environ. Monit. Assess.* **2014**, *186*, 2215–2229. [CrossRef] [PubMed]
6. Kaushal, S.S.; Mayer, P.M.; Vidon, P.G.; Smith, R.M.; Pennino, M.J.; Newcomer, T.A.; Duan, S.; Welty, C.; Belt, K.T. Land use and climate variability amplify carbon, nutrient, and contaminant pulses: A review with management implications. *JAWRA* **2014**, *50*, 585–614. [CrossRef]
7. Schoeneberger, M.M.; Bentrup, G.; Patel-Weynand, T. *Agroforestry: Enhancing Resiliency in U.S. Agricultural Landscapes under Changing Conditions*; U.S. Department of Agriculture, Forest Service: Washington, DC, USA, 2017; p. 228.
8. Fortier, J.; Truax, B.; Gagnon, D.; Lambert, F. Soil nutrient availability and microclimate are influenced more by genotype than by planting stock type in hybrid poplar bioenergy buffers on farmland. *Ecol. Engin.* **2020**, *157*, 105995. [CrossRef]
9. Truax, B.; Fortier, J.; Gagnon, D.; Lambert, F. Long-term effects of white-tailed deer overabundance, hybrid poplar genotype and planting stock type on tree growth and ecosystem services provision in bioenergy buffers. *For. Ecol. Manag.* **2021**, *480*, 118673. [CrossRef]
10. Ferrarini, A.; Serra, P.; Almagro, M.; Trevisan, M.; Amaducci, S. Multiple ecosystem services provision and biomass logistics management in bioenergy buffers: A state-of-the-art review. *Renew. Sustain. Energy Rev.* **2017**, *73*, 277–290. [CrossRef]
11. Fortier, J.; Truax, B.; Gagnon, D.; Lambert, F. Potential for hybrid poplar riparian buffers to provide ecosystem services in three watersheds with contrasting agricultural land use. *Forests* **2016**, *7*, 37. [CrossRef]
12. Schultz, R.C.; Isenhart, T.M.; Simpkins, W.W.; Colletti, J.P. Riparian forest buffers in agroecosystems—lessons learned from the Bear Creek Watershed, central Iowa, USA. *Agrofor. Syst.* **2004**, *61–62*, 35–50.
13. Truax, B.; Gagnon, D.; Lambert, F.; Fortier, J. Multiple-use zoning model for private forest owners in agricultural landscapes: A case study. *Forests* **2015**, *6*, 3614–3664. [CrossRef]
14. Fortier, J.; Truax, B.; Gagnon, D.; Lambert, F. Mature hybrid poplar riparian buffers along farm streams produce high yields in response to soil fertility assessed using three methods. *Sustainability* **2013**, *5*, 1893–1916. [CrossRef]
15. Truax, B.; Gagnon, D.; Fortier, J.; Lambert, F. Biomass and volume yield in mature hybrid poplar plantations on temperate abandoned farmland. *Forests* **2014**, *5*, 3107–3130. [CrossRef]
16. Krigstin, S.; Wetzel, S. A review of mechanisms responsible for changes to stored woody biomass fuels. *Fuel* **2016**, *175*, 75–86. [CrossRef]

17. Dzurenda, L.; Banski, A. The effect of firewood moisture content on the atmospheric thermal load by flue gases emitted by a boiler. *Sustainability* **2019**, *11*, 284. [CrossRef]

18. Nord-Larsen, T.; Bergstedt, A.; Farver, O.; Heding, N. Drying of firewood—the effect of harvesting time, tree species and shelter of stacked wood. *Biomass Bioenergy* **2011**, *35*, 2993–2998. [CrossRef]

19. United State Environmental Protection Agency. Burn Wise. Available online: https://www.epa.gov/burnwise (accessed on 9 November 2020).

20. Manzone, M. Performance evaluation of different techniques for firewood storage in Southern Europe. *Biomass Bioenergy* **2018**, *119*, 22–30. [CrossRef]

21. Kim, D.-W.; Murphy, G. Forecasting air-drying rates of small Douglas-fir and hybrid poplar stacked logs in Oregon, USA. *Int. J. For. Eng.* **2013**, *24*, 137–147. [CrossRef]

22. Visser, R.; Berkett, H.; Spinelli, R. Determining the effect of storage conditions on the natural drying of radiata pine logs for energy use. *N. Z. J. For. Sci.* **2014**, *44*, 3. [CrossRef]

23. Röser, D.; Mola-Yudego, B.; Sikanen, L.; Prinz, R.; Gritten, D.; Emer, B.; Väätäinen, K.; Erkkilä, A. Natural drying treatments during seasonal storage of wood for bioenergy in different European locations. *Biomass Bioenergy* **2011**, *35*, 4238–4247. [CrossRef]

24. Lee, J.S.; Sokhansanj, S.; Lau, A.K.; Lim, C.J.; Bi, X.T.; Basset, V.; Yazdanpanah, F.; Melin, S.O. The effects of storage on the net calorific value of wood pellets. *Can. Biosyst. Eng.* **2015**, *57*, 5–12. [CrossRef]

25. Obernberger, I.; Brunner, T.; Bärnthaler, G. Chemical properties of solid biofuels—Significance and impact. *Biomass Bioenergy* **2006**, *30*, 973–982. [CrossRef]

26. Ravishankara, A.R.; Daniel, J.S.; Portmann, R.W. Nitrous oxide ($N_2O$): The dominant ozone-depleting substance emitted in the 21st century. *Science* **2009**, *326*, 123. [CrossRef] [PubMed]

27. Heil, J.; Vereecken, H.; Brüggemann, N. A review of chemical reactions of nitrification intermediates and their role in nitrogen cycling and nitrogen trace gas formation in soil. *Eur. J. Soil Sc.* **2016**, *67*, 23–39.

28. Rodríguez-Soalleiro, R.; Eimil-Fraga, C.; Gómez-García, E.; García-Villabrille, J.D.; Rojo-Alboreca, A.; Muñoz, F.; Oliveira, N.; Sixto, H.; Pérez-Cruzado, C. Exploring the factors affecting carbon and nutrient concentrations in tree biomass components in natural forests, forest plantations and short rotation forestry. *For. Ecosyst.* **2018**, *5*, 35. [CrossRef]

29. Augusto, L.; Meredieu, C.; Bert, D.; Trichet, P.; Porté, A.; Bosc, A.; Lagane, F.; Loustau, D.; Pellerin, S.; Danjon, F.; et al. Improving models of forest nutrient export with equations that predict the nutrient concentration of tree compartments. *Ann. For. Sci.* **2008**, *65*, 808.

30. Robitaille, A.; Saucier, J.-P. *Paysages Régionaux du Québec Méridional*; Les Publications du Québec: Ste-Foy, QC, Canada, 1998; p. 213.

31. Government of Canada. Station Results—1981–2010. Climate Normals and Averages. Available online: http://climate.weather.gc.ca/climate_normals/station_select_1981_2010_e.html?searchType=stnProv&lstProvince=QC (accessed on 11 May 2020).

32. Government of Canada. Historical Climate Data. Available online: https://climate.weather.gc.ca/index_e.html (accessed on 11 May 2020).

33. Manikandan, S. Data transformation. *J. Pharm. Pharm.* **2010**, *1*, 126–127. [CrossRef]

34. Petersen, R.G. Use and misuse of multiple comparison procedures. *Agron. J.* **1977**, *69*, 205–208. [CrossRef]

35. Royston, P. An extension of Shapiro and Wilk's W test for normality to large samples. *Appl. Stat.* **1982**, *31*, 115–124. [CrossRef]

36. Chambers, J.M.; Cleveland, W.S.; Kleiner, B.; Tukey, P.A. *Graphical Methods for Data Analysis*; Wadsworth International Group: Belmont, CA, USA, 1983.

37. Spartz, J.T. A Hot Debate: Stacking Firewood Bark Up or Bark Down? Lab Note. News from the Forest Products Laboratory. USDA Forests Service. Available online: https://www.fpl.fs.fed.us/labnotes/?p=436#:~{}:text=If%20split%20wood%20is%20stored,and%20accelerate%20decay%2C%20says%20Knaebe (accessed on 19 January 2021).

38. Erber, G.; Huber, C.; Stampfer, K. To split or not to split: Feasibility of pre-storage splitting of large poplar (Populus spp. L.) fuelwood logs. *Fuel* **2018**, *220*, 817–825. [CrossRef]

39. Fortier, J.; Truax, B.; Gagnon, D.; Lambert, F. Allometric equations for estimating compartment biomass and stem volume in mature hybrid poplars: General or site-specific? *Forests* **2017**, *8*, 309. [CrossRef]

40. Wallenstein Equipment Inc. Splitter Model WX330 Horizontal Tractor. Available online: https://www.wallensteinequipment.com/ca/en/model/wx330 (accessed on 15 January 2021).

41. Balatinecz, J.; Mertens, P.; de Boever, L.; Yukun, H.; Jin, J.; van Acker, J. 10 Properties, processing and utilization. In *Poplar and Willows: Tree for the Society and the Environment*; Isebrands, J.G., Richardson, J., Eds.; CAB Interantional and FAO: Rome, Italy, 2014; pp. 527–561.

42. Stutter, M.; Kronvang, B.; Ó hUallacháin, D.; Rozemeijer, J. Current insights into the effectiveness of riparian management, attainment of multiple benefits, and potential technical enhancements. *J. Environ. Qual.* **2019**, *48*, 236–247. [CrossRef] [PubMed]

43. Carmona, R.; Nuñez, T.; Alonso, M.F. Biomass yield and quality of an energy dedicated crop of poplar (Populus spp.) clones in the Mediterranean zone of Chile. *Biomass Bioenergy* **2015**, *74*, 96–102. [CrossRef]

44. Myeong, S.-J.; Han, S.-H.; Shin, S.-J. Analysis of chemical compositions and energy contents of different parts of yellow poplar for development of bioenergy technology. *J. Korean For. Soc.* **2010**, *99*, 706–710.

45. Therasme, O.; Eisenbies, M.H.; Volk, T.A. Overhead protection increases fuel quality and natural drying of leaf-on woody biomass storage piles. *Forests* **2019**, *10*, 390. [CrossRef]

46. Nurmi, J. The storage of logging residue for fuel. *Biomass Bioenergy* **1999**, *17*, 41–47. [CrossRef]
47. Rytter, L. Nutrient content in stems of hybrid aspen as affected by tree age and tree size, and nutrient removal with harvest. *Biomass Bioenergy* **2002**, *23*, 13–25. [CrossRef]
48. Matthews, S. Effect of drying temperature on fuel moisture content measurements. *Int. J. Wildland Fire* **2010**, *19*, 800–802. [CrossRef]
49. Fortier, J.; Gagnon, D.; Truax, B.; Lambert, F. Nutrient accumulation and carbon sequestration in 6 year-old hybrid poplars in multiclonal agricultural riparian buffer strips. *Agric. Ecosyst. Environ.* **2010**, *137*, 276–287. [CrossRef]
50. Rashin, E.B.; Clishe, C.J.; Loch, A.T.; Bell, J.M. Effectiveness of timber harvest practices for controlling sediment related water quality impacts. *JAWRA* **2006**, *42*, 1307–1327. [CrossRef]

 *forests*

*Communication*

# Bacterial Canker Disease on *Populus* × *euramericana* Caused by *Lonsdalea populi* in Serbia

Milica Zlatković [1,*], Imola Tenorio-Baigorria [2], Tamás Lakatos [3], Tímea Tóth [3], András Koltay [2], Predrag Pap [1], Miroslav Marković [1] and Saša Orlović [1]

1 Institute of Lowland Forestry and Environment (ILFE), University of Novi Sad, Antona Čehova 13d, 21000 Novi Sad, Serbia; pedjapap@uns.ac.rs (P.P.); miroslavm@uns.ac.rs (M.M.); sasao@uns.ac.rs (S.O.)
2 Department of Forest Protection, NARIC Forest Research Institute (ERTI), Hegyalja utca 18, 3232 Mátrafüred, Hungary; tenorio.imola@gmail.com (I.T.-B.); koltay.andras@erti.naik.hu (A.K.)
3 NARIC Fruitculture Research Institute, Vadas-tag 2, 4244 Újfehértó, Hungary; lakatos.tamas@fruitresearch.naik.hu (T.L.); toth.timea@fruitresearch.naik.hu (T.T.)
* Correspondence: milica.zlatkovic@uns.ac.rs; Tel.: +381-21-540-383

Received: 2 September 2020; Accepted: 2 October 2020; Published: 9 October 2020

**Abstract:** *Populus* × *euramericana* (Dode) Guinier clone (cl.) "I-214" is a fast-growing interspecific hybrid between Eastern cottonwood (*P. deltoides* Bartr. ex Marsh) and European black poplar (*Populus nigra* L.). *Populus* × *euramericana* was introduced into Serbia in the 1950s and has become one of the most widely grown poplar species. In September 2019, cankers were observed on stems and branches of *P.* × *euramericana* cl. "I-214" trees in a two-year-old poplar plantation in the province of Vojvodina, Serbia. The canker tissue was soft and watery, and a colorless fluid that smelled rotten flowed from the cracks in the bark, suggesting possible bacterial disease. After two weeks, diseased trees experienced crown die-back and oozing of foamy, odorous exudates and this study aimed to identify the causal agent of the disease. Canker margins and exudates were collected from 20 symptomatic trees. The associated bacterium was isolated and identified using biochemical characteristics, phylogenetic analyses based on 16S rRNA gene sequences, and multilocus sequence analyses (MLSA) based on partial sequencing of three housekeeping genes (*gyrB*, *infB*, and *atpD*). The pathogen was identified as *Lonsdalea populi*. Pathogenicity tests were conducted on rooted cuttings of *P.* × *euramericana* cl. "I-214" in an environmental test chamber and demonstrated that the isolated bacterial strain was able to reproduce symptoms of softened, water-soaked cankers and exudation. To the best of our knowledge, this is the first report of *L. populi* causing bacterial canker disease on *P.* × *euramericana* cl. "I-214" in Serbia and in southeastern Europe (SEE). It is also the first report of a bacterial disease on hybrid poplars, including *P.* × *euramericana* in this country and in SEE. If the disease spreads into new areas, selection for *L. populi* resistance may need to be integrated into future poplar breeding programs.

**Keywords:** *Populus* × *euramericana*; *Lonsdalea populi*; canker diseases; poplar diseases; bacterial canker of poplars; die-back of poplars; MLSA

## 1. Introduction

Canadian poplar (*Populus* × *euramericana* (Dode) Guinier, syn. *Populus* × *canadensis* Moench) is a fast-growing interspecific hybrid between North American Eastern cottonwood (*Populus deltoides* Bartr. ex Marsh ♀) and European black poplar (*Populus nigra* L. ♂). It is an important tree species in many European countries. *Populus* × *euramericana* is characterized by rapid growth rates, ease of clonal propagation, coppice regeneration, high biomass production and carbon sequestration, potential for phytoremediation, and suitability for multiple industrial uses, e.g., sawn timber, veneers, and fuelwood [1–3].

In Serbia, *P. × euramericana* is the most widely grown poplar species [4]. It is cultivated on floodplains and along the riverbanks of the major Serbian lowland rivers, i.e., the Danube, Sava, Tisa, Tamiš, and Morava on hydromorphic soil types, including fluvisol, humofluvisol and humogley [5]. Although several clones and cultivars of *P. × euramericana* are used in Serbia, *P. × euramericana* clone (cl.) "I-214" is the most common, most productive, and the most economically important poplar clone in the country [4,6].

Several fungal diseases are known to affect *P. × euramericana* cl. "I-214" in Europe. These include *Dothichiza* canker caused by *Dothichiza populea*, Sacc. et Briard, *Marssonina* leaf spot caused by *Drepanopeziza brunnea* (Ellis & Everh.) Rossman & W.C. Allen, *Cytospora* canker caused by *Cytospora chrysosperma* (Pers.) Fr., *Botryosphaeria* canker caused by *Botryosphaeria dothidea* (Moug. ex Fr.) Ces. et De Not., and *Melampsora* leaf rust caused by *Melampsora* spp. Moreover, *Xanthomonas populi* (ex-Ridé 1958) Ridé and Ridé 1992 and *Lonsdalea populi* (Tóth et al. 2013) Li et al. 2017 have been reported as causal agents of bacterial canker disease on *P. × euramericana* in poplar plantations [7–9].

*Populus × euramericana* cl. "I-214" was introduced into Serbia in the 1950s. At the time of introduction this clone was shown to be highly productive and resistant to various diseases, including spring defoliation caused by *Venturia populina* (Vuill.) Fabric., *Dothichiza* canker, leaf curl caused by *Taphrina populina* Fr. (Fr.), *Melampsora* leaf rust and mosaic virus disease caused by poplar mosaic virus (PMV) [4,9]. However, during the past 70 years, *P. × euramericana* cl. "I-214" has gradually become susceptible to multiple leaf and stem diseases, i.e., *Marssonina* leaf spot, *Venturia* spring defoliation, *Melampsora* leaf rust, and *Dothichiza* and *Cytospora* stem canker [9].

In September 2019, symptoms of a bacterial canker disease were observed in a two-year-old *P. × euramericana* cl. "I-214" plantation in Vojvodina, Serbia. Affected trees initially exhibited longitudinal cracks in the bark of the stems and branches accompanied by oozing of a small amount of colorless or whitish sap. As the disease progressed, the cracks in the bark enlarged, the vascular tissues under the bark became necrotic, soft, and water-soaked and copious amounts of sticky and often foamy sap with a rotten smell flowed from the cracks. Once exposed to the air the sap gradually darkened, becoming reddish or brownish and causing staining of the tree bark (Figure 1a–d). In some cases, the infected bark peeled away from the sunken canker area exposing a creamy mass of whitish exudates with a fermentation odor and these cankers usually appeared on the bark surface of the lower trunk. In severe cases of the disease, cankers caused crown die-back and the diseased trees died within a few weeks (Figure 1d). These symptoms resembled those of a recently described bacterial canker disease of hybrid poplars in Hungary, Portugal, Spain, and China caused by *L. populi* [8,10–12]. The aim of this study was to identify the bacterium associated with the disease symptoms observed on *P. × euramericana* cl. "I-214" trees in Serbia. This was done using biochemical characteristics, phylogenetic analyses based on 16S ribosomal RNA (rRNA), multilocus sequence analyses (MLSA) of three housekeeping genes, i.e., part of the DNA gyrase subunit B (*gyrB*), translation initiation factor IF2 (*infB*) and ATP synthase subunit beta (*atpD*), and a pathogenicity test.

**Figure 1.** Bacterial canker disease on *Populus* × *euramericana* clone "I-214" caused by *Lonsdalea populi* in Serbia. (**a**) Necrotic bark with foamy sap flowing from the infection site. (**b**) Canker with cracked bark and exudates emerging from the infected stem. (**c**) Softening and darkening of the vascular part of the trunk in the cankered area. (**d**) Dead tree with exudates staining the bark. (**e**) Colonies of *L. populi* after 24 h of incubation at 30 °C on tryptone soya agar. (**f**) Water-soaked sunken canker formed on *P.* × *euramericana* rooted cutting one month after inoculation with *L. populi*. (**g**) Dark, soft, and watery wood beneath the bark of a canker formed on *P.* × *euramericana* rooted cutting one month after inoculation with *L. populi*. (**h**) Negative control showing absence of canker development.

## 2. Materials and Methods

### 2.1. Sample Collection, Isolation, and Biochemical Characterization

For pathogen isolation, stem and branch tissues showing symptoms of a bacterial canker and whitish creamy exudates were collected using sterile equipment from twenty symptomatic *P.* × *euramericana* cl. "I-214" trees grown in a poplar plantation near Glogonj, in Vojvodina, Serbia (N 44°59′; E 20°32′). Samples of cankers and exudates were placed in polyethylene bags and sterile 2 mL

Eppendorf tubes, respectively, and kept at 4 °C until isolation was undertaken. Small pieces (3–5 mm diameter) of woody tissue were cut from the canker margins, surface sterilized using 70% (v/v) ethanol for 1 min. followed by a solution of 10% (v/v) sodium hypochlorite for 1 min., and then rinsed in sterile distilled water. The tissue was macerated in 1 mL of sterile distilled water using a sterile mortar and pestle; the resulting suspension was transferred to 2 mL Eppendorf tubes, and shaken for 2 min. using ZX3 advanced vortex mixer (VELP Scientifica, Milan, Italy). The suspension was serially diluted (10-fold dilutions to $10^{-9}$) and 50 μL of each dilution was spread onto tryptone soya agar (TSA, Titan Biotech Ltd., New Delhi, India). Moreover, samples of exudates were diluted in the same manner and spread onto TSA. Petri dishes were incubated at 30 °C for 48 h (Heidolph incubator 1000, Heidolph co., Kelheim, Germany). In total, seven bacterial colony types were isolated. The prevalent colonies were similar in appearance to *L. populi*, i.e., white-ivory colored, slightly bluish on the underside, round, and slightly convex [8]. These colonies were purified by subculturing and subjected to Gram test which was performed using a non-staining method with a 3% KOH solution [13]. Gram negative bacterial strains were further examined with an Olympus BX53F light microscope (Olympus Co., Tokyo, Japan) at ×400 and ×1000 magnification using an Olympus SC50 digital camera and accompanying software. A Gram-negative strain with cell morphology like *L. populi* (cells 0.5–1 × 1–2 μm in size, short-rod-shaped, motile, occurring single, or aggregated in clumps, Figure S1) [8] was selected and named ILFE-LP1. The strain was stored in tryptone soya broth (TSB, Titan Biotech Ltd., New Delhi, India) containing 40% glycerol (v/v) at −80 °C, deposited in the culture collections of the Institute of Lowland Forestry and Environment (ILFE) and NARIC Fruitculture Research Institute and further used in this study to identify the bacterium and confirm its pathogenicity.

The analyses of biochemical characteristics of the bacterium were conducted using API 20E kit (Bio-Mérieux, Marcy L'Etoile, France) following the manufacturer's instructions and the test strip was incubated for 24 h. A type strain of *L. populi* (NY060, provided by the NARIC Forest Research Institute, Mátrafüred, Hungary) was used as a positive control and isolates were assessed twice.

*2.2. DNA Extraction, PCR Amplification, and Sequencing*

Total bacterial genomic DNA was isolated from cells harvested from culture grown for 24 h at 30 °C in TSB using a mericon DNA bacteria kit (Qiagen, Hilden, Germany) following the manufacturer's protocols. The DNA was quantified with a nanodrop (BioSpec-nano, Shimadzu Biotech, Kyoto, Japan), stored at −20 °C and diluted to the concentration of 20 ng/μL prior to use in PCR reactions. Partial 16S rRNA gene was amplified by PCR using the primers and conditions as published by [14] (Table 1). Three housekeeping genes, including *gyrB*, *infB*, and *atpD* were amplified using primers designed by [15] (Table 1). The conditions for PCR amplification of the housekeeping genes were as previously determined by [16]. The PCR products were separated by electrophoresis on 1.5% (w/v) agarose gels in 1 x TBE buffer, stained with Roti-GelStain (Carl Roth, Karlsruhe, Germany) and visualized under UV illumination. The size of the products was estimated using O'gene ruler 100bp DNA ladder (Thermo Fisher Scientific Inc., Bremen, Germany). The PCR products were purified using a PCR purification kit (QIAquick, Qiagen, Hilden, Germany). Sanger sequencing was performed by Mycrosynth (Balgach, Switzerland) using primers designed by [14,15] (Table 1).

*2.3. Phylogenetic Analyses*

Raw sequence data were examined and combined into a consensus sequence using BioEdit version 7.2.5 [17] and MEGA X [18]. Sequences were compared to those of the other *Lonsdalea* strains available in the GenBank database using BLAST and related sequences were downloaded and included in the analyses (Table S1). Multiple sequence alignments were obtained with MAFFT version 7 (on-line version) [19], checked manually for alignment errors in MEGA X [18] and corrected where necessary. The phylogenetic analyses were performed using Maximum Parsimony (MP) and Maximum Likelihood (ML) analyses. ML analyses were conducted for both 16S rRNA and the combined data set of three housekeeping genes, whereas MP analyses were run only for the combined data set. The

MP analyses and the partition homogeneity test (PHT) were conducted as described by [20] and they were performed in PAUP version 4.0b10 [21]. ML analyses were run using an online version of PhyML 3.0 [22] by applying smart model selection [23]. Bootstrap analysis was carried out using 1000 replicates [24]. Phylogenetic trees were visualized using MEGA X [19]. The DNA sequences of isolate ILFE-LP1 obtained in this study were deposited in GenBank (MT505705-16S, MT537174- *atpD*, MT559754-*gyrB*, and MT559753-*infB*, Table S1).

**Table 1.** Primers used in this study to amplify and sequence 16S rRNA gene and housekeeping genes (*gyrB*, *atpD* and *infB*) from ILFE-LP1 bacterial strain isolated from a *Populus* × *euramericana* cl. "I-214" tree with symptoms of a bacterial canker in Serbia.

| PCR Primers | Sequence | Reference |
|---|---|---|
| 16SP1 | 5′-GAAGAGTTTGATCATGGCTC-3′ | [15] |
| 16SP2 | 5′-AAGGAGGTGATCCAGCCGCA-3′ | [15] |
| *gyrB* 01-F | 5′-TAARTTYGAYGAYAACTCYTAYAAAGT-3′ | [16] |
| *gyrB* 02-R | 5′-CMCCYTCCACCARGTAMAGTT-3′ | [16] |
| *atpD* 01-F | 5′-RTAATYGGMGCSGTRGTNGAYGT-3′ | [16] |
| *atpD* 02-R | 5′-TCATCCGCMGGWACRTAWAYNGCCTG-3′ | [16] |
| *infB* 01-F | 5′-ATYATGGGHCAYGTHGAYCA-3′ | [16] |
| *infB* 02-R | 5′-ACKGAGTARTAACGCAGATCCA-3′ | [16] |
| **Sequencing Primers** | | |
| SP1 | 5′-ACCGCGGCTGCTGGCACG-3′ | [15] |
| SP2 | 5′-CTCGTTGCGGGACTTAAC-3′ | [15] |
| 16SP2 | 5′-AAGGAGGTGATCCAGCCGCA-3′ | [15] |
| *gyrB* 07-F | 5′-GTVCGTTTCTGGCCVAG-3′ | [16] |
| *gyrB* 08-R | 5′-CTTTACGRCGKGTCATWTCAC-3′ | [16] |
| *atpD* 03-F | 5′-TGCTGGAAGTKCAGCARCAG-3′ | [16] |
| *atpD* 04-R | 5′-CCMAGYARTGCGGATACTTC-3′ | [16] |
| *infB* 03-F | 5′-ACGGBATGATYACSTTCCTGG-3′ | [16] |
| *infB* 04-R | 5′-AGYTTAGATTTCTGCTGACG-3′ | [16] |

## 2.4. Pathogenicity Test

In January 2020, shoots were collected from stooled beds of *P.* × *euramericana* cl. "I-214" established at an experimental forest nursery "Kaćka forest" of ILFE in Kać, Novi Sad (N 45°17′; E 19°53′). Dormant cuttings (diameter: 14 ± 0.3 mm; length: 30 ± 0.2 cm) were prepared from the lower parts of the collected shoots and stored in polyethylene bags in a cold chamber at 4 °C for two months. In March 2020, the cuttings were first soaked in water for two days in the dark at room temperature (18 ± 2 °C). They were then surface sterilized using 70% ethanol (v/v) and their top ends were sealed using grafting wax (Savacoop, Novi Sad, Serbia) to prevent desiccation and contamination. The cuttings were placed in 3 L plastic pots containing loamy fluvisol soil [5] obtained from the "Kaćka forest" nursery. They were kept in a greenhouse at ILFE (23 ± 2 °C day temperature, 19 ± 2 °C night temperature, 60–70% humidity, 16/8h day/night cycle) for one month and watered every other day to field capacity.

Thirty cuttings that developed roots were transferred to an environmental test chamber (Sanyo, MLR-351H) for the pathogenicity test. They were arranged in a completely randomized design with ten replicates (poplar plants) per treatment. Plants were inoculated with Serbian strain ILFE-LP1 and a Hungarian type strain of *L. populi* NY060 to serve as a positive control. Negative controls were mock inoculated using sterile distilled water. Prior to inoculation, bacterial strains were cultured for 24 h at 30 °C on TSA. Inoculum was prepared in 20 mL TSB. Single bacterial colonies were transferred to

Erlenmeyer flasks and incubated at 30 °C and 180rpm for 24 h in a shaker incubator (Unimax 1010, Heidolph co., Kelheim, Germany). The number of colony forming units (CFU)/ml was determined by spread-plate technique [25] on TSA incubated at 30 °C for 24 h. Bacterial suspension in TSB was transferred to 2 mL Eppendorf tubes, centrifuged at 13,200 rpm for 10 min. and the inoculum concentration was adjusted with sterile distilled water to $10^8$ CFU/ml. Plants were first surface sterilized using 70% ethanol (v/v) and then a cork borer was used to create a 6 mm diameter wound in the middle of each plant. The bacterial suspension was injected into the wound (40 µL) using a pipette and the wound was sealed with Parafilm (Pechiney, Chicago, IL, USA) to retain moisture and protect the wound from contamination. The inoculated plants were maintained at 28 °C, with a relative humidity of 90% under a 16/8 day/night cycle [26] and watered as described above. The experiment was carried out for one month and plants were monitored every day for the appearance of symptoms. At the end of the experiment, the presence of external and internal symptoms was recorded, and the length of internal canker lesions was measured. The pathogenicity test was repeated once.

### 2.5. Statistical Analyses

Statistical analyses of pathogenicity experiment data were performed using Statistica 12.0 (StatSoft Inc., Tulsa, OK, USA). The data were checked for normality using Kolmogorov–Smirnov test and homogeneity of variances was tested with Levene's test. The analyses were further conducted using non-parametric Mann–Whitney U test ($\alpha = 0{,}05$). Because there were no significant differences in lesion lengths produced by the same strain in the two subsequent pathogenicity trials, the data from a single strain were combined for further analyses.

## 3. Results

### 3.1. Biochemical Characterization

Isolate ILFE-LP1 was positive for acetoin and citrate utilization, and negative for β-galactosidase, arginine dihydrolase, lysine decarboxylase, ornithine decarboxylase, $H_2S$, urease, tryptophan deaminase, indole, and gelatinase production. Acid was produced from D-glucose, D-mannitol, D-sucrose, and amygdalin. Nitrates were not reduced to nitrites. Biochemical characteristics of ILFE-LP1 resembled those of the type strain NY060 of *L. populi*.

### 3.2. Phylogenetic Analyses

The 16S data set contained 12 sequences and 1351 characters (1282 parsimony informative, 46 parsimony uninformative, CI = 0.9, RI = 0.8, TL = 76) and the model HKY85+I was chosen for the ML analyses (I = 0.897). The topology of the ML and MP phylogenetic trees was similar, and the ML tree is presented (Table S1, Figure S2).

The combined dataset of three housekeeping genes contained 36 sequences with *Brenneria nigrifluens* as an outgroup (Table S1). The sequence dataset contained 1833 characters (195 parsimony informative, 1638 parsimony uninformative, CI = 0.8, RI = 0.9, TL = 320). The result of the PHT test was not significant and showed that three loci can be combined (P = 0.03). The model GTR+G+I was chosen for the ML analyses (G = 0.507, I = 0.451). The MP and ML analyses produced phylogenetic trees with the similar topology and therefore, only the ML tree is shown (Figure 2). Serbian strain ILFE-LP1 formed a monophyletic clade with *L. populi* strains from Hungary, Portugal, and China within *Lonsdalea* species in the phylogenetic analyses. The separation of *L. populi* from other *Lonsdalea* species was moderately supported in the phylogenetic analyses of 16S rRNA sequences (bootstrap support = 85% ML, MP) and strongly supported in the phylogenetic analyses of a combined *atpD*, *gyrB* and *infB* dataset (bootstrap support = 81% ML, 100% MP). Although the branch lengths indicate differences in the concatenated sequences, the scale bar represents 0.001 nucleotide changes per site. Based on phylogenetic analyses, strain ILFE-LP1 isolated in this study was identified as *L. populi* (Figure 1, Figure S2).

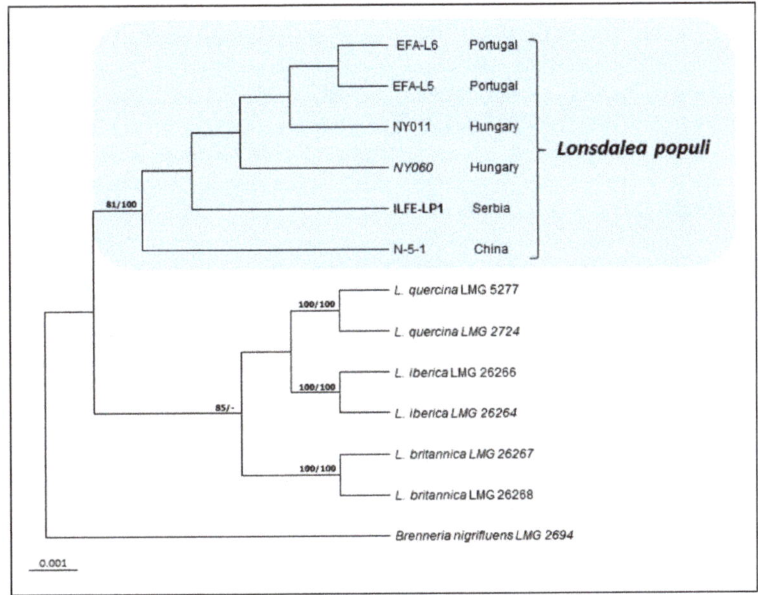


**Figure 2.** Maximum-likelihood (ML) tree resulting from ML analyses of the concatenated ***atpD, gyrB*** and ***infB*** gene sequences and showing the phylogenetic position of *Lonsdalea populi* in relation to its closely related species. The bootstrap support values (ML/MP $\geq$ 80% (maximum parsimony: MP)) are indicated at the nodes, and the scale bar represents the expected number of changes per site. The tree was rooted to *Brenneria nigrifluens*. Strain ILFE-LP1 identified in this study is shown in bold and a clade corresponding to *L. populi* is highlighted.

### 3.3. Pathogenicity Test

Four days after inoculation, the oozing of a colorless fluid and a small sunken area around the inoculation site were evident on most stems inoculated with a bacterial suspension of ILFE-LP1 and on stems of a positive control inoculated with *L. populi* NY060. The lesion gradually expanded further in the following days and oozing continued. One month after inoculation, sunken, water-soaked external cankers were visible on each stem inoculated with isolates ILFE-LP1 and *L. populi* NY060 and after the bark was removed, water-soaked, necrotic lesions were observed around the inoculation points and measured from 1.4 to 3.6 mm (average length = 1.8 mm) and 1.4 to 3.8 mm (average length = 1.9 mm), respectively (Figure 1). There were no significant differences in lesion lengths between the two strains ($p = 0.47$). No cankers formed on the stems of the control plants. *Lonsdalea populi* was successfully (100%) re-isolated from canker margins on TSB and its identity was confirmed using morphology, Gram test, PCR, and sequencing of the *atpD* gene following the procedure described above (Figure S3). *Lonsdalea populi* was never isolated from negative controls.

## 4. Discussion

This study provides the first record of *L. populi* on *P.* × *euramericana* cl. "I-214" in Serbia, and southeastern Europe (SEE). The geographic range of this Gram-negative bacterium has extended and its host association with *P.* × *euramericana* was confirmed. *Lonsdalea populi* was identified using a polyphasic approach, i.e., biochemical characteristics, phylogenetic analyses of the 16S rRNA, and MLSA of three housekeeping genes (*gyrB*, *infB* and *atpD*). The pathogenicity test confirmed that *L. populi* is the causal agent of the bacterial canker disease of *P.* × *euramericana* cl. "I-214" in Serbia.

The present study is also the first report of a bacterial disease on hybrid poplars, including *P.* × *euramericana* in Serbia, and SEE. Despite the importance of hybrid poplars, no previous research has been conducted on bacterial diseases of these trees in Serbia and SEE. Moreover, little research has been conducted on bacterial diseases of *Populus* spp. in SEE and only 'Candidatus Phytoplasma asteris'-related phytoplasmas (yellow disease phytoplasmas) were reported from Lombardy poplar (*Populus nigra* L. 'Italica') trees planted as ornamentals in Belgrade, Serbia and in Zagreb, Croatia [27,28].

The isolation of *L. populi* from *P.* × *euramericana* in Serbia is not surprising, given that it is a well-known pathogen that causes bacterial canker disease of hybrid poplars [12]. *Lonsdalea populi* has been isolated from *P.* × *euramericana* in previous studies in Hungary, Spain, Portugal, and China [8,10–12]. It has also been found associated with *P.* × *interamericana* in Spain and recently reported as a pathogen of Chinese willow (*Salix matsudana* Koidz.) causing cankers with large amounts of white sour exudates in China [11,29].

The disease symptoms (oozing cankers with water-soaked, soft wood) caused by *L. populi* observed in this study are consistent with previous reports of *Lonsdalea* canker of poplars [8,10–12]. In the current study, however, *L. populi* was isolated from two-year old *P.* × *euramericana* trees, whereas in other countries it was found on more than three-year-old trees [8,10–12]. Symptoms of a bacterial canker disease on *P.* × *euramericana* have previously also been reported associated with *Xanthomonas populi* (Ridé) Ridé and Ridé. This bacterium was a major concern in poplar-growing regions of Europe in the 1950s [30]. However, symptoms observed in this study were not typical of those caused by *X. populi* and swollen cankers with deep cracks in the bark have not been observed in the field. Moreover, *Neocosmospora solani* sensu lato has been found associated with cankers of hardwood trees, including *Populus* spp. [31]. Likewise, apart from *L. populi*, Li et al. [12] isolated *N. solani* (Mart.) L. Lombard & Crous from diseased tissues of *P.* × *euramericana* trees experiencing stem cankers in China. The authors, however, reported that *N. solani* is not an aggressive pathogen of this tree species and concluded that *L. populi* is the causal agent of a canker disease of *P.* × *euramericana* in China. Nevertheless, because disease symptoms indicated a possible bacterial infection, fungal isolations were not performed in this study, but additional research is currently being conducted to see if *L. populi* alone is causing the canker symptoms observed in the field.

To prevent the spread of the disease into new areas and plantations, in this study, a pathogenicity test was conducted using *P.* × *euramericana* rooted cuttings in an environmental test chamber under controlled conditions as described in Hou et al. [26]. Due to the high humidity to which poplars were exposed during the test (90%) to promote bacterial activity, and the fast-growing nature of *P.* × *euramericana* the experiment lasted for one month and symptoms of water-soaked cankers with exudation were successfully reproduced. Moreover, cankers of a similar size were formed when *P.* × *euramericana* was inoculated with the strain type of *L. populi* that was used as a positive control. However, oozing was not as abundant, foamy, and creamy as seen on trees in the field in an advanced stage of the disease development. This may be due to the age of the plants used for inoculation, and the duration of the test. Similarly, in a study of Li et al. [12] *L. populi* induced canker symptoms when inoculated into water-cultured excised stems in an environmental test chamber, but abundant, white, sour exudates were observed only when the test was carried out under field conditions using 3–5-year old trees. Difficulties in reproducing disease symptoms in pathogenicity trials have also been reported for other plant pathogenic bacteria, including *Lonsdalea quercina* (Hauben et al. 1999) Brady et al. 2012, *Brenneria nigrifluens* (Wilson et al. 1957) Hauben et al. 1999, and *Brenneria rubrifaciens* (Wilson et al. 1957) Hauben et al. 1999 [32–34].

The occurrence and pathogenicity of *L. populi* on *P.* × *euramericana* cl. "I-214" is of a major concern in Serbia because cl. "I-214" is the most widely grown and economically important poplar clone in the country. Intensively cultured plantations (even-aged, clonal stands) and monoclonality have already increased the vulnerability of this clone to various leaf and canker diseases [4,9]. Because the use of antibiotics for plant disease control in Serbia is prohibited [35], management options for bacterial disease problems in Serbian poplar plantations are limited. Therefore, genetic improvement programs that continuously screen new clones for disease resistance while assuring highest possible volume production could be the most promising strategy to combat *Lonsdalea* canker of poplars. Moreover, an integrated approach of disease prevention and control, focusing not only on selection and breeding for resistance, but also on biological control is needed to assure long-term sustainability of poplar plantations in Serbia.

## 5. Conclusions

To the best of our knowledge, this is the first report of *L. populi* causing bacterial canker disease on *P.* × *euramericana* cl. "I-214" in Serbia and in SEE. It is also the first record of a bacterial disease on *P.* × *euramericana* in SEE.

*Lonsdalea populi* is currently the most serious pathogen affecting *P.* × *euramericana* plantations in Europe. It is also a serious threat to Serbian poplar production. Therefore, there is a need for disease management strategies that are not only economically practical, efficient, and sustainable, but also likely to be accepted by poplar growers. If the disease spreads into new areas, selection and breeding for *Lonsdalea* canker disease resistance might be such a strategy.

**Supplementary Materials:** The following are available online at http://www.mdpi.com/1999-4907/11/10/1080/s1. Figure S1: Primers used in this study. Figure S2: Maximum-likelihood (ML) tree resulting from ML analyses of the partial 16S rRNA gene sequences (1351 bp). The bootstrap support values (ML/MP ≥ 80%) are indicated at the nodes, and the scale bar represents the number of changes. The tree was rooted to *Brenneria nigrifluens*. Figure S3: *Lonsdalea*-like colonies (marked blue) re-isolated from symptomatic tissue of *Populus* × *euramericana* clone "I-214" on tryptone soya agar. Petri dishes were incubated at 30 °C for 48 h. Table S1: Bacterial strains used for phylogenetic analyses.

**Author Contributions:** Conceptualization, M.Z. and S.O.; methodology, M.Z., I.T.-B., T.L., T.T.; investigation, M.Z., I.T.-B., T.L., T.T., P.P.; resources, S.O., T.L., A.K.; data curation, M.Z., T.L., T.T.; writing—original draft preparation, M.Z.; writing—review and editing, M.Z., S.O., T.L., I.T.-B., T.T., P.P., M.M., A.K.; visualization, M.Z., P.P., M.M.; funding acquisition, S.O., T.L. All authors have read and agreed to the published version of the manuscript.

**Funding:** This research was funded by the Ministry of Education, Science and Technological Development of the Republic of Serbia, project no. 451-03-68/2020-14/ 200197. Further funding was provided by the Ministry of Agriculture, Forestry and Fisheries of the Republic of Serbia, project no. 401-00-00058/2020-10.

**Acknowledgments:** The authors wish to acknowledge Sreten Vasić and Igor Guzina (ILFE), for technical assistance.

**Conflicts of Interest:** The authors declare no conflict of interest. The funders had no role in the design of the study; in the collection, analyses, or interpretation of data; in the writing of the manuscript, or in the decision to publish the results.

## References

1. Barrio-Anta, M.; Sixto-Blanco, H.; Canellas-Rey, D.V.I.; Castedo-Dorado, F. Dynamic growth model for I-214 poplar plantations in the northern and central plateau in Spain. *For. Ecol. Manag.* **2008**, *255*, 1167–1178. [CrossRef]
2. Eimil-Fraga, C.; Proupín-Castiñeiras, X.; Rodríguez-Añón, J.A.; Rodríguez-Soalleiro, R. Effects of shoot size and genotype on energy properties of poplar biomass in short rotation crops. *Energies* **2019**, *12*, 2051. [CrossRef]
3. Pilipović, A.; Zalesny, R.S., Jr.; Orlović, S.; Drekić, M.; Pekeč, S.; Katanić, M.; Poljaković-Pajnik, L. Growth and physiological responses of three poplar clones grown on soils artificially contaminated with heavy metals, diesel fuel, and herbicides. *Int. J. Phytoremediation* **2020**, *22*, 436–450. [CrossRef] [PubMed]
4. Orlović, S.; Pilipović, A.; Galić, Z.; Ivanišević, P.; Radosavljević, N. Results of poplar clone testing in field experiments. *Genetika* **2006**, *38*, 259–266. [CrossRef]
5. Pekeč, S.; Katanić, M.; Galović, V.; Andrašev, S.; Pilipović, A. Characteristics of some hydromorphic soils in the inundation of the middle course of the Sava river. *Topola/Poplar* **2018**, *201/202*, 45–52. [In Serbian with English summary]
6. Andrašev, S.; Bobinac, M.; Pekeč, S.; Sarić, R. Characteristics of thinning in a plantation of poplar clone I-214 with a moderate spacing 13 years after establishment. *Topola/Poplar* **2017**, *199/200*, 77–93. [In Serbian with English summary]

7. Phillips, D.H.; Burdekin, D.A. Diseases of poplar (Populus spp.). In *Diseases of Forest and Ornamental Trees*; Palgrave Macmillan: London, UK, 1992; pp. 329–357. [CrossRef]

8. Tóth, T.; Lakatos, T.; Koltay, A. *Lonsdalea quercina* subsp. populi subsp. nov., isolated from bark canker of poplar trees. *Int. J. Syst. Evol. Microbiol.* **2013**, *63*, 2309–2313. [CrossRef]

9. Pap, P.; Drekić, M.; Poljaković-Pajnik, L.; Vasić, V.; Marković, M.; Zlatković, M.; Stojanović, D.V. Monitoring and forecasting of harmful organisms in forests and plantations of Vojvodina, Serbia in 2018. *Topola/Poplar* **2018**, *201/202*, 251–274. [In Serbian with English summary]

10. Abelleira, A.; Moura, L.; Aguín, O.; Salinero, C. First report of *Lonsdalea populi* causing bark canker disease on poplar in Portugal. *Plant Dis.* **2019**, *103*, 2121. [CrossRef]

11. Berruete, I.M.; Cambra, M.A.; Collados, R.; Monterde, A.; López, M.M.; Cubero, J.; Palacio-Bielsa, A. First report of bark canker disease of poplar caused by *Lonsdalea quercina* subp. *populi* in Spain. *Plant Dis.* **2016**, *100*, 2159. [CrossRef]

12. Li, Y.; He, W.; Ren, F.; Guo, L.; Chang, J.; Cleenwerck, I.; Ma, Y.; Wang, H. A canker disease of *Populus× euramericana* in China caused by *Lonsdalea quercina* subsp. populi. *Plant Dis.* **2014**, *98*, 368–378. [CrossRef] [PubMed]

13. Suslow, T.; Schroth, M.; Isaka, M. Application of a rapid method for Gram differentiation of plant pathogenic and saprophytic bacteria without staining. *Phytopathology* **1982**, *72*, 917–918. [CrossRef]

14. Tailliez, P.; Pages, S.; Ginibre, N.; Boemare, N. New insight into diversity in the genus *Xenorhabdus*, including the description of ten novel species. *Int. J. Syst. Evol. Microbiol* **2006**, *56*, 2805–2818. [CrossRef]

15. Brady, C.; Cleenwerck, I.; Venter, S.; Vancanneyt, M.; Swings, J.; Coutinho, T. Phylogeny and identification of *Pantoea* species associated with plants, humans and the natural environment based on multilocus sequence analysis (MLSA). *Syst. Appl. Microbiol.* **2008**, *31*, 447–460. [CrossRef] [PubMed]

16. Akhurst, R.J.; Boemare, N.E.; Janssen, P.H.; Peel, M.M.; Alfredson, D.A.; Beard, C.E. Taxonomy of Australian clinical isolates of the genus *Photorhabdus* and proposal of *Photorhabdus asymbiotica* subsp. asymbiotica subsp. nov. and *P. asymbiotica* subsp. australis subsp. nov. *Int. J. Syst. Evol. Microbiol.* **2004**, *54*, 1301–1310. [CrossRef] [PubMed]

17. Hall, T.A. BioEdit: A user-friendly biological sequence alignment editor and analysis program for Windows 95/98/NT. *Nucleic Acids Symp. Ser.* **1999**, *41*, 95–98.

18. Kumar, S.; Stecher, G.; Li, M.; Knyaz, C.; Tamura, K. MEGA X: Molecular evolutionary genetics analysis across computing platforms. *Mol. Biol. Evol.* **2018**, *35*, 1547–1549. [CrossRef]

19. Katoh, K.; Rozewicki, J.; Yamada, K.D. MAFFT online service: Multiple sequence alignment, interactive sequence choice and visualization. *Brief. Bioinform.* **2019**, *20*, 1160–1166. [CrossRef]

20. Zlatković, M.; Keča, N.; Wingfield, M.J.; Jami, F.; Slippers, B. Botryosphaeriaceae associated with the die-back of ornamental trees in the Western Balkans. *A. Van Leeuw. J. Microb.* **2016**, *109*, 543–564. [CrossRef]

21. Swofford, D.L.; Sullivan, J. Phylogeny inference based on parsimony and other methods using PAUP*. In *The Phylogenetic Handbook: A Practical Approach to DNA and Protein Phylogeny*; Salemi, M., Vandamme, A.-M., Eds.; Cambidge University Press: Cambridge, UK, 2003; Volume 7, pp. 160–206.

22. Guindon, S.; Dufayard, J.F.; Lefort, V.; Anisimova, M.; Hordijk, W.; Gascuel, O. New Algorithms and Methods to Estimate Maximum-Likelihood Phylogenies: Assessing the Performance of PhyML 3.0. *Syst. Biol.* **2010**, *59*, 307–321. [CrossRef]

23. Lefort, V.; Longueville, J.E.; Gascuel, O. SMS: Smart model selection in PhyML. *Mol. Biol. Evol.* **2017**, *34*, 2422–2424. [CrossRef] [PubMed]

24. Felsenstein, J. Confidence intervals on phylogenetics: An approach using bootstrap. *Evolution* **1985**, *39*, 783–791. [CrossRef] [PubMed]

25. Koch, A.L. Growth measurement. In *Methods for General and Molecular Microbiology*, 3rd ed.; Reddy, C., Beveridge, T., Breznak, J., Marzluf, G., Schmidt, T., Snyder, L., Eds.; ASM Press: Washington, DC, USA, 2007; pp. 172–199. [CrossRef]

26. Hou, J.; Wu, Q.; Zuo, T.; Guo, L.; Chang, J.; Chen, J.; Wang, Y.; He, W. Genome-wide transcriptomic profiles reveal multiple regulatory responses of poplar to *Lonsdalea quercina* infection. *Trees* **2016**, *30*, 1389–1402. [CrossRef]

27. Mitrović, J.; Paltrinieri, S.; Contaldo, N.; Bertaccini, A.; Duduk, B. Occurrence of two 'Candidatus Phytoplasma asteris'-related phytoplasmas in poplar trees in Serbia. *Bull. Insectology* **2011**, *64*, 57–58.

28. Šeruga, M.; Škorić, D.; Botti, S.; Paltrinieri, S.; Juretić, N.; Bertaccini, A.F. Molecular characterization of a phytoplasma from the aster yellows (16SrI) group naturally infecting *Populus nigra* L. 'Italica' trees in Croatia. *For. Path.* **2003**, *33*, 113–125. [CrossRef]

29. Li, Y.; Wang, H.M.; Chang, J.P.; Guo, L.M.; Yang, X.Q.; Xu, W. A new bacterial bark canker disease of *Salix matsudana* caused by *Lonsdalea populi* in China. *Plant Dis.* **2019**, *103*, 2467. [CrossRef]

30. Nesme, X.; Steenackers, M.; Steenackers, V.; Picard, C.; Ménard, M.; Ridé, S.; Ridé, M. Differential host-pathogen interactions among clones of poplar and strains of *Xanthomonas populi* pv. populi. *Phytopathology* **1994**, *84*, 101–107. [CrossRef]

31. Boyer, M.G. A *Fusarium* canker disease of *Populus deltoides* Marsh. *Can. J. Bot.* **1961**, *39*, 1195–1204. [CrossRef]

32. Biosca, E.G.; González, R.; López-López, M.J.; Soria, S.; Montón, C.; Pérez-Laorga, E.; López, M.M. Isolation and characterization of *Brenneria quercina*, causal agent for bark canker and drippy nut of *Quercus* spp. in Spain. *Phytopathology* **2003**, *93*, 485–492. [CrossRef]

33. González, R.; López-López, M.J.; Biosca, E.G.; López, F.; Santiago, R.; López, M.M. First report of bacterial deep bark canker of walnut caused by *Brenneria (Erwinia) rubrifaciens* in Europe. *Plant Dis.* **2002**, *86*, 696. [CrossRef]

34. Wilson, E.E.; Stake, M.; Berger, J.A. Bark canker, a bacterial disease of the Persian Walnut tree. *Phytopathology* **1957**, *47*, 669–673.

35. Law on Plant Protection Products of the Republic of Serbia 41/2009-124, 17/2019-18. Available online: https://www.pravno-informacioni-sistem.rs/SlGlasnikPortal/eli/rep/sgrs/skupstina/zakon/2009/41/8/reg (accessed on 19 August 2020).

*Article*

# Biochemical and Gene Expression Analyses in Different Poplar Clones: The Selection Tools for Afforestation of Halomorphic Environments

Vladislava Galović [1,*], Marko Kebert [1], Boris M. Popović [2], Branislav Kovačević [1], Verica Vasić [1], Mary Prathiba Joseph [3,4], Saša Orlović [1] and László Szabados [4]

[1] Institute of Lowland Forestry and Environment (ILFE), University of Novi Sad, Antona Čehova 13d, 21000 Novi Sad, Serbia; kebertm@uns.ac.rs (M.K.); branek@uns.ac.rs (B.K.); vericav@uns.ac.rs (V.V.); sasao@uns.ac.rs (S.O.)

[2] Department of Field and Vegetable Crops, Faculty of Agriculture, University of Novi Sad, Trg Dositeja Obradovića 8, 21000 Novi Sad, Serbia; boris.popovic@polj.uns.ac.rs

[3] Laboratory of Plant Physiology and Biophysics, Plant Science Group, Institute of Molecular, Cell, and Systems Biology, University of Glasgow, Glasgow G12 8QQ, UK; mary.joseph@glasgow.ac.uk

[4] Laboratory of Arabidopsis Molecular Genetics, Institute of Plant Biology, Biological Research Center of the Hungarian Academy of Sciences (BRC), University of Szeged, Temesvári krt. 62, 6726 Szeged, Hungary; szabados.laszlo@brc.mta.hu

[*] Correspondence: galovic@uns.ac.rs; Tel.: +381-692-038-911

**Citation:** Galović, V.; Kebert, M.; Popović, B.M.; Kovačević, B.; Vasić, V.; Joseph, M.P.; Orlović, S.; Szabados, L. Biochemical and Gene Expression Analyses in Different Poplar Clones: The Selection Tools for Afforestation of Halomorphic Environments. *Forests* **2021**, *12*, 636. https://doi.org/10.3390/f12050636

Academic Editors: Ronald S. Zalesny Jr. and Andrej Pilipović

Received: 2 April 2021
Accepted: 10 May 2021
Published: 18 May 2021

**Publisher's Note:** MDPI stays neutral with regard to jurisdictional claims in published maps and institutional affiliations.

**Abstract:** Halomorphic soils cover a significant area in the Vojvodina region and represent ecological and economic challenges for agricultural and forestry sectors. In this study, four economically important Serbian poplar clones were compared according to their biochemical and transcriptomic responses towards mild and severe salt stress to select the most tolerant clones for afforestation of halomorphic soils. Three prospective clones of *Populus deltoides* (Bora-B229, Antonije-182/81 and PE19/66) and one of hybrid genetic background *P. nigra x P. deltoides*, e.g., *P. x euramericana* (Pannonia-M1) were hydroponically subjected to NaCl as a salt stress agent in a concentration range from 150 mM to 450 mM. Plant responses were measured at different time periods in the leaves. Biochemical response of poplar clones to salt stress was estimated by tracking several parameters such as different radical scavenging capacities (estimated by DPPH, FRAP and ABTS assays), accumulation of total phenolic content and flavonoids. Furthermore, accumulation of two osmolytes, glycine betaine and proline, were quantified. The genetic difference of those clones has been already shown by single nucleotide polymorphisms (SNPs) but this paper emphasized their differences regarding biochemical and transcriptomic salt stress responses. Five candidate genes, two putative poplar homologues of GRAS family TFs (*PtGRAS17* and *PtGRAS16*), *PtDREB2* of DREB family TFs and two abiotic stress-inducible genes (*PtP5SC1*, *PtSOS1*), were examined for their expression profiles. Results show that most salt stress-responsive genes were induced in clones M1 and PE19/66, thus showing they can tolerate salt environments with high concentrations and could be efficient in phytoremediation of salt environments. Clone M1 and PE19/66 has ABA-dependent mechanisms expressing the *PtP5CS1* gene while clone 182/81 could regulate the expression of the same gene by ABA-independent pathway. To improve salt tolerance in poplar, two putative GRAS/SCL TFs and *PtDREB2* gene seem to be promising candidates for genetic engineering of salt-tolerant poplar clones.

**Keywords:** poplar; salt stress; gene expression analyses; radical scavenger capacity; osmolytes

## 1. Introduction

Mitigation of climate change became a pivotal mission in the 21st century which employs vast scientific resources since climate change threatens some of the most important forest tree species with extinction. Rising global temperatures are expected to accelerate salinization of soil. Until now, 20% of the world's cultivated land and nearly half of irrigated land is believed to contain elevated concentrations of salt that reduce plant yield significantly below their genetic potential. Every minute, soil salinity claims about three hectares of arable land from conventional crop farming [1] and as the climate warms up, the soil continues to get saltier [2]. This implicates not only crop yield losses but also other environmental problems [3–5], in the forestry sector which is also under serious pressure [6]. This increasing tendency is occurring in Serbia at a greater frequency, especially in the agricultural region of Vojvodina. Halomorphic soil area of 106,000 hectares (5.5%) estimated by Ivanišević et al. [7] is increasing due to industrial pollution, mining and neglected irrigation [8].

One of the major consequences of various environmental stresses, including salt stress, is oxidative stress [9,10]. Oxidative stress presents an imbalance between antioxidant defense and production of reactive oxygen species (ROS), including hydrogen peroxide and oxygen-centered free radicals like superoxide, hydroxyl and hydroperoxyl radicals [11]. Plants produce many antioxidants responsible for the neutralization and detoxification of ROS. One of the largest and most diverse groups of antioxidants present in plants are polyphenols. Phenolic compounds act as antioxidants preventing the escalation of oxidative stress and protecting plants against the oxidative damage of increased ROS levels [12]. Moreover, beside antioxidant activity, polyphenols exhibit multiple roles in plant–environment interactions, including signaling and plant defense [13–15]. Many studies have confirmed that salt and drought stress provoke polyphenol accumulation in most vascular plant species [12–15] but there are still only a small number of research studies on woody plant species. Post-harvest treatment of *Ginkgo biloba* leaves with 200 mmol/L NaCl significantly increased the accumulation of flavonoids [16,17]. The effects of salt stress on polyphenol and antioxidant status of poplar are still unexplored and poorly understood.

High levels of sodium ions (Na$^+$) are toxic to plants because of their adverse effects on cellular metabolism and ion homeostasis [18–20]. Therefore, maintaining high and low levels of Na$^+$ in the cell, specifically in the cytoplasm, is essential for plants [19,21]. Various plant genetic strategies have been proposed to solve the salinity issue in the climate change context. One strategy is to find a mechanism based on biosynthesis of secondary metabolites that would counteract and balance ionic and hyperosmotic stresses and alleviate overall stress to survive these conditions [20,21]. The other strategy proposed by Shabala et al. [1] is based on targeting the mechanisms conferring Na$^+$ sequestration in external storage organs in halophytes. Also, there are strategies based on manipulation with regulatory genes as a more effective approach for developing stress-tolerant plants [22]. According to Hasegawa et al. [23], functional genomics studies of plant stress responses, particularly the identification of a core set of stress-related transcripts, are crucial for both tolerant germplasm exploitation and tolerant crop development through genetic manipulation.

Numerous scientists characterized physiological, biochemical and molecular responses of different model plants to investigate salt stress mechanisms that each of them has developed as a unique salt-adaptive feature. Hence, plants that exhibit high genetic variation for salt tolerance and biological response to salt stress are highly species, genotype and organ specific. Due to the entirely sequenced genome of poplar (*Populus trichocarpa*) [24], this species became a model plant system for molecular research in woody plant species, which significantly facilitated molecular biology research in forestry. Another poplar species that draws a lot of scientific attention is *Populus euphratica* Oliv. that is known for its tolerance to salinity, which was of great importance for large-scale afforestation on saline desert sites in China due to its ability to tolerate up to 450 mM NaCl [25].

When plants are subjected to salt stress, they can increase their stress tolerance by regulating the production of certain metabolites that can reduce damage. Examples include synthesizing osmotic adjustment substances, such as proline, glycine betaine, polyols, polyamines and some soluble proteins, as well as protective enzymes [25–27]. Extensive literature about different salt and drought effects on accumulation of free proline in various vascular but also woody plant species is available [28–33]. An overall conclusion could be that proline accumulation has a significant impact on plant tolerance to salt since most of the studies reported considerably higher amounts of this amino acid in plants subjected to salt stress [34–36]. For that reason, proline has been introduced as a universal salinity- and drought-inducible biochemical marker. Due to its osmoprotective and antioxidant properties, an increase in proline content was linked to improved salt tolerance in plants. The same can be said about proline in the context of other abiotic stress factors such as drought, extreme temperatures, UV radiation and heavy metals induced stress that consequently causes secondary stresses such as osmotic and/or oxidative stress [37,38]. Beside mentioned modes of action against environmental stresses, other functions were later identified for proline, such as its chaperone function and its important role in signaling and in modulation of the translation of proline-rich proteins, as well as redox potential buffering and regulation of different enzymes' activities and stabilization of ROS-scavenging enzymes, as reviewed by Cushman et al. [37]. It was found in *G. biloba* seedlings that the proline content decreased significantly under low NaCl concentrations (50 and 100 mmol/L) and increased significantly under higher concentrations compared to non-treated controls [16]. Intriguingly, proline accumulation was strongly suppressed in tobacco leaves when the plants were exposed to a combination of different stresses [29].

Another important salt stress biochemical marker is quaternary ammonium compound (QAC), known as glycine betaine (GB), that is a fully N-methylated derivative of glycine. GB has a strong osmoprotective action caused by dehydration injuries at the cellular level as well as during salt stress [39]. Plants that are characterized with higher amounts of GB at the organ or cell level normally represent genotypes that are more tolerant to salt than sensitive genotypes. Significantly higher levels of GB during salt stress were found in different crop plants including sugar beet (*Beta vulgaris*), spinach (*Spinacia oleracea*), barley (*Hordeum vulgare*), wheat (*Triticum aestivum*) and sorghum (*Sorghum bicolor*) under various abiotic factors [39,40]. Large embodiments of the literature are focused on the mechanisms of increased stress tolerance of plants after exogenous proline and glycine betaine application, especially concerning drought, salt, cold or high-temperature stresses of genetically engineered plants that overproduce GB and/or proline [30,40,41].

Encoding different structural and regulatory proteins, numerous genes are upregulated under stress conditions in vegetative tissue [20,30]. The candidate genes for this study belong to different metabolic pathways and have various modes of action but all play a role in the salt stress defense mechanism. Delta-1-pyrroline-5-carboxylate synthetase enzymes (P5CS), which catalyze the rate-limiting step of proline biosynthesis, are encoded by two closely related P5CS genes (*P5CS1* and *P5CS2*) in Arabidopsis. Transcription of the P5CS genes is differentially regulated by drought, salinity and abscisic acid, suggesting that these genes play specific roles in the control of proline biosynthesis [42]. According to the same authors, p5cs1-1 mutants accumulate less proline in response to salt stress than wild-type seedlings. Their roots are more hypersensitive to salt stress and have increased evidence of oxidative stress and lipid peroxidation under salt stress than the wild-type plants. Proline accumulation is thought to function as a compatible osmolyte that stabilizes membranes and subcellular components [31,32]. The SOS1 gene is one of the $Na^+$ transporters that modulate salt tolerance in plants [43]. It belongs to an overly salt-sensitive pathway and acts as a proton exchanger. Maintaining low levels of sodium ions in the cell cytosol is critical for plant growth and development. Biochemical studies suggest that $Na^+/H^+$ exchangers in the plasma membrane of plant cells contribute to cellular sodium homeostasis by transporting sodium ions out of the cell [44]. Genetic analysis has linked components of the overly salt-sensitive pathway (SOS1–3) to salt tolerance in *Arabidopsis thaliana* [45,46]. The predicted SOS1 protein sequence and comparisons of sodium ion accumulation in wild-type and SOS1 plants suggest that SOS1 is involved directly in the transport of sodium ions across the plasma membrane. SOS1 contributes to plasma membrane $Na^+/H^+$ exchange and SOS2 and SOS3 regulates SOS1 transport activity [43]. Steady-state SOS1 transcript levels increase significantly in roots and to a much lesser extent in shoots when seedlings are exposed to high levels of NaCl. This regulation by salt is mediated, at least in part, by the other identified components of the SOS pathway, SOS3 (a calcineurin B-like calcium-binding protein) and SOS2 (a serine/threonine protein kinase) [44].

Transcription factors (TFs) play important regulatory roles in targeting specific stress-related genes via binding to cis-acting elements in the DNA adjacent to the specific gene [47,48]. The DREB transcription factors play important roles in regulating abiotic stress-related genes and thereby imparting tolerance to stresses such as cold, drought and high-salt environment to the plant [49,50]. To date, many DREB genes have been identified from various plant species, and the products of these genes have been classified into six groups, termed A-1 to A-6 [49]. DREB members of different groups play diverse roles in plants. The CBF/DREB1s of the A-1 group play a critical role in cold-responsive gene expression, whereas DREB2s in group A-2 show expression under dehydration and high-salt stresses [51,52]. Meanwhile, crosstalk between CBF/DREB1 and DREB2 pathways may exist, because some DREB1 genes are also induced by osmotic stress or high-salt stress [53].

Salt and drought-inducible poplar GRAS protein SCL7 showed that this gene is potentially useful for engineering drought and salt tolerance in trees, thus focusing our attention to a GRAS/SCL transcription factor (TF) as a candidate gene for this study [48]. GRAS gene homologs have been found in *A. thaliana* and in other higher plants like black cottonwood (*P. trichocarpa*) where two putative homologous nucleotide sequences have been obtained, *PtGRAS17* and *PtGRAS16*. Moreover, poplar *GRAS16*, according to phylogenetic analyses, does not exist in the Arabidopsis genome. Characterization of these two salt stress-associated genes, by screening for nucleotide diversity (SNPs) in the coding region, proved that these four clones differ from each other [54], which leads to the assumption that those clones might also have a different response to salt stress. In our earlier research, the genetic difference of the clones was proven by SNPs but their biochemical and transcriptomic background regarding their response to salt stress will be revealed in this paper for the first time.

Investigating the polyphenol and antioxidant status as well as revealing the molecular response of different poplar clones will be helpful in developing selection strategies for improving poplar salt tolerance. Moreover, information on salt-responsive proteins/genes will be crucial for improving salt tolerance through genetic engineering techniques [22]. Although numerous genetic studies in the field of adaptability of forest species have been published, little is known about the molecular basis of this process [54]. This research contributes to revealing the molecular basis of adaptation (e.g., tolerance to abiotic stress) involving biochemical and gene expression analyses on domestic poplar clones.

The aim of this research was to employ biochemical and gene expression analyses to characterize the genetic impact on the clones in response to salt stress and evaluate reliable candidate marker genes as tools for clone selection for afforestation of the halomorphic environment.

## 2. Materials and Methods

### 2.1. Plant Material and Growth Conditions

One-year-old Populus cuttings were collected from the Gene bank of the Institute of Lowland Forestry and Environment (ILFE), Novi Sad, Serbia. The three-way random experiment was established in hydroponic culture where 4 different poplar clones were analyzed. The clones were chosen for their favorable selection characteristics like vigorous growth, straight stem, excellent rooting potential and tolerance against prevalent pests and diseases. Three clones represent *Populus deltoides* (B229, collection number B229; 182/81, collection number 182/81 and PE19/66) and one was of hybrid origin, M1, collection number M1 (*Populus x euramericana*) (Table 1). All biochemical and expression assays were done in triplicates on the leaf level. In total, 144 cuttings (36 cuttings per clone) were used for the experiment. The cuttings were randomized in 4 trays, dipped in aerated Hoagland solution and exposed to a 16 -hour photoperiod until full root and leaf development. After a month of growth and once fully developed, they were gradually subjected to salt stress with variation in NaCl concentrations (150 mM, 300 mM, 450 mM). For expression studies the time point sampling (control and treated) were 3, 8 and 24 h after NaCl treatment, while for biochemical assays were 3, 8, 12 and 24 h upon stress induction. Samples were frozen in liquid nitrogen and kept at −80 °C for further procedures.

**Table 1.** Poplar clones used in this study.

| Clone Name | Clone Collection No. | Species |
|---|---|---|
| Bora | B229 | *P. deltoides* |
| In the process | PE19/66 | *P. deltoides* |
| Pannonia | M1 | *P. x euramericana* |
| Antonije | 182/81 | *P. deltoides* |

*2.2. Biochemical Analyses*

2.2.1. Extract Preparation for Biochemical Assays

20 mg of leaf tissue were homogenized in 2 mL of 96% ethanol. The homogenate was centrifuged at 15,000 $g$ for 10 min at 4 °C and the supernatant was used for biochemical assays. All assays were measured on a spectrophotometric plate reader (MultiScan GO, Thermo Fisher Scientific, Waltham, MA, USA).

2.2.2. Flavonoid Determination

The aluminum chloride colorimetric method [55] with slight modifications was used for flavonoid determination. Briefly, 30 µL of the extract was added to 90 µL of methanol, then 6 µL 1.0 M $NaCH_3COO$ plus 6 µL 0.75 M $AlCl_3$ were added to the mixture and the volume was brought to 300 µL with water. The absorbance of the reaction mixture was measured at 415 nm. The number of total flavonoids was calculated from a calibration curve constructed in 8 points in the interval of 10–250 µg using quercetin as a standard. Results were expressed as quercetin equivalents (QE) in milligrams per gram FW of the initial sample extracted in ethanol using 10 as a dilution factor.

2.2.3. Total Phenol Content (TPC) Assay

Amounts of total phenols were determined by using the method given by Chang et al. [55]. Extracts of 25 µL, 125 µL of 0.1 M Folin–Ciocalteu reagent and 100 µL of sodium carbonate (7.5% $Na_2CO_3$) were mixed. The absorbance was read at 760 nm on a spectrophotometric plate reader (MultiScan GO, Thermo Fisher Scientific, Waltham, MA USA). The results were expressed as gallic acid equivalents (GAE) in milligrams per g of fresh weight.

2.2.4. DPPH Assay

Extracts were tested for their scavenging effect on the 2.2-diphenyl-1-picrylhydrazyl (DPPH) radical according to the method of Kim et al. [56]. 10 µL of plant extract were added to 270 µl of a 0.004% ($w/v$) solution of DPPH in 95% ethanol. The reaction mixture was shaken vigorously, and the absorbance of the remaining DPPH was measured at 520 nm after 5 min. Radical scavenger capacity (RSC) of poplar extracts against DPPH was expressed in mmol of Trolox equivalents per g fresh weight (mmol TE/g FW). The DPPH radical scavenging capacity (RSC%) was expressed as a percentage calculated by using the Equation (1):

$$RSC\ [\%] = (A_{Control} - A_{Sample})/A_{Control} * 100\% \tag{1}$$

where $A_{Control}$ stands for the absorbance of the DPPH reagent while $A_{Sample}$ represents the absorbance of DPPH reagent in the presence of the extract.

2.2.5. $ABTS^+$ Assay

Antioxidant activity of ethanolic extracts was estimated in terms of the $ABTS^+$ radical-scavenging capacity following the procedure described by Arnao [57]. 10 µL of sample were added to 290 µL of diluted ABTS$^+$ solution in microplate wells, and the absorbance was measured at 734 nm. The free radical-scavenging activity was expressed as mmol of Trolox Equivalent (TE) per gram FW sample (mmol TE g$^{-1}$ FW).

### 2.2.6. FRAP Assay

The FRAP test measures the ability of antioxidants to reduce the ferric 2.4.6-tripyridyl-s-triazine complex (Fe$^{3+}$-(TPTZ)$_2$)$^{3-}$ to the intensively blue-colored ferrous complex ((Fe$^{2+}$-(TPTZ)$_2$)$^{2-}$ in acidic medium. To perform the assay, 20 µL of leaf extract was added to 225 µL of FRAP reagent and 25 µL of water and shaken for 20 s and the absorbance was recorded at 593 nm. Ascorbic acid with concentrations from 0 to 500 µM was used as a standard. Results were expressed as ascorbic equivalents (AE) in milligrams per gram FW of the initial sample extracted in ethanol [58].

### 2.3. Biochemical Data Analysis

Tolerance index (TI) [59] was used to calculate clone response to the examined treatments, in comparison to control, for every biochemical parameter by the Equation (2):

$$TI = X_t / X_{Control} \tag{2}$$

where $X_t$ stands for the value of the examined treatment and the $X_{Control}$ stands for each value obtained for the control.

The obtained data for biochemical parameters were analyzed by repeated measures two-way analysis of variance, with clone and salt concentration as the main effects and time of measurement as the effect of repeated measures. The results of repeated measures ANOVA were used for Tukey's HSD (honestly significant difference) test. In graphs and tables presenting results of Tukey's test, treatments that have the same letter belong to the same homogenous group at the level of a = 0.05 (the same letter stands for the treatments that are not statistically significant). These tests were performed in STATISTICA 13 software [60].

### 2.4. Gene Expression Analyses

#### 2.4.1. RT-PCR and Dye-Based qPCR Analyses

Reverse transcription PCR (RT-PCR) and subsequently quantitative PCR (qPCR) were performed to test the expression level of five stress-inducible candidate genes in the controls and treated poplar leaf tissue.

Experimental part and all preparations for expression analyses, including RT-PCR, were carried out in ILFE molecular laboratory while the qPCR analyses were conducted at Biological Research Centre, Szeged, Hungary (BRC). Total RNA from each sample (control and treated) was extracted from leaves using RNeasy Mini kit (Qiagen, Hilden, Germany). The integrity of total RNA was assessed and determined together with its quality and quantity by MultiNA chip electrophoresis system (SHIMADZU, Kyoto, Japan). The RNA (1 µg) was reverse transcribed into cDNA using the AMV Reverse Transcriptase enzyme (Merck Millipore, Burlington, MA, USA). The reaction mixture for reverse transcription was prepared following the manufacturer's instruction. Integrity of cDNA was first check by BioSpec-nano Micro-volume UV-Vis Spectrophotometer (SHIMADZU, Kyoto, Japan), followed by qPCR expression check of different cDNA dilutions with the reference gene and the genes of interest. All the samples were set in triplets. The diluted cDNA (1:20) was subsequently used as template for qPCR using the primers: *PtP5CS1*-F (5′-ggcgttctcctgattgttttt-3′) and *PtP5CS1*-R (5′-gagtccattcccacttctgatt-3′); *PtSOS1*-F (5′-ttgattggaaaaactcctgctc-3′) and *PtSOS1*-R (5′-tcctgatggaatgacagcctac-3′); *PtDREB2*-F (5′-gattgttctcggggagttga-3′) and *PtDREB2*-R (5′-ccacgaaggattttctgattga-3′); *PtGRAS17*-F (5′-cttaaaaatccctctctctctctcc-3′) and *PtGRAS17*-R (5′-tctccagccaaccttcttactt-3′); *PtGRAS16*-F (5′-actatttctttagacccaacgacgac-3′) and *PtGRAS16*-R (5′-atcgcctccacaacagcc-3′); *PtActin_*F (5′-ggatattcagccccttgtctg-3′) and *PtActin_*R (5′-ttctgccccattccaacc-3′) (Table 2).

**Table 2.** List of stress-inducible candidate genes, accession numbers and corresponding primer sequences used for expression analyses.

| Stress-Induced Genes | NCBI Accession Number | | Sequences (5′→3′) | Tm (0C) | Product Size (bp) |
|---|---|---|---|---|---|
| P5CS1 | EEF01373 | PtP5CS1-F | ggcgttctcctgattgttttt | 60.48 | 84 |
| | | PtP5CS1-R | gagtccattcccacttctgatt | 59.44 | |
| SOS1 | EEF02008 | PtSOS1-F | ttgattggaaaaactcctgctc | 60.59 | 146 |
| | | PtSOS1-R | tcctgatggaatgacagcctac | 61.40 | |
| DREB2 | XM_002315114 | PtDREB2-F | gattgttctcggggagttga | 60.05 | 78 |
| | | PtDREB2-R | ccacgaaggattttctgattga | 61.33 | |
| GRAS17 (Scaffold_7) | XM_002310190 | PtGRAS17-F | cttaaaaatccctctctctctcc | 59.46 | 118 |
| | | PtGRAS17-R | tctccagccaaccttcttactt | 59.41 | |
| GRAS16 (Scaffold_5) | XM_002327770 | PtGRAS16-F | actatttctttagacccaacgacgac | 62.25 | 66 |
| | | PtGRAS16-R | atcgcctccacaacagcc | 62.10 | |
| β-Actin | XM_024591321 | PtActin_F | ggatattcagccccttgtctg | 60.90 | 141 |
| | | PtActin_R | ttctgccccattccaacc | 61.00 | |

All gene expression analyses were carried out using an Abi Prism 7900 qPCR machine (Applied Biosystems, Waltham, MA, USA). Dye-based qPCR analyses were performed using SYBR Green JumpStart™ Taq ReadyMix™ (Merck Millipore, Burlington, MA, USA). All qPCRs were performed using 12.5 µL SYBR Green JumpStart™ Taq ReadyMix™ (Merck Millipore, Burlington, MA, USA), a pair of primers for each candidate gene (0.2 µM each) and 10 µL of 1:20 diluted cDNA in a final volume of 25 µL. The qPCR protocol was as follows: 95 °C for 10 min and 40 cycles of 95 °C for 15 s, followed by 60 °C for 1 min. The technical replicates for qPCR were in triplets per each sample. The β-*Actin* gene (XM_024591321) was used as the most reliable internal control to quantify the relative transcript level of each candidate gene in each sample. The relative expression level of target genes was calculated with the $2^{-\Delta\Delta Ct}$ method [61].

### 2.4.2. Stress-Related Candidate Genes

After bioinformatics data mining, five stress-related candidate genes were selected: *PtP5SC1* (*PtP5CS2*): delta-1-pyrroline-5-carboxylate synthetase (*P. trichocarpa*), *PtSOS1*: sodium proton exchanger (*P. trichocarpa*), *PtDREB2*: dehydration, cold and high-salt stress protein (*P. trichocarpa*), *PtGRAS17*: GRAS family transcription factor (*P. trichocarpa*), *Pt-GRAS16*: GRAS family transcription factor (*P. trichocarpa*). List of stress-inducible candidate genes, accession numbers and corresponding primer sequences were shown in Table 2.

### 2.4.3. Gene Expression Data Analyses

Using the Arabidopsis Information Resource (TAIR) and National Center for Biotechnological Information (NCBI) database, candidate genes were selected to test their relative expression in stress-exposed poplar tissue. Candidate genes, *PtP5CS1*, *PtSOS1*, *PtDREB2*, *PtGRAS16*, *PtGRAS17*, activated in response to various abiotic stresses, including salt stress, were selected. A protein–protein BLAST search against the *Populus trichocarpa* genome using the *Arabidopsis thaliana* sequences was performed and related primer sequences were synthesized. After expression analysis by the qPCR method, samples were analyzed. Data analyses determined the relative expression pattern, a certain time after treatment. The relative expression level was calculated by normalizing the PCR threshold cycle number of each gene with that of the β-*Actin* reference gene. Expression pattern graphs were done in the Rggplot package (R, 2013) [62].

### 3. Results and Discussion

#### 3.1. Biochemical Analyses

The results obtained from the biochemical investigation of the salt stress effect on four black poplar clones (B229, 182/81, M1-M1 and PE19/66) are presented in Figures 1–6 and in Tables S1–S4 (Supplementary File). Figures represent the dynamic response of each clone 3, 8, 12 and 24 h after treatment with 150, 300 and 450 mM NaCl. Each parameter was expressed as the specific tolerance index (TI) [9]. Data on total phenolic content tolerance index—TPCTI—are presented in Figure 1. Figures 1–6 represents results regarding polyphenol parameters (total phenolic content, total flavonoid content and flavonoid/phenolic ratio), antioxidant capacity (DPPH radical-scavenging activity) and indicators of osmotic stress (proline content and glycine betaine content).

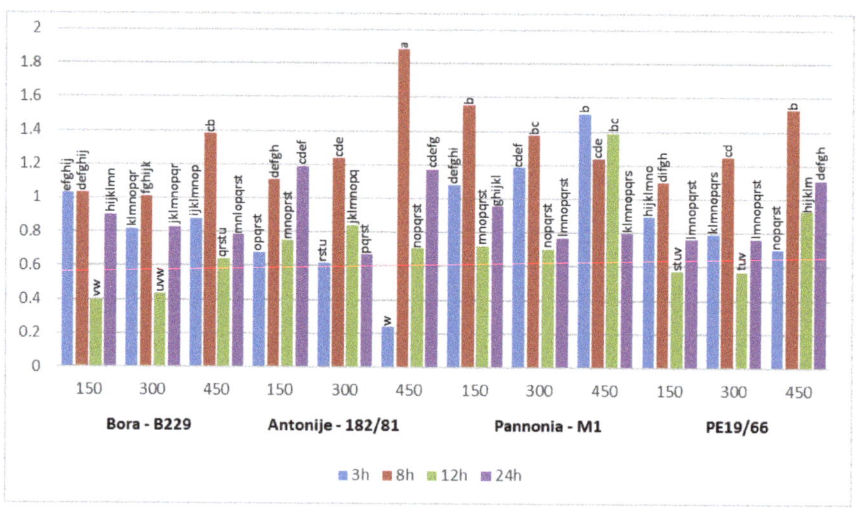

**Figure 1.** Tukey HSD test ($CI_{0.05} = 0.219$) for total phenolic content tolerance index (TPCTI) in poplar clones (Bora-B229, Antonije-182/81, Pannonia-M1 and PE19/66) 3, 8, 12 and 24 h after treatment with 150, 300 and 450 mM NaCl. Treatments that have the same letter belong to the same homogenous group at the level of $\alpha = 0.05$ (the same letter stands for the treatments that are not statistically significant). These tests were performed in STATISTICA 13 software [60].

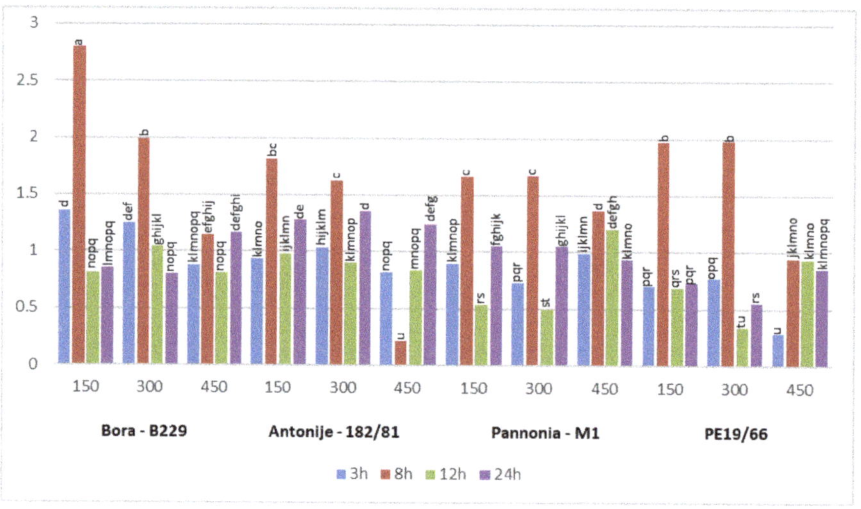

**Figure 2.** Tukey HSD test ($CI_{0.05} = 0.201$) for total flavonoid content tolerance index (TFCTI) in poplar clones (Bora-B229, Antonije-182/81, Pannonia-M1 and PE19/66) 3, 8, 12 and 24 h after treatment with 150, 300 and 450 mM NaCl. Treatments that have the same letter belong to the same homogenous group at the level of $\alpha = 0.05$ (the same letter stands for the treatments that are not statistically significant). These tests were performed in STATISTICA 13 software [60].

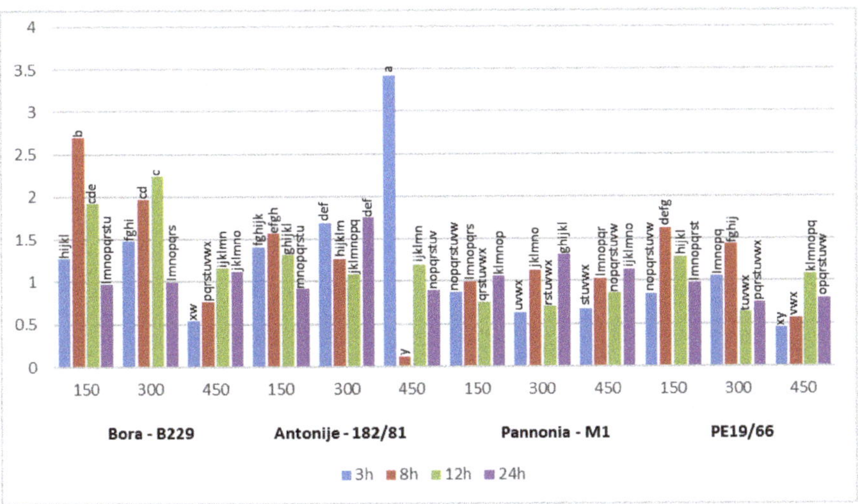

**Figure 3.** Tukey HSD test (CI$_{0.05}$ = 0.350) for the flavonoid/phenolic ratio tolerance index (FPRTI) in poplar clones (Bora-B229, Antonije-182/81, Pannonia-M1 and PE19/66) 3, 8, 12 and 24 h after treatment with 150, 300 and 450 mM NaCl. Treatments that have the same letter belong to the same homogenous group at the level of $\alpha$ = 0.05 (the same letter stands for the treatments that are not statistically significant). These tests were performed in STATISTICA 13 software [60].

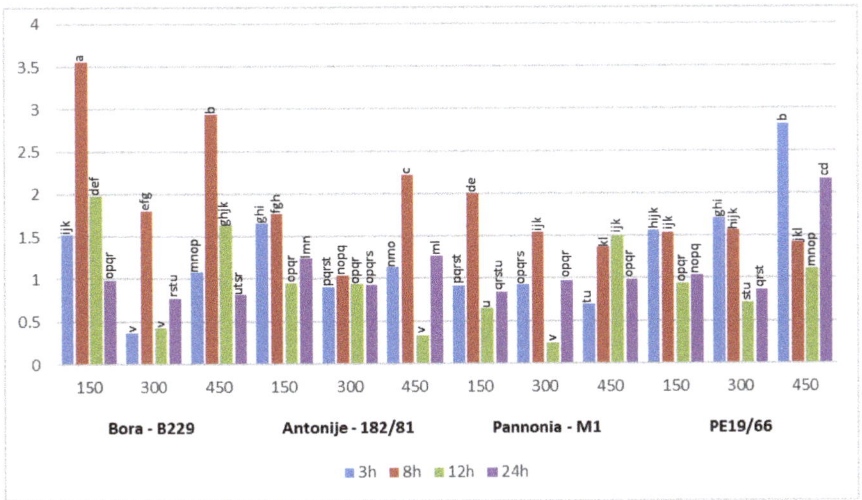

**Figure 4.** Tukey HSD test (CI$_{0.05}$ = 0.217) for DPPH-scavenging activity tolerance index (DPPHTI) in poplar clones (Bora-B229, Antonije-182/81, Pannonia-M1 and PE19/66) 3, 8, 12 and 24 h after treatment with 150, 300 and 450 mM NaCl. Treatments that have the same letter belong to the same homogenous group at the level of $\alpha$ = 0.05 (the same letter stands for the treatments that are not statistically significant). These tests were performed in STATISTICA 13 software [60].

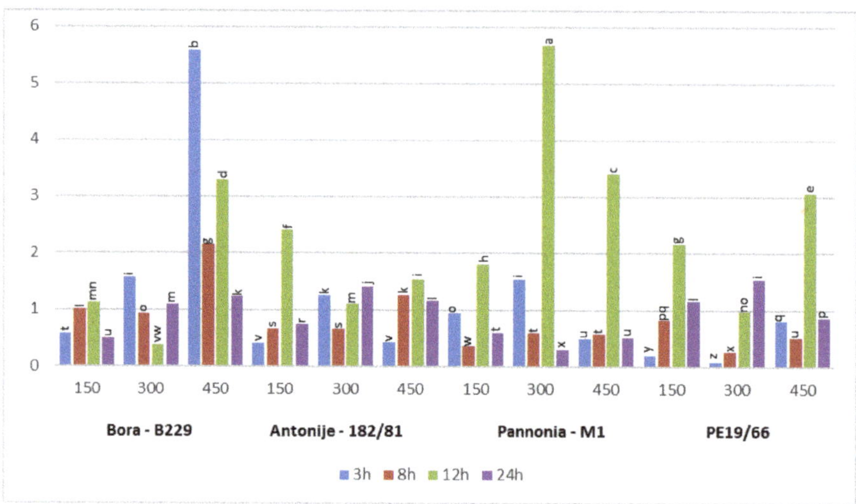

**Figure 5.** Tukey HSD test ($CI_{0.05} = 0.051$) for proline content tolerance index (PCTI) in poplar clones (Bora-B229, Antonije-182/81, Pannonia-M1 and PE19/66) 3, 8, 12 and 24 h after treatment with 150, 300 and 450 mM NaCl. Treatments that have the same letter belong to the same homogenous group at the level of $\alpha = 0.05$ (the same letter stands for the treatments that are not statistically significant). These tests were performed in STATISTICA 13 software [60].

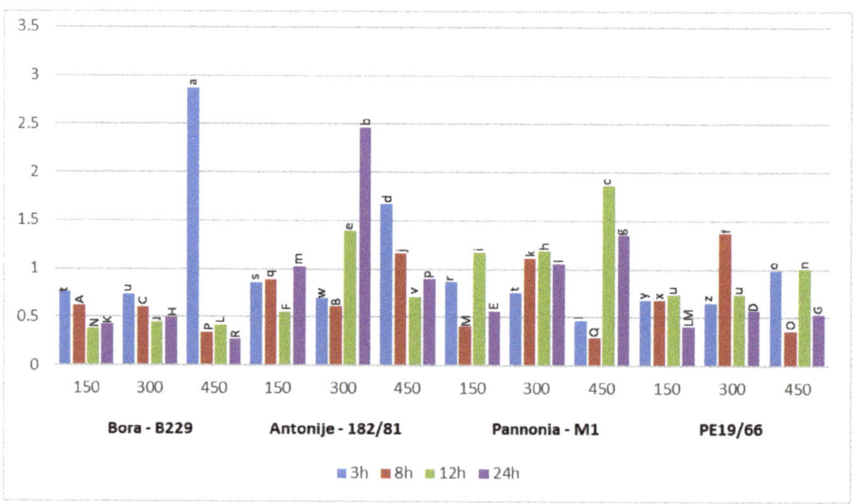

**Figure 6.** Tukey HSD test ($CI_{0.05} = 0.004$) for the glycine betaine content tolerance index (GBCTI) in poplar clones (Bora-B229, Antonije-182/81, Pannonia-M1 and PE19/66) 3, 8, 12 and 24 h after treatment with 150, 300 and 450 mM NaCl. Treatments that have the same letter belong to the same homogenous group at the level of $\alpha = 0.05$ (the same letter stands for the treatments that are not statistically significant). These tests were performed in STATISTICA 13 software [60].

The maximum of the total phenolic content tolerance index (TPCTI) was observed 8 h after the treatment was set (at all concentrations), especially in treatment with the highest concentration of salt (450 mM) in clones PE19/66 and B229. This was also observed in clones 182/81 at 300 and 450 mM and M1 at 150 and 300 mM. That parameter usually increased with the concentration in all investigated poplar clones, except for M1 where TPCTI 8 h after the treatment declined with increasing salt concentration. Twelve hours after the experiment was set, significant decrease in TPCTI was usually observed. It was followed by none significantly different or rarely higher values observed at 24 h after the treatment, except in M1 in treatments with 450 mM NaCl. The lowest TPCTI was observed three hours after the treatment with 450 mM in poplar clone 182/81, which was followed by the highest observed TPCTI (1.884) after 8 h at 450 mM NaCl. By a general comparison of all the results, it can be concluded that in all clones except M1, there was a spike of TPCTI at 8 h after the treatment, which rises with the increase in salt concentration.

Many authors have found that excessive salinity increased plant phenolic levels [10–12]. The increase in total phenolics content can be related to the increased expression of the gene for the enzyme L-phenylalanine ammonia-lyase responsible for the biosynthesis of phenolic acids [13]. Investigation of the response of two poplar hybrid clones to the high-salt stress showed that they differ in phenolic and antioxidant levels [14] which is in accordance with our findings, especially in the case of clone M1.

The results obtained by total flavonoid content measurements are expressed as total flavonoids content tolerance index (TFCTI) and flavonoids/phenolics ratio (FPRTI) are presented in Figures 2 and 3, respectively.

The highest response of TFCTI on treatments with salt concentrations of 150 and 300 mM was observed eight hours after the treatment in all clones. Clone B229 had the highest TFCTI at 150 mM and 300 mM NaCl. At the same time, clone PE19/66 had a slightly weaker reaction with the TFCTI of cca 1.97 at 150 and 300 mM NaCl, followed by clones 182/81 and M1. A significant decrease of total flavonoids was observed 12 h after treatment with concentrations of 150 and 300 mM NaCl in all clones when compared to the situation at 8 h after treatment, but at 450 mM NaCl, the response of the clones, compared to response level at 8 h post treatment, differed. In B229, TFCTI declined, in clone 182/81 it increased, while a significant decline was noticed at 8 h post treatment, and in clones M1 and PE19/66 the difference between TFCTI at 8 and 12 h was not significant.

The ratio between total flavonoids and total phenolics can be interpreted as a contribution of flavonoid to the total phenol content, and the tolerance index based on this parameter (FPRTI) can also be a valuable indicator of differences between genotypes in their reaction to abiotic stresses [63]. All the results presented in Figure 3 were obtained by comparing them to the control value normalized to 1 (100%). In clone B229, the FPRTI value increased compared to the control value at 150 mM NaCl by 170% after 8 h and by 90% after 12 h. At 300 mM NaCl, the same index increased from 97% after 8 h to 124% after 12 h. In clone 182/81, the FPRTI value dramatically increased (242%) 3 h after treatment at 450 mM NaCl compared to the control. The initial increase was followed by a significant decrease in FPRTI value (89%) compared to the control. Although the accumulation of flavonoids under the influence of the stress occurred in all clones, in clones B229 and 182/81, the flavonoid/total phenol ratio tolerance index was higher than the other two examined clones.

Polyphenols are secondary metabolites, which are not primarily connected to the processes of growth and development but are of vital importance for their ecological interactions and their role in defense mechanisms, modulating transcriptional regulation, signal transduction and hormonal regulation [15]. According to Xu et al. [16], salt stress can elicit intensive gene expression changes which affect physiological and molecular pathways, including the synthesis of flavonoids. Flavonoids are a major class of polyphenols, which play important roles in eliminating free radicals and preventing oxidation [15]. Our observation that, in some poplar genotypes, the flavonoid/total phenol ratio can increase is in accordance with published results that, despite the decrease in total phenol content in he investigated poplar genotypes, flavonoids content can be affected by stress [15]. These observations could be explained by the essential role of flavonoids among other phenolics in physiological regulation and response to stress. Vuksanović et al. [63] stressed that a high flavonoids/phenolics ratio was associated with low-intensity stress and with genotypes tolerant to acidification, while at high-intensity stress and in genotypes with low tolerance to acidification, this ratio declined due to a higher content of total phenolics and lower content of flavonoids. In that sense, the higher FPRTI observed in our study for clones B229 and 182/81 indicates their higher tolerance to salt stress than in M1 and PE 19/66.

Total antioxidant capacity was measured by determination of DPPH radical-scavenging activity. The results were presented by the analog tolerance index (DPPHTI) in Figure 4.

The highest DPPHTI was observed 8 h after the experiment set under all applied concentrations in all clones, except for PE19/66 which had its peaks already in 3 h, especially at 450 mM NaCl. The highest DPPHTI was detected in clone B229, which achieved 255% (150 mM, 8 h), 80.5% (300 mM, 8 h) and 193% (450 mM, 8 h) higher DPPH radical-scavenging activity than the analog control treatment. High initial induction (after 3 h) of DPPH radical scavengers was observed also in clone PE19/66 (182% higher than control) after the treatment with the highest salt concentration (450 mM NaCl). Considering obtained results, it can be concluded that the greatest initial antioxidant response to salt stress, according to the DPPH method, was found in genotype B229.

Diverse environmental stress factors, including salt stress, can lead to an imbalance between antioxidant defenses and the number of reactive oxygen species (ROS) causing oxidative stress [28,34]. The scavenging of reactive oxygen species (ROS) is one of the possible mechanisms of plants action against oxidative stress provoked by salt stress. The DPPH method is based on the scavenging action toward artificial 2,2-diphenyl-1-picrylhydrazyl (DPPH) [64]. A positive correlation between total phenols and DPPH radical-scavenging capacity has also been found both in the salt-stressed and control poplar plants [14] and generally in plant tissue [64]. Generally, the increase in antioxidant activity is a well-established mechanism for the stress response in plants [15,63]. However, Štajner et al. [64] suggest that an intense increase of DPPH-scavenging capacity, as well as a strong increase of total phenolic content, was not characteristic for tolerant genotypes. Instead, moderate increase of these parameters, and moderate decrease of total flavonoids and, especially, the ratio between total flavonoids and total phenolics is characteristics of tolerant genotypes. In that sense, high DPPHTI at 450 mM NaCl 24 h after the treatment suggests that genotype PE19/66 is less tolerable to high-salt stress than the other three examined clones. Results concerning proline and glycine betaine accumulation were expressed by tolerance indices based on proline content and glycine betaine content (Figures 5 and 6).

As presented in Figure 5, the highest proline content tolerance index (PCTI) was observed mostly after 12 h in all four clones by specific salt concentrations. The most intense peaks after 12 h were observed for clone B229 at 450 mM NaCl (230%), clone M1 at 300 mM NaCl (469%) and at 450 mM NaCl (242%) and clone PE19/66 at 450 mM NaCl (206%), while clone B229 showed a high peek at 450 mM NaCl 3 h after treatment. At 24 h, a decrease in the proline content tolerance index was observed with few exceptions. Concerning the overall results for proline, clones B229 and M1 seem to achieve the strongest response to salt stress, while M1 and PE 19/66 showed the lowest PCTI 24 h after the treatment.

The response of the other osmolyte, glycine betaine, depended on the salt concentration as well as clone and duration of the treatment. It was not possible to determine any general trend. It was observed that clone B229 showed the highest initial response by the glycine betaine content tolerance index (GBCTI) after 3 h, especially under 450 mM (187%), while in M1, the highest GBCTI were recorded after 12 h, especially at 450 mM NaCl (86.5%). The maximal response of clone 182/81 was after 24 h, especially under 300 mM (146%), but at 450 mM NaCl the maximum was after 3 h, while in clone PE19/66, GBCTI exceeded 1 only at 300 mM NaCl 8 h after treatment.

One of the most important strategies to enhance plant tolerance to osmotic stress, including water and salt stress, is the accumulation of osmotically active substances. Many studies have shown that the proline content in higher plants increases under different environmental stresses including drought, high salinity, UV irradiation, heavy metals, oxidative stress and in response to biotic stresses [37]. Proline has a complex effect on stress responses, including cell membrane protection from oxidative stress, by enhancing antioxidant activity, reducing $H_2O_2$ levels and facilitating growth [63]. It has been found that hybrid black poplar subjected to osmotic stress, enhances the accumulation of glycine betaine and proline as the most important strategy to resist the osmotic stress and overcome water deficit [63]. The effect of salt stress on ion concentration, proline content, antioxidant enzyme activities and gene expression has been studied in tomato cultivars as well [33]. It was revealed that plant tissue dynamically responds to the salt stress and that the proline concentration had peaks after certain time intervals (6 h and seven days) for the genotype that was more tolerant to the salt stress. Investigation of the role of glycine betaine in enhanced salinity tolerance in some plants pointed to its possible role in the maintenance of $K^+$ homeostasis, reduction of lipid peroxidation and an increase in SOD activity, a key enzyme of the reactive oxygen species-scavenging system [65].

### 3.2. Comparative Genomic Studies

To facilitate the identification of stress-related candidate genes for expression studies and future poplar clone improvement, we queried the previously published protein sequences that show homology with *P. trichocarpa* stress-related genes *PtP5CS1* (*PtP5CS2*): delta-1-pyrroline-5-carboxylate synthetase, *PtSOS1*: sodium proton exchanger, *PtDREB2*: dehydration, cold and high-salt stress protein, *PtGRAS17* and *PtGRAS16*: GRAS family transcription factor. The phylogenetic tree differentiated four distinct clades where our protein sequences showed high evolutionary relatedness with the proteins of *Populus euphratica*, which is a model plant species for salt stress studies and is of great importance for large-scale afforestation on saline desert sites (Figure 7).

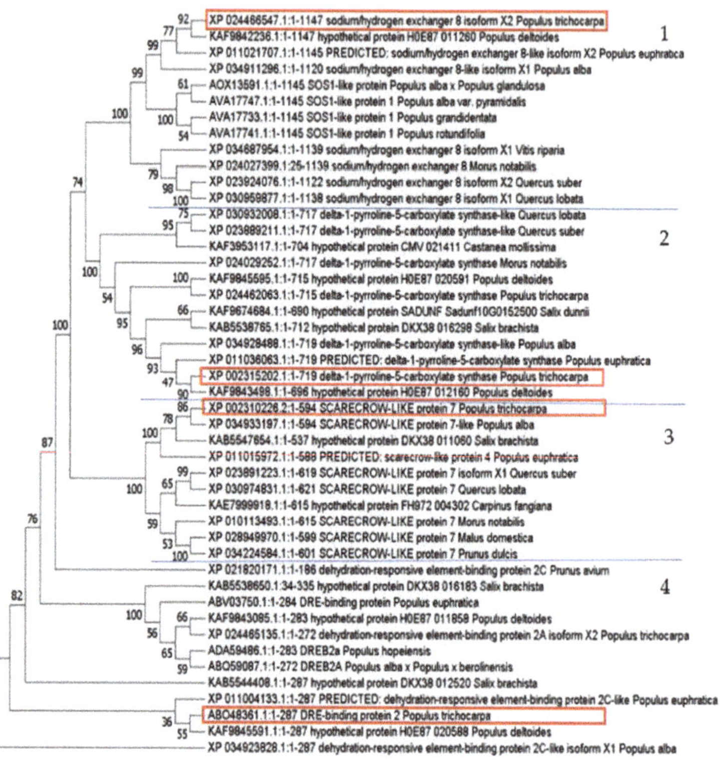

0.2

**Figure 7.** Phylogenetic analyses of protein sequences encoded by four stress-related genes of *P. trichocarpa*. The evolutionary history was inferred using the Neighbour-Joining method [66,67]. The optimal tree with the sum of branch length = 6.68 is shown. The tree is drawn to scale, with branch lengths in the same units as those of the evolutionary distances used to infer the phylogenetic tree. The evolutionary distances were computed using the Poisson correction method [68] and are in the units of the number of amino acid substitutions per site. The analysis involved 46 amino acid sequences. All ambiguous positions were removed for each sequence pair. There were a total of 1153 positions in the final dataset. Analyses were conducted in MEGA5 [69].

### 3.3. Differential Expression Pattern of Salt Stress-Related Candidate Genes

#### 3.3.1. Salt Stress Responses of *PtP5CS1*

The expression pattern of the *PtP5CS1* gene shows that the highest induction of this gene occurred in clone PE19/66 where the most transcript abundance was noticed at 300 mM NaCl after 3 h of exposure. This clone responded in a gradual expression rate, in comparison to control, in a low concentration of salt (150 mM) in early phases (3 and 8 h). The same patterns were recorded at 450 mM NaCl after 3 and 8 h. Relative expression was fivefold higher when this clone was exposed to 300 mM NaCl after 3 h of stress exposure in comparison to the M1 clone. It is interesting that clone PE19/66 after 24 h of exposure to all stressor concentrations holds the similar relative expression level.

Clone M1 and 182/81 shared similar expression values, however, clone 182/81 showed a lower induction while M1 recorded a somewhat higher and steady induction throughout the experiment. The weakest induction was noticed in the clone B229 where the *PtP5CS1* transcript level was almost equal to β-*Actin* expression (Figure 8 and Table S5 in Supplementary File). According to Fabro et al. [32], drought and salt stress differentially activate the expression of two P5CS-related genes in Arabidopsis thaliana, *AtP5CS1* and *AtP5CS2*, where *AtP5CS1* is activated by an ABA-dependent signal transduction pathway. Our results support the findings of those authors where the highest induction of the *PtP5CS1* gene was found in clone PE19/66 that was already confirmed for the same clone for the *PtRD29B* gene that was activated by an ABA-dependent pathway [70].

The expression pattern of *PtP5CS1* is supported with the biochemical data where the accumulation of proline in PE19/66 reveals a similar pattern. These findings suggest that the *PtP5CS1* gene is involved in proline accumulation during salt stress which was a confirmation of what was expected during the experiment. From examination of all gathered data, it could be concluded that the clone PE19/66 has different mechanisms expressing the *PtP5CS1* gene during salt stress through the ABA-dependent signaling pathway compared to 182/81 and M1 clones.

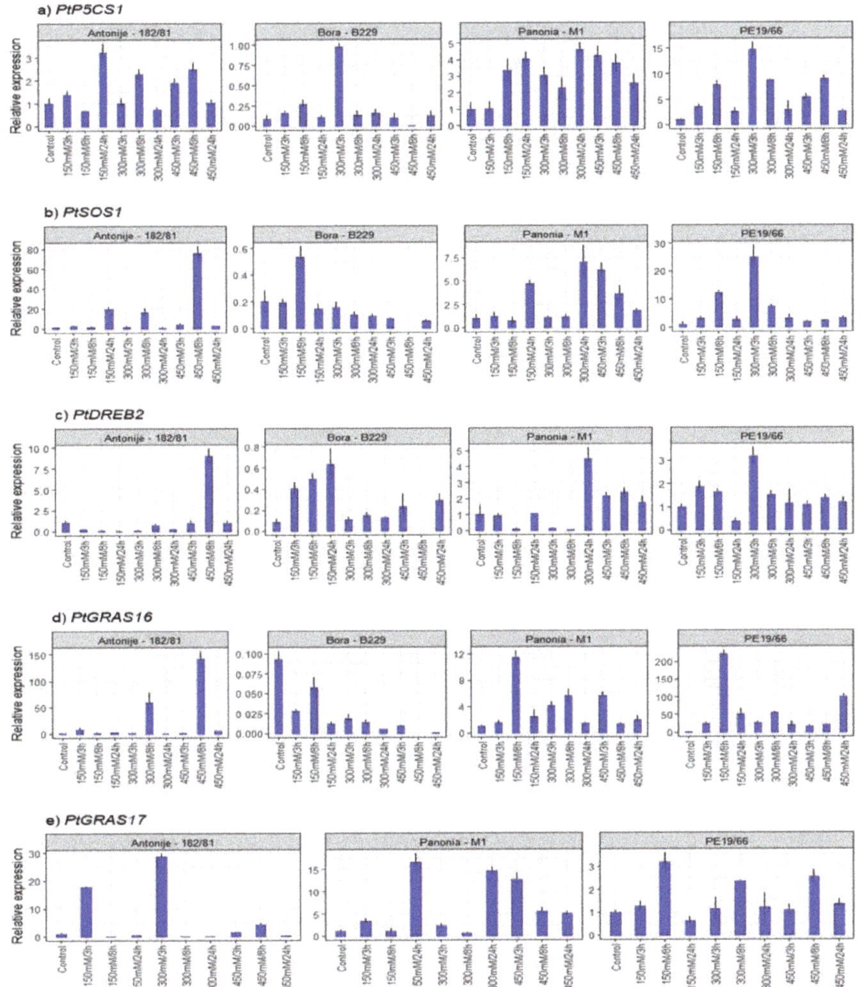

**Figure 8.** Two-step RT-qPCR results. Differential expression profiles of *PtP5CS1*, *PtSOS1*, *PtGRAS16*, *PtGRAS17*, *PtDREB2* genes induced by salt (NaCl) treatment at various concentrations (150 mM, 300 mM and 450 mM) and differing treatment durations (3 h, 8 h, 24 h).

### 3.3.2. Salt Responses of PtSOS1

This transporter gene is associated with cellular sodium homeostasis by transporting sodium ions out of the cell. Clone B229 showed the weakest expression. In the M1 clone, the *PtSOS1* gene showed a late increase (24 h) in expression at 150 mM and 300 mM and a gradual decrease at 450 mM. Clones 182/81 and PE19/66 showed abundant transcript amplification at higher salt concentrations (450 mM and 300 mM) after 8 h and 3 h, respectively. Clone PE19/66 showed a gradual increase in early expression (150 mM after 3 and 8 h) and almost no transcript amplification in high salt concentration at all time points (450 mM in 3, 8 and 24 h). Steady state of the *SOS1* transcript levels increase significantly in roots and to a much lesser extent in shoots when seedlings are exposed to high levels of NaCl [44]. This finding is supported by the same authors where they stated that this regulation by salt is mediated, at least in part, by the other identified components of the SOS pathway, SOS3 (a calcineurin B-like calcium-binding protein) and SOS2 (a serine/threonine protein kinase). Even though the *PtSOS1* gene expression pattern differed between all clones, clone M1 showed the highest expression with in more saturated stressor at 300 mM and 450 mM after 3, 8 and 24 h, while clone PE19/66 reacted earlier and expressed this gene at a low salt concentration. The latest studies revealed that *AtSOS1* is maybe involved in long-distance transport of $Na^+$ from the roots to the shoots and mediates absorption and transport of other ions ($K^+$, $Ca^{2+}$ and $H^+$), maintaining pH homeostasis in plants [71,72]. Regarding these findings we can assume that clone PE19/66 by higher induction of this gene in the leaf tissue has the possibility to make faster uptake of the salt while clone M1 responded slower and at heavier stress conditions. These give us a choice in recommendation of these two clones for phytoremediation of devastated soils and afforestation of holomorphic environment.

### 3.3.3. Salt Responses of PtDREB2

In three out of four clones, higher salt concentrations (300 mM and 450 mM) induced this regulatory gene with low transcript abundance. In clones M1 and PE19/66 were detected induction of the gene expression at 300 mM 24 h after exposure was detected, however, it was threefold lower than that observed in the 182/81 clone at 450 mM 8 h after exposure. Clone B229 reacted in the opposite manner in comparison to other clones, showing the lowest activation of the *PtDREB2* gene with a gradual increase in its transcript level at the lower concentration of the stressor agent (150 mM at 3, 8 and 24 h).

The manner of ABA-independent regulation manner of DREB genes in general and the results of higher salt stress induction of the DREB1 gene are supported by [53]. ABA-independent signaling systems have been described as pathways that mediate adaptation to stress by the activation of the CBF/DREB (cold-binding factor/dehydration-responsive element binding) regulon [73]. The most recent study [74] supports crosstalk between ABA-dependent and ABA-independent pathways in the regulation of gene expression under salt stress. From that point of view, both clones, M1 and PE19/66, had the most prominent induction level eventhough it was concluded earlier about their ABA-dependent way of salt stress regulation. Since clone 182/81 showed ABA-independent regulation of the *PtDREB2* gene in this study and *PtRD17* expression [70], it is not surprising that this clone showed low expression patterns in comparison to the other clones under investigation. In order to improve salt tolerance in poplar, due to TFs specific role as master regulators of many stress-responsive genes, *PtDREB2* transcription factor is a good candidate for validation through generation of transgenic poplar lines.

### 3.3.4. Salt Responses of PtGRAS16 and PtGRAS17

By single nucleotide screening (SNPs) of these two homologues it was recently proved the genetic differences between all four clones that were used also in this study were recently proved [54]. In clone 182/81, the *PtGRAS16* homolog showed two peaks after 8 h at the high salt concentrations (300 mM and 450 mM, respectively) but early-expressed transcripts after 8 h and at the low salt concentration (150 mM) in PE19/66 clone. This clone showed a higher expression level in comparison to the 182/81 clone. There was no considerable activation of this gene in B229 clone. Differential expression analyses revealed that the *PtGRAS17* homolog of the *PtGRAS* gene was not expressed in the B229 clone. Its highest activation values were noted in 182/81 at early stress stages (150 mM and 300 mM) 3 h after treatment and it was twofold higher in comparison to clone M1 where this clone expressed this homolog at 150 mM and 300 mM 24 h of after stress induction. In PE19/66, there was slight activation throughout the salt treatment.

By a comparison of the expression pattern between two homologues, clone 182/81 expressed *PtGRAS17* earlier (3 h after treatment at 150 mM and 300 mM) while PtGRAS16 was expressed later (8 h after treatment at 300 mM and 450 mM). Clone M1 reacted by higher induction of *PtGRAS17* at 300 mM and 450 mM while the same clone exhibit abundant amplification at 150 mM of salt stress. Additionally, it is interesting that clone PE19/66 highly expressed the *PtGRAS16* homologue almost 10 folds higher than *PtGRAS16*. All the above mentioned findings undoubtedly confirmed that those homologues can be used as a marker gene in the differentiation of clones in breeding programs. This finding is supported by Fladung et al. [75].

It is interesting that clone B229 revealed a low induction of polyphenols and showed low accumulation of flavonoid content as well as low transcription rate for all salt-responsive genes in all concentrations, and even lower levels at higher concentrations of the stress agent. These explanations can vary according to available literature. If there is a mutation of any of the components of the salt overly sensitive pathway (SOS), that could result in severely reduced salt tolerance [74], which is not possible since it will be noticed in the clone phenotype. The other assumption is that this genotype is behaving like halophyte species where, under salt stress, primary and secondary metabolites, including proline and sugar alcohols, function as osmolytes in halophytes and glycophytes [76], whereas in halophytes, prestress metabolite levels are significantly higher. This may explain its greater capacity for osmotic stress adjustment. This behavior can be compared to halophyte *P. euphratica* where the evidence for a general activation of stress-relevant genes was not detected. Because of its low transcriptional responsiveness to salinity, it was hypothesized that this species exhibits an innate activation of the stress-protective genes in comparison to sensitive poplars [77]. The evolutionary adaptation of *P. euphratica* to salt environments is linked with higher energy requirements of cellular metabolism and a loss of transcriptional regulation [77]. Therefore, we can assume that clone B229 was prone to tolerate salt environments even at high concentrations like 450 mM and could be efficient in phytoremediation of salt environments. Since the expression data for this clone are unexpectedly low for all investigated genes, it would be necessary to take additional measurements in the future work.

According to gene expression analyses, it can be assumed that clones PE19/66 and M1 clone regulate the expression of the *PtP5CS1* gene through the ABA-dependent pathway, while the clone 182/81 expressed *PtDREB2* through the ABA-independent pathway. This could be the evidence that clones have different regulation systems in gene expression and respond to salt stress differently by expressing the salt-responsive genes in different manner, which points out their genetic divergence. This knowledge could be used in breeding programs in order to improve salt tolerance by using *PtDREB2* and *PtGRAS* transcription factor genes. They could be good candidates for screening the filial generations of stress-tolerant poplars because of their role as master regulators of many stress-responsive genes.

The most important segment of this work represents accomplishing a successful strategy for approaching the climate change problem by employing biochemical and functional genomic tools using RT-qPCR technology. By testing several salt stress-related gene expressions, in this study, we broaden our knowledge about the genetic background and obtained the first applicable insights into the salt stress tolerance of different economically important poplar clones. Revealing their different biochemical and molecular response to the salt stress will enhance the possibility of phytoremediation therefore, the afforestation of halomorphic soils with these poplar clones in the Vojvodina region, Serbia.

**Supplementary Materials:** The following are available online at https://www.mdpi.com/article/10.3390/f12050636/s1; Supplementary Table S1: F-test for examined biochemical parameters in black poplar clones with different concentrations of NaCl; Supplementary Table S2: Tukey HSD test for biochemical parameters in black poplar clones with examined salt concentrations in the first 24 h; Supplementary Table S3: F-test for tolerance indices based on examined biochemical parameters in black poplar clones with different concentrations of NaCl; Supplementary Table S4: Tukey HSD test for tolerance indices of biochemical parameters in black poplar clones with examined salt concentrations in the first 24 h; Supplementary Table S5: Tukey HSD test analyses of expression data performed using $2^{-\Delta\Delta Ct}$ value distribution of *PtP5CS1*, *PtSOS1*, *PtGRAS16*, *PtGRAS17*, *PtDREB2* genes induced by salt (NaCl) treatment at various concentrations (150 mM, 300 mM and 450 mM) and differing treatment durations (3 h, 8 h, 24 h) in poplar clones (Bora-B229, Antonije-182/81, Pannonia-M1 and PE19/66). Treatments that have the same letter belong to the same homogenous group at the level of $\alpha = 0.05$ (the same letter stands for the treatments that are not statistically significant). The number 1000 stands for a missing value. These tests were performed in STATISTICA 13 software [60].

**Author Contributions:** Conceptualization of the experiment was done by V.G. Methodology was done partly by V.G. for molecular research and M.K. for biochemical research. Processing of the molecular data was carried out partly by V.G. and M.P.J. while biochemical data processing using various types of software was done by B.K. and V.V. Validation of the text was done by M.P.J. and M.K. Writing the original draft preparation was done by V.G. for the molecular part and B.M.P. for the biochemical part. Supervision was carried out by L.S. and S.O. All authors have read and agreed to the published version of the manuscript.

**Funding:** This study was funded partly by IPA Project Hungary–Serbia "Oxidative stress tolerance in plants: from models to trees (OXIT) (Project ID: HUSRB 1002/214/036) and by the Ministry of Education, Science and Technological Development of the Republic of Serbia (Project No: 451-03-9/2021-14/200197).

**Institutional Review Board Statement:** Not applicable.

**Informed Consent Statement:** Not applicable.

**Data Availability Statement:** Not applicable.

**Acknowledgments:** This is a conjoined research work of the Biological Research Center of the Hungarian Academy of Sciences, Institute of Plant Biology Laboratory of Arabidopsis Molecular Genetics University of Szeged, Hungary and the Institute of Lowland Forestry and Environment, Novi Sad, Serbia. We wish to thank Laszlo Szabados and Saša Orlović for supervision and support throughout the research work and Laura Zsigmond for assistance in bioinformatics. We would like to thank Ed Bauer from the USDA Forest Service, Northern Research Station, Institute for Applied Ecosystem Studies, Rhinelander, USA, WI for English and style editing. We thank Saša Kostić for excellent data processing in R software. We wish to thank Sreten Vasić for excellent technical assistance in the molecular laboratory in ILFE, Serbia.

**Conflicts of Interest:** The authors declare no conflict of interest.

## References

1. Shabala, S.; Bose, J.; Hedrich, R. Salt bladders: Do they matter? *Trends Plant Sci.* **2014**, *9*, 687–691. [CrossRef] [PubMed]
2. Zhou, Y.; Liu, C.; Tang, D.; Yan, L.; Wang, D.; Yang, Y.; Gui, J.; Zhao, X.-Y.; Li, L.; Tang, X.-D.; et al. The Receptor-like Cytoplasmic Kinase STRK1 Phosphorylates and Activates CatC, thereby Regulating $H_2O_2$ Homeostasis and Improving Salt Tolerance in Rice. *Plant Cell* **2018**, *30*, 1100–1118. [CrossRef]
3. Chen, S.; Polle, A. Salinity tolerance of Populus. *Plant Biol.* **2010**, *12*, 317–333. [CrossRef] [PubMed]
4. Turner, R.C.; Marshal, C. The accumulation of zinc by subcellular fractions of roots of *Agrostis tenuis* Sibth. in relation to zinc tolerance. *New Phytol.* **1972**, *71*, 671–676. [CrossRef]
5. Zorić, M.; Đukić, I.; Kljajić, L.; Karaklić, D.; Orlović, S. The possibilities for improvement of ecosystem services in Tara National Park. *Topola* **2019**, *203*, 53–63.
6. Kesić, L.; Matović, B.; Stojnić, S.; Stjepanović, S.; Stojanović, D. Climate change as a factor reducing the growth of trees in the pure Norway spruce stand (*Picea abies* (L.) *H. Karst.*) in the national park "Kopaonik". *Topola* **2016**, *197/198*, 25–34.
7. Ivanišević, P.; Galić, Z.; Rončević, S.; Kovačević, B.; Marković, M. Significance of establishment of forest tree and shrub plantations for the stability and sustainable development of ecosystems in Vojvodina. *Topola* **2008**, *181/182*, 31–40.

8. Ličina, V.; Nešić, L.; Belić, M.; Hadžić, V.; Sekulić, P.; Vasin, J.; Ninkov, J. The soils of Serbia and their degradation. *Ratar. Povrt.* **2011**, *48*, 285–290. [CrossRef]

9. Wilkins, D.A. The measurement of tolerance to edaphic factors by means of root growth. *New Phytol.* **1978**, *80*, 623–633. [CrossRef]

10. Ben Ahmed, C.; Ben Rouina, B.; Sensoy, S.; Boukhriss, M. Saline Water Irrigation Effects on Fruit Development, Quality, and Phenolic Composition of Virgin Olive Oils, Cv. Chemlali. *J. Agric. Food Chem.* **2009**, *57*, 2803–2811. [CrossRef]

11. Petridis, A.; Therios, I.; Samouris, G.; Tananaki, C. Salinity-induced changes in phenolic compounds in leaves and roots of four olive cultivars (*Olea europaea* L.) and their relationship to antioxidant activity. *Environ. Exp. Bot.* **2012**, *79*, 37–43. [CrossRef]

12. Taârit, M.B.; Msaada, K.; Hosni, K.; Marzouk, B. Physiological changes, phenolic content and antioxidant activity of *Salvia officinalis* L. grown under saline conditions. *J. Sci. Food Agric.* **2012**, *92*, 1614–1619. [CrossRef] [PubMed]

13. Hura, T.; Hura, K.; Grzesiak, S. Contents of total phenolics and ferulic acid, and PAL activity during water potential changes in leaves of maize single-cross hybrids of different drought tolerance. *J. Agron. Crop Sci.* **2008**, *194*, 104–112. [CrossRef]

14. Nguyen, K.K.; Cuellar, C.; Mavi, P.S.; LeDuc, D.; Bañuelos, G.; Sommerhalter, M. Two Poplar Hybrid Clones Differ in Phenolic Antioxidant Levels and Polyphenol Oxidase Activity in Response to High Salt and Boron Irrigation. *J. Agric. Food Chem.* **2018**, *66*, 7256–7264. [CrossRef] [PubMed]

15. Popović, B.M.; Štajner, D.; Ždero-Pavlović, R.; Tumbas-Šaponjac, V.; Čanadanović-Brunet, J.; Orlović, S. Water stress induces changes in polyphenol profile and antioxidant capacity in poplar plants (*Populus* spp.). *Plant Physiol. Biochem.* **2016**, *105*, 242–250. [CrossRef] [PubMed]

16. Xu, N.; Liu, S.; Lu, Z.; Pang, S.; Wang, L.; Wang, L.; Li, W. Gene Expression Profiles and Flavonoid Accumulation during Salt Stress in *Ginkgo biloba* Seedlings. *Plants* **2020**, *9*, 1162. [CrossRef]

17. Valifard, M.; Mohsenzadeh, S.; Kholdebarin, B.; Rowshan, V. Effects of salt stress on volatile compounds, total phenolic content and antioxidant activities of *Salvia mirzayanii*. *S. Afr. J. Bot.* **2014**, *93*, 92–97. [CrossRef]

18. Vuksanović, V.; Kovačević, B.; Kebert, M.; Milović, M.; Kesić, L.; Karaklić, V.; Orlović, S. In vitro modulation of antioxidant and physiological properties of white poplar induced by salinity. *Bull. Fac. For.* **2019**, *120*, 179–196. [CrossRef]

19. Popović, B.M.; Štajner, D.; Ždero-Pavlović, R.; Tari, I.; Csiszár, J.; Gallé, Á.; Poór, P.; Galović, V.; Trudić, B.; Orlović, S. Biochemical response of hybrid black poplar tissue culture (*Populus* × *canadensis*) on water stress. *J. Plant Res.* **2017**, *130*, 559–570. [CrossRef]

20. Niu, X.; Bressan, R.A.; Hasegawa, P.M.; Pardo, J.M. Ion Homeostasis in NaCl Stress Environments. *Plant Physiol.* **1995**, *109*, 735–742. [CrossRef]

21. Štajner, D.; Orlović, S.; Popović, B.M.; Kebert, M.; Galić, Z. Screening of drought oxidative stress tolerance in *Serbian melliferous* plant species. *Afr. J. Biotechnol.* **2011**, *10*, 1609–1614. [CrossRef]

22. Pessarakli, M.; Szabolcs, I. Soil salinity and sodicity as particular plant/crop stress factors. In *Handbook of Plant and Crop Stress*; Pessarakli, M., Ed.; Dekker: New York, NY, USA, 1999; pp. 1–16. ISBN 0824719484.

23. Hasegawa, P.M.; Bressan, R.A.; Zhu, J.-K.; Bohnert, H.J. Plant cellular and molecular responses to high salinity. *Annu. Rev. Plant Physiol. Plant Mol. Biol.* **2000**, *51*, 463–499. [CrossRef]

24. Tuskan, G.; DiFazio, S.; Bohlmann, J.; Grigoriev, I.; Hellsten, U.; Jansson, S.; Putnam, N.; Ralph, S.; Rombauts, S.; Salamov, A.; et al. The genome of black cottonwood, *Populus trichocarpa* (Torr. & Gray ex Brayshaw). *Science* **2006**, *313*, 1596–1604. [CrossRef]

25. Brosché, M.; Vinocur, B.; Alatalo, E.R.; Lamminmäki, A.; Teichmann, T.; Ottow, E.A.; Djilianov, D.; Afif, D.; Bogeat-Triboulot, M.B.; Altman, A.; et al. Gene expression and metabolite profiling of *Populus euphratica* growing in the Negev desert. *Genome Biol.* **2005**, *6*, R101. [CrossRef]

26. Sairam, R.K.; Tyagi, A. Physiology and Molecular Biology of Salinity Stress Tolerance in Plants. *Curr. Sci.* **2004**, *86*, 407–421. [CrossRef]

27. Giri, J. Glycinebetaine and abiotic stress tolerance in plants. *Plant Signal. Behav.* **2011**, *6*, 1746–1751. [CrossRef]

28. Qureshi, M.; Abdin, M.; Ahmad, J.; Iqbal, M. Effect of long-term salinity on cellular antioxidants, compatible solute and fatty acid profile of sweet Annie (*Artemisia annua* L.). *Phytochemistry* **2013**, *95*, 215–223. [CrossRef]

29. Kebert, M.; Rapparini, F.; Neri, L.; Bertazza, G.; Orlović, S.; Biondiet, S. Copper-induced responses in poplar clones are associated with genotype- and organ-specific changes in peroxidase activity and proline, polyamine, ABA, and IAA levels. *J. Plant Growth Regul.* **2017**, *36*, 131–147. [CrossRef]

30. Ashrafa, M.; Foolad, M.R. Roles of glycine betaine and proline in improving plant abiotic stress resistance. *Environ. Exp. Bot.* **2007**, *59*, 206–216. [CrossRef]

31. Székely, G.; Abrahám, E.; Cséplo, A.; Rigó, G.; Zsigmond, L.; Csiszár, J.; Ayaydin, F.; Strizhov, N.; Jásik, J.; Schmelzer, E.; et al. Duplicated P5CS genes of *Arabidopsis* play distinct roles in stress regulation and developmental control of proline biosynthesis. *Plant J.* **2008**, *53*, 11–28. [CrossRef] [PubMed]

32. Fabro, G.; Kovács, I.; Pavet, V.; Szabados, L.; Alvarez, M.E. Proline accumulation and AtP5CS2 gene activation are induced by plant-pathogen incompatible interactions in *Arabidopsis*. *Mol. Plant Microbe Interact.* **2004**, *17*, 343–350. [CrossRef]

33. Gharsallah, C.; Fakhfakh, H.; Grubb, D.; Gorsane, F. Effect of salt stress on ion concentration, proline content, antioxidant enzyme activities and gene expression in tomato cultivars. *AoB Plants* **2016**, *8*, plw055. [CrossRef]

34. Pottosin, I.; Velarde-Buendía, A.M.; Bose, J.; Zepeda-Jazo, I.; Shabala, S.; Dobrovinskaya, O. Cross-talk between reactive oxygen species and polyamines in regulation of ion transport across the plasma membrane: Implications for plant adaptive responses. *J. Exp. Bot.* **2014**, *65*, 1271–1283. [CrossRef]

35. Kavi Kishor, P.B.; Sangam, S.; Amrutha, R.N.; Laxmi, P.S.; Naidu, K.R.; Rao, K.R.S.S.; Rao, S.; Theriappan, P.; Sreenivasulu, N. Regulation of proline biosynthesis, degradation, uptake and transport in higher plants: Its implications in plant growth and abiotic stress tolerance. *Curr. Sci.* **2005**, *88*, 424–438.

36. Bray, E.A.; Bailey-Serres, J.; Weretilnyk, E. Responses to abiotic stress. In *Biochemistry & Molecular Biology of Plants*; Gruissem, W., Jones, R., Eds.; American Society of Plant Physiologists: Rockville, MA, USA, 2000; pp. 1158–1249. ISBN 0943088399.

37. Cushman, J.C.; Bohnert, H.J. Genomic approaches to plant stress tolerance. *Curr. Opin. Plant Biol.* **2000**, *3*, 117–124. [CrossRef]

38. Szabados, L.; Savouré, A. Proline: A multifunctional amino acid. *Trends Plant Sci.* **2010**, *15*, 89–97. [CrossRef]

39. Rizhsky, L.; Liang, H.; Shuman, J.; Shulaev, V.; Davletova, S.; Mittler, R. When Defense Pathways Collide. The Response of Arabidopsis to a Combination of Drought and Heat Stress. *Plant Physiol.* **2004**, *134*, 1683–1696. [CrossRef]

40. Yang, D.S.; Zhang, J.; Li, M.-X. Metabolomics Analysis Reveals the Salt-Tolerant Mechanism in *Glycine soja*. *J. Plant Growth Regul.* **2017**, *36*, 460–471. [CrossRef]

41. Rhodes, D.; Hanson, A.D. Quaternary ammonium and tertiary sulfonium compounds in higher-plants. *Annu. Rev. Plant Physiol. Plant Mol. Biol.* **1993**, *44*, 357–384. [CrossRef]

42. Seki, M.; Narusaka, M.; Ishida, J.; Nanjo, T.; Fujita, M.; Oono, Y.; Kamiya, A.; Nakajima, M.; Enju, A.; Sakurai, T.; et al. Monitoring the expression profiles of 7000 *Arabidopsis genes* under drought, cold and high salinity stresses using a full-length cDNA microarray. *Plant J.* **2002**, *31*, 279–292. [CrossRef]

43. Wang, Q.; Guan, C.; Wang, P.; Ma, Q.; Bao, A.K.; Zhang, J.L.; Wang, S.M. The Effect of AtHKT1;1 or AtSOS1 Mutation on the Expressions of $Na^+$ or $K^+$ Transporter Genes and Ion Homeostasis in Arabidopsis thaliana under Salt Stress. *Int. J. Mol. Sci.* **2019**, *20*, 1085. [CrossRef] [PubMed]

44. Qiu, Q.-S.; Guo, Y.; Dietrich, M.A.; Schumaker, K.S.; Zhu, J.-K. Regulation of SOS1, a plasma membrane Na+/H+ exchanger in *Arabidopsis thaliana*, by SOS2 and SOS3. *Proc. Natl. Acad. Sci. USA* **2002**, *99*, 8436–8441. [CrossRef] [PubMed]

45. Zhu, J.K. Salt and drought stress signal transduction in plants. *Annu. Rev. Plant Biol.* **2002**, *53*, 247–273. [CrossRef] [PubMed]

46. Shi, H.; Ishitani, M.; Kim, C.; Zhu, J.-K. The *Arabidopsis thaliana* salt tolerance gene SOS1 encodes a putative $Na^+/H^+$ antiporter. *Proc. Natl. Acad. Sci. USA* **2000**, *97*, 6896–6901. [CrossRef]

47. Wang, H.; Wang, H.; Shao, H.; Tang, X. Recent Advances in Utilizing Transcription Factors to Improve Plant Abiotic Stress Tolerance by Transgenic Technology. *Front. Plant Sci.* **2016**, *7*, 67. [CrossRef] [PubMed]

48. Ma, H.S.; Liang, D.; Shuai, P.; Xia, X.L.; Yin, W.L. The salt- and drought-inducible poplar GRAS protein SCL7 confers salt and drought tolerance in *Arabidopsis thaliana*. *J. Exp. Bot.* **2010**, *61*, 4011–4019. [CrossRef] [PubMed]

49. Dubouzet, J.G.; Sakuma, Y.; Ito, Y.; Kasuga, M.; Dubouzet, E.G.; Miura, S.; Seki, M.; Shinozaki, K.; Yamaguchi-Shinozaki, K. OsDREB genes in rice, *Oryza sativa* L., encode transcription activators that function in drought-, high-salt- and coldresponsive gene expression. *Plant J.* **2003**, *33*, 751–763. [CrossRef] [PubMed]

50. Magome, H.; Yamaguchi, S.; Hanada, A.; Kamiya, Y.; Oda, K. Dwarf and delayedflowering 1, a novel *Arabidopsis* mutant deficient in gibberellin biosynthesis because of overexpression of a putative AP2 transcription factor. *Plant J.* **2004**, *37*, 720–729. [CrossRef] [PubMed]

51. Liu, Q.; Kasuga, M.; Sakuma, Y.; Abe, H.; Miura, S.; Yamaguchi-Shinazaki, K.; Shinozaki, K. Two transcription factors, DREB1 and DREB2, with an AP2/EREBP DNA-binding domain separate two cellular signal transduction pathways in drought- and low-temperature-responsive gene expression in *Arabidopsis*. *Plant Cell* **1998**, *10*, 1391–1406. [CrossRef] [PubMed]

52. Yamaguchi-Shinozaki, K.; Shinozaki, K. Organization of cis-acting regulatory elements in osmotic- and cold-stress responsive promoters. *Trends Plant Sci.* **2005**, *10*, 88–94. [CrossRef]

53. Chen, J.; Xia, X.; Yin, W. Expression profiling and functional characterization of a DREB2-type gene from *Populus euphratica*. *Biochem. Biophys. Res. Commun.* **2009**, *378*, 483–487. [CrossRef] [PubMed]

54. Galovic, V.; Orlovic, S.; Fladung, M. Characterization of two poplar homologs of the GRAS/SCL gene, which encodes a transcription factor putatively associated with salt tolerance. *iForest Biogeosci. For.* **2015**, *8*, 780–785. [CrossRef]

55. Chang, C.; Yang, M.; Wen, H.; Chern, J. Estimation of total flavonoid content in propolis by two complementary colorimetric methods. *J. Food Drug Anal.* **2002**, *10*, 178–182. [CrossRef]

56. Kim, D.O.; Jeong, S.W.; Lee, C.Y. Antioxidant capacity of phenolic phytochemicals from various cultivars of plums. *Food Chem.* **2003**, *81*, 321–326. [CrossRef]

57. Arnao, M.B. Some methodological problems in the determination of antioxidant activity using chromogen radicals: A practical case. *Trends Food Sci. Technol.* **2000**, *11*, 419–421. [CrossRef]

58. Miller, N.; Rice-Evans, C. Factors influencing the antioxidant activity determined by the ABTS radical cation assay. *Free Radic. Res.* **1997**, *26*, 195–199. [CrossRef]

59. Benzie, I.F.F.; Strain, J.J. The ferric reducing ability of plasma as a measure of "antioxidant power": The FRAP assay. *Anal. Biochem.* **1996**, *239*, 70–76. [CrossRef] [PubMed]

60. TIBCO Software Inc. Statistica (Data Analysis Software System). Version 13. 2017. Available online: https://docs.tibco.com/products/tibco-statistica-13-3-0 (accessed on 2 April 2021).

61. Livak, K.J.; Schmittgen, T.D. Analysis of relative gene expression data using realtime quantitative PCR and the 2(−delta delta c(t)) method. *Methods* **2001**, *25*, 402–408. [CrossRef] [PubMed]

62. R Core Team. *R: A Language and Environment for Statistical Computing*; R Core Team: Vienna, Austria, 2013.

63. Vuksanovic, V.; Kovacevic, B.; Kebert, M.; Katanic, M.; Pavlovic, L.; Kesic, L.; Orlovic, S. Clone specificity of white poplar (*Populus alba* L.) acidity tolerance in vitro. *Fresenius Environ. Bull.* **2019**, *11*, 8307–8313.

64. Štajner, D.; Popović, B.M.; Ćalić-Dragosavac, D.; Malenčić, Đ.; Zdravković-Korać, S. Comparative Study on *Allium schoenoprasum* Cultivated Plant and *Allium schoenoprasum* Tissue Culture Organs Antioxidant Status. *Phyther. Res.* **2011**, *25*, 1618–1622. [CrossRef]

65. Meloni, D.A.; Martínez, C.A. Glycinebetaine improves salt tolerance in vinal (*Prosopis ruscifolia* Griesbach) seedlings. *Braz. J. Plant Physiol.* **2009**, *21*, 233–241. [CrossRef]

66. Saitou, N.; Nei, M. The neighbor-joining method: A new method for reconstructing phylogenetic trees. *Mol. Biol. Evol.* **1987**, *4*, 406–425. [CrossRef] [PubMed]

67. Felsenstein, J. Confidence limits on phylogenies: An approach using the bootstrap. *Evolution* **1985**, *39*, 783–791. [CrossRef] [PubMed]

68. Zuckerkandl, E.; Pauling, L. Evolutionary divergence and convergence in proteins. In *Evolving Genes and Proteins*; Bryson, V., Vogel, H.J., Eds.; Academic Press: New York, NY, USA, 1965; pp. 97–166. ISBN 9781483227344.

69. Tamura, K.; Peterson, D.; Peterson, N.; Stecher, G.; Nei, M.; Kumar, S. MEGA5: Molecular Evolutionary Genetics Analysis using Maximum Likelihood, Evolutionary Distance, and Maximum Parsimony Methods. *Mol. Biol. Evol.* **2011**, *28*, 2731–2739. [CrossRef] [PubMed]

70. Galović, V.; Orlović, O.; Prathiba, M.J.; Pekeč, S.; Vasić, V.; Vasić, S.; Szabados, L. Characterization of abiotic stress responsive RD29B and RD17 genes in different poplar clones. *Topola* **2020**, *206*, 13–20. [CrossRef]

71. Shi, H.Z.; Quintero, F.J.; Pardo, J.M.; Zhu, J.K. The putative plasma membrane Na+/H+ antiporter SOS1 controls long-distance Na+ transport in plants. *Plant Cell* **2002**, *14*, 465–477. [CrossRef] [PubMed]

72. Oh, D.H.; Lee, S.Y.; Bressan, R.A.; Yun, D.J.; Bohnert, H.J. Intracellular consequences of SOS1 deficiency during salt stress. *J. Exp. Bot.* **2010**, *61*, 1205–1213. [CrossRef] [PubMed]

73. Lata, C.; Prasad, M. Role of DREBs in regulation of abiotic stress responses in plants. *J. Exp. Bot.* **2011**, *62*, 4731–4748. [CrossRef] [PubMed]

74. Van Zelm, E.; Zhang, Y.; Testerink, C. Salt tolerance mechanisms of plants. *Annu. Rev. Plant Biol.* **2020**, *71*, 403–433. [CrossRef] [PubMed]

75. Fladung, M.; Buschbom, J. Identification of single nucleotide polymorphisms in different *Populus* species. *Trees* **2009**, *23*, 1199–1212. [CrossRef]

76. Flowers, T.J.; Munns, R.; Colmer, T.D. Sodium chloride toxicity and the cellular basis of salt tolerance in halophytes. *Ann. Bot.* **2015**, *115*, 419–431. [CrossRef] [PubMed]

77. Janz, D.; Behnke, K.; Schnitzler, J.P.; Kanawati, B.; Schmitt-Kopplin, P.; Polle, A. Pathway analysis of the transcriptome and metabolome of salt sensitive and tolerant poplar species reveals evolutionary adaption of stress tolerance mechanisms. *BMC Plant Biol.* **2010**, *10*, 150. [CrossRef] [PubMed]

*forests*

*Article*

# Establishment of Regional Phytoremediation Buffer Systems for Ecological Restoration in the Great Lakes Basin, USA. I. Genotype × Environment Interactions

Ronald S. Zalesny, Jr. [1,*], Andrej Pilipović [2], Elizabeth R. Rogers [1,3], Joel G. Burken [4], Richard A. Hallett [5], Chung-Ho Lin [3], Bernard G. McMahon [6], Neil D. Nelson [6], Adam H. Wiese [1], Edmund O. Bauer [1], Larry Buechel [7], Brent S. DeBauche [3], Mike Peterson [7], Ray Seegers [8] and Ryan A. Vinhal [1,3]

[1] USDA Forest Service, Northern Research Station, Institute for Applied Ecosystem Studies, Rhinelander, WI 54501, USA; elizabeth.r.rogers@usda.gov (E.R.R.); adam.wiese@usda.gov (A.H.W.); ebauer@charter.net (E.O.B.); ryan.vinhal@usda.gov (R.A.V.)

[2] Institute of Lowland Forestry and Environment, University of Novi Sad, 21000 Novi Sad, Serbia; andrejp@uns.ac.rs

[3] Center for Agroforestry, School of Natural Resources, University of Missouri–Columbia, Columbia, MO 65201, USA; LinChu@missouri.edu (C.-H.L.); debaucheb@umsystem.edu (B.S.D.)

[4] Department of Civil, Architectural and Environmental Engineering, Missouri University of Science and Technology, Civil, Architectural, and Environmental Engineering, Rolla, MO 65401, USA; burken@mst.edu

[5] USDA Forest Service, Northern Research Station, New York City Urban Field Station, Bayside, NY 11360, USA; richard.hallett@usda.gov

[6] Natural Resources Research Institute, University of Minnesota Duluth, Duluth, MN 55807, USA; bmcmahon@nrri.umn.edu (B.G.M.); nnelson2@d.umn.edu (N.D.N.)

[7] Environmental Legacy Management Group—Midwest, Waste Management, Inc., Menomonee Falls, WI 53051, USA; lbuechel@wm.com (L.B.); mpeterso2@wm.com (M.P.)

[8] Waste Management of Wisconsin, Inc., Whitelaw, WI 54247, USA; rseegers@wm.com

* Correspondence: ronald.zalesny@usda.gov

**Citation:** Zalesny, R.S., Jr.; Pilipović, A.; Rogers, E.R.; Burken, J.G.; Hallett, R.A.; Lin, C.-H.; McMahon, B.G.; Nelson, N.D.; Wiese, A.H.; Bauer, E.O.; et al. Establishment of Regional Phytoremediation Buffer Systems for Ecological Restoration in the Great Lakes Basin, USA. I. Genotype × Environment Interactions. *Forests* **2021**, *12*, 430. https://doi.org/10.3390/f12040430

Academic Editors: Francisca C. Aguiar

Received: 6 March 2021
Accepted: 30 March 2021
Published: 2 April 2021

**Publisher's Note:** MDPI stays neutral with regard to jurisdictional claims in published maps and institutional affiliations.

**Abstract:** Poplar remediation systems are ideal for reducing runoff, cleaning groundwater, and delivering ecosystem services to the North American Great Lakes and globally. We used phyto-recurrent selection (PRS) to establish sixteen phytoremediation buffer systems (phyto buffers) (buffer groups: 2017 × 6; 2018 × 5; 2019 × 5) throughout the Lake Superior and Lake Michigan watersheds comprised of twelve PRS-selected clones each year. We tested for differences in genotypes, environments, and their interactions for health, height, diameter, and volume from ages one to four years. All trees had optimal health. Mean first-, second-, and third-year volume ranged from $71 \pm 26$ to $132 \pm 39$ cm$^3$; $1440 \pm 575$ to $5765 \pm 1132$ cm$^3$; and $8826 \pm 2646$ to $10{,}530 \pm 2110$ cm$^3$, respectively. Fourth-year mean annual increment of 2017 buffer group trees ranged from $1.1 \pm 0.7$ to $7.8 \pm 0.5$ Mg ha$^{-1}$ yr$^{-1}$. We identified generalist varieties with superior establishment across a broad range of buffers ('DM114', 'NC14106', '99038022', '99059016') and specialist clones uniquely adapted to local soil and climate conditions ('7300502', 'DN5', 'DN34', 'DN177', 'NM2', 'NM5', 'NM6'). Using generalists and specialists enhances the potential for phytoremediation best management practices that are geographically robust, being regionally designed yet globally relevant.

**Keywords:** ecosystem services; multi-environmental trials (MET); phenotypic plasticity; phyto buffers; phyto-recurrent selection; phytotechnologies; poplars; *Populus*

## 1. Introduction

Ninety-five percent of the United States' surface freshwater and 20% of the world's freshwater reserve are contained within the Great Lakes Basin [1,2]. The Basin provides ecosystem services (including clean drinking water) to over 34 million people in the United States and Canada [3–5], and contributes substantially to the United States' economy, with a gross regional product (GRP) estimated at nearly 4.1 trillion USD [6]. Anthropogenic activities in the region, combined with population growth and land use changes, induce a vast number of stressors such as toxic pollution, increases in invasive species populations, and climate change [7], all of which can disturb terrestrial water cycles [8] and impact water quality of the Great Lakes and their watersheds [9,10]. Non-point sources of pollution such as landfills and similar sites contribute to watershed contamination by runoff and leakage. Although pollutant levels within landfills generally decrease over time through chemical and biological alteration and degradation [11–13], treatment of leachate and wastewaters can help mitigate soil and water contamination [14]. Sustainable, long-term restoration practices are needed to preserve and enhance ecosystem services in the Great Lakes Basin. In the early 2010s, over 1.5 billion USD were invested in restoration projects to halt or mitigate pollution of the Great Lakes ecosystem [15]. Ecological restoration techniques and technologies are more economically viable and sustainable long-term solutions than off-site treatment methods for mitigating contamination.

Ecological restoration may help reverse land degradation, improve the resilience of biodiversity, and deliver important ecosystem services [16,17]. Phytotechnologies are an ideal solution for preventing water contamination, and they include biological recovery activities that have been classified into four categories: remediation, reclamation, restoration, and rehabilitation [18]. The most common phytotechnology, phytoremediation, involves the use of plants for remediation and prevention of water contamination [19]. Often, phytoremediation is accomplished by installing vegetative covers, riparian buffers, and/or hydraulic control systems. Depending upon their type and chemical properties, contaminants are accumulated, immobilized, metabolized or volatilized [20–22].

Fast-growing trees such as *Populus* L. species and their hybrids (hereafter referred to as poplars) have been studied and used in plantation-based silvicultural systems known as short rotation woody crop (SRWC) systems for over 100 years [23,24]. The primary focus for SRWCs has been on the production of wood and wood products. More recently, poplar production systems have been implemented to provide a variety of ecosystem services [25–27], including their application in phytotechnologies for ecological clean up and restoration [28]. Phytoremediation research starting in the 1990s showed that poplars were great candidates for remediation systems because of their vigorous growth, easy vegetative propagation, well-developed root systems, and high productivity on marginal and liability lands [27,29,30]. The results of such studies indicate the potential of poplars for phytoremediation of various inorganic and organic contaminants [20,31–35]. Another factor that adds to the value of poplars in phytoremediation is that they exhibit a broad range of genetic diversity given their ability to undergo spontaneous and controlled intra- and inter-species hybridization, which thereby creates a high number of simple and complex hybrids [36].

Phytoremediation projects are often installed on sites and in soils that are not ideal for plant growth. Variability in environmental conditions affects evolutionary processes of populations or individuals within species, resulting in broadening or narrowing of their ecological niches, defined as generalists and specialists, respectively [37,38]. Poplars are ideal candidates for tree improvement by hybridization and breeding [39–41], with further aims to fully exploit poplar genomic and genetic potential and maximize wood biomass productivity from fast-growing trees worldwide. The ability of genotypes to produce different phenotypes in distinct environmental conditions is defined as phenotypic plasticity, and includes variation in the morphology, physiology, behavior, and life history of organisms [42]. This can lead to unpredictability in performance that can complicate selection of proper genotypes for phytoremediation projects. The genotype by environment ($G \times E$) interaction, in which a phenotype is a function of the genotype, the environment, and the differential phenotypic response of genotypes to site-specific edaphic and climatic conditions, is a determining factor of clonal site performance [43]. Such interaction exists when comparative performances of genotypes vary according to local site conditions, with superiority of a genotype in a certain environment converted to inferiority in another environment [44]. Genotypic variability across a variety of environments can lead to unpredictability in performance that can complicate selection of proper genotypes for commercial use. Therefore, in tree breeding, multi-environmental trials (MET) are implemented to evaluate the degree and pattern of G × E interactions, as well as to test the robustness of genotypes in different environments [45,46].

However, sometimes phenotypic plasticity can hinder expression of G × E interactions. For example, testing of hybrid poplar clones for growth performance and robustness at different sites in the Midwestern USA showed low G × E interaction and high plasticity of tested genotypes, indicating their suitability for growing on a wider spectrum of habitats [46,47]. In poplar research, G × E interactions and definitions of generalists vs specialists have been studied for various traits ranging from biomass production [48–50] and rooting ability [51], to physiological traits related to productivity [36] and drought tolerance [52], to wood properties [53,54]. Often, G × E interactions are also determined by the genomic groups of the hybrids, where components of different *Populus* species (e.g., *P. deltoides* Bartr. ex. Marsh, *P. maximowiczii* A. Henry, *P. nigra* L., *P. trichocarpa* Torr. et Gray) significantly contribute to the expression of specific traits by a given genotype, and are reflected in rooting ability and adaptability to certain climate, latitude or soil-water conditions [48,49,55,56]. Therefore, clonal testing to maximize understanding of G × E interactions is paramount in establishing successful poplar plantings for a broad range of end uses and ecosystem services, including environmental objectives such as phytoremediation. For example, optimizing G × E interactions may be most appropriate for small-scale applications with site-specific needs, while minimum G × E interactions are needed to deploy superior and robust clones at a justifiable cost for large-scale commercial systems [46].

The current study is a component of a regional phytoremediation testing network originally funded by the Great Lakes Restoration Initiative (GLRI) [2] to select and test the ability of new poplar genotypes to reduce surface runoff and prevent groundwater contamination at landfills and similar sites in the Great Lakes Basin. We implemented phyto-recurrent selection, a stepwise greenhouse- and field-based approach that combines crop and tree improvement strategies to evaluate, identify and select superior-performing clones based on multiple testing cycles [32,57]. Results of the initial screening of candidate clones in greenhouse experiments, previously published by Rogers et al. [58], were used to develop the MET testing network consisting of sixteen phytoremediation buffer systems (i.e., phyto buffers) at ten sites located in the Lake Superior (i.e., Michigan's Upper Peninsula) and Lake Michigan (i.e., eastern Wisconsin) watersheds. Our objective was to test for differences in genotypes, environments, and their interactions for health, height, diameter, and volume during early field establishment (i.e., from one to four years after planting). Our results identify poplar clones with maximum phytoremediation potential that can be established across a wide range of environmental conditions. These data are useful for clonal selection to maximize phytoremediation potential in future phytotechnologies, regardless of specific site conditions or genotypes deployed.

## 2. Materials and Methods

### 2.1. Site Description

A regional phytotechnologies network consisting of sixteen phytoremediation buffer systems (i.e., phyto buffers) was established in 2017 (×6 phyto buffers), 2018 (×5), and 2019 (×5) in the Lake Superior watershed of the Upper Peninsula of Michigan, USA and the Lake Michigan watershed of eastern Wisconsin, USA. Multiple phyto buffers were installed at some locations, resulting in ten field testing sites throughout the network (Figure 1). The sites ranged in latitude from 46.7840 to 42.8382 °N and in longitude from −89.1291 to −86.5976 °W, which is consistent with poplar productivity supplysheds in the region [47,59–61]. Site-related climate properties are shown in Table 1. Twenty-year historical monthly averages for precipitation, temperature, and drought were determined across each growing season (April to October) for the time period 2000 to 2020 and summed (precipitation) and averaged (temperature, drought) to obtain final annual values. Based on the nearest weather station to each site, precipitation (P, mm) along with maximum ($T_{max}$, °C) and minimum ($T_{min}$, °C) air temperatures were obtained from the National Oceanic and Atmospheric Administration (NOAA) National Climate Data Center (https://www.ncdc.noaa.gov/cdo-web/, accessed on 20 January 2021). Average temperature (i.e., $T_{avg} = ((T_{max} + T_{min})/2)$, °C) and the difference between maximum and minimum temperatures (i.e., $T_{diff} = T_{max} - T_{min}$, °C) were calculated for each site. Daily growing degree days ($GDD_{day}$) were calculated as the average temperature minus a base temperature of 10 °C (i.e., $GDD_{day} = T_{avg} - 10$ °C) and summed across each growing season. Average annual growing degree days for each site ($GDD_{annual}$) were estimated by averaging the seasonal GDD values from 2000 to 2020. The United States Drought Monitor (https://droughtmonitor.unl.edu/, accessed on 20 January 2021) was accessed to obtain drought index scores according to percent area within each county belonging to four drought index categories: D0 (abnormally dry), D1 (moderate drought), D2 (severe drought), D3 (extreme drought), and D4 (exceptional drought). Categories D3 and D4 were negligible and, therefore, not reported in Table 1. Buffer-specific soil properties were acquired from the USDA Natural Resources Conservation Service (NRCS) Web Soil Survey (https://websoilsurvey.sc.egov.usda.gov/, accessed on 20 January 2021) and are provided in Table 2.

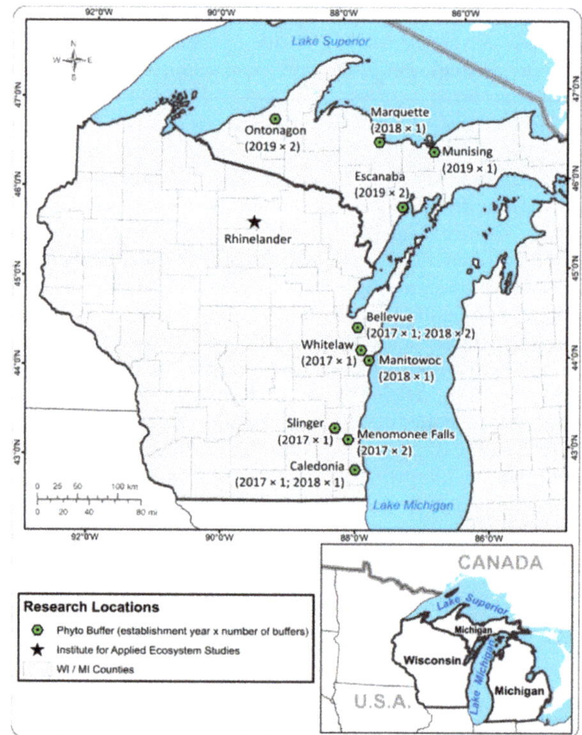


**Figure 1.** Regional phytotechnologies network consisting of sixteen phytoremediation buffer systems (i.e., phyto buffers) established in 2017 (×6 phyto buffers), 2018 (×5), and 2019 (×5) in the Lake Superior watershed of the Upper Peninsula of Michigan, USA and the Lake Michigan watershed of eastern Wisconsin, USA.

**Table 1.** Precipitation, temperature, growing degree days, and drought indices of ten field testing sites in a regional phyto technologies network consisting of sixteen phytoremediation buffer systems (i.e., phyto buffers) established from 2017 to 2019 in the Lake Superior watershed of the Upper Peninsula of Michigan, USA and the Lake Michigan watershed of eastern Wisconsin, USA.

| Site | Bellevue, WI | Caledonia, WI | Escanaba, MI | Manitowoc, WI | Marquette, MI |
|---|---|---|---|---|---|
| County | Brown | Racine | Delta | Manitowoc | Marquette |
| Buffer group (i.e., year of planting) | 2017, 2018 | 2017, 2018 | 2019 | 2018 | 2018 |
| Total number of phyto buffers | 3 | 2 | 2 | 1 | 1 |
| Annual precipitation (P) (mm) | $613 \pm 27$ | $686 \pm 36$ | $556 \pm 32$ | $614 \pm 27$ | $530 \pm 28$ |
| Average temperature ($T_{avg}$) (°C) | $15.3 \pm 0.2$ | $15.7 \pm 0.2$ | $13.6 \pm 0.2$ | $14.8 \pm 0.2$ | $13.1 \pm 0.4$ |
| Maximum temperature ($T_{max}$) (°C) | $21.1 \pm 0.2$ | $21.5 \pm 0.2$ | $20.0 \pm 0.3$ | $19.2 \pm 0.3$ | $17.3 \pm 0.4$ |
| Minimum temperature ($T_{min}$) (°C) | $9.5 \pm 0.1$ | $10.0 \pm 0.1$ | $7.2 \pm 0.2$ | $10.4 \pm 0.2$ | $8.9 \pm 0.4$ |
| Maximum–minimum temperature ($T_{diff}$) (°C) | $11.6 \pm 0.1$ | $11.5 \pm 0.1$ | $12.8 \pm 0.2$ | $15.8 \pm 0.4$ | $8.5 \pm 0.1$ |
| Annual growing degree days ($GDD_{annual}$) | $1342 \pm 27$ | $1418 \pm 31$ | $1017 \pm 24$ | $1213 \pm 37$ | $997 \pm 54$ |
| Drought index (abnormally dry) (D0) (%) | $24.0 \pm 5.3$ | $22.4 \pm 5.7$ | $31.4 \pm 6.9$ | $20.1 \pm 5.0$ | $36.5 \pm 7.7$ |
| Drought index (moderate drought) (D1) (%) | $8.0 \pm 3.1$ | $9.5 \pm 4.6$ | $10.3 \pm 3.8$ | $7.9 \pm 3.2$ | $14.7 \pm 4.8$ |
| Drought index (severe drought) (D2) (%) | $0.8 \pm 0.8$ | $4.9 \pm 3.0$ | $3.9 \pm 2.0$ | $1.2 \pm 1.1$ | $4.5 \pm 2.4$ |

| Site | Menomonee Falls, WI | Munising, MI | Ontonagon, MI | Slinger, WI | Whitelaw, WI |
|---|---|---|---|---|---|
| County | Waukesha | Alger | Ontonagon | Washington | Manitowoc |
| Buffer group (i.e., year of planting) | 2017 | 2019 | 2019 | 2017 | 2017 |
| Total number of phyto buffers | 2 | 1 | 2 | 1 | 1 |
| Annual precipitation (P) (mm) | $649 \pm 23$ | $655 \pm 25$ | $551 \pm 26$ | $653 \pm 36$ | $640 \pm 26$ |
| Average temperature ($T_{avg}$) (°C) | $15.3 \pm 0.1$ | $12.3 \pm 0.2$ | $13.4 \pm 0.2$ | $15.1 \pm 0.2$ | $14.9 \pm 0.1$ |
| Maximum temperature ($T_{max}$) (°C) | $21.2 \pm 0.2$ | $17.0 \pm 0.2$ | $19.7 \pm 0.3$ | $21.1 \pm 0.2$ | $21.0 \pm 0.2$ |
| Minimum temperature ($T_{min}$) (°C) | $9.4 \pm 0.1$ | $7.7 \pm 0.2$ | $7.1 \pm 0.2$ | $9.0 \pm 0.2$ | $8.9 \pm 0.1$ |
| Maximum–minimum temperature ($T_{diff}$) (°C) | $11.8 \pm 0.1$ | $9.3 \pm 0.1$ | $12.7 \pm 0.1$ | $12.1 \pm 0.1$ | $12.1 \pm 0.1$ |
| Annual growing degree days ($GDD_{annual}$) | $1344 \pm 26$ | $877 \pm 30$ | $1044 \pm 39$ | $1286 \pm 29$ | $1295 \pm 26$ |
| Drought index (abnormally dry) (D0) (%) | $20.9 \pm 5.4$ | $28.4 \pm 6.4$ | $37.9 \pm 8.0$ | $17.7 \pm 5.1$ | $20.1 \pm 5.0$ |
| Drought index (moderate drought) (D1) (%) | $9.1 \pm 4.4$ | $9.0 \pm 3.3$ | $15.2 \pm 6.3$ | $9.2 \pm 4.2$ | $7.9 \pm 3.2$ |
| Drought index (severe drought) (D2) (%) | $4.1 \pm 2.5$ | $3.6 \pm 2.1$ | $7.2 \pm 3.8$ | $2.7 \pm 1.7$ | $1.2 \pm 1.1$ |

Climate and drought data are means ± one standard error across each growing season (April to October) from 2000 to 2020. Climate source: National Oceanic and Atmospheric Administration (NOAA) National Climate Data Center (https://www.ncdc.noaa.gov/cdo-web/, accessed on 20 January 2021). Drought source: United States Drought Monitor (https://droughtmonitor.unl.edu/, accessed on 20 January 2021); percent of area within the county in each category (D0 to D2).

## 2.2. Clone Selection

Phyto-recurrent selection was conducted through a polycyclic evaluation process to choose superior genotypes for out planting [32,58,62]. For each phyto buffer, soils were collected in the field, brought to the USDA Forest Service, Institute for Applied Ecosystem Studies in Rhinelander, Wisconsin, USA (Figure 1), and processed (i.e., dried and sifted to remove large debris) for use as soil treatments during three greenhouse selection cycles [58]. The number of clones tested in each cycle decreased from 140 (cycle 1) to 60 (cycle 2) to 15 (cycle 3) based on multiplicative rank summation indices incorporating tree survival, growth, physiology, and health [63]. Twelve clones were selected for cycle 4 field testing, resulting in three buffer groups related to phyto buffer establishment in 2017, 2018, and 2019. Phyto buffers within buffer group years had the same twelve clones, but the twelve clones differed among the three buffer groups based on cycle 3 phyto-recurrent selection results. In the current study, separate analyses were conducted for each buffer group. Each set of buffer group clones consisted of genotypes that have: (1) been commonly used for commercial and/or research purposes in the region (i.e., Common clones), (2) a rich history of testing but are still at the experimental stage (i.e., Experimental), and (3) been bred, tested, and selected at the University of Minnesota Duluth, Natural Resources Research Institute (NRRI) and show promise for broad-ranging applications (i.e., NRRI) [46,56]. Specific clones and their genomic groups are listed in Table 3.

**Table 2.** Soil properties of sixteen phytoremediation buffer systems (i.e., phyto buffers) comprising a regional phytotechnologies network established from 2017 to 2019 in the Lake Superior watershed of the Upper Peninsula of Michigan, USA and the Lake Michigan watershed of eastern Wisconsin, USA.

| Phyto Buffer [a] | BC | BE | BW | CE | CW | EE, EW | MA | ME, MW | MQ | MU | ON, OS | SL | WH |
|---|---|---|---|---|---|---|---|---|---|---|---|---|---|
| Soil series | Manawa | Kewaunee | Bellevue | Fox | Matherton | Croswell | Hochheim | Sebewa | Schweitzer | Kalkaska | Oldman | Casco | Boyer |
| Drainage class [b] | SPD | WD | SPD | MWD | SPD | MWD | WD | PD | WD | SED | MWD | SED | WD |
| Slope (%) | 0 to 3 | 2 to 6 | 2 to 6 | 2 to 6 | 1 to 3 | 0 to 3 | 6 to 12 | 0 to 2 | 6 to 25 | 0 to 6 | 6 to 35 | 20 to 30 | 6 to 12 |
| K factor (erodibility, 0.02 to 0.69 scale) | 0.37 | 0.49 | 0.28 | 0.37 | 0.28 | 0.05 | 0.37 | 0.37 | 0.20 | 0.04 | 0.37 | 0.32 | 0.43 |
| Texture [c] | SiCL | SiCL | SiCL | L | L | S | L | L | SL | S | L | SL | SCL |
| Sand (%) | 10.1 | 13.3 | 19.8 | 39.5 | 50.1 | 87.4 | 45.4 | 37.3 | 55.9 | 94.7 | 51.4 | 54.0 | 58.2 |
| Silt (%) | 45.9 | 47.7 | 50.0 | 39.7 | 28.1 | 10.4 | 34.4 | 42.1 | 41.1 | 4.4 | 41.4 | 28.6 | 18.8 |
| Clay (%) | 44.0 | 39.0 | 30.2 | 20.8 | 21.8 | 2.2 | 20.2 | 20.6 | 3.0 | 0.9 | 7.2 | 17.4 | 23.0 |
| Organic matter (%) | 1.3 | 1.1 | 3.7 | 1.1 | 1.5 | 6.8 | 1.3 | 1.5 | 2.2 | 3.5 | 6.4 | 0.6 | 0.4 |
| Soil organic carbon (%) | 0.7 | 0.6 | 2.2 | 0.6 | 0.9 | 4.0 | 0.7 | 0.8 | 1.3 | 2.0 | 3.7 | 0.4 | 0.2 |
| pH | 7.0 | 6.6 | 7.2 | 5.8 | 6.2 | 4.9 | 7.4 | 7.0 | 4.9 | 5.0 | 4.6 | 7.4 | 6.9 |
| Bulk density (g cm$^{-3}$) | 1.43 | 1.45 | 1.48 | 1.53 | 1.52 | 1.45 | 1.51 | 1.47 | 1.47 | 1.51 | 1.36 | 1.53 | 1.61 |
| Cation exchange capacity (meq 100 g$^{-1}$) | 23.1 | 20.0 | 25.0 | 12.7 | 13.6 | na [d] | 10.8 | 17.1 | 7.4 | 1.1 | na | 13.8 | 8.1 |
| Saturated hydraulic conductivity (Ksat) ($\mu$m sec$^{-1}$) | 1.6 | 3.4 | 3.0 | 9.0 | 17.7 | 91.7 | 6.5 | 9.0 | 8.0 | 90.3 | 20.0 | 13.2 | 13.2 |
| Frost free days (#) | 160 | 160 | 135 | 173 | 150 | 130 | 145 | 152 | 115 | 130 | 110 | 169 | 140 |
| Depth to water table (cm) | >200 | >200 | 0 | 178 | 30 | 60 | >200 | 15 | >200 | >200 | 30 | >200 | >200 |
| Available water capacity (cm cm$^{-1}$) | 0.1 | 0.2 | 0.2 | 0.2 | 0.2 | 0.1 | 0.2 | 0.2 | 0.1 | 0.1 | 0.1 | 0.1 | 0.2 |
| Available water storage (cm) | 8.4 | 8.8 | 11.6 | 10.9 | 10.0 | 6.0 | 9.1 | 12.0 | 7.7 | 5.0 | 8.3 | 6.0 | 8.7 |
| Water content (15 Bar) (%) | 25.7 | 22.1 | 22.7 | 13.5 | 14.2 | 3.1 | 12.7 | 13.1 | 3.5 | 2.7 | 3.2 | 11.2 | 14.3 |
| Water content (1/3 Bar) % | 33.9 | 31.4 | 34.3 | 26.8 | 24.2 | 11.7 | 23.4 | 26.3 | 10.7 | 9.2 | 9.3 | 21.0 | 21.5 |

Source: USDA Natural Resources Conservation Service (NRCS) Web Soil Survey (https://websoilsurvey.sc.egov.usda.gov/, accessed on 20 January 2021). [a] Phyto buffers: BC: Bellevue (Central); BE: Bellevue (East); BW: Bellevue (West); CE: Caledonia (East); CW: Caledonia (West); EE: Escanaba (East); EW: Escanaba (West); MA: Manitowoc; ME: Menomonee Falls (East); MW: Menomonee Falls (West); MQ: Marquette; MU: Munising; ON: Ontonagon (North); OS: Ontonagon (South); SL: Slinger; WH: Whitelaw. [b] Drainage classes: MWD: moderately well drained; PD: poorly drained; SED: somewhat excessively drained; SPD: somewhat poorly drained; WD: well drained. [c] Textures: L: loam; S: sand; SCL: sandy clay loam; SiCL: silty clay loam; SL: sandy loam. [d] na: not available.

**Table 3.** Genomic groups, clones, buffer groups (i.e., years of planting), and clone groups for *Populus* genotypes tested in a regional phyto technologies network of sixteen phytoremediation buffer systems (i.e., phyto buffers) established from 2017 to 2019 in the Lake Superior watershed of the Upper Peninsula of Michigan, USA and the Lake Michigan watershed of eastern Wisconsin, USA.

| Genomic Group [a] | Clone | Buffer Group | | | Clone Group [b] |
|---|---|---|---|---|---|
| *P. deltoides* 'D' | 7300502 | 2017 | 2018 | | Experimental |
| *P. deltoides* × *P. maximowiczii* 'DM' | DM114 | 2017 | 2018 | 2019 | Experimental |
| *P. deltoides* × *P. nigra* 'DN' | NC14106 | 2017 | | | Experimental |
| | 99038022 | 2017 | | 2019 | NRRI |
| | 99059016 | 2017 | | | NRRI |
| | 9732-11 | | 2018 | 2019 | NRRI |
| | 9732-24 | | 2018 | 2019 | NRRI |
| | 9732-31 | | 2018 | 2019 | NRRI |
| | 9732-36 | 2017 | 2018 | 2019 | NRRI |
| | DN2 | | 2018 | 2019 | Experimental |
| | DN5 | 2017 | 2018 | | Common |
| | DN34 | 2017 | 2018 | 2019 | Common |
| | DN177 | 2017 | | 2019 | Experimental |
| *P. nigra* × *P. maximowiczii* 'NM' | NM2 | 2017 | 2018 | 2019 | Common |
| | NM5 | 2017 | 2018 | 2019 | Experimental |
| | NM6 | 2017 | 2018 | 2019 | Common |

[a] Species authorities: *P. deltoides* Bartr. Ex Marsh; *P. nigra* L.; *P. maximowiczii* A. Henry NRRI = promising genotypes bred, tested, and selected at the University of Minnesota Duluth, Natural Resources Research Institute (NRRI) for broad-ranging applications [46,56]; [b] Experimental = genotypes with a rich history of testing but that are still at the experimental stage; Common = genotypes commonly used for commercial and/or research purposes in the region.

### 2.3. Phyto Buffer Establishment and Experimental Design

Dormant, unrooted hardwood cuttings (measuring 25.4 cm in length, and containing at least one primary bud within 2.54 cm from the top of each cutting) were processed from one-year-old whips collected during the dormant season (i.e., January through March) from: the USDA Forest Service Hugo Sauer Nursery (Rhinelander, WI, USA); University of Minnesota NRRI Clonal Orchard at the North Central Research and Outreach Center Nursery (Grand Rapids, Minnesota, USA); Iowa State University Clonal Orchard at the Iowa Department of Natural Resources State Nursery (Ames, IA, USA); and Michigan State University Clonal Orchard at the Tree Research Center (Lansing, MI, USA).

Processed cuttings were stored in polyethylene bags at 5° C and, prior to planting, soaked in water to a height of 16.93 cm for 48 h in a dark room at 21 °C during May and June in 2017, 2018, and 2019 (Table S1). Prior to planting, rocks and other large obstructions were removed and the soil was tilled to a depth of 30 cm. Thus, at planting, there was no competition from weeds and/or grasses. To reduce potential competition effects over time, subsequent site maintenance throughout the study period included: (1) tilling to a depth of 30 cm, (2) hand weeding to a minimum diameter of 0.61 m around each individual tree, and (3) removing rocks and other obstructions. For each phyto buffer, a minimum of one maintenance cycle per month throughout each growing season was conducted.

At each of the six (2017) or five (2018, 2019) phyto buffers, trees were planted in a randomized complete block design (RCBD) with eight blocks and twelve clones at a spacing of 2.44 m $\times$ 2.44 m (i.e., 1680 trees ha$^{-1}$). Due to field space constraints, four blocks were planted at Slinger, Wisconsin. Clones were arranged in randomized complete blocks to minimize effects of potential environmental gradients, especially those related to runoff and soil physico-chemical properties. Two border rows were established on the perimeter of each phyto buffer to reduce potential border effects [64,65]. All phyto buffers were fenced using 2.3 m tall Trident extra strength deer fencing (Trident Enterprises, Waynesboro, PA, USA) to eliminate potential impacts from white-tailed deer (*Odocoileus virginianus* Zimmerman) browse. At the end of the first growing season, tree survival across all phyto buffers and clones was 97.1 %, with 95.5%, 96.2%, and 99.6 % survival for the 2017, 2018, and 2019 buffer groups, respectively. All trees that died were replanted with the same genotype to ensure full stocking of 1680 trees ha$^{-1}$. However, the replanted trees were not included in the analyses below.

### 2.4. Field Measurements

At the end of each growing season, tree height to the nearest 0.1 m was measured from the ground to the apical bud. Diameter was measured at 10 cm above the soil surface for one- and two-year-old trees and at breast height (i.e., DBH at 1.37 m) for three-year-old trees. All diameter estimates were determined to the nearest 0.1 cm. Tree volume (V) was calculated from height (H) and diameter (D; including one- and two-year diameter and DBH) using the model: $V = D^2 \times H$ [66]. Starting in 2020, trees of the 2017 buffer group were too tall to continue height measurements at the requisite precision noted above. As a result, 2020 DBH measurements collected from 2017 buffer group trees were used to estimate mean annual increment (MAI; Mg ha$^{-1}$ yr$^{-1}$). In this particular case, individual-tree biomass was estimated using the general model: Biomass = $10^{a_0} \times DBH^{a_1}$ while applying genomic-group specific coefficients from Headlee and Zalesny [67] (*P. deltoides* 'D': $a_0$ = -0.65, $a_1$ = 2.01; *P. deltoides* $\times$ *P. nigra* 'DN': $a_0$ = $-1.02$, $a_1$ = 2.36; *P. deltoides* $\times$ *P. maximowiczii* 'DM': $a_0$ = $-1.03$, $a_1$ = 2.33; *P. nigra* $\times$ *P. maximowiczii* 'NM': $a_0$ = $-0.50$, $a_1$ = 1.94). Individual-tree biomass estimates were then scaled to MAI using standard metric conversion factors and stocking of 1680 trees ha$^{-1}$ at four years after planting.

### 2.5. Health Assessments

Based on a modified methodology of Rogers et al. [58], individual tree health parameters were scored using a five-category qualitative scale ranging from 1 to 5, where 1 = optimal health, 2 = good health, 3 = moderate health, 4 = poor health, and 5 = dead (i.e., health score was inversely related to health). Two researchers scored each parameter to promote consistency in ratings. There were six parameters: (1) vigor, (2) defoliation, (3) leaf discoloration, (4) chlorosis, (5) leaf scorch, and (6) leaf spots. A multiplicative weighted summation index was used to calculate final health index values, with vigor receiving a coefficient of 0.25 and all other parameters having a health coefficient equal to 0.15. Health assessments were not conducted in 2020.

*2.6. Data Analysis*

Health (of all buffer groups) and MAI (of the 2017 buffer group) data were subjected to analyses of variance (ANOVA) and analyses of means (ANOM) using SAS® (PROC GLM; PROC ANOM; SAS INSTITUTE, INC., Cary, NC, USA) assuming a two-way factorial design including six (2017) or five (2018, 2019) buffers, twelve clones, and their interactions. Fisher's Least Significant Difference (LSD) was used to identify significant differences among least-squares means for main effects and interactions at $P < 0.05$.

Height and volume (of all buffer groups) and diameter (excluding 2020 diameter of 2017 buffer group trees) data were subjected to analyses of variance (ANOVA) and analyses of means (ANOM) using SAS® (PROC MIXED; PROC ANOM; SAS INSTITUTE, INC., Cary, NC, USA) assuming a three-way, repeated measures factorial design including six (2017) or five (2018, 2019) buffers, twelve clones, three (2017, 2018) or two (2019) ages, and their interactions. The ages (representing tree growth after each growing season) were analyzed as the repeated measure. To account for pseudo-replication over time, six different covariance structures (i.e., vc, cs, ar (1), toep, ante (1), un) were tested in PROC MIXED to determine which one provided the best model fit based on the lowest BIC scores. Using these covariance structures, ANOVA were conducted in PROC MIXED for all traits, and multiple comparisons analyses were conducted to identify significant differences among least-squares means for main effects and interactions as noted above.

## 3. Results

*3.1. Health*

There were significant differences in health of the 2017 buffer group trees ($\text{HEALTH}_{2017}$) among buffer and clone main effects for the first, second, and third year of growth ($P_{2017,2018,2019} < 0.0001$), yet the buffer $\times$ clone interaction governed health for all three years ($P_{2017} < 0.0001$; $P_{2018} < 0.0001$; $P_{2019} = 0.0152$) (Table S2). $\text{HEALTH}_{2017}$ of trees measured in 2017 (i.e., $\text{HEALTH}_{2017(2017)}$) ranged from $1.0 \pm 0.0$ ('NM2' and 'NM5' at Menomonee Falls (East); 'NM2' at Whitelaw; most healthy] to $1.6 \pm 0.1$ ('DM114' at Caledonia (East); least healthy), with an overall mean of $1.1 \pm 0.1$ (Figure 2). Thus, all trees were of optimal health (i.e., health index ranging from 1 to 2). The healthiest trees were grown at Whitelaw, which had 14.1% better $\text{HEALTH}_{2017(2017)}$ than at Caledonia (East), the buffer with trees exhibiting the poorest health. The range in health scores was broader for clones, with 'NM5' having 37.0% healthier trees than 'DM114'. Common clones had the healthiest and Experimental clones the least-healthiest trees, with NRRI genotypes being intermediate for $\text{HEALTH}_{2017(2017)}$. Seven buffer $\times$ clone interactions resulted in $\text{HEALTH}_{2017(2017)}$ values that were significantly greater than the overall mean ['DM114', 'DN177', and 'NC14106' at Caledonia (East); 'DM114' and 'NC14106' at Menomonee Falls (East); 'DM114' at Menomonee Falls (West); 'DM114' at Slinger]. With the exception of 'DM114' and 'DN177', all clones were generalists for $\text{HEALTH}_{2017(2017)}$, with health index scores not varying by more than 0.3 for any buffer $\times$ clone combinations. Trees grown at Bellevue (West) and Whitelaw were 38.4% healthier than those at Menomonee Falls (East) and Caledonia (East) for 'DM114'. Similarly, for 'DN177', trees at Whitelaw were 32.2% healthier than at Caledonia (East) (Figure 2). Trends in second- ($\text{HEALTH}_{2017(2018)}$) and third-year ($\text{HEALTH}_{2017(2019)}$) health of the 2017 buffer group trees were similar to $\text{HEALTH}_{2017(2017)}$ (Figures S1 and S2).

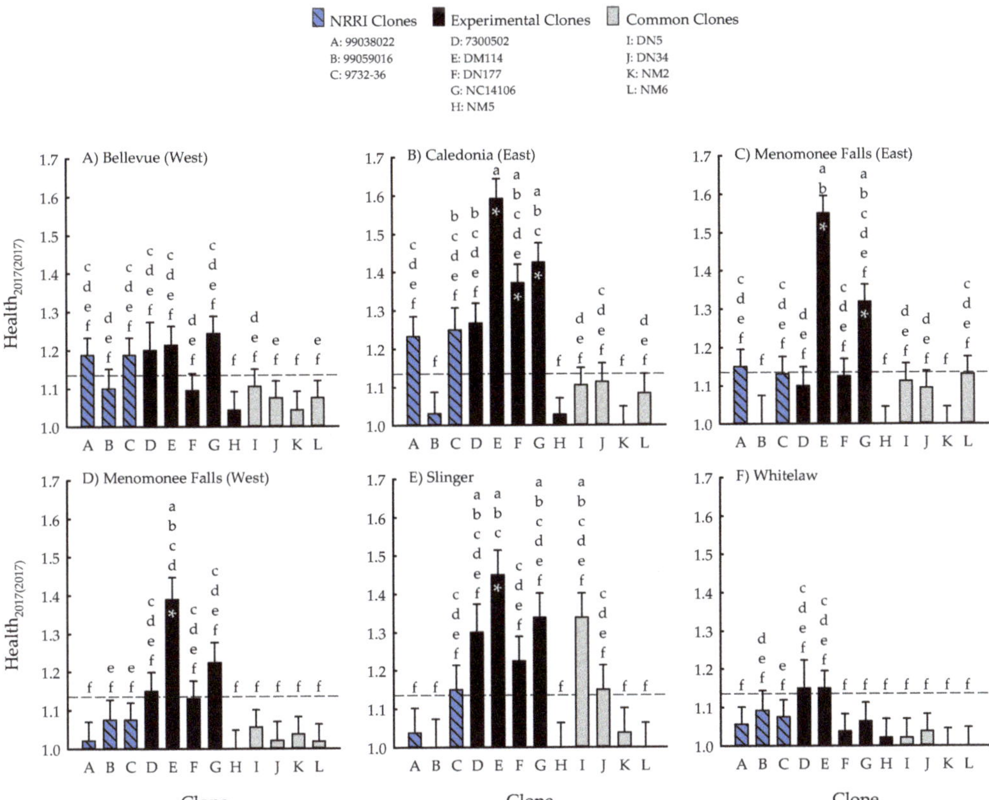



**Figure 2.** Tree health (± one standard error) determined after the 2017 growing season of twelve poplar clones tested in six phytoremediation buffer systems (i.e., phyto buffers) (**A–F**) established in 2017 (i.e., the 2017 buffer group) in the Lake Michigan watershed of eastern Wisconsin, USA. The dashed line represents the overall mean, and asterisks indicate means different than the overall mean at $P < 0.05$. Bars with different letters across all buffer × clone combinations are different at $P < 0.05$. See Materials and Methods for complete tree health definitions (1 = optimal health, 2 = good health, 3 = moderate health, 4 = poor health, and 5 = dead).

Differences among buffer and clone main effects were significant for first- ($\text{HEALTH}_{2018(2018)}$) and second-year ($\text{HEALTH}_{2018(2019)}$] health of the 2018 buffer group trees ($P_{2018} = 0.0009$ for Clone; $P_{2018,2019} < 0.0001$ for all other main effects), yet the buffer × clone interaction governed health for both years ($P_{2018} = 0.0029$; $P_{2019} < 0.0001$) (Table S2). $\text{HEALTH}_{2018(2018)}$ ranged from $1.0 \pm 0.0$ ('9732-11', '9732-24', 'DM114', 'DN2', 'NM2', 'NM5', 'NM6' at Bellevue (East); '9732-11', 'DN2' at Bellevue (Central); most healthy) to $1.5 \pm 0.1$ ('DM114' at Manitowoc; least healthy), with an overall mean of $1.1 \pm 0.1$ (Figure 3). Similar to $\text{HEALTH}_{2017(2017)}$ above, all trees were of optimal health. The healthiest trees were grown at Marquette and Bellevue (East), which had 25.5 % better $\text{HEALTH}_{2018(2018)}$ than at Manitowoc that exhibited the poorest health. The range in health scores was narrower for clones, with 'NM5' having 12.9 % healthier trees than 'DM114' that had the poorest health. Common and NRRI clones had the best overall health scores that were similar to one another yet better than Experimental genotypes. Seven buffer × clone interactions resulted in $\text{HEALTH}_{2018(2018)}$ values there were significantly greater (i.e., of poorer health) than the overall mean, with all occurring at Manitowoc: '9732-11', '9732-24', '9732-31', '9732-36', '7300502', 'DM114', 'DN2', and 'DN5'. With the exception of '7300502' and 'DM114', all clones were classified as generalists for $\text{HEALTH}_{2018(2018)}$. Trees grown at

Bellevue (East) and Marquette were 52.5% healthier than those at Manitowoc for '7300502', while those at Bellevue (East), Bellevue (Central), and Marquette were 48.8% healthier than at Manitowoc (Figure 3). Trends in second-year health of the 2018 buffer group trees (HEALTH$_{2018(2019)}$) were similar to HEALTH$_{2018(2018)}$ (Figure S3).

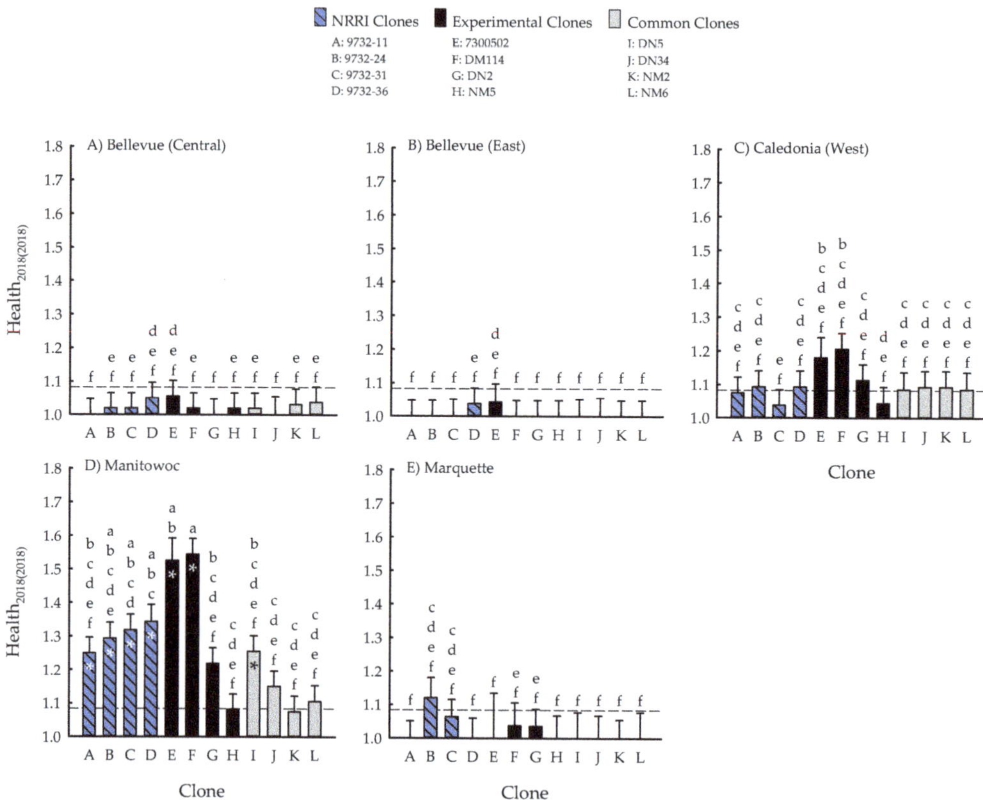



**Figure 3.** Tree health (±one standard error) determined after the 2018 growing season of twelve poplar clones tested in five phytoremediation buffer systems (i.e., phyto buffers) (**A–E**) established in 2018 (i.e., the 2018 buffer group) in the Lake Superior watershed of the Upper Peninsula of Michigan, USA and the Lake Michigan watershed of eastern Wisconsin, USA. The dashed line represents the overall mean, and asterisks indicate means different than the overall mean at $P < 0.05$. Bars with different letters across all buffer × clone combinations are different at $P < 0.05$. See Materials and Methods for complete tree health definitions (1 = optimal health, 2 = good health, 3 = moderate health, 4 = poor health, and 5 = dead).

Differences among buffer and clone main effects were significant for first-year health of the 2019 buffer group trees (HEALTH$_{2019(2019)}$) ($P_{Buffer} < 0.0001$; $P_{Clone} = 0.0159$), yet the buffer × clone interaction governed health ($P = 0.0036$) (Table S2). HEALTH$_{2019(2019)}$ ranged from $1.0 \pm 0.0$ ['9732-24', 'DM114', 'DN2', 'NM2', 'NM6' at Munising; '9732-31' at Ontonagon (North); '99038022', '9732-11', 'DM114', 'DN2', 'NM6' at Ontonagon (South); most healthy] to $1.3 \pm 0.0$ ['DM114', 'DN2' Escanaba (West); least healthy], with an overall mean of $1.1 \pm 0.0$ (Figure 4). All trees were of optimal health. The healthiest trees were grown at Ontonagon (South), which had 13.7% better HEALTH$_{2019(2019)}$ than at Escanaba (West) where trees had the poorest health. The range in health scores was narrower for clones, with 7.2% separating the healthiest trees of 'NM2' from the least healthy trees of 'DN2', 'DN34', and 'DN177'. Similar to HEALTH$_{2017(2017)}$, Common clones had the healthiest and Experimental clones the least-healthiest trees, with NRRI genotypes being intermediate for HEALTH$_{2019(2019)}$. Two buffer × clone interactions resulted in HEALTH$_{2019(2019)}$ values there were significantly greater than the overall mean ('DM114'and 'DN2' at Escanaba (West)). No clones varied by more than 0.3 health index points for any buffer × clone combinations, thus indicating that all had generalist health performance. The largest variation in clonal responses to specific buffers was where trees were 21.7% healthier at all buffers relative to Escanaba (West) for 'DM114', and where trees were 26.3% healthier at Munising and Ontonagon (South) than Escanaba (West) for 'DN2' (Figure 4).

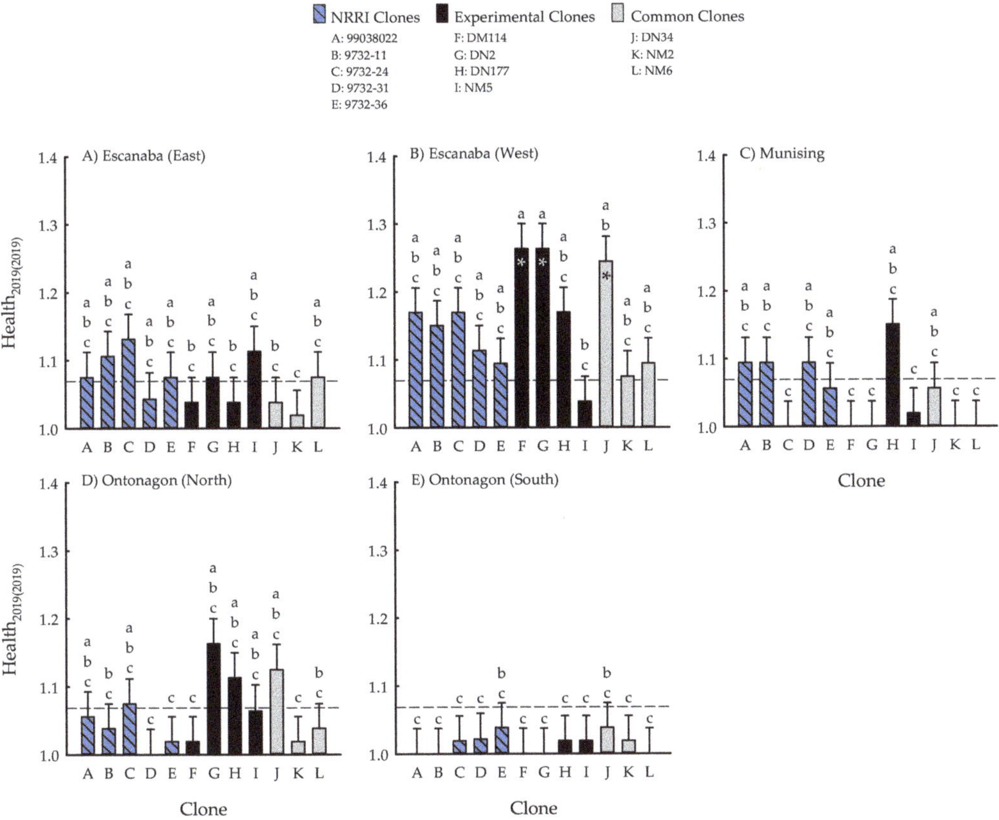


**Figure 4.** Tree health (±one standard error) determined after the 2019 growing season of twelve poplar clones tested in five phytoremediation buffer systems (i.e., phyto buffers) (**A–E**) established in 2019 (i.e., the 2019 buffer group) in the Lake Superior watershed of the Upper Peninsula of Michigan, USA. The dashed line represents the overall mean, and asterisks indicate means different than the overall mean at $P < 0.05$. Bars with different letters across all buffer × clone combinations are different at $P < 0.05$. See Materials and Methods for complete tree health definitions (1 = optimal health, 2 = good health, 3 = moderate health, 4 = poor health, and 5 = dead).

### 3.2. Biomass and Growth

Differences for buffer and clone main effects were significant for fourth-year mean annual increment of the 2017 buffer group trees (MAI$_{2017(2020)}$) ($P < 0.0001$), yet the buffer × clone interaction governed MAI$_{2017(2020)}$ ($P < 0.0001$) (Table S2). MAI$_{2017(2020)}$ ranged from 1.10 ± 0.73 ('7300502' at Whitelaw) to 7.67 ± 0.5 Mg ha$^{-1}$ yr$^{-1}$ ('NM5' at Menomonee Falls (East)), with an overall mean of 3.20 ± 0.51 Mg ha$^{-1}$ yr$^{-1}$ (Figure 5). The largest trees were grown at Menomonee Falls (West), which had 55.4% greater MAI$_{2017(2020)}$ than at Whitelaw, the buffer with the smallest trees. The range in MAI$_{2017(2020)}$ was broader for clones, with

'NM5' exhibiting 69.8% more biomass than 'NC14106' that had the smallest trees of any genotype. While $MAI_{2017(2020)}$ varied across genotypes, trends across clone groups were non-existent, with Common, Experimental, and NRRI clones performing similarly across buffer × clone combinations. Many buffer × clone interactions resulted in $MAI_{2017(2020)}$ values that were significantly greater than the overall mean, with the most notable being 'NM5' outperforming the mean at four of the six buffers: Caledonia (East), Menomonee Falls (East), Menomonee Falls (West), and Slinger. In contrast, 'NC14106' had significantly less $MAI_{2017(2020)}$ than the overall mean at three buffers: Bellevue (West), Menomonee Falls (East), and Whitelaw.

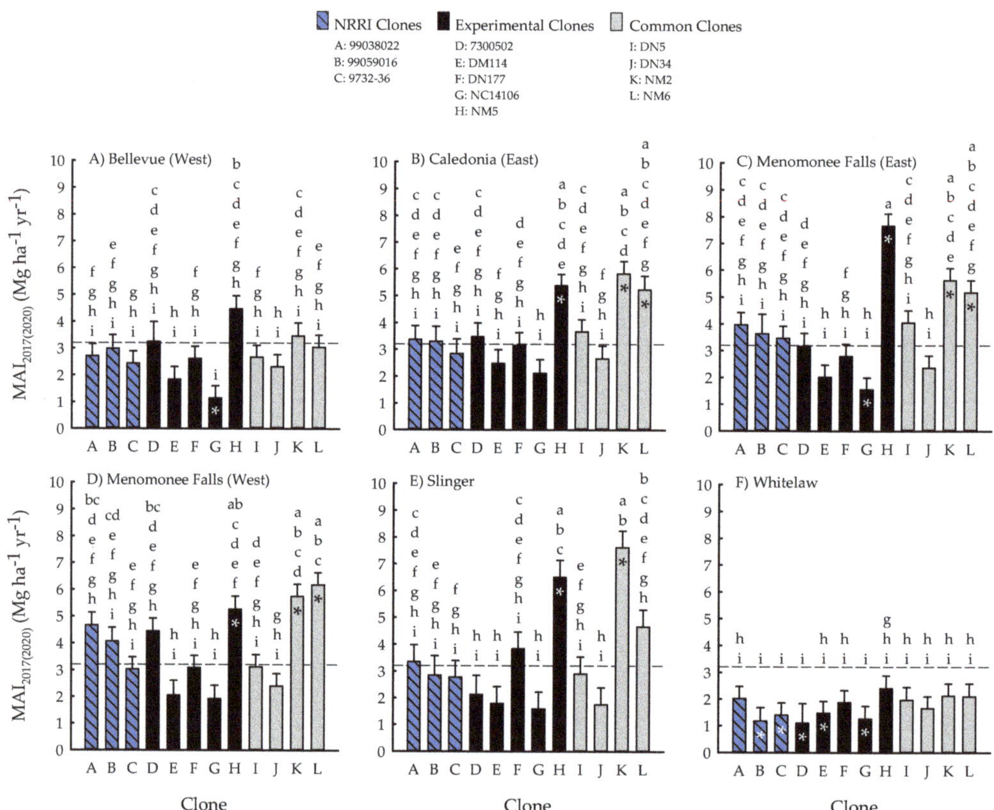



**Figure 5.** Mean annual increment $(Mg\ ha^{-1}\ yr^{-1})$ (±one standard error) determined after the 2020 growing season of twelve poplar clones tested in six phytoremediation buffer systems (i.e., phyto buffers) (**A–F**) established in 2017 (i.e., the 2017 buffer group) in the Lake Michigan watershed of eastern Wisconsin, USA. The dashed line represents the overall mean, and asterisks indicate means different than the overall mean at $P < 0.05$. Bars with different letters across all buffer × clone combinations are different at $P < 0.05$.

In addition to these changes in magnitude, there were distinct changes in $MAI_{2017(2020)}$ ranks that defined the genotypes as generalists or specialists. In particular, clones exhibited generalist $MAI_{2017(2020)}$, with four exceptions (Table 4). First, '99059016' had stable performance across five of the six buffers, with Whitelaw (i.e., the buffer with its lowest rank of eleventh) having 60.9% less $MAI_{2017(2020)}$ than Bellevue (West), where '99059016' ranked fifth. Second, '7300502' had broad variation in $MAI_{2017(2020)}$ across all six buffers, with a 75.3% reduction in $MAI_{2017(2020)}$ at Whitelaw (rank = 12) versus Bellevue (West) (rank = 3). Third, 'DN177' was a specialist because of its high $MAI_{2017(2020)}$ at Slinger (where it ranked fourth), which was 36.7% greater than Menomonee Falls (East) (rank = 9). Fourth, despite higher ranking (rank = 7) at Whitelaw relative to other buffers for 'DN34', $MAI_{2017(2020)}$ at Whitelaw was 60.6% less than that of Caledonia (East) (rank = 11), resulting in its specialist response (Table 4). However, this classification for 'DN34' should be interpreted with caution, especially given that 'DN34' ranked tenth at four buffers and eleventh at one buffer, indicating stable performance across buffers.

**Table 4.** Clonal rank for mean annual increment (MAI2017(2020)) measured after the fourth growing season of twelve poplar clones tested in six phytoremediation buffer systems (i.e., phyto buffers) established in 2017 (i.e., the 2017 buffer group) in the Lake Michigan watershed of eastern Wisconsin, USA.

| Clone | Response Group [b] | Phyto Buffer [a] | | | | | |
|---|---|---|---|---|---|---|---|
| | | BW | CE | ME | MW | SL | WH |
| 99038022 | Generalist | 6 | 6 | 5 | 4 | 5 | 4 |
| 99059016 | Specialist | 5 | 7 | 6 | 6 | 7 | 11 |
| 9732-36 | Generalist | 9 | 9 | 7 | 9 | 8 | 9 |
| 7300502 | Specialist | 3 | 5 | 8 | 5 | 9 | 12 |
| DM114 | Generalist | 11 | 11 | 11 | 11 | 10 | 8 |
| DN177 | Specialist | 8 | 8 | 9 | 8 | 4 | 6 |
| NC14106 | Generalist | 12 | 12 | 12 | 12 | 12 | 10 |
| NM5 | Generalist | 1 | 2 | 1 | 3 | 2 | 1 |
| DN5 | Generalist | 7 | 4 | 4 | 7 | 6 | 5 |
| DN34 | Specialist | 10 | 10 | 10 | 10 | 11 | 7 |
| NM2 | Generalist | 2 | 1 | 2 | 2 | 1 | 2 |
| NM6 | Generalist | 4 | 3 | 3 | 1 | 3 | 3 |

[a] BW: Bellevue (West); CE: Caledonia (East); ME: Menomonee Falls (East); MW: Menomonee Falls (West); SL: Slinger; WH: Whitelaw. [b] Generalist = clone exhibiting stable $MAI_{2017(2020)}$ across phyto buffers (i.e., minimal rank changes); Specialist = clone exhibiting exceptional $MAI_{2017(2020)}$ at one or more phyto buffers relative to the other buffers (i.e., broad rank changes).

The buffer × clone × year interaction was significant for height ($P$ = 0.0483), diameter ($P$ = 0.0018), and volume ($P$ = 0.0001) of the 2017 buffer group trees (Table S3). $VOLUME_{2017}$ of trees measured in 2017 ($VOLUME_{2017(2017)}$) ranged from 3.1 ± 54.4 ('7300502' at Whitelaw) to 459.6 ± 22.6 cm$^3$ ('7300502' at Slinger), with an overall mean of 132.0 ± 38.7 cm$^3$, while $VOLUME_{2017(2018)}$ (i.e., 2017 buffer group trees measured in 2018) ranged from 503.5 ± 1619.6 ('7300502' at Whitelaw) to 13,027.0 ± 1402.6 cm$^3$ ('NM2' at Slinger), with an overall mean of 5765.2 ± 1132.3 cm$^3$ (Table 5). $VOLUME_{2017(2019)}$ ranged from 1668.7 ± 3018.1 ('7300502' at Whitelaw) to 26,652.0 ± 1848.0 cm$^3$ ('NM5' at Menomonee Falls (East)), with an overall mean of 10,530.6 ± 2109.5 cm$^3$ (Table 5). Across all buffer × clone × year combinations, volume increased 43.7-fold from the first year to the second year after planting, and then 1.8-fold from the second year to the third year. After the first and second growing seasons, the largest trees were grown at Slinger, which had 1301 and 268% greater volume than the buffer with the smallest trees (Whitelaw), respectively. For the third growing season, trees at Menomonee Falls (East) were largest, with $VOLUME_{2017(2019)}$ being 338% greater than Whitelaw, which had the smallest trees.

The range in volume was narrower for clones, with 'DN5' having 204% greater volume than the least productive clone ('99059016') for $VOLUME_{2017(2017)}$, 'NM5' being 86% greater than '7300502' for $VOLUME_{2017(2018)}$, and 'NM5' being 114% greater than 'NC14106' for $VOLUME_{2017(2019)}$. With the exception of the first growing season where NRRI clones exhibited the least overall volume followed by Experimental and Common (most volume) genotypes, Common clones had the largest and Experimental clones the smallest trees, with NRRI genotypes being intermediate for $VOLUME_{2017(2018)}$ and $VOLUME_{2017(2019)}$. Trends in height and diameter of the 2017 buffer group trees were similar to volume (Tables S4 and S5). Furthermore, in addition to the broad variability in the magnitude of differences among buffer × clone × year combinations, the frequency and magnitude of changes in rank within and across years defined genotypes as generalists (i.e., minimal rank changes) or specialists (i.e., broad variation resulting in $\geq 5$ rank changes for at least one buffer × clone × year pair) (Table S6). In particular, the NRRI clones ('99038022', '99059016', '9732-36') were high-level generalists characterized by nearly universal stability in ranks across buffers within measurement years, less than three substantial (i.e., >5 ranks) rank changes across all three-way combinations, and moderate to high rank stability over time. Clones 'DM114' and 'NC14106' were also generalists, exhibiting moderate rank stability within years and consistent ranks over time. The remaining genotypes were specialists. Clone '7300502' had the most variability in early ranks of all clones, followed by high rank stability in later years that were not consistent over time. For example, '7300502' ranked first at Slinger in 2017 but then twelfth at this buffer in 2018 and 2019. Clones 'NM2', 'NM5', and 'NM6' were high-level specialists characterized by broad variability in ranks across buffers within measurement years, frequent substantial rank changes across all three-way combinations, and moderate stability over time. Similarly, 'DN5', 'DN34', and 'DN177' were specialists, albeit with more moderate rank variation than the 'NMx' genotypes (Table S6).

**Table 5.** Volume ($cm^3$) (±one standard error) of twelve poplar clones tested in six phytoremediation buffer systems (i.e., phyto buffers) established in 2017 (i.e., the 2017 buffer group) in the Lake Michigan watershed of eastern Wisconsin, USA. Trees were measured following the 2017, 2018, and 2019 growing seasons. Volume values with different letters within a clone column across measurement years are different at $P < 0.05$.

| | Clone | | | | | | | | | | | |
|---|---|---|---|---|---|---|---|---|---|---|---|---|
| Buffer [a] | 99038022 | | 99059016 | | 9732-36 | | 7300502 | | DM114 | | DN177 | |
| *2017 Measurement year* | | | | | | | | | | | | |
| BW | 51 ± 34 | i | 14 ± 38 | v | 26 ± 33 | i | 29 ± 54 | w | 32 ± 39 | gh | 29 ± 34 | s |
| CE | 74 ± 41 | i | 56 ± 42 | v | 101 ± 42 | i | 137 ± 38 | w | 100 ± 39 | fgh | 154 ± 36 | t |
| ME | 199 ± 36 | h | 78 ± 54 | v | 180 ± 34 | i | 128 ± 28 | w | 168 ± 38 | f | 188 ± 34 | t |
| MW | 247 ± 38 | gh | 48 ± 44 | v | 106 ± 23 | i | 203 ± 27 | w | 130 ± 44 | fg | 143 ± 41 | t |
| SL | 359 ± 64 | g | 109 ± 47 | wv | 250 ± 42 | i | 460 ± 23 | w | 279 ± 42 | e | 337 ± 47 | u |
| WH | 39 ± 33 | i | 13 ± 38 | v | 27 ± 33 | i | 3 ± 54 | w | 26 ± 35 | h | 17 ± 34 | s |
| *2018 Measurement year* | | | | | | | | | | | | |
| BW | 6412 ± 992 | de | 4859 ± 1145 | yx | 4221 ± 992 | fgh | 2323 ± 1620 | w | 3559 ± 1060 | cd | 4748 ± 992 | wv |
| CE | 5084 ± 1145 | ef | 6080 ± 1255 | yx | 5653 ± 1254 | efg | 4702 ± 1145 | xw | 5204 ± 1145 | bcd | 6303 ± 1060 | xw |
| ME | 8719 ± 992 | bcd | 8800 ± 1620 | y | 8305 ± 992 | bcde | 4599 ± 1060 | xw | 5663 ± 992 | bc | 7861 ± 992 | x |
| MW | 9423 ± 1060 | bc | 8780 ± 1145 | y | 6579 ± 992 | defg | 6418 ± 1060 | yx | 6105 ± 1254 | abc | 7842 ± 992 | x |
| SL | 8076 ± 1403 | cde | 7336 ± 1619 | yx | 7327 ± 1403 | cdef | 3520 ± 1620 | xw | 5383 ± 1403 | bcd | 9714 ± 1402 | yx |
| WH | 3286 ± 992 | f | 1454 ± 1145 | wv | 1852 ± 992 | hi | 504 ± 1620 | w | 2230 ± 992 | d | 2396 ± 992 | v |
| *2019 Measurement year* | | | | | | | | | | | | |
| BW | 7837 ± 1848 | cde | 8008 ± 2134 | yx | 7026 ± 1848 | cdefg | 6827 ± 3018 | yx | 6622 ± 1976 | abc | 9900 ± 1848 | yx |
| CE | 11,140 ± 2134 | bc | 10,301 ± 2338 | zy | 10,429 ± 2338 | abcd | 9992 ± 2134 | zy | 10,684 ± 2134 | a | 12,331 ± 1976 | y |
| ME | 17,228 ± 1848 | a | 14,235 ± 3018 | z | 15,618 ± 1848 | a | 9834 ± 1976 | zy | 9903 ± 1848 | a | 13,598 ± 1848 | y |
| MW | 17,490 ± 1976 | a | 14,740 ± 2134 | z | 11,408 ± 1849 | abc | 14,137 ± 1976 | z | 8800 ± 2338 | ab | 12,334 ± 1848 | y |
| SL | 14,022 ± 2614 | ab | 10,795 ± 3018 | zy | 12,544 ± 2614 | ab | 5379 ± 3018 | yxw | 8762 ± 2613 | abc | 20,272 ± 2614 | z |
| WH | 4834 ± 1848 | ef | 2393 ± 2134 | xw | 2770 ± 1848 | ghi | 1669 ± 3018 | w | 3909 ± 1848 | bcd | 4960 ± 1848 | xw |
| Buffer | NC14106 | | NM5 | | DN5 | | DN34 | | NM2 | | NM6 | |
| *2017 Measurement year* | | | | | | | | | | | | |
| BW | 32 ± 33 | d | 70 ± 44 | qp | 34 ± 32 | j | 27 ± 34 | s | 45 ± 36 | e | 24 ± 33 | v |
| CE | 69 ± 38 | d | 68 ± 36 | qp | 184 ± 32 | i | 108 ± 36 | ts | 159 ± 36 | e | 114 ± 38 | v |
| ME | 115 ± 33 | d | 269 ± 41 | r | 182 ± 32 | i | 190 ± 34 | ut | 121 ± 33 | e | 202 ± 33 | v |
| MW | 93 ± 38 | d | 133 ± 44 | q | 118 ± 47 | ij | 140 ± 36 | t | 169 ± 34 | e | 151 ± 33 | v |
| SL | 245 ± 48 | d | 330 ± 57 | r | 409 ± 51 | h | 276 ± 51 | u | 415 ± 47 | e | 323 ± 47 | v |
| WH | 18 ± 36 | d | 20 ± 43 | p | 42 ± 34 | j | 24 ± 34 | s | 18 ± 34 | e | 22 ± 36 | v |
| *2018 Measurement year* | | | | | | | | | | | | |
| BW | 2675 ± 992 | c | 6631 ± 1060 | ut | 5872 ± 992 | efg | 6424 ± 992 | xw | 4820 ± 1060 | d | 3399 ± 992 | w |
| CE | 3951 ± 1145 | Bc | 4398 ± 669 | ts | 6646 ± 992 | efg | 5988 ± 1060 | xw | 6071 ± 1060 | cd | 5337 ± 1145 | xw |
| ME | 4340 ± 992 | bc | 10,815 ± 992 | yv | 8521 ± 992 | cde | 7185 ± 992 | yxw | 6296 ± 992 | cd | 6517 ± 992 | x |
| MW | 5300 ± 1145 | abc | 6875 ± 1060 | vu | 7869 ± 992 | def | 8656 ± 1060 | yx | 8644 ± 992 | c | 7505 ± 992 | yx |
| SL | 5102 ± 1403 | abc | 10,229 ± 1402 | v | 6930 ± 1403 | efg | 7040 ± 1402 | xw | 13,027 ± 1403 | b | 7628 ± 1403 | yx |
| WH | 1815 ± 1060 | cd | 2129 ± 1060 | s | 3422 ± 1060 | g | 2605 ± 992 | v | 1567 ± 992 | e | 1548 ± 1060 | wv |
| *2019 Measurement year* | | | | | | | | | | | | |
| BW | 4568 ± 1848 | abc | 11,230 ± 1976 | yv | 6959 ± 1848 | defg | 7644 ± 1848 | yxw | 8281 ± 1976 | c | 6106 ± 1848 | x |
| CE | 9496 ± 2134 | a | 10,632 ± 1203 | v | 12,763 ± 1848 | ab | 9440 ± 1976 | zy | 15,705 ± 1976 | b | 11,155 ± 2134 | zy |
| ME | 7928 ± 1848 | ab | 26,652 ± 1848 | z | 17,335 ± 1848 | a | 13,068 ± 1848 | z | 15,887 ± 1848 | b | 14,848 ± 1848 | z |
| MW | 8497 ± 2134 | ab | 13,425 ± 1976 | y | 12,108 ± 1848 | bc | 9817 ± 1976 | zy | 15,186 ± 1848 | b | 13,598 ± 1848 | z |
| SL | 7720 ± 2614 | abc | 23,814 ± 2614 | z | 11,875 ± 2613 | bcd | 10,255 ± 2614 | zy | 26,238 ± 2614 | a | 13,526 ± 2614 | z |
| WH | 3344 ± 1976 | bc | 3051 ± 1976 | ts | 3982 ± 1976 | fg | 4076 ± 1848 | wv | 2812 ± 1848 | de | 2448 ± 1976 | wv |

[a] BW: Bellevue (West); CE: Caledonia (East); ME: Menomonee Falls (East); MW: Menomonee Falls (West); SL: Slinger; WH: Whitelaw.

The buffer $\times$ clone $\times$ year interaction was significant for height, diameter, and volume ($P < 0.0001$) of the 2018 buffer group trees (Table S3). $\text{VOLUME}_{2018(2018)}$ ranged from $12.2 \pm 38.4$ ('DN5' at Marquette) to $185.7 \pm 25.3$ cm$^3$ ('NM2' at Manitowoc), with an overall mean of $71.2 \pm 26.3$ cm$^3$, while $\text{VOLUME}_{2018(2019)}$ ranged from $287.8 \pm 1518.7$ ('DN5' at Marquette) to $11{,}085.0 \pm 930.0$ cm$^3$ ('NM2' at Manitowoc), with an overall mean of $3418.4 \pm 1035.2$ cm$^3$ (Table 6). $\text{VOLUME}_{2018(2020)}$ ranged from $261.0 \pm 3882.2$ ('DN5' at Marquette) to $27{,}220.0 \pm 2377.4$ cm$^3$ ('NM5' at Manitowoc), with an overall mean of $8826.0 \pm 2646.2$ cm$^3$ (Table 6). Across all buffer $\times$ clone $\times$ year combinations, volume increased 48-fold from the first year to the second year after planting, and then 2.6-fold from the second year to the third year. After the first growing season, the largest trees were grown at Caledonia (West), which had 529% greater volume than Marquette, the buffer with the smallest trees. During the second and third growing seasons, the largest trees were from Manitowoc, which had 1079 and 1744% greater volume than the buffer with the smallest trees (Marquette), respectively. As with volume for the 2017 buffer group trees, there was less variability among clones than buffers, with '9732-31' having 156% greater volume than the least productive clone ('7300502') for $\text{VOLUME}_{2018(2018)}$, 'NM5' being 163% greater than '7300502' for $\text{VOLUME}_{2018(2018)}$, and 'NM5' being 143% greater than 'DM114' for $\text{VOLUME}_{2018(2018)}$. During the first growing season, NRRI clones had the greatest overall volume, while Common clones exhibited the largest trees at two and three years after planting. For all three years, Experimental clones had the least volume. Trends in height and diameter of the 2018 buffer group trees were similar to volume (Tables S7 and S8). Moreover, as with 2017 buffer group trees, changes in magnitude and rank of 2018 buffer group clones across buffer $\times$ year combinations defined them as generalists or specialists (defined as for Table S6 above) (Table S9). Similar to 2017, the NRRI clones ('9732-11', '9732-24', '9732-31', '9732-36') were high-level generalists with universal stability in ranks, few rank changes greater than five ranks, and high rank stability over time. Clone 'DM114' was also a generalist, having only two substantial rank changes and nearly identical ranks from 2017 to 2019. All other genotypes were specialists. As in 2017, clone '7300502' had a high level of early rank variability and low to moderate rank consistency over time; clones 'NM2', 'NM5', and 'NM6' were high-level specialists with broad rank variability, numerous substantial rank changes, and moderate stability as trees aged; 'DN2', 'DN5', and 'DN34' were consistent specialists with moderate levels of rank changes and age-dependent stability (Table S9).

The buffer $\times$ clone $\times$ year interaction was significant for height ($P = 0.0079$), diameter ($P < 0.0001$), and volume ($P < 0.0001$) of the 2019 buffer group trees (Table S3). $\text{VOLUME}_{2019(2019)}$ ranged from $8.6 \pm 26.9$ ('DN177' at Ontonagon (North)) to $396. \pm 26.7$ cm$^3$ ('99038022' at Escanaba (West)), with an overall mean of $88.7 \pm 26.7$ cm$^3$, while $\text{VOLUME}_{2019(2020)}$ ranged from $92.6 \pm 612.7$ ('NM5' at Ontonagon (North)) to $8909.6 \pm 573.1$ cm$^3$ ('NM2' at Escanaba (West)), with an overall mean of $1440.4 \pm 575.1$ cm$^3$ (Table 7). Across all buffer $\times$ clone $\times$ year combinations, volume increased 16.2-fold from the first year to the second year after planting. After the first and second growing seasons, the largest trees were grown at Escanaba (West), which had 1008 and 1066% greater volume than the buffer with the smallest trees (Ontonagon (North)), respectively. For $\text{VOLUME}_{2019(2019)}$, '99038022' had 101% greater volume than the least productive clone ('DN177'), while 'NM2' was 164% greater than '9732-11' for $\text{VOLUME}_{2019(2020)}$. NRRI clones exhibited the greatest first-year volume, followed by Common and Experimental genotypes. For $\text{VOLUME}_{2019(2020)}$, ranks of NRRI, Experimental, and Common clones changed, with Common genotypes having the most $\text{VOLUME}_{2019(2020)}$ followed by Experimental and NRRI (least volume) clones. Trends in height and diameter of the 2019 buffer group trees were similar to volume (Tables S10 and S11). Furthermore, 2019 clones were classified as generalists or specialists (defined as for Table S6 above) (Table S12).

**Table 6.** Volume (cm$^3$) ($\pm$one standard error) of twelve poplar clones tested in five phytoremediation buffer systems (i.e., phyto buffers) established in 2018 (i.e., the 2018 buffer group) in the Lake Superior watershed of the Upper Peninsula of Michigan, USA and the Lake Michigan watershed of eastern Wisconsin, USA. Trees were measured following the 2018, 2019, and 2020 growing seasons. Volume values with different letters within a clone column across measurement years are different at $P < 0.05$.

| | Clone | | | | | | | | | | | |
|---|---|---|---|---|---|---|---|---|---|---|---|---|
| Buffer [a] | 9732-11 | | 9732-24 | | 9732-31 | | 9732-36 | | 7300502 | | DM114 | |
| | *2018 Measurement year* | | | | | | | | | | | |
| BC | 57 ± 28 | f | 40 ± 30 | u | 69 ± 0 | f | 26 ± 33 | t | 28 ± 21 | d | 25 ± 00 | w |
| BE | 70 ± 28 | f | 49 ± 36 | u | 59 ± 29 | f | 43 ± 28 | t | 42 ± 45 | d | 14 ± 23 | w |
| CW | 157 ± 24 | f | 139 ± 24 | u | 182 ± 24 | ef | 147 ± 24 | ut | 71 ± 30 | d | 104 ± 24 | xw |
| MA | 133 ± 20 | f | 95 ± 20 | u | 143 ± 23 | ef | 118 ± 25 | t | 33 ± 33 | d | 68 ± 22 | w |
| MQ | 16 ± 23 | f | 31 ± 30 | u | 40 ± 25 | f | 31 ± 16 | t | 18 ± 66 | d | 17 ± 28 | w |
| | *2019 Measurement year* | | | | | | | | | | | |
| BC | 2344 ± 930 | de | 2422 ± 930 | xwv | 2496 ± 930 | cd | 1522 ± 930 | vut | 867 ± 930 | cd | 1622 ± 930 | xw |
| BE | 2239 ± 930 | e | 2117 ± 930 | wv | 2068 ± 994 | de | 1966 ± 930 | wvu | 1046 ± 1074 | bcd | 1587 ± 930 | xw |
| CW | 5469 ± 930 | cd | 3901 ± 930 | yxw | 7876 ± 930 | b | 3893 ± 930 | xwv | 3455 ± 1176 | bc | 3072 ± 930 | x |
| MA | 7857 ± 930 | bc | 5378 ± 930 | yx | 7162 ± 930 | b | 5251 ± 994 | yx | 3084 ± 1315 | bc | 4340 ± 930 | yx |
| MQ | 474 ± 994 | ef | 718 ± 1176 | vu | 1002 ± 994 | def | 1068 ± 1176 | vut | 496 ± 2631 | cd | 607 ± 1316 | xw |
| | *2020 Measurement year* | | | | | | | | | | | |
| BC | 6777 ± 2377 | bc | 7454 ± 2377 | y | 7886 ± 2377 | b | 5194 ± 2377 | yxw | 4258 ± 2377 | ab | 4751 ± 2377 | zyx |
| BE | 6160 ± 2377 | bcd | 6240 ± 2377 | yx | 6031 ± 2541 | bc | 4870 ± 2377 | yxwv | 2364 ± 2745 | bcd | 3541 ± 2377 | yx |
| CW | 9917 ± 2377 | b | 7452 ± 2377 | y | 23,912 ± 2377 | a | 7554 ± 2377 | y | 11,061 ± 3007 | a | 7537 ± 2377 | zy |
| MA | 20,902 ± 2377 | a | 14,368 ± 2377 | z | 18,160 ± 2377 | a | 15,068 ± 2542 | z | 10,918 ± 3362 | a | 9297 ± 2377 | z |
| MQ | 767 ± 2542 | ef | 945 ± 3007 | wvu | 1802 ± 2542 | def | 1935 ± 3007 | wvut | 549 ± 6724 | cd | 1242 ± 3362 | xw |
| Buffer | DN2 | | NM5 | | DN5 | | DN34 | | NM2 | | NM6 | |
| | *2018 Measurement year* | | | | | | | | | | | |
| BC | 45 ± 32 | f | 45 ± 16 | u | 37 ± 20 | d | 51 ± 29 | w | 30 ± 24 | e | 25 ± 24 | w |
| BE | 73 ± 16 | f | 37 ± 22 | u | 30 ± 25 | d | 52 ± 39 | w | 27 ± 42 | e | 37 ± 23 | w |
| CW | 116 ± 24 | f | 173 ± 25 | vu | 148 ± 25 | d | 117 ± 24 | w | 130 ± 24 | de | 170 ± 25 | w |
| MA | 74 ± 23 | f | 148 ± 24 | u | 50 ± 20 | d | 136 ± 24 | w | 186 ± 25 | de | 160 ± 23 | w |
| MQ | 20 ± 25 | f | 16 ± 33 | u | 12 ± 38 | d | 22 ± 33 | w | 20 ± 30 | e | 21 ± 38 | w |
| | *2019 Measurement year* | | | | | | | | | | | |
| BC | 2796 ± 930 | e | 2584 ± 930 | w | 2780 ± 930 | cd | 3175 ± 1074 | w | 1592 ± 930 | de | 1684 ± 930 | w |
| BE | 3120 ± 930 | de | 1981 ± 930 | wv | 2170 ± 994 | cd | 2500 ± 1074 | w | 1318 ± 930 | de | 1762 ± 930 | w |
| CW | 7874 ± 930 | c | 8262 ± 994 | yx | 6381 ± 994 | b | 4564 ± 930 | xw | 6930 ± 930 | c | 5804 ± 994 | x |
| MA | 5662 ± 930 | cd | 10,362 ± 930 | y | 4625 ± 930 | bc | 6640 ± 930 | yx | 11,085 ± 930 | b | 9552 ± 930 | y |
| MQ | 555 ± 994 | ef | 380 ± 1315 | wvu | 288 ± 1519 | d | 481 ± 1315 | w | 321 ± 1074 | de | 478 ± 1519 | w |
| | *2020 Measurement year* | | | | | | | | | | | |
| BC | 6135 ± 2377 | cd | 6280 ± 2377 | yx | 7036 ± 2377 | b | 7613 ± 2745 | yx | 4692 ± 2377 | cd | 3075 ± 2377 | xw |
| BE | 6731 ± 2377 | c | 4971 ± 2378 | xw | 4073 ± 2541 | bcd | 5246 ± 2745 | yxw | 3384 ± 2377 | cde | 3010 ± 2377 | xw |
| CW | 23,055 ± 2377 | a | 25,221 ± 2542 | z | 17,328 ± 2542 | a | 9214 ± 2377 | y | 21,570 ± 2377 | a | 17,352 ± 2542 | z |
| MA | 16,079 ± 2377 | b | 27,220 ± 2377 | z | 13,630 ± 2377 | a | 16,728 ± 2377 | z | 24,636 ± 2377 | a | 22,265 ± 2377 | z |
| MQ | 1157 ± 2541 | ef | 506 ± 3362 | wvu | 261 ± 3882 | d | 1132 ± 3362 | w | 713 ± 2745 | de | 340 ± 3882 | w |

[a] BC: Bellevue (Central); BE: Bellevue (East); CW: Caledonia (West); MA: Manitowoc; MQ: Marquette.

As in previous buffer groups, 'DM114' was a high-level generalist. With the exception of '99038022' that was a high-level generalist with nearly universal rank stability, only two substantial rank changes across all three-way combinations, and moderate stability over time, the NRRI clones were specialists for the 2019 buffer group. In particular, '9732-11', '9732-24', '9732-31', and '9732-36' were moderate- to high-level specialists characterized by broad rank variability, substantial rank changes, and moderate stability over time. Similarly, the performance of 'DN2' as a generalist in the 2019 buffer group was different relative to its specialist volume in the 2018 buffer group. Across 2019 phyto buffers, 'DN2' exhibited moderate rank stability within years and consistent ranks over time. All remaining clones were moderate- ('DN34', 'DN177') and high-level ('NM2', 'NM5', 'NM6') specialists with trends for changes in magnitude and rank similar to their performance in 2017 and 2018 buffer groups (Table S12).

**Table 7.** Volume (cm$^3$) ($\pm$one standard error) of twelve poplar clones tested in five phytoremediation buffer systems (i.e., phyto buffers) established in 2019 (i.e., the 2019 buffer group) in the Lake Superior watershed of the Upper Peninsula of Michigan, USA. Trees were measured following the 2019 and 2020 growing seasons. Volume values with different letters within a clone column across measurement years are different at $P < 0.05$.

| | Clone | | | | | | | | | | | |
| --- | --- | --- | --- | --- | --- | --- | --- | --- | --- | --- | --- | --- |
| Buffer [a] | 99038022 | | 9732-11 | | 9732-24 | | 9732-31 | | 9732-36 | | DM114 | |
| *2019 Measurement year* | | | | | | | | | | | | |
| EE | 73 ± 26 | b | 49 ± 25 | y | 98 ± 26 | c | 81 ± 28 | y | 44 ± 27 | d | 47 ± 26 | y |
| EW | 396 ± 27 | b | 239 ± 27 | y | 198 ± 27 | c | 228 ± 27 | y | 312 ± 27 | cd | 215 ± 27 | y |
| MU | 89 ± 26 | b | 28 ± 26 | y | 49 ± 26 | c | 20 ± 27 | y | 47 ± 26 | d | 38 ± 26 | y |
| ON | 54 ± 27 | b | 28 ± 27 | y | 23 ± 27 | c | 29 ± 27 | y | 16 ± 27 | d | 36 ± 27 | y |
| OS | 60 ± 27 | b | 49 ± 27 | y | 56 ± 27 | c | 33 ± 28 | y | 28 ± 27 | d | 42 ± 27 | y |
| *2020 Measurement year* | | | | | | | | | | | | |
| EE | 517 ± 573 | b | 470 ± 573 | y | 1553 ± 573 | ab | 1040 ± 613 | zy | 463 ± 573 | bc | 828 ± 573 | y |
| EW | 2397 ± 573 | a | 2206 ± 573 | z | 2444 ± 573 | a | 2155 ± 573 | z | 3189 ± 573 | a | 3166 ± 573 | z |
| MU | 2075 ± 573 | a | 878 ± 573 | zy | 1709 ± 573 | ab | 950 ± 573 | zy | 1761 ± 573 | a | 2463 ± 573 | z |
| ON | 691 ± 573 | b | 205 ± 573 | y | 189 ± 573 | c | 295 ± 573 | y | 285 ± 573 | cd | 924 ± 573 | y |
| OS | 439 ± 573 | b | 785 ± 573 | zy | 805 ± 573 | bc | 494 ± 613 | y | 441 ± 573 | bcd | 806 ± 573 | y |
| Buffer | DN2 | | DN177 | | NM5 | | DN34 | | NM2 | | NM6 | |
| *2019 Measurement year* | | | | | | | | | | | | |
| EE | 45 ± 26 | b | 56 ± 26 | y | 42 ± 26 | c | 50 ± 26 | y | 43 ± 27 | b | 47 ± 26 | x |
| EW | 382 ± 27 | b | 231 ± 27 | y | 304 ± 27 | c | 278 ± 27 | y | 337 ± 27 | b | 269 ± 27 | x |
| MU | 29 ± 27 | b | 20 ± 26 | y | 45 ± 27 | c | 56 ± 27 | y | 35 ± 27 | b | 60 ± 27 | x |
| ON | 14 ± 27 | b | 09 ± 27 | y | 24 ± 29 | c | 18 ± 27 | y | 26 ± 27 | b | 29 ± 27 | x |
| OS | 33 ± 27 | b | 18 ± 27 | y | 25 ± 27 | c | 26 ± 27 | y | 26 ± 27 | b | 38 ± 27 | x |
| *2020 Measurement year* | | | | | | | | | | | | |
| EE | 531 ± 573 | b | 755 ± 573 | y | 543 ± 573 | bc | 668 ± 573 | y | 645 ± 573 | b | 510 ± 573 | x |
| EW | 2675 ± 573 | a | 2448 ± 573 | z | 7687 ± 573 | a | 1881 ± 573 | z | 8910 ± 573 | a | 6129 ± 573 | z |
| MU | 1834 ± 573 | a | 1230 ± 573 | z | 1932 ± 573 | b | 2245 ± 573 | z | 1770 ± 573 | b | 3302 ± 573 | y |
| ON | 206 ± 573 | b | 106 ± 573 | y | 93 ± 613 | c | 275 ± 573 | y | 337 ± 573 | b | 278 ± 573 | x |
| OS | 727 ± 573 | b | 305 ± 573 | y | 267 ± 573 | c | 515 ± 573 | y | 339 ± 573 | b | 662 ± 573 | x |

[a] EE: Escanaba (East); EW: Escanaba (West); MU: Munising; ON: Ontonagon (North); OS: Ontonagon (South).

## 4. Discussion and Conclusions

### 4.1. Genotype × Environment Interactions

Understanding genotype by environment (G × E) interactions is a necessary step for identifying and selecting poplar clones used for phytoremediation and associated phytotechnologies [68]. Poplar phenotypes are a function of their genotype, environment, and genotypic response to specific site conditions [49]. Phyto-recurrent selection has been used to choose superior poplar genotypes in the Midwestern United States [32,68]. Using both generalist and specialist genotypes enhances ecosystem services provided by phytoremediation applications. Deploying generalists with low G × E interactions and robust productivity across the region may be beneficial for cost and operational efficiencies [47,56], while specialists with high G × E interactions may maximize productivity, phytoremediation potential, and overall benefits of ecosystem services [26,28,49]. In the current study, sixteen phytoremediation buffer systems (i.e., multi-environmental trials (MET)) were established to evaluate trends in G × E interactions and identify generalist and specialist poplars in order to reduce runoff and clean groundwater (Figure 1).

In this study there were significant main (buffer, clone, and year) and interaction effects on tree health and growth parameters. In particular, interactions involving the buffer main effect were major factors governing clonal productivity. Buffer effects reflect tree responses to combined edaphic and local climatic conditions, and influence clonal performance traits such as: stem biomass production [69,70]; foliar fungal microbiomes [71]; leaf characteristics [72]; and diameter, height, and wood volume production [49,73]. In the current study, the broad spectrum of MET buffers with varying soil and climate conditions led to a wide range in clonal performance related to changes in both genotypic magnitude and ranks over time. More favorable climatic conditions (i.e., warmer, more precipitation; Table 1) and adequate soils for poplar cultivation (e.g., suitable texture, water-air properties, pH; Table 2) likely led to greater volume of most clones at Menomonee Falls, Slinger, Manitowoc and Caledonia. On the other hand, lower performance of clones at Marquette and Ontonagon can be attributed to less favorable climatic and soil conditions such as lower precipitation, temperature, and soil pH (i.e., more acidity). Similar results were obtained by Hansen et al. [74] who showed that soil water availability played a key role in the productivity of woody biomass plantations. In the current study, sites with irrigation or shallow water tables exhibited the greatest wood volume, which was further corroborated through poplar biomass productivity modeling in the Midwestern United States [59,60]. Overall, G × E interactions resulted in mean annual increment (MAI) of four-year-old trees from the current 2017 buffer group ranging from 1.1 to 7.8 Mg ha$^{-1}$ yr$^{-1}$ (mean = 3.2 Mg ha$^{-1}$ yr$^{-1}$), which agreed with results for similarly-aged (i.e., 3 to 5 years) poplars in the region, whose MAI ranged from 0.6 Mg ha$^{-1}$ yr$^{-1}$ to 7.1 Mg ha$^{-1}$ yr$^{-1}$ [49,75].

Multi-environmental trials are key tools for defining gains achieved through identifying genotype characteristics, their stability, and relevance of their interaction with varying environmental conditions (i.e., G × E interactions) [76]. Although G × E interactions can have a significant impact on the precision of breeding value estimates, often resulting in decreased genetic gain [53], matching superior species and clones to particular site and growing conditions has been critical in maximizing the productivity of SRWC plantations [70]. Tree age is an important factor shown to govern G × E interactions for poplars. Although Riemenschneider et al. [48] found significant G × E interactions first occurring at three years after planting, Semerci et al. [52] recorded significant G × E interactions for growth and phenology traits in one-year-old poplar clones grown on sites with different water availability in Turkey. Similarly, in the present study, all tested traits exhibited G × E interactions after the first year in all three buffer groups (e.g., for one- to four-year-old trees). In contrast to the results presented here, greenhouse phyto-recurrent selection experiments with soils from the six phyto buffers of the 2017 buffer group [Bellevue (West), Caledonia (East), Menomonee Falls (East), Menomonee Falls (West), Slinger, and Whitelaw] showed a lack of G × E interactions for root-shoot ratio and growth performance index of many poplar clones tested at the current MET. Nevertheless, there were significant G × E interactions for tree health [58]. Such differences may be attributed to variability in environmental conditions between the greenhouse and field buffers and/or the length of the experiment (i.e., months versus years).

Regardless, such results have indicated that G × E interactions in poplar clones vary during the life cycle of the trees. Zalesny and Headlee [25] found significant G × E interactions for biomass and carbon production in both 10- and 20-year-old poplar plantations, despite negligible genotypic effects on both traits for 20-year-old trees. The presence/absence of G × E interactions within clones during the production cycle also can be expressed by variability in growth patterns across clones. Netzer et al. [77] recorded that some clones had greater biomass productivity in the second half of the stand rotation, while Ghezehei et al. [78] recorded both lack and presence of significant differences in clonal productivity of four- and eight-year-old poplars.

*4.2. Generalist and Specialist Response Groups*

Phenotypic responses determine comparative genotypic performance, resulting in some clones growing well and providing higher levels of ecosystem services across a broad range of soil, climate, and/or contaminant conditions (i.e., generalists). On the other hand, specialists optimize their growth and physiological processes when subjected to specific site conditions [43,44]. We identified both generalist ('DM114', 'NC14106', '99038022', '99059016') and specialist ('7300502', 'DN5', 'DN34', 'DN177', 'NM2', 'NM5', 'NM6') clones, along with others that exhibited volume consistent with both response groups across buffers and years ('9732-11', '9732-24', '9732-31', '9732-36', 'DN2') (i.e., those that shifted from generalists to specialists as trees aged; see below) (Table S13). Classification of these clones has important practical implications in reducing uncertainties associated with field-deployment of these genotypes for multiple applications, including phytoremediation.

Changes in both magnitude and ranks across buffer × clone × year combinations defined the G × E interactions of the current study [43,44]. Different classifications were found for volume versus mean annual increment (MAI), which supports the need for long-term monitoring throughout plantation development [68]. One explanation for differences between these traits may be related to the age at which the trees were measured. As noted previously, in the Southeastern United States, clonal rankings in poplar wood volume production changed with increasing stand age [70,79]. In this study, all volume estimates were from one- to three-year-old trees, while MAI was determined for trees after their fourth growing season, the start of the mid-rotation growth stage for poplars used for phytoremediation [80]. Similar changes were also apparent when evaluating measurement years within individual buffer groups. That is, oftentimes clonal rankings dramatically changed as trees aged (e.g., '7300502' had the greatest volume at Slinger during the establishment year only to have the least volume at this buffer after two and three growing seasons). A second explanation for differences in classifications between volume and MAI may be related to individual clones expressing higher levels of genetic variation and phenotypic plasticity as they responded to highly variable and changing soil conditions both within and across growing seasons at the phyto buffers. Guet et al. [72] reported that *P. deltoides* and *P. nigra* (which were the most common species used as parents in the current study) exhibited high levels of such genetic variation and plasticity, allowing them to better adapt to site-level spatial and temporal heterogeneity. These responses could lead to a greater propensity for specialist growth performance, and may explain why some NRRI clones ('9732-11', '9732-24', '9732-31', '9732-36') shifted from generalists to specialists in the current study (Table S13). In contrast, Nelson et al. [46] identified most of these clones as being geographically robust across both latitudinal and longitudinal gradients in North America. One of these clones, '9732-36', had very consistent volume production in our 2017 and 2018 buffer groups (as well as MAI in our 2017 buffer group), yet trended towards specialist responses at phyto buffers established in 2019. This may have been due to a negligible relationship between phenotypic plasticity and G × E interactions. As in our interpretation (and its definition), Des Marais et al. [81] linked G × E interactions to changes in clonal ranking and growth performance (i.e., variance-changing interaction) [56].

This concept of variance-changing interaction also supports differences in classifying generalist and specialist clones of the current study within individual measurement years associated with buffer × clone interactions for specific buffer groups. Specifically, individual response group designations for buffer × clone × year combinations (from Tables S6, S9 and S12) may differ somewhat from final classifications listed in Table S13, given the need to assess stability and magnitude of ranks within years and over time (i.e., classifying the clones holistically). With the exception of the NRRI clones, most other genotypes from the *P. deltoides* × *P. nigra* 'DN' genomic group (with 'DN2' being the only exception), as well as the *P. deltoides* 'D' clone '7300502' and all clones of the *P. deltoides* × *P. maximowiczii* 'DM' and *P. nigra* × *P. maximowiczii* 'NM" genomic groups exhibited consistent classifications across buffer groups. Nevertheless, of particular interest was that individual clones within genomic groups (or breeding groups, for the 'DN' hybrids) performed similarly, indicating that selection of genomic groups may be effective for early phyto-recurrent selection cycles (i.e., when choosing base populations for testing). Such genomic group trends have been reported for the same or related genotypes used in other phytoremediation applications [35,82]. Overall, the preponderance of specialist clones in the current study supports the need for phyto-recurrent selection in order to match genotypes to sites for small-scale applications with location-specific requirements (i.e., see variation in ranks for 'NM5' from Tables S6, S9 and S12), as well as the parallel need for continued testing of new genetic material, such as the NRRI clones, to select robust genotypes with minimal G × E interactions that can be used for large-scale, commercial applications at a justifiable cost while providing a multitude of ecosystem services across the rural to urban continuum [46].

**Supplementary Materials:** The following are available online at https://www.mdpi.com/article/10.3390/f12040430/s1; **Table S1:** Dates of planting for each phytoremediation buffer system (i.e., phyto buffer); **Table S2:** Probability values from analyses of variance for health and mean annual increment (MAI); **Table S3:** Probability values from analyses of variance for height, diameter, and volume; **Table S4:** Height for the buffer × clone × year interaction (2017 buffer group); **Table S5:** Diameter for the buffer × clone × year interaction (2017 buffer group); **Table S6**: Volume clone rank for the buffer × clone × year interaction (2017 buffer group); **Table S7**: Height for the buffer × clone × year interaction (2018 buffer group); **Table S8**: Diameter for the buffer × clone × year interaction (2018 buffer group); **Table S9**: Volume clone rank for the buffer × clone × year interaction (2018 buffer group); **Table S10**: Height for the buffer × clone × year interaction (2019 buffer group); **Table S11**: Diameter for the buffer × clone × year interaction (2019 buffer group); **Table S12**: Volume clone rank for the buffer × clone × year interaction (2019 buffer group); **Table S13**: Final classification of clones into generalist and specialist response groups; **Figure S1**: Health for the buffer × clone interaction measured in 2018 (2017 buffer group); **Figure S2**: Health for the buffer × clone interaction measured in 2019 (2017 buffer group); **Figure S3**: Health for the buffer × clone interaction measured in 2019 (2018 buffer group).

**Author Contributions:** Conceptualization, E.O.B., L.B., J.G.B., B.S.D., R.A.H., C.-H.L., M.P., R.S., A.H.W., and R.S.Z.J.; methodology, E.O.B., J.G.B., B.S.D., R.A.H., C.-H.L., B.G.M., A.P., E.R.R., R.S., R.A.V., A.H.W., and R.S.Z.J.; validation, A.P., E.R.R., and R.S.Z.J.; formal analysis, R.S.Z.J.; investigation, E.O.B., J.G.B., B.S.D., R.A.H., C.-H.L., B.G.M., N.D.N., A.P., E.R.R., R.A.V., A.H.W., and R.S.Z.J.; resources, L.B., J.G.B., R.A.H., C.-H.L., B.G.M., N.D.N., M.P., R.S., and R.S.Z.J.; data curation, B.S.D., E.R.R., R.A.V., A.H.W., and R.S.Z.J.; writing—original draft preparation, A.P., E.R.R., and R.S.Z.J.; writing—review and editing, E.O.B., L.B., J.G.B., B.S.D., R.A.H., C.-H.L., B.G.M., N.D.N., M.P., A.P., E.R.R., R.S., R.A.V., A.H.W., and R.S.Z.J.; visualization, E.R.R., and R.S.Z.J.; supervision, J.G.B., B.S.D., C.-H.L., N.D.N., E.R.R., A.H.W., and R.S.Z.J.; project administration, J.G.B., C.-H.L., and R.S.Z.J.; funding acquisition, A.H.W. and R.S.Z.J. All authors have read and agreed to the published version of the manuscript.

**Funding:** This work was funded by the Great Lakes Restoration Initiative (GLRI; Template #738 Landfill Runoff Reduction).

**Acknowledgments:** The findings and conclusions in this publication are those of the authors and should not be construed to represent any official USDA or United States Government determination or policy. This work was funded by the Great Lakes Restoration Initiative (GLRI; Template #738 Landfill Runoff Reduction). We are grateful to S. Johnson, C.H. Perry, and N. Vrevich of the USDA Forest Service for GLRI support. We thank B. Dean (formerly), K. Corcoran, and M. Mujaddedi of the International Programs Office of the USDA Forest Service for administrative support of scientific exchanges between A. Pilipović and R. Zalesny, as well as their administration and support of International Forestry Fellows: D. Karlsson (Sweden), P. Muñoz Gomez de la Serna (Spain), and A. Peqini (Albania). We thank the following site managers for access to their field sites: J. Forney and B. Pliska (Waste Management, Inc.); K. Dorow, D. Koski, and K. McDaniel (City of Manitowoc, Wisconsin); D. Henderson (AECOM Technical Services, Inc.); B. Austin and J. Wales (Marquette County Solid Waste Management Authority); D. Pyle (Delta County Solid Waste Management Authority); and S. Coron (Great American Disposal), as well as T. Beggs (Wisconsin Department of Natural Resources) for regulatory assistance and B. Sexton (Sand County Environmental) for insight, knowledge, and guidance on sustainable phytoremediation systems. We appreciate laboratory, greenhouse, and field technical assistance from: B.A. Birr, T. Cook, S. Eddy, F. Erdmann, C. Espinosa, D. Karlsson, R. Lange, P. Manley, M. Mueller, C. Munch, P. Muñoz Gomez de la Serna, D. Nguyen, A. Peqini, M. Ramsay, E. Schmidt, J. Schutts, and M. Wagler. We are grateful to J.M. Suvada for developing Figure 1, as well as R. Klevickas and P. Bloese for providing cutting material. We thank W.L. Headlee for statistical mentoring and reviewing earlier versions of this manuscript.

**Conflicts of Interest:** The authors declare no conflict of interest.

# References

1. Fuller, K.; Shear, H.; Wittig, J. *The Great Lakes: An Environmental Atlas and Resource Book*, 3rd ed; U.S. Environmental Protection Agency and Government of Canada: Washington, DC, USA, 1995; Volume 95, p. 46, Issue 1 of EPA 905-B.
2. GLRI (Great Lakes Restoration Initiative). *Action Plan I (2010–2014)*; GLRI: Washington, DC, USA, 2010; 41p.
3. MEA (Millenium Ecosystem Assessment). *Ecosystems and Human Well-Being: Synthesis*; Island Press: Washington, DC, USA, 2005; p. 155.
4. Steinman, A.D.; Cardinale, B.J.; Munns, W.R.; Ogdahl, M.E.; Allan, J.D.; Angadi, T.; Bartlett, S.; Brauman, K.; Byappanahalli, M.; Doss, M.; et al. Ecosystem services in the Great Lakes. *J. Great Lakes Res.* **2017**, *43*, 161–168. [CrossRef]
5. Michigan Sea Grant. Available online: https://www.michiganseagrant.org (accessed on 8 April 2020).

6.    Campbell, M.; Cooper, M.J.; Friedman, K.; Anderson, W.P. The economy as a driver of change in the Great Lakes–St. Lawrence River basin. *J. Great Lakes Res.* **2015**, *41*, 69–83. [CrossRef]
7.    Allan, J.D.; McIntyre, P.B.; Smith, S.D.P.; Halpern, B.S.; Boyer, G.L.; Buchsbaum, A.; Burton, G.A.; Campbell, L.M.; Chadderton, W.L.; Ciborowski, J.J.H.; et al. Joint analysis of stressors and ecosystem services to enhance restoration effectiveness. *Proc. Natl. Acad. Sci. USA* **2012**, *110*, 372–377. [CrossRef] [PubMed]
8.    Vörösmarty, C.J.; Sahagian, D. Anthropogenic disturbance of the terrestrial water cycle. *BioScience* **2000**, *50*, 753–765. [CrossRef]
9.    Donohoe, M. Causes and health consequences of environmental degradation and social injustice. *Soc. Sci. Med.* **2003**, *56*, 573–587. [CrossRef]
10.   Wu, Q.; Zhou, H.; Tam, N.F.; Tian, Y.; Tan, Y.; Zhou, S.; Li, Q.; Chen, Y.; Leung, J.Y. Contamination, toxicity and speciation of heavy metals in an industrialized urban river: Implications for the dispersal of heavy metals. *Mar. Pollut. Bull.* **2016**, *104*, 153–161. [CrossRef] [PubMed]
11.   Cureton, P.M.; Groenevelt, P.H.; McBride, R.A. Landfill Leachate Recirculation: Effects on Vegetation Vigor and Clay Surface Cover Infiltration. *J. Environ. Qual.* **1991**, *20*, 17–24. [CrossRef]
12.   Kjeldsen, P.; Barlaz, M.A.; Rooker, A.P.; Baun, A.; Ledin, A.; Christensen, T.H. Present and Long-Term Composition of MSW Landfill Leachate: A Review. *Crit. Rev. Environ. Sci. Technol.* **2002**, *32*, 297–336. [CrossRef]
13.   Duggan, J. The potential for landfill leachate treatment using willows in the UK—A critical review. *Resour. Conserv. Recycl.* **2005**, *45*, 97–113. [CrossRef]
14.   Wong, M.; Leung, C. Landfill Leachate as Irrigation Water for Tree and Vegetable Crops. *Waste Manag. Res.* **1989**, *7*, 311–323. [CrossRef]
15.   Allan, J.D.; Smith, S.D.; McIntyre, P.B.; A Joseph, C.; E Dickinson, C.; Marino, A.L.; Biel, R.G.; Olson, J.C.; Doran, P.J.; Rutherford, E.S.; et al. Using cultural ecosystem services to inform restoration priorities in the Laurentian Great Lakes. *Front. Ecol. Environ.* **2015**, *13*, 418–424. [CrossRef]
16.   Wortley, L.; Hero, J.-M.; Howes, M.J. Evaluating Ecological Restoration Success: A Review of the Literature. *Restor. Ecol.* **2013**, *21*, 537–543. [CrossRef]
17.   Nunez-Mir, G.C.; Iannone, B.V.; Curtis, K.; Fei, S. Evaluating the evolution of forest restoration research in a changing world: A "big literature" review. *New For.* **2015**, *46*, 669–682. [CrossRef]
18.   Lima, A.T.; Mitchell, K.; O'Connell, D.W.; Verhoeven, J.; Van Cappellen, P. The legacy of surface mining: Remediation, restoration, reclamation and rehabilitation. *Environ. Sci. Policy* **2016**, *66*, 227–233. [CrossRef]
19.   Arthur, E.L.; Rice, P.J.; Rice, P.J.; Anderson, T.A.; Baladi, S.M.; Henderson, K.L.D.; Coats, J.R. Phytoremediation—An Overview. *Crit. Rev. Plant Sci.* **2005**, *24*, 109–122. [CrossRef]
20.   Burken, J.G.; Schnoor, J.L. Predictive Relationships for Uptake of Organic Contaminants by Hybrid Poplar Trees. *Environ. Sci. Technol.* **1998**, *32*, 3379–3385. [CrossRef]
21.   Cooke, J.A.; Johnson, M.S. Ecological restoration of land with particular reference to the mining of metals and industrial minerals: A review of theory and practice. *Environ. Rev.* **2002**, *10*, 41–71. [CrossRef]
22.   Chaney, R.L.; Baklanov, I.A. Phytoremediation and Phytomining. *Adv. Bot. Res.* **2017**, *83*, 189–221. [CrossRef]
23.   Dickmann, D. Silviculture and biology of short-rotation woody crops in temperate regions: Then and now. *Biomass Bioenergy* **2006**, *30*, 696–705. [CrossRef]
24.   Johnson, J.M.; Coleman, M.D.; Gesch, R.W.; Jaradat, A.A.; Mitchell, R.; Reicosky, D.C.; Wilhelm, W.W. Biomass-bioenergy crops in the United States: A changing paradigm. *Am. J. Plant Sci. Biotechnol.* **2007**, *1*, 1–28.
25.   Zalesny, R.S., Jr.; Headlee, W.L. Developing Woody Crops for the Enhancement of Ecosystem Services under Changing Climates in the North Central United States. *J. For. Environ. Sci.* **2015**, *31*, 78–90. [CrossRef]
26.   Zalesny, R.S., Jr.; Stanturf, J.A.; Gardiner, E.S.; Perdue, J.H.; Young, T.M.; Coyle, D.R.; Headlee, W.L.; Bañuelos, G.S.; Hass, A. Ecosystem Services of Woody Crop Production Systems. *BioEnergy Res.* **2016**, *9*, 465–491. [CrossRef]
27.   Zalesny, R.S., Jr.; Berndes, G.; Dimitriou, I.; Fritsche, U.; Miller, C.; Eisenbies, M.; Ghezehei, S.; Hazel, D.; Headlee, W.L.; Mola-Yudego, B.; et al. Positive water linkages of producing short rotation poplars and willows for bioenergy and phytotechnologies. *Wiley Interdiscip. Rev. Energy Environ.* **2019**, *8*, 345. [CrossRef]
28.   Zalesny, R.S., Jr.; Stanturf, J.A.; Gardiner, E.S.; Bañuelos, G.S.; Hallett, R.A.; Hass, A.; Stange, C.M.; Perdue, J.H.; Young, T.M.; Coyle, D.R.; et al. Environmental Technologies of Woody Crop Production Systems. *BioEnergy Res.* **2016**, *9*, 492–506. [CrossRef]
29.   Licht, L.A.; Isebrands, J. Linking phytoremediated pollutant removal to biomass economic opportunities. *Biomass Bioenergy* **2005**, *28*, 203–218. [CrossRef]
30.   Pilipović, A.; Orlovic, S.; Rončević, S.; Nikolić, N.; Župunski, M.; Spasojevic, J. Results of selection of poplars and willows for water and sediment phytoremediation. *J. Agric. For.* **2015**, *61*, 205–211. [CrossRef]
31.   Zalesny, R.S., Jr.; E Riemenschneider, D.; Hall, R.B. Early rooting of dormant hardwood cuttings of Populus: Analysis of quantitative genetics and genotype × environment interactions. *Can. J. For. Res.* **2005**, *35*, 918–929. [CrossRef]
32.   Zalesny, J.A.; Zalesny, R.S., Jr.; Wiese, A.H.; Hall, R.B. Choosing Tree Genotypes for Phytoremediation of Landfill Leachate Using Phyto-Recurrent Selection. *Int. J. Phytoremediat.* **2007**, *9*, 513–530. [CrossRef]
33.   Pilipović, A.; Orlović, S.; Nikolić, N.; Borišev, M.; Krstić, B.; Rončević, S. Growth and plant physiological parameters as markers for selection of poplar clones for crude oil phytoremediation. *Šumarski List* **2012**, *136*, 273–281.

34. Nikolić, N.; Zorić, L.; Cvetković, I.; Pajević, S.; Borišev, M.; Orlović, S.; Pilipović, A. Assessment of cadmium tolerance and phytoextraction ability in young Populus deltoides L. and Populus × euramericana plants through morpho-anatomical and physiological responses to growth in cadmium enriched soil. *iForest Biogeosci. For.* **2017**, *10*, 635–644. [CrossRef]
35. Zalesny, R.S., Jr.; Bauer, E.O. Genotypic variability and stability of poplars and willows grown onnitrate-contaminated soils. *Int. J. Phytoremediat.* **2019**, *21*, 969–979. [CrossRef]
36. Orlović, S.; Guzina, V.; Merkulov, L. Genetic variability in anatomical, physiological and growth characteristics of hybrid poplar (*Populus × euramericana* Dode (Guinier)) and eastern cottonwood (*Populus deltoides* Bartr.) clones. *Silvae Genet.* **1998**, *47*, 183–189.
37. Reboud, X.; Bell, G. Experimental evolution in Chlamydomonas. III. Evolution of specialist and generalist types in environments that vary in space and time. *Heredity* **1997**, *78*, 507–514. [CrossRef]
38. Kassen, R. The experimental evolution of specialists, generalists, and the maintenance of diversity. *J. Evol. Biol.* **2002**, *15*, 173–190. [CrossRef]
39. Stout, A.B.; Schreiner, E.J. Results of a Project in Hybridizing Poplars. *J. Hered.* **1933**, *24*, 217–229. [CrossRef]
40. Bradshaw, H.; Ceulemans, R.; Davis, J.; Stettler, R. Emerging Model Systems in Plant Biology: Poplar (Populus) as A Model Forest Tree. *J. Plant Growth Regul.* **2000**, *19*, 306–313. [CrossRef]
41. Isebrands, J.; Zalesny, R.S., Jr. Reflections on the contributions of Populus research at Rhinelander, Wisconsin, USA. *Can. J. For. Res.* **2021**, *51*, 139–153. [CrossRef]
42. Sommer, R.J. Phenotypic Plasticity: From Theory and Genetics to Current and Future Challenges. *Genetics* **2020**, *215*, 1–13. [CrossRef]
43. De Leon, N.; Jannink, J.-L.; Edwards, J.W.; Kaeppler, S.M. Introduction to a Special Issue on Genotype by Environment Interaction. *Crop. Sci.* **2016**, *56*, 2081–2089. [CrossRef]
44. Li, Y.; Suontama, M.; Burdon, R.D.; Dungey, H.S. Genotype by environment interactions in forest tree breeding: Review of methodology and perspectives on research and application. *Tree Genet. Genomes* **2017**, *13*, 60. [CrossRef]
45. Calleja-Rodriguez, A.; Gull, B.A.; Wu, H.X.; Mullin, T.J.; Persson, T. Genotype-by-environment interactions and the dynamic relationship between tree vitality and height in northern *Pinus sylvestris. Tree Genet. Genomes* **2019**, *15*, 36. [CrossRef]
46. Nelson, N.D.; Berguson, W.E.; McMahon, B.G.; Meilan, R.; Smart, L.B.; Gouker, F.E.; Bloese, P.; Miller, R.; Volk, T.A.; Cai, M.; et al. Discovery of Geographically Robust Hybrid Poplar Clones. *Silvae Genet.* **2019**, *68*, 101–110. [CrossRef]
47. Nelson, N.D.; Meilan, R.; Berguson, W.E.; McMahon, B.G.; Cai, M.; Buchman, D. Growth performance of hybrid poplar clones on two agricultural sites with and without early irrigation and fertilization. *Silvae Genet.* **2019**, *68*, 58–66. [CrossRef]
48. Riemenschneider, D.E.; Isebrands, J.G.; Berguson, W.E.; Dickmann, D.I.; Hall, R.B.; Mohn, C.A.; Stanosz, G.R.; Tuskan, G.A. Poplar breeding and testing strategies in the north-central U.S.: Demonstration of potential yield and consideration of future research needs. *For. Chron.* **2001**, *77*, 245–253. [CrossRef]
49. Zalesny, R.S., Jr.; Hall, R.B.; Zalesny, J.A.; McMahon, B.G.; Berguson, W.E.; Stanosz, G.R. Biomass and Genotype × Environment Interactions of Populus Energy Crops in the Midwestern United States. *BioEnergy Res.* **2009**, *2*, 106–122. [CrossRef]
50. Sixto, H.; Salvia, J.; Barrio, M.; Ciria, M.P.; Cañellas, I. Genetic variation and genotype-environment interactions in short rotation Populus plantations in southern Europe. *New For.* **2011**, *42*, 163–177. [CrossRef]
51. Zalesny, R.S., Jr.; Bauer, E.O.; Hall, R.B.; Zalesny, J.A.; Kunzman, J.; Rog, C.J.; Riemenschneider, D.E. Clonal Variation in Survival and Growth of Hybrid Poplar and Willow in an *in situ* trial on Soils Heavily Contaminated with Petroleum Hydrocarbons. *Int. J. Phytoremediat.* **2005**, *7*, 177–197. [CrossRef]
52. Semerci, A.; Guevara, C.A.; Gonzalez-Benecke, C.A. Water availability effects on growth and phenology of 11 poplar cultivars growing in semiarid areas in Turkey. *New For.* **2020**, 1–20. [CrossRef]
53. Pliura, A.; Zhang, S.; MacKay, J.; Bousquet, J. Genotypic variation in wood density and growth traits of poplar hybrids at four clonal trials. *For. Ecol. Manag.* **2007**, *238*, 92–106. [CrossRef]
54. Headlee, W.L.; Zalesny, R.S., Jr.; Hall, R.B.; Bauer, E.O.; Bender, B.; Birr, B.A.; Miller, R.O.; Randall, J.A.; Wiese, A.H.; Zalesny, R.S. Specific Gravity of Hybrid Poplars in the North-Central Region, USA: Within-Tree Variability and Site × Genotype Effects. *Forests* **2013**, *4*, 251–269. [CrossRef]
55. Cervera, M.T.; Storme, V.; Soto, A.; Ivens, B.; Van Montagu, M.; Rajora, O.P.; Boerjan, W. Intraspecific and interspecific genetic and phylogenetic relationships in the genus Populus based on AFLP markers. *Theor. Appl. Genet.* **2005**, *111*, 1440–1456. [CrossRef]
56. Nelson, N.D.; Berguson, W.E.; McMahon, B.G.; Cai, M.; Buchman, D.J. Growth performance and stability of hybrid poplar clones in simultaneous tests on six sites. *Biomass Bioenergy* **2018**, *118*, 115–125. [CrossRef]
57. Zalesny, R.S., Jr.; Bauer, E.O. Selecting and Utilizing Populus and Salixfor Landfill Covers: Implications for Leachate Irrigation. *Int. J. Phytoremediat.* **2007**, *9*, 497–511. [CrossRef]
58. Rogers, E.R.; Zalesny, R.S., Jr.; Hallett, R.A.; Headlee, W.L.; Wiese, A.H. Relationships among Root–Shoot Ratio, Early Growth, and Health of Hybrid Poplar and Willow Clones Grown in Different Landfill Soils. *Forests* **2019**, *10*, 49. [CrossRef]
59. Zalesny, R.S., Jr.; Donner, D.M.; Coyle, D.R.; Headlee, W.L. An approach for siting poplar energy production systems to increase productivity and associated ecosystem services. *For. Ecol. Manag.* **2012**, *284*, 45–58. [CrossRef]
60. Headlee, W.L.; Zalesny, R.S., Jr.; Donner, D.M.; Hall, R.B. Using a Process-Based Model (3-PG) to Predict and Map Hybrid Poplar Biomass Productivity in Minnesota and Wisconsin, USA. *BioEnergy Res.* **2012**, *6*, 196–210. [CrossRef]
61. Miller, R.O. Growth Variation Among Hybrid Poplar Varieties in Michigan, USA and the Implications for Commercial Biomass Production. *BioEnergy Res.* **2018**, *11*, 816–825. [CrossRef]

62. Zalesny, R.S., Jr.; Burken, J.G.; Hallett, R.A.; Pilipović, A.; Wiese, A.H.; Rogers, E.R.; Bauer, E.O.; Buechel, L.; DeBauche, B.S.; Henderson, D.; et al. The Great Lakes Restoration Initiative: Reducing Runoff from landfills in the Great Lakes Basin, USA. In Proceedings of the 12th Biennial Short Rotation Woody Crops Operations Working Group Conference: 2018 Woody Crops International Conference, Rhinelander, WI, USA, 22–27 July 2018.
63. Zalesny, R.S., Jr.; Burken, J.G.; Hallett, R.A.; Wiese, A.H. Using Phyto-Recurrent Selection to Reduce Impacts of Runoff from Closed Landfills in the Lake Michigan Drainage Basin, USA. In Proceedings of the 14th International Phytotechnologies Conference: Phytotechnologies–New Sustainable Solutions for Environmental Challenges, Montreal, QC, Canada, 25–29 September 2017.
64. Hansen, E.A. Root length in young hybrid *Populus* plantations: Its implications for border width of research plots. *For. Sci.* **1981**, *27*, 808–814. [CrossRef]
65. Zavitkovski, J. Small plots with unplanted plot border can distort data in biomass production studies. *Can. J. For. Res.* **1981**, *11*, 9–12. [CrossRef]
66. Kershaw, J.A.; Ducey, M.J.; Beers, T.W.; Husch, B. *Forest Mensuration*; Wiley: Hoboken, NY, USA, 2016; p. 630.
67. Headlee, W.L.; Zalesny, R.S., Jr. Allometric Relationships for Aboveground Woody Biomass Differ Among Hybrid Poplar Genomic Groups and Clones in the North-Central USA. *BioEnergy Res.* **2019**, *12*, 966–976. [CrossRef]
68. Zalesny, R.S., Jr.; Headlee, W.L.; Gopalakrishnan, G.; Bauer, E.O.; Hall, R.B.; Hazel, D.W.; Isebrands, J.G.; Licht, L.A.; Negri, M.C.; Nichols, E.G.; et al. Ecosystem services of poplar at long-term phytoremediation sites in the Midwest and Southeast, United States. *Wiley Interdiscip. Rev. Energy Environ.* **2019**, *8*, 349. [CrossRef]
69. Fortier, J.; Gagnon, D.; Truax, B.; Lambert, F. Biomass and volume yield after 6 years in multiclonal hybrid poplar riparian buffer strips. *Biomass Bioenergy* **2010**, *34*, 1028–1040. [CrossRef]
70. Ghezehei, S.B.; Nichols, E.G.; Maier, C.A.; Hazel, D.W. Adaptability of *Populus* to Physiography and Growing Conditions in the Southeastern USA. *Forests* **2019**, *10*, 118. [CrossRef]
71. Bálint, M.; Bartha, L.; O'Hara, R.B.; Olson, M.S.; Otte, J.; Pfenninger, M.; Robertson, A.L.; Tiffin, P.; Schmitt, I. Relocation, high-latitude warming and host genetic identity shape the foliar fungal microbiome of poplars. *Mol. Ecol.* **2014**, *24*, 235–248. [CrossRef] [PubMed]
72. Guet, J.; Fabbrini, F.; Fichot, R.; Sabatti, M.; Bastien, C.; Brignolas, F. Genetic variation for leaf morphology, leaf structure and leaf carbon isotope discrimination in European populations of black poplar (*Populus nigra* L.). *Tree Physiol.* **2015**, *35*, 850–863. [CrossRef] [PubMed]
73. Dhillon, G.P.S.; Singh, A.; Singh, P.; Sidhu, D.S. Field Evaluation of *Populus deltoides* Bartr. ex Marsh. at Two Sites in Indo-gangetic Plains of India. *Silvae Genet.* **2010**, *59*, 1–7. [CrossRef]
74. Hansen, E.A.; Ostry, M.E.; Johnson, W.D.; Tolsted, D.N.; Netzer, D.A.; Berguson, W.E.; Hall, R.B. *Field Performance of Populus in Short-Rotation Intensive Culture Plantations in the North-Central U.S.*; USDA Forest Service: Washington, DC, USA, 1994; Volume 320, p. 13.
75. Hansen, E. Mid-rotation yields of biomass plantations in the north central U.S. *Mid Rotat. Yields Biomass Plant. North Cent. U.S.* **1992**, *309*, 8. [CrossRef]
76. Sixto, H.; Gil, P.; Ciria, P.; Camps, F.; Sánchez, M.; Cañellas, I.; Voltas, J. Performance of hybrid poplar clones in short rotation coppice in Mediterranean environments: Analysis of genotypic stability. *GCB Bioenergy* **2013**, *6*, 661–671. [CrossRef]
77. Netzer, D.; Tolsted, D.; Ostry, M.E.; Isebrands, J.G.; Riemenschneider, D.; Ward, K. *Growth, Yield, and Disease Resistance of 7- to 12-Year-Old Poplar Clones in the North Central United States*; USDA Forest Service: Washington, DC, USA, 2002; Volume 229, p. 31.
78. Ghezehei, S.B.; Wright, J.; Zalesny, R.S., Jr.; Nichols, E.G.; Hazel, D.W. Matching site-suitable poplars to rotation length for optimized productivity. *For. Ecol. Manag.* **2020**, *457*, 117670. [CrossRef]
79. Ghezehei, S.B.; Nichols, E.G.; Hazel, D.W. Early Clonal Survival and Growth of Poplars Grown on North Carolina Piedmont and Mountain Marginal Lands. *BioEnergy Res.* **2016**, *9*, 548–558. [CrossRef]
80. Zalesny, R.S., Jr.; Zhu, J.Y.; Headlee, W.L.; Gleisner, R.; Pilipović, A.; Van Acker, J.; Bauer, E.O.; Birr, B.A.; Wiese, A.H. Ecosystem Services, Physiology, and Biofuels Recalcitrance of Poplars Grown for Landfill Phytoremediation. *Plants* **2020**, *9*, 1357. [CrossRef]
81. Marais, D.L.D.; Hernandez, K.M.; Juenger, T.E. Genotype-by-Environment Interaction and Plasticity: Exploring Genomic Responses of Plants to the Abiotic Environment. *Annu. Rev. Ecol. Evol. Syst.* **2013**, *44*, 5–29. [CrossRef]
82. Zalesny, J.A.; Zalesny, R.S.; Coyle, D.R.; Hall, R.B. Growth and biomass of Populus irrigated with landfill leachate. *For. Ecol. Manag.* **2007**, *248*, 143–152. [CrossRef]

*Article*

# Establishment of Regional Phytoremediation Buffer Systems for Ecological Restoration in the Great Lakes Basin, USA. II. New Clones Show Exceptional Promise

Andrej Pilipović [1], Ronald S. Zalesny, Jr. [2,*], Elizabeth R. Rogers [2,3], Bernard G. McMahon [4], Neil D. Nelson [4], Joel G. Burken [5], Richard A. Hallett [6] and Chung-Ho Lin [3]

1 Institute of Lowland Forestry and Environment, University of Novi Sad, 21000 Novi Sad, Serbia; andrejp@uns.ac.rs
2 USDA Forest Service, Northern Research Station, Institute for Applied Ecosystem Studies, Rhinelander, WI 54501, USA; elizabeth.r.rogers@usda.gov or errccd@mail.missouri.edu
3 School of Natural Resources, Center for Agroforestry, University of Missouri–Columbia, Columbia, MO 65201, USA; LinChu@missouri.edu
4 Natural Resources Research Institute, University of Minnesota Duluth, Duluth, MN 55807, USA; bmcmahon@nrri.umn.edu (B.G.M.); nnelson2@d.umn.edu (N.D.N.)
5 Civil, Architectural, and Environmental Engineering, Missouri University of Science and Technology, Rolla, MO 65401, USA; burken@mst.edu
6 USDA Forest Service, Northern Research Station, New York City Urban Field Station, Bayside, NY 11360, USA; richard.hallett@usda.gov
* Correspondence: ronald.zalesny@usda.gov

**Citation:** Pilipović, A.; Zalesny, R.S., Jr.; Rogers, E.R.; McMahon, B.G.; Nelson, N.D.; Burken, J.G.; Hallett, R.A.; Lin, C.-H. Establishment of Regional Phytoremediation Buffer Systems for Ecological Restoration in the Great Lakes Basin, USA. II. New Clones Show Exceptional Promise. *Forests* **2021**, *12*, 474. https://doi.org/10.3390/f12040474

Academic Editor: Richard Dodd

Received: 6 March 2021
Accepted: 10 April 2021
Published: 13 April 2021

**Publisher's Note:** MDPI stays neutral with regard to jurisdictional claims in published maps and institutional affiliations.

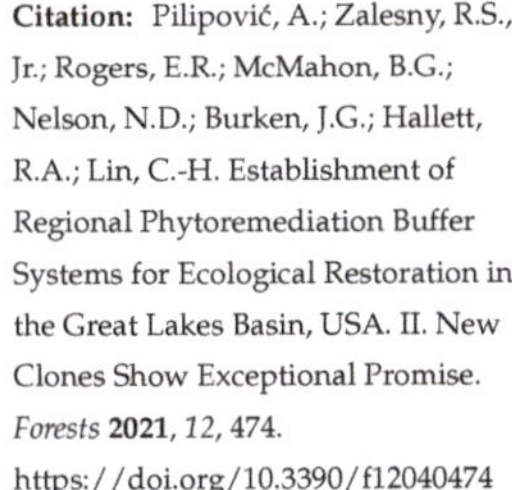

**Abstract:** Poplar tree improvement strategies are needed to enhance ecosystem services' provisioning and achieve phytoremediation objectives. We evaluated the establishment potential of new poplar clones developed at the University of Minnesota Duluth, Natural Resources Research Institute (NRRI) from sixteen phytoremediation buffer systems (phyto buffers) (buffer groups: 2017 × 6; 2018 × 5; 2019 × 5) throughout the Lake Superior and Lake Michigan watersheds. We divided clones into Experimental (testing stage genotypes) and Common (commercial and/or research genotypes) clone groups and compared them with each other and each NRRI clone (NRRI group) at the phyto buffers. We tested for differences in clone groups, phyto buffers, and their interactions for survival, health, height, diameter, and volume from ages one to four years. First-year survival was 97.1%, with 95.5%, 96.2%, and 99.6% for the 2017, 2018, and 2019 buffer groups, respectively. All trees had optimal health. Fourth-year mean annual increment of 2017 buffer group trees ranged from $2.66 \pm 0.18$ to $3.65 \pm 0.17$ Mg ha$^{-1}$ yr$^{-1}$. NRRI clones '99038022' and '9732-31' exhibited exceptional survival and growth across eleven and ten phyto buffers, respectively, for all years. These approaches advance poplar tree improvement efforts throughout the region, continent, and world, with methods informing clonal selection for multiple end-uses, including phytotechnologies.

**Keywords:** clonal selection; genotype × environment (G × E) interactions; multi-environmental trials (MET); phenotypic plasticity; phyto buffers; phyto-recurrent selection; phytotechnologies; poplars; *Populus*

## 1. Introduction

The Great Lakes Basin is one of the most important natural resources in North America, providing numerous environmental, economic, and societal benefits. Zalesny et al. [1] elaborated on these benefits, in addition to the substantial role of the Basin in provisioning freshwater and related ecosystem services to millions of people each year [2,3]. This unique water resource, however, is becoming increasingly degraded by anthropogenic activities. Legacy pollution, urban runoff and stormwater, and agricultural inputs (i.e., herbicides, pesticides, nutrients) have all contributed to declining water quality of the Basin, leading to 99% of the surface water being impaired for one or more designated use(s) [4].

Landfills, waste dumps, and similar sites have contributed to non-point source pollution, especially due to the continuous rise in waste generation and concomitant increases in landfill size [5]. Landfill leachate is a potential pollution source from municipal landfill sites that is often characterized by low biodegradability, high nitrogen content, and presence of other pollutants [6]. Leachate and associated surface runoff are often managed through proactive preventative measures or reactive remediation strategies to prevent water contamination. Phytoremediation is one potential long-term, sustainable solution for achieving runoff reduction and cleaning/filtering of water, in which plants and their associated microorganisms are used for environmental cleanup [7,8]. Pollutants are remediated by various mechanisms such as accumulation in plant tissues, plant and microbe metabolism, and volatilization [9–11]. Plant water uptake can also reduce contaminant mobility at a site [12].

Purpose-grown trees, particularly poplars (*Populus* spp.) and other short rotation woody crops (SRWCs), are well-suited to phytotechnology applications due to their ideal physiological, morphological, and genetic traits [13]. Poplars can help managers achieve remediation goals in a condensed timeframe (e.g., <20 years) based on specific silvicultural prescriptions that are matched to site and management objectives [14]. Additionally, poplar-based phytotechnologies can provide other ecosystem services such as carbon sequestration and biomass feedstocks for biofuels, bioenergy, and bioproducts [15–17]. In recent decades, poplar biomass production systems have become more important globally, given the large demand for wood combined with sustainable forest management goals. As a result, tree breeding and improvement strategies are needed now more than ever to maximize the performance of poplars for achieving specific remediation and ecosystem service objectives.

As with agronomic and horticultural crops, tree breeding and improvement began hundreds of years ago, and over time has expanded to include numerous coniferous and broadleaved species [18]. Significant results have been obtained within the *Populus* genus through spontaneous and controlled hybridization and breeding throughout the last century [19,20]. Broad genetic variation, both within and among *Populus* species, coupled with their ability to undergo successful intra- and inter-specific hybridization, in addition to the ability of some species to propagate readily from cuttings, have driven the success of poplar tree improvement [21–23]. To prove the superiority of new collections and crosses, poplar genotypes and cultivars undergo complex testing in multi-environmental trials (MET), in which phenotypic responses to different environments, defined as genotype by environment interactions (G × E), are evaluated. Similarly, METs are used to test the robustness in genotypic performance across varying site and climatic conditions [24,25]. These G × E interactions have been studied often, leading to the characterization of genotypes as generalists or specialists [26,27]. Over the years, traits of interest in poplar breeding programs have evolved from agronomic characteristics (e.g., yield, pest and disease resistance, rooting capabilities) to more contemporary traits relating to biomass production (e.g., physiological drivers of productivity and wood properties) [22] and ecosystem services [13,16].

Regional clonal development in the Midwestern United States has proliferated since the 1930s due to extensive open-pollination collections, intra- and inter-sectional hybridization, and increased interest in wood biomass production [19,20,28]. Over 100,000 poplar offspring have been created since the 1950s [14], with the majority produced by regional breeding programs at the University of Illinois (J. Jokela; B. McMahon), Iowa State University (R. Hall; B. McMahon), University of Minnesota (C. Mohn; D. Riemenschneider), and University of Minnesota Duluth (B. McMahon; W. Berguson). Clonal testing has been highly active since the 1990s [20], with multiple MET networks being established around the Midwest to monitor biomass production [29–31]. From these METs, Netzer et al. [32] showed the greatest potential of clones was for *P. deltoides* Bartr. ex Marsh × *P. nigra* L. 'DN' hybrids (a.k.a., *P.* × *euramericana* (Dode) Guinier; *P.* × *canadensis* Moench) 'DN21', 'DN154', 'DN164', 'DN170', 'DN177', and 'NE264', in addition to *P. nigra* × *P. maximowiczii* A. Henry 'NM' hybrid 'NM2'. With the exception of 'NM2', all clones were 'DN' hybrids exhibiting generalist growth performance. Another poplar clonal regional testing network was established in 1995, 1997, and 2000 across Iowa, Michigan, Minnesota, and Wisconsin [28]. This MET network initially contained 42 clones but was expanded to a total of 187 clones, most of which were from the aforementioned Midwestern breeding programs [33]. Results from these METs showed greater biomass productivity rates than any previously recorded in the region, leading Riemenschneider et al. [28] to conclude the need for continued tree improvement activities. Significant G × E interactions defined generalist ('NC14105', 'Crandon', 'NM2') and specialist ('7300501', '80 × 01015', 'NC14103') clones [33], which have since been tested for ecosystem services and environmental technologies [16,34].

The most recent poplar breeding and testing has been conducted at the University of Minnesota Duluth, Natural Resources Research Institute (NRRI) [35,36]. In parallel with traditional clonal testing of poplar productivity through evaluation of genotypic growth and stability [36], NRRI researchers have tested the application of different silvicultural measures [37] and defined geo-robust clones (i.e., extreme generalists) for establishment across broader latitudinal and longitudinal ranges [38]. A contemporary goal of this and other poplar breeding efforts is to test clones for a wide range of ecosystem services such as carbon sequestration and phytoremediation [17,39].

Poplars have been tested and deployed extensively in phytoremediation systems to remediate organic [9,40–43] and inorganic contaminants [44–47], in addition to newer classes of pollutants such as contaminants of emerging concern (CECs) [48–50]. Testing poplar clones for phytoremediation is a complex process including breeding and selection for: (1) traditional traits related to growth and productivity [51–53]; (2) tolerance of contaminants, determined by investigating physiological and metabolic processes [42,43,47,54]; and (3) phytoremediation potential exhibited by contaminant accumulation/degradation [41,55,56]. Simultaneous selection for such a broad range of breeding traits can be achieved with phyto-recurrent selection, a stepwise testing process. In this method, crop and tree improvement strategies are implemented over multiple testing cycles to identify and select clones with superior performance [14,57]. Throughout the selection process, the number of clones decreases while the number of tested parameters and cycle length increase. Selection using basic traits such as growth and root:shoot ratio is enhanced with data on additional parameters such as tree health and growth performance index [58]. Further investigation often includes greenhouse and field-testing clonal performance related to contaminant effects and accumulation, ecophysiology, and morpho-anatomical changes [45,46]. Following multiple selection cycles in the greenhouse, field validation of selected clones is a necessary step in phyto-recurrent selection. For example, testing clones used in the current study, Zalesny and Bauer [59] reported broad clonal variation across eleven-year-old trees grown for nitrate phytoremediation in the Midwestern US. These phyto-recurrent selection results further emphasize the importance of long-term phytoremediation studies in evaluating clonal performance throughout stand development [17].

As described by Zalesny et al. [1], phyto-recurrent selection was used to establish an ongoing MET testing network consisting of sixteen phytoremediation buffer systems (i.e., phyto buffers) at sites located in the Lake Superior (i.e., Michigan's Upper Peninsula) and Lake Michigan (i.e., eastern Wisconsin) watersheds. Given the potential of new genotypes in the biomass productivity networks illustrated above, our overarching objective in the current study was to test for ecological restoration potential of new clones developed at NRRI. To do so , we divided clones into Experimental (i.e., genotypes with a rich history of testing but are still at the experimental stage) and Common (i.e., genotypes commonly used for commercial and/or research purposes in the region) clone groups that we then compared with each other and each NRRI clone planted at the phyto buffers. Although Zalesny et al. [1] compared individual clones, these current comparisons are warranted because poplar clones in the Midwestern United States are often selected in groups based on stage of testing (i.e., Experimental versus Common) rather than individually, due to uncertainties with nursery production and availability of clonal material. Specifically , we tested for differences in the three clone groups (i.e., NRRI, Experimental, Common), phyto buffers (i.e., environments), and their interactions for health, height, diameter, and volume during early field establishment (i.e., from one to four years after planting). These data are useful to advance poplar tree improvement efforts throughout the region, continent, and world, informing clonal selection for multiple end-uses, including phytotechnologies.

## 2. Materials and Methods

### 2.1. Site Description

Zalesny et al. [1] provided a detailed description of the regional phytotechnologies network tested in the current study, including climate- and soil-related information. In summary, there were sixteen phytoremediation buffer systems (i.e., phyto buffers) established across ten field testing sites in 2017 ($\times$6 phyto buffers), 2018 ($\times$5), and 2019 ($\times$5) in the Lake Superior watershed of the Upper Peninsula of Michigan, USA and the Lake Michigan watershed of eastern Wisconsin, USA (Figure 1). The sites ranged in latitude from 46.7840 to 42.8382° N and in longitude from −89.1291 to −86.5976° W. Twenty-year (2000 to 2020) historical monthly averages for precipitation and temperature were obtained from the National Oceanic and Atmospheric Administration (NOAA) National Climate Data Center (https://www.ncdc.noaa.gov/cdo-web/ (accessed on 20 January 2021)) and are listed in Table 1. Table 2 provides buffer-specific soil properties that were acquired from the USDA Natural Resources Conservation Service (NRCS) Web Soil Survey (https://websoilsurvey.sc.egov.usda.gov/ (accessed on 20 January 2021)).

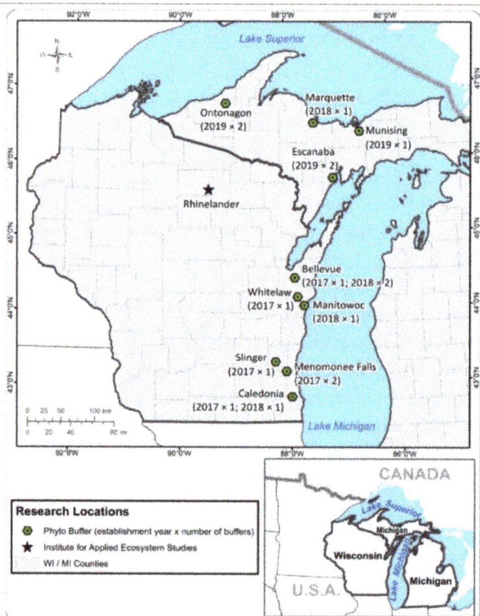


**Figure 1.** Regional phytotechnologies network consisting of sixteen phytoremediation buffer systems (i.e., phyto buffers) established in 2017 ($\times$6 phyto buffers), 2018 ($\times$5), and 2019 ($\times$5) in the Lake Superior watershed of the Upper Peninsula of Michigan, USA and the Lake Michigan watershed of eastern Wisconsin, USA. From Zalesny et al. [1].

**Table 1.** Precipitation and temperature of ten field testing sites in a regional phytotechnologies network consisting of sixteen phytoremediation buffer systems (i.e., phyto buffers) established from 2017 to 2019 in the Lake Superior watershed of the Upper Peninsula of Michigan, USA and the Lake Michigan watershed of eastern Wisconsin, USA. Adapted from Zalesny et al. [1].

| Site | Bellevue, WI | Caledonia, WI | Escanaba, MI | Manitowoc, WI | Marquette, MI |
|---|---|---|---|---|---|
| County | Brown | Racine | Delta | Manitowoc | Marquette |
| Buffer group (i.e., year of planting) | 2017, 2018 | 2017, 2018 | 2019 | 2018 | 2018 |
| Phyto buffer [a] | BC, BE, BW | CE, CW | EE, EW | MA | MQ |
| Annual precipitation (P) (mm) [b] | $613 \pm 27$ | $686 \pm 36$ | $556 \pm 32$ | $614 \pm 27$ | $530 \pm 28$ |
| Average temperature ($T_{avg}$) (°C) | $15.3 \pm 0.2$ | $15.7 \pm 0.2$ | $13.6 \pm 0.2$ | $14.8 \pm 0.2$ | $13.1 \pm 0.4$ |
| **Site** | **Menomonee Falls, WI** | **Munising, MI** | **Ontonagon, MI** | **Slinger, WI** | **Whitelaw, WI** |
| County | Waukesha | Alger | Ontonagon | Washington | Manitowoc |
| Buffer group (i.e., year of planting) | 2017 | 2019 | 2019 | 2017 | 2017 |
| Phyto Buffer | ME, MW | MU | ON, OS | SL | WH |
| Annual precipitation (P) (mm) | $649 \pm 23$ | $655 \pm 25$ | $551 \pm 26$ | $653 \pm 36$ | $640 \pm 26$ |
| Average temperature ($T_{avg}$) (°C) | $15.3 \pm 0.1$ | $12.3 \pm 0.2$ | $13.4 \pm 0.2$ | $15.1 \pm 0.2$ | $14.9 \pm 0.1$ |

[a] BC: Bellevue (Central); BE: Bellevue (East); BW: Bellevue (West); CE: Caledonia (East); CW: Caledonia (West); EE: Escanaba (East); EW: Escanaba (West); MA: Manitowoc; ME: Menomonee Falls (East); MW: Menomonee Falls (West); MQ: Marquette; MU: Munising; ON: Ontonagon (North); OS: Ontonagon (South); SL: Slinger; WH: Whitelaw. [b] Precipitation and temperature data are means ± one standard error across each growing season (April to October) from 2000 to 2020. Data source: National Oceanic and Atmospheric Administration (NOAA) National Climate Data Center (https://www.ncdc.noaa.gov/cdo-web/ (accessed on 20 January 2021)).

**Table 2.** Soil properties of sixteen phytoremediation buffer systems (i.e., phyto buffers) comprising a regional phytotechnologies network established from 2017 to 2019 in the Lake Superior watershed of the Upper Peninsula of Michigan, USA and the Lake Michigan watershed of eastern Wisconsin, USA. Adapted from Zalesny et al. [1].

| Phyto Buffer [a] | BC | BE | BW | CE | CW | EE, EW | MA | ME, MW | MQ | MU | ON, OS | SL | WH |
|---|---|---|---|---|---|---|---|---|---|---|---|---|---|
| Soil series | Manawa | Kewaunee | Bellevue | Fox | Matherton | Croswell | Hochheim | Sebewa | Schweitzer | Kalkaska | Oldman | Casco | Boyer |
| Drainage class [b] | SPD | WD | SPD | MWD | SPD | MWD | WD | PD | WD | SED | MWD | SED | WD |
| Texture [c] | SiCL | SiCL | SiCL | L | L | S | L | L | SL | S | L | SL | SCL |
| Sand (%) | 10.1 | 13.3 | 19.8 | 39.5 | 50.1 | 87.4 | 45.4 | 37.3 | 55.9 | 94.7 | 51.4 | 54.0 | 58.2 |
| Silt (%) | 45.9 | 47.7 | 50.0 | 39.7 | 28.1 | 10.4 | 34.4 | 42.1 | 41.1 | 4.4 | 41.4 | 28.6 | 18.8 |
| Clay (%) | 44.0 | 39.0 | 30.2 | 20.8 | 21.8 | 2.2 | 20.2 | 20.6 | 3.0 | 0.9 | 7.2 | 17.4 | 23.0 |
| pH | 7.0 | 6.6 | 7.2 | 5.8 | 6.2 | 4.9 | 7.4 | 7.0 | 4.9 | 5.0 | 4.6 | 7.4 | 6.9 |
| Frost free days (#) | 160 | 160 | 135 | 173 | 150 | 130 | 145 | 152 | 115 | 130 | 110 | 169 | 140 |
| Depth to water table (cm) | >200 | >200 | 0 | 178 | 30 | 60 | >200 | 15 | >200 | >200 | 30 | >200 | >200 |

Source: USDA Natural Resources Conservation Service (NRCS) Web Soil Survey (https://websoilsurvey.sc.egov.usda.gov/ (accessed on 20 January 2021)). [a] Phyto buffers: BC: Bellevue (Central); BE: Bellevue (East); BW: Bellevue (West); CE: Caledonia (East); CW: Caledonia (West); EE: Escanaba (East); EW: Escanaba (West); MA: Manitowoc; ME: Menomonee Falls (East); MW: Menomonee Falls (West); MQ: Marquette; MU: Munising; ON: Ontonagon (North); OS: Ontonagon (South); SL: Slinger; WH: Whitelaw. [b] Drainage classes: MWD: moderately well drained; PD: poorly drained; SED: somewhat excessively drained; SPD: somewhat poorly drained; WD: well drained. [c] Textures: L: loam; S: sand; SCL: sandy clay loam; SiCL: silty clay loam; SL: sandy loam.

## 2.2. Clone Selection

Rogers et al. [58] and Zalesny et al. [1] described the phyto-recurrent selection process that was used to choose genotypes for phyto buffer field establishment. Twelve clones were selected, outplanted, and tested for each of three buffer groups (i.e., with buffer groups defined as phyto buffers established in 2017 ($\times$6), 2018 ($\times$5), and 2019 ($\times$5)), and separate analyses were conducted for each buffer group for the particular set of twelve clones. Based on the objective of the current study, clones were categorized into three clone groups: (1) 'NRRI' clones that are new genotypes produced by the University of Minnesota Duluth, Natural Resources Research Institute (NRRI), in Duluth, Minnesota, USA [36,38] (these genotypes were not combined with one another and were analyzed individually, collectively representing the NRRI clone group); (2) 'Experimental' clones that have been tested broadly in the region but have not reached commercial status (combined for current analyses); and (3) 'Common' clones that have been used in decades of testing and deployment in the Midwestern United States (combined). Clones, genomic groups, and their respective clone groups are listed in Table 3.

**Table 3.** Clone groups and buffer groups (i.e., years of planting) of clones and their genomic groups for *Populus* genotypes tested in a regional phytotechnologies network of sixteen phytoremediation buffer systems (i.e., phyto buffers) established from 2017 to 2019 in the Lake Superior watershed of the Upper Peninsula of Michigan, USA and the Lake Michigan watershed of eastern Wisconsin, USA.

| Clone Group [a,b] | | |
| --- | --- | --- |
| NRRI | Experimental | Common |
| | ———— 2017 Buffer group ———— | |
| 99038022 'DN' | 7300502 'D' | DN5 'DN' |
| 99059016 'DN' | DM114 'DM' | DN34 'DN' |
| 9732-36 'DN' | NC14106 'DM' | NM2 'NM' |
| | DN177 'DN' | NM6 'NM' |
| | NM5 'NM' | |
| | ———— 2018 Buffer group ———— | |
| 9732-11 'DN' | 7300502 'D' | DN5 'DN' |
| 9732-24 'DN' | DM114 'DM' | DN34 'DN' |
| 9732-31 'DN' | DN2 'DN' | NM2 'NM' |
| 9732-36 'DN' | NM5 'NM' | NM6 'NM' |
| | ———— 2019 Buffer group ———— | |
| 99038022 'DN' | DM114 'DM' | DN34 'DN' |
| 9732-11 'DN' | DN2 'DN' | NM2 'NM' |
| 9732-24 'DN' | DN177 'DN' | NM6 'NM' |
| 9732-31 'DN' | NM5 'NM' | |
| 9732-36 'DN' | | |

[a] Genomic groups: *P. deltoides* Bartr. Ex Marsh 'D'; *P. deltoides* $\times$ *P. maximowiczii* A. Henry 'DM'; *P. deltoides* $\times$ *P. nigra* L. 'DN'; *P. nigra* $\times$ *P. maximowiczii* 'NM'; [b] 'NRRI' = promising genotypes bred, tested, and selected at the University of Minnesota Duluth, Natural Resources Research Institute (NRRI) for broad-ranging applications [36,38]; analyzed individually. 'Experimental' = genotypes with a rich history of testing but that are still at the experimental stage; analyzed as a group. 'Common' = genotypes commonly used for commercial and/or research purposes in the region; analyzed as a group.

## 2.3. Phyto Buffer Establishment and Experimental Design

Individual phyto buffers were established during May and June in 2017, 2018, and 2019 by planting 25.4 cm, dormant, unrooted hardwood cuttings that were soaked in water to a height of 16.93 cm for 48 h in a dark room at 21 °C before planting. Site preparation consisted of removing rocks and other obstructions followed by tilling to a depth of 30 cm. For site maintenance, soils were tilled to a depth of 30 cm, rocks and other obstructions were continually removed, and vegetation was removed via hand weeding to a minimum diameter of 0.61 m around each individual tree. At least one maintenance entry per month was performed at each phyto buffer throughout each growing season.

The experimental design consisted of eight randomized complete blocks (RCBD) and twelve clones per block at a spacing of 2.44 × 2.44 m (i.e., 1680 trees ha$^{-1}$). There was one exception: four blocks were planted at Slinger, Wisconsin due to space constraints. Two border rows were established on the perimeter of each phyto buffer to reduce potential border effects [60,61]. All phyto buffers were fenced using 2.3 m tall Trident extra strength deer fencing (Trident Enterprises, Waynesboro, PE, USA) to eliminate potential impacts from white-tailed deer (*Odocoileus virginianus* Zimmerman) browse. Replanting of dead trees with identical clones occurred each growing season to ensure full stocking of 1680 trees ha$^{-1}$. Analyses did not include the replanted trees.

*2.4. Field Measurements*

Tree height (to the nearest 0.1 m) and diameter (to the nearest 0.1 cm) were measured after each growing season. Height was consistently measured from the ground to the apical bud, whereas diameter measurements changed as trees aged. At one and two years after planting, diameter was measured at 10 cm above the soil surface; starting in year three, diameter at breast height (i.e., DBH at 1.37 m) was determined. Based on height (H) and diameter (D; including one- and two-year diameter and DBH), tree volume (V) was calculated using the following equation provided by Kershaw et al. [62]: $V = D^2 \times H$. After four years of growth, the 2017 buffer group trees were too tall to be measured to the nearest 0.1 m. For these trees, DBH values were used to estimate mean annual increment (MAI; Mg ha$^{-1}$ yr$^{-1}$) according to genomic-group specific coefficients from Headlee and Zalesny [63] applied in the following model: $Biomass_{Individual\ Tree} = 10^{a0} \times DBH^{a1}$. Standard metric conversion factors and the stocking of 1680 trees ha$^{-1}$ were used to scale these individual-tree values to stand-level MAI.

*2.5. Health Assessments*

Six tree health parameters were scored by two researchers to reduce variability in the ratings: (1) vigor, (2) defoliation, (3) leaf discoloration, (4) chlorosis, (5) leaf scorch, and (6) leaf spots. Scoring consisted of a five-category qualitative scale ranging from 1 to 5, where 1 = optimal health, 2 = good health, 3 = moderate health, 4 = poor health, and 5 = dead (modified from Rogers et al. [58]; i.e., health score was inversely related to health). Final health index values were calculated using a multiplicative weighted summation index with a coefficient of 0.25 for vigor and 0.15 for all other parameters. Health assessments were not conducted in 2020.

*2.6. Data Analysis*

Clone groups described above and listed in Table 3 (i.e., NRRI, Experimental, Common) were substituted for clones in Zalesny et al. [1]; otherwise, data analysis methods were the same for both studies.

As directly reported in Zalesny et al. [1], "Health (of all buffer groups) and MAI (of the 2017 buffer group) data were subjected to analyses of variance (ANOVA) and analyses of means (ANOM) using SAS® (PROC GLM; PROC ANOM; SAS INSTITUTE, INC., Cary, North Carolina, USA) assuming a two-way factorial design including six (2017) or five (2018, 2019) buffers, [five (2017), six (2018), or seven (2019) clone groups], and their interactions. Fisher's Least Significant Difference (LSD) was used to identify significant differences among least-squares means for main effects and interactions at $p < 0.05$".

As directly reported in Zalesny et al. [1], "Height and volume (of all buffer groups) and diameter (excluding 2020 diameter of 2017 buffer group trees) data were subjected to analyses of variance (ANOVA) and analyses of means (ANOM) using SAS® (PROC MIXED; PROC ANOM; SAS INSTITUTE, INC., Cary, NC, USA) assuming a three-way, repeated measures factorial design including six (2017) or five (2018, 2019) buffers, [five (2017), six (2018), or seven (2019) clone groups], three (2017, 2018) or two (2019) ages, and their interactions. The ages (representing tree growth after each growing season) were analyzed as the repeated measure. To account for pseudo-replication over time, six different covariance structures (i.e., vc, cs, ar(1), toep, ante(1), un) were tested in PROC MIXED to determine which one provided the best model fit based on the lowest Bayesian Information Criterion (BIC) scores. Using these covariance structures, ANOVA were conducted in PROC MIXED for all traits, and multiple comparisons analyses were conducted to identify significant differences among least-squares means for main effects and interactions as noted above".

## 3. Results

### 3.1. Survival

First-year survival across all phyto buffers and clones was 97.1%, with 95.5%, 96.2%, and 99.6% survival for the 2017, 2018, and 2019 buffer groups, respectively. For the 2017 buffer group, an additional 24 trees (4.5%) were replanted due to external factors not associated with direct mortality. Specifically, three trees were coppiced due to encroachment of a powerline, 12 trees were impacted by beavers, and nine trees exhibited some level of winter dieback that was not fatal. Additionally, trees at Caledonia (East) were flooded for five days during early May of the 2018 growing season. All trees survived the flood and growth may have been impacted initially, but growth reductions were not evident during end-of-year measurements. For the 2018 buffer group, 33 trees (6.9%) were impacted by external factors, with 11 trees experiencing substantial growth reductions associated with runoff of water used to cool an adjacent mulch pile, and 22 trees having deer browse and broken tops. For the 2019 buffer group, no external factors impacted tree survival. All trees that died or were impacted were replanted to ensure full stocking of 1680 trees ha$^{-1}$ in subsequent years.

For the 2017 buffer group, first-year survival ranged from 37.5 ('99059016' at Menomonee Falls (East)) to 100% (for 18 of 30 possible buffer × clone group combinations) (Table 4). There was minimal variability across phyto buffers, with survival at Whitelaw, which had the lowest number of trees alive, being 3.1% less than Bellevue (West), the buffer with the greatest survival. The variability increased for clone groups, ranging from 77.3% ('99059016') to 100% ('99038022'), although this range in survival was driven by the fact that only 37.5% of the '99059016' trees were alive at Menomonee Falls (East). The next lowest survival for all buffer × clone group combinations was 75% for '99059016' at Caledonia (East) and Slinger. Experimental and Common clone groups exhibited at least 92.5% survival at all buffers. For the 2018 buffer group, first-year survival ranged from 87.5% ('9732-36' at Marquette) to 100% (for 21 of 30 possible buffer × clone group combinations) (Table 4). Variability across phyto buffers was stable, with trees at Bellevue (East) and Marquette (the buffers with the lowest survival) exhibiting 3.1% fewer trees alive than at Bellevue (Central), which had the highest survival. The percentage of trees alive across clone groups increased 5.7% from the Common clones to three of the NRRI genotypes: '9732-11'; '9732-24'; and '9732-31'. With the exception of '9732-36' grown at Marquette, all NRRI clones exhibited 100% survival across buffers, whereas the lowest survival for the Experimental and Common clones was 90.6% at Manitowoc and Marquette, respectively. For the 2019 buffer group, first-year survival was 100% for all buffer × clone group combinations, with two exceptions (Table 4). Survival was 87.5% for '9732-31' at Escanaba (East) and Ontonagon (South).

**Table 4.** First-year survival (percentage) of three poplar clone groups tested in sixteen phytoremediation buffer systems (i.e., phyto buffers) that were established in 2017, 2018, and 2019 (i.e., buffer groups) in the Lake Superior watershed of the Upper Peninsula of Michigan, USA and the Lake Michigan watershed of eastern Wisconsin, USA.

| | | | | Clone Group [a] | | |
|---|---|---|---|---|---|---|
| | | **NRRI** | | | | |
| **Buffer** [b] | 99038022 | 99059016 | 9732-36 | Experimental | Common | Overall |
| | | | | 2017 Buffer group | | |
| BW | 100.0 | 75.0 | 100.0 | 97.5 | 100.0 | 96.9 |
| CE | 100.0 | 87.5 | 87.5 | 100.0 | 93.8 | 95.8 |
| ME | 100.0 | 37.5 | 100.0 | 100.0 | 100.0 | 94.8 |
| MW | 100.0 | 100.0 | 100.0 | 90.0 | 100.0 | 95.8 |
| SL | 100.0 | 75.0 | 100.0 | 95.0 | 100.0 | 95.8 |
| WH | 100.0 | 87.5 | 100.0 | 92.5 | 93.8 | 93.8 |
| Overall | 100.0 | 77.3 | 97.7 | 95.9 | 97.7 | 95.5 |

| | | | | Clone Group | | |
|---|---|---|---|---|---|---|
| | | **NRRI** | | | | |
| **Buffer** [c] | 9732-11 | 9732-24 | 9732-31 | 9732-36 | Experimental | Common | Overall |
| | | | | 2018 Buffer group | | | |
| BC | 100.0 | 100.0 | 100.0 | 100.0 | 100.0 | 93.8 | 97.9 |
| BE | 100.0 | 100.0 | 100.0 | 100.0 | 93.8 | 90.6 | 94.8 |
| CW | 100.0 | 100.0 | 100.0 | 100.0 | 93.8 | 96.8 | 96.8 |
| MA | 100.0 | 100.0 | 100.0 | 100.0 | 90.6 | 100.0 | 96.9 |
| MQ | 100.0 | 100.0 | 100.0 | 87.5 | 96.9 | 90.6 | 94.8 |
| Overall | 100.0 | 100.0 | 100.0 | 97.5 | 95.0 | 94.3 | 96.2 |

| | | | | Clone Group | | | | |
|---|---|---|---|---|---|---|---|---|
| | | | **NRRI** | | | | | |
| **Buffer** [d] | 99038022 | 9732-11 | 9732-24 | 9732-31 | 9732-36 | Experimental | Common | Overall |
| | | | | | 2019 Buffer group | | | |
| EE | 100.0 | 100.0 | 100.0 | 87.5 | 100.0 | 100.0 | 100.0 | 99.0 |
| EW | 100.0 | 100.0 | 100.0 | 100.0 | 100.0 | 100.0 | 100.0 | 100.0 |
| MU | 100.0 | 100.0 | 100.0 | 100.0 | 100.0 | 100.0 | 100.0 | 100.0 |
| ON | 100.0 | 100.0 | 100.0 | 100.0 | 100.0 | 100.0 | 100.0 | 100.0 |
| OS | 100.0 | 100.0 | 100.0 | 87.5 | 100.0 | 100.0 | 100.0 | 99.0 |
| Overall | 100.0 | 100.0 | 100.0 | 95.0 | 100.0 | 100.0 | 100.0 | 99.6 |

[a] 'NRRI' = promising genotypes bred, tested, and selected at the University of Minnesota Duluth, Natural Resources Research Institute (NRRI) for broad-ranging applications [36,38]. 'Experimental' = genotypes with a rich history of testing but that are still at the experimental stage. 'Common' = genotypes commonly used for commercial and/or research purposes in the region. [b] BW: Bellevue (West); CE: Caledonia (East); ME: Menomonee Falls (East); MW: Menomonee Falls (West); SL: Slinger; WH: Whitelaw. [c] BC: Bellevue (Central); BE: Bellevue (East); CW: Caledonia (West); MA: Manitowoc; MQ: Marquette. [d] EE: Escanaba (East); EW: Escanaba (West); MU: Munising; ON: Ontonagon (North); OS: Ontonagon (South).

*3.2. Health*

Buffer main effects were significant for first-year health of 2017 ($p = 0.0006$), 2018 ($p < 0.0001$), and 2019 ($p < 0.0001$) buffer group trees (Table S1). Health of 2017 buffer group trees measured in 2017 (i.e., HEALTH$_{2017(2017)}$) ranged from $1.06 \pm 0.02$ (Whitelaw; most healthy) to $1.17 \pm 0.02$ (Caledonia (East); least healthy), with an overall mean of $1.11 \pm 0.02$ (Figure 2). Thus, all trees were of optimal health (i.e., health index ranging from 1 to 2). Trees grown at Menomonee Falls (West) and Whitelaw were 4.8% to 9.1% significantly healthier than at the remaining phyto buffers, which were not different than each other. Whitelaw trees were also 4.6% healthier than the overall mean, and the mean was 4.7% healthier than those from Caledonia (East). Health of 2018 buffer group trees measured in 2018 (i.e., HEALTH$_{2018(2018)}$) ranged from $1.01 \pm 0.02$ (Bellevue (East); most healthy) to $1.28 \pm 0.02$ (Manitowoc; least healthy), with an overall mean of $1.09 \pm 0.02$ (Figure 2). Trees at Manitowoc had 15% unhealthier trees than Caledonia (West), and both of these buffers had significantly lower health than at Bellevue (Central), Bellevue (East), and Marquette, which were not different from one another. With the exception of Caledonia (West), all buffers exhibited health index scores significantly different than the overall mean, with Manitowoc being the only buffer with poorer health (i.e., by 15%). HEALTH$_{2019(2019)}$ ranged from $1.02 \pm 0.01$ (Ontonagon (South); most healthy) to $1.14 \pm 0.01$ (Escanaba (West); least healthy), with an overall mean of $1.07 \pm 0.01$ (Figure 2). Trees at Escanaba (West) were significantly less healthy than those at Escanaba (East), Munising, and Ontonagon (North), the latter of which had similar health to Ontonagon (South), which exhibited 5% greater health than the overall mean.

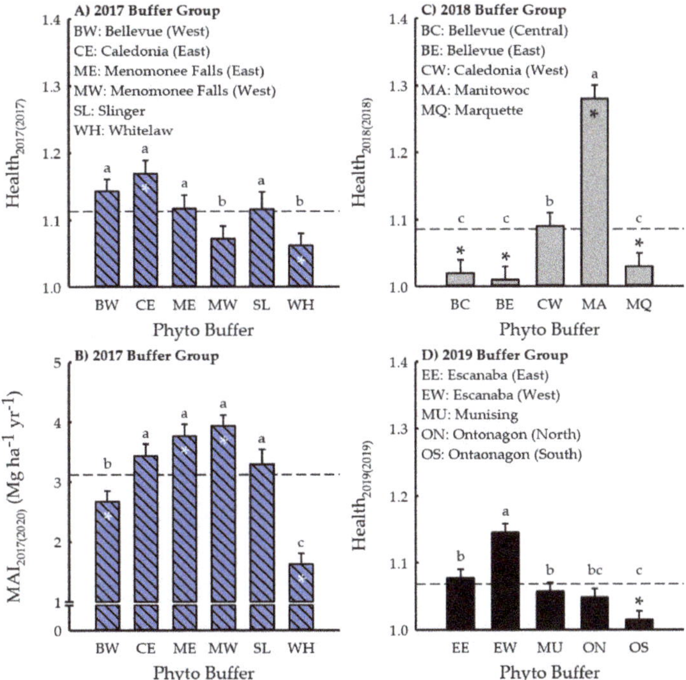


**Figure 2.** First-year tree health (**A**) and fourth-year mean annual increment (MAI) (**B**) of six phytoremediation buffers (i.e., phyto buffers) established in 2017 (i.e., the 2017 Buffer Group), in addition to the first-year tree health of five phyto buffers established in the 2018 Buffer Group (**C**) and 2019 Buffer Group (**D**) of a regional phytotechnologies network in the Lake Superior watershed of the Upper Peninsula of Michigan, USA and the Lake Michigan watershed of eastern Wisconsin, USA. Error bars represent one standard error of the mean. The dashed line represents the overall mean, and asterisks indicate means different than the overall mean at $p < 0.05$. Bars with different letters are different at $p < 0.05$. See Section 2 for complete tree health definitions (1 = optimal health, 2 = good health, 3 = moderate health, 4 = poor health, and 5 = dead).

The clone group main effect was significant for HEALTH$_{2017(2017)}$ ($p < 0.0001$) (Table S1). HEALTH$_{2017(2017)}$ ranged from $1.05 \pm 0.02$ ['99059016'; most healthy] to $1.19 \pm 0.02$ [Experimental; least healthy], with an overall mean of $1.11 \pm 0.02$ (Figure 3). The healthiest trees were from '99059016' and the Common clone group, which did not differ from one another but were 5.7% and 4% healthier than the overall mean, respectively. Although health of the Experimental trees did not differ from '9732-36', they were of 6.8% poorer health than the overall mean.

Differences among buffer and clone main effects were significant for second- and third-year health of the 2017 buffer group trees and second-year health of the 2018 buffer group trees ($p < 0.05$), yet the buffer × clone group interaction governed health for all three combinations ($p_{2017(2018)} = 0.0233$; $p_{2017(2019)} = 0.0010$; $p_{2018(2019)} = 0.0023$) (Table S1). HEALTH$_{2017(2018)}$ ranged from 1.20 ± 0.08 ('99059016' at Menomonee Falls (East); most healthy) to 1.76 ± 0.07 ('9732-36' at Slinger; least healthy), with an overall mean of 1.36 ± 0.05 (Figure 4). The healthiest trees were grown at Menomonee Falls (East), which had 17.8% better HEALTH$_{2017(2018)}$ than at Slinger, which exhibited the poorest health. The range in health scores was narrower for clone groups, with '99059016' having 10.5% healthier trees than Experimental genotypes that had the poorest health. Three buffer × clone group interactions resulted in HEALTH$_{2017(2018)}$ values that were significantly greater (i.e., of poorer health) than the overall mean: Experimental at Bellevue (West); '9732-36' and Experimental at Slinger (Figure 4). Trends in HEALTH$_{2017(2019)}$ (Figure S1) and HEALTH$_{2018(2019)}$ (Figure S2) were similar to HEALTH$_{2017(2018)}$.

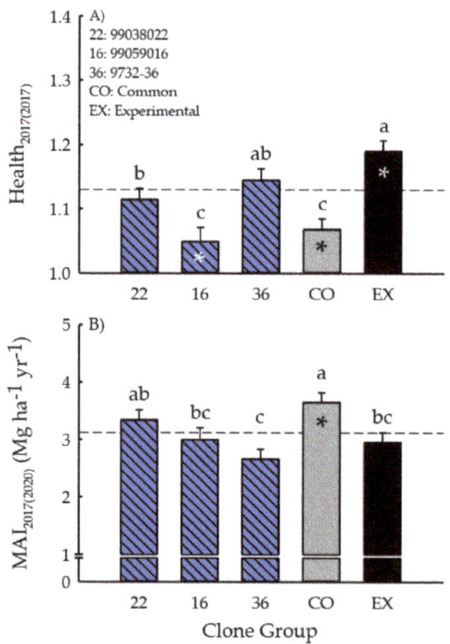


**Figure 3.** First-year tree health (**A**) and fourth-year mean annual increment (MAI) (**B**) of three clone groups (i.e., NRRI = 22, 16, 36; Common; Experimental; see Table 3 for definitions) tested in six phytoremediation buffers (i.e., phyto buffers) established in 2017 (i.e., the 2017 Buffer Group) in the Lake Michigan watershed of eastern Wisconsin, USA. Error bars represent one standard error of the mean. The dashed line represents the overall mean, and asterisks indicate means different than the overall mean at $p < 0.05$. Bars with different letters are different at $p < 0.05$. See Section 2 for complete tree health definitions (1 = optimal health, 2 = good health, 3 = moderate health, 4 = poor health, and 5 = dead).

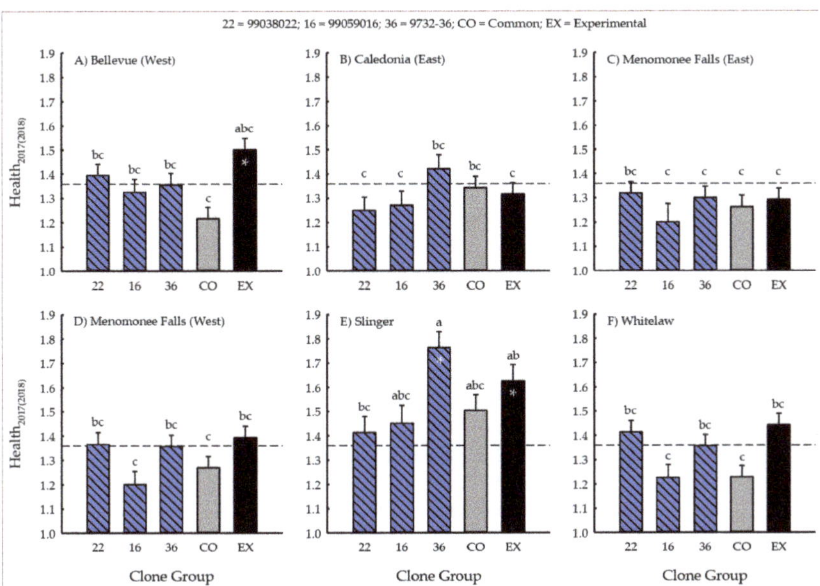



**Figure 4.** Tree health (± one standard error) determined after the 2018 growing season of three clone groups (i.e., NRRI = 22, 16, 36; Common; Experimental; see Table 3 for definitions) tested in six phytoremediation buffer systems (i.e., phyto buffers) established in 2017 (i.e., the 2017 Buffer Group) in the Lake Michigan watershed of eastern Wisconsin, USA. The dashed line represents the overall mean, and asterisks indicate means different than the overall mean at $p < 0.05$. Bars with different letters across all buffer × clone group combinations are different at $p < 0.05$. See Section 2 for complete tree health definitions (1 = optimal health, 2 = good health, 3 = moderate health, 4 = poor health, and 5 = dead).

### 3.3. Biomass and Growth

Buffer main effects were significant for mean annual increment (MAI) of 2017 buffer group trees measured in 2020 (i.e., $MAI_{2017(2020)}$) ($p < 0.0001$) (Table S1). $MAI_{2017(2020)}$ ranged from 1.62 ± 0.18 (Whitelaw) to 3.93 ± 0.18 Mg ha$^{-1}$ yr$^{-1}$ (Menomonee Falls (West)), with an overall mean of 3.12 ± 0.20 Mg ha$^{-1}$ yr$^{-1}$ (Figure 2). The largest trees were grown at Caledonia (East), Menomonee Falls (East), Menomonee Falls (West), and Slinger; these trees were at least 23.1% significantly greater than those grown at Bellevue (West), which were 64% larger than Whitelaw trees. Trees grown at Bellevue (West) and Whitelaw had significantly less biomass than the overall mean, while those at both Menomonee Falls buffers had biomass greater than the mean.

The clone group main effect was significant for $MAI_{2017(2020)}$ ($p = 0.0010$) (Table S1). $MAI_{2017(2020)}$ ranged from 2.66 ± 0.18 ('9732-36') to 3.65 ± 0.17 Mg ha$^{-1}$ yr$^{-1}$ (Common), with an overall mean of 3.12 ± 0.18 Mg ha$^{-1}$ yr$^{-1}$ (Figure 3). Trees of the Common clone group were the largest, having 16.9% more biomass than the overall mean. Whereas '99038022' had similar $MAI_{2017(2020)}$ to Common trees, this NRRI genotype was also similar in biomass to '99059016' and Experimental trees, which were not different from one another.

The buffer × year interaction was significant for height, diameter, and volume of the 2017 buffer group trees, in addition to height for the 2018 and 2019 buffer group trees ($p < 0.0001$ for all interactions) (Table S2). Across buffers, volume increased 49.4-fold from 2017 to 2018 and then 1.8-fold from 2018 to 2019. In particular, $VOLUME_{2017(2017)}$ ranged from 24.9 ± 14.8 (Whitelaw) to 280.9 ± 206.0 cm$^3$ (Slinger) (mean = 121.6 ± 17.3 cm$^3$), $VOLUME_{2017(2018)}$ ranged from 2151.2 ± 450.5 (Whitelaw) to 7929.7 ± 503.5 cm$^3$ (Menomonee Falls (East)) (mean = 6010.9 ± 499.4 cm$^3$), and $VOLUME_{2017(2019)}$ ranged from 3371.1 ± 728.5 (Whitelaw) to 15,226.0 ± 814.3 cm$^3$ (Menomonee Falls (East)) (mean = 10,656.1 ± 807.7 cm$^3$) (Figure 5). There was more variability across buffers during the first growing season than subsequent years. Within years, there was a general trend of Slinger and both Menomonee Falls buffers to have trees with the largest volume, whereas Whitelaw had the smallest trees, and Bellevue (West) and Caledonia (East) were intermediate. Trends in HEIGHT$_{2017}$ (Figure S3), DIAMETER$_{2017}$ (Figure S4), HEIGHT$_{2018}$ (Figure S5), and HEIGHT$_{2019}$ (Figure S6) were similar to VOLUME$_{2017}$ for the buffer $\times$ year interaction.

The clone group $\times$ year interaction was significant for height ($p < 0.0001$), diameter ($p = 0.0184$), and volume ($p = 0.0449$) of the 2017 buffer group trees (Table S2). Across clone groups, volume increased 29.5-fold from 2017 to 2018 and then 1.6-fold from 2018 to 2019. In particular, VOLUME$_{2017(2017)}$ ranged from 53.2 $\pm$ 21.9 ('99059016') to 161.5 $\pm$ 17.0 cm$^3$ ('99038022') (mean = 121.6 $\pm$ 15.5 cm$^3$), VOLUME$_{2017(2018)}$ ranged from 5202.6 $\pm$ 430.1 (Experimental) to 6833.2 $\pm$ 444.4 cm$^3$ ('99038022') (mean = 6010.9 $\pm$ 458.1 cm$^3$), and VOLUME$_{2017(2019)}$ ranged from 9965.7 $\pm$ 724.7 ('9732-36') to 12,092.0 $\pm$ 718.8 cm$^3$ ('99038022') (mean = 10,656.1 $\pm$ 740.7 cm$^3$) (Figure 6). Similar to the buffer $\times$ year interaction, volume of the first growing season had greater variability in clone group performance relative to years two and three. Within years, '99038022' consistently exhibited the greatest volume, although not necessarily from a statistical standpoint. In 2017, however, '99059016' had the lowest volume, which was 203.7% significantly lower than that of '99038022'. Trends in HEIGHT$_{2017}$ (Figure S7) and DIAMETER$_{2017}$ (Figure S8) were similar to those of VOLUME$_{2017}$ for the clone group $\times$ year interaction.

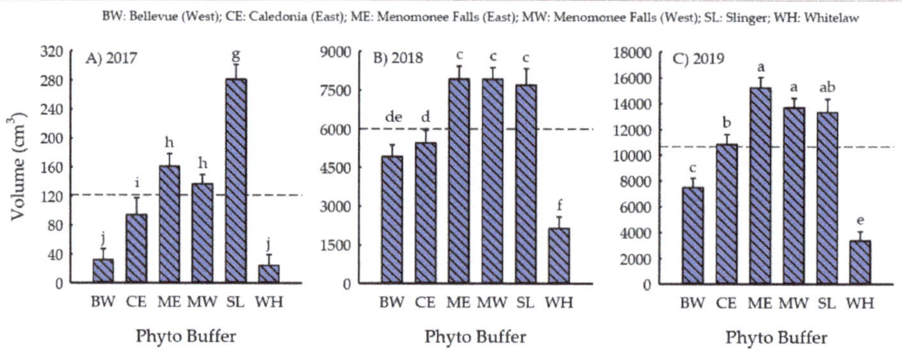


**Figure 5.** First- (**A**), second- (**B**), and third-year (**C**) volume ($\pm$one standard error) of six phytoremediation buffers (i.e., phyto buffers) established in 2017 (i.e., the 2017 Buffer Group) in the Lake Michigan watershed of eastern Wisconsin, USA. The dashed line represents the overall mean, and asterisks indicate means different than the overall mean at $p < 0.05$. Bars with different letters across all buffer $\times$ year combinations are different at $p < 0.05$.

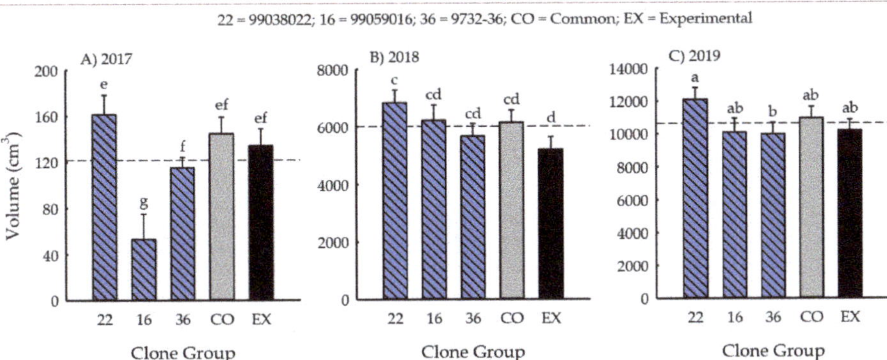



**Figure 6.** First- (**A**), second- (**B**), and third-year (**C**) volume (±one standard error) of three clone groups (i.e., NRRI = 22, 16, 36; Common; Experimental; see Table 3 for definitions) tested in six phytoremediation buffers (i.e., phyto buffers) established in 2017 (i.e., the 2017 Buffer Group) in the Lake Michigan watershed of eastern Wisconsin, USA. The dashed line represents the overall mean, and asterisks indicate means different than the overall mean at $p < 0.05$. Bars with different letters across all clone group $\times$ year combinations are different at $p < 0.05$.

The buffer $\times$ clone group $\times$ year interaction was significant for diameter ($p = 0.0036$) and volume ($p < 0.0001$) of the 2018 buffer group trees (Table S2). VOLUME$_{2018(2018)}$ ranged from 15.9 $\pm$ 28.3 ('9732-11' at Marquette) to 182.1 $\pm$ 22.1 cm$^3$ ('9732-31' at Caledonia (West)), with an overall mean of 76.5 $\pm$ 22.7 cm$^3$, whereas VOLUME$_{2018(2019)}$ ranged from 383.5 $\pm$ 825.9 (Common at Marquette) to 7975.7 $\pm$ 825.9 cm$^3$ (Common at Manitowoc), with an overall mean of 3399.7 $\pm$ 848.2 cm$^3$ (Table 5). VOLUME$_{2018(2020)}$ ranged from 632.2 $\pm$ 2129.9 (Common at Marquette) to 23,912.0 $\pm$ 2129.9 cm$^3$ ('9732-31' at Caledonia (West)), with an overall mean of 8763.9 $\pm$ 2187.2 cm$^3$ (Table 5). Across all buffer $\times$ clone group $\times$ year combinations, VOLUME$_{2018}$ increased 44.4-fold from the first year to the second year after planting, and then 2.6-fold from the second year to the third year. After the first growing season, trees with the greatest volume were grown at Caledonia (West), which had 470.3% greater volume than Marquette, the buffer with the smallest trees. For the second and third growing seasons, the largest trees were grown at Manitowoc, which had 837.1% and 1322.3% greater volume than the buffer with the smallest trees (Marquette), respectively. The range in volume was narrower for clone groups, with '9732-31' exhibiting the greatest volume in all years. For 2018, '9732-31' had 67.6% bigger trees than those of the Experimental group, which had the smallest trees. Similarly, '9732-31' produced 50.4% and 66.9% larger trees than '9732-36' in 2019 and 2020, respectively. Trends in diameter of the 2018 buffer group trees were similar to those of volume (Table S3).

**Table 5.** Volume ($cm^3$) ($\pm$one standard error) of three poplar clone groups tested in five phytoremediation buffer systems (i.e., phyto buffers) established in 2018 (i.e., the 2018 Buffer Group) in the Lake Superior watershed of the Upper Peninsula of Michigan, USA and the Lake Michigan watershed of eastern Wisconsin, USA. Trees were measured following the 2018, 2019, and 2020 growing seasons. Volume values with different letters within a clone column across measurement years are different at $p < 0.05$.

| | Clone Group [a] | | | | | | | | | | |
|---|---|---|---|---|---|---|---|---|---|---|---|
| | NRRI | | | | | | | | | | |
| Buffer [b] | 9732-11 | | 9732-24 | | 9732-31 | | 9732-36 | | Experimental | | Common | |
| | | | | | 2018 Measurement year | | | | | | | |
| BC | 57 ± 15 | f | 40 ± 21 | u | 69 ± 23 | e | 26 ± 26 | u | 36 ± 22 | v | 34 ± 28 | c |
| BE | 70 ± 22 | f | 49 ± 23 | u | 59 ± 28 | e | 43 ± 20 | u | 40 ± 22 | v | 35 ± 20 | c |
| CW | 157 ± 22 | f | 139 ± 22 | u | 182 ± 22 | e | 147 ± 22 | u | 113 ± 22 | v | 138 ± 22 | c |
| MA | 133 ± 24 | f | 95 ± 21 | u | 143 ± 22 | e | 118 ± 25 | u | 86 ± 22 | v | 133 ± 23 | c |
| MQ | 16 ± 28 | f | 31 ± 25 | u | 40 ± 23 | e | 31 ± 25 | u | 18 ± 22 | v | 18 ± 18 | c |
| | | | | | 2019 Measurement year | | | | | | | |
| BC | 2344 ± 826 | de | 2422 ± 826 | we | 2496 ± 826 | cd | 1522 ± 826 | vu | 1968 ± 826 | w | 2207 ± 826 | c |
| BE | 2239 ± 826 | e | 2117 ± 826 | wv | 2068 ± 883 | d | 1966 ± 826 | wv | 1970 ± 826 | xw | 1866 ± 826 | c |
| CW | 5469 ± 826 | cd | 3901 ± 826 | yxw | 7876 ± 826 | b | 3893 ± 826 | xwv | 5632 ± 826 | y | 5978 ± 826 | b |
| MA | 7857 ± 826 | b | 5378 ± 826 | yx | 7162 ± 826 | b | 5251 ± 883 | yx | 6185 ± 826 | y | 7976 ± 826 | b |
| MQ | 474 ± 883 | ef | 718 ± 1045 | vu | 1002 ± 883 | de | 1068 ± 1045 | vu | 602 ± 826 | wv | 383 ± 826 | c |
| | | | | | 2020 Measurement year | | | | | | | |
| BC | 6777 ± 2130 | bc | 7454 ± 2130 | y | 7886 ± 2130 | b | 5194 ± 2130 | yxw | 5356 ± 2130 | yx | 5415 ± 2130 | b |
| BE | 6160 ± 2130 | bcd | 6240 ± 2130 | yx | 6031 ± 2277 | bc | 4870 ± 2130 | yxwv | 4521 ± 2130 | yxw | 3814 ± 2130 | bc |
| CW | 9917 ± 2130 | b | 7452 ± 2130 | y | 23,912 ± 2130 | a | 7554 ± 2130 | y | 16,443 ± 2130 | z | 16,333 ± 2130 | a |
| MA | 20,902 ± 2130 | a | 14,368 ± 2130 | z | 18,160 ± 2130 | a | 15,068 ± 2277 | z | 16,444 ± 2130 | z | 19,315 ± 2130 | a |
| MQ | 767 ± 2277 | ef | 945 ± 2694 | wvu | 1802 ± 2277 | de | 1935 ± 2694 | wvu | 1249 ± 2130 | wv | 632 ± 2130 | c |

[a] 'NRRI' = promising genotypes bred, tested, and selected at the University of Minnesota Duluth, Natural Resources Research Institute (NRRI) for broad-ranging applications [36,38]. 'Experimental' = genotypes with a rich history of testing but that are still at the experimental stage. 'Common' = genotypes commonly used for commercial and/or research purposes in the region. [b] BC: Bellevue (Central); BE: Bellevue (East); CW: Caledonia (West); MA: Manitowoc; MQ: Marquette.

The buffer $\times$ clone group $\times$ year interaction was significant for diameter ($p = 0.0293$) and volume ($p < 0.0001$) of the 2019 buffer group trees (Table S2). $VOLUME_{2019(2019)}$ ranged from 16.4 ± 26.0 ('9732-36' at Ontonagon (North)) to 396.3 ± 25.8 $cm^3$ ('99038022' at Escanaba (West)), with an overall mean of 91.1 ± 25.9 $cm^3$, whereas $VOLUME_{2019(2020)}$ ranged from 189.0 ± 391.4 ('9732-24' at Ontonagon (North)) to 5639.8 ± 391.4 $cm^3$ (Common at Escanaba (West)), with an overall mean of 1294.4 ± 393.02 $cm^3$ (Table 6). $VOLUME_{2019}$ increased 14.2-fold from the first year to the second year after planting. For the first and second growing seasons, the largest trees were grown at Escanaba (West), which had 903.4% and 860.1% greater volume than the buffer with the smallest trees (Ontonagon (North)), respectively. Clone groups exhibited less variation, with '99038022' exhibiting the greatest first-year volume, which was 71.8% more than that of '9732-31', which had the smallest trees. The Common group trees produced 108.8% larger trees than '9732-11' at two years after planting. Trends in diameter of the 2019 buffer group trees were similar to those of volume (Table S4).

**Table 6.** Volume ($cm^3$) ($\pm$one standard error) of three poplar clone groups tested in five phytoremediation buffer systems (i.e., phyto buffers) established in 2019 (i.e., the 2019 Buffer Group) in the Lake Superior watershed of the Upper Peninsula of Michigan, USA. Trees were measured following the 2019 and 2020 growing seasons. Volume values with different letters within a clone column across measurement years are different at $p < 0.05$.

| | Clone Group [a] | | | | | | | | | | | | |
| | NRRI | | | | | | | | | | | | |
| Buffer [b] | 99038022 | | 9732-11 | | 9732-24 | | 9732-31 | | 9732-36 | | Experimental | | Common | |
| | | | | | | 2019 Measurement year | | | | | | | |
| EE | $73 \pm 26$ | y | $49 \pm 26$ | c | $98 \pm 26$ | cd | $81 \pm 28$ | x | $44 \pm 26$ | d | $48 \pm 26$ | c | $47 \pm 26$ | x |
| EW | $396 \pm 26$ | y | $239 \pm 26$ | bc | $198 \pm 26$ | c | $228 \pm 26$ | yx | $312 \pm 26$ | cd | $283 \pm 26$ | c | $295 \pm 26$ | x |
| MU | $89 \pm 26$ | y | $28 \pm 26$ | c | $49 \pm 26$ | d | $20 \pm 26$ | x | $47 \pm 26$ | cd | $33 \pm 26$ | c | $50 \pm 26$ | x |
| ON | $54 \pm 26$ | y | $28 \pm 26$ | c | $23 \pm 26$ | d | $29 \pm 26$ | x | $16 \pm 26$ | d | $20 \pm 26$ | c | $24 \pm 26$ | x |
| OS | $60 \pm 26$ | y | $49 \pm 26$ | c | $56 \pm 26$ | d | $33 \pm 28$ | x | $28 \pm 26$ | d | $30 \pm 26$ | c | $30 \pm 26$ | x |
| | | | | | | 2020 Measurement year | | | | | | | |
| EE | $517 \pm 391$ | y | $470 \pm 391$ | bc | $1553 \pm 391$ | ab | $1040 \pm 418$ | z | $463 \pm 391$ | c | $664 \pm 391$ | c | $607 \pm 391$ | x |
| EW | $2397 \pm 391$ | z | $2206 \pm 391$ | a | $2444 \pm 391$ | a | $2155 \pm 391$ | z | $3189 \pm 391$ | a | $3994 \pm 391$ | a | $5640 \pm 391$ | z |
| MU | $2075 \pm 391$ | z | $878 \pm 391$ | b | $1709 \pm 391$ | ab | $950 \pm 391$ | y | $1761 \pm 391$ | b | $1865 \pm 391$ | b | $2439 \pm 391$ | y |
| ON | $691 \pm 391$ | y | $205 \pm 391$ | bc | $189 \pm 391$ | cd | $295 \pm 391$ | yx | $285 \pm 391$ | cd | $332 \pm 391$ | c | $296 \pm 391$ | x |
| OS | $439 \pm 391$ | y | $785 \pm 391$ | bc | $805 \pm 391$ | bc | $494 \pm 418$ | yx | $441 \pm 391$ | cd | $526 \pm 391$ | c | $505 \pm 391$ | x |

[a] 'NRRI' = promising genotypes bred, tested, and selected at the University of Minnesota Duluth, Natural Resources Research Institute (NRRI) for broad-ranging applications [36,38]. 'Experimental' = genotypes with a rich history of testing but that are still at the experimental stage. 'Common' = genotypes commonly used for commercial and/or research purposes in the region. [b] EE: Escanaba (East); EW: Escanaba (West); MU: Munising; ON: Ontonagon (North); OS: Ontonagon (South).

## 4. Discussion and Conclusions

Selection of *Populus* and other short rotation woody crop (SRWC) species to match specific site and growing conditions is imperative for maximizing productivity [64]. The availability of appropriate genotypes can be necessary for plantation or site managers in the absence of precise site information [65]. Species of *Populus*, a genus utilized ubiquitously for environmental applications, have been bred and tested extensively for biomass production [66], especially beginning in the early 1990s with international germplasm exchanges and other cooperative tree improvement efforts between the United States and Europe [67]. Results of these testing efforts have shown great potential of new genotypes for biomass production. Building on these successful partnerships, the poplar breeding and testing program at the University of Minnesota Duluth, Natural Resources Research Institute (NRRI) has produced thousands of genotypes since the mid-1990s [35,36]. Some of these clones have been defined as geo-robust, meaning they are extreme generalists with the capability for establishment across broader latitudinal and longitudinal ranges [38]. The ecological restoration potential of a subset of these clones was tested in the current study, and clones '99038022' and '9732-31' exhibited exceptional survival and growth across eleven and ten phytoremediation buffer systems (i.e., phyto buffers), respectively, in the Lake Superior watershed of the Upper Peninsula of Michigan, USA and the Lake Michigan watershed of eastern Wisconsin, USA. Other NRRI clones showed exceptional promise at individual phyto buffers, demonstrating the value of matching individual genotypes to specific site conditions. This combination of generalist and specialist genotypes corroborated the importance of such multi-environmental trials (MET) throughout plantation development, making the current data useful for advancing poplar tree improvement efforts throughout the region, continent, and world, informing clonal selection for multiple end-uses, including phytotechnologies.

Across the United States, average annual poplar productivity of approximately 9 Mg ha$^{-1}$ yr$^{-1}$ is common, with advanced genotypes exhibiting nearly 2.5 times as much growth [68]. In the Midwestern United States, the location of the current study, a wide range of poplar biomass productivity potential has been reported. Most common stand densities of 1075 and 1736 trees ha$^{-1}$ (i.e., $3 \times 3$ and $2 \times 2$ m spacing, respectively) have resulted in mean annual increment (MAI) ranges similar to those of our study for the same age. Poplar biomass plantations with 1736 trees ha$^{-1}$ had MAI values ranging from 2.8 to 6.1 Mg ha$^{-1}$ yr$^{-1}$ at four years after planting [29,32] and 6.7 to 9.0 Mg ha$^{-1}$ yr$^{-1}$ for five-year-old trees [30]. Maximum productivity resulting from 3-PG modeling resulted in 13.0 Mg ha$^{-1}$ yr$^{-1}$ at the end of ten-year rotations [69]. Plantations of the same stand density as the current study (i.e., 1075 trees ha$^{-1}$) exhibited productivity ranging from 4.3 to 5.3 Mg ha$^{-1}$ yr$^{-1}$ at age four years [43] and 5.1 to 16.8 Mg ha$^{-1}$ yr$^{-1}$ after six years of growth [28]. Optimizing genotype $\times$ environment interactions for the best performing clones resulted in MAI values of 3.0 to 11.0 Mg ha$^{-1}$ yr$^{-1}$ for four-year-old trees [28]. Such a wide range in productivities can be attributed in part to site conditions and planting stock (i.e., rooted vs. unrooted cuttings). Effective clonal selection is integral to maximizing productivity, regardless of application (e.g., biomass for bioenergy, phytotechnologies, etc.). Productivity values in the lower part of this range have been shown for poplars grown for phytotechnologies. At phytoremediation plantations in the Midwest planted at stand densities from 434 to 4310 trees ha$^{-1}$, MAI values ranged from 4.4 to 15.5 Mg ha$^{-1}$ yr$^{-1}$ for some of the same clones as the current study ('DN5', 'DN34', 'NM2', 'NM6') [17]. However, lower productivity (0.5 to 2.5 Mg ha$^{-1}$ yr$^{-1}$) also has been reported for poplar clones 'DN5', 'NC14106', 'NM2', and 'NM6' irrigated with landfill leachate grown for two years with a stand density of 3472 trees ha$^{-1}$ [52]. These results corroborated the growth productivity of clones in the current study, for which MAI ranged from 1.6 to 3.9 Mg ha$^{-1}$ yr$^{-1}$ across all phyto buffers and clones. Considering that phyto buffers in our study were located adjacent to landfills and similar sites, clone productivity can be considered satisfactory because the presence of potential soil heterogeneity can significantly affect biomass production of poplar clones [14,42,46,47,53,57,70].

Optimal site conditions for poplar growth include deep, fertile sandy-loam to clay-loam soils with pH ranging from 5.0 to 7.5 that are well drained, but not droughty [71]. Thus, annual precipitation is another influential factor and, in the present study, all phyto buffers fit within the regional precipitation gradient range of 76.2 to 88.9 cm [30]. Site conditions at the buffers significantly affected growth and productivity of the tested clones, specifically concerning soil water availability and pH, which serve as limiting factors for poplar growth. By comparison, there was a lack of phyto buffer $\times$ clone group interaction regarding MAI at four years after planting. Such an outcome can be explained by the origin of the hybrids; NRRI clones belong to the 'DN' genomic group, whereas Control and Experimental clone groups contain clones originating from different poplar species and inter- and intra-sectional hybrids [17].

Trends in health were similar across phyto buffer groups; phyto buffer and clone group main effects governed health during the year of establishment, and in the following years, phyto buffer $\times$ clone group interactions were expressed. Such results can be explained by a stronger influence of site conditions and clone group characteristics (i.e., rooting ability) on vitality during the year of establishment, whereas the interaction of the factors evolved in subsequent years. Greenhouse experiments of Rogers et al. [58] showed a similar health response of NRRI clones '99038022' and '9732-36' compared to Experimental ('NC14106') and Common ('DN34', 'NM2', 'NM6') clones grown in soils from six of the phyto buffers of the current study (BW: Bellevue (West); CE: Caledonia (East); ME: Menomonee Falls (East); MW: Menomonee Falls (West); SL: Slinger; WH: Whitelaw). Finally, despite significant effects of phyto buffer, clone group, and their interaction, all health assessment values were within the optimal health category, with values ranging from 1.11 to 1.36 across all phyto buffer $\times$ clone group $\times$ year combinations (Figures 1–3, Figures S1 and S2; Table S1), indicating no substantial influence on clonal vitality across all sites.

As expected, the phyto buffer $\times$ clone group $\times$ year interaction for diameter and volume production of clones was significant, indicating different growth patterns of tested clones and, further, changes in annual growth increment of poplars throughout the production cycle [72]. Such an explanation could also be applied for MAI, which was lower (though not always significantly) for NRRI clones than those of the Common clone group. These results were corroborated considering volume production of the clones in the 2017 Phyto Buffer Group. NRRI clones '99059016' and '9732-36' had significantly lower wood volume than Common clones after the first year, whereas these differences were negligible after two and three years of growth. In the current study, the lack of a significant phyto buffer × clone group × year interaction for height can be explained by the fact that although height and diameter are typically positively correlated for poplars (and trees in general), this correlation is influenced by variation due to the site and G × E interactions, leading to the need for matching clones to specific site conditions [73]. In addition, different biomass allocation growth patterns (e.g., terminal vs. lateral shoot growth) among clones could have impacted the current results [64].

In general, NRRI clones showed potential for use in phytotechnologies, with high productivity exhibited for clones '99038022' and '9732-31'. Previously, NRRI clones '99038022', '99059016', '9732-11', '9732-24', and '9732-31' demonstrated high productivity for mean basal area and volume, often outperforming Common clones [36,37]. Although the productivity of NRRI clones have varied markedly across sites, the identification of geographically robust clones holds promise for efficiently meeting diverse environmental objectives [38]. Breeding and selecting clonal forest reproductive material has many advantages, including utilization of both additive and non-additive variance, resulting in larger genetic gains [35,74,75]. On the other hand, environmental factors can diminish genetic gains. According to Pliura et al. [76], the presence of a significant G × E interaction implies that: (1) a genotype's performance in a specific environment can be less accurately predicted by the overall genotypic mean, and (2) a genotype's overall performance can be less accurately predicted by the genotypic mean in a specific environment. Both of these responses can result in biased estimates and, thus, decreases in genetic gains [76].

The aforementioned results, including those of the present study, indicated that NRRI clones, which originated from a narrow range of latitudes, were well-suited to the latitudinal range of the phyto buffers. For example, 'D125' (selected from Dr. Carl Mohn's long-term *P. deltoides* program at the University of Minnesota) is the female Minnesota *P. deltoides* parent used for all $F_1$ full-sib progeny within family pedigree '9732'. In contrast, some genotypes of the Common and Experimental clone groups originated from other parts of North America and Europe, making them less adapted to certain phyto buffer site conditions. The intra-specific breeding strategy for NRRI clones uses *P. deltoides* parents of a limited geographic range (Minnesota) combined with other *Aigeiros* species (e.g., *P. nigra*) to produce progeny of increased performance [35]. The *P. nigra* component of 'DN' hybrids has produced a strong heterotic effect not exhibited in *P. trichocarpa* Torr. et Gray × *P. deltoides* 'TD' hybrids due to greater genomic relatedness between *P. deltoides* and *P. nigra* relative to poplars from the *Tacamahaca* section (e.g., *P. trichocarpa*, *P. maximowiczii* A. Henry) [77]. This genetic closeness was corroborated by mitochondrial DNA variation [78] and simple sequence repeat (SSR) markers [79]. In addition, species biology likely contributed substantially to the performance of NRRI hybrids. According to Sixto et al. [65], the plasticity of certain *Aigeiros* species enabled them to grow on a vast range of habitats (e.g., from poor, dry and stony to optimal silty or sandy loamy soils) versus *Tacamahaca* balsam poplars that preferred alluvial, fertile soils in wetter climates and higher elevations. Their results were verified by findings of positive sensitivity to increases in median temperature and negative sensitivity to increased sand content by *P. nigra* clones, with the opposite occurring for *P. trichocarpa* × *P. deltoides* hybrids [65]. Nelson et al. [37] hypothesized that the *P. nigra* male component of *P. deltoides* × *P. nigra* hybrids imparts broad adaptability to these genotypes.

Overall, in the current study, NRRI clones exhibited positive growth performance at all sixteen phyto buffers during the first four years of establishment. Their height, diameter, and volume, like those of the Common and Experimental clone groups, were influenced by site conditions, which was expected considering soil heterogeneity at the phyto buffers.

NRRI clones, the progeny of Minnesota-selected *P. deltoides* and *P. nigra*, were robust and well-adapted to the varying climate and soils at the phyto buffers. Our results corroborated previous testing of NRRI clones in more traditional SRWC production plantations [35–38], indicating their potential for use in phytotechnologies.

**Supplementary Materials:** The following are available online at https://www.mdpi.com/article/10.3390/f12040474/s1; Table S1: Probability values from analyses of variance for health and mean annual increment (MAI); Table S2: Probability values from analyses of variance for height, diameter, and volume; Table S3: Diameter for the buffer × clone group × year interaction (2018 Buffer Group); Table S4: Diameter for the buffer × clone group × year interaction (2019 Buffer Group); Figure S1: Health for the buffer × clone group interaction measured in 2019 (2017 Buffer Group); Figure S2: Health for the buffer × clone group interaction measured in 2019 (2018 Buffer Group); Figure S3: Height for the buffer × year interaction (2017 Buffer Group); Figure S4: Diameter for the buffer × year interaction (2017 Buffer Group); Figure S5: Height for the buffer × year interaction (2018 Buffer Group); Figure S6: Height for the buffer × year interaction (2019 Buffer Group); Figure S7: Height for the clone group × year interaction (2017 Buffer Group); Figure S8: Diameter for the clone group × year interaction (2017 Buffer Group).

**Author Contributions:** Conceptualization, J.G.B., R.A.H., C.-H.L. and R.S.Z.J.; methodology, J.G.B., R.A.H., C.-H.L., B.G.M., A.P., E.R.R. and R.S.Z.J.; validation, A.P., E.R.R. and R.S.Z.J.; formal analysis, R.S.Z.J.; investigation, J.G.B., R.A.H., C.-H.L., B.G.M., N.D.N., A.P., E.R.R. and R.S.Z.J.; resources, J.G.B., R.A.H., C.-H.L., B.G.M., N.D.N. and R.S.Z.J.; data curation, E.R.R. and R.S.Z.J.; writing: original draft preparation, A.P., E.R.R., and R.S.Z.J.; writing: review and editing, J.G.B., R.A.H., C.-H.L., B.G.M., N.D.N., A.P., E.R.R. and R.S.Z.J.; visualization, E.R.R. and R.S.Z.J.; supervision, J.G.B., C.-H.L., N.D.N., E.R.R. and R.S.Z.J.; project administration, J.G.B., C.-H.L. and R.S.Z.J.; funding acquisition, R.S.Z.J. All authors have read and agreed to the published version of the manuscript.

**Funding:** This work was funded by the Great Lakes Restoration Initiative (GLRI; Template #738 Landfill Runoff Reduction).

**Acknowledgments:** The findings and conclusions in this publication are those of the authors and should not be construed to represent any official USDA or United States Government determination or policy. This work was funded by the Great Lakes Restoration Initiative (GLRI; Template #738 Landfill Runoff Reduction). We are grateful to S. Johnson, C.H. Perry, and N. Vrevich of the USDA Forest Service for GLRI support. We thank B. Dean (formerly), K. Corcoran, and M. Mujaddedi of the International Programs Office of the USDA Forest Service for administrative support of scientific exchanges between A. Pilipović and R. Zalesny, as well as their administration and support of International Forestry Fellows: D. Karlsson (Sweden), P. Muñoz Gomez de la Serna (Spain), and A. Peqini (Albania). We thank the following site managers for access to and information about their field sites: L. Buechel, J. Forney, M. Peterson, B. Pliska, and R. Seegers (Waste Management, Inc.); K. Dorow, D. Koski, and K. McDaniel (City of Manitowoc, Wisconsin); D. Henderson (AECOM Technical Services, Inc.); B. Austin and J. Wales (Marquette County Solid Waste Management Authority); D. Pyle (Delta County Solid Waste Management Authority); and S. Coron (Great American Disposal), as well as T. Beggs (Wisconsin Department of Natural Resources) for regulatory assistance and B. Sexton (Sand County Environmental) for insight, knowledge, and guidance on sustainable phytoremediation systems. We appreciate laboratory, greenhouse, and field technical assistance from: E. Bauer, B. Birr, T. Cook, B. DeBauche, S. Eddy, F. Erdmann, C. Espinosa, D. Karlsson, R. Lange, P. Manley, M. Mueller, C. Munch, P. Muñoz Gomez de la Serna, D. Nguyen, A. Peqini, M. Ramsay, E. Schmidt, J. Schutts, C.P. Vinhal, M. Wagler, and A. Wiese. We are grateful to: J.M. Suvada for developing Figure 1; R. Klevickas and P. Bloese for providing cutting material; and E. Bauer, B. DeBauche, C.P. Vinhal, and A. Wiese for project leadership. We thank W. Headlee for statistical mentoring and reviewing earlier versions of this manuscript.

**Conflicts of Interest:** The authors declare no conflict of interest.

# References


1. Zalesny, R.S., Jr.; Pilipović, A.; Rogers, E.R.; Burken, J.G.; Hallett, R.A.; Lin, C.-H.; McMahon, B.G.; Nelson, N.D.; Wiese, A.H.; Bauer, E.O.; et al. Establishment of regional phytoremediation buffer systems for ecological restoration in the Great Lakes Basin, USA. I. Genotype × environment interactions. *Forests* **2021**, *12*, 430. [CrossRef]
2. Fuller, K.; Shear, H.; Wittig, J. *The Great Lakes: An Environmental Atlas and Resource Book*, 3rd ed.; U.S. Environmental Protection Agency: Chicago, IL, USA; Government of Canada: Toronto, ON, Canada, 1995; Volume 95, p. 46, Issue 1 of EPA 905-B.
3. GLRI (Great Lakes Restoration Initiative). *Action Plan I (2010–2014)*; GLRI (Great Lakes Restoration Initiative): Washington, DC, USA, 2010; p. 41.
4. United States Environmental Protection Agency (USEPA). *National Water Quality Inventory Report to Congress*; United States Environmental Protection Agency (USEPA): Washington, DC, USA, 2017; p. 21, EPA/841/R-16/011.
5. United States Environmental Protection Agency (USEPA). *National Overview: Facts and Figures on Materials, Wastes and Recycling, 2020*; United States Environmental Protection Agency (USEPA): Washington, DC, USA, 2020. Available online: https://www.epa.gov/facts-and-figures-about-materials-waste-and-recycling/national-overview-facts-and-figures-materials#Trends1960-Today (accessed on 14 January 2021).
6. Justin, M.Z.; Pajk, N.; Zupanc, V.; Zupančič, M. Phytoremediation of landfill leachate and compost wastewater by irrigation of Populus and Salix: Biomass and growth response. *Waste Manag.* **2010**, *30*, 1032–1042. [CrossRef] [PubMed]
7. Salt, D.E.; Blaylock, M.; Kumar, N.P.B.A.; Dushenkov, V.; Ensley, B.D.; Chet, I.; Raskin, I. Phytoremediation: A Novel Strategy for the Removal of Toxic Metals from the Environment Using Plants. *Nat. Biotechnol.* **1995**, *13*, 468–474. [CrossRef] [PubMed]
8. Rock, S.; Pivetz, B.; Madalinski, K.; Adam, N.; Wilson, T. *Introduction to Phytoremediation*; U.S. Environmental Protection Agency: Washington, DC, USA, 2000; p. 72, EPA/600/R-99/107 (NTIS PB2000-106690).
9. Burken, J.G.; Schnoor, J.L. Predictive Relationships for Uptake of Organic Contaminants by Hybrid Poplar Trees. *Environ. Sci. Technol.* **1998**, *32*, 3379–3385. [CrossRef]
10. Arthur, E.L.; Rice, P.J.; Rice, P.J.; Anderson, T.A.; Baladi, S.M.; Henderson, K.L.D.; Coats, J.R. Phytoremediation—An Overview. *Crit. Rev. Plant Sci.* **2005**, *24*, 109–122. [CrossRef]
11. Chaney, R.L.; Baklanov, I.A. Phytoremediation and Phytomining. *Adv. Bot. Res.* **2017**, *83*, 189–221. [CrossRef]
12. Robinson, B.H.; Bañuelos, G.; Conesa, H.M.; Evangelou, M.W.H.; Schulin, R. The Phytomanagement of Trace Elements in Soil. *Crit. Rev. Plant Sci.* **2009**, *28*, 240–266. [CrossRef]
13. Isebrands, J.G.; Aronsson, P.; Carlson, M.; Ceulemans, R.; Coleman, M.D.; Dickinson, N.M.; Dimitriou, J.; Doty, S.L.; Gardiner, E.S.; Heinsoo, K.; et al. Environmental applications of poplars and willows. *Poplars Willows Trees Soc. Environ.* **2014**, *6*, 258–336. [CrossRef]
14. Zalesny, R.S., Jr.; Bauer, E.O. Selecting and utilizing *Populus* and *Salix* for landfill covers: Implications for leachate irrigation. *Intl. J. Phytoremed.* **2007**, *9*, 497–511. [CrossRef]
15. Licht, L.A.; Isebrands, J. Linking phytoremediated pollutant removal to biomass economic opportunities. *Biomass Bioenergy* **2005**, *28*, 203–218. [CrossRef]
16. Zalesny, R.S., Jr.; Stanturf, J.A.; Gardiner, E.S.; Perdue, J.H.; Young, T.M.; Coyle, D.R.; Headlee, W.L.; Bañuelos, G.S.; Hass, A. Ecosystem services of woody crop production systems. *BioEnergy Res.* **2016**, *9*, 465–491. [CrossRef]
17. Zalesny, R.S., Jr.; Headlee, W.L.; Gopalakrishnan, G.; Bauer, E.O.; Hall, R.B.; Hazel, D.W.; Isebrands, J.G.; Licht, L.A.; Negri, M.C.; Guthrie-Nichols, E.; et al. Ecosystem services of poplar at long-term phytoremediation sites in the Midwest and Southeast, United States. *Wiley Interdiscip. Rev. Energy Environ.* **2019**, *8*, e349. [CrossRef]
18. Burdon, R.D. Forest genetics and tree breeding: Current and future signposts. In *Encyclopedia of Forest Sciences*; Burley, J., Eveno, E., Youngquist, J.A., Eds.; Elsevier Academic: Amsterdam, the Netherlands, 2004; pp. 1538–1545; ISBN 978-0-12-145160-8.
19. Stout, A.B.; Schreiner, E.J. Results of a project in hybridizing poplars. *J. Hered.* **1933**, *24*, 217–229. [CrossRef]
20. Isebrands, J.; Zalesny, J.R. Reflections on the contributions of *Populus* research at Rhinelander, Wisconsin, USA. *Can. J. For. Res.* **2021**, *51*, 139–153. [CrossRef]
21. Eckenwalder, J.E. Natural intersectional hybridization between North American species of *Populus* (Salicaceae) in sections *Aigeiros* and *Tacamahaca*. II. Taxonomy. *Can. J. Bot.* **1984**, *62*, 325–335. [CrossRef]
22. Stanton, B.J.; Neale, D.B.; Li, S. Populus Breeding: From the Classical to the Genomic Approach. In *Genetics and Genomics of Populus*; Jansson, S., Bhalerao, R., Groover, A., Eds.; Springer: New York, NY, USA, 2010; Volume 8, pp. 309–348.
23. Mahama, A.A.; Hall, R.B.; Zalesny, R.S., Jr. Differential interspecific incompatibility among *Populus* hybrids in sections *Aigeiros* Duby and *Tacamahaca* Spach. *For. Chron.* **2011**, *87*, 790–796. [CrossRef]
24. De Leon, N.; Jannink, J.L.; Edwards, J.W.; Kaeppler, S.M. Introduction to a special issue on genotype by environment interaction. *Crop Sci.* **2016**, *56*, 2081–2089. [CrossRef]
25. Calleja-Rodriguez, A.; Gull, B.A.; Wu, H.X.; Mullin, T.J.; Persson, T. Genotype-by-environment interactions and the dynamic relationship between tree vitality and height in northern Pinus sylvestris. *Tree Genet. Genomes* **2019**, *15*, 36. [CrossRef]
26. Orlović, S.; Guzina, V.; Merkulov, L. Genetic variability in anatomical, physiological and growth characteristics of hybrid poplar (*Populus* × *euramericana* Dode (Guinier)) and eastern cottonwood (*Populus deltoides* Bartr.) clones. *Silvae Genet.* **1998**, *47*, 183–190.
27. Kassen, R. The experimental evolution of specialists, generalists, and the maintenance of diversity. *J. Evol. Biol.* **2002**, *15*, 173–190. [CrossRef]

28. Riemenschneider, D.E.; Isebrands, J.G.; Berguson, W.E.; Dickmann, D.I.; Hall, R.B.; Mohn, C.A.; Stanosz, G.R.; Tuskan, G.A. Poplar breeding and testing strategies in the north-central U.S.: Demonstration of potential yield and consideration of future research needs. *For. Chron.* **2001**, *77*, 245–253. [CrossRef]

29. Hansen, E.A. *Mid-Rotation Yields of Biomass Plantations in the North Central United States. Research Paper NC-309*; U.S. Department of Agriculture, Forest Service, North Central Forest Experiment Station: St. Paul, MN, USA, 1992; p. 8. [CrossRef]

30. Hansen, E.A.; Ostry, M.E.; Johnson, W.D.; Tolsted, D.N.; Netzer, D.A.; Berguson, W.E.; Hall, R.B. *Field Performance of Populus in Short-Rotation Intensive Culture Plantations in the North-Central Research Paper NC-320*; U.S. Department of Agriculture, Forest Service, North Central Forest Experiment Station: St. Paul, MN, USA, 1994; p. 13. [CrossRef]

31. Volk, T.A.; Berguson, B.; Daly, C.; Halbleib, M.D.; Miller, R.; Rials, T.G.; Abrahamson, L.P.; Buchman, D.; Buford, M.; Cunningham, M.W.; et al. Poplar and shrub willow energy crops in the United States: Field trial results from the multiyear regional feedstock partnership and yield potential maps based on the PRISM-ELM model. *GCB Bioenergy* **2017**, *10*, 735–751. [CrossRef]

32. Netzer, D.A.; Tolsted, D.; Ostry, M.E.; Isebrands, J.G.; Riemenschneider, D.E.; Ward, K.T. *Growth, Yield, and Disease Resistance of 7- to 12-Year-Old Poplar Clones in the North Central United States. General Technical Report NC-229*; U.S. Department of Agriculture, Forest Service, North Central Research Station: St. Paul, MN, USA, 2002; p. 31. [CrossRef]

33. Zalesny, R.S., Jr.; Hall, R.B.; Zalesny, J.A.; Berguson, W.E.; McMahon, B.G.; Stanosz, G.R. Biomass and genotype × environment interactions of *Populus* energy crops in the Midwestern United States. *BioEnergy Res.* **2009**, *2*, 106–122. [CrossRef]

34. Zalesny, R.S., Jr.; Stanturf, J.A.; Gardiner, E.S.; Bañuelos, G.S.; Hallett, R.A.; Hass, A.; Stange, C.M.; Perdue, J.H.; Young, T.M.; Coyle, D.R.; et al. Environmental technologies of woody crop production systems. *BioEnergy Res.* **2016**, *9*, 492–506. [CrossRef]

35. Berguson, W.; McMahon, B.; Riemenschneider, D. Additive and Non-Additive Genetic Variances for Tree Growth in Several Hybrid Poplar Populations and Implications Regarding Breeding Strategy. *Silvae Genet.* **2017**, *66*, 33–39. [CrossRef]

36. Nelson, N.D.; Berguson, W.E.; McMahon, B.G.; Cai, M.; Buchman, D.J. Growth performance and stability of hybrid poplar clones in simultaneous tests on six sites. *Biomass Bioenergy* **2018**, *118*, 115–125. [CrossRef]

37. Nelson, N.D.; Meilan, R.; Berguson, W.E.; McMahon, B.G.; Cai, M.; Buchman, D. Growth performance of hybrid poplar clones on two agricultural sites with and without early irrigation and fertilization. *Silvae Genet.* **2019**, *68*, 58–66. [CrossRef]

38. Nelson, N.D.; Berguson, W.E.; McMahon, B.G.; Meilan, R.; Smart, L.B.; Gouker, F.E.; Bloese, P.; Miller, R.; Volk, T.A.; Cai, M.; et al. Discovery of Geographically Robust Hybrid Poplar Clones. *Silvae Genet.* **2019**, *68*, 101–110. [CrossRef]

39. Zalesny, R.S., Jr.; Berndes, G.; Dimitriou, I.; Fritsche, U.; Miller, C.; Eisenbies, M.; Ghezehei, S.; Hazel, D.; Headlee, W.L.; Mola-Yudego, B.; et al. Positive water linkages of producing short rotation poplars and willows for bioenergy and phytotechnologies. *Wiley Interdiscip. Rev. Energy Environ.* **2019**, *8*, e345. [CrossRef]

40. Lin, C.-H.; Goyne, K.W.; Kremer, R.J.; Lerch, R.N.; Garrett, H.E. Dissipation of Sulfamethazine and Tetracycline in the Root Zone of Grass and Tree Species. *J. Environ. Qual.* **2010**, *39*, 1269–1278. [CrossRef]

41. Doty, S.L.; Freeman, J.L.; Cohu, C.M.; Burken, J.G.; Firrincieli, A.; Simon, A.; Khan, Z.; Isebrands, J.G.; Lukas, J.; Blaylock, M.J. Enhanced Degradation of TCE on a Superfund Site Using Endophyte-Assisted Poplar Tree Phytoremedia. *Environ. Sci. Technol.* **2017**, *51*, 10050–10058. [CrossRef] [PubMed]

42. Pilipović, A.; Orlović, S.; Nikolić, N.; Borišev, M.; Krstić, B.; Rončević, S. Growth and plant physiological parameters as markers for selection of poplar clones for crude oil phytoremediation. *Šumarski List* **2012**, *136*, 273–281. Available online: https://www.sumari.hr/sumlist/pdf/201202730.pdf (accessed on 10 January 2021).

43. Pilipović, A.; Zalesny, R.S., Jr.; Orlović, S.; Drekić, M.; Pekeč, S.; Katanić, M.; Poljaković-Pajnik, L. Growth and physiological responses of three poplar clones grown on soils artificially contaminated with heavy metals, diesel fuel, and herbicides. *Int. J. Phytoremedia.* **2019**, *22*, 436–450. [CrossRef]

44. Gardea-Torresdey, J.L.; Peralta-Videa, J.R.; De La Rosa, G.; Parsons, J. Phytoremediation of heavy metals and study of the metal coordination by X-ray absorption spectroscopy. *Coord. Chem. Rev.* **2005**, *249*, 1797–1810. [CrossRef]

45. Nikolić, N.; Zorić, L.; Cvetković, I.; Pajević, S.; Borišev, M.; Orlović, S.; Pilipović, A. Assessment of cadmium tolerance and phytoextraction ability in young *Populus deltoides* L. and *Populus × euramericana* plants through morpho-anatomical and physiological responses to growth in cadmium enriched soil. *iFor. Biogeosci. For.* **2017**, *10*, 635–644. [CrossRef]

46. Pilipović, A.; Nikolić, N.; Orlović, S.; Petrović, N.; Krstić, B. Cadmium phytoextraction potential of poplar clones (*Populus* spp.). *Z. Naturforsch.* **2005**, *60*, 247–251.

47. Pilipović, A.; Zalesny, R.S., Jr.; Rončević, S.; Nikolić, N.; Orlović, S.; Beljin, J.; Katanić, M. Growth, physiology, and phytoextraction potential of poplar and willow established in soils amended with heavy-metal contaminated, dredged river sediments. *J. Environ. Manag.* **2019**, *239*, 352–365. [CrossRef]

48. Pierattini, E.C.; Francini, A.; Huber, C.; Sebastiani, L.; Schröder, P. Poplar and diclofenac pollution: A focus on physiology, oxidative stress and uptake in plant organs. *Sci. Total Environ.* **2018**, *636*, 944–952. [CrossRef]

49. Raffaelli, A.; Pierattini, E.C.; Francini, A.; Sebastiani, L. ESI and APCI LC-MS/MS in model investigations on the absorption and transformation of organic xenobiotics by poplar plants (*Populus alba* L.). In *Comprehensive Analytical Chemistry*; Cappiello, A., Palma, P., Eds.; Elsevier Academic: Amsterdam, The Netherlands, 2018; Volume 79, pp. 241–266; ISBN 978-0-44-463914-1.

50. Vannucchi, F.; Traversari, S.; Raffaelli, A.; Francini, A.; Sebastiani, L. Populus alba tolerates and efficiently removes caffeine and zinc excesses using an organ allocation strategy. *Plant Growth Regul.* **2020**, *92*, 597–606. [CrossRef]

51. Zalesny, R.S., Jr.; Bauer, E.O.; Hall, R.B.; Zalesny, J.A.; Kunzman, J.; Rog, C.J.; Riemenschneider, D.E. Clonal variation in survival and growth of hybrid poplar and willow in an in situ trial on soils heavily contaminated with petroleum hydrocarbons. *Intl. J. Phytoremed.* **2005**, *7*, 177–197. [CrossRef]

52. Zalesny, J.A.; Zalesny, R.S., Jr.; Coyle, D.R.; Hall, R.B. Growth and biomass of *Populus* irrigated with landfill leachate. *For. Ecol. Manage.* **2007**, *248*, 143–152. [CrossRef]

53. Zalesny, R.S., Jr.; Bauer, E.O. Evaluation of *Populus* and *Salix* continuously irrigated with landfill leachate I. Genotype-specific elemental phytoremediation. *Intl. J. Phytoremed.* **2007**, *9*, 281–306. [CrossRef] [PubMed]

54. Kebert, M.; Rapparini, F.; Neri, L.; Bertazza, G.; Orlović, S.; Biondi, S. Copper-Induced Responses in Poplar Clones are Associated with Genotype- and Organ-Specific Changes in Peroxidase Activity and Proline, Polyamine, ABA, and IAA Levels. *J. Plant Growth Regul.* **2017**, *36*, 131–147. [CrossRef]

55. Bañuelos, G.S.; Shannon, M.C.; Ajwa, H.; Draper, J.H.; Jordahl, J.; Licht, J. Phytoextraction and Accumulation of Boron and Selenium by Poplar (*Populus*) Hybrid Clones. *Int. J. Phytoremediation* **1999**, *1*, 81–96. [CrossRef]

56. Baldantoni, D.; Cicatelli, A.; Bellino, A.; Castiglione, S. Different behaviours in phytoremediation capacity of two heavy metal tolerant poplar clones in relation to iron and other trace elements. *J. Environ. Manag.* **2014**, *146*, 94–99. [CrossRef]

57. Zalesny, J.A.; Zalesny, R.S., Jr.; Wiese, A.H.; Hall, R.B. Choosing tree genotypes for phytoremediation of landfill leachate using phyto-recurrent selection. *Intl. J. Phytoremed.* **2007**, *9*, 513–530. [CrossRef]

58. Rogers, E.R.; Zalesny, J.R.S.; Hallett, R.A.; Headlee, W.L.; Wiese, A.H. Relationships among Root–Shoot Ratio, Early Growth, and Health of Hybrid Poplar and Willow Clones Grown in Different Landfill Soils. *Forest* **2019**, *10*, 49. [CrossRef]

59. Zalesny, R.S., Jr.; Bauer, E.O. Genotypic variability and stability of poplars and willows grown on nitrate-contaminated soils. *Intl. J. Phytoremed.* **2019**, *21*, 969–979. [CrossRef]

60. Hansen, E.A. Root length in young hybrid *Populus* plantations: Its implications for border width of research plots. *For. Sci.* **1981**, *27*, 808–814. [CrossRef]

61. Zavitkovski, J. Small plots with unplanted plot border can distort data in biomass production studies. *Can. J. For. Res.* **1981**, *11*, 9–12. [CrossRef]

62. Kershaw, J.A.; Ducey, M.J.; Beers, T.W.; Husch, B. *Forest Mensuration*, 5th ed.; Wiley-Blackwell: Hoboken, NJ, USA, 2016; p. 630. ISBN 978-1-118-90203-5.

63. Headlee, W.L.; Zalesny, R.S., Jr. Allometric relationships for aboveground woody biomass differ among hybrid poplar genomic groups and clones in the north-central USA. *BioEnergy Res.* **2019**, *12*, 966–976. [CrossRef]

64. Ghezehei, S.B.; Nichols, E.G.; Maier, C.A.; Hazel, D.W. Adaptability of *Populus* to physiography and growing conditions in the Southeastern USA. *Forests* **2019**, *10*, 118. [CrossRef]

65. Sixto, H.; Gil, P.M.; Ciria, P.; Camps, F.; Cañellas, I.; Voltas, J. Interpreting genotype-by-environment interaction for biomass production in hybrid poplars under short-rotation coppice in Mediterranean environments. *GCB Bioenergy* **2016**, *8*, 1124–1135. [CrossRef]

66. Clifton-Brown, J.; Harfouche, A.; Casler, M.D.; Jones, H.D.; MacAlpine, W.J.; Murphy-Bokern, D.; Smart, L.B.; Adler, A.; Ashman, C.; Awty-Carroll, D.; et al. Breeding progress and preparedness for mass-scale deployment of perennial lignocellulosic biomass crops switchgrass, miscanthus, willow and poplar. *GCB Bioenergy* **2019**, *11*, 118–151. [CrossRef]

67. Hall, R.B.; Hanna, R.D. Exchange, evaluation, and joint testing of genetic stock. *Biomass Bioenergy* **1995**, *9*, 81–87. [CrossRef]

68. Vance, E.D.; Maguire, D.A.; Zalesny, R.S., Jr. Research strategies for increasing productivity of intensively managed forest plantations. *J. For.* **2010**, *108*, 183–192. [CrossRef]

69. Headlee, W.L.; Zalesny, R.S., Jr.; Donner, D.M.; Hall, R.B. Using a process-based model (3-PG) to predict and map hybrid poplar biomass productivity in Minnesota and Wisconsin, USA. *BioEnergy Res.* **2013**, *6*, 196–210. [CrossRef]

70. Zalesny, R.S., Jr.; Bauer, E.O. Evaluation of *Populus* and *Salix* continuously irrigated with landfill leachate II. Soils and early tree development. *Intl. J. Phytoremed.* **2007**, *9*, 307–323. [CrossRef]

71. Hansen, E.A.; Netzer, D.A.; Tolsted, D.N. *Guidelines for Establishing Poplar Plantations in the North-Central U.S. Research Paper NC-363*; U.S. Department of Agriculture, Forest Service, North Central Forest Experiment Station: St. Paul, MN, USA, 1993; p. 6. [CrossRef]

72. Andrašev, S.; Rončević, S. Developmental characteristics of selected black poplar clones (section *Aigeiros* Duby). *For. J.* **2008**, *54*, 3–10.

73. Ghezehei, S.B.; Nichols, E.G.; Hazel, D.W. Early Clonal Survival and Growth of Poplars Grown on North Carolina Piedmont and Mountain Marginal Lands. *BioEnergy Res.* **2016**, *9*, 548–558. [CrossRef]

74. Mullin, T.; Park, Y. Estimating genetic gains from alternative breeding strategies for clonal forestry. *Can. J. For. Res.* **1992**, *22*, 14–23. [CrossRef]

75. Eriksson, G.; Ekberg, I.; Clapham, D. *An Introduction to Forest Genetics*; Swedish University of Agricultural Sciences (SLU): Uppsala, Sweden, 2006; p. 185. ISBN 91-576-7190-7.

76. Pliura, A.; Zhang, S.; MacKay, J.; Bousquet, J. Genotypic variation in wood density and growth traits of poplar hybrids at four clonal trials. *For. Ecol. Manag.* **2007**, *238*, 92–106. [CrossRef]

77. Cervera, M.T.; Storme, V.; Soto, A.; Ivens, B.; Van Montagu, M.; Rajora, O.P.; Boerjan, W. Intraspecific and interspecific genetic and phylogenetic relationships in the genus *Populus* based on AFLP markers. *Theor. Appl. Genet.* **2005**, *111*, 1440–1456. [CrossRef]

78. Barrett, J.W.; Rajora, O.P.; Yeh, F.C.H.; Dancik, B.P.; Strobeck, C. Mitochondrial DNA variation and genetic relationships of Populus species. *Genome* **1993**, *36*, 87–93. [CrossRef]

79. Orlović, S.; Galović, G.; Zorić, M.; Kovačević, B.; Pilipović, A.; Zoran, G. Evaluation of interspecific DNR variability in poplars using AFLP and SSR markers. *Afr. J. Biotechnol.* **2009**, *8*, 5241–5247. [CrossRef]

*forests*

Article

# Field Testing of Selected Salt-Tolerant Screened Balsam Poplar (*Populus balsamifera* L.) Clones for Use in Reclamation around End-Pit Lakes Associated with Bitumen Extraction in Northern Alberta

Yue Hu [1,*], David Kamelchuk [2], Richard Krygier [3] and Barb R. Thomas [1]

1   Department of Renewable Resources, University of Alberta, Edmonton, AB T6G 2E3, Canada; bthomas@ualberta.ca
2   Alberta-Pacific Forest Industries Inc., Boyle, AB T0A 0M0, Canada; david.kamelchuk@alpac.ca
3   Natural Resources Canada, Canadian Forest Service, Canadian Wood Fibre Centre, Edmonton, AB T6H 3S5, Canada; richard.krygier@canada.ca
*   Correspondence: yhu6@ualberta.ca

**Abstract:** For the oil sands mine sites in northern Alberta, the presence of salty process affected water, a byproduct of the hot-water bitumen extraction process, is anticipated to pose a challenge on some reconstructed landforms. The fundamental challenge when re-vegetating these sites is to ensure not only survival, but vigorous growth where plants are subjected to conditions of high electrical conductivity owing to salts in process affected water that may be contained in the substrate. Finding plants suitable for high salt conditions has offered the opportunity for Alberta-Pacific Forest Industries Inc. (Al-Pac) to investigate the potential role of using native balsam poplar (*Populus balsamifera* L.) as a key reclamation species for the oil sands region. Two years of greenhouse screening (2012 and 2013) of 222 balsam poplar clones from Al-Pac's balsam poplar tree improvement program, using process affected discharge water from an oil sands processing facility in Ft. McMurray, has suggested an opportunity to select genetically suitable native clones of balsam poplar for use in reclamation of challenging sites affected by process water. In consideration of the results from both greenhouse and field testing, there is an opportunity to select genetically suitable native clones of balsam poplar that are tolerant to challenging growing conditions, making them more suitable for planting on saline sites.

**Keywords:** oil sands reclamation; end-pit lake; balsam poplar; salt tolerance

**Citation:** Hu, Y.; Kamelchuk, D.; Krygier, R.; Thomas, B.R. Field Testing of Selected Salt-Tolerant Screened Balsam Poplar (*Populus balsamifera* L.) Clones for Use in Reclamation around End-Pit Lakes Associated with Bitumen Extraction in Northern Alberta. *Forests* **2021**, *12*, 572. https://doi.org/10.3390/f12050572

Academic Editors: Ronald S. Zalesny Jr. and Andrej Pilipović

Received: 26 March 2021
Accepted: 29 April 2021
Published: 2 May 2021

**Publisher's Note:** MDPI stays neutral with regard to jurisdictional claims in published maps and institutional affiliations.

## 1. Introduction

In Canada, the Alberta oil sands region is located in the Cold Lake, Peace River, and Athabasca regions in the North America Boreal Plain and covers approximately 142,200 km$^2$ [1]. Currently, approximately 856,000 barrels of bitumen per day (bbl day$^{-1}$) are produced in the mineable portion of the Athabasca region [1]. Surface mining for oil sands production in the Athabasca Oil Sands Region (AOSR) of Alberta has resulted in a cumulative disturbance footprint of 895 km$^2$ (until 2013), with only 0.2% of the total land base disturbed by mining being certified as reclaimed by the Government of Alberta [2]. In addition, open-pit mining leaves a reconstructed landscape of overburden dumps and tailings deposits that require reclamation, targeting self-sustaining and locally common ecosystems [3]. The process of bitumen extraction requires vast amounts of water [4] and the resultant oil sands process water (OSPW) must be contained and not returned to the region's river system owing to a current zero-discharge policy by the Alberta Environmental Protection and Enhancement Act [5]. This requirement has resulted in over a billion cubic meters of tailings water being held through various types of containment systems [6], one of which is often referred to as an 'end-pit' lake. As development expands, large areas of disturbed land in the oil sands region will require reclamation with suitable, well-chosen, native plant material. Reclamation, in the AOSR, is defined as the process to return the disturbed ecosystems to an "equivalent land capacity" as the pre-disturbance ecosystem [7]. This may involve practices such as recontouring the ground, replacing the subsoil and topsoil, revegetation, and monitoring the environmental conditions [7].

Soil salinity and sodicity in oil sands reclamation areas have been listed among the most challenging revegetation concerns [8–10]. Salt stress leads to reductions in growth, productivity, and survival in numerous plant species [11]. The stress induced by salinity on plants is the result of three mechanisms: osmotic stress due to a more negative soil water potential, accumulation of toxic ions, and disturbances in nutrient balance [11]. These effects, in turn, lead to reductions in growth, productivity, and survival in numerous plant species [11]. In the AOSR in Alberta, salinity problems associated with OSPW and exposed marine shale overburden are two major potential challenges when reclaiming upland landscapes [9]. Salt-stress is particularly detrimental for boreal woody species as most exhibit relatively low tolerance to salinity [9]. The fundamental challenge when revegetating these sites is not only to ensure survival, but to achieve growth rates appropriate to the ecosystem class even where plants are subjected to conditions of high electrical conductivity (EC) owing to salts in process affected water that is contained in the substrate.

The challenge of finding plants suitable for high salt conditions has offered the opportunity for Alberta-Pacific Forest Industries Inc. (Al-Pac) to investigate the potential role of using native balsam poplar (*Populus balsamifera* L.) as a key reclamation species for the oil sands region. Poplars are used throughout North America to reclaim sites containing heavy metals, salts, pesticides, solvents, explosives, radionuclide hydrocarbons, and landfill leachates [12–14]. Many studies have suggested that *Populus* species are tolerant to salinity and can even lower soil salinity [15–18]. For example, Euphrates poplar (*Populus euphratica* Oliv.) can grow well in soils with up to 8000 mg L$^{-1}$ salinity [15]. Moreover, Liu et al. (2001) [19] reported that white poplar (*Populus alba* L.) tolerated 2000 mg L$^{-1}$ salinity irrigation in a sandy soil in a greenhouse for two years. Poplars are also well suited for phytoremediation thanks to their ability to uptake high levels of nutrients and mineral salts, accumulate above and below ground carbon, improve soil structure and function, and reduce erosion [12,16]. Balsam poplar is a desirable species for boreal forest reclamation thanks to its fast growth and ease of vegetative propagation [20], combined with its natural role as a pioneer species. In addition, the EC tolerance range for balsam poplar is very high, ranging from 14.58 to 31.38 mS cm$^{-1}$, whereas the tolerance of white spruce (*Picea glauca* (Moench) Voss) is 8.75 to 14.92 mS cm$^{-1}$ and that of jack pine (*Pinus banksiana* Lamb.) is even lower at 1.02–6.33 mS cm$^{-1}$ [9].

Al-Pac is a pulp company that manages a 6.37 million ha forest management agreement (FMA) area in northeastern Alberta with an overlapping tenure with the oil sands region of Alberta, Canada. Balsam poplar is native to the region and has been the focus of Al-Pac's controlled parentage (tree improvement) program (CPP) (PB1-Alberta-Pacific Controlled Parentage Program plan for balsam poplar (2011)). The CPP consists of clones selected from within the FMA area (also the CPP deployment region) and outside the FMA area, with a minimum of 10 clones per provenance and 52 provenances. Approximately 520 clones were selected and have been planted on six test sites throughout the FMA area, including extreme (i.e., dry) locations, to investigate both local adaptability and potential regional adaptability under climate change. While Al-Pac is testing these trees for their reforestation potential, they are also of significant interest for their oil sands reclamation potential. After two years of greenhouse screening (2012 and 2013) of 222 balsam poplar clones from Al-Pac's program and based on their responses to varying levels of exposure to OSPW, clones were grouped into three categories (see Section 2.2.1 for details) for further field testing.

Our objective for the greenhouse study was to identify clones through screening and select genotypes that would be expected to survive and grow when used for reclamation on sites affected by OSPW. For the field trial, our objective was to test and identify balsam poplar clones, selected for salt tolerance from the greenhouse trials, exhibiting higher survival and increased growth (e.g., height and diameter) on reclamation sites compared with the following: (i) clones that did not exhibit tolerance to elevated salt levels in the greenhouse trials and (ii) a local seed zone Stream 1 wild balsam poplar cutting collection (local control).

## 2. Materials and Methods

### 2.1. Greenhouse Set-Up (2012&2013)

### 2.1.1. Plant Material and Growing Conditions

Two aeroponic greenhouse experiments were conducted in 2012 and 2013 using balsam poplar clones selected from Al-Pac's balsam poplar CPP. We selected 148 and then another 86 clones for screening in 2012 and 2013, respectively. In addition, 12 of the top clones from the 2012 experiment were retested in 2013.

All trees were propagated from 10 cm long dormant hardwood cuttings, from 1-year-old tissue, collected during the winter, prior to each experiment, from stooling beds grown at the Al-Pac mill site (54° 53′ N, 112° 51′ W, 575 m). Cuttings were stored in a chest freezer prior to commencement of the experiment. Cuttings were grown aeroponically in plastic containers filled with one of three treatment solutions and each aerated using a tubing system connected to a dedicated air compressor. The experiment was a completely randomized design with three treatments: (1) 100% reverse osmosis (RO) water (city water run through a reverse osmosis system); (2) 25% OSPW combined with 75% RO water; and (3) 50% OSPW with 50% RO water, with three replicates for each clone and water treatment combination (three containers per treatment) for a total of nine containers. Each treatment container held 80 L of solution.

The OSPW was collected directly from an outflow spout at a mine facility in Ft. McMurray, AB, into plastic jugs and transported to the Northern Forestry Centre (Natural Resources Canada, Canadian Forest Service) in Edmonton, AB, where the experiments were conducted each year. Prior to initiating the experiment, 170 mL of Hoagland's solution [21] was added to each treatment container. A near neutral pH was maintained 15 days prior to the start of the experiment and then monitored during the experiment for all water treatments by adding either phosphoric acid ($H_3PO_4$) (if higher than 7.5) or potassium hydroxide (KOH) (if lower than 6.5). Additional RO water was added to the containers at week four and week six to maintain adequate levels of liquid, compensating for water used by the plants and through evaporation. Additional OSPW was not added.

In 2012 and 2013, once pH was stabilized and after the experiment began (day 0), pH was maintained between 6.55 and 7.26 for all three water treatments for the duration of each experiment. The mean pH values for control (100% RO), 25% process water, and 50% process water were 6.91, 6.96, and 6.96, respectively, in 2012. In 2013, the mean pH values for control, 25% process water, and 50% process water were 6.96, 6.91, and 6.90, respectively. The mean electrical conductivity (EC) levels for control (100% RO), 25% process water, and 50% process water were 1.16 mS cm$^{-1}$, 2.14 mS cm$^{-1}$, and 3.28 mS cm$^{-1}$, respectively. In 2013, the mean EC levels for control, 25% process water, and 50% process water were 1.08 mS cm$^{-1}$, 2.25 mS cm$^{-1}$, and 3.31 mS cm$^{-1}$, respectively.

Containers (97 cm × 77 cm × 44 cm) used for this experiment had a cell arrangement of 11 cells long by 15 cells wide with a cell opening of 4 cm diameter into which a rubber bung, 4 cm long, was placed. Cuttings were placed into a hole in the middle of the rubber bung that fit into the cells in the lid of the container, suspending the cutting above the water. Cuttings were completely randomized for the location in each container. The experimental greenhouse had a day time temperature of approximately 24 °C and a mean night time temperature of 18 °C. Humidity was maintained at 65–85% with an 18 h photoperiod. In 2012, the cuttings were planted on 4 July (day 0) and grown until 17 August (day 44). In 2013, the cuttings were planted on 14 August (day 0) and grown until 18 October (day 65). An extended photoperiod was maintained with natural light supplemented with sodium vapor lamps at a light intensity of 250 μmol m$^{-2}$ s$^{-1}$. Each container had its own water pump that sprayed the solution into the air space where the cuttings were suspended, and the roots grew, every 30 min. In 2012, two water pumps failed (one for treatment 2 and one for treatment 3) just prior to harvest, resulting in only two replicates for each treatment at harvest, and in 2013, one water pump failed for treatment 2 at day 30. All data were collected up until the point of pump failure, which was day 22 in the 2012 experiment; therefore, only the final harvest and gas exchange data were affected.

### 2.1.2. Data Collection

In 2012, water samples were collected from the containers and analyzed three times during each experiment. Treatment water was sampled prior to planting the cuttings (day 15), near the middle of the growth period (~day 35), and at the end of the experiment (day 44, 2012 and day 65, 2013) and analyzed for basic nutrients, pH, and EC by the Biogeochemical Analytical Service Laboratory at the University of Alberta. Initial cutting diameters (mm) were measured for all trees at planting. Mortality was assessed prior to harvesting. To be considered DEAD, the cuttings were required to have had leaves emerge and then die; otherwise, a cutting that never flushed was considered a missing value. At the end of the experiment, prior to measurements and destructive sampling, a qualitative visual assessment of tree health was completed using the following scale: (1) dead tree; (2) tree was dying, leaves or stem were wilting and turning black; (3) tree appeared stressed, significant yellowing, or dropping of leaves; (4) tree showed signs of chlorosis, but otherwise looked healthy; (5) leaves were green and tree looked healthy; and (6) leaves were dark green and tree was thriving. Final height (cm) was measured from the base of the new growth (attachment of stem to cutting) or the rubber bung surface, whichever was higher, to the base of the terminal bud. Final basal stem diameter (mm) was measured at the base of the new growth (attachment of stem to cutting), or the rubber bung surface, whichever was higher. In 2012, photosynthesis rate (A) was measured using an infrared gas analyzer (LI-6400; Li-Cor, Lincoln, NE, USA) prior to the final harvest. Measurements were made on one fully expanded mature leaf per cutting between 08:00 and 13:30 with a supplied (saturating) light level of 1200 PAR. Following growth measurements, plants were destructively harvested and separated into leaf, stem, root, and original cutting components; these were oven-dried in paper bags for 72 h at 65 °C and then weighed and used to calculate biomass and root/shoot ratio.

### *2.2. Field Testing (2014–2019)*

### 2.2.1. Treatment Groups

Thirty-five balsam poplar clones from the PB1-CPP were screened and assessed for salt tolerance according to the greenhouse study [22]. Twenty-five of the selected clones were the top performing clones in the 50% process affected water treatment (treatment 1) and were selected as the 'high salt tolerant' treatment group and 10 clones that were poor performing clones in the 50% process affected water treatment were selected as a 'low salt tolerant' control group (treatment 2). There was an additional Stream 1 vegetative control lot collected from the local seed zone (CM2.2) with a minimum of 75 genotypes [23], which was not screened previously in the greenhouse (treatment 3).

### 2.2.2. Plant Material

One-year-old whips were collected from the Al-Pac mill site in February 2014, processed into 10 cm long cuttings, and placed in freezer storage at −3 °C. Cuttings were removed from freezer storage and soaked for two days in cool, fresh water prior to striking into Beaver Plastics® 512A styroblock (Beaver Plastics Ltd., Acheson, AB, Canada) containers on 4 June 2014 at Bonnyville Forest Nursery. Stream 1 control cuttings were collected in the winter of 2013/14 and kept in freezer storage until being soaked for one day and struck into 512A styroblock containers on 13 June at Bonnyville Forest Nursery.

All cuttings were grown under commercial nursery growing conditions from June to September 2014. Once they were hardened off and set bud, rooted cuttings were sorted and labeled and transported to the mine site. Planting was completed by 15 October 2014.

### 2.2.3. Testing Environment

The end-pit lakeshore used for this study is located north of Ft. McMurray (57°0′30″ N, 111°37′18″ W, 290 m a.s.l.) (Figure A2). The climate is considered a "warm-summer humid continental climate (Dfb)" according to Köppen climate classification [24]. The 20-year average (1999–2019) mean annual precipitation is 474 mm and the average temperature is 1.8 °C [24]. The site has a gentle slope running parallel to the water's edge with good nutrient condition. The former 50 to 60 m deep mine pit was largely filled with fluid fine tails (FFTs), and then capped with 4 m of process water and later 2 m of fresh water [25,26]. Given that the pore water of the FFT and the process water on top are brackish, vegetation on the shore of the lake was expected to be subjected to salty water (roughly 10% of the salinity of seawater) for the foreseeable future.

### 2.2.4. Experiment Design

The trial was a randomized block design and was planted on 15 October 2014. There were four ramets of each of 35 Al-Pac clones (25 treatment 1; 10 treatment 2) and 60 Stream I control trees (treatment 3) planted in each of three blocks on the south shore of the end-pit lake (Figure A3). Each block contained a total of 200 trees. In order to reduce within block variability, blocks were laid out with five trees running perpendicular from the lakeshore by 40 trees parallel to the lakeshore. Trees were planted 1 m apart in rows moving away from the lakeshore, with block 1 as close to the edge of the water as possible to maximize exposure to potentially saline lake water and ground water discharge from the adjacent hillside. All blocks followed the curving edge of the lake to keep them at as consistent an elevation and soil moisture condition within each block as possible. Rows up from the lake were tilled and covered with plastic mulch prior to planting. These rows were spaced approximately 3 m apart. Additional plastic mulch was placed manually between the rows to cover the entire trial area to minimize weed competition. Tree locations and identities were individually marked and mapped (Figure A3).

### 2.2.5. Growth and Survival Data Collection

Tree height (Ht) and basal root collar diameter (RCD) were measured according to protocols described in the trial measurement manual [27]. All trees were measured after installation for Ht, RCD in year 1, and diameter at breast height (DBH) was measured starting in year 2, and they were remeasured each fall in years 1, 2, 3, 4, and 5 (fall 2019). Survival was evaluated based on a visual assessment of the above ground stem from 2015 to 2019.

### 2.2.6. Tissue Nutrient Analysis

In summer 2016, two sets of leaf tissue samples were collected. Two leaves were collected from each live tree of the 35 clones and grouped into composite samples by clone and block. The Stream 1 control treatment trees had two leaves per tree collected from six randomly selected trees from each block and grouped to provide a single composite sample. A total of 108 samples were collected for the primary sample analysis. The Stream 1 control trees were chosen for the heavy metal analysis owing to the composition of this lot (i.e., 60 trees/block, minimum of 75 clones collected in the lot) representing a random, composite sample of multiple clones and collected from trees not used for the primary tissue analysis.

All 108 tissue samples were analyzed at Exova Laboratories, Surrey, British Columbia, to determine the uptake of nutrients and other compounds from the site (including boron, calcium, copper, iron, magnesium, manganese, molybdenum, phosphorous, potassium, sodium, sulfur, zinc, and nitrogen). The six Stream 1 control tree samples were also analyzed for 33 heavy metals (including aluminum, antimony, arsenic, barium, beryllium, bismuth, boron, cadmium, chromium, cobalt, copper, iron, lead, lithium, magnesium, manganese, mercury, molybdenum, nickel, phosphorus, potassium, selenium, silicon, silver, sodium, strontium, sulfur, thallium, tin, titanium, vanadium, zinc, and zirconium). The heavy metal analysis was completed as an indicator only of potential heavy metal accumulation.

### 2.3. Data Analysis

For both greenhouse and field studies, all growth and nutrient data were analyzed by two-way mixed model analysis of variance (ANOVA) using SAS 9.4 [28].

For the greenhouse study, treatment (0, 25%, or 50% OSPW) was considered a fixed effect and clone and container were considered random effects. The treatment × clone interaction was also included in the model and initial diameter of the cutting was used as a covariate. Following significant main effects analysis, multiple comparisons among means were completed using the Student–Newman–Keuls (SNK) test. We used $p \leq 0.05$ to determine significance.

For the field testing, treatments were considered a fixed effect and block considered as a random effect. Multiple comparisons among means were completed using the Student–Newman–Keuls test with $p \leq 0.05$ used to determine significance.

## 3. Results

### 3.1. Greenhouse Study

#### 3.1.1. Survival and Growth

In both 2012 and 2013, there were significant effects of treatment ($p < 0.001$, 2012; $p < 0.001$, 2013) and clone ($p < 0.001$, 2012; $p = 0.036$, 2013) on visual health assessment. The average ratings of the visual health assessment for the 2012 experiment, done at the end of the experiment, were 3.84, 3.39, and 2.81 for the control, 25%, and 50% treatments, respectively, out of a maximum score of 6. The control was significantly healthier ($p < 0.05$) than those in either the 25% or 50% treatments, which also differed from one another ($p < 0.05$). Most of the trees in the 25% and 50% treatments showed signs of chlorosis in both the younger and older leaves, and between the leaf veins. Some trees that ranked as a 3 or lower had necrotic leaf spots, were losing leaves, or in the most severe cases were dead.

In the 2013 experiment, the average visual health assessment ratings were 4.48, 3.95, and 4.38 for the control, 25%, and 50% treatments, respectively. The control and 50% treatments were significantly healthier ($p < 0.05$) than the 25% treatment; however, there was no significant difference between control and 50%. Observation of mortality showed that overall survival for the control treatment was 77.0%, the 25% solution was 80.4%, and the 50% solution was 64.9%. Survival ranged from 0 to 100%.

In both 2012 and 2013, there were significant effects of treatment and clone for final stem height, stem basal diameter and stem, and root and leaf biomass, with no clone by treatment interaction effect (Table 1). In 2012, the overall mean stem height for all clones did not differ between the control and 25% OSPW treatment, measuring on average 26 cm. Mean stem height for the 50% treatment was 30% significantly lower than both the control and 25% treatment. In 2013, the overall mean stem height for all clones in the control treatment was significantly greater than those in either the 25% or 50% treatments, which also differed from one another.

For stem basal diameter in the 2012 experiment, there was a decreasing trend from the control to the 25% to 50% treatments, with means averaging about 3.40 mm (Table 1). The trees in the 50% treatment were significantly smaller than in the control and 25% treatments. In the 2013 experiment, mean stem basal diameter did not differ between the control and 25% treatment; however, stem basal diameter in the 50% treatment was significantly lower ($p < 0.001$) than either the control or 25% treatment. Initial cutting diameter was used as a covariate for both the final stem height and final basal diameter analyses; it was not significant for the final stem height and final basal diameter in 2012 ($p = 0.06$; final height, $p = 0.052$; basal diameter), but it was significant in 2013 ($p = 0.034$; final height, $p = 0.003$; basal diameter).

**Table 1.** Mean values ($\pm$SE) for growth and biomass measurements of balsam poplar in 2012 (top) and 2013 (bottom) for all three treatments. Significant differences between treatment means within each row are indicated by different letters based on results of analysis of variance followed by post-hoc Student–Newman–Keuls (SNK) tests.

| 2012 | Treatment | | |
|---|---|---|---|
| | **Control** | **25% process water** | **50% process water** |
| Final stem height (cm) | $26.49 \pm 0.73$a | $26.74 \pm 0.80$a | $18.08 \pm 0.62$b |
| Stem basal diameter (mm) | $3.46 \pm 0.06$a | $3.40 \pm 0.07$a | $3.20 \pm 0.06$b |
| Stem biomass (g) | $0.37 \pm 0.02$a | $0.31 \pm 0.02$b | $0.16 \pm 0.01$c |
| Root biomass (g) | $0.19 \pm 0.01$a | $0.13 \pm 0.01$b | $0.09 \pm 0.01$c |
| Leaf biomass (g) | $0.86 \pm 0.04$a | $0.73 \pm 0.04$b | $0.52 \pm 0.03$c |
| Total biomass (g) | $1.42 \pm 0.08$a | $1.17 \pm 0.06$b | $0.77 \pm 0.04$c |
| **2013** | **Treatment** | | |
| | **Control** | **25% process water** | **50% process water** |
| Final stem height (cm) | $39.79 \pm 1.28$a | $33.70 \pm 1.43$b | $31.15 \pm 0.98$c |
| Stem basal diameter (mm) | $4.69 \pm 0.09$a | $4.62 \pm 0.11$a | $4.06 \pm 0.07$b |
| Stem biomass (g) | $1.06 \pm 0.07$a | $0.86 \pm 0.07$b | $0.54 \pm 0.03$c |
| Root biomass (g) | $0.60 \pm 0.04$a | $0.47 \pm 0.04$b | $0.30 \pm 0.02$c |
| Leaf biomass (g) | $1.63 \pm 0.08$a | $1.30 \pm 0.09$b | $1.04 \pm 0.05$c |
| Total biomass (g) | $3.32 \pm 0.19$a | $2.88 \pm 0.21$b | $1.90 \pm 0.10$c |

In the 2012 experiment, leaf, stem, root, and total biomass decreased significantly from the control to the 25% and 50% treatments (Table 1). In the 2012 experiment, the control treatment had the highest root/shoot ratio ($0.15 \pm 0.005$) followed by the 50% treatment ($0.14 \pm 0.005$), while the 25% treatment had the lowest root/shoot ratio ($0.12 \pm 0.006$). There was no significant difference between the control and 50% treatment, although both differed from the 25% treatment. In the 2013 experiment, the control treatment had the highest root/shoot ratio ($0.19 \pm 0.008$) followed by the 25% treatment ($0.17 \pm 0.008$) and then the 50% treatment ($0.15 \pm 0.005$).

Photosynthesis rates (A) were significantly influenced by both clone and treatment ($p < 0.001$) in 2012. Post-hoc comparisons ($p < 0.05$) showed that the control treatment had a significantly higher A than either the 25% or 50% process water treatments, which did not differ from one another. Pearson's correlation coefficient (r) between A and overall biomass was 0.435 for control, 0.434 for the 25% process water, and 0.435 for the 50% process water treatments. Significantly positive correlations ($p < 0.001$) were detected between all the treatments' total biomass and A. Despite some clones performing better under the OSPW treatments, the control water treatment plants had the highest rates of photosynthesis as well as the highest visual score for plant health at 3.84 in 2012 (vs. 3.39 in 25% and 2.81 in 50%).

Nitrogen and magnesium levels appeared to be adequate as compared with the control for all treatments (Table 2); however, there were noticeable decreasing trends with control > 25% > 50% process-affected water for iron, indicating that an iron deficiency may have been present. It was observed that the iron levels in the 25% and 50% process water samples were very low by the end of the experiment (Table 2). It is likely that the elevated phosphate levels, which were due to the addition of $H_3PO_4$ to reduce or maintain a stable pH, caused the iron to precipitate out of solution, making it unavailable to the plants. 'Rust' observed on the bottoms of both the 25% and 50% process water treatment containers supports this hypothesis.

**Table 2.** Results of analysis of the water solution (means) for the control (reverse osmosis: RO), 25% process water treatment (process $H_2O$ 25%), and 50% process water (process $H_2O$ 50%) treatments before and after the addition of Hoagland's solution (day 15), day 35, and day 44 in 2012. Values for undiluted process water (process $H_2O$ 100%) are shown for comparison. Note: TDN = total dissolved nitrogen; TDP = total dissolved phosphorus.

| | $NH_4^+$ (N µg/L) | $NO_2^+$ $NO_3$ (N µg/L) | TDN (N µg/L) | TDP (P µg/L) | Cl (mg/L) | $SO_4$ (mg/L) | Na (mg/L) | K (mg/L) | Ca (mg/L) | Mg (mg/L) | Fe (mg/L) | Al (mg/L) |
|---|---|---|---|---|---|---|---|---|---|---|---|---|
| Minimum level of detection | 2 | 1 | 10 | 3 | 0.03 | 0.04 | 0.016 | 0.009 | 0.005 | 0.01 | 0.016 | 0.004 |
| RO pre H * Day 15 | 27 | 289 | <LOD *** | <LOD | 5.51 | 31.03 | 13.05 | 0.90 | 31.75 | 7.61 | <LOD | <LOD |
| RO post H ** Day 15 | 428 | 3800 | <LOD | <LOD | 5.72 | 34.28 | 13.19 | 7.47 | 32.03 | 8.61 | 0.06 | <LOD |
| RO Day 35 | 17 | 42,867 | <LOD | 45,181 | 9.92 | 179.60 | 21.85 | 183.99 | 63.32 | 43.33 | 0.43 | <LOD |
| RO Day 44 | 43 | 37,933 | <LOD | 52,183 | 8.38 | 135.47 | 20.04 | 122.40 | 41.98 | 31.54 | 0.21 | <LOD |
| Process $H_2O$ 100% | 5787 | 16 | <LOD | <LOD | 717 | 397 | 1195 | 15 | 21 | 11 | <LOD | <LOD |
| Process $H_2O$ 25% pre H Day 15 | 2140 | 701 | <LOD | <LOD | 195.16 | 131.12 | 269.37 | 5.37 | 27.58 | 8.21 | <LOD | <LOD |
| Process $H_2O$ 25% post H Day 15 | 2390 | 2910 | <LOD | <LOD | 194.52 | 132.66 | 270.05 | 10.02 | 29.45 | 9.53 | 0.04 | <LOD |
| Process $H_2O$ 25% Day 35 | 10 | 50,567 | <LOD | 102,986 | 206 | 303 | 306 | 180 | 42 | 43 | 0.10 | 10 |
| Process $H_2O$ 25% Day 44 | 33 | 38,933 | <LOD | 109,417 | 198 | 257 | 275 | 120 | 30 | 32 | 0.06 | <LOD |
| Process $H_2O$ 50% pre H Day 15 | 4020 | 870 | <LOD | <LOD | 367.83 | 231.82 | 540.19 | 10.09 | 25.80 | 9.68 | <LOD | <LOD |
| Process $H_2O$ 25% post H Day 15 | 4060 | 2880 | <LOD | <LOD | 365.58 | 233.04 | 529.52 | 14.53 | 26.55 | 10.53 | 0.04 | <LOD |
| Process $H_2O$ 50% Day 35 | 26 | 33,933 | <LOD | 151,163 | 365.17 | 380.95 | 524.48 | 187.20 | 22.96 | 39.34 | 0.02 | <LOD |
| Process $H_2O$ 50% Day 44 | 27 | 53,467 | <LOD | 174,730 | 397.45 | 388.63 | 575.50 | 160.50 | 19.87 | 36.45 | 0.02 | <LOD |

* pre H = before addition of Hoagland's solution. ** post H = after addition of Hoagland's solution. *** LOD = limit of detection.

### 3.1.2. Clonal Variation

Fourteen clones consistently ranked in the top 30 (30 was selected as a target number of clones to ensure genetic diversity standards would be met [23]; Ne = 18 for operational deployment of native species from a CPP program onto public lands in Alberta) for stem height growth across all three treatments (Figure 1a). For stem basal diameter growth, 12 clones ranked consistently in the top 30 in all three treatments (Figure A1a). The same trends were observed for the top 12 clones in the 2013 experiment (Figure 1b; Figure A1b).

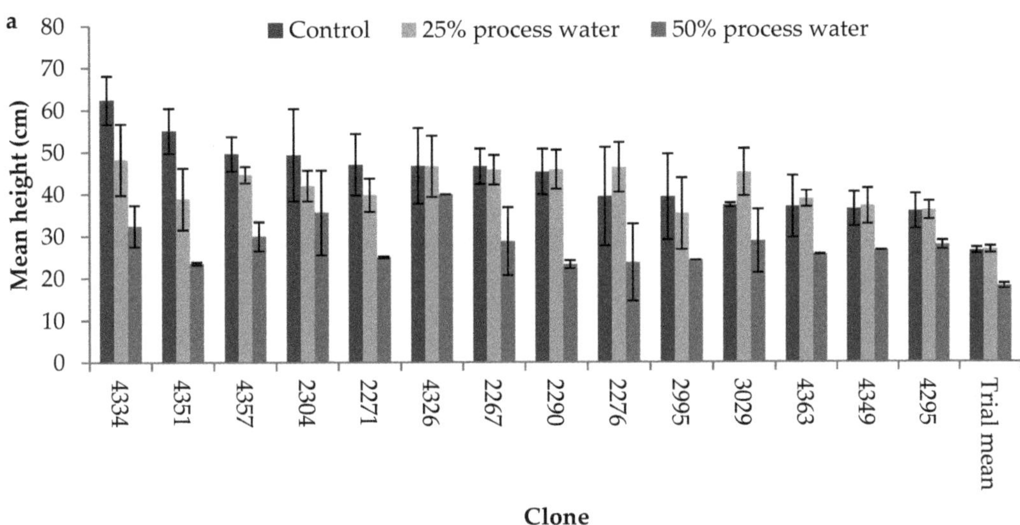


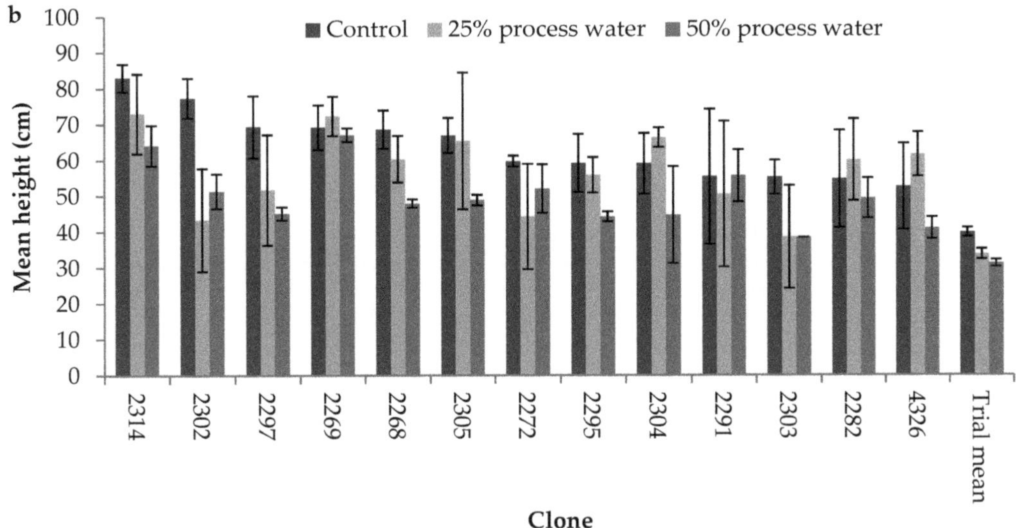


**Figure 1.** (**a**) Mean height (±SE) of 14 balsam poplar clones that ranked in the top 30 for all three water treatments as compared with the overall trial mean for 148 clones after 44 days of growth in the 2012 experiment. (**b**) Mean height (±SE) of 13 balsam poplar clones that ranked in the top 30 for all three water treatments as compared with the overall trial mean for 86 clones after 65 days of growth in the 2013 experiment.

Within the 148 clones tested in 2012, there were 38 clones that had higher total biomass in the 25% process water treatment than in the control treatment, while 24 clones had a higher mean total biomass in the 50% process water treatment than the control, and five clones ('AP2309', 'AP2453', 'AP3033', 'AP3127', and 'AP4356') showed the opposite trend with the 50% process-affected water treatment > 25% process-affected water treatment > control. Clones 'AP2453' and 'AP4356', which exhibited this reverse trend, also ranked within the top 30 clones for total biomass in all three treatments.

In the 2013 experiment, there was, again, a significant decrease in leaf, stem, root, and total biomass from the control to the 25%, and 50% treatment (Figure 2). Within the 86 clones that were tested in 2013, there were 27 clones that had higher total biomass for the 25% process water treatment than control (3.32 g $\pm$ 0.19), and 10 clones had higher total biomass in the 50% process water treatment than the control. See Appendix A Tables A1 and A2 for summary total biomass data for the 30 top performing clones in 2012 and 2013. Overall performance showed similar trends across both years for the top 10 clones (Figure 3). However, owing to the longer growth period in 2013, the biomass totals for 2013 were higher overall than those in 2012. There were, however, exceptions to this trend on an individual clone basis. In addition, some of the highest root/shoot ratios were observed in clones that had below average total biomass growth.

### 3.2. Field Testing
### 3.2.1. Survival and Growth

Overall, all trees grew well at the edge of the end-pit lake. There was no significant difference between treatments for survival in 2019, which overall remained very high at 82%, 84%, and 85% for treatments 1, 2, and 3, respectively (Table 3). However, mortality rates increased from 11% (2015) to 17% (2019), which was likely due to higher mortality in the first row of trees adjacent to the lakeshore, with some having almost eroded into the lake.

**Table 3.** Per cent (%) survival rate ($\pm$SE) of balsam poplar among treatments (1 = 25 selected tolerant clones, 2 = 10 selected control clones, 3 = Stream 1 vegetative lot clones) in the fall of each year.

| Treatment | Year (Age) | | | | |
|---|---|---|---|---|---|
| | 2015 (Age 1) | 2016 (Age 2) | 2017 (Age 3) | 2018 (Age 4) | 2019 (Age 5) |
| 1 | 87% $\pm$ 2% | 87% $\pm$ 2% | 86% $\pm$ 2% | 83% $\pm$ 3% | 82% $\pm$ 3% |
| 2 | 92% $\pm$ 2% | 90% $\pm$ 3% | 90% $\pm$ 3% | 87% $\pm$ 4% | 84% $\pm$ 4% |
| 3 | 89% $\pm$ 4% | 89% $\pm$ 4% | 89% $\pm$ 4% | 87% $\pm$ 5% | 85% $\pm$ 5% |

The average height and DBH at year five for treatments 1, 2, and 3 were 3.75, 3.58, and 3.61 m for height (Figure 4) and 27.88, 27.02, and 26.38 mm for DBH, respectively (Figure 5). However, there were no significant differences in height and DBH among treatments. Mean growth increments for both height (Figure 4) and basal RCD or DBH (Figure 5) showed similar growth trends across all three treatments from 2015 to 2019. The largest annual height increment was in 2016, which averaged approximately 1 m for all three treatments (Figure 4).

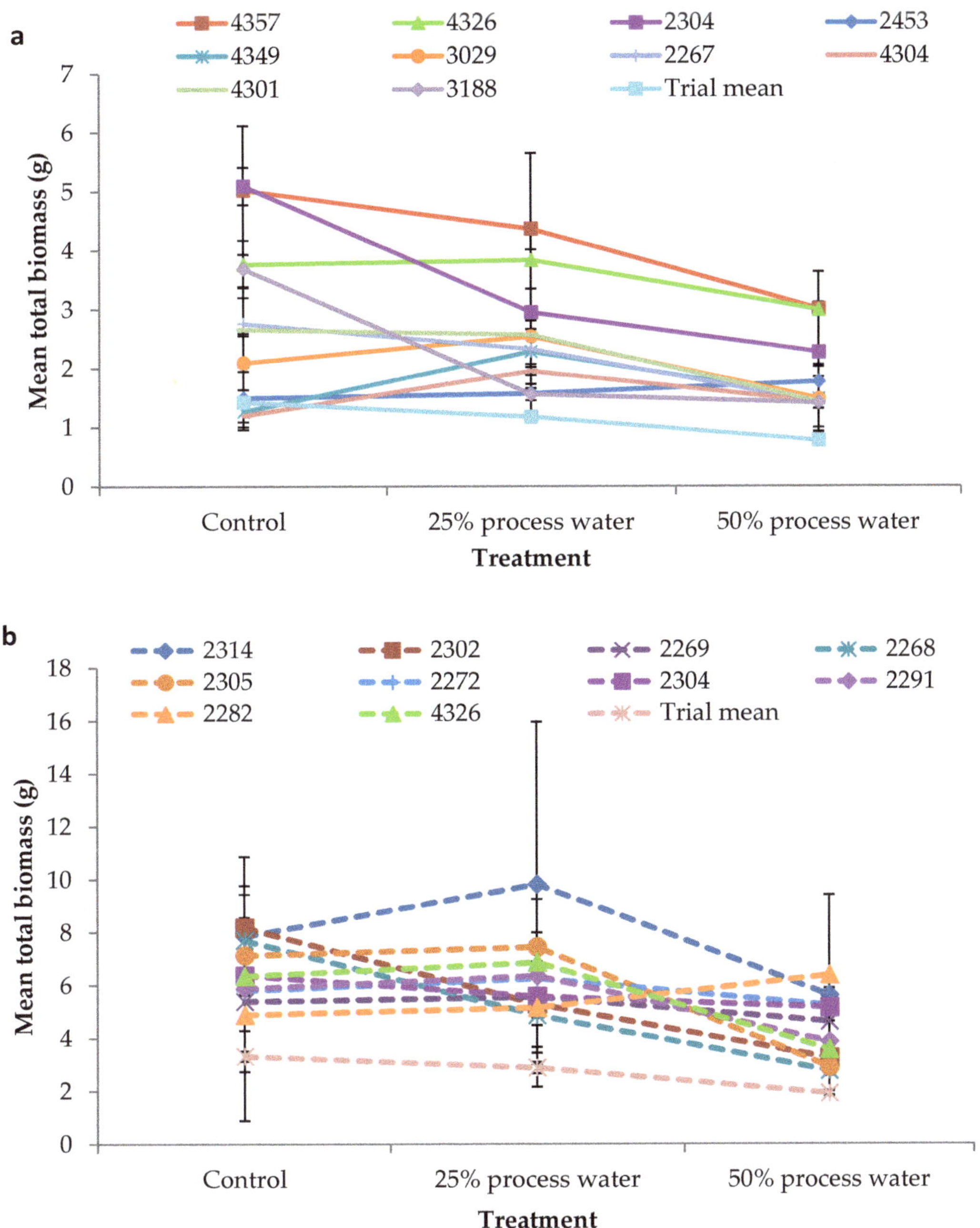

**Figure 2.** (**a**) Total biomass (mean dry weight ± SE) in each water treatment for the top 10 balsam poplar clones in the three water treatments and the overall means for all trees after 44 days of growth in the 2012 experiment (solid line). (**b**) Total biomass (mean dry weight ± SE) vs. treatment for the top 10 balsam poplar clones in the control, 25%, and 50% process water treatment solutions and the overall treatment mean after 65 days of growth in the 2013 experiment (dash line). Clones tested in both years (4326 and 2304) are in the same colour.

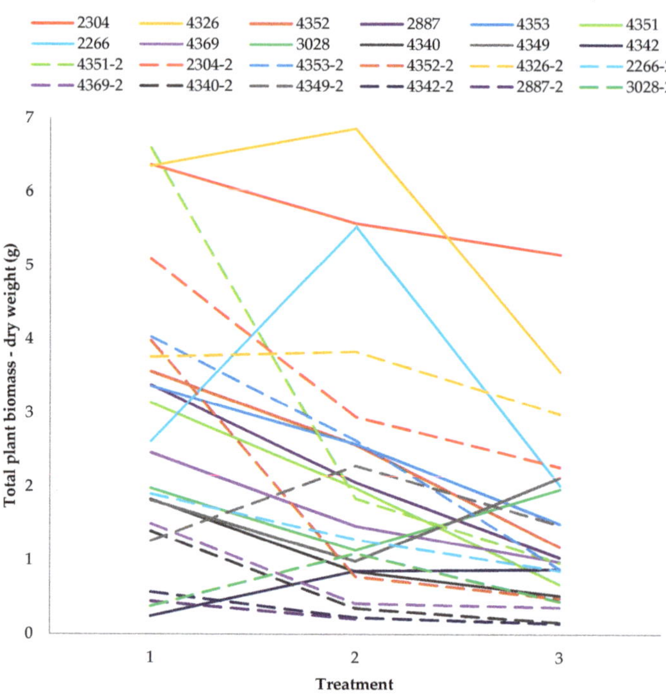

**Figure 3.** Performance (total plant biomass (g)) of the 12 balsam poplar clones grown in both 2012 and 2013 trials (2013 = solid line, 2012 = dashed line) under three treatments (1 = control; 2 = 25% process water; 3 = 50% process water).

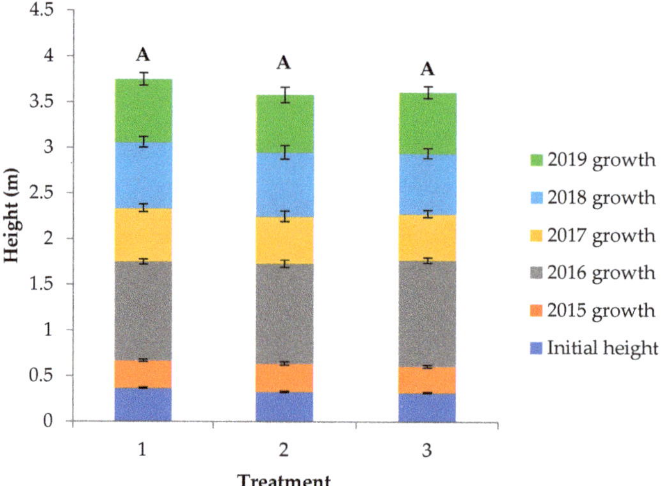

**Figure 4.** Mean initial height and mean annual height increment (±SE) (year 1 = 2015 growth; year 2 = 2016 growth; year 3 = 2017 growth; year 4 = 2018 growth; year 5 = 2019 growth) (m) (±SE) for balsam poplar trees planted in three treatments (1 = 25 selected tolerant clones, 2 = 10 selected control clones, 3 = Stream 1 veg. lot clones). Significant differences between treatment means for height are indicated by different letters at $p \leq 0.05$.

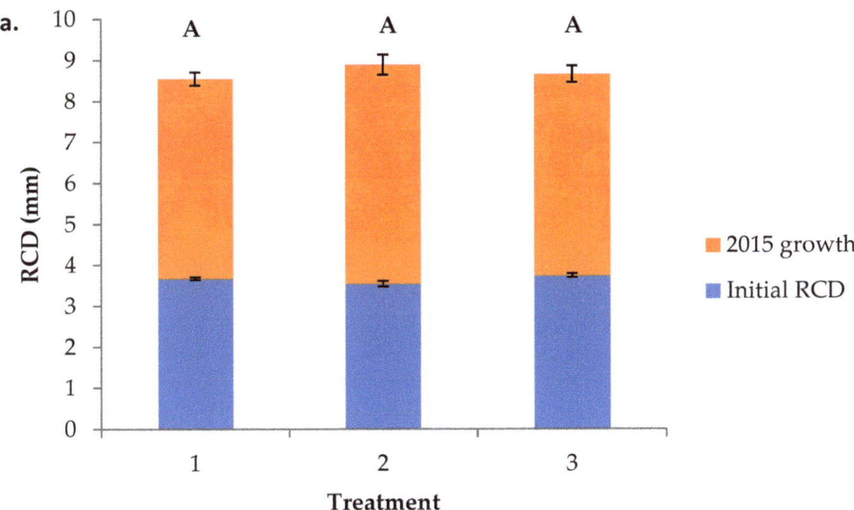

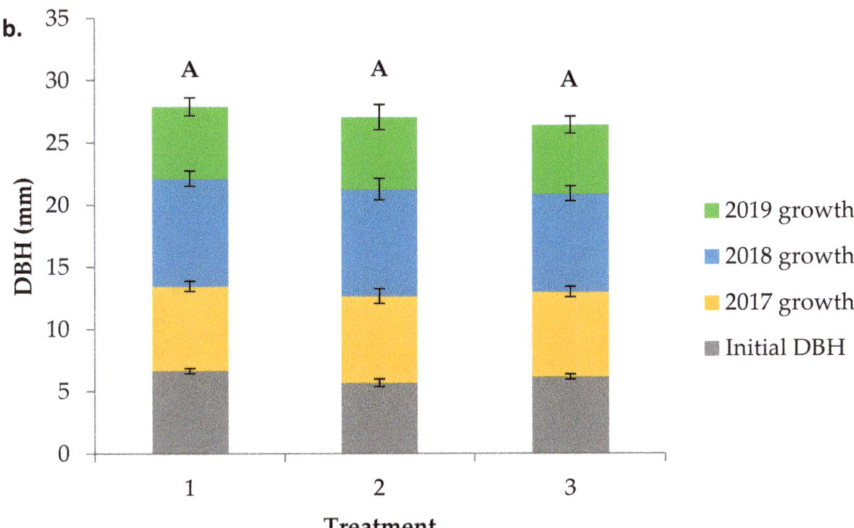

**Figure 5.** Mean initial diameter and mean annual growth increment (±SE) (year 1 = 2015 growth; year 2 = 2016 growth; year 3 = 2017 growth; year 4 = 2018 growth; year 5 = 2019 growth) (cm) (±SE) for balsam poplar trees planted in three treatments (1 = 25 selected tolerant clones, 2 = 10 selected control clones, 3 = Stream 1 veg. lot clones). (**a**) Basal root collar diameter (RCD mm) (2014–2015); (**b**) diameter at breast height (DBH mm) (2016–2019). Significant differences between treatment means for DBH are indicated by different letters at $p \leq 0.05$.

Significant differences were found among different blocks for both height and DBH growth parameters (Figure 6). The distance to the shoreline was used to determine the block design running parallel to the shore. Trees in Block 2 (10 m away to the lake edge) showed the best tree performance and this block represented the middle distance from the shoreline (between Block 1, closest to the water's edge, and Block 3, furthest up the slope from the water's edge).

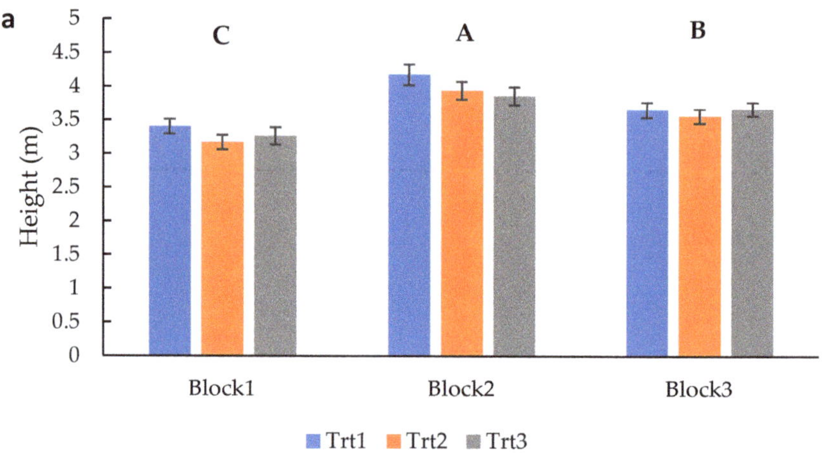

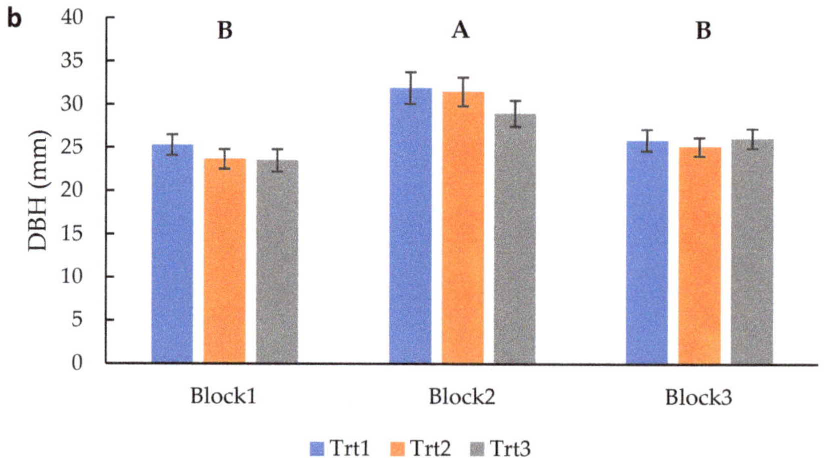

**Figure 6.** Treatment means (1 = 25 selected tolerant clones, 2 = 10 selected control clones, 3 = Stream 1 vegetative lot clones) of growth parameters for balsam poplar trees planted in three blocks (Block 1 = 5 m from the lake edge; Block 2 = 10 m from the lake edge; Block 3 = 15 m from the lake edge) in 2019. (**a**) Height (m); (**b**) DBH (mm). Significant differences between block means are indicated by different letters. Significant differences between treatment means for height or DBH in each block are indicated by different letters at $p \leq 0.05$.

Growth in height ranged from 2.5 m to more than 5 m across all clones (Figure 7a), while DBH ranged from 17 mm to more than 36 mm (Figure 7b). The Stream 1, treatment 3 clones showed average growth when compared with the 35 Stream 2 selected clones (treatment 1 + treatment 2) for both height and DBH (Figure 7a,b). Not surprisingly, there was a strong correlation (r = 0.925) between height and DBH by the fall of 2019, indicating that the taller trees also had, in general, great DBH.

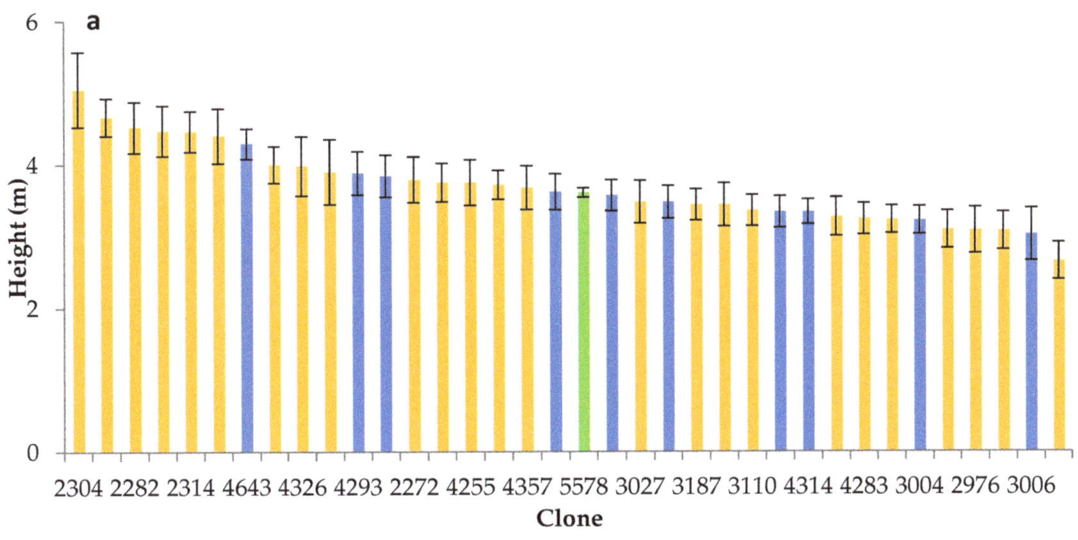

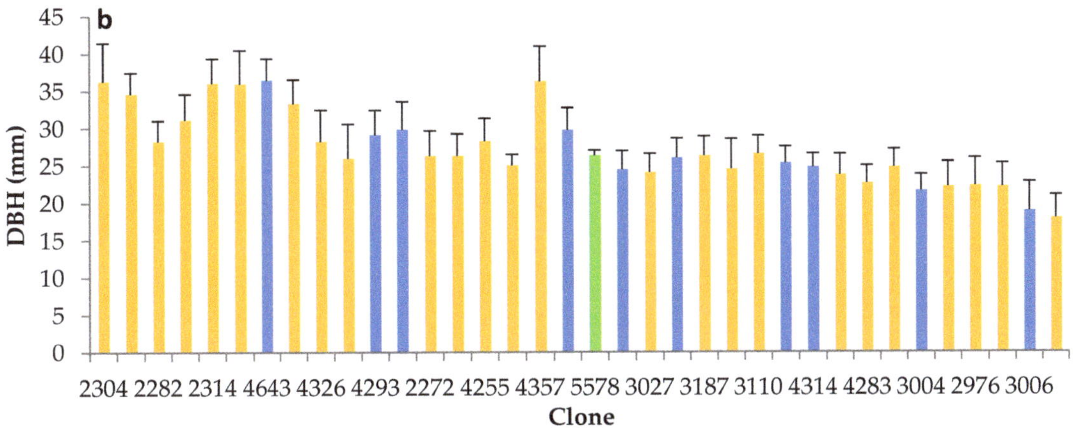

**Figure 7.** Mean growth (±SE) by balsam poplar clone (fall 2019). The yellow bars are treatment 1 clones, the dark blue bars are treatment 2 clones, and the green bar is treatment 3 (treatment 1 = 25 selected tolerant clones, treatment 2 = 10 selected control clones, treatment 3 = Stream 1 vegetative lot clones; # 5578 was the lot number for all Stream 1 clones). (**a**) Height (m); (**b**) DBH (mm).

When considering the tallest 18 clones from treatment 1 (where 25 clones were tested), which is the minimum number of clones required for unrestricted registration of a Stream 2 'lot' to be deployed operationally (i.e., Ne = 18) as determined by the government standards [23], the results showed a significant difference in height, but not DBH, when compared with treatment 2 (10 'low salt tolerant clones' from 50% process water testing) and, more importantly, the Stream 1 local 'wild' collection (treatment 3), for both height and DBH (Table 4).

**Table 4.** Growth data (height and diameter at breast height (DBH)) (±SE) of balsam poplar in 2019 for all treatments (top 18 clones from treatment 1 and all clones from treatment 2 and treatment 3). Significant differences between treatment means are indicated by different letters.

| Treatment/Lot Type | Height (m) | DBH (mm) |
|---|---|---|
| 1 (top 18 clones)/Stream 2 | 4.01 ± 0.08 [a] | 29.99 ± 0.86 [a] |
| 2 (10 control clones)/Stream 2 | 3.58 ± 0.09 [b] | 27.02 ± 1.00 [b] |
| 3 Local control/Stream 1 | 3.61 ± 0.06 [b] | 26.38 ± 0.69 [b] |

Stem volume was calculated for each tree using fall 2019 data based on the following equation: $V = A_b \times H/3$ (where V: stem volume ($cm^3$), $A_b$: basal area = $\pi \times DBH^2$ (diameter at breast height)/4 ($cm^2$), and H: height (cm)) [29] (Figure 8). Although no significant differences were found in either height or DBH, when stem volume was calculated, trees in treatment 1 (including all 25 clones) (1060.47 ± 68.24 $mm^3$) had a larger stem volume than trees in treatment 3 (814.87 ± 54.25 $mm^3$) (Figure 8). However, when considering the tallest 18 clones from treatment 1 (where 25 clones were tested), the stem volume of the top 18 clones was 1254.55 ± 86.71 $mm^3$, which is significantly greater than ($p < 0.05$) treatment 2 (893.39 ± 80.67 $mm^3$) and treatment 3.

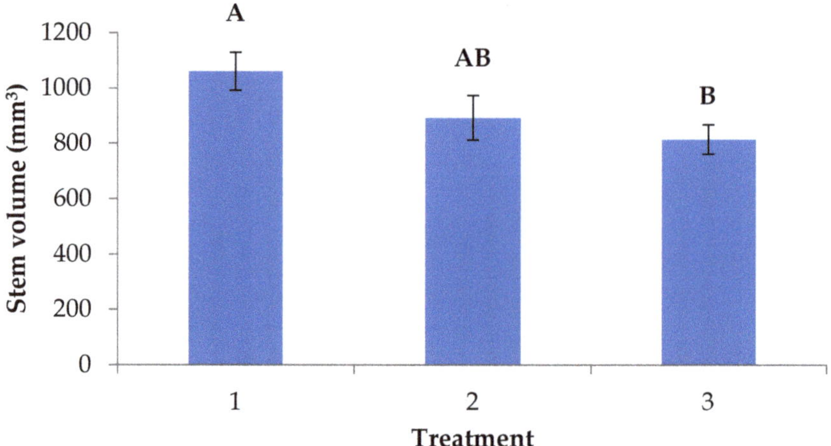


**Figure 8.** Mean stem volume ($mm^3$) (±SE) for five-year-old balsam poplar trees planted in 2019 (1 = 25 selected tolerant clones, 2 = 10 selected control clones, 3 = Stream 1 vegetative lot clones). Significant differences between treatment means for stem volume are indicated by different letters at $p \leq 0.05$.

### 3.2.2. Tissue Nutrient Analysis

Overall, nutrient analysis from bulk leaf tissue samples showed a low sodium level and a high calcium level (Table 5). There were no significant differences found for any of the nutrients by treatment except for magnesium, where treatment 3 had significantly higher levels compared with treatments 1 or 2 (Table 6). Significant differences were found among different blocks for some tissue nutrients (i.e., Cu, Fe, Mn, P, Zn, and N) for both treatments (Table 7).

**Table 5.** Mean tissue nutrient (a) and heavy metal (b) analysis (±SE) based on leaf tissue samples in two-year-old balsam poplar trees.

| | | | a. Tissue nutrient | | | | |
|---|---|---|---|---|---|---|---|
| Site | B(ug/g) | Ca (%) | Cu (ug/g) | Fe (ug/g) | Mg (%) | Mn (ug/g) | Mo (ug/g) |
| End-pit lake | 33.44 ± 0.82 | 1.05 ± 0.02 | 4.69 ± 0.12 | 136.06 ± 4.84 | 0.282 ± 0.004 | 9.65 ± 0.26 | 0.51 ± 0.01 |
| | P (%) | K (%) | S (%) | Zn (ug/g) | N (%) | Na (%) * | |
| | 0.164 ± 0.003 | 1.45 ± 0.03 | 0.36 ± 0.01 | 127.48 ± 5.51 | 1.38 ± 0.06 | - | |

| | | | b. Heavy metal ** | | | | |
|---|---|---|---|---|---|---|---|
| Site *** | Al (ug/g) | Ba (ug/g) | Cd (ug/g) | Cr (ug/g) | Co (ug/g) | Li (ug/g) | Ni (ug/g) |
| End-pit lake | 55.90 ± 14.70 | 1.87 ± 0.27 | 0.20 ± 0.00 | 0.79 ± 0.26 | 0.38 ± 0.13 | 0.45 ± 0.05 | 0.74 ± 0.11 |
| | Si (ug/g) | Sr (ug/g) | Sn (ug/g) | Ti (ug/g) | V (ug/g) | | |
| | 270.67 ± 47.10 | 11.73 ± 1.53 | 1.40 ± 0.06 | 2.30 ± 0.30 | 3.83 ± 0.12 | | |

* Na levels were all <0.01, so no statistical analysis could be completed. ** Heavy metal analysis was only conducted on Stream 1 control trees. *** Metal levels <0.5 (ug/g) are not shown.

**Table 6.** Mean leaf tissue nutrient analysis (±SE) results by treatment in two-year-old balsam poplar trees. Significant differences between treatment means at $p \leq 0.05$ are indicated by different letters. Treatment 1 = 25 selected tolerant clones, treatment 2 = 10 selected control clones, treatment 3 = Stream 1 vegetative lot clones.

| Treatment | B(ug/g) | Ca (%) | Cu (ug/g) | Fe (ug/g) | Mg (%) | Mn (ug/g) | Mo (ug/g) |
|---|---|---|---|---|---|---|---|
| 1 | 33.62 ± 1.02 | 1.04 ± 0.02 | 4.42 ± 0.13 | 139.83 ± 6.47 | 0.283 ± 0.004 [ab] | 9.56 ± 0.33 | 0.48 ± 0.04 |
| 2 | 32.88 ± 1.52 | 1.07 ± 0.04 | 5.39 ± 0.21 | 127.00 ± 6.16 | 0.275 ± 0.007 [b] | 9.67 ± 0.44 | 0.55 ± 0.04 |
| 3 | 33.70 ± 0.55 | 0.97 ± 0.05 | 4.61 ± 0.57 | 132.33 ± 14.44 | 0.32 ± 0.026 [a] | 8.70 ± 1.79 | 0.47 ± 0.03 |

| Treatment | P (%) | K (%) | S (%) | Zn (ug/g) | N (%) | Na (%) |
|---|---|---|---|---|---|---|
| 1 | 0.161 ± 0.003 | 1.46 ± 0.03 | 0.35 ± 0.01 | 119.06 ± 6.61 | 1.35 ± 0.07 | - |
| 2 | 0.172 ± 0.006 | 1.44 ± 0.04 | 0.37 ± 0.01 | 149.42 ± 9.48 | 1.43 ± 0.13 | - |
| 3 | 0.160 ± 0.011 | 1.46 ± 0.12 | 0.33 ± 0.04 | 118.47 ± 10.90 | 1.38 ± 0.31 | - |

* Na levels were all <0.01, so no statistical analysis could be completed.

**Table 7.** Block means of tissue nutrient analysis for each treatment in two-year-old balsam poplar trees. Significant differences between block means at $p \leq 0.05$ are indicated by different letters. Block 1 = 5 m from the lake edge; Block 2 = 10 m from the lake edge; Block 3 = 15 m from the lake edge.

| | | | a. Treatment 1 (25 selected tolerant clones) | | | | |
|---|---|---|---|---|---|---|---|
| Block | B (ug/g) | Ca (%) | Cu (ug/g) | Fe (ug/g) | Mg (%) | Mn (ug/g) | Mo (ug/g) |
| 1 | 34.14 ± 1.74 | 1.06 ± 0.04 | 5.19 ± 0.20 [a] | 178.44 ± 6.29 [a] | 0.270 ± 0.008 | 10.96 ± 0.69 [a] | 0.58 ± 0.03 |
| 2 | 33.65 ± 1.50 | 1.03 ± 0.05 | 4.46 ± 0.19 [b] | 123.68 ± 6.58 [b] | 0.280 ± 0.008 | 8.94 ± 0.43 [b] | 0.53 ± 0.04 |
| 3 | 33.16 ± 2.07 | 1.02 ± 0.04 | 3.65 ± 0.18 [c] | 115.36 ± 7.47 [c] | 0.300 ± 0.008 | 8.77 ± 0.47 [b] | 0.42 ± 0.04 |

| Block | P (%) | K (%) | S (%) | Zn (ug/g) | N (%) | Na (%) |
|---|---|---|---|---|---|---|
| 1 | 0.170 ± 0.007 [a] | 1.55 ± 0.04 | 0.36 ± 0.01 | 137.54 ± 9.69 [a] | 1.81 ± 0.07 [a] | - |
| 2 | 0.170 ± 0.007 [a] | 1.44 ± 0.07 | 0.36 ± 0.02 | 115.71 ± 8.26 [b] | 1.34 ± 0.09 [b] | - |
| 3 | 0.140 ± 0.005 [b] | 1.41 ± 0.07 | 0.33 ± 0.02 | 100.94 ± 7.71 [c] | 0.89 ± 0.14 [c] | - |

| | | | b. Treatment 2 (10 selected control clones) | | | | |
|---|---|---|---|---|---|---|---|
| Block | B (ug/g) | Ca (%) | Cu (ug/g) | Fe (ug/g) | Mg (%) | Mn (ug/g) | Mo (ug/g) |
| 1 | 35.65 ± 2.70 [a] | 1.09 ± 0.06 | 6.22 ± 0.38 [a] | 166.80 ± 4.50 [a] | 0.290 ± 0.010 | 10.98 ± 0.92 [a] | 0.62 ± 0.02 |
| 2 | 32.39 ± 2.29 [b] | 1.08 ± 0.09 | 5.10 ± 0.28 [b] | 112.40 ± 5.62 [b] | 0.270 ± 0.020 | 9.36 ± 0.74 [a] | 0.53 ± 0.03 |
| 3 | 30.60 ± 2.91 [b] | 1.03 ± 0.05 | 4.84 ± 0.33 [b] | 101.80 ± 6.70 [b] | 0.270 ± 0.009 | 8.67 ± 0.44 [b] | 0.57 ± 0.03 |

| Block | P (%) | K (%) | S (%) | Zn (ug/g) | N (%) | Na (%) |
|---|---|---|---|---|---|---|
| 1 | 0.180 ± 0.012 [a] | 1.48 ± 0.07 | 0.39 ± 0.02 | 183.00 ± 8.17 [a] | 1.77 ± 0.25 [a] | - |
| 2 | 0.172 ± 0.012 [a] | 1.47 ± 0.06 | 0.39 ± 0.02 | 152.00 ± 9.23 [b] | 1.56 ± 0.15 [b] | - |
| 3 | 0.160 ± 0.007 [b] | 1.36 ± 0.09 | 0.34 ± 0.03 | 113.30 ± 11.01 [c] | 0.95 ± 0.17 [c] | - |

## 4. Discussion

### 4.1. Greenhouse Study

Interest in and acceptance of poplar and willow for use in reclamation and phytoremediation have been increasing in the last 15 years [30,31]. Salinity is known to reduce water absorption and cause water stress [32]. Salts are taken up by plants and the increasing tissue ion concentration contributes to a decrease in water potentials. In addition, some plant species accumulate solutes under stress to maintain a positive water balance [33]. This osmotic adjustment allows plants to maintain turgor in saline environments. Approximately 1% of all plant species are halophytes and can complete their life cycle in relatively high saline environments, such as 200 mM NaCl or more [34]. The ability of poplars to grow on harsh sites, along with being relatively easy to propagate, through the use of cuttings, makes them an ideal candidate species for use in reclamation, more specifically with respect to this study, on reclaimed oil sands mine sites in northeastern Alberta.

In both years of the greenhouse studies, clear differences in the phenotypic growth response of the tested balsam poplar clones were observed, indicating tolerance for OSPW by native balsam poplars and clonal variability in that tolerance. These findings support the assertion that the opportunity exists to select and propagate an easily propagated native species for use in reclaiming these challenging sites. More specifically, there was a high degree of genetic (clonal) variability in survival, height, diameter, and biomass growth in response to the control, 25%, and 50% process water treatments. Most clones performed more poorly in the 25% and 50% process water solutions as compared with the control. There were several clones, however, that performed consistently better than the average for all of the traits measured across all three treatments, exhibiting desirable traits for selection from the population of clones tested. These results suggest that genetic differences in clones should be considered in the selection of genetic materials for use on reclamation sites impacted by high salt-containing tailings generated from oil sands operations.

Clone 'AP4357', tested in both years of the study, is an example of an OSPW-tolerant clone that consistently performed at or near the top for all traits measured, and in all three water treatments across both years. There were also a number of clones that performed better in the 25% process water treatment, 50% process water treatment, or both as compared with their control treatment performance. These tolerant clones appeared to actually prefer the saline conditions, which indicates that there are balsam poplar clones that are salt loving or 'halophiles'.

Significant positive correlations between height growth and rooting traits such as root length and root dry weight of poplar have been reported [35]. Thus, we believe height growth can be used as a surrogate measure of root development, which may have increased associated microbial activity in the rhizosphere. Therefore, our better-rooting clones (i.e., clones 'AP4326' and 'AP2304') may exhibit greater remedial potential. In addition, root/shoot ratios are often very useful in determining if plants have healthy root systems relative to above ground biomass [36]. Because salty soils often limit root penetration and inhibit root growth [37], clones that have higher root growth are likely going to have increased performance in saline environments. Therefore, higher root/shoot ratios would be desirable. However, one must be careful not to look at root/shoot ratios as a single trait for selection as it gives no indication of the actual growth performance of the plant. Ideally, in the selection of suitable clones from this experiment, clones that have high total above ground biomass with an above average root/shoot ratio would be considered desirable.

### 4.2. Field Testing

### 4.2.1. Survival and Growth

The initial high survival rate of all treatments at the end-pit lake (Table 3) indicated these trees were well adapted to the reclamation mine site. Additionally, survival through the first two years was high, indicating that early survival is an important indicator of later survival and growth. Owing to good water availability and sufficient nutrients throughout each growing season, trees grew very well (Figures 4 and 5). The trees also experienced little to no competition owing to the installed black plastic, which also likely made the soil warmer, although this was not measured. In this trial, distance to the shoreline had a significant impact on the trees' height and DBH (Figure 6). For trees planted adjacent to the shoreline, the trees had ready access to the water table while also being more vulnerable to shoreline erosion.

As no significant differences were found between treatment 3 (Stream 1 vegetative lot clones) and the other two plant treatment groups for height and DBH, these results suggested the growth rates of the Stream 1 vegetative lot clones might be considered acceptable when compared with the Al-Pac selected clones (salt-tolerant and controls from Al-Pac's program) on a site with 'ideal' growing conditions. However, mean stem volume showed a significant difference between treatments (treatment 1 $\geq$ treatment 2 $\geq$ treatment 3), indicating the selected Stream 2 clones from Al-Pac's CPP program, overall, grew better than the Stream 1 vegetative lot clones. In addition, analysis of the top 18 clones from the Stream 2 lot selected from the treatment 1 group showed significant differences for both height and DBH when compared with treatments 2 and 3 (Table 4). This suggests that there is potential to plant groups of selected clones that would outperform wild clonal collections of Stream 1 native balsam, with the added advantage of not being restricted to deployment only within their local seed zone, but taking advantage of the entire region associated with the balsam poplar controlled parentage program.

Interestingly, despite clones selected from treatment 2 being chosen from the group of clones that did not exhibit superior salt tolerance (i.e., greatest growth) in the initial greenhouse screening trials (2012 and 2013), they still showed a level of tolerance to salinity testing [22]. It is worth noting that the selected high tolerance clones (treatment 1) were not necessarily the tallest trees overall in the initial screening experiment; some of the clones performed well in all three treatments, while others were selected because they performed better in the 50% process water treatment than in the 25% and control treatments. Moreover, the EC level of the 50% process water in the greenhouse was between 3 and 3.6 mS cm$^{-1}$ [22]; however, the EC level of surface water from the end-pit lake was only 2.7–3.0 mS cm$^{-1}$ (salty water) [37], which suggests that the Al-Pac selected clones did not experience the same level of stress in the field as their previous screening showed they had the capability to sustain. Therefore, if the end-pit lake site was not, in fact, heavily impacted by salts, the selected high salt tolerant treatment clones that did the best in the greenhouse trial might not have had the opportunity to exhibit their superior 'salt tolerance', and thus to this point in time, looked similar to the clones represented in treatment 2 and in the Stream 1 vegetative lot.

From the current data, Stream 2 clones, selected for salt tolerance, had greater volume when compared with the Stream 1 wild lot, even though for height and DBH alone, clones from treatment 1 and treatment 2 showed growth increments that were interspersed with each other, and treatment 3 was close to the median in performance (Figure 7). As treatment 1 (Stream 2) clones showed salt tolerance in the greenhouse study and performed well in the field study, selecting and planting these trees may prove beneficial in the future if reclamation sites become more challenging. In addition, access to the Stream 2 clones from stoolbeds and/or existing trees could simplify collections while also ensuring the material can be planted over a much wider area (i.e., no seed zone restrictions) associated with the CPP.

### 4.2.2. Nutrient Analysis

Sodium and calcium were of particular interest in this experiment as they are the main drivers of potential 'salinity' conditions on reclamation sites [38]. White and Liber (2018) [37] characterized the chemical constituents in surface water from this end-pit lake, and found sodium was one of the main ions that contributed to salinity. However, foliar sodium levels in our current study were below any accurately detectable level and, as such, were reported as being <0.01% (Table 5a). Foliar calcium levels, which were in the normal value range (from 0.1% to 5%) [39], however, were higher than sodium levels in the samples (Table 5a). The low sodium levels measured may suggest that sodicity is currently not a concern, or that the poplars did not accumulate it in their leaves. The heavy metal accumulation patterns were similar to the published results under field conditions [40,41] (Table 5b), where Zn and Cd were accumulated in the leaves, indicating the phytoextraction ability of balsam poplar. The significant differences found in magnesium (Mg) levels among the three treatments indicate that the Stream 1 vegetative lot clones could be potentially used in a high Mg contaminated field. Overall, the trend from tissue nutrient analysis using block means (Table 7) showed the nutrient levels were higher in the block closest to the water. This finding indicated that a high-water table might also offer the opportunity for salt accumulation in both the immediately adjacent surrounding soil and eventually in the plant tissues that could affect tree growth. In addition, the data from the heavy metal analysis could be used as supplementary data that might be useful in the future as a source for comparison. However, there is very limited published literature that outlines the range of acceptable nutrient concentrations in balsam poplar leaves.

## 5. Conclusions

In consideration of the results obtained from both the greenhouse and field studies, there is an opportunity to select genetically suitable native clones of balsam poplar that are tolerant to challenging growing conditions, making them more suitable for planting in reclamation efforts on potentially saline sites than unselected clones or populations. The field testing indicated the potential use of selected Stream 2 clones (selected high salt-tolerant clones) from Al-Pac's balsam poplar controlled parentage program for oil sands reclamation sites in northeastern Alberta. In addition, the Stream 1 wild lot showed comparable growth performance under "ideal" conditions. However, if reclamation were being conducted on challenging, salty sites, clones from the Al-Pac selections are recommended. Furthermore, the selected salt-tolerant clones showed greater stem volumes, which indicates that they are potentially the most flexible trees as they will likely do better under higher salt conditions, and they will have a greater volume even when conditions are favourable. Balsam poplar has shown considerable genetic diversity in growth performance in this study and such results are encouraging in light of an expanding industrial energy sector footprint. Moreover, poplars are well known for their ability to tolerate salinity [17,18,42] and, therefore, screening clones for salt tolerance, and maximizing the potential use of the tree improvement program trees available through Al-Pac, while meeting government regulations for genetic diversity, could provide a significant opportunity for reclamation in the oil sands region in Alberta. Reclamation challenges are in their infancy in Alberta and adjacent regions, and selected material from native species may provide greater benefit as a source of reclamation materials than untested material to help meet those challenges.

**Author Contributions:** B.R.T. and D.K. conceived and designed the experiments. Y.H., D.K., and B.R.T. performed the experiments and analyzed the data. Y.H., B.R.T., and D.K. wrote the paper. R.K. contributed nutrient solutions and the aeroponic system in the greenhouse study. Y.H., D.K., R.K., and B.R.T. reviewed and edited the paper. All authors have read and agreed to the published version of the manuscript.

**Funding:** This research was funded by Syncrude Canada Ltd. through a contract with Alberta-Pacific Forest Industries Inc. (Al-Pac). In addition, Y.H. received funding through MITACS and Al-Pac.

**Data Availability Statement:** The data presented in this study are available on request from the corresponding author.

**Acknowledgments:** This experiment has been supported by Alberta-Pacific Forest Industries Inc. and received considerable logistic assistance from Syncrude Canada Ltd. We thank the MITACS program and Al-Pac for providing support to Y.H. to complete this project. Thanks, is also given to Nathalie Startsev, Chen Ding, and Marc Robbins for technical assistance in the greenhouse experiments. The authors wish to express their gratitude to Craig Farnden (Syncrude Canada Ltd.) for his assistance in field data collection and logistics. The authors also thank Ellen Macdonald (University of Alberta) for her valuable suggestions and assistance during the greenhouse phase of the project.

**Conflicts of Interest:** The authors declare no conflict of interest.

## Appendix A

**Table A1.** Mean total biomass (±SE, grams) of the top 30 balsam poplar clones in the 50% process water treatment with corresponding means (±SE) for the control and 25% process water treatments in 2012. Clone numbers in bold identify clones that ranked in the top 30 for all three water treatments, while underlined clone numbers indicated consistent performance in both years' experiments. Periods indicate dead plants.

| Clone | Control Mean (g) | ±SE | 25% Process Water Mean (g) | ±SE | 50% Process Water Mean (g) | ±SE |
|---|---|---|---|---|---|---|
| **4357** | 5.03 | 0.63 | 4.36 | 0.26 | 3.00 | 0.63 |
| **4326** | 3.76 | 1.65 | 3.84 | 0.25 | 2.99 | 0.85 |
| **2304** | 5.10 | 0.32 | 2.95 | 1.07 | 2.27 | 1.36 |
| 2453 | 1.49 | 0.05 | 1.57 | 1.11 | 1.77 | 0.45 |
| 4349 | 1.27 | 0.07 | 2.29 | | 1.48 | 0.68 |
| **3029** | 2.10 | 0.54 | 2.54 | 0.36 | 1.47 | |
| **2267** | 2.76 | 0.45 | 2.32 | | 1.43 | |
| 4304 | 1.20 | 1.96 | 1.96 | 0.28 | 1.43 | |
| **4301** | 2.66 | 0.46 | 2.57 | 0.81 | 1.42 | 0.55 |
| 3188 | 3.69 | 0.21 | 1.56 | 0.76 | 1.41 | 0.12 |
| **2995** | 3.26 | 0.47 | 2.60 | 0.40 | 1.39 | |
| **4363** | 2.47 | 0.15 | 2.02 | 0.11 | 1.38 | |
| 4255 | 1.92 | 0.49 | 1.02 | 0.10 | 1.36 | 0.10 |
| 4296 | 3.93 | 0.18 | 1.07 | 0.01 | 1.36 | 0.08 |
| **4334** | 5.62 | 0.15 | 2.58 | 0.47 | 1.35 | 0.07 |
| 4277 | 1.16 | 0.34 | 2.57 | 0.48 | 1.34 | 0.25 |
| 4295 | 1.63 | 0.29 | 1.26 | 0.73 | 1.32 | 0.61 |
| **2288** | 4.01 | 0.27 | 1.86 | 0.28 | 1.26 | 0.08 |
| 3110 | 1.70 | 1.24 | 0.99 | 0.09 | 1.22 | 0.08 |
| 4285 | 1.53 | 0.22 | 4.17 | | 1.19 | |
| 4315 | 0.23 | 0.71 | 2.15 | 0.24 | 1.19 | 0.43 |
| 3187 | 0.70 | 0.19 | 0.62 | 0.71 | 1.14 | . |
| 2447 | 1.63 | 0.03 | 0.24 | 0.25 | 1.13 | 0.17 |
| 4249 | 0.91 | 0.21 | 1.61 | 0.32 | 1.11 | 0.30 |
| 2976 | 1.26 | 1.18 | 0.54 | 1.81 | 1.09 | |
| 2312 | 0.41 | 0.97 | 2.47 | 0.91 | 1.08 | 0.31 |
| 4317 | 1.56 | 0.17 | 1.23 | 0.70 | 1.06 | |
| 4297 | 0.61 | | 0.36 | | 1.02 | 0.06 |
| 4356 | 0.66 | 1.09 | 0.89 | 0.02 | 1.02 | 0.12 |
| 3106 | 1.02 | 0.91 | 2.30 | 0.74 | 1.01 | 0.13 |
| **Treatment mean** | 1.42 | 0.08 | 1.17 | 0.06 | 0.77 | 0.04 |

**Table A2.** Mean total biomass (±SE, grams) of the top 30 balsam poplar clones in the 50% process water treatment with corresponding means (±SE) for the control and 25% process water treatments in 2013. Clone numbers in bold identify clones that ranked in the top 30 for all three water treatments, while underlined clone numbers indicated consistent performance in both years' experiments. Periods indicate dead plants.

| Clone | Control Mean (g) | ±SE | 25% Process Water Mean (g) | ±SE | 50% Process Water Mean (g) | ±SE |
|---|---|---|---|---|---|---|
| **2282** | 4.89 | 2.16 | 5.16 | 0.40 | 6.39 | 3.02 |
| **2314** | 7.85 | 0.56 | 9.82 | 6.16 | 5.62 | 0.27 |
| **2287** | 8.91 | 5.21 | 9.85 | 3.71 | 5.55 | 0.68 |
| **2272** | 5.80 | 0.39 | 6.26 | 0.98 | 5.24 | 0.87 |
| <u>**2304**</u> | 6.37 | 1.65 | 5.57 | 1.39 | 5.16 | 2.02 |
| **2313** | 3.92 | 1.21 | 4.81 | 2.03 | 5.01 | 0.24 |
| **2269** | 5.39 | 1.09 | 5.56 | 1.07 | 4.63 | 0.88 |
| **2291** | 5.88 | 4.99 | 6.35 | 2.91 | 3.86 | 0.98 |
| <u>**4326**</u> | 6.35 | 3.52 | 6.86 | 1.03 | 3.56 | 0.37 |
| 2301 | 3.40 | 1.95 | 2.26 | 1.63 | 3.43 | 0.72 |
| **2302** | 8.18 | 1.26 | 5.29 | 2.27 | 3.24 | 1.40 |
| **2289** | 3.67 | 1.89 | 2.97 | 1.13 | 3.14 | 1.34 |
| **2305** | 7.13 | 1.42 | 7.44 | 0.55 | 2.89 | 0.32 |
| **2300** | 6.42 | 1.91 | 5.61 | 1.34 | 2.88 | 0.47 |
| **2268** | 7.70 | 2.07 | 4.86 | 2.71 | 2.76 | 0.77 |
| 2278 | 4.85 | 1.46 | 2.81 | 1.15 | 2.63 | 0.77 |
| 3027 | 5.33 | 1.17 | 1.46 | 0.013 | 2.59 | 0.30 |
| **2307** | 4.71 | 2.63 | 3.86 | 1.47 | 2.58 | 0.64 |
| 2997 | 2.68 | 1.90 | 1.26 | 0.72 | 2.47 | 0.44 |
| 2284 | | | 1.985 | | 2.3 | |
| **2297** | 8.50 | 1.93 | 8.71 | 0.64 | 2.22 | 0.14 |
| **2295** | 4.20 | 1.55 | 3.27 | 0.35 | 2.21 | 0.32 |
| 4274 | 1.05 | 0.43 | 1.55 | 0.12 | 2.16 | |
| <u>4349</u> | 1.82 | 1.23 | 0.99 | 0.64 | 2.13 | 0.28 |
| 4283 | 3.05 | 0.64 | 0.88 | 0.51 | 2.08 | 0.41 |
| 2266 | 2.62 | 0.27 | 5.53 | 2.93 | 2.02 | 0.68 |
| **2303** | 4.93 | 0.75 | 4.16 | 0.90 | 1.99 | |
| 2293 | 3.61 | 0.47 | 3.52 | 0.87 | 1.99 | 0.26 |
| 915 | 4.27 | 0.86 | 2.70 | 1.63 | 1.97 | 0.30 |
| 3028 | 1.98 | 0.24 | 1.14 | 0.048 | 1.97 | 0.68 |
| Treatment mean | 3.32 | 0.19 | 2.88 | 0.21 | 1.90 | 0.10 |

**Table A3.** Growth variables of balsam poplar tested by ANOVA indicating the source of variation, F-value, degrees of freedom (df), and *p*-value for the 2012 and 2013 greenhouse experiments for different clones, treatments, and interaction effects. Trt = treatment.

| | | Growth Season | | | | | |
| | | 2012 | | | 2013 | | |
| Growth variable | Source of variation | F | df | *p*-value | F | df | *p*-value |
|---|---|---|---|---|---|---|---|
| Stem height (cm) | Clone | 5.26 | 144 | <0.001 | 9.24 | 85 | <0.001 |
| | Trt | 85.78 | 2 | <0.001 | 31.64 | 2 | <0.001 |
| | Clone * Trt | 1.17 | 237 | 0.091 | 0.83 | 169 | 0.929 |
| Basal diameter (mm) | Clone | 3.83 | 144 | <0.001 | 6.24 | 85 | <0.001 |
| | Trt | 7.86 | 2 | 0.004 | 23.28 | 2 | <0.001 |
| | Clone * Trt | 1.01 | 237 | 0.471 | 0.92 | 169 | 0.730 |
| Stem biomass (g) | Clone | 4.65 | 144 | <0.001 | 6.36 | 85 | <0.001 |
| | Trt | 51.64 | 2 | <0.001 | 35.82 | 2 | <0.001 |
| | Clone * Trt | 1.45 | 237 | 0.007 | 0.84 | 169 | 0.897 |
| Root biomass (g) | Clone | 4.6 | 144 | <0.001 | 4.29 | 85 | <0.001 |
| | Trt | 39.32 | 2 | <0.001 | 25.4 | 2 | <0.001 |
| | Clone * Trt | 1.23 | 237 | 0.039 | 0.75 | 169 | 0.985 |
| Leaf biomass (g) | Clone | 4.51 | 144 | <0.001 | 5.62 | 85 | <0.001 |
| | Trt | 39.56 | 2 | <0.001 | 30.31 | 2 | <0.001 |
| | Clone * Trt | 1.17 | 237 | 0.082 | 0.99 | 169 | 0.538 |
| Total biomass (g) | Clone | 4.97 | 144 | <0.001 | 5.77 | 85 | <0.001 |
| | Trt | 47.06 | 2 | <0.001 | 34.60 | 2 | <0.001 |
| | Clone * Trt | 1.28 | 237 | 0.016 | 0.87 | 169 | 0.841 |

**Table A4.** ANOVA for mean stem volume ($mm^3$) of balsam poplar in 2019 at the end-pit lake (treatment 1 = 25 selected tolerant clones, treatment 2 = 10 selected control clones, treatment 3 = Stream 1 vegetative lot clones).

| Source | DF | F Value | *p*-Value |
|---|---|---|---|
| Block | 2 | 19.95 | <0.0001 |
| Treatment | 2 | 5.03 | 0.0076 |
| Clone | 33 | 4.04 | <0.0001 |
| Error | 458 | | |
| Total | 495 | | |

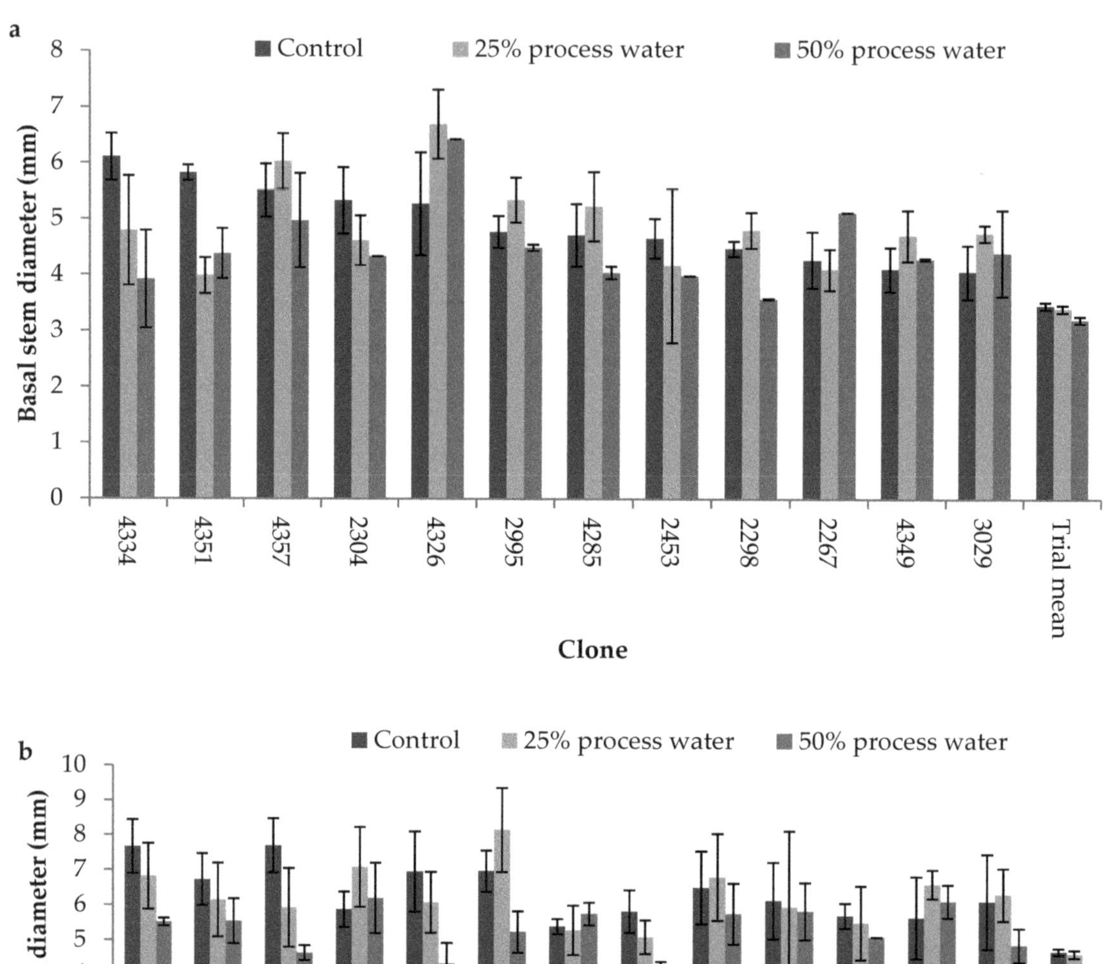

**Figure A1. (a).** Mean basal stem basal diameter (±SE, mm) of 12 balsam poplar clones that ranked in the top 30 for all three water treatments as compared with the overall trial mean for 148 clones after 44 days of growth in the 2012 experiment. **(b).** Mean basal stem basal diameter (±SE, mm) of 11 clones that ranked in the top 30 for all three water treatments as compared with the overall trial mean for 86 clones after 65 days of growth in the 2013 experiment.

**Figure A2.** Aerial view in winter of the end-pit lake showing the location (marked in red) of the trial adjacent to the lake edge.

**End-pit Lake**

North ↑

| | | | | | | | | | | | | | | | | | | | | | | | | | | | | | | | | | | | | | | | | |
|---|---|---|---|---|---|---|---|---|---|---|---|---|---|---|---|---|---|---|---|---|---|---|---|---|---|---|---|---|---|---|---|---|---|---|---|---|---|---|---|---|
| Rep 1 | 4283<br>361 | 2314<br>362 | 5578<br>363 | 5578<br>364 | 947<br>365 | 5578<br>366 | 3027<br>367 | 4643<br>368 | 3027<br>369 | 2304<br>370 | 5578<br>380 | 5578<br>379 | 4357<br>378 | 4304<br>377 | 5578<br>376 | 5578<br>375 | 4285<br>374 | 4274<br>373 | 2272<br>372 | 4296<br>371 | 2452<br>381 | 2887<br>382 | 5578<br>383 | 4643<br>384 | 4283<br>385 | 2453<br>386 | 4249<br>387 | 4255<br>388 | 4293<br>389 | 4314<br>390 | 3006<br>400 | 2997<br>399 | 4314<br>398 | 4277<br>397 | 5578<br>396 | 5578<br>395 | 3006<br>394 | 3006<br>393 | 427<br>392 | 3110<br>391 |
| | 3188<br>440 | 2314<br>439 | 4301<br>438 | 5578<br>437 | 4249<br>436 | 2452<br>435 | 4295<br>434 | 4295<br>433 | 947<br>432 | 5578<br>431 | 3027<br>430 | 4304<br>429 | 3002<br>428 | 4301<br>427 | 2997<br>426 | 5578<br>425 | 2282<br>424 | 5578<br>423 | 4277<br>422 | 3029<br>421 | 5578<br>411 | 2272<br>412 | 2304<br>413 | 4277<br>414 | 2976<br>415 | 5578<br>416 | 5578<br>417 | 2272<br>418 | 5578<br>419 | 5578<br>420 | 2976<br>410 | 4314<br>409 | 947<br>408 | 5578<br>407 | 5578<br>406 | 2314<br>405 | 5578<br>404 | 4357<br>403 | 557<br>402 | 5578<br>401 |
| | 4274<br>441 | 2887<br>442 | 4296<br>443 | 4301<br>444 | 3188<br>445 | 2995<br>446 | 4283<br>447 | 3004<br>448 | 4357<br>449 | 4301<br>450 | 3110<br>451 | 3002<br>452 | 4277<br>453 | 4285<br>454 | 2272<br>455 | 2887<br>456 | 4283<br>457 | 5578<br>458 | 5578<br>459 | 4352<br>460 | 2976<br>461 | 4296<br>462 | 2314<br>463 | 3110<br>464 | 5578<br>465 | 4295<br>466 | 5578<br>467 | 5578<br>468 | 5578<br>469 | 4326<br>470 | 5578<br>471 | 3004<br>472 | 2452<br>473 | 5578<br>474 | 4352<br>475 | 5578<br>476 | 3002<br>477 | 5578<br>478 | 299<br>479 | 3187<br>480 |
| | 5578<br>520 | 5578<br>519 | 2452<br>518 | 2453<br>517 | 3029<br>516 | 3188<br>515 | 4304<br>514 | 2453<br>513 | 3110<br>512 | 4285<br>511 | 5578<br>510 | 3187<br>509 | 3029<br>508 | 2304<br>507 | 5578<br>506 | 5578<br>505 | 5578<br>504 | 4255<br>503 | 3004<br>502 | 4255<br>501 | 2282<br>500 | 2997<br>499 | 5578<br>498 | 2304<br>497 | 5578<br>496 | 3002<br>495 | 4295<br>494 | 2997<br>493 | 3188<br>492 | 5578<br>491 | 2282<br>490 | 5578<br>489 | 4249<br>488 | 2453<br>487 | 4293<br>486 | 4352<br>485 | 5578<br>484 | 5578<br>483 | 557<br>482 | 5578<br>481 |
| | 5578<br>521 | 5578<br>522 | 3187<br>523 | 3027<br>524 | 5578<br>525 | 4293<br>526 | 4285<br>527 | 5578<br>528 | 4249<br>529 | 4326<br>530 | 4274<br>531 | 3029<br>532 | 4296<br>533 | 947<br>534 | 2995<br>535 | 4643<br>536 | 4293<br>537 | 5578<br>538 | 4352<br>539 | 2887<br>540 | 3187<br>541 | 5578<br>542 | 4255<br>543 | 5578<br>544 | 4326<br>545 | 5578<br>546 | 4326<br>547 | 3004<br>548 | 3006<br>549 | 4314<br>550 | 2995<br>551 | 4643<br>552 | 2976<br>553 | 4357<br>554 | 2282<br>555 | 5578<br>556 | 5578<br>557 | 5578<br>558 | 430<br>559 | 5578<br>560 |
| Rep 2 | 5578<br>561 | 2887<br>562 | 4301<br>563 | 4293<br>564 | 4249<br>565 | 5578<br>566 | 4326<br>567 | 3029<br>568 | 4296<br>569 | 2314<br>570 | 3187<br>571 | 5578<br>572 | 3188<br>573 | 4357<br>574 | 5578<br>575 | 5578<br>576 | 4283<br>577 | 2453<br>578 | 5578<br>579 | 2997<br>580 | 4301<br>581 | 4352<br>582 | 3006<br>583 | 3029<br>584 | 2282<br>585 | 5578<br>586 | 2452<br>587 | 3002<br>588 | 4326<br>589 | 2453<br>590 | 4357<br>591 | 2995<br>592 | 5578<br>593 | 4301<br>594 | 3110<br>595 | 5578<br>596 | 4296<br>597 | 2997<br>598 | 427<br>599 | 5578<br>600 |
| | 2995<br>640 | 5578<br>639 | 2997<br>638 | 3027<br>637 | 5578<br>636 | 5578<br>635 | 5578<br>634 | 5578<br>633 | 4314<br>632 | 5578<br>631 | 5578<br>630 | 2272<br>629 | 4277<br>628 | 2314<br>627 | 2452<br>626 | 2452<br>625 | 5578<br>624 | 5578<br>623 | 5578<br>622 | 2453<br>621 | 4643<br>620 | 3027<br>619 | 2304<br>618 | 5578<br>617 | 3188<br>616 | 3006<br>615 | 4255<br>614 | 5578<br>613 | 5578<br>612 | 5578<br>611 | 5578<br>610 | 4296<br>609 | 5578<br>608 | 4249<br>607 | 3004<br>606 | 2976<br>605 | 5578<br>604 | 4277<br>603 | 557<br>602 | 5578<br>601 |
| | 5578<br>641 | 4295<br>642 | 5578<br>643 | 4326<br>644 | 2272<br>645 | 4643<br>646 | 5578<br>647 | 4301<br>648 | 3004<br>649 | 4283<br>650 | 2997<br>651 | 4255<br>652 | 5578<br>653 | 2304<br>654 | 5578<br>655 | 3006<br>656 | 3002<br>657 | 4293<br>658 | 4274<br>659 | 5578<br>660 | 2995<br>661 | 2314<br>662 | 2453<br>663 | 4285<br>664 | 5578<br>665 | 2887<br>666 | 4255<br>667 | 4304<br>668 | 5578<br>669 | 4296<br>670 | 4277<br>671 | 4643<br>672 | 4314<br>673 | 2314<br>674 | 947<br>675 | 2976<br>676 | 4357<br>677 | 4314<br>678 | 557<br>679 | 5578<br>680 |
| | 4352<br>720 | 4643<br>719 | 4314<br>718 | 2304<br>717 | 4285<br>716 | 5578<br>715 | 3110<br>714 | 4274<br>713 | 4285<br>712 | 4249<br>711 | 3029<br>710 | 4283<br>709 | 5578<br>708 | 4295<br>707 | 4357<br>706 | 3027<br>705 | 4293<br>704 | 5578<br>703 | 4274<br>702 | 5578<br>701 | 4352<br>700 | 4304<br>699 | 4283<br>698 | 4326<br>697 | 5578<br>696 | 5578<br>695 | 2272<br>694 | 4304<br>693 | 2976<br>692 | 2976<br>691 | 5578<br>690 | 2272<br>689 | 2282<br>688 | 3187<br>687 | 947<br>686 | 5578<br>685 | 4295<br>684 | 3002<br>683 | 311<br>682 | 5578<br>681 |
| | 5578<br>721 | 4352<br>722 | 2282<br>723 | 4285<br>724 | 2304<br>725 | 5578<br>726 | 5578<br>727 | 5578<br>728 | 5578<br>729 | 2452<br>730 | 947<br>731 | 3002<br>732 | 3110<br>733 | 4304<br>734 | 4277<br>735 | 3004<br>736 | 2995<br>737 | 2887<br>738 | 2282<br>739 | 4295<br>740 | 4249<br>741 | 3188<br>742 | 3187<br>743 | 3029<br>744 | 4293<br>745 | 5578<br>746 | 5578<br>747 | 4255<br>748 | 3187<br>749 | 3006<br>750 | 3027<br>751 | 2887<br>752 | 5578<br>753 | 3188<br>754 | 5578<br>755 | 5578<br>756 | 5578<br>757 | 5578<br>758 | 947<br>759 | 3004<br>760 |
| Rep 3 | 5578<br>761 | 4357<br>762 | 5578<br>763 | 4304<br>764 | 5578<br>765 | 4293<br>766 | 4285<br>767 | 4285<br>768 | 3188<br>769 | 2304<br>770 | 4357<br>771 | 5578<br>772 | 5578<br>773 | 4277<br>774 | 4301<br>775 | 3110<br>776 | 4301<br>777 | 4293<br>778 | 5578<br>779 | 2887<br>780 | 5578<br>781 | 2314<br>782 | 5578<br>783 | 4295<br>784 | 5578<br>785 | 5578<br>786 | 4296<br>787 | 2453<br>788 | 4295<br>789 | 5578<br>790 | 3002<br>791 | 5578<br>792 | 4314<br>793 | 3002<br>794 | 5578<br>795 | 3004<br>796 | 5578<br>797 | 2997<br>798 | 464<br>799 | 3029<br>800 |
| | 4277<br>840 | 4283<br>839 | 4285<br>838 | 3188<br>837 | 4249<br>836 | 4296<br>835 | 5578<br>834 | 4352<br>833 | 2995<br>832 | 4357<br>831 | 3002<br>830 | 4314<br>829 | 2314<br>828 | 2997<br>827 | 4304<br>826 | 2452<br>825 | 4352<br>824 | 4643<br>823 | 3029<br>822 | 5578<br>821 | 5578<br>820 | 4352<br>819 | 4274<br>818 | 3006<br>817 | 947<br>816 | 5578<br>815 | 4357<br>814 | 2453<br>813 | 3110<br>812 | 4296<br>811 | 5578<br>810 | 2452<br>809 | 2282<br>808 | 5578<br>807 | 5578<br>806 | 4285<br>805 | 5578<br>804 | 5578<br>803 | 429<br>802 | 2272<br>801 |
| | 4304<br>841 | 5578<br>842 | 5578<br>843 | 4283<br>844 | 5578<br>845 | 4295<br>846 | 3110<br>847 | 5578<br>848 | 3004<br>849 | 3006<br>850 | 2995<br>851 | 947<br>852 | 5578<br>853 | 4277<br>854 | 3027<br>855 | 2976<br>856 | 5578<br>857 | 5578<br>858 | 5578<br>859 | 2887<br>860 | 5578<br>861 | 947<br>862 | 2453<br>863 | 2452<br>864 | 2272<br>865 | 3027<br>866 | 5578<br>867 | 4326<br>868 | 5578<br>869 | 4326<br>870 | 2282<br>871 | 5578<br>872 | 4255<br>873 | 5578<br>874 | 4326<br>875 | 4255<br>876 | 2314<br>877 | 3188<br>878 | 288<br>879 | 4314<br>880 |
| | 5578<br>920 | 5578<br>919 | 2997<br>918 | 5578<br>917 | 5578<br>916 | 4283<br>915 | 5578<br>914 | 3187<br>913 | 5578<br>912 | 5578<br>911 | 2976<br>910 | 2997<br>909 | 3027<br>908 | 5578<br>907 | 2453<br>906 | 2304<br>905 | 2995<br>904 | 2282<br>903 | 2282<br>902 | 5578<br>901 | 4274<br>900 | 4249<br>899 | 4326<br>898 | 3029<br>897 | 2887<br>896 | 5578<br>895 | 3187<br>894 | 4643<br>893 | 5578<br>892 | 2272<br>891 | 4293<br>890 | 3006<br>889 | 3027<br>888 | 4643<br>887 | 5578<br>886 | 4314<br>885 | 5578<br>884 | 5578<br>883 | 557<br>882 | 2272<br>881 |
| | 4249<br>921 | 4301<br>922 | 2976<br>923 | 3187<br>924 | 3006<br>925 | 5578<br>926 | 2314<br>927 | 2976<br>928 | 4283<br>929 | 5578<br>930 | 5578<br>931 | 4352<br>932 | 2995<br>933 | 4295<br>934 | 3029<br>935 | 5578<br>936 | 3187<br>937 | 4301<br>938 | 4304<br>939 | 4274<br>940 | 3004<br>941 | 4274<br>942 | 947<br>943 | 4249<br>944 | 3004<br>945 | 3002<br>946 | 5578<br>947 | 5578<br>948 | 4277<br>949 | 2304<br>950 | 3110<br>951 | 5578<br>952 | 3188<br>953 | 5578<br>954 | 4293<br>955 | 2452<br>956 | 4255<br>957 | 5578<br>958 | 425<br>959 | 2304<br>960 |

XXXX — Clone Number
xxx — Position Number

**Treatment 1** (25 selected tolerant clones): 2272, 2282, 2304, 2314, 2453, 2976, 2995, 2997, 3027, 3029, 3110, 3187, 3188, 4249, 4255, 4274, 4277, 4283, 4285, 4295, 4296, 4301, 4304, 4326, 4357

**Treatment 2** (10 selected control clones): 947, 2452, 2887, 3002, 3004, 3006, 4352, 4293, 4314, 4643

**Treatment 3** (Stream 1 vegetative lot clones): 5578

**Figure A3.** Map of the field trial site showing individual tree locations within each block (rep).

# References


1. Government of Alberta. Oil Sands Facts and Statistics. Available online: https://www.alberta.ca/oil-sands-facts-and-statistics.aspx/ (accessed on 21 December 2020).
2. Natural Resources Canada. Oil Sands: Land Use and Reclamation. Available online: https://www.nrcan.gc.ca/energy/publications/18740/ (accessed on 21 December 2020).
3. Fung, M.Y.P.; Macyk, T.M. Reclamation of Oil Sands Mining Areas. In *Reclamation of Drastically Disturbed Lands*, 2nd ed.; Barnhisel, R.I., Darmody, R.G., Lee Daniels, W., Eds.; American Society of Agronomy Monograph: Madison, WI, USA, 2000; Volume 41, p. 1082.
4. MacKinnon, M.D. Development of the tailings pond at Syncrude's oil sands plant, 1978–1987. *AOSTRA J. Res.* **1989**, *5*, 109–133.
5. Province of Alberta. Chapter E-12. In *Environmental Protection and Enhancement Act*; Revised Statues of Alberta; Alberta Queen's Printer: Edmonton, AB, Canada, 2017; p. 168.
6. Del Rio, L.F.; Hadwin, A.K.M.; Pinto, L.J.; MacKinnon, M.D.; Moore, M.M. Degradation of naphthenic acids by sediment micro-organisms. *J. Appl. Microbiol.* **2006**, *101*, 1049–1061. [CrossRef] [PubMed]
7. Province of Alberta. Chapter R-17.3. In *Responsible Energy Development Act*; Statutes of Alberta; Alberta Queens Printer: Edmonton, AB, Canada, 2014; p. 44.
8. Renault, S.; Lait, C.; Zwiazek, J.J.; MacKinnon, M.D. Effect of high salinity tailings waters produced from gypsum treatment of oil sands tailings on plants of the boreal forest. *Environ. Pollut.* **1998**, *102*, 177–184. [CrossRef]
9. Howat, D.R. *Acceptable Salinity, Sodicity and pH Values for Boreal Forest Reclamation*; Alberta Environment; Environmental Sciences Division: Edmonton, AB, Canada, 2000; p. 191.
10. Barbour, S.L.; Chanasyk, D.; Hendry, J.; Leskiw, L.; Macyk, T.; Mendoza, C.; Naeth, A.; Nichol, C.; O'Kane, M.; Purdy, B.; et al. *Soil Capping Research in the Athabasca Oil Sands Region Volume 1: Technology Synthesis*; Syncrude Canada Ltd.: Fort McMurray, AB, Canada, 2007; p. 175.
11. Munns, R.; Tester, M. Mechanisms of salinity tolerance. *Annu. Rev. Plant Biol.* **2008**, *59*, 651–681. [CrossRef]
12. Isebrands, J.G.; Karnosky, D.F. Environmental benefits of poplar culture. In *Poplar Culture in North America*; Dickmann, D.I., Isebrands, J.G., Eckenwalder, J.E., Richardson, J., Eds.; NRC-CNRC Press: Ottawa, ON, Canada, 2001; pp. 207–218.
13. Licht, L.A.; Isebrands, J.G. Linking phytoremediated pollutant removal to biomass economic opportunities. *Biomass Biol.* **2005**, *28*, 203–218. [CrossRef]
14. Richardson, J.; Cook, J.E.K.; Isebrands, J.G.; Thomas, B.R.; Van Rees, K.C.J. Poplar research in Canada—A historical perspective with a view to the future. *Can. J. Bot.* **2007**, *85*, 1136–1146. [CrossRef]
15. Ma, H.C.; Fung, L.; Wang, S.S.; Altman, A. Photosynthetic response of *Populus euphratica* to salt stress. *For. Ecol. Manag.* **1997**, *93*, 55–61. [CrossRef]
16. Gorden, J.C. Poplars: Trees of the people, trees of the future. *For. Chron.* **2001**, *77*, 217–219. [CrossRef]
17. Sixto, H.; Aranda, I.; Grau, J.M. Assessment of salt tolerance in Populus alba clones using chlorophyll fluorescence. *Photosynthetica* **2006**, *44*, 169–173. [CrossRef]
18. Imada, S.; Yamanaka, N.; Tamai, S. Effects of salinity on the growth, Na partitioning, and Na dynamics of a salt-tolerant tree, *Populus alba* L. *J. Arid Environ.* **2009**, *73*, 245–251. [CrossRef]
19. Liu, H.; Tamai, S.; Furukawa, I. The influence on the growth and biomass of *Populus alba* L. under saline irrigation in sandy soil. *Appl. For. Sci.* **2001**, *10*, 37–44.
20. DesRochers, A.; Thomas, B.R.; Butson, R. Reclamation roads and landings with balsam poplar cuttings. *For. Ecol. Manag.* **2004**, *199*, 39–50. [CrossRef]
21. Johnson, C.M.; Scout, P.R.; Broyer, T.C.; Carlton, A.B. Comparative chlorine requirements of different plant species. *Plant Soil* **1957**, *7*, 337–353. [CrossRef]
22. Thomas, B.; Kamelchuk, D.; Macdonald, E.; Hu, Y.; Krygier, R. *Screening Populus Balsamifera Clones for Use in Reclamation on Challenging Sites*; Internal Document; Alberta-Pacific Forest Industries Inc.: Boyle, AB, Canada, 2013; p. 26.
23. Alberta Agriculture and Forestry (Forestry Division) (AAF). *Alberta Forest Genetic Resource Management and Conservation Standards (FGRMS)*; Alberta Agriculture and Forestry; Government of Alberta: Edmonton, AB, Canada, 2016; p. 165.
24. Climate FortMcMurray. Available online: https://en.climate-data.org/north-america/canada/alberta/fort-mcmurray-4623/ (accessed on 25 April 2021).
25. Thomas, B.R.; Kamelchuk, D.; Dietrich, S. Native balsam poplar clones for use in reclamation of salt-impacted sites. In *COSIA Land EPA 2017 Mine Site Reclamation Research Report*; Canada's Oil Sands Innovation Alliance (COSIA): Calgary, AB, Canada, 2018; pp. 162–163.
26. Tailings. Available online: https://www.syncrude.ca/sustainability/tailings/ (accessed on 26 March 2021).
27. Alberta Agriculture and Forestry (Alberta Tree Improvement and Seed Center) (AAF). *Trial Measurement Manual: Best Practices Field Procedures for Tree Measurements in Genetic Field Trials*; Alberta Agriculture and Forestry; Government of Alberta: Edmonton, AB, Canada, 2016; p. 22.
28. SAS Institute Inc. *Base SAS®9.4 Procedures Guide: Statistical Procedures*, 2nd ed.; SAS Institute Inc.: Cary, NC, USA, 2013; p. 558.

29. Brown, K.; van den Driessche, R. Growth and nutrition of hybrid poplars over 3 years after fertilization at planting. *Can. J. For. Res.* **2002**, *32*, 226–232. [CrossRef]

30. Mirck, J.; Volk, T.A. Response of three shrub willow varieties (*Salix* spp.) to storm water treatments with different concentrations of salts. *Bioresour. Technol.* **2010**, *101*, 3484–3492. [CrossRef] [PubMed]

31. Algreen, M.; Trapp, S.; Rein, A. Phytoscreening and phytoextraction of heavy metals at Danish polluted sites using willow and poplar trees. *Environ. Sci. Pollut. Res.* **2014**, *21*, 8992–9001. [CrossRef]

32. Walker, R.R. Sodium exclusion and potassium-sodium selectivity in salt treated trifoliate orange (*Poncirus triliata*) and cleopatra mandarin (*Citrus reticulata*) plants. *Aust. J. Plant Physiol.* **1986**, *13*, 293–303.

33. Negrão, S.; Schmöckel, S.M.; Tester, M. Evaluating physiological responses of plants to salinity stress. *Ann. Bot.* **2017**, *119*, 1–11. [CrossRef] [PubMed]

34. Flowers, T.J.; Colmer, T.D. Salinity tolerance in halophytes. *New Phytol.* **2008**, *179*, 945–963. [CrossRef]

35. Zalesny, R.S.; Riemenschneider, D.E.; Hall, R.B. Early rooting of dormant hardwood cuttings of *Populus*: Analysis of quantitative genetics and genotype × environment interactions. *Can. J. For. Res.* **2005**, *35*, 918–929. [CrossRef]

36. Landhäusser, S.M.; Silins, U.; Lieffers, V.J.; Liu, W. Responses of *Populus tremuloides*, *Populus balsamifera*, *Betula papyrifera* and *Picea glauca* seedlings to low soil temperature and water-logged soil conditions. *Scand. J. For. Res.* **2003**, *18*, 391–400. [CrossRef]

37. Tisdale, S.L.; Nelson, W.L.; Beaton, J.D.; Havlin, J.L. *Soil Fertility and Fertilizers*; MacMillan Publishing Co. Inc.: New York, NY, USA, 1993; p. 648.

38. White, K.B.; Liber, K. Early chemical and toxicological risk characterization of inorganic constituents in surface water from the Canadian oil sands first large-scale end pit lake. *Chemosphere* **2018**, *211*, 745–757. [CrossRef]

39. White, P.J.; Broadley, M.R. Calcium in plants. *Ann. Bot.* **2003**, *92*, 487–511. [CrossRef] [PubMed]

40. Hu, Y.; Nan, Z.; Su, J.; Wang, N. Heavy metal accumulation by poplar in calcareous soil with various degrees of multi-metal contamination: Implications for phytoextraction and phytostabilization. *Environ. Sci. Pollut. Res.* **2013**, *20*, 7194–7203. [CrossRef] [PubMed]

41. Kacálková, L.; Tlustoš, P.; Száková, J. Phytoextraction of risk elements by willow and poplar trees. *Int. J. Phytoremediat.* **2015**, *17*, 414–421. [CrossRef]

42. Gordon, M.P.; Choe, N.; Duffy, J.; Ekuan, G.; Heilman, P.E.; Muizniecks, I.; Newman, L.A.; Ruszaj, M.; Shurtleff, B.B.; Strand, S.E.; et al. Phytoremediation of trichloroethylene with hybrid poplars. *Environ. Health Perspect.* **1998**, *106*, 1001–1004. [PubMed]

 *forests*

*Article*

# High Biomass Productivity of Short-Rotation Willow Plantation in Boreal Hokkaido Achieved by Mulching and Cutback


Qingmin Han [1],[*],[†] , Hisanori Harayama [1] , Akira Uemura [1],[†] , Eriko Ito [1] , Hajime Utsugi [2] , Mitsutoshi Kitao [1] and Yutaka Maruyama [3]

[1]  Hokkaido Research Center, Forestry and Forest Products Research Institute (FFPRI), 7 Hitsujigaoka, Toyohira, Sapporo 062-8516, Hokkaido, Japan; harahisa@ffpri.affrc.go.jp (H.H.); akirauem@ffpri.affrc.go.jp (A.U.); iter@ffpri.affrc.go.jp (E.I.); kitao@ffpri.affrc.go.jp (M.K.)

[2]  Principal Research Director, FFPRI, 1 Matsunosato, Tsukuba 305-8687, Ibaraki, Japan; utsugi@ffpri.affrc.go.jp

[3]  Department of Forest Science and Resources, Nihon University, 1866 Kameino, Fujisawa 252-8510, Kanagawa, Japan; maruyama.yutaka@nihon-u.ac.jp

[*]  Correspondence: qhan@ffpri.affrc.go.jp; Tel./Fax: +81-29-829-8220

[†]  Present address: Department of Plant Ecology, FFPRI, 1 Matsunosato, Tsukuba 305-8687, Ibaraki, Japan.



Received: 12 March 2020; Accepted: 27 April 2020; Published: 1 May 2020



**Abstract:** Weed control, which is commonly achieved by herbicides, is important in successfully establishing short-rotation coppice (SRC) of willow. In this study, we examined agricultural mulch film as a means of effective weed control and the influence of cutback practice (coppicing the first year's shoot growth in the winter following planting) on biomass production in boreal Hokkaido, Japan. One-year-old cuttings from two clones each of *Salix pet-susu* and *S. sachalinensis* were planted in double-rows at a density of 20,000 plants $ha^{-1}$. All plants were harvested three growing seasons after cutback. Average oven-dried biomass yield was 5.67 t $ha^{-1}$ $yr^{-1}$ with mulching, whereas it was 0.46 t $ha^{-1}$ $yr^{-1}$ in the unmulched control with a weed biomass of 4.13 t $ha^{-1}$ $yr^{-1}$, indicating that mulching was an effective weed control. However, weeds grew vigorously on the ground between mulch sheets and their dry biomass amounted to 0.87 t $ha^{-1}$ $yr^{-1}$. Further weeding between the mulch sheets enhanced the willow biomass yield to 10.70 t $ha^{-1}$ $yr^{-1}$ in the treatment with cutback. In contrast, cutback even reduced the willow yield when there were weeds between the mulch sheets. This negative effect of cutback on the willow yield resulted from nutrient competition with weeds; there was similar leaf nitrogen content and dry biomass per unit land area for the weeds and willows combined in the control and mulching treatments. These results suggest that growing SRC willow is feasible in boreal Hokkaido if combined with complete weed control and cutback, and is facilitated by using mulch film.

**Keywords:** cutback; mulch; *Salix*; short-rotation coppice; weed control; woody biomass


---

## 1. Introduction

Woody biomass production is an economically viable and ecologically sound solution to address increasing energy demands; it also has positive effects on reducing global atmospheric $CO_2$ and, consequently, the greenhouse effect. Willows (*Salix* spp.) are one of the best species for short-rotation coppice (SRC) in temperate climates because they are easily propagated vegetatively, they have a high yield potential in a few years, a broad genetic base, and they can re-sprout from their coppiced stools after harvesting [1]. However, despite their general plasticity, the adoption of willows as a bioenergy production system remains a challenge in respect to high yield [2]. In particular, weed control is of top priority in order to achieve a long-term high yielding SRC willow plantation [3–6]. It is extremely important because weeds have a negative effect on the SRC willows as they compete for light, water, and nutrients, and they consequently reduce the survival rate of cuttings and biomass production [2,7]. Furthermore, the response to weed competition may differ between species and clones, as do growth patterns and competitive abilities, just as soil properties and climate differ between plantations [8]. To date, willow selection and breeding have focused on high biomass production under optimum conditions, and no commercial clone has been found to have high biomass productivity in the presence of competition from weeds [2,7,8]. Therefore, development of environmentally friendly, efficient, and cheap weed control measures might be the best way forward for the establishment of SRC willow plantations.

The application of mulches in agriculture and horticulture has increased dramatically in the last two decades throughout the world. This increase is due to benefits such as suppressed weed growth, soil moisture conservation, reducing certain insect pests, higher crop yield, earlier harvests, improved fruit quality, and more efficient use of soil nutrients [9–15]. The selection of an appropriate mulching material depends on crop type, crop management practices, and climatic conditions. Of the various materials being used as mulches, plastics are most commonly used in agriculture and horticulture. However, to our knowledge, there has been little research on the use of plastic mulch in plantations [16–18], whereas a report demonstrated that plastic mulch is an excellent way of promoting willow productivity by regulating water content in the soil and increasing soil temperature and controlling weeds on the poorly drained soil [17].

Coppicing the first year's shoot growth in the winter following planting, which is hereafter called cutback, is another common practice during the establishment of SRC willow plantations. This is mainly done to promote multiple stem sprouting and to facilitate fertilization and additional weeding during the second growth season [4–6,19,20]. However, recent studies indicate that this practice may reduce biomass productivity [8,20], and is no longer recommended in Sweden [21]. The factors and mechanisms determining the effect of cutback on biomass production are not fully understood. The reduced canopy due to cutback may provide weeds with new establishment opportunities and increase the need for further weed control during the years following establishment. Consequently, willow biomass production would decrease. Therefore, more studies are needed to determine whether cutback affects the ability of willows to compete with weeds, especially at the beginning of the second growing season, since some weed species may start to grow earlier than the willow. The interaction between cutback and weeds in relation to biomass production may be also be linked to resource limitation within plantations [22–24]. In this respect, information on nitrogen acquisition and allocation between willows and weeds would provide an insight into the effect of cutback practice on willow production [23,25,26].

Compared to European and North American countries, SRC willow cultivation is still a developing field in Japan [27–29]. The Feed-in Tariff Policy for renewable energy implemented in 2012 after the Fukushima Daiichi Nuclear Power Plant accident has become a driver, and the demand for wood resources has increased rapidly in recent years. Considering the above factors and their influence on the fast establishment of SRC willow plantations and long-term biomass productivity, we evaluated plastic mulch as a method of weed control. We further tested cutback and its interaction with weed control on the biomass production of *Salix pet-susu* and *S. sachalinensis* in the first harvest cycle, three growing seasons after cutback. There are no commercial willow cultivars in Japan, and these two species were selected because they are widely distributed in Northeastern region of Honshu and Hokkaido [30].

## 2. Materials and Methods

### 2.1. Study Site and Experimental Design

This study was carried out in a former grassland located near the Sanru river in Shimokawa town, Hokkaido (44°25′ N, 142°42′ E). The parent material of the soil is quaternary alluvial deposits with rounded or subrounded rock fragments covering the surface [31]. Soil nitrate nitrogen averaged from surface to 40 cm depth was about 4.6 mg per kg soil [31]. During the period 1978–2016, mean annual precipitation was 918.4 mm, mean diurnal mean, maximum, and minimum temperatures were 5.1 °C, 10.3 °C, and −0.5 °C, respectively, obtained at a nearby meteorological station (44°18′ N, 142°38′ E, 143 m a.s.l.; Japanese Bureau of Meteorology). Average days with snow cover depth more than 3 cm at the same station amounted to 150 during the period 1984–2016. Further details about precipitation and temperature during the period 2013–2016 (the first harvest cycle) are presented in Figure S1.

To control weeds, plastic mulch film (150 cm width and 0.021 mm thickness; Sunshat, C.I. TAKIRON Corporation, Tokyo, Japan) was laid using an agricultural machine, following the crushing of surface rock fragments to 20 cm soil depth using a mechanical stone crusher in autumn of 2012 (Figure 1a). We selected this polyethylene mulch because of its excellent strength and weather resistance. In May 2013, one-year-old 20 cm long unrooted cuttings—the length most commonly recommended in most European and northern American countries [3–5,21,32,33]—were planted in a double-row arrangement (Figure 1a). The distances were 1.5 m between the double rows, 0.5 m between rows, and 0.5 m between cuttings within a row, resulting in a density of 20,000 cuttings per hectare. This system is used worldwide in order to facilitate the management of sites using farm machinery [3–5]. Each clone (either P81 and P82 from *S. pet-susu* or S27 and S67 from *S. sachalinensis*) was planted in separate mulched rows next to each other (Figure 1b). In addition, a row without mulch was included as a natural control to confirm the effect of weed competition on biomass productivity. An electric fence was installed around the site to protect the willows from predation by deer and rabbits.

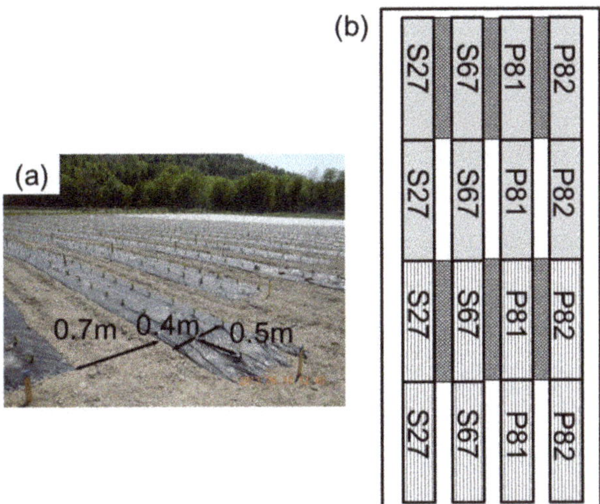


**Figure 1.** Experimental design: (**a**) double row planting of 20 cm long cuttings through mulch films, (**b**) layout of cutback and weed control treatments of two clones of *Salix sachalinensis* (S27, S67) and *S. pet-susu* (P81, P82) within a block. Vertical lines represent cutback and grey columns represent non-cutback treatments. White areas between the mulched areas represent weed-control treatments and black cross-hatching represent non-weed-control treatments, respectively.

The cutback treatment was conducted over half of the site in order to examine its effect on biomass production of SRC willow. In the winter following planting (the end of 2013), the year's shoots were cut back using pruning shears in order to stimulate development of multiple stems on each plant in the next growing season. Although no weeds grew on soils covered by mulch, weeds grew vigorously on the ground between the mulch sheets in the first growing season. Therefore, half of this site was sprayed with herbicides (Roundup Maxload, Tokyo, Nissan Chemical Corporation) in May 2014 after weeds stated growing. A month later, we confirmed that all weeds had died.

Therefore, four treatments were established in a five-block design (Figure 1b) before the second growing season: with or without cutback and/or weed control. There were 20 cuttings in each treatment within a block.

### 2.2. Allometric Equations

Twelve sprouted stems from *S. sachalinensis* were cut down to 0.03 m above the ground in August 2015 to enable us to generate allometric equations based on the main-axis cutting method [34]. Immediately after felling, the total height ($H$) and basal diameter at 0.03 m ($D_0$) of each stem were measured. $D_0$ ranged from 13.85 to 41.15 mm, and $H$ ranged from 221 to 445 cm. Each stem was sealed in a plastic bag and stored in an air-conditioned car until taken back to the laboratory. The main stems were divided into 0.50 m lengths. The length and basal diameter ($D$) of each stem section was measured and organs attached to the main stem were classified into leaves and branches. All samples were then oven-dried at 75 °C for 72 h and weighed. Allometric equations were estimated for biomass of stem, branch, and foliage of each stem in terms of $D_0$, $H$, and $D$ (Equations in Supplementary Material).

### 2.3. Growth Measurements and Harvesting

In November 2016 after leaf senescence, three growing seasons after the cutback treatment, all stems of both *S. sachalinensis* and *S. pet-susu* were harvested. Fresh biomass of aboveground organs including branches and stems from each cutting were weighed immediately. Sub-samples of branches and stems were taken back to the laboratory for biomass production determination, expressed in oven-dried tonnes per hectare per year (t ha$^{-1}$ yr$^{-1}$). The ratio of dry biomass to fresh biomass was 0.5132, 0.5326, 0.5268, and 0.5474 for P81, P82, S27, and S67, respectively. Neither fertilization nor pesticide were applied during the first harvest cycle.

In order to examine the effect of cutback and weed control on the biomass increment process each year before harvesting, the $D_0$ and $H$ from each cutting of *S. sachalinensis* were measured in November 2014. All stems were revisited in November 2015 to measure $D_0$ and $H$, as well as diameter at the height of the lowest branch ($D$). The number of sprouted stems from each cutting of *S. sachalinensis* was counted in November 2014.

### 2.4. Nitrogen Analysis

For nitrogen analysis, mature leaves from the aforementioned twelve sprouted stems used to develop the allometric equations were sampled. On the same occasion, weeds from the ground between mulches in treatments without weed control were harvested in three plots (50 cm × 70 cm) from each treatment.

All samples from leaves of both willow and weed were dried to constant mass at 75 °C and ground to a fine powder in a steel ball mill (MM400; Retsch, Haan, Germany). Total nitrogen concentration of the fine powder derived from each sample was measured after combustion in a CHN Analyzer (Vario Max CN, Elementar, Hanau, Germany).

### 2.5. Statistical Analyses

The effects of cutback, weeding, and species and their interactions on all dependent variables were analyzed by a two way or three way repeated-measures analysis of variance (ANOVA) using SigmaPlot 13.0 (Systat Software Inc., San Jose, CA, USA). We tested for normality and variance homogeneity at $p < 0.05$, and all variables were log-transformed when needed. The effect of the block on all variables was not significant at $p < 0.05$, and thus, these results are not shown. The Holm–Sidak method procedures were used for pairwise multiple comparison when significant treatment effects were revealed.

## 3. Results

Mulch had a significant effect on biomass production in the first harvest cycle, three growing seasons after cutback (Table 1). Without mulch, weeds grew fast and its height could reach up to 120 cm in August, which resulted in 69% of unrooted cuttings of willow dying. Therefore, the average biomass production of willows was 0.46 t ha$^{-1}$ yr$^{-1}$, which was even lower than weed production (4.13 t ha$^{-1}$ yr$^{-1}$). In contrast, willow biomass production increased to 5.67 t ha$^{-1}$ yr$^{-1}$ when mulches were used to control weeds, although weeds between the rows of willows growing in the ground not covered by mulch produced about 0.87 t ha$^{-1}$ yr$^{-1}$. Both total plant biomass and leaf nitrogen content had similar values per unit land area in the control and mulched treatment.

**Table 1.** Annual average biomass production and leaf nitrogen content of both weeds and willow in land under natural conditions (no weed control) or plastic mulch. Values shown are mean ± SE.

| Treatment | Plant Type | Biomass (t ha$^{-1}$ yr$^{-1}$) | Leaf Nitrogen (%) | Leaf Nitrogen (kg ha$^{-1}$ yr$^{-1}$) |
|---|---|---|---|---|
| Natural conditions | Weeds | 4.13 ± 0.15 | 1.36 ± 0.07 | 56.22 ± 0.16 |
|  | Willow | 0.46 ± 0.05 | na | na |
| Mulch | Weeds | 0.87 ± 0.11 | 1.84 ± 0.08 | 15.40 ± 1.27 |
|  | Willow | 5.67 ± 0.28 | 2.39 ± 0.10 | 57.29 ± 0.22 |

With further weed control on the ground between mulches, willow biomass production increased to 11.43 and 8.84 t ha$^{-1}$ yr$^{-1}$ in the cutback treatment in *S. sachalinensis* and *S. pet-susu*, respectively (Figure 2). Without weed control, in contrast, cutback had no effect on biomass production in *S. pet-susu* and actually reduced yield by about 32% in *S. sachalinensis*. In addition, cutback and weed control had a significant interaction with respect to productivity (Figure 2, Table S1). Since the canopy of the willows closed during the second growing season after cutback, further weed control was not necessary.

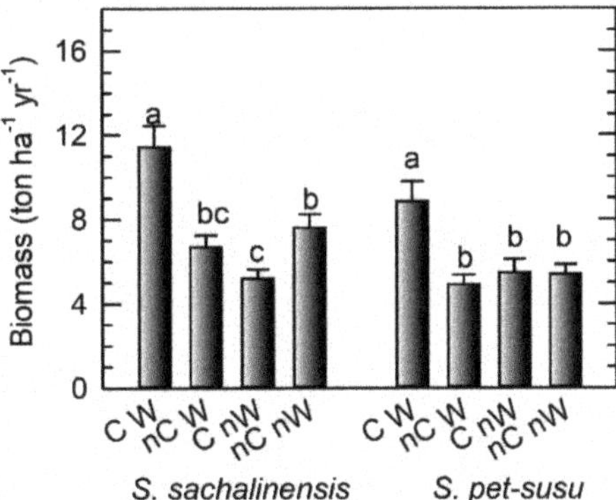


**Figure 2.** Comparison of dry biomass production in the first harvest cycle after three growing seasons: C and nC represent with and without cutback; W and nW represent with and without weed control, respectively. The significance values of the factorial analysis are shown in Table S1. Different letters indicate significant differences in the corresponding values from the same species at $p < 0.05$. Values shown are mean ± SE from two clones of the same species.

Cutback practice even reduced biomass increase without weed control on the ground between the mulch sheets (Figure 3, Table S2). This negative effect was especially obvious in branches and stems when comparing pairs without weed control in the first growing season (Figure 3b,c). In the second growing season, the negative effect of cutback became insignificant in leaves and branches, but was still significant in stems. Stem biomass increase in the second growing season exhibited a three-fold increase compared to the first growing season in all treatments.

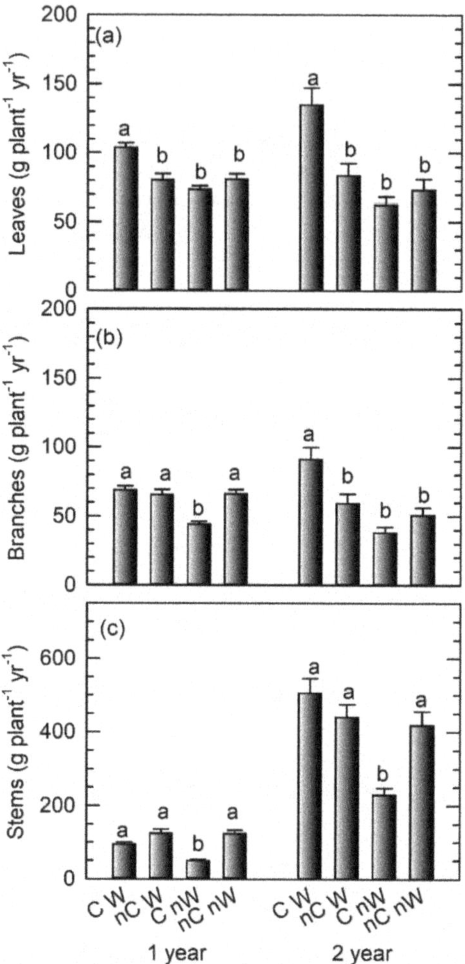



**Figure 3.** Annual biomass increments of (**a**) leaves, (**b**) branches, and (**c**) stems of *Salix sachalinensis* in the first and second growing seasons after cutback. C and nC represent with and without cutback; W and nW represent with and without weed control, respectively. The significance values of the factorial analysis are shown in Table S2. Different letters indicate significant differences in the corresponding values in a single year at $p < 0.05$. Values shown are mean ± SE from two clones of the same species.

Cutback promoted the formation of multiple stems (Figure 4, Table S1). Interestingly, a significant interaction between cutback and weed control was observed with respect to the number of stems per cutting, indicating that weed control is important for cutback to enhance productivity.

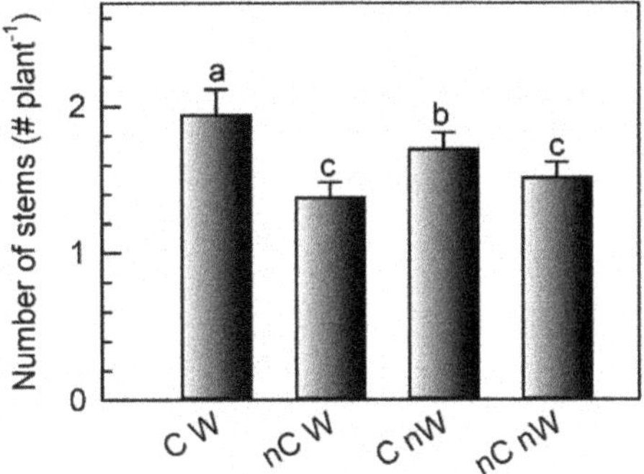

**Figure 4.** Number of stems sprouting from a single stool, counted after one season of growth since cutback in *Salix sachalinensis*. C and nC represent with and without cutback; W and nW represent with and without weed control, respectively. The significance values of the factorial analysis are shown in Table S1. Different letters indicate significant differences in the corresponding values at $p < 0.05$. Values shown are mean ± SE from two clones of the same species.

## 4. Discussion

Weed control is commonly achieved by one or two applications of herbicide during land preparation and planting for establishing an SRC willow plantation [3–6]. This study demonstrated that agricultural mulches are also an effective measure to control weeds in the boreal region studied, which has been less studied in forest plantations in comparison with agriculture and horticulture [16–18]. With complete weed suppression combined with cutback, annual average dry biomass production amounted to 10.70 t $ha^{-1}$ $yr^{-1}$ in the first harvest cycle, three growing seasons after cutback. However, cutback had no or even a negative effect on biomass production in the presence of weed competition, as found in previous studies [2,8]. These results suggest that cutback should only be conducted when weeds are absent.

### 4.1. Weed Control by Agricultural Mulch Film

In spite of mulching, weeds grew vigorously on the ground between mulch sheets. This adversely affected the growth of willows and as a result the biomass production in the first harvest cycle was reduced by 32% in *S. sachalinensis*. These results suggest that weeds on the ground between mulches also need to be controlled before cutback. In previous studies, the extent to which weeds influenced willow growth was reported to depend on both willow species and weed species [8,25]. In the present study, yield difference between *S. sachalinensis* and *S. pet-susu* provides further evidence for this conclusion. In addition, we found that total leaf nitrogen content per unit land area from weeds and willows combined was similar in the unweeded control and the mulched treatment. Thus, total biomass production per unit land area in the control was similar to that in the mulched treatment. These results suggest that competition for nutrients with weeds is the main cause of yield decrease at this site. Weed-related growth reduction may also differ depending on the plantation in terms of different soil nutrient limitation [2,35].

No weeds were observed in the second growing season after cutback because the willow canopy had closed, meaning that weed control conducted once during the establishment stage of a SRC plantation is sufficient. However, the negative effect of cutback on the biomass increment of stems remained as previously reported by Albertsson et al. [8], suggesting that weed control during the establishment year is extremely important in order to obtain a high yield SRC willow plantation [3–6]. Even though weed-related growth reduction decreases over time, the economic returns during the expected life cycle of a plantation will probably be much less if weeds are not controlled [8]. Yields over subsequent harvest cycles should be monitored, however, to validate this statement.

Mulches used in this study remained in good condition after the first harvest. Removal of the plastic is time-consuming (about 16 h/ha) [9]. In the long run, mulches are broken into pieces, some pieces being buried during soil preparation for a new crop and some remaining on the soil surface. The buried pieces are more difficult to decompose since they are less affected by light and high temperatures, creating serious soil problems whose environmental repercussion has not been fully evaluated. Considering that no weeds grow in the second growing season after cutback, biodegradable plastic mulch in organic production may be an environmentally friendly alternative to control weeds that deserves further research [9].

*4.2. Cutback*

When weeds between mulch sheets were not controlled, cutback had no effect on biomass production in *S. pet-susu* and actually reduced yield by about 32% in the first harvest cycle in *S. sachalinensis.* This negative effect has recently been reported in other willow species [8,20]. Therefore, the commonly recommended practice of cutback during establishment of a willow plantation should only be undertaken when there is sufficient weed control in place. A recent study showed that competition with weeds resulted in reduced willow growth at lower levels of fertilizer application compared to higher levels [7]. In this study, there was little difference in total leaf biomass and leaf nitrogen per land area between the unweeded control and the mulched treatment, indicating the same demand for nitrogen between weeds and willow. This limited nitrogen supply for willows in the presence of weeds resulted in lower leaf and branch increments in the cutback treatment and reduced biomass yield in the first harvest cycle.

*S. sachalinensis,* which had relatively higher growth reduction in the cutback and no weeding treatment compared to *S. pet-susu*, exhibited higher biomass production under optimum conditions (complete weed control). Similar results were reported previously for 10 commercial breeding clones in southern Sweden [8]. The relatively low biomass production even without weed competition may partly explain the low growth reduction with weed competition. In this respect, commercial breeding has been mainly aimed at the goal of high biomass production, and this is mostly considered under optimum environmental conditions. Willow species with high biomass productivity may have high nutrient demands and, thus, their productivity may decrease when exposed to competition from weeds. Therefore, further studies on physiological traits, such as nutrient and water acquisition and their use efficiency, canopy structure and development, and biomass allocation between aboveground organs and roots of different species and clones, are critical for the effective management of SRC plantations [22,25,36,37]. All these factors can be manipulated through genetic improvement and silvicultural practice [38].

Cutback increased the number of stems produced by each stool in *S. sachalinensis*, in agreement with previous studies [20]. In our previous study, the number of sprouted stems of the same species produced in a nursery in Sapporo was higher than in this study [27]. This difference is probably related to differences in climatic conditions because the mean annual temperature in Sapporo is 2 °C higher than at the current site, indicating that local climatic factors should be taken into account when planning to establish an SRC willow plantation and considering biomass yield and economic return. The number of stems from the experimental willows in this study overlapped the lower end of commercial cultivars in other countries [4–6,17,20,33], indicating the importance of breeding for commercial cultivars to achieve the highest possible yields that are also competitive with weeds [38].

The average oven dry biomass production recorded for the two willows in this study falls within the range of European and North American countries, which is 7.0–18.0 t ha$^{-1}$ yr$^{-1}$ [2,20,21,39–41]. Yields depend on cultivars, site-specific conditions (soil fertility, climate), and type and intensity of management (fertilization, weeding, and protection from blight and insect). For example, fertilized willow SRC plantations had on average 38% higher yield than non-fertilized SRC plantations in Sweden [2]. In addition, it is generally found that yields are higher from the second harvest cycle [19,39,40]. However, nutrient export in harvested biomass over multiple rotations will require soil nutrient amendments to maintain the SRC willow productivity that deserves further research [41,42].

## 5. Conclusions

This study demonstrated that agricultural mulches are an effective measure for controlling weeds during the establishment of an SRC willow plantation. However, weeds between mulch sheets had negative effects on yield, especially in treatments with cutback, which was resulted from nutrient competition with weeds. Cutback practice should be used only when weeds are completely controlled. With complete weed control, 10.70 t dry biomass per hectare was achieved three years after cutback, indicating the feasibility of successfully establishing an SRC willow plantation in boreal Hokkaido, Japan.

**Supplementary Materials:** Supplementary Materials can be found online at http://www.mdpi.com/1999-4907/11/5/505/s1.

**Author Contributions:** Conceptualization, Q.H., H.H., A.U. and H.U.; Methodology, Q.H., H.H., A.U. and H.U.; Formal Analysis, Q.H.; Investigation, Q.H., H.H., A.U., E.I., M.K., Y.M. and H.U.; Writing—Original Draft Preparation, Q.H.; Writing—Review & Editing, Q.H., H.H., A.U., E.I., M.K., Y.M. and H.U.; Project Administration, Q.H., A.U. and H.U.; Funding Acquisition, Q.H. and H.U. All authors have read and agreed to the published version of the manuscript.

**Acknowledgments:** The authors thank Haruka Yamamoto for the biomass measurements, Yuji Takahashi of Shimokawa town for maintenance the experimental site, and Oji Holdings Co. for offering the willow cuttings. The authors thanks also the editor and the two anonymous reviewers for their valuable comments. This work was supported by an FFPRI Grant.

**Conflicts of Interest:** The authors declare no conflict of interest.

## References

1. Volk, T.A.; Verwijst, T.; Tharakan, P.J.; Abrahamson, L.P.; White, E.H. Growing fuel: A sustainability assessment of willow biomass crops. *Front. Ecol. Environ.* **2004**, *2*, 411–418. [CrossRef]

2. Dimitriou, I.; Rosenqvist, H.; Berndes, G. Slow expansion and low yields of willow short rotation coppice in Sweden; implications for future strategies. *Biomass Bioenerg.* **2011**, *35*, 4613–4618. [CrossRef]

3. *Manual for SRC Willow Growers*; Lantmännen Agroenergi: Stockholm, Sweden, 2012; Available online: http://www.voederbomen.nl/wordpress/wp-content/uploads/2012/08/ManualSRCWillowGrowers.pdf (accessed on 22 March 2020).

4. Abrahamson, L.P.; Volk, T.A.; Smart, L.B.; Cameron, K.D. *Shrub Willow Biomass Producser's Handbook*; State University of New York College of Environmental Science and Forestry (SUNY-ESF): Syracuse, NY, USA, 2010; Available online: https://www.esf.edu/willow/documents/ProducersHandbook.pdf (accessed on 22 March 2020).

5. Snowdon, K.; Mclvor, I.; Nicholas, I. *Energy Farming with Willow in New Zealand*; IEA Bioenergy Tasks: Paris, France, 2010.

6. Caslin, B.; Finnan, J.; Johnston, C.; McCracken, A.; Walsh, L. *Short Rotation Coppice Willow Best Practice Guidelines*. 2010. Available online: https://www.ifa.ie/wp-content/uploads/2013/10/Willow-Best-Practic-Guide-Teagasc-2010.pdf (accessed on 22 March 2020).

7. Edelfeldt, S.; Lundkvist, A.; Forkman, J.; Verwijst, T. Establishment and early growth of willow at different levels of weed competition and nitrogen fertilization. *Bioenerg. Res.* **2016**, *9*, 763–772. [CrossRef]

8. Albertsson, J.; Verwijst, T.; Hansson, D.; Bertholdsson, N.O.; Åhman, I. Effects of competition between short-rotation willow and weeds on performance of different clones and associated weed flora during the first harvest cycle. *Biomass Bioenerg.* **2014**, *70*, 364–372. [CrossRef]

9. Kasirajan, S.; Ngouajio, M. Polyethylene and biodegradable mulches for agricultural applications: A review. *Agron. Sustain. Dev.* **2012**, *32*, 501–529. [CrossRef]

10. Brodhagen, M.; Peyron, M.; Miles, C.; Inglis, D.A. Biodegradable plastic agricultural mulches and key features of microbial degradation. *Appl. Microbiol. Biotechnol.* **2015**, *99*, 1039–1056. [CrossRef]

11. Greer, L.; Dole, J.M. Aluminum foil, aluminium-painted, plastic, and degradable mulches increase yields and decrease insect-vectored viral diseases of vegetables. *Horttechnology* **2003**, *13*, 276–284. [CrossRef]

12. Shimazaki, M.; Nesumi, H. A Method for High-Quality Citrus Production Using Drip Fertigation and Plastic Sheet Mulching. *Jarq-Jpn. Agric. Res. Q.* **2016**, *50*, 301–306. [CrossRef]

13. Haapala, T.; Palonen, P.; Korpela, A.; Ahokas, J. Feasibility of paper mulches in crop production: A review. *Agric. Food Sci.* **2014**, *23*, 60–79. [CrossRef]

14. Tarara, J.M. Microclimate modification with plastic mulch. *HortScience* **2000**, *35*, 169–180. [CrossRef]

15. Adhikari, R.; Bristow, K.L.; Casey, P.S.; Freischmidt, G.; Hornbuckle, J.W.; Adhikari, B. Preformed and sprayable polymeric mulch film to improve agricultural water use efficiency. *Agric. Water Manag.* **2016**, *169*, 1–13. [CrossRef]

16. Riege, D.A.; Sigurgeirsson, A. Facilitation of afforestation by *Lupinus nootkatensis* and by black plastic mulch in south-west Iceland. *Scand. J. For. Res.* **2009**, *24*, 384–393. [CrossRef]

17. Labrecque, M.; Teodorescu, T.I.; Babeux, P.; Cogliastro, A.; Daigle, S. Impact of herbaceous competition and drainage conditions on the early productivity of willows under short-rotation intensive culture. *Can. J. For. Res.* **1994**, *24*, 493–501. [CrossRef]

18. Heiska, S.; Tikkanen, O.P.; Rousi, M.; Julkunen-Tiitto, R. Bark salicylates and condensed tannins reduce vole browsing amongst cultivated dark-leaved willows (*Salix myrsinifolia*). *Chemoecology* **2007**, *17*, 245–253. [CrossRef]

19. Guidi, W.; Pitre, F.E.; Labrecque, M. Short-rotation coppice of willows for the production of biomass in Eastern Canada. In *Biomass—Now Sustainable Growth and Use*; Matovic, M.D., Ed.; IntechOpen: London, UK, 2013. [CrossRef]

20. Zamora, D.S.; Apostol, K.G.; Wyatt, G.J. Biomass production and potential ethanol yields of shrub willow hybrids and native willow accessions after a single 3-year harvest cycle on marginal lands in central Minnesota, USA. *Agrofor. Syst.* **2014**, *88*, 593–606. [CrossRef]

21. Verwijst, T.; Lundkvist, A.; Edelfeldt, S.; Albertsson, J. *Development of Sustainable Willow Short Rotation Forestry in Northern Europe*; IntechOpen: London, UK, 2013; p. 44392.

22. Rytter, R.-M. Biomass production and allocation, including fine-root turnover, and annual N uptake in lysimeter-grown basket willows. *For. Ecol. Manag.* **2001**, *140*, 177–192. [CrossRef]

23. Weih, M. Evidence for increased sensitivity to nutrient and water stress in a fast-growing hybrid willow compared with a natural willow clone. *Tree Physiol.* **2001**, *21*, 1141–1148. [CrossRef]

24. Weih, M.; Bonosi, L.; Ghelardini, L.; Ronnberg-Wastljung, A.C. Optimizing nitrogen economy under drought: Increased leaf nitrogen is an acclimation to water stress in willow (*Salix* spp.). *Ann. Bot.* **2011**, *108*, 1347–1353. [CrossRef]

25. Brereton, N.J.B.; Pitre, F.E.; Shield, I.; Hanley, S.J.; Ray, M.J.; Murphy, R.J.; Karp, A. Insights into nitrogen allocation and recycling from nitrogen elemental analysis and $^{15}$N isotope labelling in 14 genotypes of willow. *Tree Physiol.* **2014**, *34*, 1252–1262. [CrossRef]

26. Vigl, F.; Rewald, B. Size matters?—The diverging influence of cutting length on growth and allometry of two Salicaceae clones. *Biomass Bioenerg.* **2014**, *60*, 130–136. [CrossRef]

27. Han, Q.; Harayama, H.; Uemura, A.; Ito, E.; Utsugi, H. The effect of the planting depth of cuttings on biomass of short rotation willow. *J. For. Res.* **2017**, *22*, 131–134. [CrossRef]

28. Utsugi, H.; Uemura, A.; Ishihara, M.; Takahashi, Y.; Ohara, M.; Hashida, K.; Makino, R.; Nojiri, M.; Magara, K.; Ikeda, T.; et al. *Biomass Utilization from a Field of Willow*; FFPRI: Sapporo, Japan, 2011; Available online: http://www.ffpri.affrc.go.jp/hkd/research/documents/h23biomass.pdf. (In Japanese)

29. Maruyama, Y.; Mori, S.; Kitao, M.; Tobita, H.; Koike, T. Effects of fertilization on photosynthetic traits and yield of two willow species. *Jpn. J. For. Environ.* **2002**, *44*, 71–75. (In Japanese)

30. Niiyama, K. The role of seed dispersal and seedling traits in colonization and coexistence of Salix species in a seasonally flooded habitat. *Ecol. Res.* **1990**, *5*, 317–331. [CrossRef]

31. Ito, E.; Uemura, A.; Harayama, H.; Han, Q.; Utsugi, H. Root system development and partitioning of nitrogen in short-rotation willow. *Boreal For. Res.* **2015**, *63*, 47–50. (In Japanese)

32. Rossi, P. Length of cuttings in establishment and production of short-rotation plantations of *Salix* 'Aquatica'. *New For.* **1999**, *18*, 161–177. [CrossRef]

33. Verwijst, T.; Lundkvist, A.; Edelfeldt, S.; Forkman, J.; Nordh, N.-E. Effects of clone and cutting traits on shoot emergence and early growth of willow. *Biomass Bioenerg.* **2012**, *37*, 257–264. [CrossRef]

34. Chiba, Y. Plant form analysis based on the pipe model theory. II. Quantitative analysis of ramification in morphology. *Ecol. Res.* **1990**, *6*, 21–28. [CrossRef]

35. Weih, M.; Nordh, N.-E. Characterising willows for biomass and phytoremediation: Growth, nitrogen and water use of 14 willow clones under different irrigation and fertilisation regimes. *Biomass Bioenerg.* **2002**, *23*, 397–413. [CrossRef]

36. Tharakan, P.J.; Volk, T.A.; Nowak, C.A.; Ofezu, G.J. Assessment of canopy structure, light interception, and light-use efficiency of first year regrowth of shrub willow (*Salix* sp.). *Bioenerg. Res.* **2008**, *1*, 229–238. [CrossRef]

37. Weih, M.; Rönnberg-Wästjung, A.-C. Shoot biomass growth is related to the vertical leaf nitrogen gradient in Salix canopies. *Tree Physiol.* **2007**, *27*, 1551–1559. [CrossRef] [PubMed]

38. Hanley, S.J.; Karp, A. Genetic strategies for dissecting complex traits in biomass willows (*Salix* spp.). *Tree Physiol.* **2014**, *34*, 1167–1180. [CrossRef] [PubMed]

39. Aylott, M.J.; Casella, E.; Tubby, I.; Street, N.R.; Smith, P.; Taylor, G. Yield and spatial supply of bioenergy poplar and willow short-rotation coppice in the UK. *New Phytol.* **2008**, *178*, 358–370. [CrossRef] [PubMed]

40. Wang, Z.; MacFarlane, D.W. Evaluating the biomass production of coppiced willow and poplar clones in Michigan, USA, over multiple rotations and different growing conditions. *Biomass Bioenerg.* **2012**, *46*, 380–388. [CrossRef]

41. Amichev, B.Y.; Hangs, R.D.; Konecsni, S.M.; Stadnyk, C.N.; Volk, T.A.; Bélanger, N.; Vujanovic, V.; Schoenau, J.J.; Moukoumi, J.; Van Rees, K.C.J. Willow Short-Rotation Production Systems in Canada and Northern United States: A Review. *Soil Sci. Soc. Am. J.* **2014**, *78*, S168–S182. [CrossRef]

42. Hangs, R.D.; Schoenau, J.J.; Van Rees, K.C.J.; Bélanger, N.; Volk, T.; Jensen, T. First rotation biomass production and nutrient cycling within short-rotation coppice willow plantations in Saskatchewan, Canada. *Bioenerg. Res.* **2014**, *7*, 1091–1111. [CrossRef]

 *forests*

*Article*

# Estimation of Yield Loss Due to Deer Browsing in a Short Rotation Coppice Willow Plantation in Northern Japan

**Hisanori Harayama** [1,*], **Qingmin Han** [2], **Makoto Ishihara** [1], **Mitsutoshi Kitao** [1], **Akira Uemura** [2], **Shozo Sasaki** [1], **Takeshi Yamada** [1,†], **Hajime Utsugi** [2] **and Yutaka Maruyama** [3]

[1]  Hokkaido Research Center, Forestry and Forest Products Research Institute (FFPRI), 7 Hitsujigaoka, Toyohira, Sapporo 062-8516, Hokkaido, Japan; makolin@ffpri.affrc.go.jp (M.I.); kitao@ffpri.affrc.go.jp (M.K.); shozos@ffpri.affrc.go.jp (S.S.); kenchan@ffpri.affrc.go.jp (T.Y.)

[2]  FFPRI, 1 Matsunosato, Tsukuba 305-8687, Ibaraki, Japan; qhan@ffpri.affrc.go.jp (Q.H.); akirauem@ffpri.affrc.go.jp (A.U.); utsugi@ffpri.affrc.go.jp (H.U.)

[3]  Department of Forest Science and Resources, Nihon University, 1866 Kameino, Fujisawa 252-8510, Kanagawa, Japan; maruyama.yutaka@nihon-u.ac.jp

*  Correspondence: harahisa@ffpri.affrc.go.jp

†  Current address: FFPRI, 1 Matsunosato, Tsukuba, Ibaraki 305-8687, Japan.

Received: 18 June 2020; Accepted: 24 July 2020; Published: 26 July 2020

**Abstract:** Deer browsing is a major factor causing significant declines in yield in short rotation coppice (SRC) willow, but the resultant yield loss is difficult to estimate because it requires extensive investigation, especially when the standard yield is unknown. We investigated a simple method for estimating yield loss due to deer browsing. We enclosed an experimental SRC willow plantation in Hokkaido, northern Japan, planted with 12 clones, with an electric fence; deer browsing did, however, occur in the first summer of the second harvest cycle. We counted the number of sprouting stems and deer-browsed stems per plant and, after three years, the yield of each clone was analyzed using a generalized linear model with the above two parameters for the numbers of stems as explanatory variables. The model explained the yield of 11 out of the 12 clones, and estimated that browsing of a single stem per plant could reduce yield to 80%. Losses due to deer browsing were estimated to be as much as 6.0 oven dry ton ha$^{-1}$ yr$^{-1}$. The potential yield in the absence of deer browsing ranged from 2.2 to 7.5 oven dry ton ha$^{-1}$ yr$^{-1}$ among clones, and was significantly positively correlated with the estimated yield loss due to deer browsing. Our results suggest that a generalized linear model can be used to estimate the yield loss due to deer browsing from a simple survey, and deer browsing could significantly reduce willow biomass yield from the clones we studied, and thus countermeasures to control deer browsing are therefore necessary if sufficient willow biomass yield is to be produced.

**Keywords:** deer browsing; *Salix*; short rotation coppice; woody biomass; yield loss

---

## 1. Introduction

Global warming, which is mainly attributable to greenhouse gases such as $CO_2$ [1], has recently led to increased efforts to utilize woody biomass energy from managed forests, which are treated as carbon neutral, in Europe and the United States [2,3], as well as in Japan [4]. Short rotation coppice (SRC) willow is the most successful woody biomass energy crop in the cool temperate regions of Europe and North America [5–10]; under favorable conditions, it typically yields a biomass of over 10 oven dry ton (odt) ha$^{-1}$ yr$^{-1}$, with a harvest rotation interval of 2–5 years. This capacity is associated with the coppicing ability of willow, which produces multiple stems and vigorous post-harvesting regrowth from the stump, resulting in repeated biomass yields of 20–50 odt ha$^{-1}$ every 2–5 years over a plantation life span of 20–25 years, with no need for replanting [11,12]. SRC willow can be established on marginal land that is generally unsuitable for food crops [13–15]. In addition to producing biomass for use in the bioenergy and biofuel industries, SRC willow provides positive ecosystem services, such as carbon sequestration in the roots, biodiversity, and water quality [16,17]. However, there are countries or regions where SRC willow has not become established commercially, such as Japan, despite the existence of suitable environmental conditions. This is because of risks and uncertainties concerning its production, management, and marketing, which need to be resolved if the use of SRC willow is to be promoted [18].

One cause of significant biomass loss in SRC willow plantations is the extensive damage due to browsing by large animals, such as deer [19–21]. Smaller SRC willow plantations are more susceptible to deer browsing, which is therefore a greater obstacle to commercial viability, especially in the early stages of SRC willow introduction, when plantations tend to be small [19]. The construction of fences as animal deterrents is effective but expensive [12,20]; in order to evaluate its cost effectiveness, it is therefore important to be able to estimate the yield loss due to deer browsing. In areas where commercial SRC willow has been widely developed, such as in northern Europe, potential yield by region for each willow variety is known; yield loss due to deer browsing can therefore be easily estimated by comparing the browsed actual yield with potential yield in the absence of browsing. In contrast, in areas where SRC willow is underdeveloped and potential yield is not known, estimating yield loss from deer browsing is difficult and requires more extensive investigation by, for example, individually estimating shoot losses based on shoot bite diameters using an allometry equation between shoot diameter and biomass [22].

The use of woody biomass for energy production has progressed in Japan since the Great East Japan Earthquake in 2011 [4,23], but SRC willow is still not commercially cultivated. Hokkaido, the northernmost of Japan's four main islands, is potentially suitable for growing SRC willow because of its flat terrain and cool temperate climate with sufficient rainfall. Additionally, its cold winter temperatures increase the demand for local heat and/or combined heat and power plants, which are typical destinations of harvested SRC willow [19]. In fact, in experimental plantations in Hokkaido, willow yields of more than 10 odt ha$^{-1}$ yr$^{-1}$ can be obtained with proper cultivation management [24,25]. Meanwhile, Yezo-sika deer (*Cervus nippon yesoensis*) density has increased rapidly in Hokkaido since the late 1980s [26], and the species is currently considered overabundant [27]. Since the late 1990s, Yezo-sika deer damage to agriculture and forestry has been severe [26], with recent economic impacts estimated at approximately 4 billion yen per year [28]. In general, susceptibility to deer damage varies among willow clones and is one of the key properties considered in selecting cultivars in Europe and the United States [29–31]; the selection of appropriate willow cultivars is still under development in Japan, and thus there is limited information on the susceptibility of willow clones to deer damage as well as their potential yield.

Our main objectives were to estimate: (1) the amount of yield loss due to deer browsing in willow clones with unknown potential yields, using a simple method counting the number of browsed and non-browsed stems; and (2) the yield of each clone in the absence of deer browsing—that is, potential yield.

## 2. Materials and Methods

### 2.1. Study Site

The study was conducted at an experimental SRC willow plantation in Shimokawa town, Hokkaido prefecture, northern Japan (44°25′ N, 142°42′ E, 249 m a.s.l., Figure 1a). This plantation was established to modify and improve SRC willow management procedures [24] and to select high yield clones. The parent material of soil in the study site was quaternary alluvial sediments, with rounded and sub-rounded fragments covering the surface [32]. The study site was 56 m × 60 m, almost level, and was plowed and harrowed in late October 2012. In May 2013, one-year old cuttings (length, approximately 20 cm) of *Salix pet-susu* Kimura and *S. sachalinensis* F.Schmidt clones ($n$ = 7 and 5, respectively) were planted by hand in a double-row system at a density of 20,000 cuttings ha$^{-1}$; the distance between and within double rows was 1.5 and 0.5 m, respectively, and plant spacing within rows was 0.5 m (Figure 1b). The cuttings were collected from experimental plantations in Shimokawa town and Sapporo city (Figure 1a) a few days before planting and refrigerated in plastic bags. Since willow clones appropriate for SRC were still under selection and no commercial cuttings existed in Japan, there was no guarantee the cuttings used in this study were high-yielding. All planting rows were sheeted with a plastic mulch (width, 1.5 m, and thickness, 0.021 mm; Sunshat, C.I. TAKIRON Corporation, Tokyo, Japan) to suppress weed growth. However, we did not conduct additional weeding between adjacent plastic mulches, except for half of the S.I.27, S.I.67, P.I.81, and P.I.82 clones, resulting in lower yields due to weed overgrowth in the first harvest rotation [24]. A total of 28 double rows (approximately 60 m long) were planted, each comprising plants of a single clone; the rows were randomly arranged (Figure 1c). Almost all the willow plants were cut back in November 2013 to promote sprouting, except for half of the S.I.27, S.I.67, P.I.81, and P.I.82 clones, so that the impact of the cut back procedure on biomass yield could be evaluated [24]. All plants were harvested in November 2016 using chain and hand saws, and then coppiced in spring 2017. The study site was enclosed by an electric fence, except in the snow period (November to April) to avoid destruction of the fence by snow coverage, usually over 1 m.

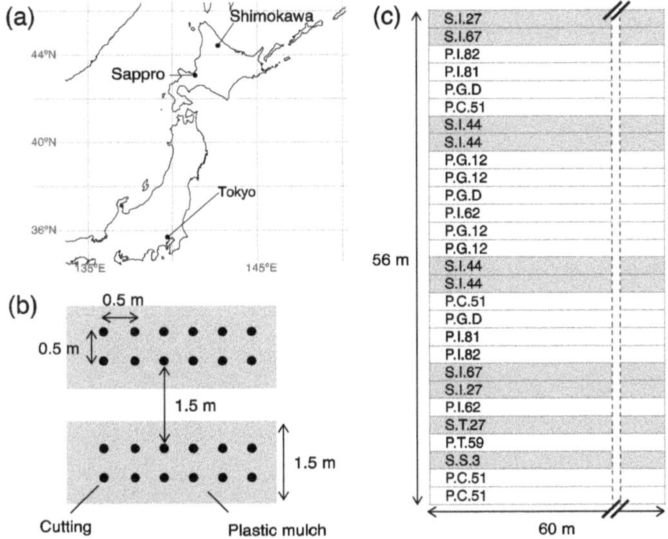


**Figure 1.** Study site and experimental design. Location of the study site in Shimokawa town, Japan (**a**), planting design with double-row system (**b**), and arrangement of willow clones in the 28 rows (**c**). Clone names beginning with "P" (white rows) and "S" (gray rows) in panel (**c**) represent *Salix pet-susu* and *S. sachalinensis*, respectively.

### 2.2. Deer Browsing Damage to Sprouting Willow Stems

In August 2017, the first year of sprouting, we found deer browsing damage to willow stems following a problem with the electric fence. Similar willow damage during growing season was previously reported to be due to browsing damage by Yezo-sika deer in a study using sensor cameras at the same study area [33]. In November 2017, we investigated the impact on the plants. An initial visual inspection suggested that browsing damage differed significantly among double rows containing different clones, and that damage to individual plants in a single-clone double-row was generally the same throughout the 60-m length (Figure 2a). Therefore, in each of the 28 double rows, the total numbers of stems and sprouting stems browsed by deer were counted for each plant positioned 5 m from the edge, that is, typically 20 plants per double row, with the exception of some rows that included dead plants (we did not find any plants that had died due to deer damage at the time of the browsing investigation). The number of plants investigated for each clone ranged from 34 to 108. The percentage of browsed stems per plant (%) was calculated by dividing the number of browsed stems by the total number of stems per plant. With the exception of August 2017 and the snow period, electric fencing protection from deer was functional, and there was no evidence of extensive deer browsing in other periods during the study.

**Figure 2.** Willow stem tips browsed by Yezo-sika deer in the experimental short rotation coppice. Planting rows of a clone with minimal browsing (white inverted triangles) adjacent to rows of a severely browsed clone (black inverted triangles) (**a**). Willow stems broken by Yezo-sika deer while browsing willow stem tips (**b**).

*2.3. Willow Yield Measurement*

In late October 2019, three years after the first harvest, we harvested all the willow plants, except for those that were sampled for fresh mass to dry mass conversion. We did this using a sugarcane harvester (UT-120K, Uotani-tekko Inc., Nara, Japan), which cuts willows into billets approximately 20 cm long and harvests them into a harvesting bag. We harvested each 60-m single-clone double row into a single bag, which was then immediately weighed using a digital crane scale (resolution: 0.5 kg; 1ACBP-K, Shuzui Scales Co., Ltd., Aichi, Japan). The fresh yield (fresh t ha$^{-1}$ yr$^{-1}$) was determined by subtracting the weight of the harvesting bag (11.5 kg) and dividing by the planted area in a double row (0.011 ha) and the rotation period of three years. To obtain the oven-dry mass yield from the fresh yield in the bags, we harvested nine typical willow plants using a saw before harvesting with the sugarcane harvester and weighed each plant (1.63 fresh kg per plant on average, 14.63 fresh kg in total). We then transported the sample plants to the laboratory, dried them at 70 °C to constant mass, and reweighed them. The oven-dry yield of each clone (odt ha$^{-1}$ yr$^{-1}$) was calculated by multiplying the fresh yield mean of each of the 12 clones by the mean of the ratio of dry to fresh weight of the sample plants.

*2.4. Data Analyses*

All statistical analyses were performed using R version 3.6.2 [34]. We estimated willow yield and yield loss due to deer browsing from generalized linear models (GLMs) with a gamma distribution and log link function. The actual browsed yield of each clone was analyzed by a GLM in which the objective variable was the mean dry yield for each clone, and the explanatory variables were, for each clone, the mean number of stems and mean number of browsed stems per plant. We also analyzed the same variables with GLMs with a gamma distribution and inverse link function and a Gaussian distribution and identity link function. We compared the three models using Akaike's information criterion (AIC) (Table S1); the GLM with a gamma distribution and log link function had the smallest AIC, and so was adopted as the willow yield estimation model. We determined that a clone (P.I.82) was an outlier from the residual plot, the quantile-quantile plot, and scale-location plot using the *plot* function in R (Figure S1). We reanalyzed the GLM for the dataset excluding the outlier clone, and estimated the GLM coefficients. The coefficient of determinations for each of the GLMs ($r^2_{GLM}$) [35] for all 12 clones and for the 11 clones excluding the outlier was calculated using the *req* function of the "rsq" package in R. The GLM excluding an outlier clone was used to simulate willow yield and yield loss when the number of stems per plant varied from 1 to 10, and the percentage of browsed stems per plant varied from 0% to 100%. In addition, we estimated the non-browsed yield by converting the number of browsed stems in the GLM to zero for each clone in the experimental plot. Yield loss by browsing was estimated by subtracting the estimated browsed yield from the estimated unbrowsed yield, that is, 0% of browsed stems per plant.

Standard major axis (SMA) regressions were performed to analyze bivariate relationships between willow yield and other variables across clones using the "smart" package in R. Interspecific differences between *S. pet-susu* and *S. sachalinensis* in the number of stems and number of browsed stems and the percentage of browsed stems per plant were analyzed by a Mann–Whitney U test using the *wilcox.test* function in R, and the differences in actual yield, estimated non-browsed yield, and estimated yield loss from browsing were analyzed by a Student t-test using the *t.test* function in R. The level of statistical significance was set at $p = 0.05$.

## 3. Results

### 3.1. Deer Browsing and Willow Yield across Clones

Yezo-sika deer typically browsed the tips of willow stems (Figure 2a). In addition, deer broke some stems in order to access the tips (Figure 2b). The number of sprouting stems, the extent of deer browsing, and actual willow yield varied largely among clones. Clonal means for the number of stems per plant ranged from 4.8 for P.I.81 to 8.1 for S.I.27, across clones (Figure 3a). The clonal means of the number of stems browsed by deer per plant ranged from 0.61 for S.T.27 to 8.1 for S.I.27 across clones (Figure 3b). The clonal means of the percentage of browsed stems per plant ranged from 11% for P.C.51 to 99% for S.I.27, across clones (Figure 3c). The actual yield of each clone ranged from 1.05 odt ha$^{-1}$ yr$^{-1}$ for clone S.T.27 to 4.26 odt ha$^{-1}$ yr$^{-1}$ for P.I.82 (Figure 3). There was no significant relationship observed between actual yield and the number of stems per plant, number of browsed stems per plant, and percentage of browsed stems per plant across clones (Figure 3). There were no significant interspecific differences in the number of stems per plant (Mann–Whitney U test; $p = 0.53$), browsed stems per plant ($p = 0.53$), percentage of browsed stems per plant ($p = 0.34$), and actual yield (t-test; $p = 0.08$).

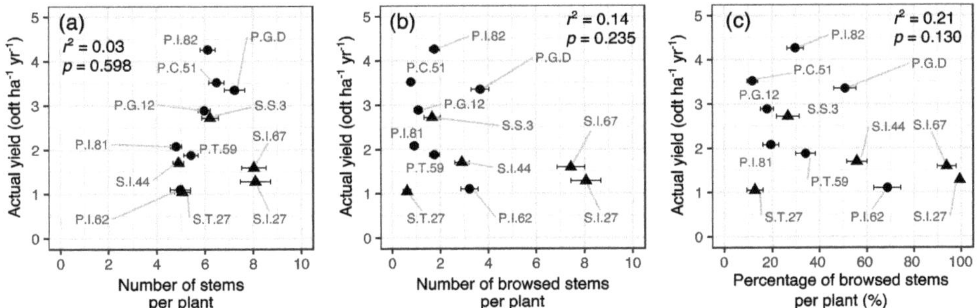


**Figure 3.** Relationships of the means of number of stems per plant (**a**), number of browsed stems per plant (**b**), and percentage of browsed stems per plant (**c**) to actual willow yield across 12 willow clones. Clone names, coefficients of determination ($r^2$), and level of significance ($p$) of the standardized major axis regression are shown in each panel. Circles represent *Salix pet-susu* clones and triangles *S. sachalinensis* clones; odt, oven dry ton. Error bars represent standard errors of the mean ($n = 34$–$108$).

### 3.2. Estimation of Yield Loss Caused by Deer Browsing

A simple GLM incorporating the number of stems and number of browsed stems could accurately estimate actual willow yields with deer browsing (Table 1, Figure 4a). In the GLM calculated for all 12 clones including an outlier, $r^2_{GLM}$ was 0.43, indicating a moderate predictive power (Table 1). When we removed the outlier clone from the analysis (Figure S1), $r^2_{GLM}$ increased to 0.74, indicating substantial predictive power (Table 1). In addition, for the model excluding an outlier, the $r^2$ value in the SMA regression between the actual and estimated yield was 0.74 ($p < 0.001$), and the regression line almost overlapped the 1:1 relationship line (Figure 4a). Yield was underestimated for the outlier clone P.I.82. Clone P.I.82, which had the highest actual yield among the willow clones studied, was estimated by the GLM to have a predicted yield of more than 1 odt ha$^{-1}$ yr$^{-1}$ lower than actual yield. The numbers of stems and browsed stems per plant had a significantly positive (approximately 1.5-fold) and negative (approximately 0.8-fold) effect on yield for each clone in both models, including and excluding an outlier clone (Table 1). The GLM simulation excluding an outlier clone predicted that yields increased with increasing number of stems per plant, but decreased significantly with the percentage of browsed stems per plant (Figure 4b). The effect of browsing damage on the yield was large; for example, even if 25% of the number of stems per plant were browsed, yield decreased to at least 61% of the non-browsed yield (Figure 4b). The GLM simulation also predicted that yield loss by deer browsing increases with the estimated yield and percentage of browsed stems per plant. For example, with a standard target yield of 10 odt ha$^{-1}$ yr$^{-1}$ for SRC willow, yield losses caused by deer browsing were estimated at approximately 4, 6, 7, and 8 odt ha$^{-1}$ yr$^{-1}$ if 25%, 50%, 75%, and 100% of the stems per plant were browsed (Figure 4c).

**Table 1.** A summary of the generalized linear models (GLMs) of estimated willow yield under deer browsing conditions for each clone. A gamma distribution with a log-link function was used for the models. One model includes all 12 clones studied, and another model comprised 11 clones, excluding an outlier. The models incorporate the number of stems and browsed stems per plant as explanatory variables. The numbers in parentheses in the "Estimate" column represent the exponential value of the estimate. SE is the standard error.

| Factor | Estimate | SE | *t* Value | *p* Value |
|---|---|---|---|---|
| GLM for 12 clones (including an outlier) [1] | | | | |
| Intercept | −1.113 (0.33) | 0.616 | −1.808 | 0.104 |
| Number of stems per plant | 0.411 (1.51) | 0.118 | 3.488 | 0.007 |
| Number of browsed stems per plant | −0.223 (0.80) | 0.055 | −4.037 | 0.003 |
| GLM for 9 clones (excluding an outlier) [2] | | | | |
| Intercept | −1.009 (0.36) | 0.519 | −1.943 | 0.088 |
| Number of stems per plant | 0.373 (1.45) | 0.100 | 3.724 | 0.006 |
| Number of browsed stems per plant | −0.199 (0.82) | 0.048 | −4.186 | 0.003 |

[1] Residual deviance: 0.891 on 9 degrees of freedom, $r^2_{GLM}$ = 0.43; [2] Residual deviance: 0.619 on 8 degrees of freedom, $r^2_{GLM}$ = 0.74.

When the GLM was applied to actual data from the experimental plantation, willow yield was estimated to increase to 2.2–7.5 odt ha$^{-1}$ yr$^{-1}$, assuming an absence of deer browsing, from 1.1–4.3 odt ha$^{-1}$ yr$^{-1}$ with deer browsing (Figure 5a). The estimated yield without deer browsing was not significantly correlated with actual yield across clones. The yield loss due to deer browsing was estimated to range from 0.1 odt ha$^{-1}$ yr$^{-1}$ for S.T.27 to 6.0 odt ha$^{-1}$ yr$^{-1}$ for S.I.27, and was significantly correlated with estimated yield under non-browsing conditions across clones (Figure 5b) and the percentage of browsed stems per plant (Figure 5c). There was no significant interspecific difference in the estimated yield without browsing ($p = 0.27$) and estimated yield loss due to browsing ($p = 0.15$). In P.I.82, with the maximum actual yield among the clones studied, although an average of 1.7 stems (i.e., 30% of the average 6.1 sprouting stems per individual) were subject to browsing, the potential yield without browsing was estimated to be approximately 4 odt ha$^{-1}$ yr$^{-1}$, which was almost the same as the actual yield with browsing (Figure 5a).

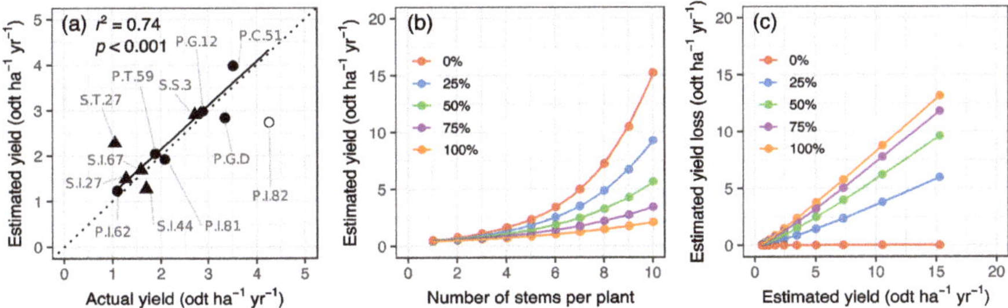

**Figure 4.** Relationship between actual and estimated willow yield with reference to deer browsing, using a generalized linear model (GLM) (**a**), and GLM simulations for estimated yield with browsing (**b**) and estimated yield loss due to browsing (**c**), assuming 0%, 25%, 50%, 75%, and 100% of browsed stems per plant. In panel (**a**), circles and triangles represent *Salix pet-susu* and *S. sachalinensi* clones, respectively; and an open symbol represents the clone that were determined to be an outlier based on the GLM residuals. Clone names are shown. The coefficient of determination ($r^2$), level of significance ($p$), and regression line of the standardized major axis analysis for 11 clones excluding an outlier clone are also shown in panel (**a**). The dotted line represents a 1:1 relationship. Loess regressions were applied in panels (**b**) and (**c**). odt, oven dry ton.

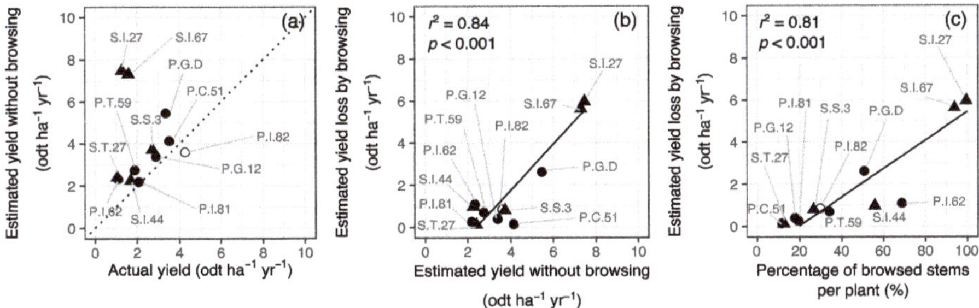

**Figure 5.** Relationships between actual and estimated yield without deer browsing (**a**), between estimated yield without browsing and estimated yield loss due to browsing (**b**), and the percentage of browsed stems per plant and estimated yield loss due to browsing (**c**) across willow clones. An open symbol represents the clone that was determined to be an outlier based on the GLM residuals. Clone names are shown. The coefficient of determination ($r^2$), level of significance ($p$), and significant regression line of the standardized major axis analyses for the 11 clones excluding an outlier clone are shown. The dotted line in panel (**a**) represents a 1:1 relationship. odt, oven dry ton.

## 4. Discussion

### 4.1. Yield Loss Caused by Deer Browsing

Although the actual willow yield with browsing could not be estimated by a single stem parameter (Figure 3), we were able to estimate the yield for 11 out of 12 clones using a simple GLM that incorporated both the number of sprouting stems and browsed stems as explanatory variables (Figure 4a). The high predictive accuracy of this model, despite the lack of stem diameter and/or height data, suggests that growth in each browsed or unbrowsed stem was comparable among the 11 clones during three years of rotation. As a result, it was possible to predict the yield only by the data on the numbers of stems. These results suggest that when an SRC willow plantation is damaged by deer browsing, the biomass loss can be easily estimated using a GLM by counting the number of browsed and unbrowsed stems per plant and measuring the yield in places where the degree of browsing damage is different in the plantation. This approach will be particularly useful in areas where SRC willow is in the developmental stage and standard yields are unknown. In addition, the fact that the number of browsed stems in the first year of rotation significantly explained yield in the final year of rotation in the GLM (Table 1) would indicate that browsing damage, which occurred only once in the first year of rotation, affected biomass increase during the three-year harvest intervals and was not offset during that period. Although the influence of deer on the SRC willow has sometimes been underestimated [21], these results indicate that deer-browsing control can be very important for the success of the SRC willow plantation. On the other hand, yield of P.I.82, which had the highest actual yield among the 12 clones studied, was greatly underestimated by the GLM (Figure 4a). This could possibly result from greater biomass per stem and/or a smaller reduction in biomass per browsed stem in this clone than in the other clones. The stem of P.I.82 might be taller than that of the other clones at the time of browsing damage and thus be less affected by browsing damage per stem, although we did not measure the height.

The model simulation showed that a single browsing event (which occurred during the first summer of our study period) resulted in a significant yield loss over three years of a harvest cycle; the potential for willow yield to be halved if 25% of the number of stems per plant that were browsed by deer was also shown (Figures 4 and 5). This high percentage of yield loss, even when the percentage of browsed stems per plant was small, is attributable not only to direct biomass loss due to deer browsing, but also to the indirect negative effects of browsing damage on biomass increase. In a conifer crop tree, *Pseudotsuga menziesii*, deer browsing and weed overgrowth inhibited the crop tree height growth, whereas deer browsing on broadleaf competitors resulted in a slightly higher tree height in comparison with that in the plot which excluded deer under high weed suppression by herbicides, indicating potential interaction between competing vegetation and browsing [36]. Because willow growth is very susceptible to weeds [24,37], and willow had a higher preference for deer than weeds at the present study site [33], the reason for the indirect negative effect of deer browsing on willow biomass in this study could be that deer browsing damage delayed the closure of the willow canopy, resulting in weed overgrowth, thus suppressing the growth of the willow susceptible to weeds [24,38]. Because our results are based on only one browsing event that occurred on a small plantation, future studies need to examine how the results of the simulation vary depending on various conditions and situations of SRC willow plantation, such as the extent of weed growth, plantation size, browsing frequency, browsing timing, etc.

*4.2. Potential Yield of Studied Willow Clones*

Only two of the 12 clones studied have a known potential yield in the absence of browsing damage in the study area. The two clones, S.I.27 and S.I.67, yielded approximately 5–11 odt ha$^{-1}$ yr$^{-1}$ in the first harvest rotation (average of approximately 8 odt ha$^{-1}$ yr$^{-1}$) with no deer browsing at the same plantation in this study [24]. We estimated the potential yield to be 7.5 and 7.3 odt ha$^{-1}$ yr$^{-1}$ for S.I.27 and S.I.67, respectively, in the second rotation using the GLM (Figure 5a). SRC willow yields are generally greater in the second than in the first rotation [39–41], but, in our study, more yield data from the second rotation for these clones were included for areas not weeded between mulches and therefore with potentially reduced yields compared with the first rotation [24]; therefore, the potential GLM yield estimates for these two clones would be reasonable in comparison to the first rotation yield. It might therefore be reasonable to assume that the estimated potential yields by the GLM of the other nine clones other than the outlier are also probably close to their actual potential yield in the absence of browsing, although we were not able to compare the estimated potential yield with the first rotation yield due to the lack of the first rotation data. Future verification with actual potential yield that was reliably protected against deer browsing will allow us to clarify the accuracy of the estimated potential yield from the model in this study.

In this study, clones with the highest estimated yield tended to show results of greater yield loss due to deer browsing (Figure 5b). In general, there is a trade-off between plant growth and defense against herbivory because of limited resources [42,43]. Therefore, it would be difficult and time consuming to search for and breed super willow clones that achieve both high-yielding and low damage from deer browsing; though susceptibility to browsing animals has already been used as one of the traits for selecting willow commercial variety in Europe [29]. Compared with Europe and North America, hunting is less popular in Japan, where hunters are aging and decreasing in number [44]. Therefore, it will be potentially difficult to use deer hunting as a major management tool for maintaining deer browsing at low levels in SRC willow plantations in Hokkaido, Japan. The potential yields of most clones were estimated to be low ($<5$ odt ha$^{-1}$ yr$^{-1}$), even considering that the overgrowth of weeds among mulches could have negatively affected willow yields. If commercial SRC willow plantations are to be promoted in Hokkaido, Japan, the selection of high-yielding clones that are less susceptible to deer browsing, together with the development of inexpensive measures protecting against deer entry to plantations, will therefore be desirable. Further research about various protective measures against deer browsing, including the introduction of physical barriers such as fencing and/or repellents, establishment of large plantations, and deer population management in cooperation with local hunters [20,21,45,46], should lead to a higher willow yield by reducing yield loss due to deer browsing and improved profit forecasts at SRC willow plantations for the studied willow clones in the study site area.

## 5. Conclusions

This study demonstrated that the GLM analysis, which incorporates two easily measured parameters—the number of stems per plant and the number of browsed stems by deer—can estimate the biomass loss due to deer browsing in an SRC willow plantation, which cannot be estimated by a single parameter. The GLM also can be used to predict yield in the absence of browsing damage, although it needs further validation in the future. In the 12 clones studied, there was a trade-off between potential biomass yield and susceptibility to browsing damage, suggesting that the importance of preventing deer browsing to obtain adequate biomass yield in an SRC willow plantation.

**Supplementary Materials:** The followings are available online at http://www.mdpi.com/1999-4907/11/8/809/s1, Table S1. Akaike's Information Criterions (AICs) of the generalized linear models predict willow yield with deer browsing. Figure S1: Residual plot, quantile-quantile plot, and scale-location plot of the generalized linear model used to determine an outlier clone.

**Author Contributions:** Conceptualization, H.H.; methodology, H.H. and S.S.; formal analysis, H.H., S.S., and T.Y.; investigation, H.H., Q.H., M.I., M.K., S.S., T.Y., H.U., and Y.M.; data curation, H.H. and T.Y.; writing—original draft preparation, H.H.; writing—review and editing, H.H., Q.H., M.I., M.K., A.U., S.S., T.Y., H.U., and Y.M.; project administration, H.H., Q.H., A.U., and H.U.; funding acquisition, H.H., Q.H., and H.U. All authors have read and agreed to the published version of the manuscript.

**Funding:** This research was funded by Forestry and Forest Products Research Institute (FFPRI) grants.

**Acknowledgments:** We thank our collaborators from Shimokawa town and the Shimokawa forest association, especially Yuji Takahashi and Toshio Yamamoto, for assistance with the establishment of this trial, maintenance and harvesting, and also Oji Holdings Cor. for offering willow cuttings, and Uotani-tekko Inc. for assistance with the harvest. We also thank Haruka Yamamoto, Eriko Ito, Tomomasa Amano, Naoyuki Furuya, Toshimitsu Nagasawa, Megumi Yasumoto, Masayoshi Takahasi, and Kiyohiko Fujimoto for their assistance in data collection.

**Conflicts of Interest:** The authors declare no conflict of interest.

## References


1.  IPCC. Global Warming of 1.5 °C. An IPCC Special Report on the Impacts of Global Warming of 1.5 °C Above Pre-Industrial Levels and Related Global Greenhouse Gas Emission Pathways, in the Context of Strengthening the Global Response to the Threat of Climate Change, Sustainable Development, and Efforts to Eradicate Poverty. 2018. Available online: https://www.ipcc.ch/sr15/download/ (accessed on 25 July 2020).
2.  Environmental Protection Agency. *EPA's Treatment of Biogenic Carbon Dioxide Emissions from Stationary Sources that Use Forest Biomass for Energy Production*; United States Environmental Protection Agency: Washington, DC, USA, 2018. Available online: https://www.epa.gov/air-and-radiation/epas-treatment-biogenic-carbon-dioxide-emissions-stationary-sources-use-forest (accessed on 25 July 2020).
3.  European Union. *Directive 2009/28/EC of the European Parliament and of the Council of 23 April 2009 on the Promotion of the Use of Energy from Renewable Sources and Amending and Subsequently Repealing Directives 2001/77/EC and 2003/30/EC*; European Union: Brussels, Belgium, 2009.
4.  Japan Woody Bioenergy Association. *Wood Biomass Energy Data Book 2018*; Japan Woody Bioenergy Association: Tokyo, Japan, 2019; p. 25. Available online: https://www.jwba.or.jp/app/download/13159696992/Data+book+2018+%28English+version%29.pdf?t=1523342683 (accessed on 25 July 2020).
5.  Ericsson, K.; Rosenqvist, H.; Nilsson, L.J. Energy crop production costs in the EU. *Biomass Bioenergy* **2009**, *33*, 1577–1586. [CrossRef]
6.  Amichev, B.Y.; Hangs, R.D.; Konecsni, S.M.; Stadnyk, C.N.; Volk, T.A.; Bélanger, N.; Vujanovic, V.; Schoenau, J.J.; Moukoumi, J.; Van Rees, K.C.J. Willow short-rotation production systems in Canada and northern United States: A review. *Soil Sci. Soc. Am. J.* **2014**, *78*, S168–S182. [CrossRef]
7.  Lindegaard, K.N.; Adams, P.W.; Holley, M.; Lamley, A.; Henriksson, A.; Larsson, S.; von Engelbrechten, H.G.; Esteban Lopez, G.; Pisarek, M. Short rotation plantations policy history in Europe: Lessons from the past and recommendations for the future. *Food Energy Secur* **2016**, *5*, 125–152. [CrossRef] [PubMed]
8.  Aylott, M.J.; Casella, E.; Tubby, I.; Street, N.R.; Smith, P.; Taylor, G. Yield and spatial supply of bioenergy poplar and willow short-rotation coppice in the UK. *New Phytol.* **2008**, *178*, 358–370. [CrossRef]
9.  Fabio, E.S.; Volk, T.A.; Miller, R.O.; Serapiglia, M.J.; Gauch, H.G.; Van Rees, K.C.J.; Hangs, R.D.; Amichev, B.Y.; Kuzovkina, Y.A.; Labrecque, M.; et al. Genotype × environment interaction analysis of North American shrub willow yield trials confirms superior performance of triploid hybrids. *GCB Bioenergy* **2017**, *9*, 445–459. [CrossRef]
10. Verwijst, T.; Lundkvist, A.; Edelfeldt, S.; Albertsso, J. Development of sustainable willow short rotation forestry in northern Europe. In *Biomass Now—Sustainable Growth and Use*; Matovic, M.D., Ed.; IntechOpen: London, UK, 2013. [CrossRef]
11. Townsend, P.A.; Haider, N.; Boby, L.A.; Heavey, J.P.; Miller, T.A. *A Roadmap for Poplar and Willow to Provide Environmental Services and to Build the Bioeconomy*; Washington State University: Washington, DC, USA, 2018; p. 37.
12. Caslin, B.; Finnan, J.; Johnston, C.; McCracken, A.; Walsh, L. *Short Rotation Coppice Willow Best Practice Guidelines*; Teagasc and Agri-Food and Biosciences Institute: Carlow, Ireland, 2015. Available online: https://www.teagasc.ie/media/website/publications/2011/Short_Rotation_Coppice_Best_Practice_Guidelines.pdf (accessed on 26 July 2020).
13. Fabio, E.S.; Smart, L.B. Effects of nitrogen fertilization in shrub willow short rotation coppice production—A quantitative review. *GCB Bioenergy* **2018**, *10*, 548–564. [CrossRef]
14. Mehmood, M.A.; Ibrahim, M.; Rashid, U.; Nawaz, M.; Ali, S.; Hussain, A.; Gull, M. Biomass production for bioenergy using marginal lands. *Sustain. Prod. Consum.* **2017**, *9*, 3–21. [CrossRef]
15. Styles, D.; Thorne, F.; Jones, M.B. Energy crops in Ireland: An economic comparison of willow and Miscanthus production with conventional farming systems. *Biomass Bioenergy* **2008**, *32*, 407–421. [CrossRef]
16. Milner, S.; Holland, R.A.; Lovett, A.; Sunnenberg, G.; Hastings, A.; Smith, P.; Wang, S.; Taylor, G. Potential impacts on ecosystem services of land use transitions to second-generation bioenergy crops in GB. *Glob Chang. Biol. Bioenergy* **2016**, *8*, 317–333. [CrossRef]
17. Vanbeveren, S.P.P.; Ceulemans, R. Biodiversity in short-rotation coppice. *Renew. Sustain. Energy Rev.* **2019**, *111*, 34–43. [CrossRef]

18. Schulz, V.; Gauder, M.; Seidl, F.; Nerlich, K.; Claupein, W.; Graeff-Hönninger, S. Impact of different establishment methods in terms of tillage and weed management systems on biomass production of willow grown as short rotation coppice. *Biomass Bioenergy* **2016**, *85*, 327–334. [CrossRef]

19. Dimitriou, I.; Rosenqvist, H.; Berndes, G. Slow expansion and low yields of willow short rotation coppice in Sweden; implications for future strategies. *Biomass Bioenergy* **2011**, *35*, 4613–4618. [CrossRef]

20. Dimitriou, I.; Rutz, D. *Sustainable Short Rotation Coppice: A Handbook*; Rutz, D., Ed.; WIP Renewable Energies: Munich, Germany, 2015.

21. Landgraf, D.; Brunner, J.; Helbig, C. The impact of wild animals on SRC in Germany—A widely underestimated factor. In Proceedings of the BENWOOD: Short rotation forestry and Agroforestry: An Exchange of Experience between CDM Countries and Europe, Conference Proceedings, Barolo, Italy, 20–22 June 2011; pp. 133–140.

22. Bergström, R.; Guillet, C. Summer browsing by large herbivores in short-rotation willow plantations. *Biomass Bioenergy* **2002**, *23*, 27–32. [CrossRef]

23. Forestry Agency Japan. *State of Japan's Forests and Forest Management—3rd Country Report of Japan to the Montreal Process*; Forestry Agency Japan: Tokyo, Japan, 2019; p. 153. Available online: https://www.maff.go.jp/e/policies/forestry/attach/pdf/index-8.pdf (accessed on 25 July 2020).

24. Han, Q.; Harayama, H.; Uemura, A.; Ito, E.; Utsugi, H.; Kitao, M.; Maruyama, Y. High biomass productivity of short-rotation willow plantation in boreal Hokkaido achieved by mulching and cutback. *Forests* **2020**, *11*, 505. [CrossRef]

25. Han, Q.; Harayama, H.; Uemura, A.; Ito, E.; Utsugi, H. The effect of the planting depth of cuttings on biomass of short rotation willow. *J. For. Res.* **2017**, *22*, 131–134. [CrossRef]

26. Kaji, K.; Saitoh, T.; Uno, H.; Matsuda, H.; Yamamura, K. Adaptive management of sika deer populations in Hokkaido, Japan: Theory and practice. *Popul. Ecol.* **2010**, *52*, 373–387. [CrossRef]

27. Shirozu, T.; Soga, A.; Morishita, Y.K.; Seki, N.; Ko-Ketsu, M.; Fukumoto, S. Prevalence and phylogenetic analysis of Cryptosporidium infections in Yezo sika deer (Cervus nippon yesoensis) in the Tokachi sub-prefecture of Hokkaido, Japan. *Parasitol. Int.* **2020**, *76*, 102064. [CrossRef]

28. Hokkaido Goverment. Results of Wildlife Damage Survey. (In Japanese). 2020. Available online: http://www.pref.hokkaido.lg.jp/file.jsp?id=853297 (accessed on 25 July 2020).

29. Caslin, B.; Finnan, J.; McCracken, A. *Willow Varietal Identification Guide*; Teagasc & AFBI: Carlow, Ireland, 2012.

30. Smart, L.B.; Volk, T.A.; Lin, J.; Kopp, R.F.; Phillips, I.S.; Cameron, K.D.; White, E.H.; Abrahamson, L.P. Genetic improvement of shrub willow (*Salix* spp.) crops for bioenergy and environmental applications in the United States. *Unasylva* **2005**, *56*, 51–55.

31. Lindegaard, K.; Barker, J.H.A. Breeding willows for biomass. *Asp. Appl. Biol.* **1996**, *49*, 1–9.

32. Ito, E.; Uemura, A.; Harayama, H.; Han, Q.; Utsugi, H. Root system development and partitioning of nitrogen in short-rotation willow. (In Japanese). *Boreal For. Res.* **2015**, *63*, 47–50.

33. Ishihara, M.; Matsuura, Y. The actual condition of animal damage in the willow plantation of Shimokawa town : A tendency of feeding damage and photography frequency by sensor camera about Yezo deer in willow growth period. (In Japanese). *Boreal For. Res.* **2014**, *62*, 83–86.

34. R Core Team. *R: A Language and Environment for Statistical Computing, 3.6.2*; R Foundation for Statistical Computing: Vienna, Austria, 2019.

35. Zhang, D. A coefficient of determination for generalized linear models. *Am. Stat.* **2016**, *71*, 310–316. [CrossRef]

36. Stokely, T.D.; Betts, M.G.; Macinnis-Ng, C. Deer-mediated ecosystem service versus disservice depends on forest management intensity. *J. Appl. Ecol.* **2019**, *57*, 31–42. [CrossRef]

37. Helby, P.; Rosenqvist, H.; Roos, A. Retreat from Salix—Swedish experience with energy crops in the 1990s. *Biomass Bioenergy* **2006**, *30*, 422–427. [CrossRef]

38. Bullard, M.J.; Mustill, S.J.; Carver, P.; Nixon, P.M.I. Yield improvements through modification of planting density and harvest frequency in short rotation coppice *Salix* spp.—2. Resource capture and use in two morphologically diverse varieties. *Biomass Bioenergy* **2002**, *22*, 27–39. [CrossRef]

39. Amichev, B.Y.; Volk, T.A.; Hangs, R.D.; Bélanger, N.; Vujanovic, V.; Van Rees, K.C.J. Growth, survival, and yields of 30 short-rotation willow cultivars on the Canadian Prairies: 2nd rotation implications. *New For.* **2018**, *49*, 649–665. [CrossRef]

40. Wickham, J.; Rice, B.; Finnan, J.; McConnon, R. *A Review of Past and Current Research on Short Rotation Coppice in Ireland and Abroad*; COFORD: Dublin, Ireland, 2010; p. 36.

41. Larsen, S.U.; Jørgensen, U.; Kjeldsen, J.B.; Lærke, P.E. Effect of fertilisation on biomass yield, ash and element uptake in SRC willow. *Biomass Bioenergy* **2016**, *86*, 120–128. [CrossRef]

42. Coley, P.D.; Bryant, J.P.; Chapin, F.S., 3rd. Resource availability and plant antiherbivore defense. *Science* **1985**, *230*, 895–899. [CrossRef]

43. Herms, D.A.; Mattson, W.J. The dilemma of plants: To grow or defend. *Q. Rev. Biol.* **1992**, *67*, 283–335. [CrossRef]

44. Iijima, H. The Effects of Landscape Components, Wildlife Behavior and Hunting Methods on Hunter Effort and Hunting Efficiency of Sika Deer. *Wildl. Biol.* **2017**, *2017*. [CrossRef]

45. Charles, J.G.; Nef, L.; Allegro, G.; Collins, C.M.; Delplanque, A.; Gimenez, R.; Höglund, S.; Jiafu, H.; Larsson, S.; Luo, Y.; et al. Insect and other pests of poplars and willows. In *Poplars and Willows: Trees for Society and the Environment*; Isebrands, J.G., Richardson, J., Eds.; CABI: Wallingford, UK, 2014; pp. 459–526. [CrossRef]

46. Uno, H.; Kaji, K.; Tamada, K. Sika Deer Population Irruptions and Their Management on Hokkaido Island, Japan. In *Sika Deer: Biology and Management of Native and Introduced Populations*; McCullough, D.R., Takatsuki, S., Kaji, K., Eds.; Springer: Tokyo, Japan, 2009; pp. 405–419. [CrossRef]

 *forests*

*Article*

# Intensive Mechanical Site Preparation to Establish Short Rotation Hybrid Poplar Plantations—A Case-Study in Québec, Canada


**Nelson Thiffault** [1,2,*] , **Raed Elferjani** [2] , **François Hébert** [3] , **David Paré** [4] **and Pierre Gagné** [5]

1    Canadian Wood Fibre Centre, Canadian Forest Service, Natural Resources Canada, 1055 du P.E.P.S., P.O. Box 10380, Sainte-Foy Stn., Québec, QC G1V 4C7, Canada
2    Réseau Reboisement Ligniculture Québec, Université TÉLUQ, 5800 Saint-Denis, Bureau 1105, Montréal, QC H2S 3L5, Canada; Raed.Elferjani@teluq.ca
3    Ministère des Forêts, de la Faune et des Parcs du Québec, Direction de la Protection des Forêts, 5700 4e avenue Ouest, Québec, QC G1H 6R1, Canada; Francois.Hebert@mffp.gouv.qc.ca
4    Laurentian Forestry Centre, Canadian Forest Service, Natural Resources Canada, 1055 du P.E.P.S., P.O. Box 10380, Sainte-Foy Stn., Québec, QC G1V 4C7, Canada; david.pare@canada.ca
5    Chaire Industrielle de Recherche sur la Construction Écoresponsable en Bois, Université Laval, 2425 de la Terrasse, Québec, QC G1V 0A6, Canada; pierre.gagne@sbf.ulaval.ca
*    Correspondence: nelson.thiffault@canada.ca; Tel.: +1-418-454-1976



Received: 19 June 2020; Accepted: 17 July 2020; Published: 21 July 2020



**Abstract:** Because they generate more wood per area and time, short rotation plantations are likely to play an increasing role in meeting the global increase in the demand for wood fiber. To be successful, high-yield plantations require costly intensive silviculture regimes to ensure the survival and maximize yields. While hybrid poplar (*Populus* spp.) is frequently used in intensive, short rotation forestry, it is particularly sensitive to competition and resource levels. Mechanical site preparation is thus of great importance to create microsites that provide sufficient light levels and adequate soil water and nutrient availability. We conducted an experiment in Québec (Canada) to compare two intensive site preparation treatments commonly used to establish hybrid poplar. We compared the effects of double-blade site preparation (V-blade), mounding and a control on hybrid poplar growth and nutritional status four growing seasons after planting on recently harvested forested sites. We also evaluated the effects of site preparation and planted poplar on inorganic soil N. Our results confirmed general positive effects of site preparation on the early growth of hybrid poplar clones. After four growing seasons, survival was higher in the mounding treatment (99%) than in the V-blade (91%) and the control (48%). Saplings planted in the V-blade and in the mounding treatments had mean diameters that were respectively 91% and 155% larger than saplings planted in the control plots. Saplings were 68% taller in the mounding treatment than the control plots, but differences between the V-blade and controls were not significant. We did not detect significant effects of site preparation or the presence of planted hybrid poplar on soil inorganic N. Sapling foliar nutrient concentrations were not influenced by the site preparation treatments. Based on these results, mounding appears to be a good management approach to establish hybrid poplar plantations under the ecological conditions we have studied, as it is less likely to cause erosion because of the localized nature of the treatment. However, these environmental benefits need to be balanced against economic and social considerations.

**Keywords:** intensive silviculture; *Populus maximowiczii* × *P. deltoides* × *P. trichocarpa*; fast-growing tree species; severe soil disturbance; foliar nutrition; soil inorganic N

## 1. Introduction

Fast-growing plantations established using performant plant material and managed over short rotations are increasingly being considered as a solution to meet the global increase in the demand for wood fiber. Moreover, when integrated into functional zoning, high-yield plantations contribute to offsetting timber production losses related to extensive management and conservation areas [1]. They also provide an opportunity to produce wood fiber over shorter periods of time than in natural forests or less intensively-managed plantations, thus reducing the time crop trees are exposed to biotic and abiotic risk factors. The annual increase in area of planted forests worldwide was 1.2% between 2010–2015, which is half of the rate estimated to be needed to meet the six billion cubic meters of the global wood demand in 2050 [2,3].

To be successful and meet their production objectives, high-yield plantations require intensive silviculture regimes to ensure the survival and maximize the growth of planted seedlings. After harvesting of boreal and sub-boreal forest ecosystems, thick soil organic horizons and competition for resources by fast-growing herb and shrub species limit root growth of planted seedlings or cuttings and jeopardize successful establishment [4]. Mechanical site preparation is thus frequently used to create a sufficient number of suitable planting microsites that provide adequate light levels as well as proper soil water and nutrient availability, while limiting waterlogged conditions [5]. Hybrid poplar (*Populus* spp.) is frequently used in intensive, short rotation forestry [6–12]; this species offers fast growth rates and high productivity [13]. However, adequate site preparation is mandatory to ensure survival and promote early growth, as the species is particularly sensitive to competition and resource levels [14,15].

We conducted an experiment in Québec, a Canadian province characterized by the abundance of its forest resources, which extend over 760,000 km$^2$ and represent 2% of the world's forests. With 40% of the world's certified forest area within its border, Canada has the largest area of third-party independently certified forests in the world [16], and provincial forest acts, including the one in force in Québec, are based on the principles of sustainable forest management [17]. Forestry activities must thus protect ecosystem functions. For example, management activities have to ensure that site productivity is maintained over the long-term with minimal or no impacts on soil and water quality, even under intensive silviculture regimes. Additionally, in Québec, herbicide use is restricted on private lands and prohibited on public lands [18], which creates a further challenge for the establishment of hybrid poplar, a species known to be highly sensitive to root competition [19].

In this context, we compared two intensive mechanical site preparation treatments commonly used to establish hybrid poplar in Québec, namely double-blade site preparation and mounding. Blading creates series of ~3 m-wide furrows oriented parallel to each other, in which the mineral soil is entirely exposed. This technique displaces a large volume of soil and reduces competing vegetation but might negatively affect soil physical and microbial characteristics [20] and favor N leaching from the site. Mounding, on the other hand, creates ~0.7 m$^3$ mounds composed of bare mineral soil deposited on top of inverted organic material; mounds are dispersed on the site, approx. 3 m from each other. This technique generates greater cost than blading (~1500 CAN\$ ha$^{-1}$ for mounding vs. ~1200 CAN\$ ha$^{-1}$ for blading), because it requires more costly equipment and more machine time. Mounding is generally effective in enhancing soil drainage and aeration and in increasing soil temperature, thus favoring high survival and early growth of planted seedlings [21,22], including for fast growing species like hybrid poplar [19]. Although both blading and mounding create a severe soil disturbance at the microsite scale, blading results in greater site disturbance when considered at the stand scale [23–25]. The use of this technique is thus raising concerns as it can promote soil erosion, especially at the early stages of plantation establishment [26]. However, because of its fast growth rates and high nutritional needs, hybrid poplar can significantly influence soil nutrient dynamics [27] and potentially act as a buffer of site preparation effects on soils.

Our objective was to compare the effects of two site preparation treatments (plus a control) on hybrid poplar growth and nutritional status four growing seasons after planting on recently harvested forested sites. We also compared site preparation and planted poplar effects on inorganic soil N as an indicator of site potential productivity. We predicted that sapling height and diameter at breast height would be greater in site prepared plots compared to control conditions, but that blading and mounding would result in similar sapling dimensions. We however posited that blading would result in lower concentrations of soil inorganic N, compared to the other treatments, with correspondingly lower foliar macronutrient concentrations in saplings planted in these plots than in the other treatments. Finally, we also tested the effect planting hybrid poplar on soil inorganic N, and expected that the presence of hybrid poplars would have a positive effect on soil inorganic N by buffering out the negative site preparation impact (significant site preparation × plantation interaction).

## 2. Materials and Methods

### 2.1. Study Site and Experimental Design

The experiment was implemented between September 2012 and June 2013 at four sites located approximately 11 km east of the municipality of Saint-Pascal, Québec (Canada) and 4 km south of the municipality of Rivière-Bleue, Québec (Canada). More specifically, the four sites were located at 47°5048200 N, 69°6610600 W; 47°4801800 N, 69°6539300 W; 47°4714900 N, 69°0555400 W; and 47°4728400 N, 69°0535300 W, respectively. The study sites are within the balsam fir (*Abies balsamea* L. [Mill])–yellow birch (*Betula alleghaniensis* Britt.) bioclimatic domain as described in the Québec ecological classification system [28]. In this region, climate is humid-continental with a mean annual temperature of 3.1 °C (± 0.9 °C) and total annual precipitation of 1012 mm (28% fall as snow) [29]. The altitude varies between 240–360 m above sea level. Based on the Canadian System of Soil Classification [30], soils are Gleysolic Podzols and were formed from a moderate (0.5–1 m) to deep (>1 m) coarse glacial till deposit (47% sand, 34% silt, 19% clay; pH = 5.2). After harvesting, mesic sites in this region are typically invaded by fast growing, light demanding shrub and tree species such as *Rubus idaeus* L., *Prunus pensylvanica* L., *Acer spicatum* Lam., and *Populus tremuloides* Michx [31].

We established a complete block 3 × 2 split-plot design with four replicated blocks (one block per site). Each block was divided into three main plots of approx. 700 m$^2$, each of which was randomly selected to receive one of three site preparation treatments: Two mechanical site preparation treatments plus a control. The first treatment, performed between September and November 2012, was a double-blade scarification as described and illustrated in Hébert et al. [24]: A first scarification pass was executed by a "V-blade" attached in front of a bulldozer that created a 2.7 m wide × 0.47 m deep furrow that exposed the mineral soil. Then, a second pass with a smaller "V-blade" (back hoe) mounted on the back of the same bulldozer was done to increase furrow depth at its center. Furrows were laid out in parallel strips within the plots (Figure 1a). The second treatment was mounding site preparation performed with a 0.8 m$^3$ bucket mounted on an excavator that dug the soil to collect the mineral layers. Afterwards, the content of the bucket was inverted next to the hole, and created a 0.3 m high × 1.5 m wide × 1.5 m long mound (Figure 1b). Mounds were positioned at about 3 m from each other within each treated main plot. The third main plot of each block was left unprepared to serve as a control (Figure 1c).

After site preparation, each main plot was divided into two sub-plots, each of which was randomly selected to be either planted with a hybrid poplar clone or left unplanted. Following provincial guidelines, we selected a clone of *Populus maximowiczii* × *P. deltoides* × *P. trichocarpa* [32] for plantation in the selected sub-plots. Planting was performed in May 2013, as described by Hébert et al. [24]. In summary, we planted one-year-old unrooted cuttings of approximately 100 cm in height, 30 cm deep in the soil at a spacing of approx. 4 m × 3 m (~833 stems ha$^{-1}$) using a metallic rod to create a planting hole. In the double-blade scarification treatment, the cuttings were planted at the hinge position of the scarification furrows [24]. In the mounding treatment, we planted cuttings at the top of the mound. Based on operational practices in this region, we fertilized all planted cuttings in May 2014 with 460 g of a 15–30–5 $NH_4$–$NO_3$ granular fertilizer, within a radius of 50 cm around the cutting base.

Granular fertilizer was also applied in the unplanted sub-plots so that the presence of hybrid poplars was the only factor differentiating the planted from the unplanted treatment. The resulting design was a 3 × 2 split-plot with three levels of site preparation (V-blade, Mounding, Control), and two levels of plantation (with plantation, without plantation), replicated four times (i.e., on four different sites).

**Figure 1.** Example of main plots with the site preparation treatments that were tested in the study. The "V-blade" treatment resulted in 2.7 m wide × 0.47 m deep furrows that exposed the mineral soil (**a**). Mounding site preparation consisted in 0.3 m high × 1.5 m wide × 1.5 m long inverted mounds adjacent to the holes created by the excavator (**b**). Control plots consisted in unprepared soil conditions (**c**).

## 2.2. Sapling Measurements

Cuttings were identified and mapped following planting. In September 2016, we measured diameter at breast height (DBH, at 1.3 m) of all saplings, and measured total height on trees selected for foliar analyses (see below). We assessed tree survival based on initial measurements.

## 2.3. Foliar and Soil Nutrients

In each planted sub-plot of every block, we selected five planted trees for foliar sampling in September 2016 (end of the fourth growing season since planting). The selected trees were located in the four corners and near the centre of the sub-plot, respecting a buffer of one planted tree from the sub-plot borders. On each selected tree, we collected 10 leaves that we grouped to form one composite sample per tree. Leaves were collected randomly from the base of the tree crown up to a maximum height of about 2 m. Samples were kept cold until further analyzes. They were then oven-dried at 60 °C and ground to pass a 0.5 mm mesh. Using 200 mg sub-samples, we analyzed concentration of total N through direct combustion at 1350 °C and analysis using a TruMac CN Elemental Analyzer (LECO Corporation, St-Joseph, MI, USA). We used sub-samples of 90–110 mg for $H_2SO_4$–$H_2O_2$ digestion [33] (6 mL 18 M $H_2SO_4$; 4 mL 9.8 M $H_2O_2$), and measured P, K, Ca, and Mg concentrations by inductively coupled plasma analysis (Thermo Jarrel-Ash-ICAP 61E, Thermo Fisher Scientific, Waltham, MA, USA).

In each sub-plot, we collected four soil cores at the end of the 2016 growing season using a 4.8 cm diameter metal cylinder inserted down to the first 15 cm of mineral soil. In planted sub-plots, sampling spots were located at the base of the four corner-saplings that were sampled for foliar analyses. In unplanted sub-plots, we collected samples in the four corners in microsites that would have been adequate for planting. Soil samples were dried to 5% mass based moisture content and sieved at 2 mm. Five grams of soil ($\pm$0.02 g) were extracted with 50 mL 2M KCl solution [34] and inorganic N ($NH_4$–N and $NO_3$–N) was analysed by spectrophotometry (QuikChem R8500 Series 2, Lachat Instruments, Milwaukee, WI, USA) after 30 min. agitation and filtration.

## 2.4. Statistical Analyses

Data regarding DBH, height and foliar nutrients were analyzed using linear mixed models with site preparation as a fixed effect and the block as a random effect following a fully randomized block design. Results regarding inorganic soil N were analyzed using a linear mixed model with site preparation, plantation, and site preparation $\times$ plantation interaction as fixed effects and the block as a random effect following a split-plot design. An $\alpha$ threshold of 0.05 was used to identify significant effects. Normality and homoscedasticity were verified for all data using visual distribution of data and by analysis of residues. Natural logarithmic transformations were made when necessary. Comparisons between site preparation treatments were assessed with Tukey's a posteriori mean comparison tests. All statistical analyses were performed with the *nlme* and *emmeans* packages of R, version 3.5.3 [35–37].

## 3. Results

We found a significant effect of treatments on height and DBH as measured 4 growing seasons after planting ($p \leq 0.006$; Figure 2). Saplings were 68% taller in the mounding treatment than the control plots ($p = 0.005$), but height in V-blade treated plots was not different from height in control conditions ($p = 0.116$; Figure 2a). Sapling height was statistically equivalent in plots treated with mounding and V-blade ($p = 0.065$). Saplings in the V-blade and in the mounding treated plots had a DBH that was respectively 91% and 155% larger than saplings planted in the control plots ($p \leq 0.021$; Figure 2b). DBH was equivalent between the mounding and V-blade treatments ($p = 0.195$). Survival was higher in the mounding (99%), compared to 91% in the V-blade and 48% in the control treatments.

(a) 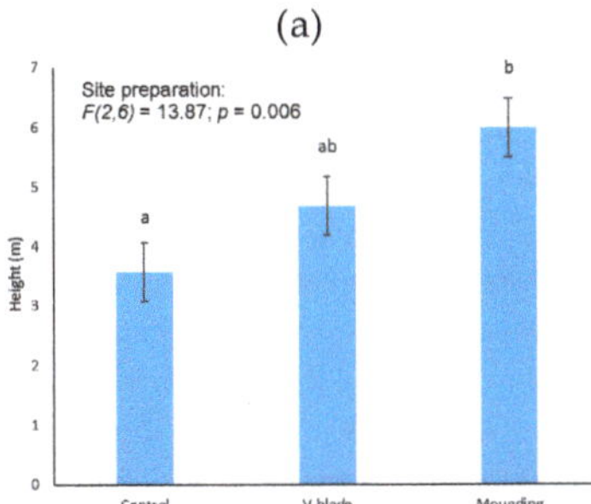


(b) 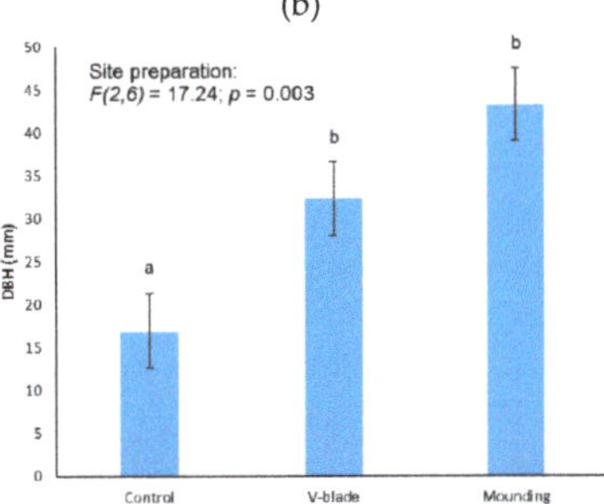


**Figure 2.** Effects of mechanical site preparation treatments on hybrid poplar sapling height (**a**) and diameter at breast height (DBH) (**b**) after four growing seasons, along with results from the linear mixed models (values in parentheses are the numerator and denominator degrees of freedom). *p*-value for the random block effect was <0.001 for both variables. For a given variable, values with similar letters are not significantly different at $\alpha = 0.05$ based on Tukey HSD pairwise comparisons. Data are presented as means $\pm$ standard error.

None of the foliar nutrient concentrations differed between site preparation treatments four growing seasons after planting (Table 1). Soil inorganic N was equivalent between site preparation treatments and was not influenced by the presence/absence of planted hybrid poplars (Table 1).

**Table 1.** Synthesis of the linear mixed model results for hybrid poplar foliar nutrition and soil inorganic N, as measured four growing seasons after planting.

| Variable | Mean (SE) | Site Preparation (SP) | | Plantation (P) | | SP × P | |
|---|---|---|---|---|---|---|---|
| | | $F_{(2,6)}$ [a] | *p*-Value [b] | $F_{(1,9)}$ [a] | *p*-Value [b] | $F_{(2,9)}$ [a] | *p*-Value [b] |
| Foliar N (g·kg$^{-1}$) | 12.4 ± 0.7 | 3.2 | 0.112 | – | – | – | – |
| Foliar P (g·kg$^{-1}$) | 2.0 ± 0.2 | 4.7 | 0.058 | – | – | – | – |
| Foliar K (g·kg$^{-1}$) | 15.3 ± 10.3 | 0.0 | 0.993 | – | – | – | – |
| Foliar Ca (g·kg$^{-1}$) | 14.0 ± 2.3 | 2.3 | 0.183 | – | – | – | – |
| Foliar Mg (g·kg$^{-1}$) | 2.2 ± 0.4 | 0.2 | 0.815 | – | – | – | – |
| Soil inorganic N (mg·kg$^{-1}$) | 3.1 ± 0.2 | 2.6 | 0.155 | 3.9 | 0.080 | 2.6 | 0.132 |

[a] Presented as *F (numerator degrees of freedom, denominator degrees of freedom)*. [b] *p*-values for the random block effect varied from < 0.001 to 0.048, depending on variables.

## 4. Discussion

Our results confirm the global positive effects of site preparation on the survival and the early growth of hybrid poplar clones, a fast growing species that is sensitive to nutrient deficiencies and competition by vegetation after planting, especially on forested sites [15,18]. Previous studies in temperate and boreal regions have reported positive effects of mechanical site preparation on species like *Populus* spp., *Pine* spp., and *Picea* spp. [38–41]. The objective of mechanical site preparation is to create adequate microsites for seedling rooting and early growth. Soil disturbance affects substrate temperature, moisture and density (porosity), and ultimately nutrient availability and light exposure ([20] and references therein). The advantages of mounding over blading might be related to the microtopography created by the treatment. Mounding is not recommended on drought-prone sites because mounds are susceptible to drying, which might compromise cutting survival, particularly during the first season after planting. However, soil elevation creates a well-drained, aerated and uncompact substrate that provides better rooting conditions, an advantage given the high annual precipitation regime in our study region (1012 mm). On the other hand, cuttings planted in furrows created by blading might suffer from soil water saturation and root hypoxia, particularly if the water table is high [39,42]. Site preparation can indeed modify rooting patterns; root distribution is more even on mounds than in furrows created by disc trenching [43].

Site preparation such as soil inversion or ploughing has been reported to increase [18,44] or decrease [45,46] soil N pool or mineralized N concentration. In our study, conducted on relatively rich boreal mixedwood sites of Eastern Canada, we did not measure any significant effect of site preparation on soil inorganic N. Similarly, sapling foliar nutrient concentrations were not influenced by the treatments either, although foliar nutrient concentrations suggest that N is below the optimum range while other nutrients appeared to be near the optimum [47]. The effects of site preparation are highly dependent on soil characteristics, including management history, surficial deposit and climate. Benefits from site preparation on nutrient availability and subsequently, on planted cutting nutritional status, are less likely to occur on sites with thin humus relative to boreal forest sites with thick humus layers [48,49]. We expected that the presence of hybrid poplar would, by itself, influence soil inorganic N. Indeed, because of the quality of its litter, the presence of *Populus tremuloides* in boreal stands has been shown to improve cation exchangeable capacity and pH of the forest floor [50]. We posit that we have not detected these effects on the sites we have studied because litter input is still too low to influence soil physical and chemical properties and because tree growth appears to be mainly N limited. Longer-term monitoring of the planted/unplanted plots of this experiment will allow testing the impacts of hybrid poplar on soil properties and disentangle them from those of site preparation.

## 5. Conclusions

Both of the mechanical site preparation treatments we have tested in this study significantly increased the radial growth of planted hybrid poplar, but only mounding had a significant effect on sapling height growth relative to control conditions. These effects could not be explained by improved soil nutrient availability or foliar nutritional status and might be the results of improved root growth, water status or abiotic or biotic interactions not studied here. Based on these results, mounding, with a higher survival rate, appears as a good management approach to establish hybrid poplar plantations under the ecological conditions studied here. Although the differences with V-blading were small in terms of sapling growth, mounding is less likely to cause erosion and nutrient lixiviation than blading because of the localized nature of the treatment. These environmental benefits need, however, to be balanced against economic and social considerations in the context of sustainable forest management. For example, apart for being more costly, the mounding treatment creates a rough microtopography that can be a challenge for vegetation management operations using motor-manual brushsaws as an alternative to chemical herbicides [51], contrary to V-blading that creates regular corridors in which workers can easily circulate.

**Author Contributions:** Conceptualization, N.T., R.E., F.H., D.P. and P.G.; methodology, N.T., F.H., D.P. and P.G.; writing—original draft preparation, N.T., R.E. and F.H.; writing—review and editing, N.T., R.E., F.H., D.P. and P.G. All authors have read and agreed to the published version of the manuscript.

**Funding:** This research was funded by Cascades, the Canadian Forest Service of Natural Resources Canada, with the participation of the Ministère des Forêts, de la Faune et des Parcs du Québec (Laboratoire de chimie organique et inorganique), the Conférence régionale des élus du Bas-Saint-Laurent, the Groupement forestier de Kamouraska, the Groupement forestier du Témiscouata and the Réseau Ligniculture Québec (now the Réseau Reboisement Ligniculture Québec).

**Acknowledgments:** We thank Clarence Dubé for his support in the establishment of the experiment. We are indebted to Sébastien Dagnault, Fanny Michaud, Olivier Jeffrey, Evelyne Gaillard and Ulysse Rémillard for their technical support, as well as to the staff of the laboratories at the Direction de la recherche forestière and the Laurentian Forestry Centre. We thank Michael Hoepting and two anonymous reviewers for providing constructive advice on an earlier version of this work.

**Conflicts of Interest:** The authors declare no conflict of interest. The funders had no role in the design of the study; in the collection, analyses, or interpretation of data; in the writing of the manuscript, or in the decision to publish the results.

## References

1. Paquette, A.; Messier, C. The role of plantations in managing the world's forests in the Anthropocene. *Front. Ecol. Environ.* **2010**, *8*, 27–34. [CrossRef]
2. Barua, S.K.; Lehtonen, P.; Pahkasalo, T. Plantation vision: Potentials, challenges and policy options for global industrial forest plantation development. *Int. For. Rev.* **2014**, *16*, 117–127. [CrossRef]
3. Payn, T.; Carnus, J.-M.; Freer-Smith, P.; Kimberley, M.; Kollert, W.; Liu, S.; Orazio, C.; Rodriguez, L.; Silva, L.N.; Wingfield, M.J. Changes in planted forests and future global implications. *For. Ecol. Manag.* **2015**, *352*, 57–67. [CrossRef]
4. Balandier, P.; Collet, C.; Miller, J.H.; Reynolds, P.E.; Zedaker, S.M. Designing forest vegetation management strategies based on the mechanisms and dynamics of crop tree competition by neighbouring vegetation. *Forestry* **2006**, *79*, 3–27. [CrossRef]
5. Thiffault, N.; Jobidon, R.; Munson, A.D. Comparing large containerized and bareroot conifer stock on sites of contrasting vegetation composition in a non-herbicide scenario. *New For.* **2014**, *45*, 875–891. [CrossRef]
6. Goehing, J.; Henkel-Johnson, D.; Macdonald, S.E.; Bork, E.W.; Thomas, B.R. Spatial partitioning of competitive effects from neighbouring herbaceous vegetation on establishing hybrid poplars in plantations. *Can. J. For. Res.* **2019**, *49*, 595–605. [CrossRef]
7. Fortier, J.; Truax, B.; Gagnon, D.; Lambert, F. Abiotic and biotic factors controlling fine root biomass, carbon and nutrients in closed-canopy hybrid poplar stands on post-agricultural land. *Sci. Rep.* **2019**, *9*, 6296. [CrossRef]
8. Thomas, B.R.; Schreiber, S.G.; Kamelchuk, D.P. Impact of planting container type on growth and survival of three hybrid poplar clones in central Alberta, Canada. *New For.* **2016**, *47*, 815–827. [CrossRef]
9. Grenke, J.S.J.; Macdonald, S.E.; Thomas, B.R.; Moore, C.A.; Bork, E.W. Relationships between understory vegetation and hybrid poplar growth and size in an operational plantation. *For. Chron.* **2016**, *92*, 469–476. [CrossRef]
10. Mamashita, T.; Larocque, G.R.; DesRochers, A.; Beaulieu, J.; Thomas, B.R.; Mosseler, A.; Major, J.; Sidders, D. Short-term growth and morphological responses to nitrogen availability and plant density in hybrid poplars and willows. *Biomass Bioenerg.* **2015**, *81*, 88–97. [CrossRef]

11.  Masse, S.; Marchand, P.P.; Bernier-Cardou, M. Forecasting the deployment of short-rotation intensive culture of willow or hybrid poplar: Insights from a Delphi study. *Can. J. For. Res.* **2014**, *44*, 422–431. [CrossRef]

12.  Larocque, G.R.; DesRochers, A.; Larchevêque, M.; Tremblay, F.; Beaulieu, J.; Mosseler, A.; Major, J.E.; Gaussiran, S.; Thomas, B.R.; Sidders, D.; et al. Research on hybrid poplars and willow species for fast-growing tree plantations: Its importance for growth and yield, silviculture, policy-making and commercial applications. *For. Chron.* **2013**, *89*, 32–41. [CrossRef]

13.  Stanturf, J.A.; van Oosten, C. Operational poplar and willow culture. In *Poplars and Willow: Trees for Society and the Environment*; Isebrands, J.G., Richardson, J., Eds.; FAO: Rome, Italy, 2014; pp. 200–257.

14.  Bilodeau-Gauthier, S.; Paré, D.; Messier, C.; Bélanger, N. Juvenile growth of hybrid poplars on acidic boreal soil determined by environmental effects of soil preparation, vegetation control, and fertilization. *For. Ecol. Manag.* **2011**, *261*, 620–629. [CrossRef]

15.  McCarthy, R.; Rytter, L.; Hjelm, K. Effects of soil preparation methods and plant types on the establishment of poplars on forest land. *Ann. For. Sci.* **2017**, *74*, 1–12.

16.  Forest Products Association of Canada. *Forest Certification in Canada. The Programs, Similarities and Achievements*; Forest Products Association of Canada (FPAC): Ottawa, ON, Canada, 2011.

17.  Québec, Sustainable Forest Development Act. Chapter A-18.1. In Les Publications du Québec: Québec, QC, Canada. 2020. Available online: http://legisquebec.gouv.qc.ca/en/ShowDoc/cs/A-18.1 (accessed on 10 June 2020).

18.  Bilodeau-Gauthier, S.; Paré, D.; Messier, C.; Bélanger, N. Root production of hybrid poplars and nitrogen mineralization improve following mounding of boreal Podzols. *Can. J. For. Res.* **2013**, *43*, 1092–1103. [CrossRef]

19.  Messier, C.; Coll, L.; Poitras-Larivière, A.; Bélanger, N.; Brisson, J. Resource and non-resource root competition effects of grasses on early- versus late-successional trees. *J. Ecol.* **2009**, *97*, 548–554. [CrossRef]

20.  Löf, M.; Dey, D.C.; Navarro, R.M.; Jacobs, D.F. Mechanical site preparation for forest restoration. *New For.* **2012**, *43*, 825–848. [CrossRef]

21.  Pennanen, T.; Heiskanen, J.; Korkama, T. Dynamics of ectomycorrhizal fungi and growth of Norway spruce seedlings after planting on a mounded forest clearcut. *For. Ecol. Manag.* **2005**, *213*, 243–252.

22.  Bolte, A.; Löf, M. Root spatial distribution and biomass partitioning in *Quercus robur* L. seedlings: The effects of mounding site preparation in oak plantations. *Eur. J. For. Res.* **2010**, *129*, 603–612. [CrossRef]

23.  Buitrago, M.; Paquette, A.; Thiffault, N.; Bélanger, N.; Messier, C. Early performance of planted hybrid larch: Effects of mechanical site preparation and planting depth. *New For.* **2015**, *46*, 319–337.

24.  Hébert, F.; Bachand, M.; Thiffault, N.; Paré, D.; Gagné, P. Recovery of plant community functional traits following severe soil perturbation in plantations: A case-study. *Int. J. Biodiv. Sci. Ecosyst. Serv. Manag.* **2016**, *12*, 116–127. [CrossRef]

25.  Sutton, R.F. Mounding site preparation: A review of European and North American experience. *New For.* **1993**, *7*, 151–192. [CrossRef]

26.  Figueiredo, T.; Fonseca, F.; Martins, A. Soil loss and run-off in young forest stands as affected by site preparation technique: A study in NE Portugal. *Eur. J. For. Res.* **2011**, *131*, 1747–1760. [CrossRef]

27.  Truax, B.; Gagnon, D.; Lambert, F.; Fortier, J. Riparian buffer growth and soil nitrate supply are affected by tree species selection and black plastic mulching. *Ecol. Eng.* **2017**, *106*, 82–93. [CrossRef]

28.  Saucier, J.P.; Robitaille, A.; Grondin, P. Cadre bioclimatique du Québec. In *Manuel de Foresterie*, 2nd ed.; Doucet, R., Côté, M., Eds.; Ordre des Ingénieurs Forestiers du Québec, Éditions Multimondes: Québec, QC, Canada, 2009; pp. 186–205.

29.  Environment Canada Canadian Climate Normals 1981–2010 Station Data. Available online: http://climate.weather.gc.ca/climate_normals/index_e.html (accessed on 10 June 2020).

30.  Soil Classification Working Group. *The Canadian System of Soil Classification*, 3rd ed.; 1646; Agriculture and Agri-Food Canada: Ottawa, ON, Canada, 1998; p. 187.

31.  Jobidon, R. *Autécologie de Quelques Espèces deCcompétition d'Importance pour la Régénération Forestière au Québec. Revue de Littérature 117*; Ministère des Ressources Naturelles, Direction de la Recherche Forestière: Québec. QC, Canada, 1995; p. 180.

32.  Réseau Ligniculture Québec. *Le Guide de Populiculture au Québec*; Réseau Ligniculture Québec: Québec, QC, Canada, 2011; p. 124.

33.  Thomas, R.R.; Sheard, R.W.; Moyer, J.R. Comparison of conventional and automated procedures for nitrogen, phosphorus and potassium, analysis of plant material using a single digestion. *Agron. J.* **1967**, *59*, 240–243. [CrossRef]

34. Carter, M.; Gregorich, E. *Soil Sampling and Methods of Analysis*, 2nd ed.; CRC Press: Boca Raton, FL, USA, 1993; p. 1262.

35. Pinheiro, J.; Bates, D.; DebRoy, S.; Sarkar, D. R Core Team nlme: Linear and Nonlinear Mixed Effects Models R package version 3.1-140. 2019. Available online: https://cran.r-project.org/package=nlme (accessed on 10 June 2020).

36. R Core Team. *R: A Language and Environment for Statistical Computing*; R Foundation for Statistical Computing: Vienna, Austria, 2017; Available online: https://www.r-project.org/ (accessed on 10 June 2020).

37. Lenth, R. Emmeans: Estimated Marginal Means, aka Least-Squares Means. R package version 1.3.5.1. 2019. Available online: https://CRAN.R-project.org/package=emmeans (accessed on 10 June 2020).

38. Fraser, E.C.; Landhäusser, S.M.; Lieffers, V.J. The effects of mechanical site preparation and subsequent wildfire on trembling aspen (*Populus tremuloides* Michx.) regeneration in central Alberta, Canada. *New For.* **2003**, *25*, 67–81. [CrossRef]

39. Simard, M.; Lecomte, N.; Bergeron, Y.; Bernier, P.Y.; Paré, D. Forest productivity decline caused by successional paludification of boreal soils. *Ecol. Appl.* **2007**, *17*, 1619–1637. [CrossRef]

40. Hébert, F.; Boucher, J.-F.; Walsh, D.; Tremblay, P.; Côté, D.; Lord, D. Black spruce growth and survival in boreal open woodlands 10 years following mechanical site preparation and planting. *Forestry* **2014**, *87*, 277–286. [CrossRef]

41. Henneb, M.; Valeria, O.; Thiffault, N.; Fenton, N.J.; Bergeron, Y. Effects of mechanical site preparation on microsite availability and growth of planted black spruce in Canadian paludified forests. *Forests* **2019**, *10*, 670. [CrossRef]

42. Sikström, U.; Hökkä, H. Interactions between soil water conditions and forest stands in boreal forests with implications for ditch network maintenance. *Silva Fenn.* **2016**, *50*, 1416. [CrossRef]

43. Celma, S.; Blate, K.; Lazdiņa, D.; Dūmiņš, K.; Neimane, S.; Štāls, T.A.; Štikāne, K. Effect of soil preparation method on root development of P. sylvestris and P. abies saplings in commercial forest stands. *New For.* **2018**, *50*, 283–290. [CrossRef]

44. Ryan, D.F.; Huntington, T.G.; Martin, C.W. Redistribution of soil nitrogen, carbon and organic matter by mechanical disturbance during whole-tree harvesting in northern hardwoods. *For. Ecol. Manag.* **1992**, *49*, 87–99. [CrossRef]

45. Sutinen, R.; Gustavsson, N.; Hänninen, P.; Middleton, M.; Räisänen, M.L. Impact of mechanical site preparation on soil properties at clear-cut Norway spruce sites on mafic rocks of the Lapland Greenstone Belt. *Soil Tillage Res.* **2019**, *186*, 52–63. [CrossRef]

46. Aleksandrowicz-Trzcińska, M.; Drozdowski, S.; Studnicki, M.; Żybura, H. Effects of site preparation methods on the establishment and natural-regeneration traits of scots pines (*Pinus sylvestris* L.) in northeastern Poland. *Forests* **2018**, *9*, 717. [CrossRef]

47. Lteif, A.; Whalen, J.K.; Bradley, R.L.; Camiré, C. Diagnostic tools to evaluate the foliar nutrition and growth of hybrid poplars. *Can. J. For. Res.* **2008**, *38*, 2138–2147. [CrossRef]

48. Thiffault, N.; Jobidon, R. How to shift unproductive *Kalmia angustifolia-Rhododendron groenlandicum* heath to productive conifer plantation. *Can. J. For. Res.* **2006**, *36*, 2364–2376. [CrossRef]

49. Thiffault, N.; Jobidon, R.; Munson, A.D. Performance and physiology of large containerized and bare-root spruce seedlings in relation to scarification and competition in Québec (Canada). *Ann. For. Sci.* **2003**, *60*, 645–655. [CrossRef]

50. Légaré, S.; Paré, D.; Bergeron, Y. Influence of aspen on forest floor properties in black spruce-dominated stands. *Plant Soil* **2005**, *275*, 207–220. [CrossRef]

51. Thiffault, N.; Roy, V. Living without herbicides in Québec (Canada): Historical context, current strategy, research and challenges in forest vegetation management. *Eur. J. For. Res.* **2011**, *130*, 117–133. [CrossRef]

 *forests*

*Article*

# Nutrient Contribution of Litterfall in a Short Rotation Plantation of Pure or Mixed Plots of *Populus alba* L. and *Robinia pseudoacacia* L.

Isabel González [1], Hortensia Sixto [1], Roque Rodríguez-Soalleiro [2] and Nerea Oliveira [1,*]

[1] Forest Research Centre, National Institute for Agriculture and Food Research and Technology (INIA-CIFOR), Crta, de la Coruña km 7.5, E-28040 Madrid, Spain; isabelgz@inia.es (I.G.); sixto@inia.es (H.S.)

[2] Sustainable Forest Management Group, University of Santiago de Compostela, C/Benigno Ledo s/n, E-27002 Lugo, Spain; roque.rodriguez@usc.es

* Correspondence: oliveira.nerea@inia.es

Received: 6 October 2020; Accepted: 23 October 2020; Published: 25 October 2020

**Abstract:** This study aims to quantify the potential contribution of nutrients derived from leaf litter in a short rotation coppice plantation which includes monocultures of the species *Populus alba* (PA) and *Robina pseudoacacia* (RP) as well as a mixture of 50PA:50RP, in the middle of the rotation. The *P. alba* monoculture was that which provided the most leaf litter (3.37 mg ha$^{-1}$ yr$^{-1}$), followed by the 50PA:50RP mixture (2.82 mg ha$^{-1}$ yr$^{-1}$) and finally the *R. pseudoacacia* monoculture (2.55 mg ha$^{-1}$ yr$^{-1}$). In addition to producing more litterfall, leaves were shed later in the *P. alba* monoculture later (December) than in the *R. pseudoacacia* monoculture (October) or the mix (throughout the fall). In terms of macronutrient supply per hectare, the contributions derived from leaf litter were higher for K, P and Mg in the case of *P. alba* and for N in *R. pseudoacacia*, the mix presenting the highest Ca content and intermediate concentrations for the rest of the nutrients. In addition, other factors such as C:N or N:MO ratios, as well as the specific characteristics of the soil, can have an important impact on the final contribution of these inputs. The carbon contribution derived from leaf fall was higher in the *P. alba* monoculture (1.5 mg ha$^{-1}$ yr$^{-1}$), intermediate in the mixed plot (1.3 mg ha$^{-1}$ yr$^{-1}$) and slightly lower for the *R. pseudoacacia* monoculture (1.3 mg ha$^{-1}$ yr$^{-1}$). Given these different strategies of monocultures with regard to the dynamism of the main nutrients, species mixing would appear to be suitable option to achieve a potential reduction in mineral fertilization in these plantations.

**Keywords:** short rotation coppice (SRC); biomass; white poplar; black locust; monocultures; mixture; leaf litter

## 1. Introduction

Forest plantations of fast-growing species under a short rotation coppice system (SRC) can contribute to the supply of biomass for use in bioenergy and bioproducts within the context of the bioeconomy [1]. *Salicaceae* (poplars and willows) are suitable species for this purpose due to their high productivity, ease of vegetative multiplication and ample availability of genetic material [2], resulting in crops of this family of genotypes being common in many areas of the world [3]. To a lesser extent, other species have been considered for SRC [4,5]. Among the latter, *Robinia pseudoacacia* L., of de *Fabaceae* family, is also a fast-growing species, with a certain degree of drought tolerance, capable of sprouting from the stumps and with a high nitrogen fixing capacity [6]. For these reasons, *Robinia* is considered suitable for cultivation in SRC in some areas of Europe [7,8], although it is also considered an invasive species introduced into Europe in the 17th century [9].

Most SRC plantations are established as monocultures with a single species, although the possibility of mixed stands has also been explored [10–12]. Mixed plantations, in addition to increasing genetic variability and favoring tolerance to certain stresses, can also provide productive benefits based on complementary or facilitation strategies [13].

One of the main internal flows of the continuous vegetation–soil–fauna dynamic in forestry or agroforestry ecosystems is leaf litter, the subsequent decomposition process, and the consequent incorporation of organic matter and nutrients, needed for growth, into the soil [14–16]. Leaf litter is therefore part of a key mechanism of recycling and redistribution of nutrients. Litterfall quantification allows the potentiality as regards the degree of annual return of nutrients to the soil to be assessed. The quantity of litterfall will depend on different factors such as the genotype, the climatic factors such as temperature and light, fertility and degree of soil moisture, type of management or age of the plot, among others [17–20].

In plantations with fast-growing species, the leaf litter plays an important role within the nutrient cycle, allowing the replacement of a high percentage of mineral nutrients to the soil [21–24]. Soil fertility and nutrient recycling is one of the main concerns in relation to sustainability [25] and the assessment of forest plantations on agricultural soils is therefore pertinent. In the specific case of deciduous *Populus* spp., it is estimated that around 88% of N, 83% P and 78% K are returned through leaf litter in mature plantations [26]. Other authors, however, report lower return rates of between 20 and 40% and suggest that the rate depends highly on the genotype [27]. In general, nutrient cycling in poplar stands is considered efficient, with no significant loss of nutrients according to Meiresonne et al. [28] and returns via leaf litter are also often rich in basic cations [29,30]. Data on these nutrient returns in the specific case of poplar growing in short rotation coppice point to between 60–80% of the nutrients absorbed being returned annually through litterfall [31]. However, in SRC plantations with another fast-growing species such as eucalyptus, Guo et al. [32] reported rates of return of around 24% for N, although in this case it is an evergreen species.

Leaf litter decomposition rate is also important for determining how nutrients enter the soil and will also determine the amount of organic matter which accumulates. This rate is controlled by both biotic and abiotic factors, with the chemical composition of the leaf litter (especially N and P concentration and C:N ratio) being one of the main influencing factors [33–35]. The variability of the chemical composition of the leaf litter will be determined, among other factors, by the efficiency of the reabsorption and relocation of nutrients at species level and by their interspecific variation [36].

Additionally, to quantify the contribution of SRC to carbon content it is necessary to determine the sources of variation in C concentration and to address potential sequestration in these farming systems [37]. In recent years, studies have pointed to the high potential contribution of plantations with fast-growing species to the global carbon budget [38,39]. Moreover, such plantations have been proposed as part of a strategy to mitigate global warming in the short term [40,41].

In forest plantations with fast-growing species, and specifically in the case of high density, short rotation plantations for biomass production, this evaluation is necessary not only to promote more sustainable management, but also in order to value the ecosystem services associated with them. Furthermore, it is necessary to evaluate the potential adverse effects on forest soils associated with the implementation of intensive plantations [42–44] on highly managed agricultural land, since these plantations are established on this type of land in many countries. Hence, the aims of this work are (i) to quantify the annual production of leaf litter and its composition in pure and mixed plots of high density, short rotation coppice (SRC) under Mediterranean conditions; and (ii) to determine the nutrient dynamics in this type of plantation and the potential impact on the soil. We hypothesized that mixed plantations improve the quality of the leaf litter with respect to monocultures, and consequently increase soil fertility.

## 2. Material and Methods

### 2.1. Study Area

The experimental plantation is located in the center of the Iberian Peninsula (40°28′ N, 3°22′ W) at an elevation of 595 m, average mean temperature 15.3 °C (mean absolute maximum 28.8 °C and mean absolute minimum of 3.1 °C), with annual precipitation of 281 mm. The main edaphic features are detailed in Table 1.

**Table 1.** Average values ± standard deviation of physical-chemical parameters and concentrations of assimilable Phosphorus and interchangeable elements (Potassium, Calcium, Magnesium and Sodium) on the surface horizon of the soil (0–30 cm).

| Parameters | | Before Plantation |
|---|---|---|
| Texture | | Sandy-loam |
| pH H$_2$O | | 8.65 ± 0.10 |
| EC | mS cm$^{-1}$ | 0.59 ± 0.24 |
| Carbonates | % | 10.08 ± 1.00 |
| N | g kg$^{-1}$ | 0.81 ± 0.08 |
| OM | % | 0.68 ± 0.08 |
| C:N | | 23.77 ± 1.97 |
| P | mg kg$^{-1}$ | <4 |
| K | cmol kg$^{-1}$ | 0.30 ± 0.05 |
| Ca | cmol kg$^{-1}$ | 38.71 ± 0.87 |
| Mg | cmol kg$^{-1}$ | 4.71 ± 0.44 |
| Na | cmol kg$^{-1}$ | 0.73 ± 0.30 |

Granulometric analysis was recorded by the Bouyoucus method; The pH and electrical conductivity (EC) was potentiometrically determined; Total nitrogen (N) was calculated according to Kjeldahl modified method; the soil organic matter (OM) was calculated by Walkley–Black method; carbonates was calculated according to Bernard calcimeter method; assimilable P was calculated according to Olsen; and the concentrations of exchangeable Ca, Mg, Na and K were determined by extracting with ammonium acetate (1N) and subsequently analyzing using ICP-OES (Optima 5300 DV, Perkin-Elmer, Massachusetts, MA, USA).

The plantation was established in 2012 with the aim of evaluating biomass yield under different species compositions. Different mixing ratios of two fast-growing species were tested under a high-density (10,000 trees ha$^{-1}$, spacing 2.5 m × 0.40 m) short rotation coppice system. The two species were *Populus alba* L., genotype '111PK', and *Robinia pseudoacacia* L., genotype 'Nyirsegy', and the trial included *P. alba* monoculture (PA), *R. pseudoacacia* monoculture (RP) and a mixture of both at a ratio of 50PA:50RP. The 50PA:50RP mix of both species was done tree by tree within the row and between rows. The figure design and the results of this research are described in Oliveira et al. [12]. The plantation was established using stem cuttings (unrooted in the case of *P. alba* and rooted for *R. pseudoacacia*) having followed the standard soil preparation procedure described for SRC plantations [45]. Since the Mediterranean climate is characterized by severe summer drought, the plantation was irrigated from June to September using a drip application system. No fertilization treatment was applied.

The experimental design included three blocks, each containing the *P. alba* and *R. pseudoacacia* monocultures as well as the 50PA:50RP mix of both species. Each block and plot contained 64 trees in total. Further details on the experimental design are given in Oliveira et al. [12].

### 2.2. Litterfall Collection

Litterfall samples were collected from September to December in the 1st vegetative period of the 2nd rotation (R4S1, where R is the root age and S is the stool age), being the value of basal area of basal diameters (BA) and the height (H) of the species the following: the *P. alba* monoculture (BA = 23.94 cm$^2$ and H = 5.46 m); the *R. pseudoacacia* monoculture (BA = 11.73 cm$^2$ and H = 3.92 m) and the mix (BA = 20.75 cm$^2$ and H = 4.89 m for PA and BA = 14.80 cm$^2$ and H = 4.50 m for RP). Twelve litterfall traps (perforated plastic boxes with a surface area of 0.17 m$^2$ and a height of 23 cm) were randomly placed in the rows of each block and plot and within the row, equidistant between two trees. The final number of traps was thirty-six.

The monthly accumulated litterfall was taken to the laboratory where leaves were separated from the rest, which included twigs, bark, seeds, shoots, and other released components. The leaves were then dried at 65 °C to constant weight and finally weighed. The leaf litter contribution in each subplot was calculated by adding the results for the different traps. The calculation per unit area was performed by dividing the sum of the total dry weight of the different fractions by the area of the trap, extrapolating the result obtained to one hectare. The unit to express the contribution of leaf litter is therefore mg ha$^{-1}$ yr$^{-1}$ in dry matter.

Prior to the abscission of the leaf (end of August), when it was probable that the translocation of nutrients from the leaves to the reserve organs had not yet begun [2,46], fresh green leaves were collected from the trees in the same blocks and plots where the traps had been placed for further analysis.

*2.3. Foliar Nutrient Analysis*

The following analyses were performed on both the green leaves and senescent leaves collected over 3 months: Total C and N by dry combustion using an elemental analyzer (CNS-2000, LECO, St Joseph, MI, USA); and P, K, Ca and Mg were determined by optical emission spectroscopy using ICP-OES (Optima 5300 DV, Perkin-Elmer, Massachusetts, MA, USA) after wet digestion of the sample with nitric acid in a closed microwave system (Ethos plus, Milestone, Sorisole, Italy).

The percentage of nutrient resorption efficiency (NRE; hereafter retranslocation) between the two types of leaves (green and senescent) was calculated according to the following Equation [47]:

$$NRE = (Nu_{green} - Nu_{senescent})/Nu_{green} \times 100 \tag{1}$$

where $Nu_{green}$ is the nutrient concentration in the green leaf and $Nu_{senescent}$ is the concentration in senescent leaf.

The nutrient use efficiency index was determined according to Vitousek [48], through the relationship between dry mass and nutrient concentration ratio of leaf litter.

*2.4. Data Analysis*

A multivariate analysis of variance was performed to assess the effect of the treatments (plot type and sampling time) on leaf litter production and the chemical composition. A one-way ANOVA was performed when evaluating the effect of a single factor. Fisher's Least Significant Difference (LSD) test was used to establish those means that are significantly different.

A non-parametric analysis was performed when the assumptions of the one-way ANOVA test were not met, using the Kruskal–Wallis test in these cases.

We worked with weighted annual averages according to weight fraction at subplot level when analyzing data related to leaf nutrients due to the variability in both concentration and input over the sampling period. The software package used was the R statistical program.

## 3. Results and Discussion

*3.1. Leaf Litter Supply*

The most representative fraction of litterfall in all plots of the plantation (both monocultures and 50PA:50RP mixture) corresponded to leaves (around 98%); therefore, we will refer to this component from now on, with leaf litter being understood as all the leaves falling into the litterfall traps. However, according to Medina-Villar et al. [49], the leaf percentages for both species growing in the riparian ecosystem were lower (69%), probably because the litterfall trap contents comprised the entire annuity.

Leaf detachment in deciduous species mainly takes place throughout the fall. Abscission can occur at any time during this period, depending on various factors such as weather and edaphic conditions (water stress and soil fertility) but also on the species [50]. In our study, under the same soil and climate conditions, the maximum leaf litter values for the *P. alba* monoculture were reached in December, while leaf shedding in the *R. pseudoacacia* monoculture occurred earlier, reaching maximum values in October (Figure 1). This fact is in accordance with observed differences in phenology, as winter buds are formed earlier in *R. pseudoacacia* (early September) and later in *P. alba* (late October). This finding has previously been reported by Medina-Villar et al. [49] for natural stands in a study area proximate to that of the present study and supports previous findings by González-Muñoz et al. [51] and Castro-Díez et al. [52]. Furthermore, it may be attributable to differences in the strategies for minimizing the energy expenditure required to keep tissues alive when the temperatures fall [53].

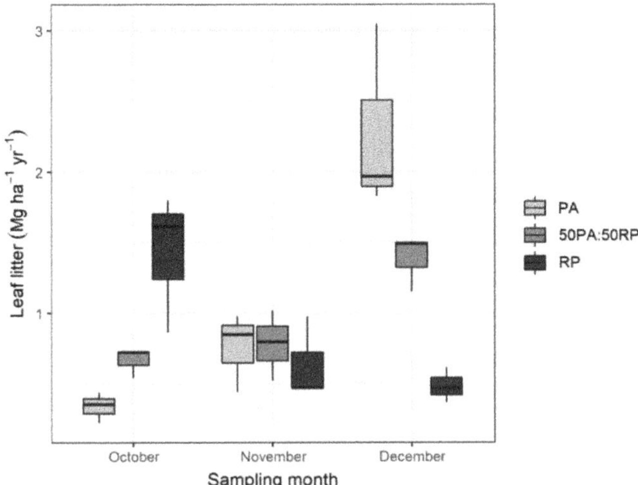

**Figure 1.** Monthly contribution (mg ha$^{-1}$ yr$^{-1}$) of the leaf litter in the *P. alba* (PA) and *R. pseudoacacia* (RP) monocultures, and the 50PA:50RP mixture.

Since a greater amount of the leaves in the total leaf litter of the mixture corresponded to *P. alba* (60 % PA to 40% RP) (Figure 2), and as this species sheds most of its leaves in December (82%), this was the month in which leaf litter in the mixture reached a maximum.

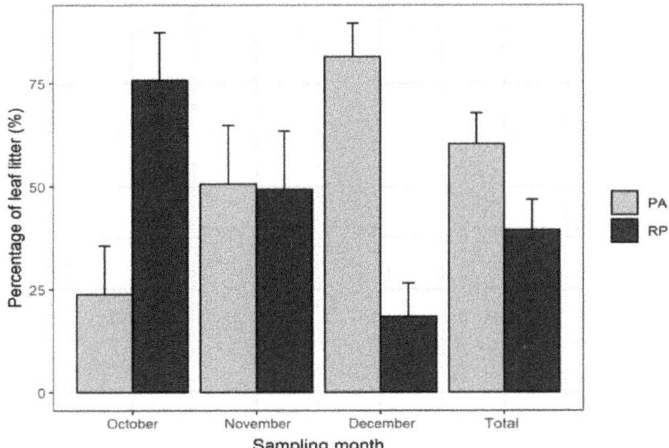

**Figure 2.** Percentage contribution of *P. alba* (PA) and *R. pseudoacacia* (RP) to the total leaf litter of the 50PA:50RP mixture.

The total leaf litter expressed in Mg per hectare in the different plots is shown in Table 2. Although the amounts of leaf litter among plots were not significantly different ($p$-value = 0.2160), the annual inflow of leaves was more than 20% higher in the *P. alba* monoculture compared to *R. pseudoacacia* and the 50PA:50RP mixture. This trend contrasts with that described by Medina-Villar et al. [49], who reported greater leaf litter for *R. pseudoacacia* and pointed to generally higher growth rates due to the invasive character of this species compared to native species [51,52]. These conflicting findings may be due to the rapid growth rate of the *P. alba* genotype in our case compared to *R. pseudoacacia* over two rotations of 3 years [12,54,55].

**Table 2.** Leaf litter total annual weight in *P. alba* or *R. pseudoacacia* monocultures and 50PA:50RP mixture plantations.

| Plots | Leaf Litterfall $(mg\ ha^{-1}\ yr^{-1})$ |
|---|---|
| *P. alba* monoculture | $3.37 \pm 0.79$ |
| 50PA:50RP mixture | $2.82 \pm 0.37$ |
| *R. pseudoacacia* monoculture | $2.55 \pm 0.16$ |

In contrast, the leaf litter production recorded in the *P. alba* monoculture ($3.37$ mg ha$^{-1}$ yr$^{-1}$) was similar to that obtained by Guenon et al. [56], who reported $3.1$ mg ha$^{-1}$ yr$^{-1}$ in SRC plantations of *Populus deltoides* × *P. nigra*, although in that case the planting density was lower (7200 tree ha$^{-1}$). However, other authors have reported higher values for the same species growing in SRC plantations ($5.3$ mg ha$^{-1}$ yr$^{-1}$) [57]. The amount of leaf litter was much lower for both species ($0.77$ mg ha$^{-1}$ yr$^{-1}$ in *P. alba* and $1.02$ mg ha$^{-1}$ yr$^{-1}$ in *R. pseudoacacia*) in the riparian ecosystems described by Medina-Villar et al. [49], which is probably because of the lower tree density and the lower growth rate. The leaf litter in *R. pseudoacacia* plantations found by Tateno et al. [58] was around $3.8$ mg ha$^{-1}$ yr$^{-1}$, which is higher than the amounts obtained in the present study ($2.55$ mg ha$^{-1}$ yr$^{-1}$), despite having a lower planting density. In the mixed plantation, leaf litter accounted for $2.82$ mg ha$^{-1}$ yr$^{-1}$, with this value being between that of the two monocultures although closer to that for the *R. pseudoacacia* monoculture, despite the greater contribution *P. alba* leaves.

*3.2. Foliar Nutrient Concentration and Retranslocation Rate in Green Leaves and Senescent Leaves*

3.2.1. Macronutrients and C

Leaf N concentration was significantly higher in green leaves compared to senescent leaves in all test plots ($p$-value = 0.0002 for PA; $p$-value = 0.0001 for 50PA:50RP and $p$-value = 0.0004 for RP) (Figure 3). This result was expected, since nitrogen resorption from senescent leaves at the end of the growing season is a key function in plants [59].

In green leaves, the concentration of N did not differ significantly between the different plots ($p$-value = 0.217). However, in absolute terms, the N concentration was higher in the *R. pseudoacacia* monoculture, the values for the mixed plantation being intermediate and the lowest values being those for *P. alba*, although still greater than $25$ g kg$^{-1}$, which is considered the threshold for nutrient-demanding broadleaves [60]. A higher concentration of N in the green leaves of a *P. deltoides* L.—*Alnus glutinosa* (L). Gaertn mixture in comparison to the monoculture of *P. deltoides* was also observed by Koupar et al. [61].

The N efficiency index showed non-significant differences ($p$-value = 0.0582), although a higher mean efficiency value was observed for *P. alba* ($357.36$) in relation to that of *R. pseudoacacia* ($171.38$). The 50PA:50RP mixture showed an intermediate ratio ($240.27$) that did not differ significantly from monocultures. The low efficiency of *R. pseudoacacia*, which may be attributable to its N$_2$-fixing character, has been previously reported by González -Muñoz et al. [51].

The average concentration of N in senescent leaves was also significantly higher in the *R. pseudoacacia* monoculture and the 50PA:50RP mixture in comparison to the *P. alba* monoculture

($p$-value = 0.0258), with the 50PA:50RP mixture presenting intermediate concentrations (Figure 3). The concentrations detected in senescent leaves are in line with those described by Lee et al. [62] for *R. pseudoacacia* leaf litter (19.9 g kg$^{-1}$), Cotrufo et al. [57] in relation to *P. alba* (9.6 g kg$^{-1}$) or Das and Chaturvedi [26] for *P. deltoides* (11.4 g kg$^{-1}$). Similar trends, although with notably higher concentrations, are mentioned by Medina-Villar et al. [49]. However, Koupar et al. [61] found higher concentrations of N in senescent leaves in mixed *Populus* and *Alnus* plantations than in their respective monocultures.

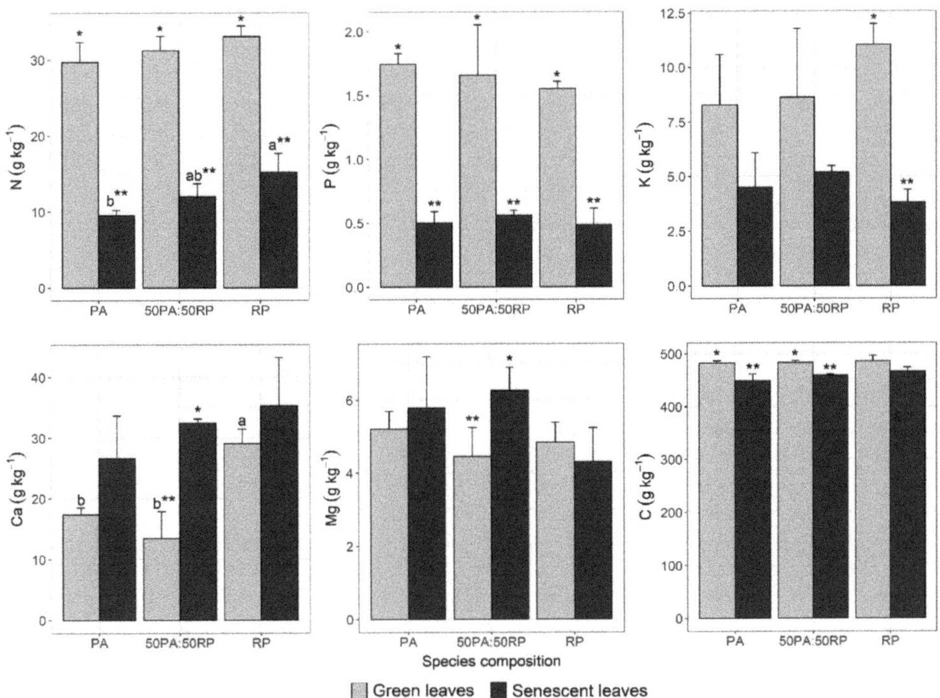


**Figure 3.** Weighted annual averages according to weight fraction and their standard deviation for macronutrients and C in green and senescent leaves of *P. alba* (PA) and *R. pseudoacacia* (RP) monocultures and 50PA:50RP mixture. The significance between plot types for green leaves, and also for senescent leaves, is shown by letters; and the significance between green and senescent leaves within the same plot (PA, 50PA:50RP and RP respectively) is shown with asterisks. Both letters and asterisks are only shown when significant differences were found.

Similarly, in relation to P, the concentration in green leaves was significantly higher than in senescent leaves in all plots ($p$-value < 0.0001 for PA; $p$-value = 0.0091 for 50PA:50RP; and $p$-value = 0.0002 for RP) (Figure 3). In the case of green leaves (50PA:50RP mixture and monocultures) no significant differences among species were observed for P concentration ($p$-value = 0.6340), although absolute values were higher in *P. alba*. In senescent leaves, no significant differences were detected ($p$-value = 0.613), the highest concentrations corresponding to the 50PA:50RP mixture and the lowest to the *R. pseudoacacia* monoculture (Figure 3).

There were no significant differences in P use efficiency ($p$-value = 0.2881), with *P. alba* presenting the highest absolute value (6801), followed by *R. pseudoacacia* (5585) and finally the mixed plantation (5078).

In this study, the P concentrations in senescent leaves were lower than those detected in other studies. Lee et al. [62] obtained mean values of 0.63 g kg$^{-1}$ in *R. pseudoacacia* or ranges from 1.14 g kg$^{-1}$ to 1.37 g kg$^{-1}$ in *Populus* spp. [26,63].

The stoichiometric N:P ratio in green leaves, widely used as an indicator of probable N:P deficiency, showed values above 16 in our study, which is the upper threshold identified by Aerts and Chapin [64] to indicate P deficiency, meaning that both species are far from displaying N deficiency. The leaves in our plots showed ratios close to normality in the *P. alba* monoculture (17.02). However, this proportion was 19.67 in the 50PA:50RP mixture and 21.34 in the case of the *R. pseudoacacia* monoculture, which could imply a progressive loss of fertility as regards P in these soils, since the leaf decomposition provides the main supply of this nutrient [65,66].

$N_2$-fixing species such as *R. pseudoacacia* may have more demand for P than non-fixing species and this element may be the most limiting for its growth [62]. Cao and Chen [67] also reported that P was more limiting than N for mature *R. pseudoacacia* plantations.

Regarding K concentration, even though the concentration in green leaves is always higher than that of senescent leaves in absolute terms, significant differences were only found in the *R. pseudoacacia* monoculture (*p*-value = 0.0004) (Figure 3), whereas no significant differences were found in the *P. alba* monoculture (*p*-value = 0.08) or the 50PA:50RP mixture (*p*-value = 0.132). No significant differences were detected among the plots (monocultures and 50PA:50RP mixture) (green leaves: *p*-value = 0.352; and senescent leaves: *p*-value = 0.311). This may be because K is a highly mobile element, both in plants and in the soil [68], which is reflected in a high variability of the concentration detected in the leaves in all plots.

K concentration in senescent leaves reported in the literature for *Populus* spp. ranges widely from 1.2 to 10.8 g $kg^{-1}$ [26,27,63,69,70], the K concentration found in this study presenting intermediate values. Less information appears to be available for K concentration in senescent *R. pseudoacacia* leaves, although Lee et al. [62] report levels of around 10.97 g $kg^{-1}$, which is more than twice our values.

The greater difference in K detected between senescent and green leaves of *R. pseudoacacia* (7.22 mg $g^{-1}$) in comparison to *P. alba* (3.77 mg $g^{-1}$) could indicate greater importance of retranslocation as compared to recirculation via leaf litter.

According to the literature, the optimal range of NPK in green leaves for *Populus* species is 17–30 g $kg^{-1}$ for N, 1.0–4.4 g $kg^{-1}$ for P and 7–20 g $kg^{-1}$ for K [71–76]. Narrower optimal ranges are established for site-demanding broadleaved [60] and more specifically for *Populus* [77] species, with 18–25 g $kg^{-1}$ for N, 1.8–3.0 g $kg^{-1}$ for P and 12–20 g $kg^{-1}$ for K. The concentrations of N obtained in the green leaves of all our plots were above the optimal range, while in the case of P, they were very close to the lower limit. In the case of K, the levels indicate deficiency. Sardans et al. [78] found green-leaf NPK ranges for *P. alba* at 41 different study points of 26.8–31.2, 1.96–2.08 and 4.8–26.5 g $kg^{-1}$ respectively. The values for the *P. alba* monoculture plots were within this range for N and K in our study, although in the case of P, the concentrations were lower.

The green-leaf NPK concentration reported by Ozbucak et al. [36] for *R. pseudoacacia* ranged from 20.0–44.2, 0.60–2.47 and 2.1–12 g $kg^{-1}$, the concentrations detected in our plots being within these ranges. For the same species, Sardans et al. [78] reported N and P ranges within those defined by Ozbucak et al. [36] (35.2–44.2 and 1.94–2.48 g $kg^{-1}$, respectively), although much higher for K (14.3–20.1 g $kg^{-1}$), the values obtained in this study being below those ranges for P and K, and very close to the lower limit in the case of N. However, if we take into account the ranges for optimum nutrition of demanding broadleaves [60,77], the concentrations obtained in this study for *R. pseudoacacia* are below the critical levels for P and K. In contrast, in the case of N, the concentrations obtained were above the optimal range, as expected, since *R. pseudoacacia* is an $N_2$-fixing species.

The low concentrations of K obtained in green leaves in this study could be due to the antagonistic relationship between Ca and K, given the high concentrations of interchangeable Ca in the soil (38.71 cmol $kg^{-1}$) (Table 1). This could cause less absorption of K by the plant due to a lower presence of this element in the soil solution as both elements compete for plasma membrane absorption sites [77,79].

As regards Ca, the absolute values of the mean concentrations were higher in senescent leaves than in green leaves in the monocultures. However, the differences between the two types of leaves were only significant in the 50PA:50RP mixture (*p*-value = 0.0877 for PA; *p*-value = 0.0018 for 50PA:50RP and $p$-value = 0.263 for RP) (Figure 3), which would indicate that a greater amount of Ca is returned to the soil in the mixed plot in comparison to the monocultures plots. This finding is consistent with that of Sayyad et al. [80] in pure and mixed stands of *Populus deltoides* and *Alnus subcordata*. The increase in the concentration in the senescent leaf is due to the low mobility of Ca, which is not an element retranslocated by plants. This low mobility causes Ca to be immobilized once assimilated, accumulating in structural components of the leaf such as membranes, cell walls and vacuoles [77]. Tzvetkova and Petkova [81] also found that Ca concentrations for *R. pseudoacacia* increased in the leaves that fall later.

In senescent leaves, no significant differences were observed between the different plots (both monocultures and the mixture) ($p$-value = 0.287), while significant differences were found in green leaves between the *R. pseudoacacia* monoculture and both the *P. alba* monoculture and the mixture ($p$-value = 0.0017).

Although no significant differences were detected in the Ca use-efficiency index ($p$-value = 0.1479), higher values were observed in the *P. alba* monoculture (132.87) than in the *R. pseudoacacia* monoculture (71.06), with mixed plots presenting an intermediate value (87.18).

In the case of Mg, as for Ca, the concentrations in absolute values were higher in senescent leaves than in green leaves for the *P. alba* monoculture and 50PA:50RP plots, these differences being significant for the mixture ($p$-value = 0.538 for PA and $p$-value = 0.0378 for 50PA:50PR) (Figure 3). However, in the *R. pseudoacacia* monoculture, despite no significant differences being detected ($p$-value = 0.437), the mean concentration of Mg was slightly higher in green leaves than in senescent leaves. This may be because Mg is an element with partial mobility, which, in addition to its involvement in photosynthesis, is a cofactor of numerous enzymatic activities, among which is the nitrogenase activity involved in the fixation of $N_2$ [82], and therefore this $N_2$-fixing species, with a higher photosynthetic activity [83], may have a greater requirement for this element. Sayyad et al. [80] found no significant differences in Mg concentrations between green and senescent leaves for the monoculture and *Populus* and *Alnus* mixture plantations. However, in *R. pseudoacacia* plantations, Tzvetkova and Petkova [81] observed a slight decrease in Mg concentrations in leaves that fall later, in agreement with our findings (from 5.15 g kg$^{-1}$ in October to 2.37 g kg$^{-1}$ in December).

No significant differences were found in the concentration of Mg between the different plots either for green leaves ($p$-value = 0.390) or senescent leaves ($p$-value = 0.132), although in terms of absolute values, the mean concentration of Mg was lower in the senescent leaves in the *R. pseudoacacia* monoculture. As in this study, Sayyad et al. [80] found lower Mg concentrations, both in green and senescent leaves in the $N_2$-fixing species.

Harvey and Van den Driessche [63] also reported higher concentrations of Ca and Mg in senescent leaves than in green leaves for *Populus* (15.80 and 4.78 g kg$^{-1}$ vs. 9.65 and 3.14 g kg$^{-1}$), these concentrations being lower than those obtained in this study.

Laganière et al. [70] and Yanai et al. [69] found ranges of between 10.8 and 18.9 g kg$^{-1}$ for Ca and between 1.8 and 2.7 g kg$^{-1}$ for Mg in senescent leaves of *Populus*, both of which are lower than the amounts found in this study.

In the case of green leaves, Sardans et al. [78] reported a range of between 21.3–48.5 g kg$^{-1}$ for Ca and between 1.9–7.2 g kg$^{-1}$ for Mg in *P. alba* and Martín-García et al. [84] reported values for the 'I-214' genotype of more than 26.4 g kg$^{-1}$ for Ca and lower than 3.6 g kg$^{-1}$ for Mg. However, Elferjani [85] reported values for hybrid poplar genotypes of between 7.9–12 g kg$^{-1}$ for Ca and 2.0–2.8 g kg$^{-1}$ for Mg, these values again being lower than those obtained in this study.

According to the literature, the optimal Ca and Mg in green leaves of *Populus* species ranges from 3–17 g kg$^{-1}$ for Ca and 1.4–4.0 g kg$^{-1}$ for Mg [71–76]. However, Bergmann [77] established narrower optimal ranges for *Populus* of 3–15 g kg$^{-1}$ for Ca and of 2.0–3.0 g kg$^{-1}$ for Mg. Hence, our Ca values for *Populus* are at the upper limit of the optimal range and above the optimal range in the case of Mg.

As regards *R. pseudoacacia*, Sardans et al. [78] reported concentration ranges in green leaves of 12.3–22.7 g kg$^{-1}$ for Ca and 2.09–2.69 g kg$^{-1}$ for Mg, the mean concentrations found in this study being above those ranges for both Ca and Mg. With respect to other demanding broadleaved species such as *Fraxinus excelsior*, Bergmann [77] established optimal ranges of 3.0–15.0 g kg$^{-1}$ for Ca and 2.0–4.0 g kg$^{-1}$ for Mg. As for less demanding broadleaved species such as beech, Stefan et al. [86] established ranges of 4–8 g kg$^{-1}$ for Ca and 1–1.5 g kg$^{-1}$ for Mg. In our study, the *R. pseudoacacia* monoculture presents concentrations above the optimal nutritional range.

The high levels of Ca and Mg measured in this study may be due to the high concentrations of these elements in the exchange complex, which, together with the basic pH, would facilitate the absorption of these nutrients by the plant (Table 1). To this fact, it should be added that high concentrations of Mg can be caused by low levels of K according to Kirkby and Mengel [87], a circumstance that would occur in the studied plantations.

The mean C values in green and senescent leaves were not significantly different between plots (*p*-value = 0.863 and *p*-value = 0.136, respectively) (Figure 3), nor were there significant differences between green and senescent leaves in the *R. pseudoacacia* monoculture (*p*-value = 0.0122 for PA; *p*-value = 0.0007 for 50PA:50RP and *p*-value = 0.0659 for RP). These values were very similar to those described for senescent poplar leaves by Cotrufo et al. [57] of around 435 g kg$^{-1}$ or to the C values in green leaves of *R. pseudoacacia* reported by Cao and Chen [67] of around 480 g kg$^{-1}$. Leaves account for an important fraction soil C contribution [88].

### 3.2.2. Retranslocation

Regarding retranslocation, as expected, there was a significant reduction in NPK concentrations in senescent leaves compared to green leaves in all plots, which indicates significant internal recycling of nutrients. Figure 4 shows the percentage of retranslocation using the weighted annual averages according to weight fraction value for senescent leaves as there are three harvesting dates.

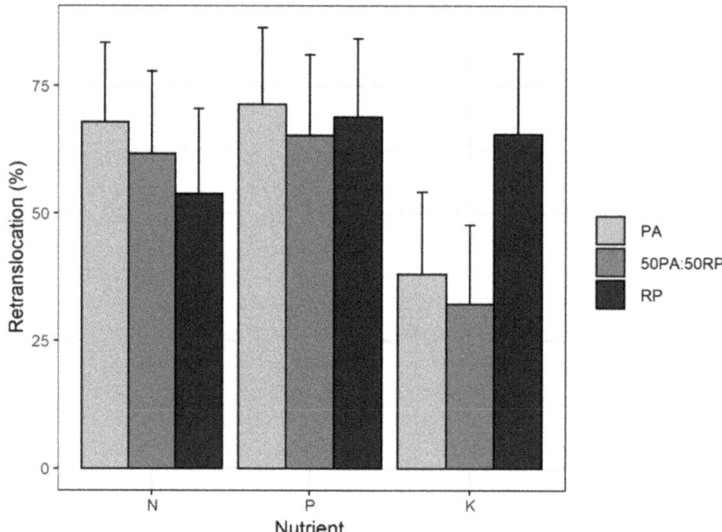


**Figure 4.** Average values and standard deviation of the percentage of efficient retranslocation of nitrogen, phosphorus and potassium in *P. alba* (PA) and *R. pseudoacacia* (RP) monocultures and 50PA:50RP mixture.

The rate of N retranslocation (N-*re*) was significantly lower in the *R. pseudoacacia* monoculture compared to the *P. alba* monoculture (*p*-value = 0.05). Higher green leaf nutrient content is normally assumed to be associated with lower retranslocation efficiency [89,90]. No significant differences were detected for P and K in the different plots (*p*-value = 0.5866 and *p*-value = 0.2521, respectively).

However, in absolute values, the rate of P retranslocation (P-*re*) was higher in *P. alba* and higher for K-*re* in *R. pseudoacacia*. Ozbucak et al. [36] reported similar trends in the retranslocation of P and K in *R. pseudoacacia*, but found the opposite for N. The higher rate of P-*re* (71.36%) with respect to the rate of N-*re* (67.87%) in all plots could indicate, as suggested by Ozbucak et al. [36], that P is a limiting nutrient in the study area. In fact, the values detected in the soil at the time of plantation were below 4 mg kg$^{-1}$, which would indicate deficiency according to Andrades and Martínez [91].

Aerts [59] states that for deciduous species in general, efficient retranslocation ranges from 40 to 75% in the case of N and from 30 to 70% for P. Hence, the rates of N-*re* and P-*re* in the three studied plots (monocultures and 50PA:50RP mixture) would be within those ranges or proximate to them. However, in the case of *R. pseudoacacia*, lower NPK retranslocation values (45, 45 and 60% respectively) were reported by Ozbucak et al. [36]. According to Salehi et al. [27], NPK in poplar hybrids ranges were 12–19%, 32–40% and 21–24%, respectively, while Das and Chaturvedi [26] detected higher limits for *P. deltoides* of around 52–54%, 40–46% and 46–47% for NPK, respectively. In the present study, the retranslocation percentage in *P. alba* was higher than 50% for all nutrients except for K.

This variability in the retranslocation rates could be due to the influence of the physical-chemical properties of the soil, as well as to the specific nutritional requirements of the plant in the retranslocation process, as suggested by Fife et al. [92]. Low percentages of retranslocation suggest a less conservative strategy, which implies a greater dependence on the circulation of nutrients from the soil [93]. In this study, where the soil has high levels of N, Ca and Mg fertility (Table 1), the retranslocation values obtained would therefore indicate a more conservative strategy as regards P in both species, as well as for K in the case of *R. pseudoacacia*, the plants being less dependent on the nutrient dynamics of the soil. Moreover, with the soil pH levels found in this study (around 8.65), formation of insoluble $CaHPO_4$ and P fixation could be expected.

### 3.3. Temporal Variability of Nutrients in Senescent Leaves

Table 3 shows the evolution over time of the NPK, Ca and Mg concentrations over the months in which leaf shedding occurred. The N concentration of *P. alba* senescent leaves was greater in October than in November or December (*p*-value = 0.0134). However, in *R. pseudoacacia*, the opposite trend was observed, with no significant differences being detected (*p*-value = 0.103). In the case of the 50PA:50RP mixture, although there were no significant differences (*p*-value = 0.0792), the N concentration decreased as the sampling period progressed. This may be due to the lower percentage of *R. pseudoacacia* leaves as the sampling period progressed, caused by differences in the time of abscission of each leaf. With regard to P, no significant variation was observed over the months (*p*-value = 0.302 for PA, *p*-value = 0.966 for 50PA:50RP and *p*-value = 0.304 for RP). K concentration did not vary significantly in any of the plots as the sampling period progressed (*p*-value = 0.6 for PA, *p*-value = 0.21 for the 50PA:50RP mixture and *p*-value = 0.0514 for RP).

As for Ca and Mg, which are less mobile nutrients that tend to accumulate in the leaves, the variation in the concentrations in senescent leaves in all plots in the case of Mg and in *R. pseudoacacia* for Ca, followed the opposite pattern to that of more mobile nutrients which are retranslocated, with concentrations being higher in the months of greatest leaf fall. In the *P. alba* monoculture, Ca concentrations were very similar between the fall months of leaf shedding, while in the case of the mixture, the concentrations were significantly lower in the month with the highest litterfall production (*p*-value = 0.0007).

The different phenology of leaf fall can influence the loss of nutrients in senescent leaves and therefore modify the chemical composition. Thus, species with earlier leaf fall tend to retranslocate nutrients more quickly to compensate for the greater amount of initial leaf loss, especially in the case of N [94]. These authors found lower nutrient contents coinciding with the period of greatest leaf fall. This finding is in agreement with our observations in the case of N, which showed lower concentrations in the period of greatest leaf-fall (October in the case of the *R. pseudoacacia* monoculture, November and December for the *P. alba* monoculture and December for the 50PA:50RP mixture). However, in the subsequent months, we observed a decrease in retranslocation in *R. pseudoacacia*, coinciding with an increase in leaf concentration (Figure 4). This may be due to the higher rates of N-fixation in autumn and spring reported for this species [95]. The increase in N concentration with leaf age has also been reported by López et al. [96] for another $N_2$-fixing species, *Alnus glutinosa*.

**Table 3.** Monthly averages ± standard deviation of NPK concentrations in senescent leaves collected in October, November and December in *P. alba* and *R. pseudoacacia* monoculture plots and mixed (50PA:50RP) plots.

| Nutrient | | Sampling Month | *P. alba* Monoculture | 50PA:50RP Mixture | *R. pseudoacacia* Monoculture |
|---|---|---|---|---|---|
| N | g kg$^{-1}$ | October | 12.87 ± 1.18 * | 14.63 ± 0.79 * | 14.57 ± 3.91 |
| | | November | 8.45 ± 1.65 [c] ** | 11.64 ± 1.85 [b] *; ** | 14.86 ± 1.10 [a] |
| | | December | 9.33 ± 1.00 [b] ** | 11.01 ± 2.09 [b] ** | 19.11 ± 0.75 [a] |
| P | g kg$^{-1}$ | October | 0.65 ± 0.20 | 0.56 ± 0.07 | 0.56 ± 0.19 |
| | | November | 0.56 ± 0.01 [a] | 0.55 ± 0.06 [a] | 0.40 ± 0.07 [b] |
| | | December | 0.45 ± 0.14 | 0.56 ± 0.06 | 0.45 ± 0.04 |
| K | g kg$^{-1}$ | October | 4.81 ± 1.42 | 6.37 ± 2.03 | 4.54 ± 1.76 |
| | | November | 5.40 ± 0.40 | 5.59 ± 0.28 | 4.60 ± 1.01 |
| | | December | 4.12 ± 2.11 | 4.40 ± 0.39 | 1.91 ± 0.28 |
| Ca | g kg$^{-1}$ | October | 32.75 ± 15.42 | 38.81 ± 0.89 * | 40.77 ± 10.04 |
| | | November | 28.74 ± 0.64 | 32.16 ± 2.29 ** | 34.06 ± 10.92 |
| | | December | 24.02 ± 8.87 | 29.73 ± 0.70 ** | 25.86 ± 5.93 |
| Mg | g kg$^{-1}$ | October | 4.82 ± 1.91 | 5.54 ± 0.91 | 5.15 ± 0.51 * |
| | | November | 6.40 ± 0.13 [a] | 6.16 ± 0.30 [a] | 4.09 ± 1.27 [b] * |
| | | December | 5.65 ± 1.92 [a] | 6.63 ± 0.64 [a] | 2.37 ± 0.17 [b] ** |

The significance between plots for each sampling month is shown by letters, and the comparison of means between samples for the same plot is represented by asterisks. Both letters and asterisks are only shown when significant differences were found.

In relation to P and K, the lowest concentrations also coincided with the period of greatest leaf fall in the *P. alba* monoculture. However, in the *R. pseudoacacia* monoculture, the lowest concentrations of P and K do not coincide with the time of greatest leaf fall. Specifically, the low concentrations of K obtained in the month of December could also be the result of leachate of this extremely mobile nutrient in leaves that have remained longer on the tree.

The decrease in NPK rates with leaf age has also been reported by other authors in deciduous species [97].

### 3.4. Stoichiometric C:N Ratio in Senescent Leaves

The average C:N ratio values (Table 4) were significantly higher in the *P. alba* monoculture compared to the *R. pseudoacacia* monoculture (*p*-value = 0.0390). The increase in the ratio was greater in November and December compared to October, both in the *P. alba* monoculture (*p*-value = 0.0665) as well as in the 50PA:50RP mixture (*p*-value = 0.0665), although it was not significantly different among months. The opposite was observed for *R. pseudoacacia*, although again there were no significant differences (*p*-value = 0.3012), probably due to the different monthly contribution of leaf fall and therefore of the organic matter, which was higher in October for this species, unlike *P. alba* and the 50PA:50RP mixture, both of which present the highest leaf fall in December.

**Table 4.** Average values ± standard deviation of C:N ratio in the three months of leaf litter harvesting for the plots studied.

| | Sampling Month | *P. alba* Monoculture | 50PA:50RP Mixture | *R. pseudoacacia* Monoculture |
|---|---|---|---|---|
| C:N | October | 35.35 ± 3.73 | 31.17 ± 1.35 | 32.67 ± 7.57 |
| | November | 50.76 ± 3.47 | 40.46 ± 6.70 | 31.50 ± 1.74 |
| | December | 49.33 ± 5.27 | 42.58 ± 7.71 | 25.69 ± 0.98 |
| | Weighted mean | 47.20 ± 2.38 [a] | 38.83 ± 5.19 [ab] | 31.20 ± 4.36 [b] |

Significant differences between plots are shown by letters. Letters are only shown when significant differences were found.

N and P are important elements of the litter and have a strong influence on the decomposition rate due to the high demand for them by decomposer microorganisms [98–100], with the initial stoichiometric C:N being a good predictor of the initial decomposition rate [101,102].

Leaf litter with a high C:N ratio normally displays a slower decomposition rate and immobilizes more N [57,103,104]. The lower average total values of C:N in the leaf litter of the *R. pseudoacacia* monoculture and in the 50PA:50RP mixed plots reflects a greater amount of N mobilized in the leaf litter, which may be evidenced by a higher initial decomposition rate than that of the leaf litter of the *P. alba* monoculture. Hirschfeld et al. [105] reported significantly lower C:N ratios in *R. pseudoacacia* (24.2) than in other North American broadleaves, such as white ash and sugar maple (43.8 and 83.2 respectively). In a *P. alba* short rotation plantation, Cotrufo et al. [57] reported an average C:N value of 43, similar to that found in this study.

The leaf litter of mixed plantations decomposes more rapidly than that of monocultures as the leaf litter mixture leads to massive decomposition by increasing the loss of mass as well as by promoting the abundance and quality of soil decomposers [106]. This statement is partially supported by the results obtained in this study, in which the 50PA:50RP mixture showed a higher initial decomposition rate than that of the *P. alba* monoculture, but not of the *R. pseudoacacia* monoculture, suggesting that species could be an influential factor. Lee et al. [62] found a higher rate of leaf litter decomposition in a mixed plot of *Quercus mongolica* and *R. pseudoacacia* than in a monoculture of *Q. mongolica*.

### 3.5. Nutrient Supply to the Soil Derived from Leaf Litter

The return to the soil of C, P and K derived from the leaf litter, expressed in kg per ha, was greater in the *P. alba* monoculture and in the 50PA:50RP mixture (Table 5), mainly because of the greater production of foliar biomass in these plots. The *R. pseudoacacia* monoculture, however, despite lower leaf litter production, presented the highest N contributions as a result of the higher N concentrations in senescent leaves.

**Table 5.** Carbon, Nitrogen, Phosphorus, Potassium, Calcium and Magnesium total annual weighted average contributions in leaf litter for *P. alba* and *R. pseudoacacia* monocultures, and 50PA:50RP mixture plantations.

| Plot | C | N | P | K | Ca | Mg |
|---|---|---|---|---|---|---|
| | kg ha$^{-1}$ yr$^{-1}$ | | | | | |
| *P. alba* monoculture | 1507.39 ± 312.62 | 31.97 ± 6.46 | 1.71 ± 0.69 | 15.84 ± 9.12 | 90.29 ± 34.98 | 19.76 ± 8.07 |
| 50PA:50RP mixture | 1296.91 ± 174.66 | 33.62 ± 4.41 | 1.58 ± 0.23 | 14.58 ± 1.63 | 91.53 ± 10.13 | 17.62 ± 2.37 |
| *R. pseudoacacia* monoculture | 1186.66 ± 63.74 | 38.56 ± 5.83 | 1.22 ± 0.27 | 9.66 ± 0.95 | 90.72 ± 26.38 | 10.97 ± 2.61 |

Das and Chaturvedi [26] found N contributions of 30.8–41.9 kg ha$^{-1}$ yr$^{-1}$, P of 4.3–5 kg ha$^{-1}$ yr$^{-1}$ and K of 19.4–20.1 kg ha$^{-1}$ yr$^{-1}$ in two 3-year-old poplar plantations. These values are similar to those obtained in our study for N, but higher than those for P and K.

In a *P. deltoides* × *P. nigra* plantation of the same age, Guénon et al. [56] found NPK contributions of 48.0, 6.2 and 10.8 kg ha$^{-1}$ yr$^{-1}$, respectively, which are greater than our values for N and P but lower for K. The higher P contributions might be explained by the higher P content of the soil and greater availability due to a strongly alkaline pH. The low P contributions in all the plots in our study will probably be reflected in a low availability of this nutrient in the soil.

In a study conducted in a riparian stand with the same species composition, *P. alba* and *R. pseudoacacia*, Medina-Villar et al. [49] reported lower N (8.8 and 15.3 kg ha$^{-1}$ yr$^{-1}$) and P (1.06 and 0.49 kg ha$^{-1}$ yr$^{-1}$) inputs, which is probably due to a much lower level of litter production compared to the levels found in our study.

Although the contribution of leaf litter was greater in *P. alba* monocultures, the 50PA:50RP mixture contributed the most in terms of Ca, which reflects the higher mean concentration of Ca in senescent leaves. In the case of Mg, the *R. pseudoacacia* monoculture presented the lowest returns because of the lower amount of leaf litter coupled with the lower concentration of Mg in senescent leaves.

In poplar SRC plantations, Guénon et al. [56] measured Ca and Mg contributions of 80 and 22 kg ha$^{-1}$, respectively, consistent with the findings of this study, although with slightly lower Ca. However, Perala and Alban [107] reported lower contributions of Ca and Mg for *P. tremuloides* (60 and 8 kg ha$^{-1}$, respectively).

The total C content was higher in the *P. alba* monoculture due to the greater content of leaf litter. Therefore, the greatest contribution of C returned to the soil through leaf litter was measured in this plot (1507 kg C ha$^{-1}$ yr$^{-1}$), followed by the 50PA:50RP mixture (1297 kg C ha$^{-1}$ yr$^{-1}$), and finally the *R. pseudoacacia* monoculture contributed the least amount (1187 kg C ha$^{-1}$ yr$^{-1}$). In a *P. deltoides* × *P. nigra* plantation of the same age as that of this study, Guénon et al. [56] measured C contributions of 1400 kg ha$^{-1}$ yr$^{-1}$, similar to those found in this study.

## 4. Conclusions

Different patterns in the amount and time of leaf fall were observed among the *P. alba* and the *R. pseudoacacia* monocultures and the mixed plantation. The soil conditions at the study site are representative of areas with high Ca and Mg availability, relatively poor in organic matter and total N, and strongly alkaline pH. As the *P. alba* monoculture provides the highest amount of litter and obtained the highest K, P and Mg from the litter, a mixed design is advantageous for the species *R. pseudoacacia* in relation to these nutrients. Even so, *P. alba* could also benefit from a mixed design derived from the higher concentration of N that *R. pseudoacacia* provides. Therefore, the mixed planting seems advantageous as a result of the different strategies shown by the two species in terms of the amount of litter and the dynamics of the main nutrients. The contribution of N derived from leaf fall in the *P. alba* monoculture is dependent on a high rate of N mineralization, which could be affected in the medium term by the higher C:N rates. The lower rates of total N and soil OM, lower N concentrations in green leaves and higher N resorption in poplar may indicate less stability in N nutrition. However, as a N$_2$-fixing species, *R. pseudoacacia* shows less N resorption, more N returned to the soil in the form of litterfall, and a lower C:N rate. The most limiting nutrients were P and K. Both species showed high retranslocation ratios for P, but poplar relied more on internal cycling for P and N, whereas *R. pseudoacacia* had higher rates of K resorption and therefore more internal cycling is required.

Although mixing the species does not increase biomass yield or net contribution of nutrients, it may be a good strategy to reduce future needs for mineral fertilization, that it is a common practice in SRC plantations, given the differences between the two monocultures in terms of processing the main nutrients. Determining the potential foliar contribution to the pool of nutrients in the soil as well as the dynamics would appear to be of importance for the sustainable management of plantations in short rotation. However, more conclusive results would require the study of leaf litter decomposition dynamics and the final incorporation of nutrients into the soil.

**Author Contributions:** Conceptualization, N.O. and H.S.; Data curation, I.G. and H.S.; Formal analysis. I.G. and N.O.; Funding acquisition, H.S; Investigation, N.O. and H.S.; Methodology, H.S., I.G. and R.R.-S.; Project administration, H.S. and R.R.-S.; Resources, N.O., H.S., and I.G.; Supervision, H.S. and R.R.-S.; Validation, H.S.; Writing—original draft, I.G., N.O. and H.S.; Writing—review and editing, I.G., H.S., R.R.-S, and N.O. All authors have read and agreed to the published version of the manuscript.

**Funding:** This research was funded by MINECO (Spain) through the framework of the project's RTA 2008-00025-C02-01 and RTA2017-00015-CO2 co-financed with funds from FEDER.

**Acknowledgments:** The authors wish to thank José Pablo de la Iglesia and Ana Parras for their careful monitoring of the plantation. We also thank Biopolar and Alasia Clones for providing the plant material. We are also grateful to Adam Collins for his English language review of the manuscript.



## References

1. European Commision. *Review of the 2012 European Bioeconomy Strategy*; European Commision: Brussels, Belgium, 2018.

2.    Farmer, R.J. The genecology of *Populus*. In *Biology of Populus and Its Implications for Management and Conservation*; Stettler, R.F., Bradshaw, H.D., Jr., Heilman, P.H., Hinckley, T.M., Eds.; NRC Research Press, National Research Council of Canada: Ottawa, ON, Canada, 1996; pp. 33–55.

3.    FAO. The International Commission on Poplars and Other Fast-Growing Trees Sustaining People and the Environment (IPC): International register of cultivars of *Populus* L. 2016. Available online: http://www.fao.org/forestry/ipc/69637@204274/en/ (accessed on 23 September 2020).

4.    Aravanopoulos, F.A. Breeding of fast growing forest tree species for biomass production in Greece. *Biomass Bioenergy* **2010**, *34*, 1531–1537. [CrossRef]

5.    Oliveira, N.; Sixto, H.; Cañellas, I.; Rodríguez-Soalleiro, R.; Pérez-Cruzado, C. Productivity model and reference diagram for short rotation biomass crops of poplar grown in Mediterranean environments. *Biomass Bioenergy* **2015**, *72*, 309–320. [CrossRef]

6.    Straker, K.C.; Quinn, L.D.; Voigt, T.B.; Lee, D.K.; Kling, G.J. Black Locust as a bioenergy feedstock: A review. *Bioenergy Res.* **2015**, *8*, 1117–1135. [CrossRef]

7.    Quinkenstein, A.; Wöllecke, J.; Böhm, C.; Grünewald, H.; Freese, D.; Schneider, B.U.; Hüttl, R.F. Ecological benefits of the alley cropping agroforestry system in sensitive regions of Europe. *Environ. Sci. Policy* **2009**, *12*, 1112–1121. [CrossRef]

8.    Slazak, A.; Freese, D. Phosphorus sorption kinetics in reclaimed lignite mine soils under different age stands of *Robinia pseudoacacia* L. in northeast Germany. *Appl. Environ. Soil Sci.* **2015**, *6*, 1–19. [CrossRef]

9.    Vítková, M.; Müllerová, J.; Sádlo, J.; Pergl, J.; Pyšek, P. Black locust (*Robinia pseudoacacia*) beloved and despised: A story of an invasive tree in Central Europe. *For. Ecol. Manag.* **2017**, *384*, 287–302. [CrossRef]

10.   Bergante, S.; Facciotto, G.; Minotta, G. Identification of the main site factors and management intensity affecting the establishment of Short-Rotation-Coppices (SRC) in Northern Italy through stepwise regression analysis. *Cent. Eur. J. Biol.* **2010**, *5*, 522–530. [CrossRef]

11.   Marron, N.; Priault, P.; Gana, C.; Gérant, D.; Epron, D. Prevalence of interspecific competition in a mixed poplar/black locust plantation under adverse climate conditions. *Ann. For. Sci.* **2018**, *75*, 1–12. [CrossRef]

12.   Oliveira, N.; del Río, M.; Forrester, D.I.; Rodríguez-Soalleiro, R.; Pérez-Cruzado, C.; Cañellas, I.; Sixto, H. Mixed short rotation plantations of *Populus alba* and *Robinia pseudoacacia* for biomass yield. *For. Ecol. Manag.* **2018**, *410*, 48–55. [CrossRef]

13.   Forrester, D.I.; Bauhus, J. A Review of Processes Behind Diversity—Productivity Relationships in Forests. *Curr. For. Rep.* **2016**, *2*, 45–61. [CrossRef]

14.   Facelli, J.M.; Pickett, S.T.A. Plant litter: Its dynamics and effects on plant community structure. *Bot. Rev.* **1991**, *57*, 1–32. [CrossRef]

15.   Maguire, D.A. Branch mortality and potential litterfall from Douglas-fir trees in stands of varying density. *For. Ecol. Manag.* **1994**, *70*, 41–53. [CrossRef]

16.   Chapin, F.S.; Matson, P.A.; Mooney, H.A. *Principles of Terrestrial Ecosystem Ecology*; Springer: New York, NY, USA, 2002; pp. 1–17.

17.   Aragão, L.E.O.C.; Malhi, Y.; Metcalfe, D.B.; Silva-Espejo, J.E.; Jiménez, E.; Navarrete, D.; Almeida, S.; Costa, A.C.L.; Salinas, N.; Phillips, O.L.; et al. Above—and below-ground net primary productivity across ten Amazonian forests on contrasting soils. *Biogeosciences* **2009**, *6*, 2759–2778. [CrossRef]

18.   El-Ramedy, H.R.; Alshaal, T.A.; Amer, M.; Domokos-Szabolcsy, E.; Elhawat, N.; Prokisch, J.; Fari, M. Soil Quality and Plant Nutrition. In *Sustainable Agricultural Reviews 14*; Springer: Cham, Switzerland, 2014; pp. 345–447. [CrossRef]

19.   Krishna, M.P.; Mohan, M. Litter decomposition in forest ecosystems: A review. *Energy Ecol. Environ.* **2017**, *2*, 236–249. [CrossRef]

20.   Giweta, M. Role of litter production and its decomposition, and factors affecting the processes in a tropical forest ecosystem: A review. *J. Ecol. Environ.* **2020**, *44*, 1–9. [CrossRef]

21.   Singh, K.P.; Tripathi, S.K. Litterfall, litter decomposition and nutrient release patterns in four native tree species raised on coal mine spoil at Singrauli, India. *Biol. Fertil. Soils* **1999**, *29*, 371–378. [CrossRef]

22.   Kumar, B.M. Litter dynamics in plantation and agroforestry systems of the tropics-a review of observations and methods. In *Ecological Basis of Agroforestry*; Batish, D.R., Kohli, R.K., Jose, S., Singh, H.P., Eds.; CRC Press: Boca Raton, FL, USA, 2008; pp. 181–209. [CrossRef]

23.   Eslamdoust, J.; Sohrabi, H. Carbon storage in biomass, litter, and soil of different native and introduced fast-growing tree plantations in the South Caspian Sea. *J. For. Res.* **2018**, *29*, 449–457. [CrossRef]

24. De la Cruz, A.C. Dinámica de Nutrientes en Parcelas Experimentales de *Populus x Euramericana* (Dode) Guinier 'I-214'. Ph.D. Thesis, Universidad Politécnica de Madrid, Madrid, Spain, 2005. Available online: http://oa.upm.es/328/ (accessed on 15 August 2020).

25. Gruhn, P.; Goletti, F.; Yudelman, M. *Integrated Nutrient Management, Soil Fertility, and Sustainable Agriculture: Current Issues and Future Challenges*; Food Agriculture and Environment Discussion Paper IFRPI 2020 Vision Brief; International Food Policy Research Institute: Washington, DC, USA, 2000.

26. Das, D.K.; Chaturvedi, O.P. Structure and function of *Populus deltoides* agroforestry systems in eastern India: 2. Nutrient dynamics. *Agrofor. Syst.* **2005**, *65*, 223–230. [CrossRef]

27. Salehi, A.; Ghorbanzadeh, N.; Salehi, M. Soil nutrient status, nutrient return and retranslocation in poplar species and clones in northern Iran. *IForest* **2013**, *6*, 336–341. [CrossRef]

28. Meiresonne, L.; De Schrijver, A.; De Vos, B. Nutrient cycling in a poplar plantation (*Populus trichocarpa x Populus deltoides* 'Beaupré') on former agricultural land in northern Belgium. *Can. J. For. Res.* **2007**, *37*, 141–155. [CrossRef]

29. De Vries, W.; Vel, E.; Reinds, G.J.; Deelstra, H.; Klap, J.M.; Leeters, E.E.J.M.; Hendriks, C.M.A.; Kerkvoorden, M.; Landmann, G.; Herkendell, J.; et al. Intensive monitoring of forest ecosystems in Europe 1. Objectives, set-up and evaluation strategy. *For. Ecol. Manag.* **2003**, *174*, 77–95. [CrossRef]

30. Stark, H.; Nothdurft, A.; Block, J.; Bauhus, J. Forest restoration with *Betula* ssp. and *Populus* ssp. nurse crops increases productivity and soil fertility. *For. Ecol. Manag.* **2015**, *339*, 57–70. [CrossRef]

31. Berthelot, A.; Ranger, J.; Gelhaye, D. Nutrient uptake and immobilization in a short-rotation coppice stand of hybrid poplars in north-west France. *For. Ecol. Manag.* **2000**, *128*, 167–179. [CrossRef]

32. Guo, L.B.; Sims, R.E.H.; Horne, D.J. Biomass production and nutrient cycling in *Eucalyptus* short rotation energy forests in New Zealand: II. Litter fall and nutrient return. *Biomass Bioenerg* **2006**, *30*, 393–404. [CrossRef]

33. Prescott, C.E.; Vesterdal, L.; Pratt, J.; Venner, K.H.; De Montigny, L.M.; Trofymow, J.A. Nutrient concentrations and nitrogen mineralization in forest floors of single species conifer plantations in coastal British Columbia. *Can. J. For. Res.* **2000**, *30*, 1341–1352. [CrossRef]

34. Rahajoe, J.S. The Role of Litter Production and Decomposition of Dominant Tree Species on the Nutrient Cycle in Natural Forest with Various Substrate Conditions. Ph.D. Thesis, Hokkaido University, Sapporo, Japan, 2003.

35. Teklay, T. Decomposition and nutrient release from pruning residues of two indigenous agroforestry species during the wet and dry seasons. *Nutr. Cycl. Agroecosyst.* **2007**, *77*, 115–126. [CrossRef]

36. Özbucak, T.B.; Kutbay, H.G.; Kilic, D.; Korkmaz, H.; Bilgin, A.; Yalçin, E.; Apaydin, Z. Foliar resorption of nutrients in selected sympatric tree species in gallery forest (black sea region). *Pol. J. Ecol.* **2008**, *56*, 227–237.

37. Arevalo, C.B.M.; Bhatti, J.S.; Chang, S.X.; Sidders, D. Land use change effects on ecosystem carbon balance: From agricultural to hybrid poplar plantation. *Agric. Ecosyst. Environ.* **2011**, *141*, 342–349. [CrossRef]

38. Brown, S.; Lugo, A.E.; Chapman, J. Biomass of tropical tree plantations and its implications for the global carbon budget. *Can. J. For. Res.* **1986**, *16*, 390–394. [CrossRef]

39. Rytter, R.M. The potential of willow and poplar plantations as carbon sinks in Sweden. *Biomass Bioenergy* **2012**, *36*, 86–95. [CrossRef]

40. Calfapietra, C.; Gielen, B.; Galema, A.N.J.; Lukac, M.; De Angelis, P.; Moscatelli, M.C.; Ceulemans, R.; Scarascia-Mugnozza, G. Free-air $CO_2$ enrichment (FACE) enhances biomass production in a short-rotation poplar plantation. *Tree Physiol.* **2003**, *23*, 805–814. [CrossRef] [PubMed]

41. Pérez-Cruzado, C.; Mohren, G.M.J.; Merino, A.; Rodríguez-Soalleiro, R. Carbon balance for different management practices for fast growing tree species planted on former pastureland in southern Europe: A case study using the $CO_2$ Fix model. *Eur. J. For. Res.* **2012**. [CrossRef]

42. Barrette, M.; Leblanc, M.; Thiffault, N.; Paquette, A.; Lavoie, L.; Bélanger, L.; Bujold, F.; Côté, L.; Lamoureux, J.; Schneider, R.; et al. Issues and solutions for intensive plantation silviculture in a context of ecosystem management. *For. Chron.* **2014**, *90*, 732–747. [CrossRef]

43. Thiffault, E.; Barrette, J.; Paré, D.; Titus, B.D.; Keys, K.; Morris, D.M.; Hope, G. Developing and validating indicators of site suitability for forest harvesting residue removal. *Ecol. Indic.* **2014**, *43*, 1–18. [CrossRef]

44. Thiffault, N.; Elferjani, R.; Hébert, F.; Paré, D.; Gagné, P. Intensive mechanical site preparation to establish short rotation hybrid poplar plantations-A case-study in Quebec, Canada. *Forests* **2020**, *11*, 785. [CrossRef]

45. Sixto, H.; Hernández-Garasa, M.J.; Ciria, P.; Carrasco, J.E.; Cañellas, I. *Manual de Cultivo de Populus spp. para la Producción deBiomasa con Fines Energéticos*; Ministerio de Ciencia e Innovación, Instituto Nacional de Investigación y Tecnología Agraria y Alimentaria (INIA): Madrid, Spain, 2010.

46. Stettler, R.F.; Bradshaw, H.D., Jr.; Heilman, P.E.; Hinckley, T.M. *Biology of Populus and Its Implications for Management and Conservation*; Stettler, R.F., Bradshaw, H.D., Jr., Heilman, P.E., Hinckley, T.M., Eds.; NRC Research Press, National Research Council of Canada: Ottawa, ON, Canada, 1996.

47. Huang, J.; Wang, X.; Yan, E. Leaf nutrient concentration, nutrient resorption and litter decomposition in an evergreen broad-leaved forest in eastern China. *For. Ecol. Manag.* **2007**, *239*, 150–158. [CrossRef]

48. Vitousek, P.M. Litterfall, nutrient cycling, and nutrient limitation in tropical forests. *Ecology* **1984**, *65*, 285–298. [CrossRef]

49. Medina-Villar, S.; Castro-Díez, P.; Alonso, A.; Cabra-Rivas, I.; Parker, I.M.; Pérez-Corona, E. Do the invasive trees, *Ailanthus altissima* and *Robinia pseudoacacia*, alter litterfall dynamics and soil properties of riparian ecosystems in Central Spain? *Plant Soil* **2015**, *396*, 311–324. [CrossRef]

50. Hobbie, S.E. Plant species effects on nutrient cycling: Revisiting litter feedbacks. *Trends Ecol. Evol.* **2015**, *30*, 357–363. [CrossRef]

51. González-Muñoz, N.; Castro-Díez, P.; Parker, I.M. Differences in nitrogen use strategies between native and exotic tree species: Predicting impacts on invaded ecosystems. *Plant Soil* **2013**, *363*, 319–329. [CrossRef]

52. Castro-Díez, P.; Valle, G.; González-Muñoz, N.; Alonso, Á. Can the life-history strategy explain the success of the exotic trees *Ailanthus altissima* and *Robinia pseudoacacia* in Iberian floodplain forests? *PLoS ONE* **2014**, *9*, 30–32. [CrossRef] [PubMed]

53. Navarro, C.M.; Pérez-Ramos, I.M.; Marañón, T.; Mediterráneo, U.B. Aporte de hojarasca al suelo en un bosque mediterraneo. *Almoraima* **2004**, *31*, 119–128.

54. Oliveira, N.; Rodríguez-Soalleiro, R.; Pérez-Cruzado, C.; Cañellas, I.; Sixto, H.; Ceulemans, R. Above- and below-ground carbon accumulation and biomass allocation in poplar short rotation plantations under Mediterranean conditions. *For. Ecol. Manag.* **2018**, *410*, 48–55. [CrossRef]

55. Sixto, H.; Gil, P.; Ciria, P.; Camps, F.; Sánchez, M.; Cañellas, I.; Voltas, J. Performance of hybrid poplar clones in short rotation coppice in Mediterranean environments: Analysis of genotypic stability. *GCB Bioenergy* **2014**, *6*, 661–671. [CrossRef]

56. Guénon, R.; Bastien, J.C.; Thiébeau, P.; Bodineau, G.; Bertrand, I. Carbon and nutrient dynamics in short-rotation coppice of poplar and willow in a converted marginal land, a case study in central France. *Nutr. Cycl. Agroecosyst.* **2016**, *106*, 293–309. [CrossRef]

57. Cotrufo, M.F.; De Angelis, P.; Polle, A. Leaf litter production and decomposition in a poplar short-rotation coppice exposed to free air $CO_2$ enrichment (POPFACE). *Glob. Chang. Biol.* **2005**, *11*, 971–982. [CrossRef]

58. Tateno, R.; Tokuchi, N.; Yamanaka, N.; Du, S.; Otsuki, K.; Shimamura, T.; Xue, Z.; Wang, S.; Hou, Q. Comparison of litterfall production and leaf litter decomposition between an exotic black locust plantation and an indigenous oak forest near Yan'an on the Loess Plateau, China. *For. Ecol. Manag.* **2007**, *241*, 84–90. [CrossRef]

59. Aerts, R. Nitrogen Partitioning between Resorption and Decomposition Pathways: A Trade-Off between Nitrogen Use Efficiency and Litter Decomposibility? *Oikos* **1997**, *80*, 603–606. [CrossRef]

60. Bonneau, M. Le diagnostic foliaire. *Rev. For. Fr.* **1988**, *40*, 19–28. [CrossRef]

61. Koupar, S.A.M.; Hosseini, S.M.; Modirrahmati, A.; Tabari, M.; Golchin, A.; Rad, F.H. Effect of pure and mixed plantations of *Populus deltoides* with *Alnus subcordata* on growth, nutrition and soil properties: A case study of Foman Region, Iran. *Asian J. Chem.* **2011**, *23*, 5261–5265. [CrossRef]

62. Lee, Y.C.; Nam, J.M.; Kim, J.G. The influence of black locust (*Robinia pseudoacacia*) flower and leaf fall on soil phosphate. *Plant Soil* **2011**, *341*, 269–277. [CrossRef]

63. Harvey, H.P.; Van Den Driessche, R. Poplar nutrient resorption in fall or drought: Influence of nutrient status and clone. *Can. J. For. Res.* **1999**, *29*, 1916–1925. [CrossRef]

64. Aerts, R.; Chapin, F.S. The Mineral Nutrition of Wild Plants Revisited: A Re-evaluation of Processes and Patterns. *Adv. Ecol. Res.* **1999**, *30*, 1–67. [CrossRef]

65. Porta, J.; López-Acevedo, M.; Roquero, C. *Edafología para la Agricultura y el Medio Ambiente*; Prensa, M., Ed.; Mundi-Prensa: Madrid, Spain, 2006; ISBN 84-7114-468-9.

66. Pritchett, W.L.; Fisher, R.F. *Properties and Management of Forest Soils*, 2nd ed.; John and Wiley and Sons: Hoboken, NJ, USA, 1987.

67. Cao, Y.; Chen, Y. Coupling of plant and soil C:N:P stoichiometry in black locust (*Robinia pseudoacacia*) plantations on the Loess Plateau, China. *Trees* **2017**, *31*, 1559–1570. [CrossRef]

68. Hyvärinen, A. Deposition on forest soils—Effect of tree canopy on throughfall. In *Acidification in Finland*; Springer: Berlin/Heidelberg, Germany, 1990; pp. 199–213.

69. Yanai, R.D.; Arthur, M.A.; Acker, M.; Levine, C.R.; Park, B.B. Variation in mass and nutrient concentration of leaf litter across years and sites in a northern hardwood forest. *Can. J. For. Res.* **2012**, *42*, 1597–1610. [CrossRef]

70. Laganière, J.; Paré, D.; Bradley, R.L. How does a tree species influence litter decomposition? Separating the relative contribution of litter quality, litter mixing, and forest floor conditions. *Can. J. For. Res.* **2010**, *40*, 465–475. [CrossRef]

71. Côté, B.; Camiré, C. Tree growth and nutrient cycling in dense plantings of hybrid poplar and black alder. *Can. J. For. Res.* **1987**, *17*, 516–523. [CrossRef]

72. McLennan, D.S. Spatial variation in black cottonwood (*Populus trichocarpa*) foliar nutrient concentrations at seven alluvial sites in coastal British Columbia. *Can. J. For. Res.* **1990**, *20*, 1089–1097. [CrossRef]

73. Makeshin, F.; Rehfuess, K.F.; Ruesch, I. Short-rotation plantations of poplars and willows on formerly arable land-sites, nutritional-status, biomass production, and ecological effects. *Forstwiss. Cent.* **1989**, *108*, 125–143. [CrossRef]

74. Rytter, L.; Ericsson, T. Leaf nutrient analysis in *Salix viminalis* (L.) energy forest stands growing on agricultural land. *Z. Pflanzenernähr. Bodenkd.* **1993**, *156*, 349–356. [CrossRef]

75. Jug, A.; Hofmann-Schielle, C.; Makeschin, F.; Rehfuess, K.E. Short-rotation plantations of balsam poplars, aspen and willows on former arable land in the Federal Republic of Germany. II. Nutritional status and bioelement export by harvested shoot axes. *For. Ecol. Manag.* **1999**, *121*, 67–83. [CrossRef]

76. Heinsdorf, D. Ergebnisse von Minersldungungsveruchen zu Androscoggin-Papplen auf Kippsanden ausgekohlter Braunkohlentagebaue im Bezirk Cottbus. *Beitr. Forstwirtsch.* **1985**, *19*, 34–40.

77. Bergmann, W. *Nutritional Disorders of Plants: Developments, Visual and Analytical Diagnosis*; Gustav Fischer Verlag Jena: New York, NY, USA, 1992.

78. Sardans, J.; Janssens, I.A.; Alonso, R.; Veresoglou, S.D.; Rillig, M.C.; Sanders, T.G.M.; Carnicer, J.; Filella, I.; Farré-Armengol, G.; Peñuelas, J. Foliar elemental composition of European forest tree species associated with evolutionary traits and present environmental and competitive conditions. *Glob. Ecol. Biogeogr.* **2015**, *24*, 240–255. [CrossRef]

79. Malavolta, E. *Avaliação do Estado Nutricional das Plantas: Princípios e Aplicações*, 2nd ed.; Potafos: Piracicaba, Brazil, 1997.

80. Sayyad, E.; Hosseini, S.M.; Mokhtari, J.; Mahdavi, R.; Jalali, S.G.; Akbarinia, M.; Tabari, M. Comparison of growth, nutrition and soil properties of pure and mixed stands of *Populus deltoides* and *Alnus subcordata*. *Silva Fenn.* **2006**, *40*, 27–35. [CrossRef]

81. Tzvetkova, N.; Petkova, K. Bioaccumulation of heavy metals by the leaves of *Robinia pseudoacacia* as a bioindicator tree in industrial zones. *J. Environ. Biol.* **2015**, *36*, 59–63. [PubMed]

82. Fernández-Pascual, M.; María, N.; Felipe, M. Fijación biológica del nitrógeno: Factores limitantes. In Proceedings of the Ciencia y Medio Ambiente—Segundas Jornadas Científicas sobre Medio Ambiente del CCMA-CSIC, Madrid, Spain, 16–17 April 2002; pp. 195–202.

83. Dawson, J.O.; Gordon, J.C. Nitrogen Fixation in Relation to Photosynthesis in *Alnus glutinosa*. *Bot. Gaz.* **1979**, *140*, 70–75. Available online: www.jstor.org/stable/2474206 (accessed on 5 August 2020). [CrossRef]

84. Martín-García, J.; Merino, A.; Diez, J.J. Relating visual crown conditions to nutritional status and site quality in monoclonal poplar plantations (*Populus* × *euramericana*). *Eur. J. For. Res.* **2012**, *131*, 1185–1198. [CrossRef]

85. Elferjani, R.; DesRochers, A.; Tremblay, F. Effects of mixing clones on hybrid poplar productivity, photosynthesis and root development in northeastern Canadian plantations. *For. Ecol. Manag.* **2014**, *327*, 157–166. [CrossRef]

86. Stefan, K.; Fürst, A.; Hacker, R.; Bartels, U. *Forest Foliar Condition in Europe. Results of Large-Scale Foliar Chemistry Surveys (Survey 1995 and Data from Previous Years)*; European Commission—United Nations/Economic Commission for Europe: Brussels, Belgium, 1997.

87. Kirkby, E.A.; Mengel, K. The role of magnesium in plant nutrition. *Z. Pflanzenernähr. Bodenkd.* **1976**, *139*, 209–222. [CrossRef]

88. Gordon, W.S.; Jackson, R.B. Nutrient concentrations in fine roots. *Ecology* **2000**, *81*, 275–280. [CrossRef]

89. Yuan, Z.Y.; Li, L.H.; Han, X.G.; Huang, J.H.; Jiang, G.M.; Wan, S.Q.; Zhang, W.H.; Chen, Q.S. Nitrogen resorption from senescing leaves in 28 plant species in a semi-arid region of northern China. *J. Arid Environ.* **2005**, *63*, 191–202. [CrossRef]

90. Kobe, R.K.; Lepczyk, C.A.; Iyer, M. Resorption efficiency decreases with increasing green leaf nutrients in a global data set. *Ecology* **2005**, *86*, 2780–2792. [CrossRef]

91. Andrades, M.; Martinez, E. *Fertilidad del Suelo y Parámetros que la Definen*, 3rd ed.; Universidad la Rioja-Servicio Publicaciones: La Rioja, Spain, 2014; ISBN 978-84-695-9286-1.

92. Fife, D.N.; Nambiar, E.K.S.; Saur, E. Retranslocation of foliar nutrients in evergreen tree species planted in a Mediterranean environment. *Tree Physiol.* **2008**, *28*, 187–196. [CrossRef] [PubMed]

93. Townsend, A.R.; Cleveland, C.C.; Asner, G.P.; Bustamante, M.M.C. Controls over foliar N:P ratios in tropical rain forests. *Ecology* **2007**, *88*, 107–118. [CrossRef]

94. Niinemets, Ü.; Tamm, Ü. Species differences in timing of leaf fall and foliage chemistry modify nutrient resorption efficiency in deciduous temperate forest stands. *Tree Physiol.* **2005**, *25*, 1001–1014. [CrossRef]

95. Hernández, M.C.P. *Entradas Biológicas de Nitrógeno en un Bosque Ripario*; Universidad Complutense de Madrid, Servicio de Publicaciones: Madrid, Spain, 1996.

96. López, E.S.; Pardo, I.; Felpeto, N. Seasonal differences in green leaf breakdown and nutrient content of deciduous and evergreen tree species and grass in a granitic headwater stream. *Hydrobiologia* **2001**, *464*, 51–61. [CrossRef]

97. Mendes, A.D.R.; de Oliveira, L.E.M.; do Nascimento, M.N.; Reis, K.L.; Bonome, L.T.D.S. Concentração e redistribuição de nutrientes minerais nos diferentes estádios foliares de seringueira. *Acta Amaz.* **2012**, *42*, 525–532. [CrossRef]

98. Yang, Y.S.; Guo, J.F.; Chen, G.S.; Xie, J.S.; Cai, L.P.; Lin, P. Litterfall, nutrient return, and leaf-litter decomposition in four plantations compared with a natural forest in subtropical China. *Ann. For. Sci.* **2004**, *61*, 465–476. [CrossRef]

99. Fang, S.; Li, H.; Xie, B. Decomposition and nutrient release of four potential mulching materials for poplar plantations on upland sites. *Agrofor. Syst.* **2008**, *74*, 27–35. [CrossRef]

100. Li, L.J.; Zeng, D.H.; Yu, Z.Y.; Fan, Z.P.; Yang, D.; Liu, Y.X. Impact of litter quality and soil nutrient availability on leaf decomposition rate in a semi-arid grassland of Northeast China. *J. Arid Environ.* **2011**, *75*, 787–792. [CrossRef]

101. Moro, M.J.; Domingo, F. Litter decomposition in four woody species in a Mediterranean climate: Weight loss, N and P dynamics. *Ann. Bot.* **2000**, *86*, 1065–1071. [CrossRef]

102. Brady, N.C.; Weil, R.R. *The Nature and Properties of Soils*; Prentice Hall: Upper Sadle River, NJ, USA, 2008.

103. Berg, B.; Staaf, H. Leaching, accumulation and release of nitrogen in decomposing forest litter. *Ecol. Bull.* **1981**, *33*, 163–178.

104. Manzoni, S.; Jackson, R.B.; Trofymow, J.A.; Porporato, A. The global stoichiometry of litter nitrogen mineralization. *Science* **2008**, *231*, 684–686. [CrossRef]

105. Hirschfeld, J.R.; Finn, J.T.; Patterson, W.A. Effects of *Robinia pseudoacacia* on leaf litter decomposition and nitrogen mineralization in a northern hardwood stand. *Can. J. For. Res.* **1984**, *14*, 201–205. [CrossRef]

106. Gartner, T.B.; Cardon, Z.G. Decomposition dynamics in mixed-species leaf litter. *Oikos* **2004**, *104*, 230–246. [CrossRef]

107. Perala, D.A.; Alban, D.H. *Rates of Forest Floor Decomposition and Nutrient Turnover in Aspen, Pine, and Spruce Stands on Two Soils*; Research Paper NC-227; North Central Forest Experiment Station, USDA Forest Service: Saint Paul, MN, USA, 1982.

**Publisher's Note:** MDPI stays neutral with regard to jurisdictional claims in published maps and institutional affiliations.

 *forests*

*Review*

# Poplar Short Rotation Coppice Plantations under Mediterranean Conditions: The Case of Spain

Nerea Oliveira [1,*], César Pérez-Cruzado [2], Isabel Cañellas [1], Roque Rodríguez-Soalleiro [2] and Hortensia Sixto [1]

[1] Forest Research Centre, National Institute for Agriculture and Food Research and Technology (INIA-CIFOR), Crta. de la Coruña km 7.5, E-28040 Madrid, Spain; canellas@inia.es (I.C.); sixto@inia.es (H.S.)

[2] Sustainable Forest Management Group, University of Santiago de Compostela, C/Benigno Ledo s/n, E-27002 Lugo, Spain; cesar.cruzado@usc.es (C.P.-C.); roque.rodriguez@usc.es (R.R.-S.)

* Correspondence: oliveira.nerea@inia.es

Received: 15 November 2020; Accepted: 14 December 2020; Published: 17 December 2020

**Abstract:** Developing a circular bioeconomy based on the sustainable use of biological resources, such as biomass, seems to be the best way of responding to the challenges associated with global change. Among the many sources, short rotation forest crops are an essential instrument for obtaining quality biomass with a predictable periodicity and yield, according to the areas of cultivation. This review aims to provide an overview of available knowledge on short rotation coppice *Populus* spp. plantations under Mediterranean conditions and specifically in Spain, in order to identify not only the status, but also the future prospects, for this type of biomass production. The analysis of available information was conducted by taking into consideration the following aspects: Genetic plant material; plantation design, including densities, rotation lengths and the number of rotations, and mixtures; management activities, including irrigation, fertilization, and weed control; yield prediction; biomass characterization; and finally, an evaluation of the sustainability of the plantation and ecosystem services provided. Despite advances, there is still much to be done if these plantations are to become a commercial reality in some Mediterranean areas. To achieve this aim, different aspects need to be reconsidered, such as irrigation, bearing in mind that water restrictions represent a real threat; the specific adaptation of genetic material to these conditions, in order to obtain a greater efficiency in resource use, as well as a greater resistance to pests and diseases or tolerance to abiotic stresses such as drought and salinity; rationalizing fertilization; quantifying and valuing the ecosystem services; the advance of more reliable predictive models based on ecophysiology; the specific characterization of biomass for its final use (bioenergy/bioproducts); technological improvements in management and harvesting; and finally, improving the critical aspects detected in environmental, energy, and economic analyses to achieve profitable and sustainable plantations under Mediterranean conditions.

**Keywords:** biomass; *Populus*; SRC (Short Rotation Coppice); short rotation woody crops; sustainability; Mediterranean conditions; management; review

## 1. Introduction

The challenges associated with climate change, along with the changing paradigm for both economic development and the energy model, have been crystallized into a Green Deal for Europe [1]. In this context, the use, production, and utilization of biomass are undoubtedly some of the main issues, very much related to the need to redirect the linear economy towards a circular bioeconomy based on the use of sustainable biological resources [2]. The Innovation Strategy for Sustainable Growth: A Bioeconomy for Europe [3] and its later revision [4] establish the bioeconomy as the general framework and key factor for achieving green, sustainable growth in Europe. In harmony with the

European strategy, while also taking into account the national possibilities, the Spanish Strategy for the Bioeconomy [5] and the current draft Law on Climate Change and Energy Transition [6] have been published.

Development in this area must take into consideration the potential of natural resources, including biomass. The options of using existing biomass in forests, agricultural residues, or the production of new biomass from crops planted specifically for this purpose are all seen as key to the development of the bioeconomy [3], not only for economic reasons, but also with environmental and social considerations in mind. In Spain, the Strategic Research and Innovation Agenda in relation to biomass has also recently been presented [7]. In this context, biomass is a highly valued resource for both bioenergy production and bioproducts, thus contributing towards addressing the abovementioned global challenges.

There are various sources of forest biomass, which can be differentiated into the following: (i) That derived from forest management activities associated with timber exploitation; (ii) residues resulting from silvicultural operations apart from timber exploitation; (iii) biomass derived from the forestry industry; and lastly, (iv) dedicated forest crops specifically designed for the production of woody biomass.

The first category includes logging residues and small diameter or crooked logs which are used for energy production (pellets, wood chips, or firewood). The first and third categories are the type of biomass mainly used in Spain today. The Spanish Renewable Energy Association (APPA) has repeatedly pointed to the underutilization of wood resources at a national level, as only 41% of the annual wood increment is currently being used, which is notably below the average of 60–70% for Europe [8].

However, if the amount of wood resources has been constantly increasing since the 1960's, what is the point of developing specific forestry crops? Firstly, these crops are seen as being of particular importance in terms of their potential contribution to the efficient diversification of biomass sources [9–11]. Secondly, of all biomass sources, forest crops specifically designed for woody biomass production are those most readily managed in terms of both time and space, with a predictable periodicity and yield. Thirdly, biomass from forest crops can contribute to generating and stabilizing the biomass market.

The aim of this study is to assess the advances made in poplar short rotation coppice (SRC) plantations over recent years, focusing on specific advances in Mediterranean conditions under irrigation (Spain) within the global context and to identify the areas of research where progress still needs to be made. The current state-of-the-art in biomass production from poplar SRC under Mediterranean conditions is addressed, highlighting both the strengths and weaknesses.

Therefore, this paper is structured around the following: (i) A global vision of SRC plantations focusing on poplar as one of the most suitable species in Mediterranean areas; (ii) the state-of-the-art of these plantations at a global level, while focusing on the progress made in Spain; and finally, (iii) conclusions drawn under Mediterranean conditions, with a particular emphasis on the weaknesses identified and the short- and long-term lines of research needed to address them.

## 2. Short Rotation Forest Crops for the Production of Biomass: The *Populus* Genus

Despite the different dedicated energy crops, the current agenda [7] only considers herbaceous and woody lignocellulosic crops for more diverse uses, which also include bioproducts.

Fast growing species are used in SRC, employing intensive or semi-intensive techniques [12], with coppicing cycles of between 2 and 8 years until stool productivity declines, which normally occurs after 15 years [13,14], depending on the site quality.

Biomass from SRC may become essential as an addition to the biomass provided by forests, contributing towards meeting the demands of European industry and assuring market stability [15]. This is probably linked to the need to find spatiotemporal complementarity in biomass resources, contributing to matching the supply with the demand; this circumstance is already a commercial reality in many parts of the world [16]. The suitability of such biomass is also linked to the intrinsic characteristics of its production and management [17], such as the abundance of improved, highly adaptable genetic material; high rate of successful rooting; good juvenile growth; and its resprouting capacity, among others [18–20]. Other traits associated with its end use are also deemed to be important, such as a low chlorine (Cl) and sulfur (S) content, low ash content, and high lignin content [21–24], among others.

SRC have been found to provide ecosystem services seldom sufficiently quantified and contrasted, such as air cleaning; the control of erosion or flooding in certain areas [25–27]; mitigation of the effects of climate change through carbon fixing in foliar or root biomass fractions [14,28,29]; increases of the biodiversity in agricultural environments [30–33]; and even soil decontamination [34–36], including mining reclamation [37,38]. From a social perspective, woody crops contribute towards the creation of employment in rural areas, given that these crops provide an opportunity to make use of poor, marginal, or surplus agricultural land [39]. Finally, the use of biomass from SRC helps to reduce the pressure on natural forests by providing a raw material much demanded by society and therefore by industry.

Despite this, to achieve sustainability in the implementation of SRC, it is important to consider the impact of land use changes in areas where these plantations compete with agricultural crops for land, as well as aspects related to water consumption in areas with limited water resources. Therefore, it is necessary to define the limitations in order to guarantee the sustainability of these crops [40]. Many of these aspects can probably be dealt with through the use of biotechnology to achieve improvements, or by using circular economy techniques such as wastewater reuse [41–43].

In Spain, the interest in producing biomass from SRC dates back to the mid-80's [44–46], coinciding to a large extent with the crisis in the oil sector. However, it was around 2000 when initiatives to increase sources of renewable energy at a global scale, particularly in Europe, provided the impetus to explore possibilities for the production of biomass as a renewable resource. The climatic, edaphic, and demographic characteristics of some parts of the country are suitable for the cultivation of SRC, with an expected high productivity [47] exceeding that obtained in other European countries. However, because of the Mediterranean climatic conditions, such plantations are only viable with the use of irrigation. In areas with an Atlantic climate, the conditions for SRC are also suitable, with limitations such as the orographic characteristics, which may complicate the intensive silviculture applied in this type of plantation, or the limitations associated with certain species due to soil acidity.

Although many studies have focused on the selection of vegetal material best adapted to given areas of Spain, there is still a long way to go with regards to identifying the interaction between the genotype and the environment, which is a determining factor in the success or failure of SRC.

There are many woody species potentially cultivable for biomass production. In general, they are fast growing broadleaf species with a high re-sprouting capacity.

In Europe, the *Salicaceae* family (*Populus* spp. and *Salix* spp.) presents the greatest developments on an industrial level. Plantations based on species and hybrids of *Populus* are well-established in both central and southern Europe, with examples in Germany [48], the United Kingdom [13], the Czech Republic [49], Bulgaria [50], Serbia [51], Poland [52], and France [53], as well as in Mediterranean regions, mainly Italy [54] and Spain [55]. In northern Europe, where *Salix* spp. has been the predominant species for this purpose, poplar cultivation is beginning to attract interest, with different trials being established to evaluate its potential [56,57].

To a large extent, *Populus* species are at the forefront of biomass production because of the highly efficient breeding programs in different countries, many of which are located in Europe [58]. These programs are favored due to the very broad genetic base with which to work in terms of traits linked to cultivation and wood properties [59,60], but also due to knowledge of the genome sequence [61] and the relatively short breeding cycles. Other factors include a high capacity for vegetative reproduction and their rapid growth rate.

The importance of cultivating *Salicaceae* for biomass production at a global scale is reflected by the abundance of information gathered over the last decade by the International Poplar Commission (statutory body within the FAO), which has specifically covered the subject in one of its working groups [62]. The International Catalogue of Base Materials of *Populus* for obtaining forest reproductive material contains 358 entries, among which several are specifically referenced for biomass production ('Boiano-4', 'Baldo', 'Hunneghem', and 'Raspalje'). Many more are at preliminary stages for inclusion and have been put forward for this use ('AF2', 'AF8', 'A4A', 'Monviso', 'Muur', 'Orion', 'Oudenberg', 'Sirio', and 'Vesten'). Within the framework of the EU-POP project, more than 17 genotypes are currently being characterized as biomass producers in a multi-environment trial in which ten European countries are taking part, including Spain [63].

In addition to the genotypes that have been catalogued (or are in the cataloguing phase) for this purpose, there are others that have been identified for their wood production, but which also seem to possess suitable characteristics for the production of biomass. Figure 1 shows the main genotypes that are being tested or planted for commercial purposes in some European countries.

**Figure 1.** Some of the main genotypes planted for biomass purposes in each country are represented on a European map.

In Spain, the main species of interest at a commercial or pre-commercial level are those belonging to the genera *Populus* spp. and *Eucalyptus* spp. The cultivation of species and hybrids of *Populus* for industrial wood (veneer) is well-established in many areas of the country [64]. Although poplar production plantations in Spain represent less than 1% of the tree-covered forest area (approximately 100,000 ha), in some provinces, these plantations account for more than 50% of the harvested wood, which is around 40% of the economic value of roundwood cuttings. In the province of Castilla y Leon, for example, poplar is the forest species with the highest economic value [65,66]. However, regarding poplar in SRC, there is only a token presence in Spain, with it occupying around 50 ha at its peak. The potential land at a national level corresponds to irrigated agricultural marginal land for food production with the edaphoclimatic requirements for the species [47]. Currently, the irrigated agricultural land in Spain is 3.8 Mha [67].

The genotypes for planting should be included in the European Catalogue of Base Material for *Populus* reproductive material. There is also a Spanish National Catalogue of Base Material for forest reproductive material of the *Populus* genus in qualified and controlled categories comprising 24 commercial quality genotypes (Table 1). Some of this material, which is well-adapted to the specific Spanish Mediterranean conditions, could also be of interest for biomass production. Most of the genotypes already tested or currently being tested are listed in Figure 1, and include materials from both the European and Spanish lists, as well as some that have not yet been catalogued. The potential of *Populus* is explained to a large degree by the availability of material adapted to existing conditions, along with appropriate knowledge relating to the management of the species in many parts of the country [68,69].

**Table 1.** Genotypes included in the National Catalogue of Base Materials for the production of forest reproductive materials related to the genus *Populus* L.

| Parentage | Section | Genotypes |
|---|---|---|
| *P.* × *canadensis* Mönch | D × N | Aigeiros | '2000 Verde', 'Agathe F' [a], 'E-298' [a], 'Branagesi', 'B-1M', 'Canadá Blanco', 'Dorskamp', 'Flevo', 'Guardi', 'I-214' [b], 'Campeador' [b], 'I-454/40', 'Luisa Avanzo', 'MC', 'Triplo' |
| *P. deltoides* W. Bartram ex Marshall | D | Aigeiros | 'Lux', 'Viriato' |
| *P.* × *generosa* Henry | T × D | Tacamahaca × Aigeiros | 'Beaupre', 'Boelare', 'Raspalje', 'Unal', 'USA 49-177' |
| *P.* × *generosa* Henry × *P. alba* L. | (T × D) × A | (Tacamahaca × Aigeiros) × Populus | 'I-114/69' |
| *P. nigra* L. | N | Aigeiros | 'Tr 56/75', 'Bordils', 'Lombardo leones' |

D is *P. deltoides*; N is *P. nigra*; T is *P. trichocarpa*; A is *P. alba*; [a] 'Agathe F' = 'E-298'; and [b] 'I-214' = 'Campeador'. Order of 24/06/1992, Order APA/544/2003 of 06/032003, Resolution of 07/07/2006 of the Dirección General de Agricultura, and Resolution of 07/11/2011 of the Dirección General de Recursos Agrícolas y Ganaderos.

In relation to *Eucalyptus*, the cultivation of some species and hybrids for the production of biomass has been significant in certain areas of the country, reaching production values of 14.6 Mg ha$^{-1}$ year$^{-1}$ in the case of *Eucalyptus globulus* Labill. and 21.5 Mg ha$^{-1}$ year$^{-1}$ for *Eucalyptus nitens* (Deane and Maiden) Maiden [70] in Atlantic areas without irrigation, whereas the development of similar yields in southern Spain would need irrigation.

Some genotypes of *Eucalyptus* exhibit the additional advantage of presenting an acceptable degree of tolerance to drought conditions [71,72], although the high levels of production are associated with scenarios where droughts do not occur or where irrigation is used. One of the main differences with respect to poplar is that *Eucalyptus* is an evergreen species, so harvested trees include twigs and even leaves, yielding biomass with a high ash content.

Furthermore, trials with other potentially usable species have taken place in recent years, leading to differing results. In this regard, at a Mediterranean scale, there have been experimental trials with *Ulmus pumila* L. [73], *Robinia pseudoacacia* L. [74–77], hybrids of *Salix* spp. [78–81], *Platanus* × *hispanica* Mill. ex Münchh [77], and different cultivars of *Paulownia* spp. [82,83], with the latter displaying severe adaptation problems to the climatic conditions of the Mediterranean area (early frosts and flooding) [84]. Although some of these may be of interest in terms of adding diversity to the area of crop development, perhaps the main limitation of most of these species stems from the fact that there is a lack of improved genetic material, which is necessary for allowing their use in different environments.

## 3. Lines of Progress: State of the Art

Sustainable improvement in biomass production from poplar SRC crops requires advances to be made in different aspects, which have been summarized in Figure 2. An overall vision of the current situation of each of these aspects is provided in this section.

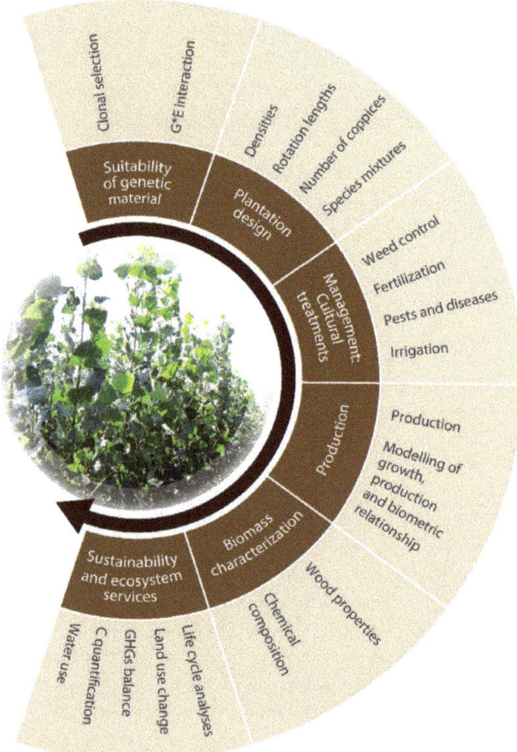


**Figure 2.** Lines of action for improving short rotation coppice-managed woody crops.

Literature searches for the combination of the terms *Populus* or poplar, and short rotation or energy crop, were conducted using Web of Science (WOS, Core collection of Web of Science), which is one of the main journal databases. There were no restrictions in terms of the year of publication or language. Figure 3 shows the number of documents per country (including only the top 30 countries) found in WOS, totaling 2185 documents. From these searches, documents from Spain were filtered and it was found that some of them were not properly classified in the database. Out of the 112 documents relating to Spain (ranked 10th) found in the WOS, only 83 were included.

In this review, we took into account not only the documents found in this database, but also other scientific publications belonging to Journal Citations Report (JCR) indexed journals, as well as numerous pieces of available gray literature. All this information was identified by tracing back papers cited in the references of the identified studies and reviewing publications by scientists who have worked or are currently working on poplar short rotation in Spain. In any case, given the abundance of existing information and the difficulties associated with the use of diverse terminology, there may be certain literature that we are not aware of, and therefore has not been considered in this review.

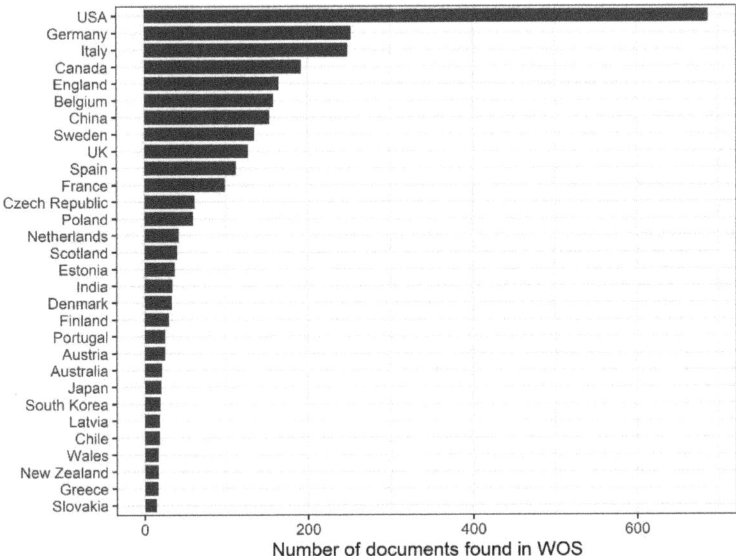


**Figure 3.** Ranking of the number of publications per country (including only the top 30 countries) found in the Web of Science (WOS) journal database, for the combination of the terms *Populus* or poplar, and short rotation or energy crop, in July 2020.

A classification of all the information available at a national level was performed according to the six categories detailed in Figure 2. This information was also broken down according to the availability of literature, separating the so-called gray literature from that contained in science journals and books (Figure 4). In Spain, biomass characterization and sustainability (mainly energy, economic, and environmental analysis) are the lines of research which have been explored the most based on scientific publications, although, if we include gray bibliography, then the most explored lines are production, modeling, and genetic material. Approximately 50.5% of the information evaluated corresponds to gray literature, 4.5% to books, and the rest to scientific publications (45%).

*3.1. Suitability of Genetic Material*

Clonal Selection

Poplar is the model tree for genetic studies and is the furthest ahead in terms of biological knowledge and genetic resources [58]. Using the best adapted material when developing plantations helps to ensure the efficient use of site resources and therefore higher levels of production. Phenotypic plasticity refers to the capacity of an organism to alter its characteristics in response to environmental conditions [85]. The genotype by environment interaction (G*E) is evidenced by the instability of phenotypic correlations derived from drastic differences in biomass production found in poplar plantations [86–88].

The G*E interaction makes poplar clonal recommendations more difficult, although it can also provide an opportunity to maximize production in these plantations by matching the most suitable material to specific site conditions. Genotypic stability is understood as the capacity of a cultivar to produce in accordance with the productive potential of each environment, that is, without straying from the behavior expected for the average genotypic value [89]. However, the strategies may differ, and this can be observed in different poplar SRC plantations; these strategies include attempting to find a broad adaptation or optimizing the adaptation to specific site conditions, hence the necessity to characterize the material [90–92]. This characterization not only has an impact on crop management, but also has repercussions for improvement strategies. In Spain, poplar material destined for biomass production has been characterized in recent years through the analysis of this interaction [55,77,93–95].

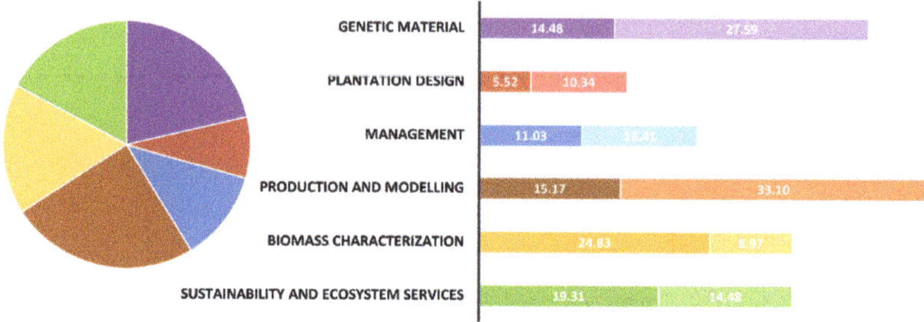


**Figure 4.** Percentages of Spanish publications in each of the six categories established. The light color corresponds to the gray literature on each subject.

With regards to the biomass, the development of genotypes aimed at achieving increased production, a greater tolerance to pests and diseases, and specific feedstock properties is ongoing through breeding programs in the countries spearheading the research, with the most prominent of these being the USA, Canada, Italy, France, Germany, and Sweden [58].

Traditional approaches based on recurrent selection as a way to increase hybrid vigor [54,96] are now supported by genomics and phenomic techniques which contribute to accelerating these processes [97,98]. For example, tools such as genome editing using Crispr/Cas9 technology [99–101], the development of transgenic plants [102], complete genome association analysis [103], and new phenotypic tools [104], are now leading to important advances in genetic improvement of the *Populus* genus. A relevant review of the progress towards an improved biomass quality and resilience of production using both traditional approaches and new technology is provided by Clifton-Brown et al. [58], specifying the breeding programs in which each of these techniques has been used. New avenues of research have also been opened up through considering epigenetic inheritance, also termed soft inheritance. This relates to inheritable changes to the gene expression induced by environmental changes [105,106].

In Spain, although progress in this line of research is limited by the lack of specific breeding programs, progress has been made in the development and testing of different transformed poplar lines expressing a pine glutamine synthetase (GS) transgene involved in N assimilation [107]; the CsDML gene that induces bud formation needed for the survival of the apical meristem under the harsh conditions of winter [108]; and a gene related to ABI3/VP1 1 (CsRAV1) to increase syllepsis, and consequently biomass production, in commercial elite trees [109,110]. The biomass potential of the different available genotypes has also been tested, with multiple clonal trials involving traditional genotypes [95,111–116], as well as new genotypes specifically selected for biomass [77,117]. The results have revealed hybrids like 'Orion' and 'AF2' as site-specific genotypes, while 'I-214' and 'Monviso' have been demonstrated to adapt well to a wide variety of scenarios. Likewise, the ability of different genotypes to produce biomass in areas degraded by mining is also being tested, with some genotypes, such as 'Raspalje' and 'AF2', displaying acceptable yields [118]. In general, a strong correlation has been identified between the response to environmental variables and the taxonomic background, with the euramerican genotypes such as '2000 Verde' responding better to lower latitudes, higher mean temperatures of the vegetative period, and a lower sandy soil content, unlike interamerican genotypes such as 'Unal' [93].

*3.2. Plantation Design*

The design of a plantation is also a highly influential factor in stand management and the optimization of production [119]. Many different types of design have been tested in Spain, combining different plantation densities [120], tree distributions [121], and rotation lengths [122]. The possibility of mixing different species or different varieties of the same species has also been put forward as a way to optimize production.

### 3.2.1. Densities

A broad range of initial stocking rates have been employed to define the optimum plantation density, from 1000 cuttings ha$^{-1}$ up to more than 40,000 cuttings ha$^{-1}$ [123,124]. Even higher densities have been contemplated, albeit less frequently, as in the case of DeBell et al. [125]. In Spain, a wide range of densities in multiple environments have been tested [46,126,127]. In terms of the yield, densities of around 20,000–25,000 cuttings ha$^{-1}$ were the most productive in the first year, but the effect of competition reduced the differences between these and less stocked plantations as rotation approached three years [120]. Other factors, such as the costs associated with the higher densities due to the greater number of cuttings, increased demand for water, and increased consumption of consumables and time required for operations such as irrigation or harvesting [128], along with impediments to mechanization of the plantation, mean that densities in excess of 15,000 cuttings ha$^{-1}$ are not recommended [120].

Furthermore, the position of the stools for a given density is also a matter of debate. Therefore, it is common for trials to include both single-row layouts [78,92] and double rows [13,129]. No significant differences in yield were found between the two layouts [130], although single-row designs tend to facilitate management operations. In Spain, both layouts have been tested [77,131], although a single row is more commonly used.

The density and distribution of stools should be directly linked to the rotation lengths applied, along with the maintenance and harvesting processes, always attempting to keep the production cost per biomass unit obtained as low as possible.

### 3.2.2. Rotation Lengths and Number of Coppices

The rotation is a factor closely linked to the density, as well as being influenced by the environment. Different rotations have been employed in the management of poplar SRC, from yearly to longer rotations for other productive objectives [132–135]. In recent years, rotations of 2 to 4 years have been the most frequently employed, with the final harvesting age depending on the genotype and site conditions [90,129,136].

In Spain, rotations from 2 to 9 years have been tested [110,117,128,137], although few studies have compared different rotation lengths. Hernández Garasa et al. [138] determined that the maximum volume production occurred between 3 and 4 years, depending on the genotype planted. The suitability of a 3- or 4-year rotation length for different genotypes under poor site conditions was evaluated by Oliveira et al. [139], who noted that biomass production increases in most genotypes from 74% to 166% when the rotation length is extended by 1 year. However, some genotypes, such as 'AF2' and 'Dorskamp', presented no significant differences. These results are consistent with those obtained under average site conditions, where the maximum volume production was not reached with a 3-year rotation length, except in the cases of certain genotypes, such as 'A2A', 'A4A', and 'Lombardo Leones' [140]. Therefore, the rotation length chosen clearly depends on both the genotype and the site conditions, with longer rotation lengths being advisable when the conditions are not optimal. Further studies are needed in order to optimize the rotation length for specific sites, genotypes, and management conditions.

Another question that has sparked controversy is whether or not to coppice after the first establishment year to encourage multiple shoot growth [141,142]. This is a common practice in some countries, such as England [143]. In Spain, the use of management techniques both with or without coppicing after the first establishment year has also been studied [144–146], although no apparent advantage was found with regard to the additional coppice after the first establishment year, either in terms of biomass yield or quality.

Opinions also vary regarding how many times coppicing should take place [129,135], given that the lifecycle is clearly related not only to the rotation length, but also to the genotype and its interaction with the environment and the cultural practices applied, with all of these factors conditioning the useful life of the stool [68]. However, studies such as Štochlová et al. [147] suggest that five or six 3-year rotations are possible, although only one commercial clone—an interspecific hybrid of *P. maximowiczii* Rupr. × *P. trichocarpa* Torr & A. Gray—was capable of biomass productivity compatible with the economic feasibility of cultivation. Therefore, it is important to take into account the factors limiting the duration of SRC crops. In Spain, several studies carried out over three successive rotations of 3 years each point to a decrease in biomass production during the third rotation cycle, suggesting the end of its productive cycle [148,149]. However, not all genotypes show the same behavior, with the genotypes 'I-214' and 'Monviso' presenting the highest yields during the third rotation [150].

### 3.2.3. Species Mixtures

There has been a sharp rise of interest in mixed forest plantations over recent decades because of the perceived benefits, not only for the environment and ecology, but also, although not always, for the yield, as a result of resource-use efficiency and facilitation. Although applying these designs in SRC plantations has attracted interest, there is still little information about it. Few examples exist of mixed plantations of poplar with other species in SRC. Some are mixtures of *Populus* spp. with *Alnus* spp. [151,152], the euramerican genotype 'AF2' with *Ulmus* spp. and *Platanus* × *hispanica* [153], and *Populus* spp. with *Robinia* spp. [154–156]. The mixture of *Populus alba* L.-*Robinia pseudoacacia* in SRC under Mediterranean conditions showed encouraging results in the first rotation [76], but they were not as promising in the second [157]. A similar design mixing *R. pseudoacacia* with the euramerican genotype 'Dorskamp' in central France demonstrated interspecific competition in the mixture as the preponderant interaction, resulting in higher mortality and lower biomass production than the two monocultures [155]. However, this mixture would appear to be advantageous given the different strategies shown by the two species in terms of the amount of litter and the dynamics of the main nutrients [158]. Therefore, although mixing the species does not increase the biomass yield, it may provide a good strategy for reducing future requirements for nitrogen addition (with the consequent ecological and economic implications), given the differences between the two monocultures in terms of processing the main nutrients [158]. Furthermore, both species—*P. alba* and *R. pseudoacacia*—are considered to be relatively tolerant to possible drought scenarios [111,159].

In any case, the current plantations are only experimental and their implementation on a commercial scale would involve restrictions in the spatial distribution as a result of the complex establishment and the harvesting requirements [76]. These difficulties often lead to mixing the species in lines or groups, thus losing the facilitation effect, although other benefits, such as the reduction of biotic and abiotic damage, landscape effects, and other environmental benefits, are retained.

Apart from mixtures of different species, complementarity is also explored in mixtures of varieties within the same genus [160,161]. Some examples in this regard include plurivarietal plantations of *P.* × *generosa* Henry [162], other hybrid groups of *Populus* [163], and different varieties of *Salix* spp. [164,165], where the aim is to attain a greater tolerance to biotic and abiotic stresses while also increasing productivity.

### 3.3. Management: Cultural Treatments

There are many aspects of plantation management which have an impact on production.

### 3.3.1. Control of Competing Vegetation

Competition from weeds in the initial crop establishment stage is one of the main reasons for plantation failure. It is not only competition for water and nutrients, but also for light and space, which is crucial in the establishment stage. Deficits of these requirements can render the plantation unviable [68]. Therefore, weed control is considered a necessary practice, with treatments also being necessary during the first establishment year and after each coppicing. These treatments may involve both chemical and/or mechanical techniques [166–168].

Currently, the application of specific herbicides for poplar cultivation is very limited due to the European regulation on the sustainable use of pesticides and their commercialization (Directive 2009/128/CE of the European parliament; Regulation (CE) n° 1107/2009 of the European parliament). In Spain, only six formulations are registered for use with hardwoods (RD 971/2014), but only one of them is specific for *Populus*.

### 3.3.2. Fertilization

The use of fertilizer in SRC plantations is a subject which attracts debate. There are many examples where fertilization provides no benefits [59,90,169–171] and others where there are positive effects on production [145,172–177].

The lack of response to fertilization may be due to the fact that soil fertility is optimal or to an inadequate assessment of the limiting nutrients. The high N requirements of poplar [178,179] mean that in locations where the soil is poor in organic matter, it is beneficial to apply fertilizer to increase the yield [180]. However, the excessive use of nitrogen fertilizers as part of conventional practice is increasingly being questioned because of environmental risks [181–183] and the high economic costs. Therefore, the use of alternatives to improve the nutritional status, such as designs with mixed plantations that include nitrogen fixing species [155] or alternative fertilization through the use of sewage sludge or waste water, have gained prominence in recent years [184–186]. Besides not having to use traditional fertilizers, these materials are purified and/or reused and the plantations thus act as phytoremediators [187]. All these alternative fertilization techniques have also been tested in poplar SRC plantations in Spain in recent years, although all in experimental plantations [42,188,189].

In all cases, it is recommended that soil analysis be conducted prior to planting in order to optimize the use, where required, of fertilizers [190].

### 3.3.3. Control of Pests and Diseases

Newly emerging pests and diseases are one of the main problems facing agriculture in the 21st century due to the presence of extreme climatic conditions [191]. In poplar plantations, this risk has increased sharply as a result of the expansion of monoclonal plantations, with only a small number of different genotypes planted. There has been clear progress made in this area and it continues to be one of the main objectives in the breeding programs of the genus.

Examples of the most serious pests and diseases affecting the leaves are those caused by *Melampsora* spp. (rusts), *Marssonina brunnea* (Ell. et Ev.) Magn., *Phloeomyzus passerinii* Sign. (woolly poplar aphid), *Venturia populina* (Vuill.) Fabr., *Chrysomela populi* L. (red poplar leaf beetle), and *Leucoma salicis* L. (white satin moth), whilst those affecting the stems and trunks are *Mycosphaerella populorum* Peck (stem canker), *Cryptorhynchus lapathi* L. (poplar and willow borer), *Paranthrene tabaniformis* Rott. (dusky clearwing), and *Sesia apiformis* L., among others, although these are not as relevant to the SRC crop [192–197].

In Spain, although phytosanitary problems in SRC crops have not occurred very frequently, the presence of rust has been observed in rust-prone genotypes when grown for timber production. *Chrysomela populi* L. has been detected in plantations in the northern half of the country, as well as *Corythucha ciliata* (Say) in the center of the peninsula, necessitating the timely application of phytosanitary products.

In any case, the sustainability of these plantations depends on the use of genetic material which is tolerant or resistant to these types of stress [68,198], as well as resorting to biological control in the case of certain pests.

### 3.3.4. Irrigation

Given the marked hygrophilic nature of the *Populus* genus [199], the availability of water in the soil is one of the limiting factors for its cultivation [200]. Due to the Mediterranean climatic conditions, it is necessary to irrigate SRC plantations in Spain during the summer drought season [201–203]. Due to limitations on water use at a global scale [204], the irrigation of SRC crops is viewed with caution in areas where this practice is necessary. This has led to changes in the way irrigation is applied, moving away from flood irrigation towards more efficient systems such as drip irrigation, although much more can still be done to increase the technical efficiency [205].

Water restrictions generally lead to production losses [206,207], although a high variability has been found in response to drought conditions. For this reason, the identification of genotypes with a greater water-use efficiency through different methodological approaches is undoubtedly of interest [208], seeking to combine materials with a high productivity and greater water-use efficiency [209]. This is especially true in the Mediterranean area [77,210–213], although accepting a certain loss of production may be advisable in these scenarios [205].

Highly productive genotypes such as 'AF2' and 'Monviso' have exhibited the greatest water-use efficiency under optimal conditions, although under restrictive water conditions, they have presented a similar water-use efficiency to that of the less productive genotypes. The strategy followed by all of them to improve the intrinsic water-use efficiency seems to be linked to stomatal control, rather than differences in the rate of photosynthesis [205].

In addition to the implications in terms of sustainability, the economic implications must also be taken into account, bearing in mind that irrigation is one of the limiting factors when assessing the profitability of these plantations under Mediterranean conditions. It has been calculated that the costs associated with irrigation, which include the irrigation system, maintenance, and the annual costs of water and electricity, account for 30% of the total costs over a whole cycle of 12 years [214].

The reusing of water from different sources represents a new approach in the context of SRC [187,188], with a solid background in the past [215,216]. In any case, it is necessary to increase the amount of research into the breeding of plants for production under conditions of water scarcity at a global level, especially in regions that suffer from water restrictions.

These alternatives, together with the improvement in irrigation techniques, could provide solutions to ensure that viable production is attained in areas where water use must be minimized, such as in Mediterranean environments.

### 3.4. Production

Short rotation coppice plantations (SRC) provide a viable alternative for the production of quality lignocellulosic biomass [217]. The biomass produced in SRC is characterized by a predictable periodicity and yield, depending on the area of cultivation. The main challenge with regard to these crops is to achieve a high level of sustainable production while maximizing benefits; that is, combining economic viability with environmental sustainability.

According to a review by Sixto et al. [68] concerning biomass production in this type of plantation using poplar genotypes, a large quantity of literature exists on the different clonal productivity under a range of environments. Table 2 presents some examples of biomass production obtained in different European countries, ranging widely from 1 to 24 Mg dm (dry matter) ha$^{-1}$ year$^{-1}$, depending on the site characteristics, the genetic material, the design, and the management scheme. Table 3 presents the production obtained in Spain under Mediterranean conditions, ranging widely from 1 to 37 Mg dm ha$^{-1}$ year$^{-1}$. The largest reported productions appear to be those associated with Mediterranean irrigated environments. Despite this, the average potential production at a national level in Spain is estimated to be around 15.3 Mg dm ha$^{-1}$ year$^{-1}$ for plantations with standard management schemes [47], although there is also high variability, depending on the previously mentioned factors [46,55,77,145,218].

**Table 2.** Different examples of the biomass yield (Mg dm ha$^{-1}$ year$^{-1}$) per genotype, age, and rotation in poplar short rotation plantations in Europe (excluding Spain).

| Species and Poplar Hybrids | Genotype | Density | Age | Country | Yield (Mg dm ha$^{-1}$ Year$^{-1}$) by Rotation | | | Reference |
| --- | --- | --- | --- | --- | --- | --- | --- | --- |
| | | | | | 1st | 2nd | 3rd | |
| *P. deltoides* × *P. trichocarpa* | 'IBW2' | | | | 2.5 | 1.6 | | |
| *P. nigra* | 'Wolterson' | | | | 8.1 | 9.7 | | |
| | 'Columbia River' | | | | 7.8 | | | |
| *P. trichocarpa* | 'Fritzi Pauley' | 10,000 | 4 | Belgium | 8.1 | 8.2 | | [19,219] |
| | 'Trichobel' | | | | 8.4 | 8.2 | | |
| *P.* × *canadensis* | 'Gibecq' | | | | 1.6 | 3.5 | | |
| *P.* × *generosa* | 'Hazendans' | | | | 10.8 | 3 | | |
| | 'Hoogvorst' | | | | 10.1 | 8.2 | | |
| *P. deltoides* × *P.* × *generosa* | 'Grimminge' | | | | 4.3 | 11.7 | 8.4 | |
| *P. nigra* | 'Brandaris' | | | | 1.4 | 7.0 | 8.7 | |
| *P. nigra* × *P. maximowiczii* | 'Wolterson' | | | | 2.7 | 11.7 | 14.6 | |
| *P. trichocarpa* × *P. maximowiczii* | 'Bakan' | | | | 4.9 | 14.3 | 18.1 | |
| | 'Skado' | | | | 5.7 | 16.8 | 20.8 | |
| | 'Ellert' | 8000 | 2 (In the 3rd rot the age is 3 year) | Belgium | 3.3 | 11.1 | 19.5 | [136] * |
| | 'Hees' | | | | 6.5 | 15.5 | 26.0 | |
| | 'Koster' | | | | 2.8 | 9.5 | 14.1 | |
| *P.* × *canadensis* | 'Muur' | | | | 3.9 | 12.4 | 17.1 | |
| | 'Oudenberg' | | | | 3.7 | 13.4 | 14.0 | |
| | 'Robusta' | | | | 2.5 | 8.1 | 15.6 | |
| | 'Vesten | | | | 4.7 | 12.6 | 16.1 | |
| *P. maximowiczii* × *P. trichocarpa* | 'NE-42' | 2222 | 3 (2nd rot: 4 year) 3 (2nd rot: 2 year) | Czech Republic | 1.0–1.4 8.3 | 9.4–9.8 15.4 | 9.1–11.4 18.9 | [220] |

Table 2. *Cont.*

| Species and Poplar Hybrids | Genotype | Density | Age | Country | Yield (Mg dm ha$^{-1}$ Year$^{-1}$) by Rotation | | | Reference |
|---|---|---|---|---|---|---|---|---|
| | | | | | 1st | 2nd | 3rd | |
| *P. balsamifera* × *P. tremula* | 'P-524' | | | | 8.1 | | | |
| *P. maximowiczii* × *P. berolinensis* | 'P-494' | | | | 10.2 | | | |
| *P. nigra* | *P. nigra* | 10,000 | 6 | Czech Republic | 2.6 | | | [221] |
| *P. nigra* × *P. maximowiczii* | 'J-104' | | | | 11.9 | | | |
| | 'J-105' | | | | 13.9 | | | |
| *P.* × *generosa* | 'P-473' | | | | 9.7 | | | |
| | 'Beaupré' | | 3 | | 1.57–11.13 | | | |
| | | | 4 | | 2.73–12.7 | | | |
| | 'Boelare' | | 3 | | 1.37–10.53 | | | |
| *P.* × *generosa* | | 3030 | 4 | France | 2.85–11.6 | | | [222] |
| | 'Hunnegem' | | 3 | | 12.1 | | | |
| | | | 4 | | 13.68 | | | |
| | 'Raspalje' | | 3 | | 2.2 | | | |
| | | | 4 | | 3.63 | | | |
| *P.* × *generosa* | 'Beaupré' | | | | 6.1 | | | |
| | 'Rap' | | | | 5.8 | | | |
| | 'Max 1' | | | | 3.3 | | | |
| *P. nigra* × *P. maximowiczii* | 'Max 3' | 8333 | 8 | Germany | 3.5 | | | [133] * |
| | 'Max 4' | | | | 3 | | | |
| *P. maximowiczii* × *P. trichocarpa* | 'Androscoggin' | | | | 3.3 | | | |
| | 'Hybride 275' | | | | 4.2 | | | |
| *P. trichocarpa* | 'Muhle Larsen' | | | | 3.7 | | | |
| *P. maximowiczii* × *P. nigra* | 'Max 1' | 8890 | 2 | Germany | 1.0–1.59 | | | [223] |
| | | 9250 | | | 6.81 | | | |
| *P. maximowiczii* × *P. nigra* | 'Max 3' | 17,778 | 4 | Germany | 8.6 | | | [48] |
| *P. maximowiczii* × *P. trichocarpa* | 'Androscoggin' | | | | 10.5 | | | |
| *P. maximowiczii* × *P. nigra* | 'Max 4' | 11,000 | 2 | Germany | 5.4–6.3 | | | [224] |

**Table 2.** *Cont.*

| Species and Poplar Hybrids | Genotype | Density | Age | Country | Yield (Mg dm ha$^{-1}$ Year$^{-1}$) by Rotation | | | Reference |
| --- | --- | --- | --- | --- | --- | --- | --- | --- |
| | | | | | 1st | 2nd | 3rd | |
| *P. deltoides* | 'Baldo' | 5747 | | | 4.75 | | | |
| | | 10,000 | | | 17.5 | | | |
| | 'Dvina' | 10,000 | | | 11.8 | | | |
| | 'Lambro' | 10,000 | | | 9.5 | | | |
| | 'Lena' | 10,000 | | | 14.2 | | | |
| | 'Lux' | 5714 | | | 7.05 | | | |
| | | 8333 | | | 4.3 | | | |
| | 'Oglio' | 10,000 | | | 14.1 | | | |
| *P. × canadensis* | 'BL-Costanzo' | 5714 | 2 | Italy | 3.25 | | | [90] |
| | | 7142 | | | 4.45 | | | |
| | 'Cima' | 5714 | | | 4.55 | | | |
| | | 7142 | | | 5.7 | | | |
| | 'I-214' | 5747 | | | 4.35 | | | |
| | | 10,000 | | | 8.4 | | | |
| | 'Luisa Avanzo' | 5714 | | | 5.55 | | | |
| | | 7142 | | | 6.25 | | | |
| | 'Orion' | 5747 | | | 4.46 | | | |
| | | 10,000 | | | 17.0 | | | |
| *P. deltoides* | 'Lux' | 10,000 | 2 | Italy | 12.38 | 6.99 | 4.99 | [225] |
| | | | 3 | | | 7.07 | 14.53 | |
| *P. deltoides* | 'Lux' | 10,000 | 1 | Italy | | | 16.4 | [135] |
| | | | 2 | | 22.5 | | | |
| | | | 3 | | 24.3 | | | |
| *P. deltoides* | 'Lux' | | | | 3.31–9.33 | 2.17–18.85 | | |
| *P. maximowiczii × P. nigra* | 'AF10' | | | | 14.84 | 20.74 | | |
| | 'AF2' | | | | 5.66–15.77 | 7.54–17.01 | | |
| *P. × canadensis* | 'I-214' | 5900 | 2 | Italy | 3.63–11.87 | 4.2–18.44 | | [226] |
| | 'Sirio' | | | | 5.65 | 8.55 | | |
| *P. × generosa × P. nigra* | 'AF6' | | | | 5.76–15.04 | 5.79–17.07 | | |
| | 'Monviso' | | | | 6.79–17.92 | 9.31–23.55 | | |
| *P. × generosa × P. trichocarpa* | 'AF8' | | | | 4.75–14.93 | 5.04–24.05 | | |

**Table 2.** *Cont.*

| Species and Poplar Hybrids | Genotype | Density | Age | Country | Yield (Mg dm ha$^{-1}$ Year$^{-1}$) by Rotation | | | Reference |
|---|---|---|---|---|---|---|---|---|
| | | | | | 1st | 2nd | 3rd | |
| *P.* × *canadensis* | 'AF2' | | | | 16 | 17 | 17 | |
| | 'I-214' | | | | 11.5 | 18 | 13.5 | |
| *P.* × *generosa* × *P. nigra* | 'AF6' | 6061 | 2 | Italy | 15 | 17.5 | 15 | [92] * |
| | 'Monviso' | | | | 17.5 | 23.5 | 17.5 | |
| *P.* × *generosa* × *P. trichocarpa* | 'AF8' | | | | 15 | 24 | 19 | |
| *P. maximowiczii* × *P. trichocarpa* | 'NE-42' | | | | 8 | | | |
| *P. trichocarpa* × *P. trichocarpa* | 'Fritzi Pauley' | | | | 8.1 | | | |
| | 'AF2' | | | | 4.1 | | | |
| | 'Albelo' | | | | 4 | | | |
| *P.* × *canadensis* | 'Degrosso' | 1333 | 7 | Poland | 6.8 | | | [52] |
| | 'Koster' | | | | 5.6 | | | |
| | 'Polargo' | | | | 4.3 | | | |
| *P.* × *generosa* × *P. trichocarpa* | 'AF8' | | | | 2.5 | | | |
| *P. nigra* × *P. maximowiczii* | 'Max 5' | 11,110 | 3 | Poland | 7.8 | | | [227] |
| | 'Columbia River' | | | | 6.71 | 6.62 | | |
| *P. trichocarpa* | 'Fritzi Pauley' | | | | 8.59 | 8.24 | | |
| | 'Trichobel' | | | | 9.08 | 9.59 | | |
| *P. trichocarpa* × *P. balsamifera* | 'Balsam Spire' | | | | 7.24 | 7.03 | | |
| | 'Gaver' | | | | 6.58 | 5.58 | | |
| *P.* × *canadensis* | 'Ghoy' | | | | 6.45 | 5.77 | | |
| | 'Gibecq' | 10,000 | 3 | United Kingdom | 5.7 | 4.73 | | [13] |
| | 'Beaupré' | | | | 7.34 | 4.87 | | |
| | 'Boelare' | | | | 6.23 | 4.2 | | |
| *P.* × *generosa* | 'Hazendans' | | | | 7.23 | 7.56 | | |
| | 'Hoogvorst' | | | | 8.84 | 8.12 | | |
| | 'Raspalje' | | | | 6.69 | 4.66 | | |
| | 'Unal' | | | | 7.55 | 5.25 | | |
| | | 10,000 | 5 | | 14.42 | | | |
| *P. balsamifera* × *P. trichocarpa* | 'Balsam Spire' | 4444 | 5 | United Kingdom | 11.20 | | | [228] |
| | | 10,000 | 4 | | 13.98 | | | |

* Estimated values from graphs.

**Table 3.** Different examples of the biomass yield (Mg dm ha$^{-1}$ year$^{-1}$) per genotype, age, and rotation in poplar short rotation plantations in Spain under Mediterranean conditions.

| Species and Poplar Hybrids | Genotype | Density | Age | Location | Yield (Mg dm ha$^{-1}$ Year$^{-1}$) by Rotation | | | Reference |
|---|---|---|---|---|---|---|---|---|
| | | | | | 1st | 2nd | 3rd | |
| *P.* × *canadensis* | 'AF2' | | | | 24.72 | | | |
| | 'Ballottino' | 13,333 | 3 | Aragón | 25.08 | | | [229] * |
| | 'I-214' | | | | 28.64 | | | |
| *P.* × *generosa* × *P. nigra* | 'Monviso' | | | | 26.91 | | | |
| *P.* × *canadensis* | 'Campeador' | | | | 12.10 | | | |
| | 'Dorskamp' | 10,000 | 5 | Soria | 6.3 | | | [230] |
| *P.* × *generosa* | 'Boelare' | | | | 16.6 | | | |
| | 'Raspalje' | | | | 19.8 | | | |
| *P.* × *canadensis* | 'Campeador' | | | | 9.3 | | | |
| | 'I-214' | 10,000 | | | 9.9 | | | |
| | 'Boelare' | | 6 | | 12.9 | | | |
| | | 2500 | | | 8.0 | | | |
| | | 5000 | | Soria | 9.2 | | | [126,127] |
| *P.* × *generosa* | 'Raspalje' | 10,000 | 6 (2nd–3rd:5year) | | 16.5 | 17.2 | 14.8 | |
| | | 10,000 | | | 9.8 | | | |
| | | 20,000 | 3 | | 12.7 | | | |
| | | 40,000 | | | 15.6 | | | |
| *P.* × *canadensis* | 'AF2' | | | | 0.84 | | | |
| | 'I-214' | | | | 2.96 | | | |
| | 'Beaupré' | | | | 3.07 | | | [231] Without irrigation |
| *P.* × *generosa* | 'Raspalje' | 8000 | 7 | Lugo | 3.1 | | | in a transition zone |
| | 'Unal' | | | | 1.97 | | | between Mediterranean |
| | 'AF6' | | | | 1.14 | | | and Atlantic climate |
| *P.* × *generosa* × *P. nigra* | 'Monviso' | | | | 0.93 | | | |
| *P. trichocarpa* | 'Trichocarpa' | | | | 3.93 | | | |
| *P.* × *canadensis* | 'AF2' and 'I-214' | 33,333 | 3 | Soria | 7–12 | 2–9.5 | 1.5–4.5 | [145] * |
| *P.* × *canadensis* | 'I-214' | 10,000 | 3 | Asturias | 9 | 11 | 6 | [148] * Without irrigation in mining zone |
| *P.* × *canadensis* | 'I-214' | 37,037 | 2 | Salamanca | 12.41 | 9.99 | 24.83 | [232] |

**Table 3.** *Cont.*

| Species and Poplar Hybrids | Genotype | Density | Age | Location | Yield (Mg dm ha$^{-1}$ Year$^{-1}$) by Rotation | | | Reference |
|---|---|---|---|---|---|---|---|---|
| | | | | | 1st | 2nd | 3rd | |
| P. × canadensis | 'AF2' | | | | 15.53 | | | |
| | 'I-214' | | | | 10.3 | | | |
| | 'Guardi' | 13,333 | 3 | Aragón | 9.03 | | | [233] *, [234] |
| P. deltoides | 'Viriato' | | | | 14 | | | |
| P. × generosa | 'Unal' | | | | 9.23 | | | |
| P. × generosa × P. nigra | 'Monviso' | | | | 13.13 | | | |
| P. × canadensis | 'I-214' | 33,333 | 2 | Spain | 15.1 | | | [112] |
| Populus spp. | 4 genotypes | 13,333 | 3 | Granada | 13.7 | | | |
| Populus spp. | 3 genotypes | 19,700 | 4 | Soria | 12.0 | | | |
| Populus spp. | 4 genotypes | 13,333 | 3 | Zamora | 7.7 | | | [235] |
| Populus spp. | 4 genotypes | 33,333 | 3 | León | 6.9 | | | |
| Populus spp. | 4 genotypes | 20,000 | 3 | Navarra | 16 | | | |
| P. nigra | 'Lombardo leones' | | | | 9.1 | 17.1 | | |
| | 'A4A' | | | | 20.0 | 12.2 | | |
| | 'A2A' | | | | 11.8 | 15.3 | | |
| P. × canadensis | 'I-214' | 17,316 | 3 | Soria | 22.0 | 30.1 | | [236] |
| | 'Luisa Avanzo' | | | | 17.3 | 27.1 | | |
| | 'MC' | | | | 16.8 | 18.4 | | |
| | '2000 Verde' | | | | 13.64 | 14.25 | | |
| P. deltoindes | 'Baldo' | | | | 8.25–19.07 | 8.59–29.94 | | |
| | 'Ballottino' | | | Gerona, | 8.25–15.94 | 11.45–19.20 | | |
| P. × canadensis | 'I-214' | 10,000 | 3 | Madrid, | 5.45–16.16 | 11.09–37.27 | | [77,80] |
| | 'Orion' | | | Soria | 12.53–22.65 | 12.70–34.56 | | |
| | 'Oudenberg' | | | | 9.43–14.27 | 9.66–19.96 | | |
| | | | | Barcelona | 4.2 | | | |
| | | | | Gerona | 5.7 | | | |
| | | | | Gerona | 11.6 | | | |
| | | | | Granada | 18.3 | | | |
| | | | | León | 4.7 | | | |
| P. × canadensis | 'I-214' | 6666-33,333 | 3 | León | 5.8 | | | [47] |
| | | | | León | 15.2 | | | |
| | | | | Madrid | 13.2 | | | |
| | | | | Navarra | 18.5 | | | |
| | | | | Soria | 8.1 | | | |
| | | | | Soria | 14.2 | | | |
| | | | | Zamora | 8.2 | | | |

**Table 3.** *Cont.*

| Species and Poplar Hybrids | Genotype | Density | Age | Location | Yield (Mg dm ha$^{-1}$ Year$^{-1}$) by Rotation | | | Reference |
|---|---|---|---|---|---|---|---|---|
| | | | | | 1st | 2nd | 3rd | |
| | 'I-214' | | | | 12.4 | | | |
| | 'Campeador' | | 4 | | 10.0 | | | |
| | 'Canada Blanco' | | | | 8.4 | | | |
| *P.* × *canadensis* | 'I-214' | 5000 | 2 | Madrid | 10.90 | | | [44] |
| | 'I-214' | | 3 | | 10.10 | | | |
| | 'I-214' | | 5 | | 17.30 | | | |
| | 'I-214' | 10,000 | 4 | | 16.9 | | | |
| *P.* × *canadensis* | 'AF2' | 13,333 | 3 | Granada | 17.1 | | | [237] |
| | 'AF2' | | | | 16.75 | | | |
| | 'Guardi' | | | | 16.14 | | | |
| *P.* × *canadensis* | 'I-214' | | | | 15.14 | | | |
| | 'MC' | | | | 15.01 | | | |
| | '2000 Verde' | 33,333 | 3 | León | 10.05 | | | [122] |
| *P.* × *generosa* | 'Unal' | | | | 11.38 | | | |
| | 'USA 49-177' | | | | 15.92 | | | |
| *P.* × *generosa* × *P. nigra* | 'Monviso' | | | | 15.16 | | | |
| | 'Pegaso' | | | | 7.37 | | | |
| | 'AF2' | | | | 16.8 | | | |
| | 'Guardi' | | | | 12.44 | | | |
| *P.* × *canadensis* | 'I-214' | | | León, | 12.21 | | | |
| | 'MC' | 33,333 | 3 | Gerona, Madrid, | 15.32 | | | [55] |
| | '2000 Verde' | | | Soria | 15.38 | | | |
| *P.* × *generosa* | 'Unal' | | | | 11.13 | | | |
| | 'Monviso' | | | | 14.47 | | | |
| *P.* × *generosa* × *P. nigra* | 'Pegaso' | | | | 6.65 | | | |
| *P. deltoides* | 'Viriato' | | | | 9.55 | 18.20 | 12.82 | |
| | 'AF2' | | | | 13.47 | 16.66 | 13.99 | |
| *P.* × *canadensis* | 'Ballottino' | 13,333 | 3 | Granada | 9.48 | 16.75 | 12.27 | [150] |
| | 'I-214' | | | | 9.11 | 13.44 | 15.81 | |
| *P.* × *generosa* × *P. nigra* | 'Monviso' | | | | 12.29 | 11.57 | 13.92 | |

**Table 3.** *Cont.*

| Species and Poplar Hybrids | Genotype | Density | Age | Location | Yield (Mg dm ha$^{-1}$ Year$^{-1}$) by Rotation | | | Reference |
|---|---|---|---|---|---|---|---|---|
| | | | | | 1st | 2nd | 3rd | |
| | 'AF2' | | | | 15.59 | | | |
| | 'A4A' | | | | 23.80 | 22.12 | | |
| *P.* × *canadensis* | 'Guardi' | | | | 15.59 | 6.43 | | |
| | 'I-214' | | | | 23.80 | 18.51 | | |
| | 'MC' | 20,000 | 3 (2nd rot: 2 year) | Navarra | 11.21 | 26.51 | | [238] |
| | 'Triplo' | | | | 18.07 | 32.92 | | |
| *P. deltoides* | 'Viriato' | | | | 16.68 | 38.18 | | |
| *P.* × *generosa* | 'Beaupré' | | | | 21.29 | 15.29 | | |
| | 'Unal' | | | | 25.86 | 12.15 | | |
| *P.* × *generosa* × *P. nigra* | 'Monviso' | | | | 15.41 | | | |
| | 'Pegaso' | | | | 13.99 | | | |
| | | 15,000 | | | 18.20 | 14.46 | | |
| *P.* × *canadensis* | 'I-214' | 20,000 | 3 (2nd rot:2 year) | Navarra | 18.68 | 18.51 | | [239] |
| | | 25,000 | | | 23.36 | 14.65 | | |
| | | 33,333 | | | 18.85 | 17.13 | | |

* Estimated values from graphs.

Modeling of Growth, Production, and Biometric Relationships in SRC Poplar Plantations

The use of estimation models to predict available biomass is becoming more frequent as research into this type of plantation progresses, not only because of the costs involved in carrying out direct estimates [240], but also due to the greater flexibility and the possibility of extrapolating the results to larger scales when the models are based on a wide range of empirical data.

Allometric models which relate the tree diameter or another easily measurable variable to the biomass are those most commonly used in forest inventories or ecological studies [241,242]. These models have also been employed in SRC plantations to estimate the available biomass [243–245]. These predictions are particularly important when evaluating the economic viability of a crop [47,246]. Apart from the choice of regression model, the assumptions that underlie the regression procedures, and the data transformations used during the procedures [247], we are confronted with a large number of factors which create uncertainty and may affect the final results [248]. Simple equations which are valid for a wide range of conditions and plant material are usually preferred. There are several examples used for poplar SRC plantations [18,30,78,136,171,226,249–253]. The precision of such allometric equations is generally sufficient for stem biomass components [254]. However, other examples of equations exist that include more predictive variables in the models, which can improve the precision of the estimations [11,47,205,255–258]. Although specific genotype-level models provide the greatest precision, in cases where genotype identification is complicated, models for genotypes which are taxonomically close or taxonomic-specific models are often used. However, under Mediterranean conditions at least, Oliveira et al. [252] proves that the genetic origin does not explain the similarities in biomass allometry among genotypes, so these approaches are not always advisable.

Over the last decade, considerable advances have been made with regards to modeling to estimate biomass from SRC poplar plantations. Not only have advances been made in allometric models, but also much effort has been channeled towards modeling other aspects of growth in these plantations, such as the leaf architecture [259], root production [260–262], and the use of process models with this aim [263–267], in addition to other management tools derived from models [47,268–270].

In recent years, considerable progress has been made in the development of specific predictive models for SRC plantations under Mediterranean conditions. Improvements in the model development methodology to achieve more robust biomass predictions have been made [248,271], and the suitability of biomass models for local populations, as well as their performance for different sample sizes, have also been evaluated [253]. Different models have been developed, including a dynamic, whole-stand model for 'I-214' poplar genotype plantations in the northern and central plateau in Spain [272], along with individual tree models and general models for estimating both the above- and below-ground biomass in poplar SRC plantations under Mediterranean conditions [252,262].

Other management tools derived from models, such as maps of possible zones of production [47,273,274], estimations derived from the use of new technologies [270,275,276], and tools such as reference diagrams, which are particularly useful for both planning and managing this type of crop [277], have also been developed in recent years.

### 3.5. Biomass Characterization

The use of lignocellulosic biomass is not limited to the production of bioenergy. In recent years, a wide range of bioproduct-related options have sprung up, such as biopolymers, bioplastics, and sugar fermentation bioproducts, among others [278–280].

Poplar lignocellulosic biomass is mainly composed of cellulose (42–49%), hemicellulose (16–23%), and lignin (21–29%) [22,233,279]. Its biomass can supply raw material for processes of thermochemical or microbiological conversion. Worthy of note among the former is combustion, either for domestic or agro-industrial applications, but also for industrial cogeneration or co-firing [281,282]. Thermochemical processes also include gasification in downdraft gasifiers or fluidized beds [283], slow pyrolysis for the production of biochar [284], and fast pyrolysis to produce biocrude oil [285]. With regards to the biochemical processes, the conversion to second-generation bioethanol is the most studied process [286–288], although lignocellulosic biomass can also be a source of biobutanol [289,290] and other second-generation biofuels [22,279,291].

In Spain, several studies have addressed the characterization of poplar biomass for thermal use [148,231,292,293] as biofuels [294–297] or for new bioproducts [298,299]. The lower heating value, which is used to calculated the available energy, is usually around 18–20 MJ kg$^{-1}$ in poplar wood on a dry basis [300,301]. Values found in Spain are within a range very similar to those obtained for other European countries, although the characteristics vary, depending on the genotype and the age, ranging from 17.61 to 18.74 MJ kg$^{-1}$ [137,302,303].

The humidity content can be as high as 48–50% at winter harvesting [231,303], but poplar biomass has shown a good ability for air drying [303]. The specific density is known to be low [304–307], thus deriving in low bulk density chips (150–260 kg m$^{-3}$, [231]), corresponding to low energy densities. Densification to produce pellets is therefore an option.

The ash content, which is negatively related to the energy value and associated with the risk of boiler corrosion [308], varies broadly (1–4%), depending on genotype and site conditions [117,137,231,303]. The fouling and slagging risk derived from ash compositions are known to be very low [308–310]. The presence of nitrogen (N), which is related to NOx emissions, is low (0.5%) in poplar biomass [148,231,311], along with chlorine (Cl, <0.02%) and sulfur (S, <0.04%), which are corrosive [145,312,313].

The combination of a high volatile matter content (much higher than 80%, [137,303,314]), softness of the wood (which is easy to grind), and a low lignin content makes poplar from SRC a promissory feedstock for gasifiers. It can also be used for the generation of bioethanol [287,315].

Lignocellulosic biomass has many other uses in the context of the bioeconomy. For example, cellulose is used in the manufacturing of cosmetics, textiles, and pharmaceutical products, among others [291,297]. Lignin can potentially be used as a raw material in the manufacturing of products with high value added, such as vanillin, biopolymers in petrochemistry, and biopesticides, as well as a material for soil enrichment [291]. Hyd-Poplar lignin could be used for the production of flame-retardant materials [299]. For example, Martín-Sampedro et al. [234] and Ibarra et al. [233] identified the genotype 'Viriato' as being very promising for use in the production of biofuels, as well as in other value-added products, all of which point to the suitability of poplar raw material for different uses in the context of the bioeconomy.

### 3.6. Sustainability and Ecosystem Services

The importance of the sustainability of crops destined for biomass production has been highlighted when evaluating their future development. During the last decade, the European Commission has carried out different analyses and much effort has been dedicated to defining criteria and indicators of sustainability (EC, Directive 2009/28/EC, 2009; EU Parliament Resolution 2013; Directive 2018/2001/EC; European Parliament 2017). The World Bioenergy Association [40] has drawn up a document which includes 24 voluntary sustainability indicators related to bioenergy in general. This document is particularly important since it represents the only multilateral initiative with a broad consensus among the different governments and international organizations, providing a framework for future policy development. Spain has been part of this association since 2008 through the Institute for Diversification and Energy Saving (IDAE). Despite this consensus, there are many different approaches to tackling these studies, sometimes because of geographical differences, which often leads to approaches that are not always homogeneous. Examples of these initiatives include the analyses conducted by Dallemand et al. [316] and Dimitriou and Rutz [317], among many others.

In any case, the criteria and indicators must be based on scientific evidence and contain specifications for each production model or source of biomass, as these vary considerably. In this regard, there are numerous aspects which need to be addressed, such as life cycle and water cycle analyses, soil quality, and erosion control, among many others [318]. The environmental impacts of establishment, harvesting, and transportation have been considered negligible when the crop is grown on marginal land in central Europe [319]. Sustainability is also necessary from economic and energetic perspectives. With regards to the latter, the adaptation would appear to be favorable. However, this may not be the case with regard to the economic viability given the current prices of biomass and absence of subsidies [320,321]. In Spain, several life cycle analyses including not only environmental, but also energetic and even economic, issues [128,237,322–325] have been carried out in order to identify the most relevant factors for these crops under Mediterranean conditions. These analyses concluded that fertilization, transport, and irrigation are some of the most influential factors [237,288,326,327]. The price of biomass, the price of land rent, harvesting, and irrigation have been identified as the most influential factors from an economic perspective in Mediterranean environments [214,237,328]. The economic viability of poplar SRC under irrigated Mediterranean conditions can be achieved either by ensuring optimum productivity or through an increase in the market prices, associated with more diversified energy use or bioproducts, along with the quantification of ecosystem services, which currently do not have a market price [214]. In addition to this, improvements in clonal selection and irrigation technology, as well as the employment of other irrigation methods, such as making use of reused water, will be essential if these plantations are to be profitable.

A review by Li et al. [329] highlights the necessity to identify sustainable sources of bioenergy. With this purpose in mind, the authors identified poplar crops among the five lignocellulosic crops with the greatest future potential. They compiled and described biomass yield information across a whole range of locations and countries, and concluded that these crops, which only account for 3% of the bioenergy in Europe [330], can be grown under a wide range of climatic conditions, thus allowing direct competition with food crops to be avoided. Moreover, it has been determined that woody crops cultivated in short rotation and herbaceous lignocellulosic crops emit between 40% and 99% less $NO_2$ than traditional crops and consequently have lower fertilization requirements and a greater N use efficiency. They also sequester carbon in the biomass that remains in the soil ($0.44$ Mg C ha$^{-1}$ year$^{-1}$) when they are planted on disused agricultural land, although the balance is not positive when they are planted on former pastureland. Other associated ecosystem benefits include an increased biodiversity (phytodiversity and zoological diversity) [32,331], erosion control and soil conservation [332], improvements in the water quality [333], and the role of the crop in phytoremediation [334–336]. In the case of Spain, the potential of the genus in the phytoremediation of soils has been evaluated through examples of the restoration of coal mining areas in northern Spain [131,337,338], or water phytodepuration [42,187,216,339]. Research has also focused on nutrient fluxes, evaluating the role of annual leaf litter in soil fertility throughout the rotation [340]; the quantification of accumulated carbon in both above- and below-ground fractions (reaching values of around $6.5$ Mg C ha$^{-1}$ year$^{-1}$ in the above-ground woody biomass, around $1.0$ Mg C ha$^{-1}$ year$^{-1}$ in the below-ground biomass [262], and around $2.5$ Mg C ha$^{-1}$ year$^{-1}$ in the case of the litter [341]), as well as the economic implications of $CO_2$ capture [342]. The impact of these forestry crops on the landscape of the agricultural environments where they are grown has not yet been evaluated in our country. Their effects on soil, biodiversity, and its function of mitigating diffuse pollution are examples of aspects that should be considered.

Among the main concerns in Europe in terms of the sustainability of crops destined for biomass production and therefore SRC plantations, are those linked to land use change. According to some studies, the balance of greenhouse gases and C in the soil indicates that bioenergy as a whole plays a role in the mitigation of climate change [343]. However, other studies question this affirmation [330], or point to a minimum cultivation period, beyond which the C balance becomes positive [344–347].

The impact on water resources must also be rigorously assessed [332,348]. In the Mediterranean area, the need for irrigation is perceived as one of the main weaknesses of these plantations. Different options aimed at minimizing or achieving more efficient irrigation in Mediterranean environments have been mentioned in Section 3.3.4. The application of wastewater in poplar plantations not only constitutes an attractive method for producing biomass through the regeneration of wastewater, but also implies a reduction or suspension of fertilizer application [42,187,188]. Hence, this approach not only provides a sustainable way to minimize water use, but also takes advantage of the ability of these plantations as vegetation filters.

Improving the sustainability of SRC crops as contributors to the biomass pool will probably require global decisions that take into account local specifications.

## 4. Current Status and Future Prospects in Spain

The availability of abandoned agricultural land, the lack of economic alternatives in rural areas, and the possibilities for complementarity afforded by SRC crops in terms of supply along with the ecosystem services associated with their establishment on agriculture land, are factors favoring the implementation of SRC forest crops [349–351]. As with other sources of biomass, it is important to analyse the biomass produced from forest crops and assess its sustainability and the best methods for producing it. However, the positive and negative aspects of this type of biomass production, indeed of biomass production in general, must always be taken into account and there will always be a certain degree of controversy surrounding its viability. Biomass from forest crops in short rotation has attracted a lot of interest in Spain, probably due to the regulations regarding electricity production in 2004 (RD 436/2004) and later in 2007 (RD 661/2007). At that time, many large and medium-sized companies considered establishing forest crops in SRC. However, the amendment to this regulation in 2013 removed the incentives at a time when many aspects associated with the establishment of crops in Mediterranean environments had still not been clarified, the ecosystem services had not been assessed, and the economic viability was far from guaranteed. This situation resulted in a declining interest in the sector, which continues today.

Despite the expected gradual increase in the contribution of renewable energy to the final gross energy consumption, in 2017, Spain was still 2.5% points from its 2020 national target [352]. Of the total contribution of renewable energy, biomass accounts for 13% [353]. With regards to the amount of biomass destined for bioproducts, as far as we know, there are no statistics for the country given that there are only a residual number of biorefineries producing such bioproducts [280]. Therefore, many aspects must still be resolved before the economic viability of renewable biomass resources can be determined. For example, the price of biomass in Spain is lower than in neighboring countries, so part of the biomass is exported. This would appear to be problematic in terms of sustainability. Moreover, no stable, predictable market for biomass exists in Spain.

In relation to biomass from forest crops in short rotation, many advances have been made in Spain over recent years, which have been described in this article. These advances include the following: (i) Increased knowledge of the genetic material in relation to the environment, with a better understanding of the adaptability of genotypes, including their water-use efficiency; (ii) maximizing production based on densities, genotypes, and management practices; (iii) evaluating the sustainability of plantations from different perspectives (environmental, energetic, and economic); (iv) exploring alternative designs, such as mixed plantations; (v) implementing plantations in marginal areas previously used for mining; (vi) quantifying accumulated carbon; (vii) specific predictive models and management tools for SRC plantations under Mediterranean conditions; (viii) physiological and molecular characterization of attributes, mainly those that are relevant to cultivation in marginal zones; (ix) chemical composition and pyrolytic behavior; and (x) the use of biotechnology to develop new materials, which should lead to increased production in the future. The role of these forest crops in water purification has also recently emerged as a matter of interest.

Despite these advances, we still have a long way to go to make SRC plantations a commercial reality in Spain. It is also known that biomass, including dedicated crops, would generate significant economic returns in rural areas [354]. This may be especially relevant in a country like Spain, where a large part of the country suffers from high levels of depopulation, although some aspects that hinder the short-term economic viability have also been identified, such as those previously mentioned regarding land rental, the need for irrigation, and the technological development of harvesting machinery at reasonable local prices. However, other issues need to be addressed in relation to crop management, such as (i) the use of irrigation, which should only be resorted to in areas where it is sustainable, by modernizing the systems, using more water-use efficient plant material or by using recycled water; (ii) the continual updating of genetic material adaptations to the site, testing of new materials, and making use of new technologies (genome editing, linkage maps, etc.), in order to produce material that is resistant to pests and diseases, tolerant to drought or a high salinity, or can adapt more effectively to specific soil and climate conditions; (iii) exploring new plantation designs; although the preliminary results of using mixtures are not particularly encouraging, many different alternatives remain to be explored; and (iv) rationalizing fertilization by evaluating the inputs derived from leaf litter and the exports of nutrients from wood, as well as by using alternative fertilizer inputs, such as those derived from sewage sludge.

Other questions should be explored based on the need to (v) quantify and value ecosystem services in terms of increased biodiversity in the agricultural landscape; carbon accumulation in each of the biomass fractions, both those that are extracted, as well as those that remain in the soil, such as foliar and root biomass; and their additional role in phytoremediation, or (vi) advance predictive modeling for Mediterranean conditions by combining the best features of empirical and process models, the advantages of which have been well-documented. The inclusion of ecophysiological variables enabling the prediction of individual tree-level biometry under different conditions, along with improvements in determining the most important variables, are some of the future objectives of modeling SRC plantations under Mediterranean conditions. In addition, as water restrictions represent a real threat, this factor is also being considered in the development of models that enable the simulation of different climate change scenarios. Regarding (vii) the characterization of biomass, although many studies have already addressed this aspect, a more in-depth knowledge will be required as the final products and their intended uses become more defined. Identifying the requirements of each of the final products will be essential in order to determine the most suitable genotypes for each crop. Given that Spain is a diverse country, it is also necessary (viii) to analyse where and how this biomass is produced at a national scale.

Improving the critical aspects detected in environmental, energetic, and economic analyses is essential for achieving profitable and sustainable plantations under Mediterranean conditions. Biomass produced ad hoc through plantations under SRC systems may be of interest in many areas of the Mediterranean, providing a further option which could contribute to the development of a circular bioeconomy while also generating important environmental services.

**Author Contributions:** Conceptualization, N.O. and H.S.; funding acquisition, H.S., R.R.-S., and I.C.; investigation, N.O., H.S., C.P.-C., R.R.-S., and I.C.; methodology, H.S. and N.O.; project administration, H.S., R.R.-S., and I.C.; supervision, H.S., I.C., C.P.-C., and R.R.-S.; writing—original draft, N.O. and H.S.; writing—review and editing, N.O., C.P.-C., I.C., R.R.-S., and H.S. All authors have read and agreed to the published version of the manuscript.

**Funding:** This research was funded by the Ministry of Science and Innovation (Spain) through the framework of the INIA projects RTA2017-00015-CO2, co-financed with funds from FEDER.

**Acknowledgments:** The authors wish to thank Adam Collins for his English language review of the manuscript.

**Conflicts of Interest:** The authors declare no conflict of interest.

## References

1. European Commission. *The European Green Deal*; European Commission: Brussels, Belgium, 2019; p. 24.
2. Hetemäki, L.; Hanewinkel, M.; Muys, B.; Ollikainen, M.; Palahí, M.; Trasobares, A. *Leading the Way to a European Circular Bioeconomy Strategy*; Science to Policy 5; European Forest Institute: Joensuu, Finland, 2017; Volume 5.
3. European Commission. *Innovating for Sustainable Growth: A Bioeconomy for Europe*; European Commision: Brussels, Belgium, 2012.
4. European Commission. *Review of the 2012 European Bioeconomy Strategy*; European Commission: Brussels, Belgium, 2018.

5.   MINECO. *Estrategia Española de Bioeconomía: Horizonte 2030*; Ministerio de Economía y Competitividad, Secretaría de Estado de Investigación, Desarrollo e Innovación: Madrid, Spain, 2015; p. 46.

6.   MITECO. *Law on Climate Change and Energy Transition*; Ministerio para la Transición Ecológica y el Reto Demográfico: Madrid, Spain, 2020.

7.   Bioplat. *Agenda Estratégica de Investigación e Innovación. Biomasa y Bioeconomía*; Ministerio de Ciencia e Innovación, Gobierno de España: Madrid, Spain, 2020.

8.   Sancho, A. *Día de la Bioenergía en España: 28 días de Independencia Energética*; Bioenergy International: Stockholm, Sweden, 2018.

9.   Gardiner, E.S.; Ghezehei, S.B.; Headlee, W.L.; Richardson, J.; Soolanayakanahally, R.Y.; Stanton, B.J.; Zalesny, R.S., Jr. The 2018 Woody Crops International Conference, Rhinelander, Wisconsin, USA, 22–27 July 2018. *Forests* **2018**, *9*, 693. [CrossRef]

10.  Broeckx, L.; Verlinden, M.; Vangronsveld, J.; Ceulemans, R. Importance of crown architecture for leaf area index of different *Populus* genotypes in a high-density plantation. *Tree Physiol.* **2012**, *32*, 1214–1226. [CrossRef] [PubMed]

11.  Huber, J.A.; May, K.; Hülsbergen, K.-J. Allometric tree biomass models of various species grown in short-rotation agroforestry systems. *Eur. J. For. Res.* **2017**, *136*, 75–89. [CrossRef]

12.  Weih, M. Intensive short rotation forestry in boreal climates: Present and future perspectives. *Can. J. For. Res. Rev. Can. Rech. For.* **2004**, *34*, 1369–1378. [CrossRef]

13.  Aylott, M.J.; Casella, E.; Tubby, I.; Street, N.R.; Smith, P.; Taylor, G. Yield and spatial supply of bioenergy poplar and willow short-rotation coppice in the UK. *New Phytol.* **2008**, *178*, 358–370. [CrossRef]

14.  Berhongaray, G.; Verlinden, M.S.; Broeckx, L.S.; Janssens, I.A.; Ceulemans, R. Soil carbon and belowground carbon balance of a short-rotation coppice: Assessments from three different approaches. *Glob. Chang. Biol. Bioenergy* **2016**, *9*, 299–313. [CrossRef]

15.  European Commission. *State of Play on the Sustainability of Solid and Gaseous Biomass Used for Electricity, Heating and Cooling in the EU (SWD 259 Final)*; European Commission: Brussels, Belgium, 2014.

16.  Dimitriou, I.; Berndes, G.; Englund, O.; Murphy, F. *Lignocellulosic Crops in Agricultural Landscapes: Production Systems for Biomass and Other Environmental Benefits–Examples, Incentives, and Barriers*; IEA Bioenergy: Paris, France, 2018.

17.  Dickmann, D.I. An overview of the genus *Populus*. In *Poplar Culture in North America*; Dickmann, D.I., Isebrands, J.G., Eckenwalder, J.E., Richardson, J., Eds.; NRC Research Press: Ottawa, ON, Canada, 2001; pp. 1–42.

18.  Al Afas, N.; Marron, N.; Van Dongen, S.; Laureysens, I.; Ceulemans, R. Dynamics of biomass production in a poplar coppice culture over three rotations (11 years). *For. Ecol. Manag.* **2008**, *255*, 1883–1891. [CrossRef]

19.  Laureysens, I.; Pellis, A.; Willems, J.; Ceulemans, R. Growth and production of a short rotation coppice culture of poplar. III. Second rotation results. *Biomass Bioenergy* **2005**, *29*, 10–21. [CrossRef]

20.  Ceulemans, R.; McDonald, A.; Pereira, J. A comparison among eucalypt, poplar and willow characteristics with particular reference to a coppice, growth-modelling approach. *Biomass Bioenergy* **1996**, *11*, 215–231. [CrossRef]

21.  Patel, B.; Gami, B. Biomass characterization and its use as solid fuel for combustion. *Iran. J. Energy Environ.* **2012**, *3*, 123–128. [CrossRef]

22.  Zhang, X.; Tu, M.; Paice, M. Routes to Potential Bioproducts from Lignocellulosic Biomass Lignin and Hemicelluloses. *Bioenergy Res.* **2011**, *4*, 246–257. [CrossRef]

23.  Surendran Nair, S.; Kang, S.; Zhang, X.; Miguez, F.E.; Izaurralde, R.C.; Post, W.M.; Dietze, M.C.; Lynd, L.R.; Wullschleger, S.D. Bioenergy crop models: Descriptions, data requirements, and future challenges. *Glob. Chang. Biol. Bioenergy* **2012**, *4*, 620–633. [CrossRef]

24.  Rodrigues, A.; Loureiro, L.; Nunes, L.J.R. Torrefaction of woody biomasses from poplar SRC and Portuguese roundwood: Properties of torrefied products. *Biomass Bioenergy* **2018**, *108*, 55–65. [CrossRef]

25.  Schultz, R.; Isenhart, T.; Simpkins, W.; Colletti, J. Riparian forest buffers in agroecosystems–lessons learned from the Bear Creek Watershed, central Iowa, USA. *Agrofor. Syst.* **2004**, *61*, 35–50. [CrossRef]

26.  Tufekcioglu, A.; Raich, J.W.; Isenhart, T.M.; Schultz, R.C. Fine root dynamics, coarse root biomass, root distribution, and soil respiration in a multispecies riparian buffer in Central Iowa, USA. *Agrofor. Syst.* **1998**, *44*, 163–174. [CrossRef]

27. Zalesny, R.S., Jr.; Stanturf, J.A.; Gardiner, E.S.; Perdue, J.H.; Young, T.M.; Coyle, D.R.; Headlee, W.L.; Banuelos, G.S.; Hass, A. Ecosystem services of woody crop production systems. *Bioenergy Res.* **2016**, *9*, 465–491. [CrossRef]

28. Coleman, M.D.; Friend, A.L.; Kern, C.C. Carbon allocation and nitrogen acquisition in a developing *Populus deltoides* plantation. *Tree Physiol.* **2004**, *24*, 1347–1357. [CrossRef]

29. Grigal, D.F.; Berguson, W.E. Soil carbon changes associated with short-rotation systems. *Biomass Bioenergy* **1998**, *14*, 371–377. [CrossRef]

30. Baum, S.; Bolte, A.; Weih, M. Short rotation coppice (SRC) plantations provide additional habitats for vascular plant species in agricultural moisac landscapes. *Bioenerg. Res.* **2012**, *5*, 573–583. [CrossRef]

31. Christian, D.P.; Niemi, G.J.; Hanowski, J.M.; Collins, P. Perspectives on biomass energy tree plantations and changes in habitat for biological organisms. *Biomass Bioenergy* **1994**, *6*, 31–39. [CrossRef]

32. Vanbeveren, S.P.P.; Ceulemans, R. Biodiversity in short-rotation coppice. *Renew. Sustain. Energy Rev.* **2019**, *111*, 34–43. [CrossRef]

33. Ferrarini, A.; Serra, P.; Almagro, M.; Trevisan, M.; Amaducci, S. Multiple ecosystem services provision and biomass logistics management in bioenergy buffers: A state-of-the-art review. *Renew. Sustain. Energy Rev.* **2017**, *73*, 277–290. [CrossRef]

34. Rockwood, D.; Naidu, C.; Carter, D.; Rahmani, M.; Spriggs, T.; Lin, C.; Alker, G.; Isebrands, J.; Segrest, S. Short-rotation woody crops and phytoremediation: Opportunities for agroforestry? In *New Vistas in Agroforestry. A Compendium for the 1st World Congress of Agroforestry*; Kluwer Academic Publisher: Dordrecht, The Netherlands; Boston, MA, USA; New York, NY, USA, 2004; pp. 51–63.

35. Dipesh, K.; Will, R.E.; Hennessey, T.C.; Penn, C.J. Evaluating performance of short-rotation woody crops for bioremediation purposes. *New For.* **2015**, *46*, 267–281. [CrossRef]

36. Pilipović, A.; Zalesny, R.S., Jr.; Orlović, S.; Drekić, M.; Pekeč, S.; Katanić, M.; Poljaković-Pajnik, L. Growth and physiological responses of three poplar clones grown on soils artificially contaminated with heavy metals, diesel fuel, and herbicides. *Int. J. Phytoremediat.* **2020**, *22*, 436–450. [CrossRef] [PubMed]

37. Zalesny, R.S., Jr.; Stanturf, J.A.; Gardiner, E.S.; Bañuelos, G.S.; Hallett, R.A.; Hass, A.; Stange, C.M.; Perdue, J.H.; Young, T.M.; Coyle, D.R. Environmental technologies of woody crop production systems. *Bioenergy Res.* **2016**, *9*, 492–506. [CrossRef]

38. Evangelou, M.W.; Conesa, H.M.; Robinson, B.H.; Schulin, R. Biomass production on trace element–contaminated land: A review. *Biomass Bioenergy* **2012**, *29*, 823–839. [CrossRef]

39. IRENA. *Global Renewables Outlook: Energy Transformation 2050*; International Renewable Energy Agency: Abu Dhabi, UAE, 2020; p. 212.

40. GBEP. *The Global Bioenergy Partnership Sustainability Indicators for Bioenergy*; Climate, Energy and Tenure Division, Food and Agriculture Organization of the United Nations (FAO): Rome, Italy, 2011.

41. Marron, N. Agronomic and environmental effects of land application of residues in short-rotation tree plantations: A literature review. *Biomass Bioenergy* **2015**, *81*, 378–400. [CrossRef]

42. de Miguel, A.; Meffe, R.; Leal, M.; González-Naranjo, V.; Martínez-Hernández, V.; Lillo, J.; Martín, I.; Salas, J.J.; Bustamante, I.d. Treating municipal wastewater through a vegetation filter with a short-rotation poplar species. *Ecol. Eng.* **2014**, *73*, 560–568. [CrossRef]

43. Zalesny, R.S., Jr.; Stanturf, J.A.; Evett, S.R.; Kandil, N.F.; Soriano, C. Opportunities for woody crop production using treated wastewater in Egypt. I. Afforestation strategies. *Int. J. Phytoremediat.* **2011**, *13*, 102–121. [CrossRef]

44. San Miguel, A.; Montoya, J. Resultados de los primeros 5 años de produccion de talleres de chopo en rotacion corta (2–5 anõs). In *Anales del Instituto Nacional de Investigaciones Agrarias. Serie Forestal*; Instituto Nacional de Investigaciones Agrarias: Madrid, Spain, 1984; pp. 73–91.

45. Marcos, F. Cultivos energéticos forestales. In Proceedings of the V Conferencia sobre Planificación, Ahorro y Alternativas Energéticas, Zaragoza, Spain, 1985.

46. Ciria, M.P.; Mazon, P.; Carrasco, J. Productivity of poplar grown on short rotation in Spain: Influence of the rotation age and plant density. In Proceedings of the 2nd biomass conference of the Americas: Energy, Environment, Agriculture, and Industry, Portland, OR, USA, 21–24 August 1995; pp. 235–243.

47. Pérez-Cruzado, C.; Sanchez-Ron, D.; Rodríguez-Soalleiro, R.; Hernández, M.; Sánchez-Martín, M.; Cañellas, I.; Sixto, H. Biomass production assessment from *Populus* spp. short-rotation irrigated crops in Spain. *Glob. Chage Biol. Bioenergy* **2013**, *6*, 312–326. [CrossRef]

48. Huber, J.A.; May, K.; Siegl, T.; Schmid, H.; Gerl, G.; Hülsbergen, K.J. Yield potential of tree species in organic and conventional short-rotation agroforestry systems in Southern Germany. *Bioenergy Res.* **2016**, *9*, 955–968. [CrossRef]

49. Havlíčková, K.; Weger, J.; Zánová, I. Short rotation coppice for energy purposes—Economy conditions and landscape functions in the Czech Republic. In *Proceedings of ISES World Congress 2007*; Springer: Berlin/Heidelberg, Germany, 2008; pp. 2482–2487.

50. Stankova, T.; Gyuleva, V.; Tsvetkov, I.; Popov, E.; Velinova, K.; Velizarova, E.; Dimitrov, D.N.; Hristova, H.; Kalmukov, K.; Dimitrova, P. Aboveground dendromass allometry of hybrid black poplars for energy crops. *Ann. For. Res.* **2016**, *59*, 61–74. [CrossRef]

51. Pilipović, A.; Orlović, S.; Kovačević, B.; Galović, V.; Stojnić, S. Selection and Breeding of Fast Growing Trees for Multiple Purposes in Serbia. In *Forests of Southeast Europe Under a Changing Climate*; Springer: Berlin/Heidelberg, Germany, 2019; pp. 239–249.

52. Niemczyk, M.; Kaliszewski, A.; Jewiarz, M.; Wróbel, M.; Mudryk, K. Productivity and biomass characteristics of selected poplar (*Populus* spp.) cultivars under the climatic conditions of northern Poland. *Biomass Bioenergy* **2018**, *111*, 46–51. [CrossRef]

53. Benbrahim, M.; Gavaland, A.; Gauvin, J. Growth and yield of mixed polyclonal stands of *Populus* in short-rotation coppice. *Scand. J. For. Res.* **2000**, *15*, 605–610. [CrossRef]

54. Nardin, F.; Alasia, F. Use of selected fast growth poplar trees for a woody biomass production die along Po valley. In Proceedings of the 2nd World Conference on Biomass for Energy, Industry and Climate Protection, Rome, Italy, 10–14 May 2004; pp. 247–249.

55. Sixto, H.; Gil, P.; Ciria, P.; Camps, F.; Sánchez, M.; Cañellas, I.; Voltas, J. Performance of hybrid poplar clones in short rotation coppice in Mediterranean environments: Analysis of genotypic stability. *Bioenegy* **2013**, *6*, 661–671. [CrossRef]

56. Nordborg, M.; Berndes, G.; Dimitriou, I.; Henriksson, A.; Mola-Yudego, B.; Rosenqvist, H. Energy analysis of poplar production for bioenergy in Sweden. *Biomass Bioenergy* **2018**, *112*, 110–120. [CrossRef]

57. Tullus, A.; Rosenvald, K.; Lutter, R.; Kaasik, A.; Kupper, P.; Sellin, A. Coppicing improves the growth response of short-rotation hybrid aspen to elevated atmospheric humidity. *Forest Ecol. Manag.* **2020**, *459*, 117825. [CrossRef]

58. Clifton-Brown, J.; Harfouche, A.; Casler, M.D.; Dylan Jones, H.; Macalpine, W.J.; Murphy-Bokern, D.; Smart, L.B.; Adler, A.; Ashman, C.; Awty-Carroll, D. Breeding progress and preparedness for mass-scale deployment of perennial lignocellulosic biomass crops switchgrass, miscanthus, willow and poplar. *Glob. Chang. Biol. Bioenergy* **2019**, *11*, 118–151. [CrossRef]

59. Scholz, V.; Ellerbrock, R. The growth productivity, and environmental impact of the cultivation of energy crops on sandy soil in Germany. *Biomass Bioenergy* **2002**, *23*, 81–92. [CrossRef]

60. Zhang, S.Y.; Yu, Q.; Chauret, G.; Koubaa, A. Selection for both growth and wood properties in hybrid poplar clones. *For. Sci.* **2003**, *49*, 901–908. [CrossRef]

61. Tuskan, G.A.; Difazio, S.; Jansson, S.; Bohlmann, J.; Grigoriev, I.; Hellsten, U.; Putnam, N.; Ralph, S.; Rombauts, S.; Salamov, A. The genome of black cottonwood, *Populus trichocarpa* (Torr. & Gray). *Science* **2006**, *313*, 1596–1604. [CrossRef] [PubMed]

62. FAO. Working Party on Sustainable Livelihoods, Land-Use, Products and Bioenergy. Available online: http://www.fao.org/forestry/ipc/69631/en/ (accessed on 15 September 2020).

63. Glas, D.; Schimer, R. *Prüfung von Pappelsorten aus Anderen EU-Staaten für Kurzumtriebsplantagen (EU-POP)*; Bundesministerium für Ernährung und Landwirtschaft: Teisendorf, Germany, 2018.

64. MAPA. Estadísticas Comisión Nacional del Chopo. Available online: https://www.mapa.gob.es/es/desarrollo-rural/temas/politica-forestal/comision-nacional-del-chopo/estadisticas.aspx (accessed on 20 September 2020).

65. ProPopulus. *La Escasez de Chopo Amenaza a la Industria Española de Contrachapado*; ProPopulus; Nature, Society, Future; The European Poplar Initiative: Spain, 2019; Available online: http://propopulus.eu/es/la-escasez-de-chopo-amenaza-a-la-industria-espanola-de-contrachapado/ (accessed on 10 December 2020).

66. Rueda, J.; García Caballero, J.L.; Cuevas, Y.; García-Jiménez, C.; Villar, C. *Cultivo de Chopos en Castilla y León*; Consejería de Fomento y Medio Ambiente, Junta de Castilla y León: Valladolid, Spain, 2019.

67. ESYRCE. *Encuesta Sobre Superficies y Rendimientos de Cultivos. Informe Sobre Regadíos en España*; Subsecretaría de Agricultura, Pesca y Alimentación y Subdirección General de Análisis, Coordinación y Estadística: Madrid, Spain, 2019.

68. Sixto, H.; Hernández, M.; Barrio, M.; Carrasco, J.; Cañellas, I. Plantaciones del género *Populus* para la producción de biomasa con fines energéticos: Revisión. *For. Syst.* **2007**, *16*, 277–294.

69. Corcuera, L.; Maestro, C.; Notivol, E. La ecofisiología como herramienta para la selección de clones más adaptados y productivos en el marco de una selvicultura clonal con chopos. *Invest. Agrar. Sist. Recur. For.* **2005**, *14*, 394–407. [CrossRef]

70. Pérez-Cruzado, C.; Merino, A.; Rodríguez-Soalleiro, R. A management tool for estimating bioenergy production and carbon sequestration in *Eucalyptus globulus* and *Eucalyptus nitens* grown as short rotation woody crops in north-west Spain. *Biomass Bioenerg* **2011**, *35*, 2839–2851. [CrossRef]

71. Pita, P.; Soria, F.; Canas, I.; Toval, G.; Pardos, J. Carbon isotope discrimination and its relationship to drought resistance under field conditions in genotypes of Eucalyptus globulus Labill. *Forest Ecol. Manag.* **2001**, *141*, 211–221. [CrossRef]

72. Valdés, A.E.; Irar, S.; Majada, J.P.; Rodríguez, A.; Fernández, B.; Pagès, M. Drought tolerance acquisition in Eucalyptus globulus (Labill.): A research on plant morphology, physiology and proteomics. *J. Proteom.* **2013**, *79*, 263–276. [CrossRef] [PubMed]

73. Perez, I.; Perez, J.; Carrasco, J.; Ciria, P. Siberian elm responses to different culture conditions under short rotation forestry in Mediterranean areas. *Turk. J. Agric. For.* **2014**, *38*, 652–662. [CrossRef]

74. Manzone, M.; Bergante, S.; Facciotto, G. Energy and economic sustainability of woodchip production by black locust (*Robinia pseudoacacia* L.) plantations in Italy. *Fuel* **2015**, *140*, 555–560. [CrossRef]

75. Böhm, C.; Quinkenstein, A.; Freese, D. Yield prediction of young black locust (*Robinia pseudoacacia* L.) plantations for woody biomass production using allometric relations. *Ann. For. Res.* **2011**, *54*, 215–227.

76. Oliveira, N.; del Río, M.; Forrester, D.I.; Rodríguez-Soalleiro, R.; Pérez-Cruzado, C.; Cañellas, I.; Sixto, H. Mixed short rotation plantations of *Populus alba* and *Robinia pseudoacacia* for biomass yield. *For. Ecol. Manag.* **2018**, *410*, 48–55. [CrossRef]

77. Sixto, H.; Cañellas, I.; van Arendonk, J.; Ciria, P.; Camps, F.; Sánchez, M.; Sánchez-González, M. Growth potential of different species and genotypes for biomass production in short rotation in Mediterranean environments. *For. Ecol. Manag.* **2015**, *354*, 291–299. [CrossRef]

78. Rosso, L.; Facciotto, G.; Bergante, S.; Vietto, L.; Nervo, G. Selection and testing of *Populus alba* and *Salix* spp. as bioenergy feedstock: Preliminary results. *Appl. Energy* **2013**, *102*, 87–92. [CrossRef]

79. Karp, A.; Shield, I. Bioenergy from plants and the sustainable yield challenge. *New Phytol.* **2008**, *179*, 15–32. [CrossRef] [PubMed]

80. Oliveira, N.; Ciria, P.; Camps, F.; Sánchez-González, M.; Sánchez, M.; Cañellas, I.; Sixto, H. Assessing genotypes in second rotation for lignocellulosic biomass production. In Proceedings of the 23rd European Biomass Conference and Exhibition (EUBCE), Viena, Austria, 1–4 June 2015.

81. Sixto, H.; Calvo, R.; Sánchez, M.M.; Arrieta, J.A.; Otero, J.M.; Salvia, J.; Cañellas, I. Fine scale site variation correlated to growth in a *Salicaceae* plantations (*Salix* and *Populus*) during the first vegetative period. In Proceedings of the 24th International Poplar Commission. Improving lives with poplars and willows, Dehradun, India, 30 October–2 November 2012.

82. Zuazo, V.H.D.; Bocanegra, J.A.J.; Torres, F.P.; Pleguezuelo, C.R.R.; Martínez, J.R.F. Biomass yield potential of *Paulownia* trees in a semi-arid Mediterranean environment (Spain). *Int. J. Renew. Energy Res.* **2013**, *3*, 789–793.

83. García-Morote, F.A.; López-Serrano, F.R.; Martínez-García, E.; Andrés-Abellán, M.; Dadi, T.; Candel, D.; Rubio, E.; Lucas-Borja, M.E. Stem biomass production of *Paulownia elongata* × *P. fortunei* under low irrigation in a semi-arid environment. *Forests* **2014**, *5*, 2505–2520. [CrossRef]

84. Hernández Garasa, M.J.; Sixto, H.; Ciria, P.; Carrasco, J.; Cañellas, I. *Paulownia* plantations for bioenergy in Spain. In Proceedings of the 16th European Biomass confernce and Exhibition, Valencia, Spain, 2–6 May 2008.

85. Bradshaw, A.D. Evolutionary significance of phenotypic plasticity in plants. *Adv. Genet.* **1965**, *13*, 115–155.

86. Zalesny, R.S., Jr.; Hall, R.B.; Zalesny, J.A.; McMahon, B.G.; Berguson, W.E.; Stanosz, G.R. Biomass and genotype × environment interactions of *Populus* energy crops in the Midwestern United States. *Bioenergy Res.* **2009**, *2*, 106–122. [CrossRef]

87. Mohn, C.; Randall, W. Interaction of cottonwood clones with site and planting year. *Can. J. For. Res.* **1973**, *3*, 329–332. [CrossRef]

88. Orlovic, S.; Guzina, V.; Krstic, B.; Merkulov, L. Genetic variability in anatomical, physiological and growth characteristics of hybrid poplar (*Populus* × *euramericana* Dode (Guinier)) and eastern cottonwood (*Populus deltoides* Bartr.) clones. *Silvae Genet.* **1998**, *47*, 183–189.

89. Becker, H.; Leon, J. Stability analysis in plant breeding. *Plant Breed.* **1988**, *101*, 1–23. [CrossRef]
90. Bergante, S.; Facciotto, G.; Minotta, G. Identification of the main site factors and management intensity affecting the establishment of Short-Rotation-Coppices (SRC) in Northern Italy through stepwise regression analysis. *Cent. Eur. J. Biol.* **2010**, *5*, 522–530. [CrossRef]
91. Li, Y.; Suontama, M.; Burdon, R.D.; Dungey, H.S. Genotype by environment interactions in forest tree breeding: Review of methodology and perspectives on research and application. *Tree Genet. Genomes* **2017**, *13*, 60. [CrossRef]
92. Sabatti, M.; Fabbrini, F.; Harfouche, A.; Beritognolo, I.; Mareschi, L.; Carlini, M.; Paris, P.; Scarascia-Mugnozza, G. Evaluation of biomass production potential and heating value of hybrid poplar genotypes in a short-rotation culture in Italy. *Ind. Crop. Prod.* **2014**, *61*, 62–73. [CrossRef]
93. Sixto, H.; Gil, P.M.; Ciria, P.; Camps, F.; Cañellas, I.; Voltas, J. Interpreting genotype-by-environment interaction for biomass production in hybrid poplars under short-rotation coppice in Mediterranean environments. *Glob. Chang. Biol. Bioenergy* **2016**, *8*, 1124–1135. [CrossRef]
94. Hernández, P.M.G.; Voltas, J.; González, M.S.; Martín, M.M.S.; Camps, F.; Ciria, M.P.; Cañellas, I.; Sixto, H. Análisis de estabilidad en genotipos híbridos de *"Populus"* L. para producción de biomasa. *Cuad. Soc. Esp. Cienc. For.* **2012**, 105–110.
95. Sixto, H.; Salvia, J.; Barrio, M.; Ciria, M.P.; Canellas, I. Genetic variation and genotype-environment interactions in short rotation *Populus* plantations in southern Europe. *New For.* **2011**, *42*, 163–177. [CrossRef]
96. Stanton, B. Poplar hybridisation and clonal evaluation: James River, Lower Columbia River Fibber Farm, Westport Research Station. In Proceedings of the International Poplar Symposium, Seattle, WA, USA, 20–25 August 1995; pp. 114–115.
97. Fladung, M.; Polak, O. Ac/Ds-transposon activation tagging in poplar: A powerful tool for gene discovery. *BMC Genom.* **2012**, *13*, 61. [CrossRef]
98. Marchadier, H.; Sigaud, P. Los álamos en la investigación biotecnológica. *Unasylva* **2005**, *56*, 38–39.
99. Fan, D.; Liu, T.; Li, C.; Jiao, B.; Li, S.; Hou, Y.; Luo, K. Efficient CRISPR/Cas9-mediated targeted mutagenesis in *Populus* in the first generation. *Sci. Rep.* **2015**, *5*, 12217. [CrossRef]
100. Zhou, X.; Jacobs, T.B.; Xue, L.J.; Harding, S.A.; Tsai, C.J. Exploiting SNPs for biallelic CRISPR mutations in the outcrossing woody perennial *Populus* reveals 4-coumarate: CoA ligase specificity and redundancy. *New Phytol.* **2015**, *208*, 298–301. [CrossRef]
101. Liu, T.; Fan, D.; Ran, L.; Jiang, Y.; Liu, R.; Luo, K. Highly efficient CRISPR/Cas9-mediated targeted mutagenesis of multiple genes in *Populus*. *Yi Chuan* **2015**, *37*, 1044–1052. [CrossRef] [PubMed]
102. Pilate, G.; Allona, I.; Boerjan, W.; Déjardin, A.; Fladung, M.; Gallardo, F.; Häggman, H.; Jansson, S.; Van Acker, R.; Halpin, C. Lessons from 25 years of GM tree field trials in Europe and prospects for the future. In *Biosafety of Forest Transgenic Trees*; Springer: Berlin/Heidelberg, Germany, 2016; pp. 67–100.
103. Porth, I.; Klapšte, J.; Skyba, O.; Hannemann, J.; McKown, A.D.; Guy, R.D.; DiFazio, S.P.; Muchero, W.; Ranjan, P.; Tuskan, G.A. Genome-wide association mapping for wood characteristics in *Populus* identifies an array of candidate single nucleotide polymorphisms. *New Phytologist.* **2013**, *200*, 710–726. [CrossRef] [PubMed]
104. Ludovisi, R.; Tauro, F.; Salvati, R.; Khoury, S.; Mugnozza Scarascia, G.; Harfouche, A. UAV-based thermal imaging for high-throughput field phenotyping of black poplar response to drought. *Front. Plant Sci.* **2017**, *8*, 1681. [CrossRef] [PubMed]
105. Gömöry, D.; Hrivnák, M.; Krajmerová, D.; Longauer, R. Epigenetic memory effects in forest trees: A victory of "Michurinian biology"? *Cent. Eur. For. J.* **2017**, *63*, 173–179. [CrossRef]
106. Le Gac, A.-L.; Lafon-Placette, C.; Chauveau, D.; Segura, V.; Delaunay, A.; Fichot, R.; Marron, N.; Le Jan, I.; Berthelot, A.; Bodineau, G. Winter-dormant shoot apical meristem in poplar trees shows environmental epigenetic memory. *J. Exp. Bot.* **2018**, *69*, 4821–4837. [CrossRef]
107. Jing, Z.P.; Gallardo, F.; Pascual, M.B.; Sampalo, R.; Romero, J.; De Navarra, A.T.; Cánovas, F.M. Improved growth in a field trial of transgenic hybrid poplar overexpressing glutamine synthetase. *New Phytol.* **2004**, *164*, 137–145. [CrossRef]
108. Conde, D.; Moreno-Cortés, A.; Dervinis, C.; Ramos-Sánchez, J.M.; Kirst, M.; Perales, M.; González-Melendi, P.; Allona, I. Overexpression of DEMETER, a DNA demethylase, promotes early apical bud maturation in poplar. *Plant Cell Environ.* **2017**, *40*, 2806–2819. [CrossRef]

109. Moreno-Cortés, A.; Hernández-Verdeja, T.; Sánchez-Jiménez, P.; González-Melendi, P.; Aragoncillo, C.; Allona, I. CsRAV1 induces sylleptic branching in hybrid poplar. *New. Phytol.* **2012**, *194*, 83–90. [CrossRef]

110. Moreno-Cortés, A.; Ramos-Sánchez, J.M.; Hernández-Verdeja, T.; González-Melendi, P.; Alves, A.; Simões, R.; Rodrigues, J.C.; Guijarro, M.; Canellas, I.; Sixto, H. Impact of RAV1-engineering on poplar biomass production: A short-rotation coppice field trial. *Biotechnol. Biofuels* **2017**, *10*, 110. [CrossRef]

111. Sixto, H. *Populus alba* clones and their hybrids growing under water deficit conditions for biomass production in short rotation forestry. In Proceedings of the VI International Poplar Symposium (IUFRO), Vancouver, BC, Canada, 20–28 July 2014.

112. Marcos, F.; Izquierdo, I.; Gracia, R.; Godino, M.; Ruiz, J.; Vilegas, S. *Estudio de plantaciones energéticas de chopo (Populus × euramericana cv. I-214) a turnos muy cortos*; Documento interno del Departamento de Ingeniería Forestal ETSI de Montes: Madrid, Spain, 2002.

113. San Miguel, A.; San Miguel, J.; Yagüe, S. Tallares de chopo a turno corto. In Proceedings of the 19ª Sesión de la Comisión Internacional del Álamo, Zaragoza, Spain, 22–25 September 1992; pp. 143–156.

114. Sixto, H.; Grau, J.M.; García-Baudín, J.M. Assessment of the effect of broad-spectrum pre-emergence herbicides in poplar nurseries. *Crop Prot.* **2001**, *20*, 121–126. [CrossRef]

115. Sixto, H.; Barrio, M.; Aranda, I. Evaluación de criterios para la selección de clones de chopo como productores de biomasa. In Proceedings of the XVII Reunión de la Sociedad Española de Fisiología Vegetal y X Congreso Hispano-Luso de Fisiología Vegetal Alcalá de Henares, Madrid, Spain, 18–21 September 2007.

116. Sixto, H.; Grau, J.; Alba, N.; Alia, R. Response to sodium chloride in different species and clones of genus *Populus* L. *Forestry* **2005**, *78*, 93–104. [CrossRef]

117. Álvarez-Álvarez, P.; Pizarro, C.; Barrio-Anta, M.; Cámara-Obregón, A.; Bueno, J.L.M.; Álvarez, A.; Gutiérrez, I.; Burslem, D.F. Evaluation of tree species for biomass energy production in Northwest Spain. *Forests* **2018**, *9*, 160. [CrossRef]

118. Castaño Díaz, M. Producción de Biomasa Procedente de Cultivos Energéticos Leñosos en Terrenos Ociosos de Minería en el PRINCIPADO de Asturias. Ph.D. Thesis, Biología de Organismos y Sistemas, Departamento de, University of Oviedo, Oviedo, Spain, 2018.

119. Fiala, M.; Bacenetti, J. Economic, energetic and environmental impact in short rotation coppice harvesting operations. *Biomass Bioenergy* **2012**, *42*, 107–113. [CrossRef]

120. Cañellas, I.; Huelin, P.; Hernández, M.; Ciria, P.; Calvo, R.; Gea-Izquierdo, G.; Sixto, H. The effect of density on Short Rotation *Populus* spp. plantations in the Mediterranean area. *Biomass Bioenergy* **2012**, *46*, 645–652. [CrossRef]

121. Hernández Garasa, M.; Montes Pita, F. Evolución del índice de área foliar en plantaciones de chopo con fines energéticos. Comparación entre métodos semidirectos e indirectos de estimación. In Proceedings of the 5° Congreso Forestal Español, Ávila, Spain, 21–25 September 2009.

122. Sixto, H.; Rueda, J. Evaluación de genotipos para la producción de biomasa con fines energéticos en la Comunidad de Castilla y León. In Proceedings of the 5° Congresos Forestales, Ávila, Spain, 21–25 September 2009.

123. Klasnja, B.; Orlovic, S.; Drekic, M.; Markovic, M. Energy production from short rotation poplar plantations. In Proceedings of the 7th International Symposium on Interdisciplinary Regional Research, Hungary, Serbia and Montenegro, Hunedoara, Romania, 25–25 September 2003; pp. 161–166.

124. Fiala, M.; Bacenetti, J.; Scaravonati, A.; Bergonzi, A. Short rotation coppice in northern Italy: Comprenhensive Sustainability. In Proceedings of the 18th European Biomass Conference and Exhibition, Lyon, France, 3–7 May 2010; pp. 342–348.

125. DeBell, D.S.; Clendenen, G.W.; Zasadat, J.C. Growing *Populus* biomass: Comparison of woodgrass versus wider-spaced short-rotation systems. *Biomass Bioenergy* **1993**, *4*, 305–313. [CrossRef]

126. Ciria, M.; González, E.; Carrasco, J. The effect of fertilization and planting density on biomass productivity of poplar harvested after three-years rotation. In Proceedings of the 12th European Conference and Technology Exhibition on Biomass for Energy, Industry and Climate Protection, Amsterdam, The Netherlands, 17–21 June 2002; pp. 28–286.

127. Ciria, P. Efecto del Turno de Corta y de la Densidad de Plantación Sobre la Productividad de Diversos Clones de Chopo en Condiciones de Corta Rotación. Ph.D. Thesis, Universidad Politécnica de Madrid, Madrid, Spain, 1999.

128. Sevigne, E.; Gasol, C.M.; Brun, F.; Rovira, L.; Pagés, J.M.; Camps, F.; Rieradevall, J.; Gabarrell, X. Water and energy consumption of *Populus* spp. bioenergy systems: A case study in Southern Europe. *Renew. Sustain. Energy Rev.* **2011**, *15*, 1133–1140. [CrossRef]

129. Dillen, S.Y.; Djomo, S.N.; Al Afas, N.; Vanbeveren, S.; Ceulemans, R. Biomass yield and energy balance of a short-rotation poplar coppice with multiple clones on degraded land during 16 years. *Biomass Bioenergy* **2013**, *56*, 157–165. [CrossRef]

130. Friedrich, E. *Anbautechnische Untersuchungen in Forstlichen Schnellwuchsplantagen und Demonstration des Leistungsvermögens Schnellwachsender Baumarten*; Landwirtschaftsverlag Münster: Leipzig, Germany, 1999.

131. Menéndez, J.; Loredo Pérez, J.L. Biomass production in surface mines: Renewable energy source for power plants. *WSEAS Trans. Environ. Dev.* **2018**, *14*, 205–211.

132. Martín-García, J.; Merino, A.; Diez, J.J. Relating visual crown conditions to nutritional status and site quality in monoclonal poplar plantations (*Populus × euramericana*). *Eur. J. For. Res.* **2012**, *131*, 1185–1198. [CrossRef]

133. Bungart, R.; Hüttl, R.F. Growth dynamics and biomass accumulation of 8-year-old hybrid poplar clones in a short-rotation plantation on a clayey-sandy mining substrate with respect to plant nutrition and water budget. *Eur. J. For. Res.* **2004**, *123*, 105–115. [CrossRef]

134. Truax, B.; Fortier, J.; Gagnon, D.; Lambert, F. Planting density and site effects on stem dimensions, stand productivity, biomass partitioning, carbon stocks and soil nutrient supply in hybrid poplar plantations. *Forests* **2018**, *9*, 293. [CrossRef]

135. O Di Nasso, N.N.; Guidi, W.; Ragaglini, G.; Tozzini, C.; Bonari, E. Biomass production and energy balance of a 12 year old short rotation coppice poplar stand under different cutting cycles. *Bioenergy* **2010**, *2*, 89–97. [CrossRef]

136. Vanbeveren, S.P.P.; Ceulemans, R. Genotypic differences in biomass production during three rotations of short-rotation coppice. *Biomass Bioenergy* **2018**, *119*, 198–205. [CrossRef]

137. Gómez-Martín, J.M.; Castaño-Díaz, M.; Cámara-Obregón, A.; Álvarez-Álvarez, P.; Folgueras-Díaz, M.B.; Diez, M.A. On the chemical composition and pyrolytic behavior of hybrid poplar energy crops from northern Spain. *Energy Rep.* **2020**, *6*, 764–769. [CrossRef]

138. Garasa, M.J.H.; Cañellas, I.; Viscasillas, E.; García, A.; Carrasco, J.E.; Sixto, H. Biomass yield in a short rotation clonal poplar trial. In Proceedings of the 18th European Biomass Conference and Exhibitions, Lyon, France, 3–7 May 2010.

139. Oliveira, N.; De la Iglesia, J.P.; Viscasillas, E.; Bachiller, A.; Parras, A.; González, I.; Grau, J.M.; Otero, J.M.; Cañellas, I.; Sixto, H. Adecuación de genotipos para la producción de biomasa en la meseta septentrional. In Proceedings of the II Simposio del Chopo, Valladolid, Spain, 17–19 October 2018.

140. Oliveira, N. Evaluación de la productividad clonal en biomasa, en un cultivo energético experimental de chopo instalado por INIA-CIFOR en Almazán (Soria). Master's Thesis, University of Santiago de Compostela, Lugo, Spain, 2012.

141. Willebrand, E.; Verwijst, T. Population dynamics of willow coppice systems and their implications for management of short-rotation forests. *Forest Chron.* **1993**, *69*, 699–704. [CrossRef]

142. Kopp, R.; Abrahamson, L.; White, E.; Volk, T.; Nowak, C.; Fillhart, R. Willow biomass production during ten successive annual harvests. *Biomass Bioenergy* **2001**, *20*, 1–7. [CrossRef]

143. Tubby, I.; Armstrong, A. Establishment and Management of Short Rotation Coppice. 2002. Available online: https://www.cabdirect.org/cabdirect/abstract/20023172097 (accessed on 15 November 2020).

144. Sixto, H.; Hernández, M.J.; de Miguel, J.; Cañellas, I. *Red de Parcelas de Cultivos Leñosos en Alta Densidad y Turno Corto*; Instituto Nacional de Investigación y Tecnología Agraria y Alimentaria, Ministerio de Economía y Competitividad: Madrid, Spain, 2013; p. 188.

145. Fernández, M.J.; Barro, R.; Pérez, J.; Losada, J.; Ciria, P. Influence of the agricultural management practices on the yield and quality of poplar biomass (a 9-year study). *Biomass Bioenergy* **2016**, *93*, 87–96. [CrossRef]

146. Sixto, H.; Ciria, P.; Rueda, J.; Perez, J.; García Caballero, J.; Mazón, P.; Montoto, J.L.; Cañellas, I. Comparison of two cutting scenarios and clone response in the first rotation period of a poplar crop for biomass energy production. In Proceedings of the 17th European Biomass Conference and Exhibition, Hamburg, Germany, 28 June–3 July 2009.

147. Štochlová, P.; Novotná, K.; Costa, M.; Rodrigues, A. Biomass production of poplar short rotation coppice over five and six rotations and its aptitude as a fuel. *Biomass Bioenergy* **2019**, *122*, 183–192. [CrossRef]

148. Fernández, M.J.; Barro, R.; Pérez, J.; Ciria, P. Production and composition of biomass from short rotation coppice in marginal land: A 9-year study. *Biomass Bioenergy* **2020**, *134*, 105478. [CrossRef]

149. Oliveira, N.; Otero, J.M.; Pasalodos, M.; Cañellas, I.; Pereira-Espinel, J.; Rodríguez-Soalleiro, R.; Sixto, H. Analisis de los costes de cultivos forestales de turno corto de chopo en España para produccion de bioenergia. In Proceedings of the XII Congreso de Economía Agraria, Lugo, Spain, 4–6 September 2019.

150. Sixto, H.; Grau, J.M.; Eugenio, M.E.; Ibarra, D.; Hernandez, J.J.; Monedero, E.; Oliveira, N.; Otero, J.M.; Cañellas, I. Evolution of biomass production with poplar hybrids in southern Europe. In Proceedings of the 7th International Poplar Symposium. New Bioeconomies: Exploring the role of Salicaceae, Buenos Aires, Argentina, 28 October–4 November 2018.

151. Dawson, J.O.; Hansen, E.A. Effect of *Alnus glutinosa* on hybrid populus growth and soil nitrogen concentration in a mixed plantation. In *Intensive Plantation Culture: 12 Years Research*; Hansen, E.A., Ed.; Gen. Tech. Rep. NC-91; U.S. Department of Agriculture, Forest Service, North Central Forest Experiment Station: St. Paul, MN, USA, 1983; pp. 29–34.

152. Radwan, M.; DeBell, D. Nutrient relations in coppiced black cottonwood and red alder. *Plant Soil* **1988**, *106*, 171–177. [CrossRef]

153. Pelleri, F.; Plutino, M.; Manetti, M.C.; Sansone, D.; Bergante, S.; Castro, G.; Moya, J.F.; Chiarabaglio, P.M.; Urban, M.I. Plantaciones policíclicas mixtas: Nogales, chopos y biomasa de rotación corta. In Proceedings of the II Simposio del Chopo, Valladolid, Spain, 17–19 October 2018.

154. Gana, C.; Epron, D.; Gérant, D.; Maillard, P.; Plain, C.; Priault, P.; Marron, N. How does poplar / black locust mixture influence growth and functioning of each species in a short rotation plantation? In Proceedings of the VI International Poplar Symposium, Vancouver, BC, Canada, 20–23 July 2014.

155. Marron, N.; Priault, P.; Gana, C.; Gérant, D.; Epron, D. Prevalence of interspecific competition in a mixed poplar/black locust plantation under adverse climate conditions. *Ann. For. Sci.* **2018**, *75*, 23. [CrossRef]

156. Oviedo, P.A.; Gutiérrez, J.A.; Martín, R.T.; Martínez, M.F. Cultivo mixto de especies forestales de turno corto (*Robinia pseudoacacia* y *Populus* × *euroamericana* clon AF2) con fines energéticos. *Cuad. Soc. Esp. Cienc. For.* **2016**, 17–30. [CrossRef]

157. Oliveira, N.; Del Río, M.; Rodríguez-Soalleiro, R.; Pérez-Cruzado, C.; Cañellas, I.; Sixto, H. Mixed short rotation plantation of *Populus alba* and *Robinia pseudoacacia* for biomass yield over the course of two rotations. In Proceedings of the 7th International Poplar Symposium, Buenos Aires, Argentina, 28 October–4 November 2018.

158. González, I.; Sixto, H.; Rodríguez-Soalleiro, R.; Oliveira, N. Nutrient Contribution of Litterfall in a Short Rotation Plantation of Pure or Mixed Plots of *Populus alba* L. and *Robinia pseudoacacia* L. *Forests* **2020**, *11*, 1133. [CrossRef]

159. Grünewald, H.; Böhm, C.; Quinkenstein, A.; Grundmann, P.; Eberts, J.; von Wühlisch, G. *Robinia pseudoacacia* L.: A lesser known tree species for biomass production. *Bioenergy Res.* **2009**, *2*, 123–133. [CrossRef]

160. DeBell, D.S.; Harrington, C.A. Deploying genotypes in short-rotation plantations: Mixtures and pure cultures of clones and species. *For. Chron.* **1993**, *69*, 705–713. [CrossRef]

161. DeBell, D.S.; Harrington, C.A. Productivity of *Populus* in monoclonal and polyclonal blocks at three spacings. *Can. J. For. Res.* **1997**, *27*, 978–985. [CrossRef]

162. Miot, S.; Frey, P.; Pinon, J. Varietal mixture of poplar clones: Effects on infection by *Melampsora larici-populina* and on plant growth. *Eur. J. For. Pathol.* **1999**, *29*, 411–423. [CrossRef]

163. Elferjani, R.; DesRochers, A.; Tremblay, F. Effects of mixing clones on hybrid poplar productivity, photosynthesis and root development in northeastern Canadian plantations. *For. Ecol. Manag.* **2014**, *327*, 157–166. [CrossRef]

164. Dillen, M.; Vanhellemont, M.; Verdonckt, P.; Maes, W.H.; Steppe, K.; Verheyen, K. Productivity, stand dynamics and the selection effect in a mixed willow clone short rotation coppice plantation. *Biomass Bioenergy* **2016**, *87*, 46–54. [CrossRef]

165. Schweier, J.; Arranz, C.; Nock, C.A.; Jaeger, D.; Scherer-Lorenzen, M. Impact of increased genotype or species diversity in Short Rotation Coppice on biomass production and wood characteristics. *Bioenergy Res.* **2019**, *12*, 497–508. [CrossRef]

166. Broeckx, L.S.; Verlinden, M.S.; Ceulemans, R. Establishment and two-year growth of a bio-energy plantation with fast-growing *Populus* trees in Flanders (Belgium): Effects of genotype and former land use. *Biomass Bioenergy* **2012**, *42*, 151–163. [CrossRef]

167. Morhart, C.; Sheppard, J.; Seidl, F.; Spiecker, H. Influence of different tillage systems and weed treatments in the establishment year on the final biomass production of short rotation coppice poplar. *Forests* **2013**, *4*, 849. [CrossRef]

168. Albertsson, J.; Hansson, D.; Bertholdsson, N.O.; Åhman, I. Site-related set-back by weeds on the establishment of 12 biomass willow clones. *Weed Res.* **2014**, *54*, 398–407. [CrossRef]

169. Heilman, P.E.; Xie, F.-G. Effects of nitrogen fertilization on leaf area, light interception, and productivity of short-rotation *Populus trichocarpa* × *Populus deltoides* hybrids. *Can. J. For. Res.* **1994**, *24*, 166–173. [CrossRef]

170. Ceulemans, R.; Deraedt, W. Production physiology and growth potential of poplars under short-rotation forestry culture. *For. Ecol. Manag.* **1999**, *121*, 9–23. [CrossRef]

171. Liberloo, M.; Calfapietra, C.; Lukac, M.; Godbold, D.; Luo, Z.-b.; Polle, A.; Hoosbeek, M.R.; Kull, O.; Marek, M.; Raines, C. Woody biomass production during the second rotation of a bio-energy *Populus* plantation increases in a future high $CO_2$ world. *Glob. Chang. Biol.* **2006**, *12*, 1094–1106. [CrossRef]

172. Van Veen, J.; Breteler, H.; Olie, J.; Frissel, M. Nitrogen and energy balance of a short-rotation poplar forest system. *Neth. J. Agri. Sci.* **1981**, *29*, 163–172. [CrossRef]

173. Coleman, M.; Tolsted, D.; Nichols, T.; Johnson, W.D.; Wene, E.G.; Houghtaling, T. Post-establishment fertilization of Minnesota hybrid poplar plantations. *Biomass Bioenergy* **2006**, *30*, 740–749. [CrossRef]

174. Stanturf, J.A.; Van Oosten, C.; Netzer, D.A.; Coleman, M.D.; Portwood, C.J. Ecology and silviculture of poplar plantations. In *Poplar Culture in North America*; Dickmann, D.I., Isebrands, J.G., Eckenwalder, J.E., Richardson, J., Eds.; NRC Research Press: Ottawa, ON, Canada, 2002; pp. 153–206.

175. Yan, X.-L.; Dai, T.-F.; Zhao, D.; Jia, L.-M. Combined surface drip irrigation and fertigation significantly increase biomass and carbon storage in a *Populus* × *euramericana* cv. Guariento plantation. *J. For. Res.* **2016**, *21*, 280–290. [CrossRef]

176. Euring, D.; Löfke, C.; Teichmann, T.; Polle, A. Nitrogen fertilization has differential effects on N allocation and lignin in two *Populus* species with contrasting ecology. *Trees* **2012**, *26*, 1933–1942. [CrossRef]

177. Coleman, M.D.; Dickson, R.E.; Isebrands, J. Growth and physiology of aspen supplied with different fertilizer addition rates. *Physiol. Plant.* **1998**, *103*, 513–526. [CrossRef]

178. Paris, P.; Mareschi, L.; Sabatti, M.; Tosi, L.; Scarascia-Mugnozza, G. Nitrogen removal and its determinants in hybrid *Populus* clones for bioenergy plantations after two biennial rotations in two temperate sites in northern Italy. *iForest* **2015**, *8*, 657–665. [CrossRef]

179. Rennenberg, H.; Wildhagen, H.; Ehlting, B. Nitrogen nutrition of poplar trees. *Plant Biol.* **2010**, *12*, 275–291. [CrossRef] [PubMed]

180. Wang, Y.; Xi, B.; Bloomberg, M.; Moltchanova, E.; Li, G.; Jia, L. Response of diameter growth, biomass allocation and N uptake to N fertigation in a triploid *Populus tomentosa* plantation in the North China Plain: Ontogenetic shift does not exclude plasticity. *Eur. J. Forest Res.* **2015**, *134*, 889–898. [CrossRef]

181. Gruber, N.; Galloway, J.N. An Earth-system perspective of the global nitrogen cycle. *Nature* **2008**, *451*, 293. [CrossRef] [PubMed]

182. Erisman, J.W.; van Grinsven, H.; Leip, A.; Mosier, A.; Bleeker, A. Nitrogen and biofuels; an overview of the current state of knowledge. *Nut. Cycl. Agroecosyst.* **2010**, *86*, 211–223. [CrossRef]

183. Heller, M.C.; Keoleian, G.A.; Volk, T.A. Life cycle assessment of a willow bioenergy cropping system. *Biomass Bioenergy* **2003**, *25*, 147–165. [CrossRef]

184. Sugiura, A.; Tyrrel, S.; Seymour, I.; Burgess, P. Water Renew systems: Wastewater polishing using renewable energy crops. *Water Sci. Technol.* **2008**, *57*, 1421–1428. [CrossRef]

185. Werner, A.; McCracken, A. The use of short rotation coppice poplar and willow for the bioremediation of sewage effluent. *Asp. Appl. Biol.* **2008**, 317–324.

186. Dimitriou, I.; Rosenqvist, H. Sewage sludge and wastewater fertilisation of Short Rotation Coppice (SRC) for increased bioenergy production—Biological and economic potential. *Biomass Bioenergy* **2011**, *35*, 835–842. [CrossRef]

187. Martínez-Hernández, V.; Leal, M.; Meffe, R.; De Miguel, Á.; Alonso-Alonso, C.; De Bustamante, I.; Lillo, J.; Martín, I.; Salas, J. Removal of emerging organic contaminants in a poplar vegetation filter. *J. Hazard. Mater.* **2018**, *342*, 482–491. [CrossRef] [PubMed]

188. Pradana, R.; Sixto, H.; González, B.; Demaria, I.; González, I.; Martínez-Hernández, V.; de Bustamante, I. Poplar Vegetation Filters for the beer industry: Wastewater treatment combined with biomass production. In Proceedings of the XI Congreso Ibérico de Gestión y Planificación del Agua, Madrid, Spain, 3–5 September 2020.

189. Paniagua, S.; Escudero, L.; Escapa, C.; Coimbra, R.N.; Otero, M.; Calvo, L.F. Effect of waste organic amendments on *Populus* sp biomass production and thermal characteristics. *Renew. Energy* **2016**, *94*, 166–174. [CrossRef]

190. Sixto, H.; Garasa, M.H.; Ciria, P.C.; Garcia, J.C.; de Vinas, I.C.R. Manual de Cultivo de *Populus* spp. Para la Producción de Biomasa Con Fines Energéticos. 2010. Available online: https://agris.fao.org/agris-search/search.do?recordID=XF2015034695 (accessed on 15 November 2020).

191. Kollert, W. La populicultura en Europa y España. In Proceedings of the II Simposio del chopo, Valladolid, Spain, 17–19 October 2018.

192. Newcombe, G.; Ostry, M.; Hubbes, M.; Périnet, P.; Mottet, M.-J. Poplar diseases. In *Poplar Culture in North America*; Dickmann, D.I., Isebrands, J.G., Eckenwalder, J.E., Richardson, J., Eds.; NRC Research Press: Ottawa, ON, Canada, 2001; pp. 249–276. [CrossRef]

193. Coyle, D.R.; Nebeker, T.E.; Hart, E.R.; Mattson, W.J. Biology and management of insect pests in North American intensively managed hardwood forest systems. *Annu. Rev. Entomol.* **2005**, *50*, 1–29. [CrossRef] [PubMed]

194. Duplessis, S.; Major, I.; Martin, F.; Séguin, A. Poplar and pathogen interactions: Insights from *Populus* genome-wide analyses of resistance and defense gene families and gene expression profiling. *Crit. Rev. Plant Sci.* **2009**, *28*, 309–334. [CrossRef]

195. Broderick, N.A.; Vasquez, E.; Handelsman, J.; Raffa, K.F. Effect of clonal variation among hybrid poplars on susceptibility of gypsy moth (*Lepidoptera*: *Lymantriidae*) to *Bacillus thuringiensis* subsp. *kurstaki*. *J. Econ. Entomol.* **2010**, *103*, 718–725. [CrossRef]

196. Steenackers, J.; Steenackers, M.; Steenackers, V.; Stevens, M. Poplar diseases, consequences on growth and wood quality. *Biomass Bioenergy* **1996**, *10*, 267–274. [CrossRef]

197. Soria, S. *Patología de las Choperas*; Caja Rural del Duero. Colegio de Ingenieros de Montes: Valladolid, Spain, 1992; pp. 61–75.

198. Nordman, E.E.; Robison, D.J.; Abrahamson, L.P.; Volk, T.A. Relative resistance of willow and poplar biomass production clones across a continuum of herbivorous insect specialization: Univariate and multivariate approaches. *For. Ecol. Manag.* **2005**, *217*, 307–318. [CrossRef]

199. Isebrands, J.G.; Richardson, J. *Poplars and Willows: Trees for Society and the Environment*; The Food and Agricultural Organization of the United Nations (FAO) and CAB International: Rome, Italy, 2014.

200. Bloemen, J.; Fichot, R.; Horemans, J.A.; Broeckx, L.S.; Verlinden, M.S.; Zenone, T.; Ceulemans, R. Water use of a multi-genotype poplar short-rotation coppice from tree to stand scale. *Glob. Chang. Biol. Bioenegy* **2016**, *9*, 370–384. [CrossRef]

201. Padró, A.; Orensanz, J. *El Chopo y su Cultivo*; Ministerio de Agricultura, Pesca y Alimentación (MAPA): Madrid, Spain, 1987.

202. Sixto, H.; Ruiz, V.; Grau, J.; Montoto, J. Primeros resultados de un ensayo de riego en vivero de planta de chopo. 1. In Proceedings of the 1er Simposio del Chopo, Zaragoza, Spain, 9–11 May 2001; pp. 159–166.

203. Sixto, H.; Grau, J.; González-Antoñanzas, F. Populicultura: *Populus* spp. e híbridos. Capítulo IV. In *Compendio de Selvicultura Aplicada en España*; Serrada, R., Montero, G., Reque, J.A., Eds.; Instituto Nacional de Investigación y Tecnología Agraria y Alimentaria, Ministerio de Educación y Ciencia: Madrid, Spain, 2008.

204. IPCC. Climate Change 2007: Impacts, adaptation and vulnerability. In *Contribution of Working Group II to the Fourth Assessment Report of the Intergovernmental Panel on Climate Change*; Cambridge University Press: Cambridge, UK, 2007; p. 976.

205. González-González, B.D.; Oliveira, N.; González, I.; Cañellas, I.; Sixto, H. Poplar biomass production in short rotation under irrigation: A case study in the Mediterranean. *Biomass Bioenergy* **2017**, *107*, 198–206. [CrossRef]

206. Hennig, A.; Kleinschmit, J.R.; Schoneberg, S.; Löffler, S.; Janßen, A.; Polle, A. Water consumption and biomass production of protoplast fusion lines of poplar hybrids under drought stress. *Front. Plant Sci.* **2015**, *6*, 330. [CrossRef]

207. Liang, Z.-S.; Yang, J.-W.; Shao, H.-B.; Han, R.-L. Investigation on water consumption characteristics and water use efficiency of poplar under soil water deficits on the Loess Plateau. *Colloid Surface B* **2006**, *53*, 23–28. [CrossRef] [PubMed]

208. Viger, M.; Smith, H.; Cohen, D.; Dewoody, J.; Trewin, H.; Steenackers, M.; Bastien, C.; Taylor, G. Adaptive mechanisms and genomic plasticity for drought tolerance identifed in European black poplar (*Populus nigra* L.). *Tree Physiol.* **2016**, *36*, 909–928. [CrossRef] [PubMed]

209. Niemczyk, M.; Hu, Y.; Thomas, B.R. Selection of Poplar Genotypes for Adapting to Climate Change. *Forests* **2019**, *10*, 1041. [CrossRef]

210. Garasa, M.J.H. Respuesta Anatómico Fisiológicas Frente a Estrés Hídrico en Plantaciones de Especies de Crecimiento Rápido Para la Producción de Biomasa. Ph.D. Thesis, Universidad Politécnica de Madrid, Madrid, Spain, 2015.

211. Navarro, A.; Facciotto, G.; Campi, P.; Mastrorilli, M. Physiological adaptations of five poplar genotypes grown under SRC in the semi-arid Mediterranean environment. *Trees* **2014**, *28*, 983–994. [CrossRef]

212. Navarro, A.; Portillo-Estrada, M.; Arriga, N.; Vanbeveren, S.P.; Ceulemans, R. Genotypic variation in transpiration of coppiced poplar during the third rotation of a short-rotation bio-energy culture. *Glob. Chang. Biol. Bioenergy* **2018**, *10*, 592–607. [CrossRef] [PubMed]

213. Navarro, A.; Portillo-Estrada, M.; Ceulemans, R. Identifying the best plant water status indicator for bio-energy poplar genotypes. *Glob. Chang. Biol. Bioenergy* **2020**, *12*, 426–444. [CrossRef]

214. Fuertes, A.; Oliveira, N.; Cañellas, I.; Sixto, H.; Rodríguez-Soalleiro, R. An economic overview of *Populus* spp. in Short Rotation Coppice systems under Mediterranean conditions: An assessment tool for decision-making. *Sustain. Energy Rev.* **2020**. under review.

215. de Bustamante, I. Filtros verdes. Un sistema para la depuración y reutilización de aguas residuales. *Tecnoambiente* **1998**, *79*, 73–75.

216. de Miguel, Á.; Martínez-Hernández, V.; Leal, M.; González-Naranjo, V.; de Bustamante, I. Del residuo al recurso: Reutilización de aguas para riego de biocombustibles-From waste to resource: Reuse of water for biofuel irrigation. *Water* **2012**, *79*, 68–70.

217. Makeschin, F. Short rotation forestry in Central and Northern Europe-introduction and conclusions. *For. Ecol. Manag.* **1999**, *121*, 1–7.

218. Ciria, M.; Mazón, M.; Carrasco, J. Poplar productivity evolution on short rotation during three consecutive cycles in extreme continental climate. In Proceedings of the 2nd World Conference on Biomass for Energy, Industry and Climate Protection, Rome, Italy, 10–14 May 2004.

219. Laureysens, I.; Deraedt, W.; Indeherberge, T.; Ceulemans, R. Population dynamics in a 6-year old coppice culture of poplar. I. Clonal differences in stool mortality, shoot dynamics and shoot diameter distribution in relation to biomass production. *Biomass Bioenerg* **2003**, *24*, 81–95. [CrossRef]

220. Benetka, V.; Novotná, K.; Štochlová, P. Biomass production of *Populus nigra* L. clones grown in short rotation coppice systems in three different environments over four rotations. *iForest* **2014**, *7*, 233–239. [CrossRef]

221. Trnka, M.; Trnka, M.; Fialová, J.; Koutecky, V.; Fajman, M.; Zalud, Z.; Hejduk, S. Biomass production and survival rates of selected poplar clones grown under a short-rotation system on arable land. *Plant Soil Environ.* **2008**, *54*, 78–88. [CrossRef]

222. Berthelot, A. Mélange de clones en taillis à courtes rotations de peuplier: Influence sur la productivité et l'homogénéité des produits récoltés. *Can. J. For. Res.* **2001**, *31*, 1116–1126. [CrossRef]

223. Hartmann, L.; Lamersdorf, N. Site conditions, initial growth and nutrient and litter cycling of newly installed short rotation coppice and agroforestry systems. In *Bioenergy from Dendromass for the Sustainable Development of Rural Areas*; Manning, D.B., Bemmann, A., Bredemeier, N., Lamersdorf, N., Ammer, C., Eds.; Wiley-VCH Verlag GmbH & Co. KGaA: Weinheim, Germany, 2015.

224. Kern, J.; Germer, S.; Ammon, C.; Balasus, A.; Bischoff, W.-A.; Schwarz, A.; Forstreuter, M.; Kaupenjohann, M. Environmental effects over the first $2\frac{1}{2}$ rotation periods of a fertilised poplar short rotation coppice. *Bioenergy Res.* **2018**, *11*, 152–165. [CrossRef]

225. Cabrera, A.; Tozzini, C.; Espinoza, S.; Santelices, R.; Bonari, E. Cálculo del balance energético de una plantación de *Populus deltoides* clon Lux con fines energéticos en un sitio con ambiente mediterráneo. *Bosque* **2014**, *35*, 133–139. [CrossRef]

226. Paris, P.; Mareschi, L.; Sabatti, M.; Pisanelli, A.; Ecosse, A.; Nardin, F.; Scarascia-Mugnozza, G. Comparing hybrid Populus clones for SRF across northern Italy after two biennial rotations: Survival, growth and yield. *Biomass Bioenergy* **2011**, *35*, 1524–1532. [CrossRef]

227. Stolarski, M.J.; Krzyżaniak, M.; Szczukowski, S.; Tworkowski, J.; Bieniek, A. Short rotation woody crops grown on marginal soil for biomass energy. *Pol. J. Environ. Stud.* **2014**, *23*, 1727–1739.

228. Proe, M.F.; Griffiths, J.H.; Craig, J. Effects of spacing, species and coppicing on leaf area, light interception and photosynthesis in short rotation forestry. *Biomass Bioenergy* **2002**, *23*, 315–326. [CrossRef]

229. Baraza, C. Cultivos demostrativos de chopos (*Populus* spp.) a altas densidades con fines energéticos realizados en Aragón. Primeros datos y resultados. In Proceedings of the 6° Congreso Forestal Español, Vitoria, Spain, 10–14 June 2013.

230. Ciria, M.; González, E.; Mazon, P.; Carrasco, J. Influence of the rotation age and plant density on the composition and quality of poplar biomass. Biomass for energy and the environment. In Proceedings of the 9th European Bioenergy Conference and 1st European Energy from Biomass Technology Exhibition, Copenhagen, Denmark, 24–27 June 1996; pp. 968–973.

231. Eimil-Fraga, C.; Proupín-Castiñeiras, X.; Rodríguez-Añón, J.A.; Rodríguez-Soalleiro, R. Effects of Shoot Size and Genotype on Energy Properties of Poplar Biomass in Short Rotation Crops. *Energies* **2019**, *12*, 2051. [CrossRef]

232. Godino, M. Análisis de los Cultivos de *Populus × euroamericana*, Clon I-214, a Turnos Muy Cortos con Fines Energéticos. Ph.D. Thesis, Universidad Politécnica de Madrid, Madrid, Spain, 2005.

233. Ibarra, D.; Eugenio, M.E.; Cañellas, I.; Sixto, H.; Martín-Sampedro, R. Potential of different poplar clones for sugar production. *Wood Sci. Technol.* **2017**, *51*, 669–684. [CrossRef]

234. Martín-Sampedro, R.; Eugenio, M.E.; Cañellas, I.; Sixto, H.; Ibarra, D. Characterization of different poplar clones for sugar production. In Proceedings of the 1st International Workshop on Biorefinery of Lignocellulosic Materials, Cordoba, Spain, 9–12 June 2015.

235. Djomo, S.N.; Ac, A.; Zenone, T.; De Groote, T.; Bergante, S.; Facciotto, G.; Sixto, H.; Ciria Ciria, P.; Weger, J.; Ceulemans, R. Energy performances of intensive and extensive short rotation cropping systems for woody biomass production in the EU. *Renew. Sustain. Energy Rev.* **2015**, *41*, 845–854. [CrossRef]

236. Oliveira, N.; Pérez-Cruzado, C.; Sixto, H.; Cañellas, I.; Rodrígez-Soalleiro, R. Modelización alométrica para la estimación de biomasa en un cultivo energético experimental de chopo. In Proceedings of the II Encuentro Juventud Investigadora. Universidad de Santiago de Compostela, Santiago de Compostela, Spain, 29–31 January 2014.

237. San Miguel, G.; Corona, B.; Ruiz, D.; Landholm, D.; Laina, R.; Tolosana, E.; Sixto, H.; Cañellas, I. Environmental, energy and economic analysis of a biomass supply chain based on a poplar short rotation coppice in Spain. *J. Clean. Prod.* **2015**, *94*, 93–101. [CrossRef]

238. Valbuena-Castro, J.; Oliveira, N.; Rodríguez-Soalleiro, R.; Sixto, H.; Cañellas, I. Biomass productivity assessment at a clonal forest of *Populus* spp. at Valtierra (Navarra, Spain). In Proceedings of the Xth Young Researchers Meeting on Conservation and Sustainable Use of Forest Systems, Valsaín, Spain, 25–26 January 2016.

239. Valbuena-Castro, J.; Oliveira, N.; Rodríguez-Soalleiro, R.; Cañellas, I.; Sixto, H. Assessment of biomass productivity and effect of planting density in a short rotation coppice poplar plantation in the north of Spain. In Proceedings of the 25th Session of the International Poplar Commission, Berlin, Germany, 12–16 September 2016.

240. Picard, N.; Saint-André, L.; Henry, M. *Manual for Building Tree Volume and Biomass Allometric Equations: From Field Measurement to Prediction*; Food and Agricultural Organization of the United Nations and Centre de Coopération Internationale en Recherche Agronomique pour le Développement, Montpellier: Rome, Italy, 2012; p. 215.

241. Peichl, M.; Arain, M.A. Allometry and partitioning of above- and belowground tree biomass in an age-sequence of white pine forests. *For. Ecol. Manag.* **2007**, *253*, 68–80. [CrossRef]

242. Shaiek, O.; Loustau, D.; Trichet, P.; Meredieu, C.; Bachtobji, B.; Garchi, S.; Aouni, M.H.E. Generalized biomass equations for the main aboveground biomass components of maritime pine across contrasting environments. *Ann. For. Sci.* **2011**, *68*, 443–452. [CrossRef]

243. Verwijst, T.; Lundkvist, A.; Edelfeldt, S.; Albertsson, J. Development of sustainable willow short rotation forestry in Northern Europe. In *Biomass Now-Sustainable Growth and Use*; Matovic, D.M.D., Ed.; BoD–Books on Demand: Croatia, 2013; pp. 479–502. Available online: https://books.google.com.hk/books?hl=zh-CN&lr=&id=WdmgDwAAQBAJ&oi=fnd&pg=PA479&dq=Development+of+sustainable+willow+short+rotation+forestry+in+Northern+Europe&ots=BJwT6IWByQ&sig=TZ6h3-l6e8g3YMKL1hdp77ywZ2w&redir_esc=y&hl=zh-CN&sourceid=cndr#v=onepage&q=Development%20of%20sustainable%20willow%20short%20rotation%20forestry%20in%20Northern%20Europe&f=false (accessed on 15 November 2020). [CrossRef]

244. Mosseler, A.; Major, J.; Labrecque, M.; Larocque, G. Allometric relationships in coppice biomass production for two North American willows (*Salix* spp.) across three different sites. *For. Ecol. Manag.* **2014**, *320*, 190–196. [CrossRef]

245. Verlinden, M.S.; Broeckx, L.S.; Ceulemans, R. First vs. second rotation of a poplar short rotation coppice: Above-ground biomass productivity and shoot dynamics. *Biomass Bioenergy* **2015**, *73*, 174–185. [CrossRef]

246. Cannell, M.; Willett, S. Shoot growth phenology, dry matter distribution and root: Shoot ratios of provenances of *Populus trichocarpa*, *Picea sitchensis* and *Pinus contorta* growing in Scotland. *Silvae Genet.* **1976**, *25*, 49–59.

247. Ruark, G.A.; Martin, G.L.; Bockheim, J.G. Comparison of constant and variable allometric ratios for estimating *Populus tremuloides* biomass. *For. Sci.* **1987**, *33*, 294–300. [CrossRef]

248. Oliveira, N.; Rodríguez-Soalleiro, R.; Hernández, M.J.; Cañellas, I.; Sixto, H.; Pérez-Cruzado, C. Improving biomass estimation in a *Populus* short rotation coppice plantation. *For. Ecol. Manag.* **2017**, *391*, 194–206. [CrossRef]

249. Di Matteo, G.; Sperandio, G.; Verani, S. Field performance of poplar for bioenergy in southern Europe after two coppicing rotations: Effects of clone and planting density. *iForest* **2012**, *5*, 224–229. [CrossRef]

250. Rock, J. Suitability of published biomass equations for aspen in Central Europe-Results from a case study. *Biomass Bioenergy* **2007**, *31*, 299–307. [CrossRef]

251. Coyle, D.R.; Coleman, M.D. Forest production responses to irrigation and fertilization are not explained by shifts in allocation. *For. Ecol. Manag.* **2005**, *208*, 137–152. [CrossRef]

252. Oliveira, N.; Rodríguez-Soalleiro, R.; Pérez-Cruzado, C.; Cañellas, I.; Sixto, H. On the genetic affinity of individual tree biomass allometry in poplar short rotation coppice. *Bioenergy Res.* **2017**, *10*, 525–535. [CrossRef]

253. Pérez-Cruzado, C.; Fehrmann, L.; Magdon, P.; Cañellas, I.; Sixto, H.; Kleinn, C. On the site-level suitability of biomass models. *Environ. Modell. Softw.* **2015**, *73*, 14–26. [CrossRef]

254. Zianis, D.; Muukkonen, P.; Mäkipää, R.; Mencuccini, M. *Biomass and Stem Volume Equations for Tree Species in Europe*; Tammer-Paino Oy: Tampere, Finland, 2005.

255. Dowell, R.C.; Gibbins, D.; Rhoads, J.L.; Pallardy, S.G. Biomass production physiology and soil carbon dynamics in short-rotation-grown *Populus deltoides* and *P. deltoides* × *P. nigra* hybrids. *For. Ecol. Manag.* **2009**, *257*, 134–142. [CrossRef]

256. Johansson, T.; Karacic, A. Increment and biomass in hybrid poplar and some practical implications. *Biomass Bioenergy* **2011**, *35*, 1925–1934. [CrossRef]

257. Zabek, L.M.; Prescott, C.E. Biomass equations and carbon content of aboveground leafless biomass of hybrid poplar in Coastal British Columbia. *For. Ecol. Manag.* **2006**, *223*, 291–302. [CrossRef]

258. Ben Brahim, M.; Gavaland, A.; Cabanettes, A. Generalized allometric regression to estimate biomass of *Populus* in short-rotation coppice. *Scand. J. For. Res.* **2000**, *15*, 171–176. [CrossRef]

259. Casella, E.; Sinoquet, H. A method for describing the canopy architecture of coppice poplar with allometric relationships. *Tree Physiol.* **2003**, *23*, 1153–1170. [CrossRef]

260. Berhongaray, G.; Janssens, I.A.; King, J.S.; Ceulemans, R. Fine root biomass and turnover of two fast-growing poplar genotypes in a short-rotation coppice culture. *Plant Soil* **2013**, *373*, 269–283. [CrossRef] [PubMed]

261. Berhongaray, G.; Verlinden, M.S.; Broeckx, L.S.; Ceulemans, R. Changes in belowground biomass after coppice in two *Populus* genotypes. *For. Ecol. Manag.* **2015**, *337*, 1–10. [CrossRef]

262. Oliveira, N.; Rodríguez-Soalleiro, R.; Pérez-Cruzado, C.; Cañellas, I.; Sixto, H.; Ceulemans, R. Above- and below-ground carbon accumulation and biomass allocation in poplar short rotation plantations under Mediterranean conditions. *For. Ecol. Manag.* **2018**, *428*, 57–65. [CrossRef]

263. Amichev, B.Y.; Johnston, M.; Van Rees, K.C.J. Hybrid poplar growth in bioenergy production systems: Biomass prediction with a simple process-based model (3PG). *Biomass Bioenergy* **2010**, *34*, 687–702. [CrossRef]

264. Headlee, W.L.; Zalesny, R.S., Jr.; Donner, D.M.; Hall, R.B. Using a process-based model (3-PG) to predict and map hybrid poplar biomass productivity in Minnesota and Wisconsin, USA. *BioEnergy Res.* **2013**, *6*, 196–210. [CrossRef]

265. Hart, Q.J.; Prilepova, O.; Merz, J.R.; Bandaru, V.; Jenkins, B.M. Modeling poplar growth as a short rotation woody crop for biofuels. UC Davis: Institute of Transportation Studies: EScholarship. University of California. 2014. Available online: https://escholarship.org/uc/item/1cc1p27b (accessed on 15 November 2020).

266. Hart, Q.J.; Tittmann, P.W.; Bandaru, V.; Jenkins, B.M. Modeling poplar growth as a short rotation woody crop for biofuels in the Pacific Northwest. *Biomass Bioenergy* **2015**, *79*, 12–27. [CrossRef]

267. Beringer, T.; Lucht, W.; Schaphoff, S. Bioenergy production potential of global biomass plantations under environmental and agricultural constraints. *Glob. Chang. Biol. Bioenergy* **2011**, *3*, 299–312. [CrossRef]

268. Rodrigues, A.; Gonçalves, A.B.; Casquilho, M.; Gomes, A.A. A GIS-based evaluation of the potential of woody short rotation coppice (SRC) in Portugal aiming at co-firing and decentralized co-generation. *Biomass Bioenergy* **2020**, *137*, 105554. [CrossRef]

269. Djomo, S.N.; De Groote, T.; Gobin, A.; Ceulemans, R.; Janssens, I.A. Combining a land surface model with life cycle assessment for identifying the optimal management of short rotation coppice in Belgium. *Biomass Bioenergy* **2019**, *121*, 78–88. [CrossRef]

270. Castaño-Díaz, M.; Álvarez-Álvarez, P.; Tobin, B.; Nieuwenhuis, M.; Afif-Khouri, E.; Cámara-Obregón, A. Evaluation of the use of low-density LiDAR data to estimate structural attributes and biomass yield in a short-rotation willow coppice: An example in a field trial. *Ann. For. Sci.* **2017**, *74*, 69. [CrossRef]

271. Sánchez-González, M.; Durbán, M.; Lee, D.-J.; Cañellas, I.; Sixto, H. Smooth additive mixed models for predicting aboveground biomass. *J. Agric. Biol. Environ. Stat.* **2017**, *22*, 23–41. [CrossRef]

272. Barrio-Anta, M.; Sixto-Blanco, H.; Viñas, I.C.-R.D.; Castedo-Dorado, F. Dynamic growth model for I-214 poplar plantations in the northern and central plateaux in Spain. *Forest Ecol. Manag.* **2008**, *255*, 1167–1178. [CrossRef]

273. de Ron, D.S.; Pérez-Cruzado, C.; González-Estevez, V.; Rodríguez-Soalleiro, R.; Cañellas, I.; Sixto, H. Visor cartográfico sobre producción de biomasa de chopo (*Populus* spp.) con fines energéticos en España. In Proceedings of the 6° Congreso Forestal Español, Vitoria-Gasteiz, Spain, 10–14 June 2013.

274. Gasol, C.M.; Gabarrell, X.; Rigola, M.; González-García, S.; Rieradevall, J. Environmental assessment:(LCA) and spatial modelling (GIS) of energy crop implementation on local scale. *Biomass Bioenergy* **2011**, *35*, 2975–2985. [CrossRef]

275. Andújar, D.; Fernández-Quintanilla, C.; Dorado, J. Matching the best viewing angle in depth cameras for biomass estimation based on poplar seedling geometry. *Sensors* **2015**, *15*, 12999–13011. [CrossRef] [PubMed]

276. Andújar, D.; Rosell-Polo, J.R.; Sanz, R.; Rueda-Ayala, V.; Fernández-Quintanilla, C.; Ribeiro, A.; Dorado, J. A LiDAR-based system to assess poplar biomass. *Gesunde Pflanz.* **2016**, *68*, 155–162. [CrossRef]

277. Oliveira, N.; Sixto, H.; Cañellas, I.; Rodríguez-Soalleiro, R.; Pérez-Cruzado, C. Productivity model and reference diagram for short rotation biomass crops of poplar grown in Mediterranean environments. *Biomass Bioenergy* **2015**, *72*, 309–320. [CrossRef]

278. Isikgor, F.H.; Becer, C.R. Lignocellulosic biomass: A sustainable platform for the production of bio-based chemicals and polymers. *Polym. Chem.* **2015**, *6*, 4497–4559. [CrossRef]

279. Sannigrahi, P.; Ragauskas, A.J.; Tuskan, G.A. Poplar as a feedstock for biofuels: A review of compositional characteristics. *Biofuel. Bioprod. Biorefin.* **2010**, *4*, 209–226. [CrossRef]

280. Bioplat. *Manual Sobre las Biorrefinerías en España*; Bioplat, Suschem y Agencia estatal de investigación del Ministerio de Economía, Industría y Competitividad: Madrid, Spain, 2017.

281. Tharakan, P.J.; Volk, T.A.; Abrahamson, L.P.; White, E.H. Energy feedstock characteristics of willow and hybrid poplar clones at harvest age. *Biomass Bioenerg* **2003**, *25*, 571–580. [CrossRef]

282. Obernberger, I.; Brunner, T.; Bärnthaler, G. Chemical properties of solid biofuels-significance and impact. *Biomass Bioenergy* **2006**, *30*, 973–982. [CrossRef]

283. Elder, T.; Groom, L.H. Pilot-scale gasification of woody biomass. *Biomass Bioenergy* **2011**, *35*, 3522–3528. [CrossRef]

284. Kloss, S.; Zehetner, F.; Dellantonio, A.; Hamid, R.; Ottner, F.; Liedtke, V.; Schwanninger, M.; Gerzabek, M.H.; Soja, G. Characterization of slow pyrolysis biochars: Effects of feedstocks and pyrolysis temperature on biochar properties. *J. Environ. Qual.* **2012**, *41*, 990–1000. [CrossRef] [PubMed]

285. Bridgwater, T. Challenges and opportunities in fast pyrolysis of biomass: Part I. *Johnson Matthey Technol. Rev.* **2018**, *62*, 118–130. [CrossRef]

286. Raud, M.; Kikas, T.; Sippula, O.; Shurpali, N.J. Potentials and challenges in lignocellulosic biofuel production technology. *Renew. Sustain. Energy Rev.* **2019**, *111*, 44–56. [CrossRef]

287. Moreno, A.D.; Alvira, P.; Ibarra, D.; Tomás-Pejó, E. Production of ethanol from lignocellulosic biomass. In *Production of Platform Chemicals from Sustainable Resources*; Springer: Berlin/Heidelberg, Germany, 2017; pp. 375–410.

288. González-García, S.; Dias, A.C.; Clermidy, S.; Benoist, A.; Bellon Maurel, V.; Gasol, C.M.; Gabarrell, X.; Arroja, L. Comparative environmental and energy profiles of potential bioenergy production chains in Southern Europe. *J. Clean. Prod.* **2014**, *76*, 42–54. [CrossRef]

289. Martínez, M.; Duret, X.; Minh, D.P.; Nzihou, A.; Lavoie, J. Conversion of lignocellulosic biomass in biobutanol by a NOVEL thermal process. *Int. J. Energy Prod. Manag.* **2019**, *4*, 298–310. [CrossRef]

290. Zhang, Y.; Xia, C.; Lu, M.; Tu, M. Effect of overliming and activated carbon detoxification on inhibitors removal and butanol fermentation of poplar prehydrolysates. *Biotechnol. Biofuels* **2018**, *11*, 178. [CrossRef]

291. Krzyżaniak, M.; Stolarski, M.J.; Warmiński, K. Life cycle assessment of poplar production: Environmental impact of different soil enrichment methods. *J. Clean. Prod.* **2019**, *206*, 785–796. [CrossRef]

292. Díaz-Ramírez, M.; Nogués, F.S.; Royo, J.; Rezeau, A. Combustion Behavior of Novel Energy Crops in Domestic Boilers: Poplar and Brassica Experiences. In *Alternative Energies*; Springer: Berlin/Heidelberg, Germany, 2013; pp. 27–45.

293. Díaz-Ramírez, M.C. *Grate-Fired Energy Crop Conversion: Experiences with Brassica Carinata and Populus sp.*; Springer: Berlin/Heidelberg, Germany, 2015.

294. Díaz-Ramírez, M.; Boman, C.; Sebastián, F.; Royo, J.; Xiong, S.; Boström, D. Environmental performance of three novel opportunity biofuels: Poplar, brassica and cassava during fixed bed combustion. In *Herbaceous Platns: Cultivation Methods, Grazing and Environmental Impacts*; Wallner, F., Ed.; Nova Science Publishers: Hauppauge, NY, USA, 2013.

295. Santos, J.I.; Fillat, Ú.; Martín-Sampedro, R.; Ballesteros, I.; Manzanares, P.; Ballesteros, M.; Eugenio, M.E.; Ibarra, D. Lignin-enriched fermentation residues from bioethanol production of fast-growing poplar and forage sorghum. *Bioresources* **2015**, *10*, 5215–5232. [CrossRef]

296. Paniagua, S.; García-Pérez, A.I.; Calvo, L.F. Biofuel consisting of wheat straw–poplar wood blends: Thermogravimetric studies and combustion characteristic indexes estimation. *Biomass Convers. Biorefin.* **2019**, *9*, 433–443. [CrossRef]

297. Serrano, D.; Coronado, J.M.; Melero, J.A. Conversion of cellulose and hemicellulose into platform molecules: Chemical routes. In *Biorefinery: From Biomass to Chemicals and Fuels*; Aresta, M., Dibenedetto, A., Dumeignil, F., Eds.; The Gruyter: Berlin/Boston, Germany, 2012; pp. 123–140.

298. Berrueco Martínez, B.; Langa Morales, E.; Maestro Tejada, C.; Mainar Fernández, A.M.; Urieta Navarro, J.S. Potencial de las yemas de chopo como fuente de compuestos bioactivos: Actividad antioxidante de sus extractos supercríticos. In Proceedings of the 5° Congreso Forestal Español, Ávila, Spain, 21–25 September 2009.

299. Martín-Sampedro, R.; Santos, J.I.; Fillat, Ú.; Wicklein, B.; Eugenio, M.E.; Ibarra, D. Characterization of lignins from *Populus alba* L. generated as by-products in different transformation processes: Kraft pulping, organosolv and acid hydrolysis. *Int. J. Biol. Macromol.* **2019**, *126*, 18–29. [CrossRef]

300. McKendry, P. Energy production from biomass (part 1): Overview of biomass. *Bioresour. Technol.* **2002**, *83*, 37–46. [CrossRef]

301. Kenney, W.A.; Sennerby-Forsse, L.; Layton, P. A review of biomass quality research relevant to the use of poplar and willow for energy conversion. *Biomass* **1990**, *21*, 163–188. [CrossRef]

302. Fernández, M.J.; Ciria, P.; Barro, R.; Losada, J.; Pérez, J.; Sixto, H.; Carrasco, J.E. Quality of the biomass produced in Short Rotation Coppices of poplar, willow, black locust and sycamore in two different spanish locations. *Biomass Resour.* **2013**, 134–139. [CrossRef]

303. Monedero, E.; Hernández, J.J.; Cañellas, I.; Otero, J.M.; Sixto, H. Thermochemical and physical evaluation of poplar genotypes as short rotation forestry crops for energy use. *Energy Convers. Manag.* **2016**, *129*, 131–139. [CrossRef]

304. Pliura, A.; Zhang, S.; MacKay, J.; Bousquet, J. Genotypic variation in wood density and growth traits of poplar hybrids at four clonal trials. *For. Ecol. Manag.* **2007**, *238*, 92–106. [CrossRef]

305. Marcos, F.; García, R.; García, F.; Godino, M.; Relova, I.; Villegas, S. Caracterización energética de la biomasa de chopo (*Populus* x *euramericana* I-214). In Proceedings of the 4° Congreso Forestal Español, Zaragoza, Spain, 26–30 September 2005.

306. Marcos, F.; Godino, M.; García, F.; Izquierdo, I.; Villegas, S. Tallares de chopo I-214 a turnos muy cortos. In Proceedings of the 4° Congreso Forestal Español, Zaragoza, Spain, 26–30 September 2005.

307. Marcos, F.; Villegas, S.; García, F.; Godino, M. Caracterización energética de la biomasa de chopo (Populus x euramericana I-214) en turnos muy cortos. *Rev. Energ.* **2006**. Available online: file:///C:/Users/MDPI/AppData/Local/Temp/16541-Texto%20del%20art%C3%ADculo-16533-1-10-20140611-1.pdf (accessed on 15 November 2020).

308. Llorente, M.J.F.; Laplaza, J.M.M.; Cuadrado, R.E.; García, J.E.C. Ash behaviour of lignocellulosic biomass in bubbling fluidised bed combustion. *Fuel* **2006**, *85*, 1157–1165. [CrossRef]

309. Bartolomé, C.; Gil, A. Ash deposition and fouling tendency of two energy crops (cynara and poplar) and a forest residue (pine chips) co-fired with coal in a pulverized fuel pilot plant. *Energy Fuel* **2013**, *27*, 5878–5889. [CrossRef]

310. Díaz-Ramírez, M.; Boman, C.; Sebastián, F.; Royo, J.; Xiong, S.; Boström, D. Ash characterization and transformation behavior of the fixed-bed combustion of novel crops: Poplar, brassica, and cassava fuels. *Energy Fuel* **2012**, *26*, 3218–3229. [CrossRef]

311. Díaz-Ramírez, M.; Sebastián, F.; Royo, J.; Rezeau, A. Influencing factors on $NO_X$ emission level during grate conversion of three pelletized energy crops. *Appl. Energy* **2014**, *115*, 360–373. [CrossRef]

312. Cai, J.; He, Y.; Yu, X.; Banks, S.W.; Yang, Y.; Zhang, X.; Yu, Y.; Liu, R.; Bridgwater, A.V. Review of physicochemical properties and analytical characterization of lignocellulosic biomass. *Renew. Sustain. Energy Rev.* **2017**, *76*, 309–322. [CrossRef]

313. Vassilev, S.V.; Baxter, D.; Andersen, L.K.; Vassileva, C.G. An overview of the chemical composition of biomass. *Fuel* **2010**, *89*, 913–933. [CrossRef]

314. Bartolomé, C.; Gil, A. Emissions during co-firing of two energy crops in a PF pilot plant: Cynara and poplar. *Fuel Process. Technol.* **2013**, *113*, 75–83. [CrossRef]

315. González-García, S.; Moreira, M.T.; Feijoo, G.; Murphy, R.J. Comparative life cycle assessment of ethanol production from fast-growing wood crops (black locust, eucalyptus and poplar). *Biomass Bioenergy* **2012**, *39*, 378–388. [CrossRef]

316. Dallemand, J.F.; Petersen, J.E.; Karp, A. *Short Rotation Forestry, Short Rotation Coppice and Perennial Grasses in the European Union: Agro-Environmental Aspects, Present Use and Perspectives*; JRC Scientific and Technical Reports: Luxembourg, 2007; pp. 106–111.

317. Dimitriou, I.; Rutz, D. *Sustainability Criteria and Recommendations for Short Rotation Woody Crops*; IEE Project SRCplus: Munich, Germany, 2014.

318. Zalesny, R.S., Jr.; Donner, D.M.; Coyle, D.R.; Headlee, W.L. An approach for siting poplar energy production systems to increase productivity and associated ecosystem services. *For. Ecol. Manag.* **2012**, *284*, 45–58. [CrossRef]

319. Schweier, J.; Molina-Herrera, S.; Ghirardo, A.; Grote, R.; Díaz-Pinés, E.; Kreuzwieser, J.; Haas, E.; Butterbach-Bahl, K.; Rennenberg, H.; Schnitzler, J.P. Environmental impacts of bioenergy wood production from poplar short-rotation coppice grown at a marginal agricultural site in Germany. *Bioenergy* **2017**, *9*, 1207–1221. [CrossRef]

320. El Kasmioui, O.; Ceulemans, R. Financial analysis of the cultivation of poplar and willow for bioenergy. *Biomass Bioenergy* **2012**, *43*, 52–64. [CrossRef]

321. Manzone, M.; Bergante, S.; Facciotto, G. Energy and economic evaluation of a poplar plantation for woodchips production in Italy. *Biomass Bioenergy* **2014**, *60*, 164–170. [CrossRef]

322. Gasol, C. Environmental and Economic Integrated Assessment of Local Energy Crops Production in Southern Europe. Ph.D. Thesis, Universitat Autònoma de Barcelona, Barcelona, Spain, 2009.

323. González-García, S.; Gasol, C.M.; Gabarrell, X.; Rieradevall, J.; Moreira, M.T.; Feijoo, G. Environmental profile of ethanol from poplar biomass as transport fuel in Southern Europe. *Renew. Energy* **2010**, *35*, 1014–1023. [CrossRef]

324. Blanc, S.; Gasol, C.M.; Martínez-Blanco, J.; Muñoz, P.; Coello, J.; Casals, P.; Mosso, A.; Brun, F. Economic profitability of agroforestry in nitrate vulnerable zones in Catalonia (NE Spain). *Span. J. Agric. Res.* **2019**, *17*, 16. [CrossRef]

325. Butnar, I.; Rodrigo, J.; Gasol, C.M.; Castells, F. Life-cycle assessment of electricity from biomass: Case studies of two biocrops in Spain. *Biomass Bioenergy* **2010**, *34*, 1780–1788. [CrossRef]

326. Gasol, C.M.; Gabarrell, X.; Anton, A.; Rigola, M.; Carrasco, J.; Ciria, P.; Rieradevall, J. LCA of poplar bioenergy system compared with *Brassica carinata* energy crop and natural gas in regional scenario. *Biomass Bioenergy* **2009**, *33*, 119–129. [CrossRef]

327. Relaño, R.L.; Esteban, E.T.; Rodríguez, S.J.H. Economic and Life Cycle Assessment of integrated wood and chips harvesting from hybrid poplar plantations in the Genil Valley (Spain). Comparison with chips harvesting from Poplar SRCs. In Proceedings of the From Theory to Practice: Challenges for Forest Engineering, Warsaw, Poland, 4–7 September 2016; p. 249.

328. Gasol, C.M.; Martínez, S.; Rigola, M.; Rieradevall, J.; Anton, A.; Carrasco, J.; Ciria, P.; Gabarrell, X. Feasibility assessment of poplar bioenergy systems in the Southern Europe. *Renew. Sustain. Energy Rev.* **2009**, *13*, 801–812. [CrossRef]

329. Li, W.; Ciais, P.; Makowski, D.; Peng, S. A global yield dataset for major lignocellulosic bioenergy crops based on field measurements. *Sci. Data* **2018**, *5*, 180169. [CrossRef] [PubMed]

330. Don, A.; Osborne, B.; Hastings, A.; Skiba, U.; Carter, M.S.; Drewer, J.; Flessa, H.; Freibauer, A.; Hyvönen, N.; Jones, M.B. Land-use change to bioenergy production in Europe: Implications for the greenhouse gas balance and soil carbon. *Glob. Chang. Biol. Bioenergy* **2012**, *4*, 372–391. [CrossRef]

331. Dimitriou, I.; Rutz, D. *Sustainable Short Rotation Coppice: A Handbook*; Rutz, D., Ed.; WIP Renewable Energies: Munich, Germany, 2015.

332. Rowe, R.L.; Street, N.R.; Taylor, G. Identifying potential environmental impacts of large-scale deployment of dedicated bioenergy crops in the UK. *Renew. Sustain. Energy Rev.* **2009**, *13*, 271–290. [CrossRef]

333. Griffiths, N.A.; Rau, B.M.; Vaché, K.B.; Starr, G.; Bitew, M.M.; Aubrey, D.P.; Martin, J.A.; Benton, E.; Jackson, C.R. Environmental effects of short-rotation woody crops for bioenergy: What is and isn't known. *GCB-Bioenergy* **2019**, *11*, 554–572. [CrossRef]

334. Pilipović, A.; Orlović, S.; Rončević, S.; Nikolić, N.; Župunski, M.; Spasojević, J. Results of selection of poplars and willows for water and sediment phytoremediation. *Agric. For.* **2015**, *61*, 205. [CrossRef]

335. Zalesny, R.S., Jr.; Headlee, W.L.; Gopalakrishnan, G.; Bauer, E.O.; Hall, R.B.; Hazel, D.W.; Isebrands, J.G.; Licht, L.A.; Negri, M.C.; Nichols, E.G. Ecosystem services of poplar at long-term phytoremediation sites in the Midwest and Southeast, United States. *Willey Interdiscip. Rev. Energy Environ.* **2019**, *8*, e349. [CrossRef]

336. Pandey, V.C.; Bajpai, O.; Singh, N. Energy crops in sustainable phytoremediation. *Renew. Sustain. Energy Rev.* **2016**, *54*, 58–73. [CrossRef]

337. Menéndez, J.; Loredo, J. Use of Reclaimed Surface Mines for Heat and Power Production. *Int. J. Renew. Energy Res.* **2018**, *3*, 2367–9123.

338. Castaño-Díaz, M.; Barrio-Anta, M.; Afif-Khouri, E.; Cámara-Obregón, A. Willow short rotation coppice trial in a former mining area in Northern Spain: Effects of clone, fertilization and planting density on yield after five years. *Forests* **2018**, *9*, 154. [CrossRef]

339. de Bustamante, I.; Lillo, J.; Hernández, J.; Leal, M.; Meffe, R.; de Santiago, A.; Martínez-Hernández, V. El chopo como materia prima e instrumento medioambiental. In Proceedings of the II Simposio del chopo, Valladolid, Spain, 17–19 October 2018.

340. Rodríguez-Soalleiro, R.; Eimil-Fraga, C.; Gómez-García, E.; García-Villabrille, J.D.; Rojo-Alboreca, A.; Muñoz, F.; Oliveira, N.; Sixto, H.; Pérez-Cruzado, C. Exploring the factors affecting carbon and nutrient concentrations in tree biomass components in natural forests, forest plantations and short rotation forestry. *For. Ecosyst.* **2018**, *5*, 35. [CrossRef]

341. Oliveira, N.; Fuertes, A.; González, I.; de la Iglesia, J.P.; Cañellas, I.; Rodríguez-Soalleiro, R.; Sixto, H. Cultivo de chopo para la producción de biomasa: Estrategia de mitigación mediante la acumulación de carbono en las diferentes fracciones. In Proceedings of the VIII Remedia Workshop. Economía Circular como Catalizador de la Sostenibilidad Medioambiental del Sector Primario Español, Elche, Spain, 22–23 September 2020.

342. Rodríguez, F.; Serrano, L.; Aunós, A. El papel del chopo como sumidero de $CO_2$ atmosférico. In Proceedings of the 4° Congreso Forestal Español, Zaragoza, Spain, 26–30 September 2005.

343. Djomo, S.N.; Kasmioui, O.E.; Ceulemans, R. Energy and greenhouse gas balance of bioenergy production from poplar and willow: A review. *Glob. Chang. Biol. Bioenergy* **2011**, *3*, 181–197. [CrossRef]

344. Pérez-Cruzado, C.; Mohren, G.M.; Merino, A.; Rodríguez-Soalleiro, R. Carbon balance for different management practices for fast growing tree species planted on former pastureland in southern Europe: A case study using the $CO_2$ Fix model. *Eur. J. For. Res.* **2012**, *131*, 1695–1716. [CrossRef]

345. Pérez-Cruzado, C.; Mansilla-Salinero, P.; Rodríguez-Soalleiro, R.; Merino, A. Influence of tree species on carbon sequestration in afforested pastures in a humid temperate region. *Plant Soil* **2012**, *353*, 333–353. [CrossRef]

346. Verlinden, M.S.; Broeckx, L.S.; Zona, D.; Berhongaray, G.; De Groote, T.; Serrano, M.C.; Janssens, I.A.; Ceulemans, R. Net ecosystem production and carbon balance of an SRC poplar plantation during its first rotation. *Biomass Bioenergy* **2013**, *56*, 412–422. [CrossRef]

347. Ceulemans, R.; Horemans, J.A.; Vanbeveren, S.P.P. Short-rotation crops reduce greenhouse gas emissions and contribute to climate change mitigation. In Proceedings of the 7th International Poplar Symposium. New Bioeconomies: Exploring the potential role of Salicaceae, Buenos Aires, Argentina, 28 October–4 November 2018.

348. Folch, A.; Ferrer, N. The impact of poplar tree plantations for biomass production on the aquifer water budget and base flow in a Mediterranean basin. *Sci. Total Environ.* **2015**, *524*, 213–224. [CrossRef]

349. MINETAD. *La Energía en España 2016*; Ministerio de Energía, Turismo y Agenda Digital: Madrid, Spain, 2017; p. 342.

350. Unión por la Biomasa. *Balance Socioeconómico de las Biomasas en ESPAÑA 2017–2021*; Unión por la Biomasa: Spain, 2018; p. 30. Available online: https://www.interempresas.net/Energia/Articulos/219726-Union-por-Biomasa-presenta-\T1\textquorightBalance-socioeconomico-de-biomasas-en-Espana-2017-2021\T1\textquoteright.html (accessed on 15 November 2020).

351. IDAE. *Biomasa: Cultivos Energéticos*; Instituto para la Diversificación y Ahorro de la Energía: Madrid, Spain, 2007.

352. Eurostat. *Renewable Energy Statistics*; European Commission: Brussels, Belgium, 2018.

353. WBA. *Global Bioenergy Statistics*; World Bioenergy Association: Stockholm, Sweden, 2019; p. 58.

354. APPA. *Estudio del Impacto Macroeconómico de las ENERGÍAS Renovables en España*; Asociación de Empresas de Energías Renovables: Spain, 2018; Available online: https://www.appa.es/wp-content/uploads/2018/10/Estudio_del_impacto_Macroeconomico_de_las_energias_renovables_en_Espa%C3%B1a_2017.pdf (accessed on 15 November 2020).

 *forests*

*Article*

# Biomass Yield of 37 Different SRC Poplar Varieties Grown on a Typical Site in North Eastern Germany

**Dirk Landgraf** [1,*], **Christin Carl** [1] and **Markus Neupert** [2]

[1]  Forestry and Ecosystem Management, Erfurt University of Applied Sciences, Leipziger Strasse 77, 99085 Erfurt, Germany; christin.carl@fh-erfurt.de
[2]  ECODIV URA INRAE/EA 1293, Rouen Normandy University, Place Emile Blondel, 76821 Mont-Saint-Aignan, France; markus.neupert1@univ-rouen.fr
*  Correspondence: dirk.landgraf@fh-erfurt.de; Tel.: +49-361-6700-295

Received: 5 August 2020; Accepted: 24 September 2020; Published: 28 September 2020

**Abstract:** A total of 37 different poplar varieties were grown in a randomized mini-rotation short rotation coppice (SRC) (harvest every three years) on a light sandy soil under continental climatic conditions in the south of the Federal State of Brandenburg, Germany. Along with well-known poplar varieties, newly bred ones that have not yet been approved for commercial use were selected for this study. Survival rates were determined after the first growing season in 2013 as well as at the first and second harvests in 2015 and 2018. Furthermore, the number of shoots, plant height, diameter at breast height, dry matter content and biomass yield of the varieties were recorded. After the second rotation period, only seven poplar varieties yielded more than 11 $t_{adm}$ ha$^{-1}$ y$^{-1}$ and can be recommended for commercial use. However, many varieties only reached about 8 $t_{adm}$ ha$^{-1}$ y$^{-1}$, and six varieties even had less than 4 $t_{adm}$ ha$^{-1}$ y$^{-1}$, among them newly bred varieties. Given the changing climate conditions, the cultivation of these varieties in SRC is not recommended. Our data also show that the biomass yield of several varieties decreased from the first to the second harvests. Since the survival rates were high and no damage by pest species was observed, the site-specific yield capacities of the individual clones are assumed to be the cause for this.

**Keywords:** bioenergy plantation; woody biomass; *Populus*; renewable energy

## 1. Introduction

The use of renewable raw materials is a sustainable and regionally sensible alternative to the continuing use of fossil raw materials. Thus, the European Union supports the transition to a low-carbon energy economy and has set a 27% target for the total share of energy from renewable sources by 2030 [1]. Fast-growing tree species play a particularly important and sustainable role as renewable raw materials in different land use systems, e.g., short rotation coppices (SRC) and agroforestry systems (AFS) [2–6]. They help reduce $CO_2$ emissions by substituting fossil fuels or the production of biofuels and thereby help to mitigate climate change [7].

SRC crops are defined as high-density plantations of fast-growing trees, managed in different rotation times. Commonly, three different types of rotation periods are distinguished: mini-rotation represents an interval of 2–4 years, midi-rotation of 5–10 years and maxi-rotation of 11–20 years [1,8–10]. Depending on the production target and the associated rotation variant, as well as the particular site conditions, different tree species are used in Europe. From a legal point of view, several species that are capable of resprouting can be grown in the aforementioned farming systems [11]. For a successful management of these types of plantations on a commercial scale, it is essential to maximize the benefits by combining economic viability with environmental sustainability. Over time, three tree species have become the most common ones used for the fast production of woody biomass. Black locust

(*Robinia pseudoacacia* L.) is mainly grown on poor, sandy sites with little rainfall in southern regions of Europe [12–14], whereas willow (*Salix* spec.) and its varieties are preferred in the north, in much more humid and colder regions [15–17]. However, the most common tree species in European SRC and AFS is poplar (*Populus* spec.) and its varieties. There are numerous examples of scientific studies, but their practical and commercial use in different rotation types has also been examined in many different regions in Europe [2,18–22].

In Germany, farmers mainly use sites with low yield expectations to establish SRC and AFS. Such sites are especially abundant in the Federal State of Brandenburg, in which 2000 ha, that is about one third of Germany's SRC, are located [23]. In recent years, the interest in fast-growing tree species, especially in poplar, has increased and new varieties have been brought onto the market. However, these often originated from Southern Europe, e.g., Italy, where climatic and site conditions are more favorable than in Germany, for example, due to higher temperatures, a longer vegetation period, less or no frost periods and a better water supply. For this reason, it was of particular interest to study the growth of the new Italian AF poplar varieties, as well as the new Matrix poplar varieties from Germany under the specific conditions occurring in the Federal State of Brandenburg, and compare them to older varieties.

The aim of this study was to record and compare the major growth parameters of 37 poplar varieties over a period of six years, including two harvests. These included (I) the survival rate, (II) the resprouting capacity, (III) the plant height, (IV) the diameter at breast height, (V) the dry matter content and (VI) the biomass yield. Finally, the suitability of these varieties for SRC and AFS was evaluated.

## 2. Materials and Methods

### 2.1. Site Description, Experimental Design and Plant Material

The study site is located near Großthiemig in the northeastern part of Germany (latitude N 51°23′52.9″, longitude 13°40′11.6″ E, 43 m a. s. l.). The local climate is mainly a typical inland climate but with a noticeable transition to the continental climate. The mean annual temperature is 8.6 °C with an average annual precipitation of 561 mm [24]. The soil has a sandy texture and consists of 4.8% of clay, 2.9% of silt and 92.3% of sand with a lightly acidic to neutral pH of 5.9 and an organic matter content of about 1.5%. According to Hartmann [25], the soil type is a Gleyic Cambisol.

In the spring of 2012, a randomized field trial including the 37 poplar varieties was established. Most varieties were typical varieties available on the German market in 2012 and bought from the P & P forest nursery (Eitelborn, Germany [E 7°42′ N 50°22′]). Newly bred varieties were obtained from the Italian breeder Alasia Franco Vivai (AFV) (Savigliano, Italy [E 7°38′ N 44°36′]), and the Thünen Institute (TI) (Großhansdorf, Germany [E 10°15′ N 53°39′]; 4 × Göttingen variety bred by the University of Göttingen, Germany, with material originating from INRA Bordeaux, France). Table 1 provides an overview of all poplar varieties included in this study.

Except for the varieties from TI, unrooted cuttings with a length of 20 cm and a minimal diameter of 1 cm were used and placed into the soil manually and evenly with the ground. The three TI varieties (P1, 4 × Göttingen and Esch 5) were delivered as rooted plants in pots, which were removed before planting. Every poplar variety was established in three plots with 33 individuals each, with a random distribution of plots. In total, 32 rows with a row spacing of 1.5 m were created. The distance between plants within a row was 0.5 m. Thus, a theoretical number of 13.333 poplars per hectare were planted. To ensure the success of the establishment, two mechanical weed control measures were carried out during the first growing season.

**Table 1.** Poplar varieties included in this study and, if known, their parentage (*n* = 99 per variety).

| Variety | Parentage | Sex | Source |
|---|---|---|---|
| Androscoggin | *P. maximowiczii* × *P. trichocarpa* | m | |
| Fritzi Pauley | *P. trichocarpa* | f | |
| Harff | *P. × euramericana* | f | |
| Heidemij | *P. × euramericana* | m | |
| I 214 | *P. deltoides* × *P. nigra* | f | |
| Isières | *P. × euramericana* | m | |
| Jacometti 78 B | *P. × euramericana* | f | |
| Koltay | *P. × euramericana* | m | |
| Kopecky | *P. × euramericana* | m | |
| Matrix 24 | *P. maximowiczii* × *P. trichocarpa* | | |
| Matrix 49 | *P. maximowiczii* × *P. trichocarpa* | | P & P |
| Max 1 | *P. nigra* × *P. maximowiczii* | f | |
| Max 3 | *P. nigra* × *P. maximowiczii* | f | |
| Max 4 | *P. nigra* × *P. maximowiczii* | f | |
| Monviso | *P. generosa* × *P. nigra* | f | |
| Muhle Larsen | *P. trichocarpa* | f | |
| NE 42 * | *P. maximowiczii* × *P. trichocarpa* | m | |
| Pannonia | *P. × euramericana* | f | |
| Rochester | *P. maximowicii* × *P. nigra* | f | |
| Robusta | *P. × euramericana* | m | |
| Weser 6 | *P. trichocarpa* | | |
| AF 2 | *P. × canadensis* Moench | m | |
| AF 6 | *P. generosa* × *P. nigra* A. Henry | f | |
| AF 8 | *P. trichocarpa* × *P. × generosa* | f | |
| AF 13 | *P. × canadensis* | | |
| AF 15 | *P. deltoides* × *P. nigra* | | |
| AF 16 | *P. × canadensis* | | |
| AF 17 | *P. deltoides* × *P. nigra* | | AFV |
| AF 18 | *P. deltoides* × *P. nigra* | | |
| AF 19 | *P. deltoides* × *P. nigra* | | |
| AF 20 | *P. deltoides* × *P. nigra* | | |
| AF 24 | *P. deltoides* × *P. nigra* | | |
| AF 27 | *P. deltoides* × *P. nigra* | | |
| AF 28 | *P. deltoides* × *P. nigra* | | |
| P1 | *P. × canescens* | | TI |
| 4 × Göttingen | *P. × canescens* (tetraploid) | | (UG/INRA |
| Esch 5 | *P. tremula* × *P. tremuloides* | | 717-1B4) |

* syn. Hybride 275.

## 2.2. Determination of Growth Parameters

After the first growing season, the survival rate was determined for all poplar varieties in January 2013. This was repeated at the time of the first harvest in January 2015 and the second harvest in January 2018. At each harvest, the number of shoots per stool was counted, and plant height and diameter at breast height (DBH) were measured with a diameter measurement tape at a height of 1.3 m on each individual shoot. The dry matter content (DMC) of each variety was determined by harvesting three trees of each plot, that is nine trees per variety in total. A sample was taken from each of these trees from the lower, middle and upper sections. These samples were grounded and weighed. Fresh weight after harvest and dry weight after dehydration were measured using an electronic scale with a spring balance (precision ± 1 g). Dry weight was determined by taking a representative, plot-based subsample of each sample and drying it for 48 h in the laboratory at 103.5 °C (DIN 52183 [24]) until a constant weight was reached [25,26]. This subsample was used to estimate the total value of dry woody biomass for each sample by creating a power function for each poplar variety [27] and calculating the dry woody biomass per shoot by means of the DBH. The Y = ax^b power function best fit the curve when using the method of least squares. Once the biomass of each shoot was calculated, we were able to estimate the biomass of each plot. The biomass of each tree is the sum of the biomass of all its shoots, and the plot biomass is therefore the sum of the biomass of all its trees. Thus, the survival rate of each plot was also taken into account. Using the number of trees per hectare, we converted the average total biomass per plot into the biomass per hectare (tons absolute dry biomass, $t_{adm}$ ha$^{-1}$). The dry matter biomass yield (DBY) ($t_{adm}$ ha$^{-1}$ y$^{-1}$) resulted from dividing the biomass per hectare by the number of years between each harvest.

### 2.3. Data Analysis

The collected data were analyzed for normality using the Shapiro–Wilk test ($\alpha = 0.05$). None of the data of the growth parameters followed a Gaussian distribution, so the differences between poplar varieties were analyzed using the non-parametric Kruskal–Wallis test ($\alpha = 0.05$). Fisher's least significant difference procedure was used as post hoc test. Correlation between the six variables was tested using Pearson's product moment coefficient ($\alpha = 0.05$). In addition, we tested the correlation between parameters and the actual plot position. No evidence for spatial autocorrelation was found. All analyses were carried out using R 3.6.3 [28] and the packages *tidyverse* [29] and *ggcorrplot* [30].

## 3. Results

### 3.1. Survival Rate

Out of the 37 poplar varieties, 26 varieties had a survival rate of ≥90% and 33 varieties of ≥80%. Thus, the overall survival rate after the first growing season was very high (Table 2). Understandably, the three TI varieties, which came in pots, did particularly well with survival rates between 93% and 100%. However, with up to 98%, for Max 4, similar survival rates were also achieved with unrooted cuttings. With 56% and 57%, the AF 20 and AF 19 varieties had particularly low survival rates. At the first harvest in January 2015, the survival rate of some varieties had drastically decreased. The reduction was most severe for AF 2, AF 24 and AF 15 with a decrease of 53%, 49% and 42%. Even though not as many AF 19 and AF 20 individuals had died, the survival rate was nevertheless reduced to a very low 37% and 17%, respectively. For other varieties, the decrease in individuals was not as drastic but still remarkable. For example, the survival rate of the Hungarian varieties Kopecky, Koltay and Pannonia was reduced by 35%, and 29% for the latter. In contrast, Isières, Max 1 and Max 4 performed very well with a decrease in the survival rate of only 1%, 2% and 3%. In comparison to the first rotation period, there were only marginal decreases in the survival rate of a few varieties.

**Table 2.** Survival rate of the poplar varieties at the end of the first growing season (2013), after three years (2015) and after six years (2018).

| | Survival Rate [%] | | |
|---|---|---|---|
| **Variety** | **2013** | **2015** | **2018** |
| Max 4 | 98 | 95 | 95 |
| Fritzi Pauley | 84 | 80 | 80 |
| Max 3 | 95 | 83 | 83 |
| AF 18 | 96 | 84 | 84 |
| AF 16 | 77 | 72 | 72 |
| AF 17 | 95 | 86 | 86 |
| Matrix 24 | 91 | 72 | 67 |
| Isières | 98 | 97 | 97 |
| Max 1 | 97 | 95 | 95 |
| AF 28 | 81 | 52 | 52 |

Table 2. *Cont.*

| Variety | Survival Rate [%] | | |
|---|---|---|---|
| | 2013 | 2015 | 2018 |
| AF 13 | 90 | 54 | 54 |
| Weser 6 | 96 | 87 | 87 |
| Matrix 49 | 89 | 81 | 81 |
| AF 27 | 95 | 62 | 62 |
| Pannonia | 93 | 64 | 63 |
| Monviso | 93 | 61 | 61 |
| Rochester | 95 | 82 | 82 |
| Koltay | 91 | 56 | 52 |
| P1 | 100 | 96 | 96 |
| Heidemij | 94 | 61 | 61 |
| AF 24 | 94 | 45 | 45 |
| NE 42 | 96 | 87 | 87 |
| AF 8 | 83 | 52 | 52 |
| Harff | 94 | 80 | 76 |
| Robusta | 90 | 75 | 75 |
| AF 19 | 57 | 37 | 37 |
| Kopecky | 96 | 61 | 60 |
| Jacometti 78 B | 88 | 72 | 72 |
| AF 2 | 96 | 43 | 43 |
| AF 6 | 85 | 46 | 46 |
| 4 × Göttingen | 98 | 85 | 85 |
| Androscoggin | 83 | 59 | 59 |
| I 214 | 96 | 62 | 62 |
| AF 15 | 91 | 49 | 48 |
| Muhle Larsen | 73 | 52 | 52 |
| Esch 5 | 93 | 74 | 74 |
| AF 20 | 56 | 17 | 17 |

## 3.2. Resprouting Capacity

In the first rotation period from 2012 to 2015, about 99% of the established poplars had only one shoot. After the first harvest, the number of shoots per stool increased and averaged 1.59 shoots for all poplar varieties (Figure 1). The highest median with 2.59 shoots per stool was recorded for Max 3, followed by Max 4 (2.46) and Heidemij (2.12). While these three varieties did not show any peak values regarding the number of shoots per stool, AF 17 had a maximum number of 13 shoots, followed by Matrix 24 with 11 shoots and Rochester, Isières, Weser 6 and I 214 with 10 shoots. The statistical analysis highlights the high number of shoots of Max 3 and Max 4, whereas, in general, no distinct significant differences were recorded for individual varieties or groups of varieties.

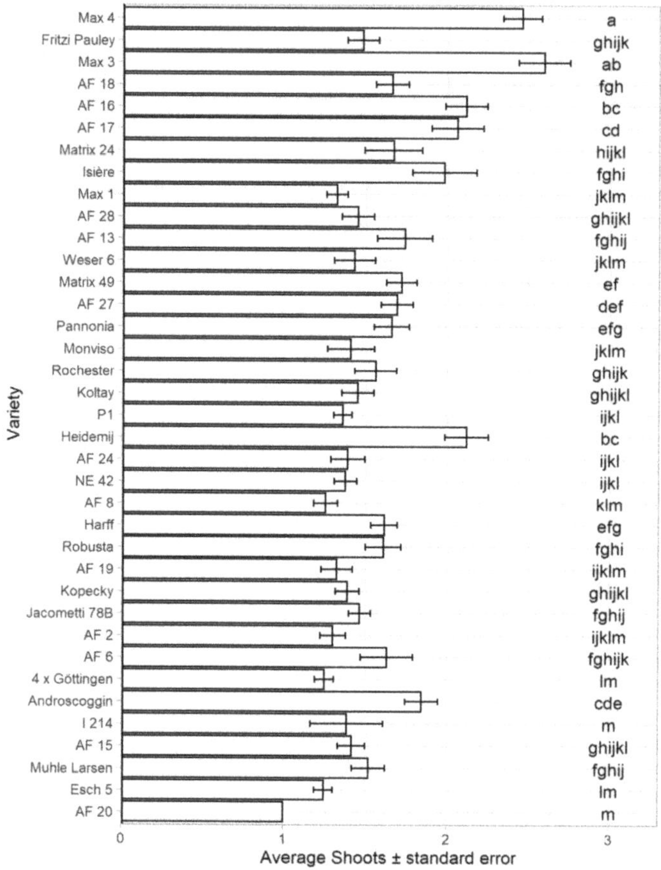

**Figure 1.** Average (±standard error) number of shoots of the poplar varieties in the winter of 2018, six years after planting and three years after the first harvest. Statistical analysis was carried out using the non-parametric Kruskal–Wallis test ($\alpha$ = 0.05, $n \in$ [17;95]) paired with Fisher's least significant difference procedure to highlight statistically different varieties; the letters on the right-hand side of the figure represent the statistical groups to which each variety belongs.

*3.3. Plant Height*

At the end of the first growing season, the 37 poplar varieties reached an average height of 1.60 m. The greatest plant heights were achieved by AF 17 with 2.35 m, followed by I 214 (2.20 m) and Max 4 (1.98 m). In contrast, Jacometti 78 B (1.12 m), Rochester (1.11 m) and AF 2 (1.09 m) showed only a weak height growth in the year of establishment. After three years, in 2015, a median height of 6.29 m was recorded for all poplar varieties. AF 13 (7.95 m), Weser 6 (7.76 m) and Fritzi Pauley (7.75 m) were the most successful ones in terms of height gain, while 4 × Göttingen (4.04 m), AF 20 (4.78 m) and Robusta (5.00 m) had achieved the lowest plant height after three growing seasons (Figure 2). In the second rotation period from 2015 to 2018, the average height of the poplar varieties only increased by 0.19 m to 6.48 m. Compared to the first rotation period, 20 varieties showed an increase in height growth at the end of the second rotation period, and 17 varieties a decrease. AF 19 had the greatest increase in height, which was 1.83 m higher than at the end of the first rotation period, followed by AF 13 and AF 24. The lowest plant height after two rotation periods was recorded for Esch 5 (4.49 m), Muhle Larsen (5.12 m) and AF 15 (5.34 m).

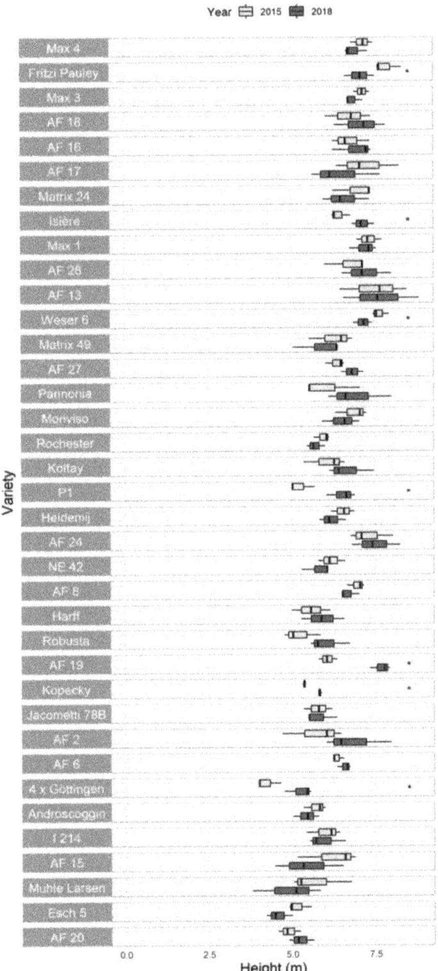



**Figure 2.** Plant height of the poplar varieties in the winters of 2015 and 2018. For each variety, the difference between the height in 2015 and in 2018 was analyzed using the non-parametric Kruskal–Wallis test ($\alpha = 0.05$, $n = 3$). Statistically significant differences are marked with an asterisk.

## 3.4. Diameter at Breast Height (DBH)

The DBH of the 37 poplar varieties reached an average of 1.0 cm after the first growing season. AF 19 achieved the highest DBH (1.2 cm) followed by AF 13 and AF 17 (both 1.1 cm). Esch 5, Muhle Larsen and Rochester had the lowest DBH with 0.6 cm. By the end of the first rotation period in 2015, the varieties achieved an average DBH of 3.3 cm. The greatest increases in DBH were recorded for AF 13 with 4.7 cm, followed by Fritzi Pauley (4.4 cm) and Max 1 with 4.3 cm. After the second rotation period in 2018, the average DBH decreased by 0.9 cm to 2.4 cm. This decrease applied to all poplar varieties. The greatest DBH at the end of the second rotation period was reached by Max 1 with 4.7 cm, followed by AF 19 and AF 13 with 4.4 cm. The lowest increase in DBH was 1.1 cm for Esch 5, followed by Muhle Larsen (1.6 cm) and Androscoggin (1.7 cm) (Figure 3).

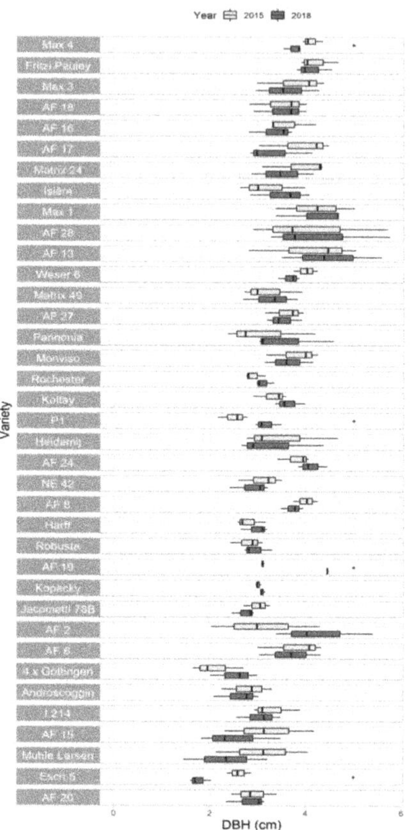



**Figure 3.** Diameter at breast height of the poplar varieties in the winters of 2015 and 2018. For each variety, the difference between the diameter in 2015 and in 2018 was analyzed using the non-parametric Kruskal–Wallis test ($\alpha = 0.05$, $n = 3$). Statistically significant differences are marked with an asterisk.

### 3.5. Dry Matter Content (DMC)

After three years, in 2015, the DMC averaged 45.2%, with large differences among the 37 poplar varieties. The greatest DMC was recorded for Rochester with 50.7%, followed by NE 42 (49.5%) and Androscoggin (48.0%). AF 28 had the lowest DMC with 42.2%, followed by Koltay (43.3%) and Max 1 (43.4%). With 44.5%, the average DMC after the second rotation period was similar to that in the first rotation period. Again, Androscoggin (54.6%), NE 42 (49.5%) and Rochester (50.7%) reached the highest DMC values. The lowest DMC was recorded for AF 27 with 38.1%, followed by AF 17 (40.5%) and AF 2 (40.2%).

### 3.6. Dry Matter Biomass Yield (DBY)

After the first rotation period, in 2015, Fritzi Pauley reached the highest yield with 14.6 $t_{adm}$ $ha^{-1}$ $y^{-1}$, followed by Weser 6 (14.3 $t_{adm}$ $ha^{-1}$ $y^{-1}$), Max 3 (12.4 $t_{adm}$ $ha^{-1}$ $y^{-1}$) and Max 4 (12.0 $t_{adm}$ $ha^{-1}$ $y^{-1}$) (Figure 4). Moderate yields between 6.8 and 6.9 $t_{adm}$ $ha^{-1}$ $y^{-1}$ were recorded for Pannonia, AF 27 and Matrix 49. AF 20 (2.4 $t_{adm}$ $ha^{-1}$ $y^{-1}$), 4 × Göttingen (2.8 $t_{adm}$ $ha^{-1}$ $y^{-1}$) and AF 19 (3.2 $t_{adm}$ $ha^{-1}$ $y^{-1}$) had the lowest yields. After the second rotation period, in 2018, the highest yield was found for Max 4 with 15.8 $t_{adm}$ $ha^{-1}$ $y^{-1}$. Fritzi Pauley (14.5 $t_{adm}$ $ha^{-1}$ $y^{-1}$) and Max 3 (13.3 $t_{adm}$ $ha^{-1}$ $y^{-1}$) reached only slightly lower yields. With 8.1 $t_{adm}$ $ha^{-1}$ $y^{-1}$, P1, Pannonia and Koltay showed moderate yields. The lowest yields were recorded for AF 20 (0.8 $t_{adm}$ $ha^{-1}$ $y^{-1}$), followed by Esch 5 (1.2 $t_{adm}$ $ha^{-1}$ $y^{-1}$)

and AF 15 (2.5 $t_{adm}$ $ha^{-1}$ $y^{-1}$). Between the first and the second rotation periods, 19 varieties were able to increase their yield, whereas 18 varieties showed a yield reduction. The increases were greatest for AF 18 (4.8 $t_{adm}$ $ha^{-1}$ $y^{-1}$), AF 16 (4.1 $t_{adm}$ $ha^{-1}$ $y^{-1}$) and Max 4 (3.8 $t_{adm}$ $ha^{-1}$ $y^{-1}$). The greatest decreases were recorded for Weser 6 (−5.4 $t_{adm}$ $ha^{-1}$ $y^{-1}$), Max 1 (−4.1 $t_{adm}$ $ha^{-1}$ $y^{-1}$) and Esch 5 (−3.0 $t_{adm}$ $ha^{-1}$ $y^{-1}$).

The statistical analysis does not show any distinct significant differences of individual varieties but allows for dividing the varieties into the groups of high-yielding (a–c) and low-yielding (i–k) varieties (Figure 5).

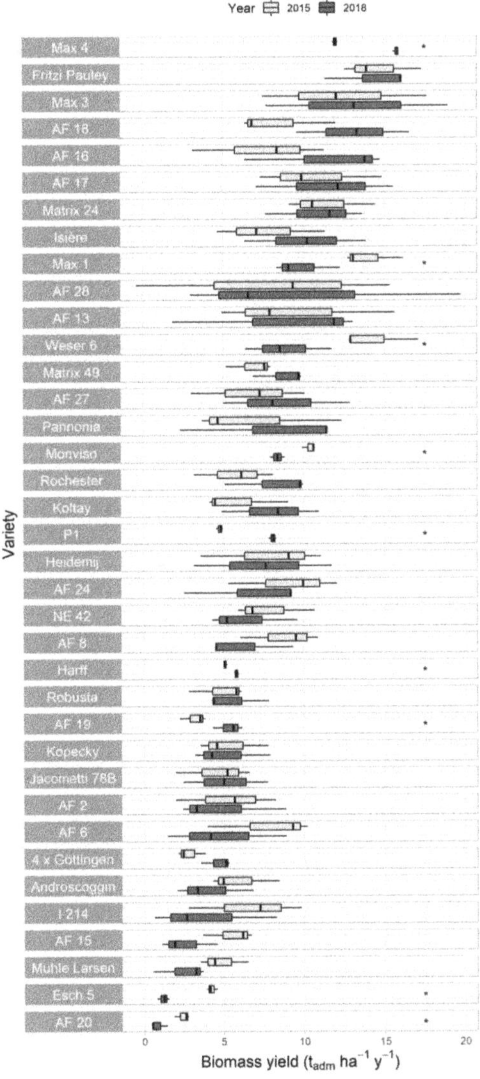


**Figure 4.** Biomass yield of the poplar varieties in the winters of 2015 and 2018. For each variety, the difference between the biomass yield in 2015 and in 2018 was analyzed using the non-parametric Kruskal–Wallis test ($\alpha = 0.05$, $n = 3$). Statistically significant differences are marked with an asterisk.

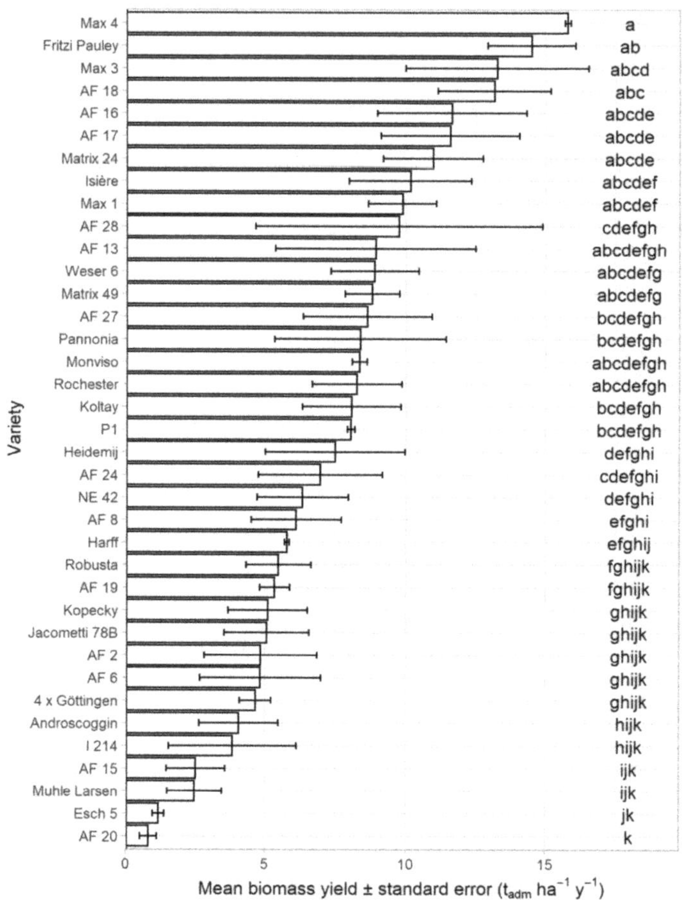



**Figure 5.** Average (±standard error) biomass yield of the poplar varieties in the winter of 2018, six years after planting and three years after the first harvest. Statistical analysis was carried out using the non-parametric Kruskal–Wallis test ($\alpha = 0.05$, $n \in [17;95]$) paired with Fisher's least significant difference procedure to highlight statistically different varieties; the letters on the right-hand side of the figure represent the statistical groups to which each variety belongs. All presented results are in order of the biomass yield presented in this figure.

Slightly positive correlations between the survival rate and the number of resprouting shoots (0.27), as well as the number of resprouting shoots and the dry matter biomass yield (0.38), were calculated (Figure 6). Furthermore, a strong correlation between plant height and DBH (0.73) was found in this study. This is also affirmed by the close correlations between DBH and biomass yield (0.75), and between plant height and biomass yield (0.69) (Figure 6).

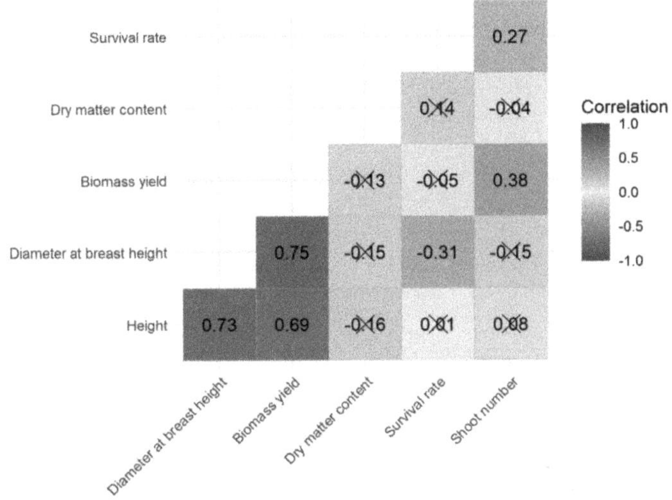



**Figure 6.** Correlation matrix of all recorded growth parameters of the poplar varieties in the winter of 2018, six years after planting and three years after the first harvest, using Pearson's product moment coefficient. Non-significant correlations ($\alpha \geq 0.05$) are crossed out.

## 4. Discussion

The economic success of an SRC is determined by the natural site conditions (e.g., soil texture, soil nutrient content, precipitation, ground water level, solar radiation and temperature), which are key factors to be considered during the planning process of an SRC [31], as well as the conditions that arise from, and can be influenced by, the plantation management itself, in particular during the establishment of an SRC, such as site preparation (tillage, herbicide application, etc.) [32], timing of planting (weather conditions, soil water status), planting material (quality of cuttings), planting techniques (setting and depth of planting) and weed control [1,31,32].

The most important and easily visible factor showing a successful management of these parameters is the survival rate of the established plants. In general, our data show very high survival rates at the end of the first growing season when compared with the literature [32]. Since high-quality propagation material and professional cultivation methods were used, the relatively low survival rates of AF 20 (56%), AF 19 (57%) and Muhle Larsen (73%) seem to be variety-specific characteristics [33]. The survival rate of all varieties decreased from the end of the first growing season in 2013 to the first harvest in 2015. The different extent of plant mortality can be assumed to be a reaction of the different poplar varieties to the specific site conditions, since neither major biotic or abiotic damage events occurred, and a proper plantation management was ensured. In the second rotation period from 2015 to 2018, there were only marginal decreases in the survival rate of a few varieties (e.g., −5% for Matrix 24 and −4% for Harff). Overall, the survival rates remained at the same level. Thus, it can be concluded that the variety-specific adaptation to the site is completed at the time of the harvest after the first rotation period.

The number of resprouting shoots after a harvest depends on the poplar variety, the planting density of the selected SRC rotation system and the site conditions [23,25,26,34]. The highest-yielding varieties in this study very often belonged to the most robust species. However, a tripling or quadrupling of the number of shoots as described by Röhle et al. [23] was not observed. The correlations between the survival rate and the number of resprouting shoots, as well the number of resprouting shoots and the dry matter biomass yield, lead to the conclusion that the greater the average number of shoots, the greater the dry matter biomass yield. This also means that the resprouting capacity of poplars does not only compensate the loss of individual plants in SRC but can also increase the yield. These results are consistent with those of Paris et al. [35].

The productivity of poplars is strongly correlated with various stem and leaf traits such as plant height, diameter, total leaf area and individual leaf area, and has been studied multiple times [36–41]. The results from our study confirm these literature results. Even the strong correlation between plant height and DBH found in our study has been described by several authors and ultimately led to various allometric functions to determine the biomass of an SRC by measuring a few, but crucial, parameters.

Aylott et al. [42,43]), Gielen and Ceulemans [44] and Cochard et al. [45] reported biomass yields between 2.0 and 9.6 $t_{adm}$ ha$^{-1}$ yr$^{-1}$ from 16 poplar varieties in England and Wales. Similar results were reported from across Europe (e.g., Rae et al. [46]) and North America (e.g., Dillen et al. [47]). To generate an economically justified income from an SRC, a yield of at least 8 $t_{adm}$ ha$^{-1}$ y$^{-1}$ needs to be achieved [46]. Since the market prices for SRC biomass fluctuate greatly depending on location and time, a reasonable average financial yield cannot be specified. However, the biomass yield of the poplar varieties included in this study can be assessed with regard to the specific site conditions, which also include the occurrence of extreme, climate change-related weather events in recent years. From the first to the second rotation periods, only approximately 50% of the poplar varieties showed increases in biomass yield (e.g., Max 4, AF 18, AF 16), whereas varieties such as Weser 6, Max 1 and AF 6 had decreasing yields. This is opposed to various publications, which reported an increasing yield over the first ten years [48]. Except for Fritzi Pauley, the six highest-yielding varieties increased their yield from the first to the second rotation periods. With 11.5 to 15.6 $t_{adm}$ ha$^{-1}$ y$^{-1}$, these varieties were able to achieve very good yields under the given site conditions. Many varieties had yields of around 8 $t_{adm}$ ha$^{-1}$ y$^{-1}$. Some of these varieties were able to increase their yield in the second rotation period, others were not. Since no active management measures were carried out during the second rotation period and no major biotic damage was recorded, it must be assumed that the determined yields represent the variety-specific yields at the study site. The varieties that had yields below 4 $t_{adm}$ ha$^{-1}$ y$^{-1}$ in the first rotation period showed a significant yield decrease in the second rotation period, which can be attributed to a reduction in all growth parameters described above (survival rate, number of shoots per stool, plant height, DBH). In conclusion, these varieties are not suitable for SRC in this location.

Looking at the biomass yield of the harvest in 2018, it is noticeable that the poplar varieties can be divided into two groups, that is into high-yielding and low-yielding varieties. The high-yielding varieties include breeds with *P. deltoides* × *P. nigra* as well as with *P. trichocarpa* × *P. maximowiczii* (i.e., Max 3, Max 4, Matrix 24). The latter is confirmed by other studies [19,49–51]. The very good performance of the *P. deltoides* × *P. nigra* varieties on this low-yield and quite dry site was not expected and, to our knowledge, has not yet been described in the literature. Benetka et al. [49] reported below-average yields of *P. nigra* varieties on low-yielding locations in the Czech Republic. Based on our study, this can be confirmed for other *P. deltoides* × *P. nigra* varieties (e.g., Robusta, Jacometti 78B, AF 2, AF 6, I 214). These varieties are known to be susceptible to fungal diseases, in particular poplar leaf rust, which can have a significant negative impact on the growth of these varieties [52,53]. The very poor performance of the AF 15 and AF 20 varieties was due to a very high plant mortality rate. Since no biotic damage (neither fungal nor insect damage) was found on either variety, we assume that they were not able to cope with the specific site conditions.

Furthermore, the poor performance of the three TI varieties was surprising. Despite the advantage of having been delivered as rooted plants in pots, they did not exhibit a satisfying growth. While P 1 still reached the standard yield with approximately 8 $t_{adm}$ ha$^{-1}$ y$^{-1}$ in the second harvest, 4 × Göttingen and Esch 5 only achieved below-average results with just under 5 and 1 $t_{adm}$ ha$^{-1}$ y$^{-1}$, respectively. Since the plant mortality rates of these varieties were low and no major biotic damage was noted, the negative performance is most likely also attributed to the site conditions.

## 5. Conclusions

Our data suggest that the selection of poplar varieties based on the particular site is extremely important for the economic success of an SRC. On the continental-influenced light sandy soil of the study site, only eight out of 37 poplar varieties showed economically sufficient growth in a mini-rotation SRC and produced a biomass yield of more than 10 $t_{adm}$ $ha^{-1}$ $y^{-1}$. This included older varieties such as Fritzi Pauley, Max 3 and Max 4, as well as newer breeds from Italy, such as AF 16, AF 17 and AF 18, and from Germany, such as Matrix 24. These varieties can be recommended for commercial use in SRC under the specific site conditions. Many varieties had a yield of around 8 $t_{adm}$ $ha^{-1}$ $y^{-1}$, with some varieties increasing and some decreasing their yield from the first to the second rotation period. Given the current climate change prognosis, we advise against the cultivation of these varieties. Six varieties did not even reach 4 $t_{adm}$ $^{ha-1}$ $y^{-1}$ in the second harvest. This included the old varieties Androscoggin and Muhle Larsen, as well as the new AF 15 and AF 20 varieties from Italy and the Esch 5 variety from TI.

Another data collection to verify these conclusions is planned for 2021 at the end of the third rotation period.

**Author Contributions:** D.L. initiated and planned the project. C.C. and M.N. performed the statistical analyses and modeled the data. D.L., C.C. and M.N. wrote the manuscript and contributed by revising the manuscript. All authors have read and agreed to the published version of the manuscript.

**Funding:** This research was funded by the German Federal Ministry of Education and Research under the grant number 0315972G for the first three years. In the following years it was financially supported by the registered association "Biomasse Schraden e. V."

**Acknowledgments:** We want to sincerely thank the employees of the P&P Forest Service GmbH for their practical help in the field, as well as Christian Steinke, Karoline Schwandt and Jan Zimmermanns for their assistance during data collection. Our personal thanks went to Christiane Helbig for her proofreading of the text.

**Conflicts of Interest:** The authors declare no conflict of interest.

## References

1. European Commission. A Policy Framework for Climate and Energy in the Period from 2020 up to 2030. Communication from the Commission to the European Parliament, the Council, the European Economic and Social Committee and the Committee of the Regions Brussels 2014: European Commission. Available online: http://eur-lex.europa.eu/legal-content/EN/TXT/PDF/?uri=CELEX:52014DC0015&from=EN (accessed on 4 February 2020).
2. Niemczyk, M.; Kaliszewski, A.; Jewiarz, M.; Wróbel, M.; Mudryk, K. Productivity and biomass characteristics of selected poplar (Populus spp.) cultivars under the climatic conditions of northern Poland. *Biomass Bioenergy* **2018**, *111*, 46–51. [CrossRef]
3. Tripathi, A.M.; Pohanková, E.; Fischer, M.; Orság, M.; Trnka, M.; Klem, K.; Marek, M.V. The Evaluation of Radiation Use Efficiency and Leaf Area Index Development for the Estimation of Biomass Accumulation in Short Rotation Poplar and Annual Field Crops. *Forests* **2018**, *9*, 168. [CrossRef]
4. Hauk, S.; Wittkopf, S.; Knoke, T. Analysis of commercial short rotation coppices in Bavaria, southern Germany. *Biomass Bioenergy* **2014**, *67*, 401–412. [CrossRef]
5. Hauk, S.; Gandorfer, M.; Wittkopf, S.; Müller, U.K.; Knoke, T. Ecological diversification is risk reducing and economically profitable—The case of biomass production with short rotation woody crops in south German land-use portfolios. *Biomass Bioenergy* **2017**, *98*, 142–152. [CrossRef]
6. Schweier, J.; Arranz, C.; Nock, C.A.; Jaeger, D.; Scherer-Lorenzen, M. Impact of Increased Genotype or Species Diversity in Short Rotation Coppice on Biomass Production and Wood Characteristics. *Bioenergy Res.* **2019**, *12*, 497–508. [CrossRef]
7. Bacenetti, J.; Bergante, S.; Facciotto, G.; Fiala, M. Woody biofuel production from short rotation coppice in Italy: Environmental-impact assessment of different species and crop management. *Biomass Bioenergy* **2016**, *94*, 209–219. [CrossRef]
8. Agostini, F.; Gregory, A.S.; Richter, G.M. Carbon Sequestration by Perennial Energy Crops: Is the Jury Still Out? *Bioenergy Res.* **2015**, *8*, 1057–1080. [CrossRef] [PubMed]

9.  Landgraf, D.; Setzer, F. *Kurzumtriebsplantagen: Holz vom Acker—So Geht's*; DLG-Verlag: Frankfurt, Germany, 2012; p. 71. ISBN 978-3-7690-2005-2. (In Germany)

10. Berhongaray, G.; Verlinden, M.S.; Broeckx, L.S.; Janssens, I.A.; Ceulemans, R. Soil carbon and belowground carbon balance of a short-rotation coppice: Assessments from three different approaches. *GCB Bioenergy* **2016**, *9*, 299–313. [CrossRef]

11. Ferrè, C.; Comolli, R. Comparison of soil $CO_2$ emissions between short-rotation coppice poplar stands and arable lands. *iForest Biogeosci. For.* **2018**, *11*, 199–205. [CrossRef]

12. European Commission. Commission Regulation (EC) No 1120/2009 of 29 October 2009 Laying Down Detailed Rules for the Implementation of the Single Payment Scheme Provided for in Title III of Council Regulation (EC) No 73/2009 Establishing Common Rules for Direct Support Schemes for Farmers under the Common Agricultural Policy and Establishing Certain Support Schemes for Farmers. 2009. Available online: https://op.europa.eu/de/publication-detail/-/publication/45b679d0-d9b9-400a-904b-8df1724e00e4/language-en (accessed on 4 February 2020).

13. Carl, C.; Biber, P.; Veste, M.; Landgraf, D.; Pretzsch, H. Key drivers of competition and growth partitioning among Robinia pseudoacacia L. trees. *For. Ecol. Manag.* **2018**, *430*, 86–93. [CrossRef]

14. Nicolescu, V.-N.; Buzatu-Goanță, C.; Bakti, B.; Keserű, Z.; Antal, B.; Rédei, K. Black locust (*Robinia pseudoacacia* L.) as a multi-purpose tree species in Hungary and Romania: A review. *J. For. Res.* **2018**, *29*, 1449–1463. [CrossRef]

15. Vítková, M.; Sádlo, J.; Roleček, J.; Petřík, P.; Sitzia, T.; Müllerová, J.; Pyšek, P. Robinia pseudoacacia-dominated vegetation types of Southern Europe: Species composition, history, distribution and management. *Sci. Total Environ.* **2020**, *707*, 134857. [CrossRef]

16. Larsen, S.U.; Jørgensen, U.; Lærke, P.E. Biomass yield, nutrient concentration and nutrient uptake by SRC willow cultivars grown on different sites in Denmark. *Biomass Bioenergy* **2018**, *116*, 161–170. [CrossRef]

17. Stolarski, M.J.; Olba-Zięty, E.; Rosenqvist, H.; Krzyżaniak, M. Economic efficiency of willow, poplar and black locust production using different soil amendments. *Biomass Bioenergy* **2017**, *106*, 74–82. [CrossRef]

18. Rogers, E.; Zalesny, J.R.S.; Hallett, R.A.; Headlee, W.L.; Wiese, A.H. Relationships among Root–Shoot Ratio, Early Growth, and Health of Hybrid Poplar and Willow Clones Grown in Different Landfill Soils. *Forests* **2019**, *10*, 49. [CrossRef]

19. Rebola-Lichtenberg, J.; Schall, P.; Annighöfer, P.; Ammer, C.; Leinemann, L.; Polle, A.; Euring, D. Mortality of Different Populus Genotypes in Recently Established Mixed Short Rotation Coppice with *Robinia pseudoacacia* L. *Forests* **2019**, *10*, 410. [CrossRef]

20. Verlinden, M.S.; Broeckx, L.; Bulcke, J.V.D.; Van Acker, J.; Ceulemans, R. Comparative study of biomass determinants of 12 poplar (Populus) genotypes in a high-density short-rotation culture. *For. Ecol. Manag.* **2013**, *307*, 101–111. [CrossRef]

21. González-González, B.D.; Oliveira, N.; González, I.; Cañellas, I.; Sixto, H. Poplar biomass production in short rotation under irrigation: A case study in the Mediterranean. *Biomass Bioenergy* **2017**, *107*, 198–206. [CrossRef]

22. Oliveira, N.; Rodríguez-Soalleiro, R.; Hernández, M.J.; Cañellas, I.; Sixto, H.; Pérez-Cruzado, C. Improving biomass estimation in a Populus short rotation coppice plantation. *For. Ecol. Manag.* **2017**, *391*, 194–206. [CrossRef]

23. Schlepphorst, H.; Hartmann, H.; Murach, D. Yield of fast-growing tree species in northeast Germany: Results of research with experimental plots (2006 to 2015). *Appl. Agric. For. Res.* **2017**, *2*, 93–110. [CrossRef]

24. Oliveira, N.; Rodríguez-Soalleiro, R.; Pérez-Cruzado, C.; Cañellas, I.; Sixto, H.; Ceulemans, R. Above- and below-ground carbon accumulation and biomass allocation in poplar short rotation plantations under Mediterranean conditions. *For. Ecol. Manag.* **2018**, *428*, 57–65. [CrossRef]

25. Goetsch, S.; Peek, R. Vergleichende Feuchtebestimmung an Bohrkernen und geschnittenen Holzproben nach DIN 52 183 Holz als Roh-und Werkstoff. *Eur. J. Wood Prod.* **1987**, *45*, 300. [CrossRef]

26. Hartmann, K.-U. Entwicklung eines Ertragsschätzers für Kurzumtriebsbestände aus Pappel. Ph.D. Thesis, Technische Universität Dresden, Dresden, Germany, 12 August 2010. Available online: https://nbn-resolving.org/urn:nbn:de:bsz:14-qucosa-62853 (accessed on 19 February 2020).

27. Ali, M.S.W. Modelling of Biomass Production Potential of Poplar in Short Rotation Plantations on Agricultural Lands of Saxony, Germany. Ph.D. Thesis, Technische Universität Dresden, Dresden, Germany, 3 March 2009. Available online: https://nbn-resolving.org/urn:nbn:de:bsz:14-ds-1237199867841-24821 (accessed on 19 February 2020).

28. Kramer, H.; Akça, A. *Leitfaden zur Waldmesslehre*; J.D. Sauerländer Verlag: Bad Orb, Germany, 2008; ISBN 9783793908807.

29. R-Core-Team. R: A language and Environment for Statistical Computing. (3.6.3). R Foundation for Statistical Computing. 2020. Available online: https://www.r-project.org/ (accessed on 23 September 2020).

30. Wickham, H.; Averick, M.; Bryan, J.; Chang, W.; D'Agostino McGowan, L.; François, R.; Grolemund, G.; Hayes, A.; Henry, L.; Hester, J.; et al. Welcome to the tidyverse. *J. Open Source Softw.* **2017**, *4*, 1686. [CrossRef]

31. Kassambra, A. Ggcorrplot: Visualization of a Correlation Matrix Using "ggplot2" (0.1.3). 2019. Available online: https://cran.r-project.org/package=ggcorrplot (accessed on 23 September 2020).

32. Hartmann, L.; Lamersdorf, N. Site conditions, initial growth and nutrient and litter cycling of newly installed short rotation coppice and agroforestry systems. In *Bioenergy from Dendromass for the Sustainable Development of Rural Areas*; Butler Manning, D., Bemmann, A., Bredemeier, M., Lamersdorf, N., Ammer, C., Eds.; Wiley-VCH: Weinheim, Germany, 2015; pp. 123–138.

33. Shifflett, S.D.; Hazel, D.W.; Nichols, E.G. Sub-Soiling and Genotype Selection Improves Populus Productivity Grown on a North Carolina Sandy Soil. *Forests* **2016**, *7*, 74. [CrossRef]

34. Friedrich, E. Anbautechnische Untersuchungen in forstlichen Schnellwuchsplantagen und Demonstration des Leistungsvermögens schnellwachsender Baumarten. In *Modellvorhaben "Schnellwachsende Baumarten"*; Hofmann, M., Ed.; Zusammenfassender Abschlussbericht; Landwirtschaftsverlag GmbH: Münster, Germany, 1999; pp. 19–150. Available online: https://www.gbv.de/dms/bs/toc/303247398.pdf (accessed on 23 September 2020).

35. Röhle, H.; Horn, H.; Müller, M.; Skibbe, K. Site-based yield estimation and biomass calculation in short rotation coppice plantations. In *Bioenergy from Dendromass for the Sustainable Development of Rural Areas*; Butler Manning, D., Bemmann, A., Bredemeier, M., Lamersdorf, N., Ammer, C., Eds.; Wiley-VCH: Weinheim, Germany, 2015; pp. 173–186.

36. Paris, P.; Mareschi, L.; Sabatti, M.; Pisanelli, A.; Ecosse, A.; Nardin, F.; Scarascia-Mugnozza, G. Comparing hybrid Populus clones for SRF across northern Italy after two biennial rotations: Survival, growth and yield. *Biomass Bioenergy* **2011**, *35*, 1524–1532. [CrossRef]

37. Larson, P.R.; Isebrands, J.G. The Plastochron Index as Applied to Developmental Studies of Cottonwood. *Can. J. For. Res.* **1971**, *1*, 1–11. [CrossRef]

38. Ceulemans, R.; Deraedt, W. Production physiology and growth potential of poplars under short-rotation forestry culture. *For. Ecol. Manag.* **1999**, *121*, 9–23. [CrossRef]

39. Scarascia-Mugnozza, G.E.; Ceulemans, R.; Heilman, P.E.; Isebrands, J.G.; Stettler, R.F.; Hinckley, T.M. Production physiology and morphology of Populus species and their hybrids grown under short rotation. II. Biomass components and harvest index of hybrid and parental species clones. *Can. J. For. Res.* **1997**, *27*, 285–294. [CrossRef]

40. Marron, N.; Villar, M.; Dreyer, E.; DeLay, D.; Boudouresque, E.; Petit, J.-M.; Delmotte, F.M.; Guehl, J.-M.; Brignolas, F. Diversity of leaf traits related to productivity in 31 Populus deltoides × Populus nigra clones. *Tree Physiol.* **2005**, *25*, 425–435. [CrossRef]

41. Monclus, R.; Dreyer, E.; Delmotte, F.M.; Villar, M.; DeLay, D.; Boudouresque, E.; Petit, J.-M.; Marron, N.; Bréchet, C.; Brignolas, F. Productivity, leaf traits and carbon isotope discrimination in 29 Populus deltoides × P. nigra clones. *New Phytol.* **2005**, *167*, 53–62. [CrossRef]

42. Aylott, M.J.; Casella, E.; Tubby, I.; Street, N.R.; Smith, P.; Taylor, G. Yield and spatial supply of bioenergy poplar and willow short?rotation coppice in the UK. *New Phytol.* **2008**, *178*, 358–370. [CrossRef]

43. Aylott, M.J.; Casella, E.; Tubby, I.; Street, N.R.; Smith, P.; Taylor, G. Corrigendum. *New Phytol.* **2008**, *178*, 897. [CrossRef]

44. Gielen, B.; Ceulemans, R. The likely impact of rising atmospheric $CO_2$ on natural and managed Populus: A literature review. *Environ. Pollut.* **2001**, *115*, 335–358. [CrossRef]

45. Cochard, H.; Casella, E.; Mencuccini, M. Xylem vulnerability to cavitation varies among poplar and willow clones and correlates with yield. *Tree Physiol.* **2007**, *27*, 1761–1767. [CrossRef] [PubMed]

46. Rae, A.M.; Street, N.R.; Robinson, K.M.; Harris, N.; Taylor, G. Five QTL hotspots for yield in short rotation coppice bioenergy poplar: The Poplar Biomass Loci. *BMC Plant Biol.* **2009**, *9*, 23. [CrossRef] [PubMed]

47. Dillen, S.; Djomo, S.N.; Al Afas, N.; Vanbeveren, S.P.; Ceulemans, R. Biomass yield and energy balance of a short-rotation poplar coppice with multiple clones on degraded land during 16 years. *Biomass Bioenergy* **2013**, *56*, 157–165. [CrossRef]

48. Landgraf, D.; Bärwolff, M.; Burger, F.; Pecenka, R.; Hering, T.; Schweier, J. *Produktivität, Management und Nutzung von Agrarholz, In Agrarholz—Schnellwachsende Bäume in der Landwirtschaft Biologie—Ökologie— Management*; Veste, M., Böhm, C., Eds.; Springer Spektrum: Berlin, Germany, 2018; ISBN 978-3-662-49930-6. [CrossRef]

49. Biertümpfel, A.; Hering, T. Energieholzanbau in Thüringen—Erträge und Praxiserfahrungen. In *Erfurter Tagung Schnellwachsender Baumarten*; Landgraf, D., Ed.; 2017; ISSN 2567-8922. Available online: https://www.researchgate.net/publication/326774259_1_ERFURTER_TAGUNG_Schnellwachsende_Bau marten_--_Etablierung_Management_und_Verwertung_vom_1611_-_17112017 (accessed on 8 May 2020).

50. Nielsen, U.B.; Madsen, P.; Hansen, J.K.; Nord-Larsen, T.; Nielsen, A.T. Production potential of 36 poplar clones grown at medium length rotation in Denmark. *Biomass Bioenergy* **2014**, *64*, 99–109. [CrossRef]

51. Benetka, V.; Novotná, K.; Štochlová, P. Biomass production of Populus nigra L. clones grown in short rotation coppice systems in three different environments over four rotations. *iForest Biogeosci. For.* **2014**, *7*, 233–239. [CrossRef]

52. Verlinden, M.S.; Broeckx, L.; Ceulemans, R. First vs. second rotation of a poplar short rotation coppice: Above-ground biomass productivity and shoot dynamics. *Biomass Bioenergy* **2015**, *73*, 174–185. [CrossRef]

53. Steenackers, V.; Strobl, S.; Steenackers, M. Collection and distribution of poplar species, hybrids and clones. *Biomass* **1990**, *22*, 1–20. [CrossRef]

*Article*

# Growth Rates of Poplar Cultivars across Central Asia

Niels Thevs [1,2,*] , Steffen Fehrenz [3], Kumar Aliev [1], Begaiym Emileva [4], Rinat Fazylbekov [5], Yerzhan Kentbaev [6], Yodgor Qonunov [7], Yosumin Qurbonbekova [8], Nurgul Raissova [5], Muslim Razhapbaev [9] and Sovietbek Zikirov [10]

[1]  World Agroforestry, Bishkek 720001, Kyrgyzstan; K.Aliev@cgiar.org
[2]  Gesellschaft für Internationale Zusammenarbeit (GIZ), 53113 Bonn, Germany
[3]  Dendroquant—Agricultural and Valuable Wood, 34359 Hannoversch Münden, Germany; steffen.fehrenz@dendroquant.com
[4]  Leibniz Institute of Agricultural Development in Transition Economies, 06120 Halle, Germany; emileva@iamo.de
[5]  Kazakh National Research Institute for Plant Protection and Quarantine, Almaty 050010, Kazakhstan; fazylbekov@gmail.com (R.F.); nraissova@gmail.com (N.R.)
[6]  Faculty of Forestry, Kazakh National Agricultural University, Almaty 050001, Kazakhstan; kentbayev@mail.ru
[7]  Mountain Societies Development Support Program, Khorog 736000, Tajikistan; yodgor.qonunov@akdn.org
[8]  School of Arts and Sciences, University of Central Asia, Khorog 736000, Tajikistan; yosumin.qurbonbekova@gmail.com
[9]  Forestry Research Center, Academy of Sciences of Kyrgyzstan, Bishkek 720001, Kyrgyzstan; mrajapbaev@yandex.ru
[10]  Spatial Planning and Development Public Foundation, Osh 723500, Kyrgyzstan; sz1.evb1@gmail.com
*   Correspondence: n.thevs@cgiar.org; Tel.: +49-17635406673

**Citation:** Thevs, N.; Fehrenz, S.; Aliev, K.; Emileva, B.; Fazylbekov, R.; Kentbaev, Y.; Qonunov, Y.; Qurbonbekova, Y.; Raissova, N.; Razhapbaev, M.; et al. Growth Rates of Poplar Cultivars across Central Asia. *Forests* **2021**, *12*, 373. https://doi.org/10.3390/f12030373

Academic Editor: Ronald S. Zalesny Jr.

Received: 30 January 2021
Accepted: 18 March 2021
Published: 20 March 2021

**Publisher's Note:** MDPI stays neutral with regard to jurisdictional claims in published maps and institutional affiliations.



**Abstract:** *Research Highlights:* Despite a long tradition of using poplars as wood source across Central Asia, recent international breeding developments have not penetrated that region yet. This study therefore explored growth performance of 30 local and international poplar cultivars. *Background and Objectives:* The Central Asian countries are forest poor countries, which need to cover the domestic wood demand through costly imports. Therefore, fast growing trees, such as poplars, are gaining increasing attention as option to grow wood domestically. The most common cultivars date back to Soviet Union times. As recent breeding developments have not reached the region, this study aims at investigate the growth performance of a number of newly developed poplar cultivars. *Materials and Methods:* The investigated cultivars were planted as cuttings across nine sites in Kyrgyzstan, Kazakhstan, and Tajikistan between 2018 and 2020. *Results:* Under warm climate conditions, i.e., low elevations, *P. deltoides x nigra* hybrids attained highest stem volumes and biomass yields, up to 16.9 t/ha*a after two years, followed by *P. nigra x maximoviczii* hybrids. One of the *P. deltoides x nigra* hybrids reached a tree height of 10.5 m after three years. On higher elevations, e.g., in the Pamirs and in Naryn, *P. maximoviczi x trichocarpa* hybrids and *P. trichocarpa* cultivars grew faster than the former hybrids. *Conclusions:* The cultivars explored in this study should be included into plantations or agroforestry systems that are being established, provided that land users are able to thoroughly control weeds and ensure nutrient and water supply. If sufficient weed control, nutrient supply, or water supply cannot be ensured, then land users should opt for local cultivars (e.g., Mirza Terek) or the *P. nigra x maximoviczii* hybrids or *P. trichocarpa*, in order to avoid failure.

**Keywords:** fast growing trees; poplar hybrids; poplar clones; tree height; DBH; stem volume; yield; agroforestry; Kyrgyzstan; Kazakhstan; Tajikistan

## 1. Introduction

Poplars are a major agroforestry tree across Central Asia and increasingly gain attention for fast growing tree woodlots and plantations [1]. Traditionally, poplars were planted along field borders and irrigation ditches to gain wood as construction material, without occupying much space of adjacent crop fields. During Soviet Union times, those poplar rows along field borders were propagated as tree wind breaks to reduce wind speed, improve the micro climate, and help to increase crop yields [2–5]. Thereby, the effects on the microclimate and crop yields were the main target rather than the wood resources potentially provided from such tree wind breaks [6–8]. After disintegration of the Soviet Union, a large share of those tree wind breaks was cut down primarily for fuel wood and secondarily for timber, as the energy supply system had broken down in the course of the disintegration of the Soviet Union [1]. Now, in parts of Central Asia, e.g., in the Ferghana Valley, poplars are being planted as tree wind breaks and plantations to gain wood as a resource and as an additional income source [9]. This is in line with policy programs or strategies of the Central Asian countries, e.g., the Green Economy Program in Kyrgyzstan, the recent Strategy to Develop the Agriculture in the Republic of Uzbekistan 2020–2030, or under the Kazakhstan 2050 strategy as reviewed by [1].

During Soviet Union times, *P. bolleana*, *P. nigra*, and *P. deltoides* cultivars were brought into Central Asia to be planted in agroforestry systems and plantations [10]. Though, today the most widely distributed poplar in agroforestry systems and plantations across Central Asia is *Populus nigra* var. *pyramidalis*, in particular the local cultivar Mirza Terek, according to own field observations and personnel communications across the region. This *Populus nigra* var. *pyramidalis* originated from an area comprising Afghanistan, the Western Pamirs, and the Western Tianshan. Standardized Mirza Terek planting material was brought from Sotchi into Central Asia in 1952 [11]. Until today, Mirza Terek is the typical poplar that is planted across most parts of Central Asia, while only in the northern part of Kyrgyzstan and SE Kazakhstan, *P. alba* is planted as well. Starting in the 1980s and 1990s, the knowledge on the physiology and genetics of poplars has been increasing (e.g., [12] and further literature there), which resulted in the development of new cultivars. Globally, a number of new poplar cultivars have been developed and released during the last decades through public breeding programs and developments by private businesses as listed by [13] and reviewed by [14]. In Asia, breeding programs have been carried out and are ongoing in China, India, Japan, Korea, and Kazakhstan. In Russia, poplar research and breeding are being carried out in a number of institutes in the European part of Russia. So far, these recent breeding developments of new cultivars have not entered Central Asia on a larger scale.

Against this background, this study aims at addressing that gap by starting to plant and monitor the survival and growth rates of a number of poplar cultivars across the different climates and elevations of Central Asia to take genotype × location interactions into account [15]. This study sees itself as a first step to explore which of the cultivars or parentages, which are listed in Table 1 below, might be selected for further more systematic in-depth field trials. Yet, as a report of work in progress this study is able to report results on the survival rates after planting, growth rates immediately after planting, and during the youth development of the planted cultivars.

**Table 1.** List of poplar cultivars, corresponding parent species, and distribution across sites. Parentages refer to the following species and hybrids: PN—*P. nigra* (section Aigeiros), PD—*P. deltoides* (section Aigeiros), PT—*P. trichocarpa* (section Tacamahaca), PA—*P. alba* (section Populus), Psi—*P. simonii* (section Tacamahaca), PPa—*P. pamirica* (section Tacamahaca), PDN—*P. x canadensis* (intra-sectional), PMT—*P. maximoviczii x trichocarpa* (intra-sectional), PNM—*P. nigra x maximoviczii* (inter-sectional), and PLfND—*P. laurifolia x canadensis* (inter-sectional). The geographical location, elevation, and climate zone are given in Table 2.

| Cultivar | Parentage | Almaty | Bishkek I | Bishkek II | Jalalabad | Osh | Lavar | Tup | Khorog | Naryn |
|---|---|---|---|---|---|---|---|---|---|---|
| Mirza Terek | PN | | | • | • | • | • | • | | • |
| Pyramidalis [1] | PN | | | | | | | | • | |
| Samsun | PD | | | • | • | | | • | | |
| 89M060 | PD | | | • | • | | | • | | |
| Oudenberg | PDN | • | • | • | • | • | • | • | • | • |
| Orion | PDN | • | • | • | • | • | • | • | • | • |
| H-8 | PDN | • | | • | • | • | • | • | • | |
| H-11 | PDN | • | | • | • | • | • | • | • | |

**Table 1.** *Cont.*

| Cultivar | Parentage | Almaty | Bishkek I | Bishkek II | Jalalabad | Osh | Lavar | Tup | Khorog | Naryn |
|---|---|---|---|---|---|---|---|---|---|---|
| H-17 | PDN | • | • | • | • | • | • | • | • | • |
| H-33 | PDN | • | • | • | • | • | • | • | • | • |
| Tiepolo | PDN | | | • | • | | | • | | |
| Bellini | PDN | | | • | • | | | • | | |
| Veronese | PDN | | | • | • | | | • | | |
| Vesten | PDN | • | • | | | • | | | | • |
| Kazakhstani | PLfND | • | | • | • | • | | • | | |
| Kyzyl-Tan | PLfND | | | | | | • | | | |
| H-275 | PMT | • | • | • | • | • | • | • | | • |
| Matrix-11 | PMT | • | • | • | • | • | • | • | | • |
| Matrix-49 | PMT | • | | | | | | | | |
| Matrix-24 | PMT | • | | | | | | | | |
| Fastwood 1 | PMT | | | • | • | | | • | • | |
| Fastwood 2 | PMT | | | • | • | | | • | • | |
| Max-3 | PNM | • | • | • | • | • | • | • | • | • |
| Max-4 | PNM | • | • | | | • | | | | • |
| Max-1 | PNM | • | | | | | | | | |
| Muhle Larsen | PT | | | • | • | | • | • | • | |
| Fritzi Pauley | PT | • | • | • | • | • | • | • | • | • |
| Trichobel | PT | • | • | • | • | • | • | • | • | • |
| Ozolin | PA | | | • | • | | | • | | |
| *P. pamirica* | PPa | | | | | | | | • | |
| *P. simonii* | PSi | | | • | • | | | • | | |

[1] Possibly this is also Mirza Terek.

**Table 2.** List of sites with their geographical location, elevation, and climate zone.

| Site Name | Country | Geographical Position | Elevation [m a.s.l.] | Climate Zone [1] |
|---|---|---|---|---|
| Almaty | Kazakhstan | 43.18° N 76.87° E | 1014 | Dfa |
| Bishkek I and II | Kyrgyzstan | 42.92° N 74.62° E | 701 | Dsa |
| Jalalabad | Kyrgyzstan | 40.94° N 72.97° E | 779 | Dsa |
| Osh | Kyrgyzstan | 40.54° N 72.89° E | 1022 | Dsa |
| Lavar | Kazakhstan | 43.57° N 78.09° E | 572 | BSk |
| Tup | Kyrgyzstan | 42.8° N 78.49° E | 1771 | Dfb |
| Khorog | Tajikistan | 37.46° N 71.61° E | 2183 | BSk |
| Naryn | Kyrgyzstan | 41.42° N 75.74° E | 1938 | BSk |

[1] Climate zone after www.climate-data.org (accessed on 16 March 2021): Dfa, Dsa, Dfb—humid and hot continental climate, BSk—cold semiarid climate.

## 2. Materials and Methods

### 2.1. Planting Material

In total, 31 cultivars were planted on nine sites in 2018, 2019, and 2020, as listed in Table 1 underneath. This set of cultivars included *P. nigra* (PN, section Aigeiros), *P. deltoides* (PD, section Aigeiros), *P. trichocarpa* (PT, section Tacamahaca), *P. alba* (PA, section Populus), and *P. simonii* (Psi, section Tacamahaca), *P. pamirica* (PPa, section Tacamahaca) cultivars as well as *P. x canadensis* (PDN, intra-sectional), *P. maximoviczii x trichocarpa* (PMT, intra-sectional), *P. nigra x maximoviczii* (PNM, inter-sectional) and *P. laurifolia x canadensis* (PLfND, inter-sectional) hybrids.

All the PT, PMT, and PNM cultivars were purchased from Wald21, Germany, while the PDN cultivars H-8, H-11, H-17, H-33, Orion, Oudenberg, and Vesten were purchased from Biopoplar, Italy. The cultivars Samsun, 89M060, Tiepolo, Bellini, Ozolin, and Veronese were obtained from the Academy of Sciences in Uzbekistan. The remaining cultivars were obtained locally from project partners. All the planting material was planted in spring as cuttings with a length of 20 cm.

### 2.2. Study Sites

The study sites were distributed across Central Asia from SE Kazakhstan over Kyrgyzstan, including the Ferghana Valley, into Tajikistan (Figure 1), with the aim to capture different growth rates of the cultivars across the climate zones and elevation range relevant for the potential use of those poplar cultivars in agroforestry or plantations [15]. The final selection of the particular sites here and their different management, as explained below, are partly owed to the availability of sites and resources.

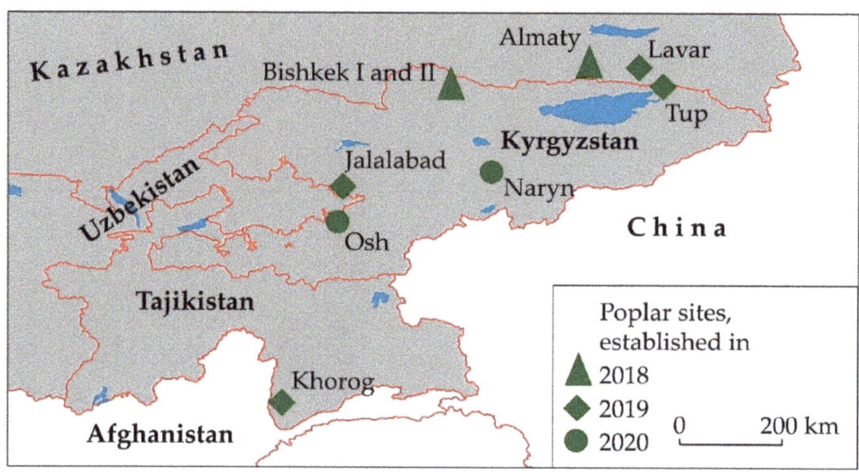


**Figure 1.** Location of the poplar testing sites.

The sites Almaty, Bishkek I and II, and Lavar represent the hot continental climate (Tables 2 and 3) with a pronounced rainy season in spring along the northern slopes of the Tianshan Mountains with its agricultural areas there. The sites Jalalabad and Osh represent the Ferghana Valley with a hot continental climate, too, but warmer and with a longer growing season compared to the former three sites (Tables 2 and 3). The Ferghana Valley is the area with the highest population density and a major agricultural region of whole Central Asia. The three latter sites, Tup, Khorog, and Naryn, represent higher elevations with colder climates and a shorter growing season compared to the former sites (Tables 2 and 3). Khorog and Naryn, in addition to their high elevation, are considered semiarid (Tables 2 and 3).

**Table 3.** Climate features at the sites (www.weatherbase.com, accessed on 16 March 2021).

| Site | Average January Temperature [°C] | Average July Temperature [°C] | Annual Precipitation [mm] | First Month of the Year with >5 °C |
|---|---|---|---|---|
| Almaty | −4.7 | 23.8 | 570 | April: 11.5 °C |
| Bishkek | −2.6 | 24.9 | 452 | March: 5.3 °C |
| Jalalabad | −1 | 25 | 430 | March: 8 °C |
| Osh | −3.4 | 25.1 | 378 | March: 6.9 °C |
| Lavar | −9.4 | 23.1 | 198 | April: 14.9 °C |
| Tup | −10.7 | 18.5 | 423 | April: 7.6 °C |
| Khorog | −6 | 22 | 260 | April: 10 °C |
| Naryn | −16 | 16 | 300 | April: 7 °C |

On each site, at least one soil profile was drilled down to 100 cm. The soils of all sites are silt and loam dominated through the first 50 cm of the soil profiles. Below 50 cm, the soil profiles contain sand and loam in Bishkek I and Jalalabad, but continue with silty horizons under the other sites. All those sites are well drained. On the site Bishkek I, there is a gradient from sandy loam to silty loam with the humus content also increasing along that gradient. The soil profile of the site Bishkek II is silty clay throughout the first 100 cm, which is not well drained. The soil in Lavar is slightly saline. According to [13], the former sites offer good to very good soil properties for poplars, while Bishkek II and Lavar offer fair to poor conditions due to their soil texture and salinity, respectively.

## 2.3. Site Management

This planting experiment was started in 2018 with the sites Almaty and Bishkek I. In 2019, the sites Khorog, Lavar, Bishkek II, Jalalabad, and Tup were started. Finally, in 2020 the sites in Osh and Naryn were added. The planting schemes and site management differed according to land, water, fertilizer, and labor availability for each of the sites (Tables 4 and 5). The site area Almaty was part of a nursery, while the two site areas in Bishkek had been unused land before starting this experiment. The sites in Tup, Jalalabad, Naryn, and Lavar belong to nursery areas of forestry enterprises and research institutes. The sites in Khorog and Osh were cropland before using them as sites within this study.

**Table 4.** Planting dates, planting schemes, and number of cuttings initially planted per site.

| Site | Planting Date | Planting Scheme | Total Number of Cuttings | Cuttings Planted per ha |
|---|---|---|---|---|
| Almaty | 18 April 2018 | 0.8 m × 0.2 m | 242 | 62,500 |
| Bishkek I | 7 April 2018 | 0.6 m × 0.6 m to 1.2 m × 0.6 m | 309 | 27,700 to 13,800 |
| Bishkek II | 6 April 2019 | 1 m × 0.6 m | 743 | 16,700 |
| Jalalabad | 10 April 2019 | 0.7 m × 1.5 m | 580 | 9500 |
| Osh | 27 May 2020 | 0.7 m × 0.6 m | 535 | 23,800 |
| Lavar | 29 March 2019 | 1.6 m × 0.6 m | 284 | 10,400 |
| Tup | 16 April 2019 | 1.4 m × 0.6 m | 828 | 11,900 |
| Khorog | 20 March 2019 | 0.5 m × 0.2 m | 352 | 100,000 |
| Naryn | 9 May 2020 | 1 m × 0.4 m | 379 | 25,000 |

**Table 5.** Water supply, weed control, and plant nutrition by site. Ammophos contained 12% N and 52% $P_2O_5$. The NPK fertilizer contained 25.6% N, 6.6% $P_2O_5$, and 21% $K_2O$.

| Site | Water Supply | Weed Control | Plant Nutrition |
|---|---|---|---|
| Almaty | Manually by water can: Every 2–3 days from mid-May through September. | Site was covered with geo-textile. | None |
| Bishkek I | Drip irrigation: Every 2–3 days from mid-May to mid-September, 3rd season once per week from June to mid-September. | Manual: 1st season: every 3 weeks until July. 2nd season: once in May and June. | 1st season: 3 g Ammophos and 6 g NPK fertilizer per tree. 2nd season: 16 g NPK fertilizer per tree. |
| Bishkek II | Drip irrigation: 1st season every 2–3 days from mid-May to mid-September, 2nd season no irrigation. | Manual: 1st season: every 3 weeks until July. 2nd season: no weed control. | 1st season: 16 g NPK fertilizer per tree. 2nd season: no fertilizer. |
| Lavar | Furrow irrigation: 1st season: once in April, once per week in May, twice per week end of July to mid-Sep. 2nd season: every 2 weeks. | Manual: Once per month from April to July | none |
| Jalalabad | Furrow irrigation: 1st season: once in April, once per week in May, twice per week end of July to mid-September 2nd season: every 2 weeks during summer. | Once per month from April to July. Manual in April, May, and July, herbicide in June. | 1st season: 3 g Ammophos and 6 g NPK fertilizer per tree. 2nd season: none. |

**Table 5.** *Cont.*

| Site | Water Supply | Weed Control | Plant Nutrition |
|---|---|---|---|
| Osh | Furrow irrigation: Twice per month from June to September. | Once per month from April to June. Manual in April, May, herbicide in June. | Manure before planting. 3 g Ammophos and 6 g NPK fertilizer per tree. |
| Tup | Furrow irrigation: Once per week from July to September | Mowing and manual: 1st season: 22 April and once per week during July. 2nd season: once per month. | 1st and 2nd season: 3 g Ammophos and 6 g NPK fertilizer per tree. |
| Khorog | Furrow irrigation: Once per week from June to August. | Manual: Twice per week from June to August. | 1st season: 3 g Ammophos per tree. |
| Naryn | Flood irrigation: Every 10 days in June, July, and August. | Manual: Once per month in June, July, and August. | None |

Leaf beetles (*Chrysomela populi*) were observed to feed on the leaves of the poplars on the sites Bishkek I and Tup during their second and third season and were treated once per season with the pesticide Doxin 100 EC from Dogal Agro, Turkey. With regard to weeds, the sites in Jalalabad and Osh suffered from a higher weed coverage throughout compared to the other site, despite the use of herbicides.

On the site Almaty, all trees were cut on October 2019, except for one tree per cultivar. In Bishkek I, all trees were cut in March 2020, except for three trees per cultivar. Therefore, Almaty and Bishkek I offer data for two seasons with a higher number of trees per cultivar, but data for the third season are only based on small numbers per cultivar.

*2.4. Data Collection and Analysis*

During the first season, the tree heights were measured at least three times during the season. These tree height measurements were continued through the second season. At the end of the second season, in addition to heights, the basal diameter, the diameters at 1 m, at 1.30 m (DBH), and at 2 m were measured to calculate the stem volume and stem biomass of each individual tree. Thereby, all trees were measured, including those on boundary rows of the sites.

The stem volumes were calculated as the sum of the volumes of the following stem sections: basis—1 m, 1 m to 1.30 m, 1.30 m to 2 m, and 2 m to the tip of the stem (or from a lower cross section to the tip of the stem for trees smaller than 2 m). The tree height was taken as stem height. The volumes of the former sections were calculated as follows:

$$V_{section} = \frac{A_{bottom} + A_{top}}{2} l$$

with $V_{section}$—volume of the given section, $A_{bottom}$ and $A_{top}$—cross section areas at the bottom and top of the given section, $l$ – length of the given section.

The volume of the top section was calculated as a cone volume:

$$V_{section} = \frac{1}{3} \pi r^2 l$$

with $V_{section}$—volume of the given section, $r$—radius of the cross section at the basis of this section, $l$ – length of the given section.

The yield of dry woody biomass (BM) was calculated as follows:

$$BM = V \delta D$$

with *V*—stem volume, $\delta$—wood density, and *D*—tree density (number of plants per hectare).

For data analysis, mean and standard deviations were calculated by site and cultivar and analyzed by an analysis of variance in SPSS (Tukey-2 post hoc test), $\alpha < 0.05$; for significant differences between cultivars by sites.

## 3. Results

The survival rates at the end of the first season were highest on the site Bishkek I with rates between 73% to 100%, followed by the sites Almaty, Lavar, and Naryn (Table S1). Thereby, on the latter three sites most cultivars had high survival rates of 70% and more, but a limited number of cultivars showed very low survival rates of below 30%, which reduced the overall survival rates per site compared to Bishkek I. For example, H-17 and H-33 had survival rates of 60% and more in Lavar and Bishkek I, but only 31% in Almaty. Also, Fritzi-Pauley had survival rates of 70% and more in Bishkek I, Almaty, and Naryn, but none of the Fritzi-Pauley cuttings survived in Lavar. On the site Bishkek I, most of the trees that did not survive were the ones planted in the sandy loam part of the site.

The survival rates on the sites Bishkek II, Jalalabad, and Tup exhibited high variability between cultivars. On those three sites, while only 10% or less of the cuttings of Orion, H-8, H-11, H-17, H-33, and Ozolin survived until the end of the first season, the cultivar Samsun showed a survival rate of 70% and more across those three sites. Other cultivars showed different survival rates across those three sites, such as Fastwood 2 with survival rates of 76%, 40%, and 30% in Bishkek II, Jalalabad, and Tup, respectively.

Survival rates by cultivar, when examined among plots, was highest for the cultivars Samsun, Max-4, and Max-3. The locally used cultivars Mirza Terek, Pyramidalis, and Pamir Poplar had survival rates of 19–100%, 64%, and 14%, respectively.

Most trees that had survived until end of the first season also survived the winter and the following second season (Table S2). On the sites Almaty, Bishkek I, Khorog, Lavar, and Tup the survival rates from the first to the second season were 90% and more by cultivar. In Jalalabad, none of the H-275, Matrix-11, Fastwood 1, and Oudenberg trees survived the winter and following second season. Less than 20% of Kazakhstani and Fritzi-Pauley survived until end of the second season. On the site Bishkek II, less than 10% of H-275, Matrix-11, Fastwood 1, Fastwood 2, Muhle-Larsen, Trichobel, and Fritzi-Pauley survived from the end of the first through the end of the second season. In addition, none of the *P. pamirica* survived the winter between the first and the second season.

The trees, all cultivars pooled together, grew tallest on the sites Bishkek I and Almaty at the end of the first and second season (Table 6). The smallest trees at the end of the first and second season were found in Khorog, Bishkek II, and Lavar.

**Table 6.** Means $\pm$ standard deviations of tree heights [m] of all trees per site at the end of the first and second growing season. Letters here indicate groups of sites that do not differ significantly at $\alpha < 0.05$.

| Site | Tree Height [m] at the End of the 1st Season | Tree Height [m] at the End of the 2nd Season |
|---|---|---|
| Almaty | $254 \pm 75$ a | $424 \pm 169$ b |
| Bishkek I | $260 \pm 86$ a | $550 \pm 106$ a |
| Bishkek II | $73 \pm 43$ e | $142 \pm 55$ e |
| Jalalabad | $167 \pm 68$ b | $334 \pm 63$ c |
| Osh | $104 \pm 42$ d | |
| Lavar | $48 \pm 28$ f | $79 \pm 55$ f |
| Tup | $131 \pm 45$ c | $281 \pm 61$ d |
| Khorog | $74 \pm 45$ e | $169 \pm 66$ e |
| Naryn | $107 \pm 35$ d | |

The tree heights at the end of the first season for the sites Bishkek I and Almaty are shown in Figures 2 and 3 (dark grey bars), where the cultivars grew into the highest trees compared across sites. On both sites, *P. x canadensis* (PDN) hybrids grew highest, followed by the *P. nigra x maximoviczii* (PNM) hybrids Max-3 and -4 as well as the *P. trichocarpa* (PT) cultivars Fritzi-Pauley and Trichobel. Of the PNM hybrids in Almaty, Max-3 attained heights comparable with the PDN hybrids. The *P. maximoviczii x trichocarpa* (PMT) hybrids ranked lowest regarding tree height at the end of the first season in these two sites.

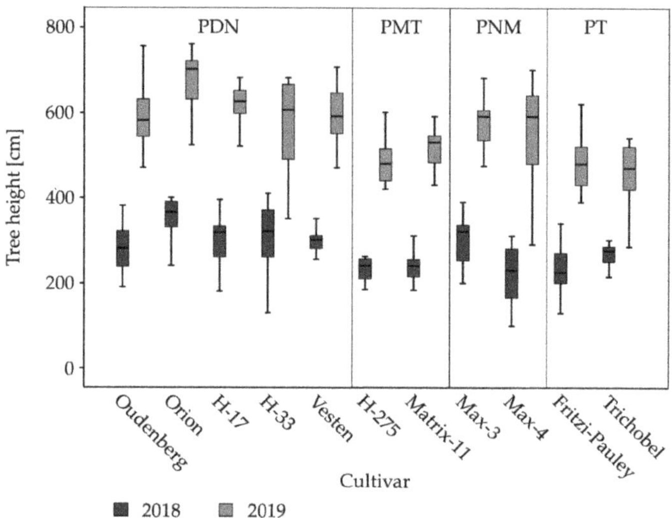

**Figure 2.** Box plots of tree heights in Bishkek I at the end of the growing seasons 2018 (dark grey) and 2019 (light grey). Parentages: PDN—*P. deltoides x nigra*, PMT—*P. maximoviczii x trichocarpa*, PNM—*P. nigra x maximoviczii*, PT—*P. trichocarpa*.

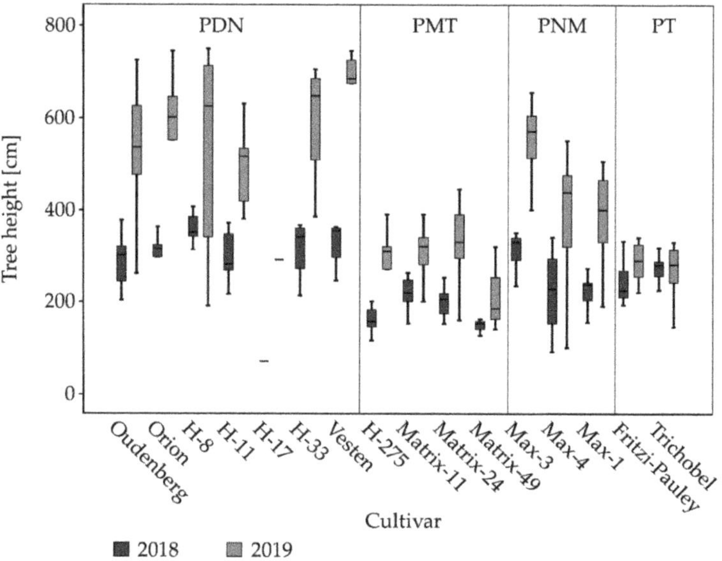

**Figure 3.** Box plots of tree heights in Almaty at the end of the growing seasons 2018 (dark grey) and 2019 (light grey). Parentages: PDN—*P. deltoides x nigra*, PMT—*P. maximoviczii x trichocarpa*, PNM—*P. nigra x maximoviczii*, PT—*P. trichocarpa*.

At the end of the second season (Figures 2 and 3, light grey bars), the PDN hybrids continued to grow highest and attained the largest stem volumes, followed by the PNM hybrids. In contrast to the first season, the PT cultivars ranked lowest with regard with tree heights and stem volumes on the site Bishkek I and similarly with the PMT hybrids in Almaty. The DBH values at the end of the second season behaved like the tree heights, with the PDN hybrids attaining the largest DBH, followed by PNM, PMT, and PT (Figures 4 and 5).

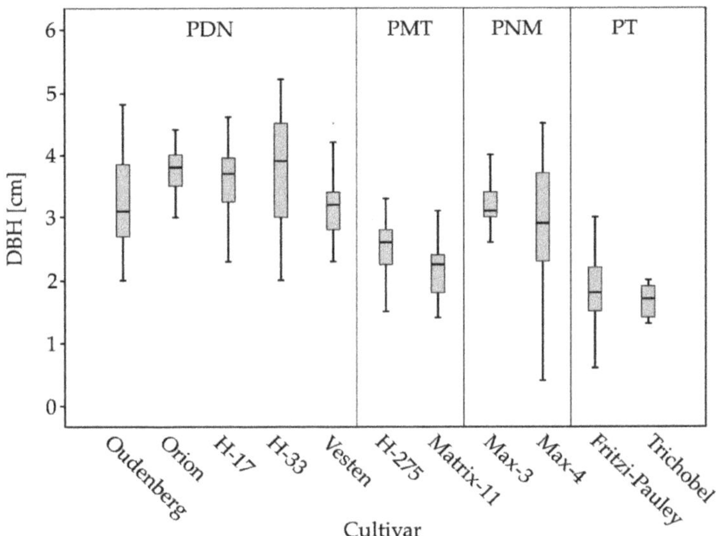

**Figure 4.** Box plots of DBH in Bishkek I at the end of the growing season 2019. Parentages: PDN—*P. deltoides x nigra*, —*P. maximoviczii x trichocarpa*, PNM—*P. nigra x maximoviczii*, PT—*P. trichocarpa*.

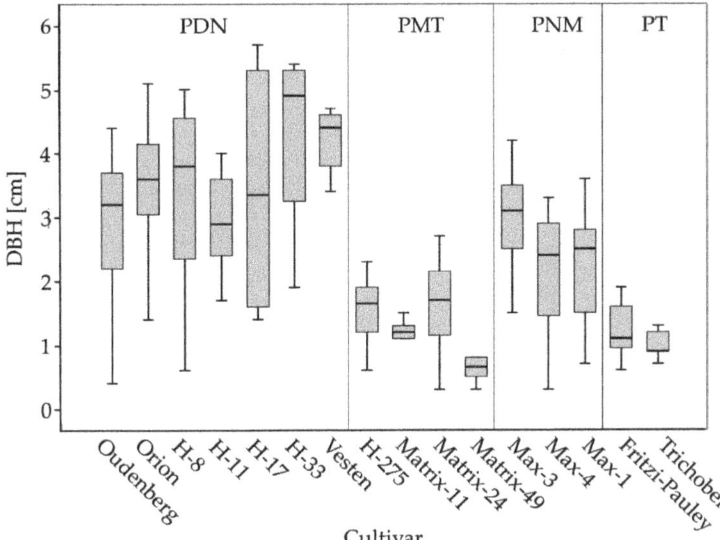

**Figure 5.** Box plots of DBH in Almaty at the end of the growing season 2019. Parentages: PDN—*P. deltoides x nigra*, PMT—*P. maximoviczii x trichocarpa*, PNM—*P. nigra x maximoviczii*, PT—*P. trichocarpa*.

Among the trees, which were carried on through the third season (2020), H-33, Orion, and H-17 (all PDN hybrids) clearly reached the largest stem volumes, tree heights, and DBH in Bishkek I (Figure 6, Table 7) as was already at the end of the second season (Table 8). At the end of the third season, H-33 had an average stem volume of 47.4 dm³ at an average tree height of 10.2 m, and DBH 10.5 cm (Table 7). Max-4 and Max-3 attained average stem volumes of 16.1 dm³ and 12.3 dm³ and average tree heights of 8.1 m and 8 m, respectively, which placed Max-4 and Max-3 among the PDN hybrids Oudenberg and Vesten (Table 7).

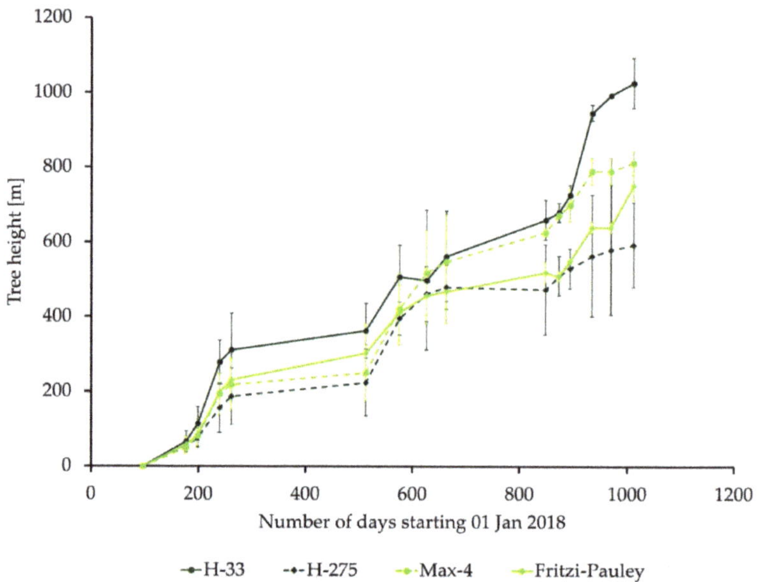


**Figure 6.** Mean tree heights by parentage groups, expressed by one cultivar per parentage (H-33—PDN, H-275—PMT, Max-4—PNM, and Fritzi-Pauley—PT) from planting time in April 2018 to end of the third season (Oct. 2020) on the site Bishkek I.

**Table 7.** Means ± standard deviations of tree heights [m], and DBH [cm] and stem volumes [dm³] at the end of the growing season 2020 for the site Bishkek I. From all cultivars, except for Trichobel, three trees were measured. Only one Trichobel survived until end of the growing season 2020.

| Cultivar | Tree Height [m] | DBH [cm] | Stem Volume [dm³] |
|---|---|---|---|
| Oudenberg | 8 ± 0.1 | 5.2 ± 0.5 | 9.9 ± 2.3 |
| Orion | 9.5 ± 0.05 | 7.5 ± 0.4 | 21.4 ± 2.2 |
| H-33 | 10.2 ± 0.5 | 10.5 ± 1.5 | 47.4 ± 12.7 |
| H-17 | 8.9 ± 0.7 | 7.2 ± 0.6 | 20.5 ± 5.1 |
| Vesten | 9.3 ± 0.2 | 6.2 ± 0.2 | 14.7 ± 1.6 |
| Max-3 | 8.5 ± 0.4 | 5.9 ± 0.9 | 12.6 ± 3.9 |
| Max-4 | 8.1 ± 0.3 | 6.8 ± 0.2 | 16.1 ± 1.3 |
| H-275 | 5.9 ± 0.9 | 4.2 ± 1.3 | 6.1 ± 3.2 |
| Matrix-11 | 7.1 ± 0.3 | 4.8 ± 0.5 | 8.3 ± 2.3 |
| Fritzi-Pauley | 7.5 ± 0.3 | 3.7 ± 0.6 | 5.7 ± 2 |
| Trichobel | 3.8 | 1.8 | 0.9 |

**Table 8.** Means ± standard deviations of stem volumes [dm$^3$] after two growing seasons for the sites Bishkek I and II, Almaty, Jalalabad and Tup. Sample sizes, minima, and maxima are listed in Table S3. Letters indicate groups of values that do not differ significantly at $\alpha < 0.05$. The last row contains means ± standard deviations by sites with all cuttings pooled together by site. Letters here indicate groups of sites that do not differ significantly at $\alpha < 0.05$.

| Cultivar | Volume [dm$^3$]. Average ± Standard Deviation | | | | |
|---|---|---|---|---|---|
|  | Almaty (2019) | Bishkek I (2019) | Bishkek II (2020) | Jalalabad (2020) | Tup (2020) |
| Mirza Terek |  |  | 0.14 ± 0.03 [a] | 1.8 ± 1.2 [a] | 0.7 ± 0.5 [b] |
| Samsun |  |  | 0.22 ± 0.31 [a] | 1.7 ± 0.8 [a] | 1.2 ± 1 [ab] |
| 89M060 |  |  | 0.24 ± 0.15 [a] | 1.7 ± 0.7 [a] | 1.2 ± 0.6 [ab] |
| Oudenberg | 2.5 ± 1.7 [bcd] | 2.9 ± 1.5 [bc] | 0.18 ± 0.14 [a] |  | 2.4 ± 0.6 [a] |
| Orion | 3.4 ± 2.2 [abc] | 4.1 ± 1.1 [a] |  |  |  |
| H-8 | 3.4 ± 2.5 [abc] |  |  |  |  |
| H-11 | 2.2 ± 1.2 [bcd] |  |  | 1.4 ± 0.3 [a] |  |
| H-17 | 4.1 ± 4 [ab] | 3.7 ± 1.3 [ab] | 0.07 ± 0.03 [a] |  |  |
| H-33 | 5.6 ± 3.4 [a] | 4.6 ± 2.6 [a] | 0.06 ± 0.01 [a] | 0.6 ± 0.06 [a] |  |
| Tiepolo |  |  | 0.07 ± 0.01[a] | 0.9 ± 0.3 [a] |  |
| Bellini |  |  | 0.19 ± 0.17 [a] |  |  |
| Veronese |  |  | 0.14 ± 0.11 [a] |  | 1.5 ± 0.8 [ab] |
| Vesten | 5.1 ± 2 [a] | 2.7 ± 1.3 [bc] |  |  |  |
| Kazakhstani |  |  |  |  | 1.4 ± 0.6 [ab] |
| H-275 | 0.7 ± 0.4 [d] | 1.8 ± 0.7 [cd] |  |  | 1.7 ± 0.8 [ab] |
| Matrix-11 | 0.3 ± 0.2 [d] | 1.4 ± 0.6 [d] |  |  | 2.2 ± 1.2 [a] |
| Matrix-49 | 0.3 ± 0.5 [d] |  |  |  |  |
| Matrix-24 | 0.8 ± 0.5 [d] |  |  |  |  |
| Fastwood 1 |  |  |  |  | 1.7 ± 0.7 [ab] |
| Fastwood 2 |  |  |  | 0.9 ± 0.3 [a] | 2.6 ± 0.9 [a] |
| Max-3 | 2.4 ± 1.3 [bcd] | 2.7 ± 1 [bc] | 0.34 ± 0.34 [a] | 2 ± 0.8 [a] | 1.7 ± 0.8 [ab] |
| Max-4 | 1.1 ± 0.9 [d] | 2.7 ± 1.6 [bc] |  |  |  |
| Max-1 | 1.3 ± 0.8 [cd] |  |  |  |  |
| Fritzi-Pauli | 0.5 ± 0.3 [d] | 1 ± 0.5 [d] | 0.13 ± 0.05 [a] | 0.6 ± 0.3 [a] | 1.2 ± 0.4 [ab] |
| Trichobel | 0.3 ± 0.2 [d] | 0.8 ± 0.3 [d] |  | 1.4 ± 0.7 [a] | 1.2 ± 0.4 [ab] |
| Ozolin |  |  | 0.12 ± 0.12 [a] |  |  |
| P. simonii |  |  |  |  | 1.6 ± 0.6 [ab] |
| All cuttings | 1.9 ± 2 [b] | 2.6 ± 1.7 [a] | 0.2 ± 0.2 [c] | 1.5 ± 0.8 [b] | 1.4 ± 0.8 [b] |

In Almaty and Bishkek I, Max-1, Max-3, and Max-4 developed leaves two to three weeks earlier than the PDN hybrids. In 2019 and 2020, a number of those trees were bent down by rain or late snow falls, but reverted back to upright trees during summer.

Annual biomass yields extrapolated to a hectare basis are listed for Bishkek I in Table 9. Only the PDN and PNM hybrids attained biomass values above the threshold of 8 t/ha*a, which was set for being economically viable. Among the PDN hybrids, H-33 ranked first with 14.8 t/ha*a (at a wood density of 0.35 t/m$^3$ and tree density of 18,500 per ha), followed by Orion and H-17. Thereby, survival rates of 100% were assumed.

**Table 9.** Yield expectation table [t/ha*a] for different tree densities between 10,000 and 20,000 trees per hectare and wood densities between 0.3 to 0.4 t/m$^3$ for each cultivar for the site Bishkek 1. The stem volumes used to calculate yields were measured at the end of the growing season 2019 as listed in Table 8. The values for wood densities were taken from own observations and center around the value of 0.35 t/m$^3$, which is given by FAO (http://www.fao.org/3/j2132s/J2132S16.htm, assessed on 10 January 2021). Grey boxes mark economically viable values of more than 8 t biomass per hectare and year. The total of 18,500 plants per hectare represents the real planting density of this site. For the calculations here, a survival rate of 100% was assumed for all cultivars.

| Trees per ha | | 10,000 | | | 15,000 | | | 18,500 | | | 20,000 | | |
| --- | --- | --- | --- | --- | --- | --- | --- | --- | --- | --- | --- | --- | --- |
| Wood Density in t/m$^3$ | | 0.3 | 0.35 | 0.4 | 0.3 | 0.35 | 0.4 | 0.3 | 0.35 | 0.4 | 0.3 | 0.35 | 0.4 |
| Cultivar | Stem Volume [dm$^3$] | | | | | | | | | | | | |
| Oudenberg | 2.9 | 4.4 | 5.1 | 5.9 | 6.6 | 7.7 | 8.8 | 8.1 | 9.5 | 10.9 | 8.8 | 10.3 | 11.7 |
| Orion | 4.1 | 6.2 | 7.2 | 8.3 | 9.3 | 10.8 | 12.4 | 11.5 | 13.4 | 15.3 | 12.4 | 14.4 | 16.5 |
| H-33 | 4.6 | 6.8 | 8.0 | 9.1 | 10.3 | 12.0 | 13.7 | 12.6 | 14.8 | 16.9 | 13.7 | 15.9 | 18.2 |
| H-17 | 3.7 | 5.5 | 6.5 | 7.4 | 8.3 | 9.7 | 11.1 | 10.2 | 12.0 | 13.7 | 11.1 | 12.9 | 14.8 |
| Vesten | 2.7 | 4.1 | 4.8 | 5.5 | 6.2 | 7.2 | 8.2 | 7.6 | 8.9 | 10.2 | 8.2 | 9.6 | 11.0 |
| Max-3 | 2.7 | 4.0 | 4.6 | 5.3 | 6.0 | 7.0 | 8.0 | 7.4 | 8.6 | 9.8 | 8.0 | 9.3 | 10.6 |
| Max-4 | 2.7 | 4.1 | 4.8 | 5.4 | 6.1 | 7.1 | 8.2 | 7.5 | 8.8 | 10.1 | 8.2 | 9.5 | 10.9 |
| H-275 | 1.8 | 2.7 | 3.1 | 3.6 | 4.0 | 4.7 | 5.4 | 5.0 | 5.8 | 6.6 | 5.4 | 6.3 | 7.2 |
| Matrix-11 | 1.4 | 2.1 | 2.5 | 2.8 | 3.2 | 3.7 | 4.2 | 3.9 | 4.6 | 5.2 | 4.2 | 4.9 | 5.6 |
| Fritzi-Pauley | 1 | 1.5 | 1.7 | 2.0 | 2.2 | 2.6 | 3.0 | 2.8 | 3.2 | 3.7 | 3.0 | 3.5 | 4.0 |
| Trichobel | 0.76 | 1.1 | 1.3 | 1.5 | 1.7 | 2.0 | 2.3 | 2.1 | 2.5 | 2.8 | 2.3 | 2.7 | 3.0 |

In Jalalabad, with a climate warmer than Bishkek and Almaty, also the PDN hybrids were among the highest trees at the end of the first season, e.g., H-33 with an average tree height of 2.2 m. The tallest trees, though, were Veronese with an average height of 2.6 m followed by the *P. deltoides* cultivar Samsun with an average height of 2.3 m (Table S4). On the other side, the *P. deltoides* cultivar 89M060 and the PDN hybrid H-11 only attained 1.2 m and 1.6 m, respectively, as average tree heights (Table S4). By the end of the second season, Max-3 (PNM) and Mirza Terek (PN) became the tallest and largest trees with average tree heights of 3.8 m and 3.5 m and average stem volumes of 2 dm$^3$ and 1.8 dm$^3$, respectively (Table S5). By end of the second season, the average stem volume of Max-3 (2 dm$^3$), which was lower, but still the same range, than the corresponding stem volumes of Max-3 in Almaty (2.4 dm$^3$) and Bishkek I (2.7 dm$^3$). H-11 only attained an average stem volume of 1.4 dm$^3$, which was only about two thirds of the corresponding stem volume of H-11 in Almaty at the end of the second season (Table 8 and Table S3).

H-33, which clearly attained the largest stem volume in Bishkek I and Almaty (Table 8) and was among the tallest in Jalalabad at the end of the first season, yielded the smallest stem volume in Jalalabad at the end of the second season with only 0.58 dm$^3$ (Tables S3 and S5).

In Osh, which has a similar climate as Jalalabad, the trees remained smaller at the end of first season compared to Jalalabad (Table S4). In Osh, the PDN hybrids H-8 and H-17 grew highest at the end of the first season (average tree heights of 1.4 m and 1.3 m, respectively), while other PDN hybrids were among the smallest with Vesten being the smallest with 0.7 m. Like in Bishkek I and Almaty, the PMT hybrids H-275 and Matrix-11 were among the smallest cultivars as well. Kazakhstani and the locally used Mirza Terek grew to average tree heights of 1.2 m and 1.1 m, respectively, which was almost as high as H-8 and H-17 (Table S4).

The site Lavar, which is saline, trees grew smaller across cultivars compared with nearby Almaty or Bishkek I, but also compared to Jalalabad and Osh, e.g., H-11, the tallest cultivar in Lavar in the middle of the second season, grew in average 1.80 m tall, which falls short more than 1 m compared to H-11 in Almaty already at the end of the first season. When comparing the different cultivars with each other on the site Lavar, the PDN hybrids also grew tallest (e.g., H-11 with 1 m average tree height) and the PMT hybrids H-275 and

Matrix-11 being among the smallest with 52 cm and 26 cm, respectively (Table S4). Max-3, which was among the tallest cultivars in Bishkek I and Almaty at the end of the second season, remained among the smallest cultivars in Lavar with an average height of 26 cm at the end of the second season. The locally developed cultivars Kazakhstani and Kyzyl-Tan also reached average heights of 26 cm only (Table S5).

On the site Bishkek II, the tree heights at the end of the first season were in a similar range as in Osh (Table S4), but less than half as tall as on the neighboring Bishkek I site or in Almaty (Tables 7 and 8). Though, the ranking between the cultivar groups was similar as in Bishkek I, Almaty, Jalalabad, or Lavar, with PD-cultivar 89M060, and the PDN hybrids Veronese, Oudenberg, H-33, Tiepolo, Bellini, and H-11 (average tree heights between 1 m and 1.6 m) being the tallest and Matrix-11, H-275, and the PT cultivars Muhle-Larsen and Trichobel being the smallest with average tree heights below 51 cm (Table S4). Mirza Terek, Fritzi-Pauley, Fastwood 1 and Fastwood 2 attained medium average tree heights with 90 cm, 89 cm, 71 cm, and 67 cm, respectively. In contrast to the other PDN hybrids, H-17 and H-8 remained small with 70 cm and 51 cm height. During the second season, Fritzi-Pauley showed the highest increment in terms of height, from 0.5 m to 1.8 m, while Max-3 attained the largest stem volume at the end of second season (0.34 dm$^3$). With regard to stem volume, the PD cultivar 89M060 and the PDN hybrids Samsun, Bellini, and Oudenberg ranked behind Max-3, followed by the locally used Mirza Terek (Table 8). Like in Jalalabad, but in stark contrast to neighboring Bishkek I, H-33 remained the smallest with only 0.06 dm$^3$ stem volume (Tables S3 and S5). Most of the trees across all parentages showed signs of phosphorus deficiency, which became visible through yellowish and partly reddish leaves at the top of the stems, while basal leaves remained green. These signs did not completely disappear after fertilizer application. On the neighboring Bishkek I site, such signs of phosphorus deficiency became visible as well, but disappeared quickly after fertilizer application.

In contrast to the previous sites, the PMT hybrids (Fastwood 1, Fastwood 2, and Matrix-11) were the tallest (in Khorog) or among the tallest (in Tup) cultivars on the two sites Khorog and Tup at the end of the first season. The average tree height of Fastwood 1 was 1.6 m in Tup and 1.3 m in Khorog at the end of the first season. At the end of the second season, Fastwood 1 and 2 were still the tallest or among the tallest cultivars with 3.2 m and 3.7 m in Tup and 2.4 m and 2.3 m in Khorog, respectively). In Tup, Fastwood 2 attained the largest stem volume (2.6 dm$^3$), followed by Oudenberg (PDN), Matrix-11 (PMT), and Max-3 (PNM) as listed in Table 7. The stem volume of Oudenberg of 2.3 dm$^3$ is lower than, but still in the range of the corresponding stem volumes in Almaty (2.5 dm$^3$) and Bishkek I (2.9 dm$^3$) at the end of their second seasons (Table S3 and Tables 8 and 9). The other PDN hybrids that had been planted in Tup had smaller stem volumes and were less tall than the PMT hybrids in Tup. In Khorog, by end of the second season the PDN hybrid Orion grew into trees as tall as Fastwood 1 and 2, while the PDN hybrids H-17 and H-11 remained among the smallest trees. On both sites, the locally used cultivars, Mirza Terek and *P. nigra* (*pyramidalis*) were among the smallest trees by end of the first and second season.

In Naryn, the third site under a cold semiarid climate, the results regarding tree heights at the end of the first season differed from the two former sites Tup and Khorog, as Orion, a PDN hybrid, grew highest (1.5 m), followed by Max-4 (PNM) with 1.2 m, and Fritzi-Pauley (PT) with 1.2 m. The PDN hybrids Oudenberg, Vesten, H-8, and Mirza Terek remained the shortest trees with average heights of 71 cm, 76 cm, 92 cm, and 60–70 cm, respectively (Table S4).

## 4. Discussion

The survival rates at the end of the first season in Bishkek I and Almaty are similar to survival rates reported by [16] and further literature there. The sites established in 2019 partly showed very low survival rates, in particular the PDN cultivars purchased from Italy on the sites Jalalabad, Bishkek II, and Tup. That can be explained by the long storage time between delivery and planting, as survival rates of those cultivars were higher in

Khorog and Lavar, where planting took place two to three weeks earlier than on the former sites. The high survival rates from the first season through winter into the second season is confounded by the findings by [16].

The annual stem biomass yields as calculated in Table 9 for Bishkek I are the same range as the annual yields published by [17] for a PDN hybrid in the Po Valley in Italy. The stem biomass yields of the PDN hybrids H-33, Orion, and H-17 after two years in Bishkek I are in the same range as *P. x canadensis* with biomass yields of 12.9 to 13.6 t/ha*a as reported by [18] from Idaho and California, but lower than NPP of 12 to 24 t/ha*a for *P. x canadensis* as listed by [13]. The lower values of Bishkek I compared to [13] can be explained as follows: in this study, stem biomass yields are presented so that branch and leaf biomass, which are a part of NPP, are ignored here. Secondly, the soil in Bishkek I had a low nutrient status, in particular with regard to phosphorus, as was visible during the first season. The fertilizer application listed for Bishkek I in Table 5 met the phosphorus requirements (20–36 kg P/ha*a), but not the nitrogen (182–246 kg N/ha*a) and potassium requirements (113–171 kg K/ha*a) as given by [13]. The tree heights and DBH values after the third season in Bishkek I are in the same range as the tree heights of 8.7–9.7 m and DBH of 6.3–7.8 cm, which were reported by [19] for PD and PDN cultivars from North Carolina. In Bishkek I, Max-3 reached an annual biomass yield which was in the same range as the 10.6 t/ha*a as published by [20] for PNM hybrid studied in Quebec, Canada. At the end of the second season, the hybrid Kazakhstani planted in Tup attained tree heights in the same range, 1.5–3.2 m, as that hybrid attained after four to five years on a site in northern Kazakhstan [21]. The slower growth in northern Kazakhstan can be explained by the colder climate and shorter growing season there compared to Tup. According to [22] local *P. nigra* var *pyramidalis* reached tree heights of 18–20 m and DBH of 11–13 cm after 20 years, while Max-3 and -4 and the PDN hybrids in Bishkek I reached DBH averages of 5.2 cm and more after three years. Therefore, it is to expect that the letter cultivars will need a shorter time to reach those heights and DBH as given for the local *P. nigra* var *pyramidalis*.

The finding that PMT and PT cultivars performed better under a colder climate than most of the PDN cultivars was confirmed by [23] on a site in Poland. There, PMT (H-275) and Frtzi-Pauley performed better than PDN hybrids. [24] reported much better growth of PDN cultivars in Italy compared to a site in Northern France. [25] studied growth and biomass yields of PDN and PNM hybrids along an elevation and climate gradient in Quebec, Canada. There, the growth and biomass yields decreased with elevation and a cooler climate across all cultivars, but the PDN hybrids exhibited the steepest decrease. The PDN hybrid attained almost the same stem volume as PNM on the site on the lowest elevation, but quickly fell onto the last rank with an increasing elevation. Next to the elevation, soil fertility impacted more strongly on growth of the PDN hybrids than on the PNM hybrid [26]. This is in line with the data of this study, as Max-3 performed better than the PDN hybrids in Jalalabad, where conditions were less favorable (less water was available and weed prevalence was higher) than in Bishkek I or Almaty.

Max-3 and Max-4 in Bishkek I reached similar tree heights, DBH, and biomass yields as Max- and Max-4 on a plantation in NE Germany published by [16]. Though, Fritzi-Pauley performed much better in that study in NE Germany compared to Bishkek I and also Almaty. Fritzi-Pauley in Almaty and Bishkek I grew slightly higher than the tree height of 2.1 m as published by [27] for Northern France after the first season. While Fritzi-Pauley grew taller than the PDN hybrids in Northern France, Fritzi-Pauley was significantly smaller than the PDN hybrids in Almaty and Bishkek I underlining that PDN hybrids have a higher potential than PT cultivars under hot continental climates.

The overall smaller tree heights and stem volumes in Lavar, the saline site, are in line with a study on growth rates of different poplar cultivars on saline soils in North Dakota [21]. There, the PDN and PNM cultivars were least affected, which reflects the results of this study regarding the PDN cultivars on the site Lavar. In contrast to [28], the PNM hybrid Max-3 remained among the smallest in Lavar.

The tree data from Bishkek II are significantly smaller than from the neighboring site Bishkek I. This stark difference can be explained by the soil properties, as the clayey texture, coupled with bad drainage, impacts negatively on poplar growth [13,29]. At the end of the second season, the pure species Fritzi-Pauley and Mirza Terek showed the largest increments, while a number of the otherwise high performing PDN hybrids (H-8, H-11, H-33) remained very small. This is particularly noteworthy, as those two cultivars grew, although the site was not irrigated throughout the whole second season. This can be explained by the better root development of the pure species compared to the PDN hybrids [12].

The lower growth performance in Jalalabad and Osh compared to Bishkek I and Almaty can be explained by the higher weed coverage on the former sites compared to the latter. This is in line with [30], who found a negative correlation between weed prevalence and stem wood production for poplars in Saskatchewan, Canada. The weeds outcompete the cuttings and trees for water, in particular during the first season after planting. Under such more adverse conditions, Mirza Terek and the PNM hybrids (Max-3) attained average tree heights in Osh and highest values in Jalalabad, while the otherwise high performing PDN hybrids (e.g., H-33) perform poorly.

The wide range of tree densities between sites might partly explain the differences with regard to grow rates between Khorog and Naryn. By end of the first season, trees in Naryn grew slightly higher compared to Khorog, though Khorog received fertilizer and weeds were controlled more intensively. The high tree density of 100,000 trees/ha in Khorog, as opposed to 25,000 trees/ha in Naryn, might have led to competition between trees, which impacted on their grow. In contrast, tree heights, dbh, and stem volumes were similar on the two sites Almaty and Bishkek I, despite the huge difference in tree densities, which were 62,500 in Almaty versus 13,800 to 27,700 trees per hectare in Bishkek I. Possibly the much better nutrient and water supply on those two sites outweighed effects of tree density at least in this young stage of tree development.

The leaf beetles did not impact on growth performance, as those beetles were controlled rapidly. Trees in Bishkek I, which was affected by pests, and in Almaty, which was not affected, attained similarly high growth rates. The two sites in Bishkek and Tup, which suffered from those beetles, were located close to other woodlands, which according to field observations harbored those beetles.

The preceding paragraphs described qualitatively relationships between the different environments of the sites, such as soil properties, climate, and weed prevalence, and resulting growth rates of the poplar cultivars included in this study. Initially, this study aimed at addressing genotype x location interactions. Though, different soil properties, e.g., the higher clay content of Bishkek II in comparison to the neighboring site Bishkek I, and differences in site management, which resulted in varying degrees of weed prevalence, water supply, and plant nutrition made it necessary to discuss the results in the light of genotype x environment interactions [15]. The design of this study, in particular the absence of control plots and replicates on the sites, e.g., with regard to different fertilizer doses or water supply, is a clear weakness of this study and needs to be addressed in further studies on local and promising new cultivars to be able to systematically describe genotype x environment interactions. These interactions are important for sound cultivar recommendations, in particular if cultivars are to be recommended under sub-optimal conditions where land users cannot guarantee optimal site management. Despite its weakness, this study is able to provide basic information to land users in the region Central Asia with regard to which group of cultivars has most promising potentials in the lower elevation versus high elevation areas. Furthermore, this study highlights to land users the most urgent operations, which are weed control, water supply, and the ability to react to nutrient deficiencies and pests, that need to be ensured to be able to tap the potentials of high yielding poplar cultivars.

## 5. Conclusions

This study investigated the growth rates of a number of poplar cultivars, *P. nigra* (PN), *P. deltoides* (PD), *P. trichocarpa* (PT), *P. alba* (PA), and *P. simonii* cultivars as well as *P. x canadensis* (PDN), *P. maximoviczii x trichocarpa* (PMT), *P. nigra x maximoviczii* (PNM), and *P. laurifolia x canadensis* (PLfND) hybrids, on experimental sites, ranging from hot continental climate to cold semiarid climate on higher elevations in Central Asia. The PDN hybrids, in particular H-33 and H-17, followed by Orion, Vesten, and Oudenberg grew tallest and yielded the highest biomass on the sites under a hot continental climate, provided that weeds were controlled thoroughly, sufficient water and plant nutrients were available, and pests were controlled. If these favorable conditions were not met, those high performing cultivars would not be able to unfold their potential. Under poor soil and management conditions, traditional and obviously more robust cultivars, as Fritzi-Pauley, Mirza Terek, or Max-3 perform better than the high yielding PDN cultivars. This implies that land users who wish to attain high growth rates and biomass yields from poplars should use high yielding PDN hybrids, but have to ensure intensive site management through the first and second season to tap the potential of those cultivars. If such intensive site management cannot be ensured, the more robust cultivars, as Fritzi-Pauley, Mirza Terek, or Max-3 should be used to avoid failure. Even if intensive site management can be ensured, Max-3 and Max-4 could be considered to be included into plantations of agroforestry systems, in order to reduce the risk of failure.

On sites on higher elevation and under a colder climate, the PMT and PT cultivars performed better than the PDN hybrids. In particular, H-8, H-11, H-17, and H-33 grew smaller than the PMT and PT cultivars, while Oudenberg and Orion only grew slightly smaller than the PMT and PT cultivars. In addition, the locally used cultivars performed worse on the higher elevation sites than the new cultivars investigated through this study.

In general, poplar-based agroforestry systems and plantations can increase their productivity by including cultivars investigated in this study. Furthermore, new cultivars, such as *P. trichocarpa x maximoviczii* or *P. deltoides x trichocarpa*, should be tested in Central Asia as well.

**Supplementary Materials:** The following are available online at https://www.mdpi.com/1999-490 7/12/3/373/s1, Table S1: survival rates [%] by sites and cultivars at the end of the first season, Table S2: survival rates [%] by sites and cultivars from the first season to the second season, Table S3: Tables of stem volumes at the end of the second growing season. The columns indicate the cultivar, average $\pm$ standard deviation, N—number of measured trees, and Min and Max—minima and maxima of tree heights. Letters indicate significant differences at $\alpha < 0.05$, Table S4: tables of tree heights at the end of the first growing season. The columns indicate the cultivar, average $\pm$ standard deviation, N—number of measured trees, and Min and Max—minima and maxima of tree heights. Letters indicate significant differences at $\alpha < 0.05$, Table S5: tables of tree heights at the end of the second growing season. The columns indicate the cultivar, average $\pm$ standard deviation, N—number of measured trees, and Min and Max—minima and maxima of tree heights. Letters indicate significant differences at $\alpha < 0.05$.

**Author Contributions:** Conceptualization, N.T. and S.F.; methodology, N.T., S.F., K.A. and Y.K.; investigation, N.T., S.F., K.A., B.E., R.F., Y.K., Y.Q. (Yodgor Qonunov), Y.Q. (Yosumin Qurbonbekova), N.R., M.R., S.Z.; writing—original draft preparation, N.T.; writing—review and editing, S.F.; visualization, N.T. and S.F.; project administration, N.T.; funding acquisition, N.T. All authors have read and agreed to the published version of the manuscript.

**Funding:** This work received financial support from the German Federal Ministry for Economic Cooperation and Development (BMZ) commissioned and administered through the Deutsche Gesellschaft für Internationale Zusammenarbeit (GIZ) Fund for International Agricultural Research (FIA), grant number: 81236142. Niels Thevs' position as an integrated expert at World Agroforestry was co-funded by BMZ as part of the Center for International Migration and Development (CIM) program. The sites in Khorog and Osh were funded by internal funds from World Agroforestry's Genebank program.

**Institutional Review Board Statement:** Not applicable.

**Informed Consent Statement:** Not applicable.

**Acknowledgments:** We wish to thank all colleagues within the partner institutions, who worked on the sites and maintained them in the course of this study.

**Conflicts of Interest:** The authors declare no conflict of interest.

# References

1. UNECE. *Forest Landscape Restoration in the Caucasus and Central Asia—Challenges and Opportunities, Background Paper for the Ministerial Roundtable on Forest Landscape Restoration in the Caucasus and Central Asia (21–22 June 2018, Astana, Kazakhstan)*; UNECE: Geneva, Switzerland, 2019.
2. Thevs, N.; Gombert, A.J.; Strenge, E.; Lleshi, R.; Aliev, K.; Emileva, B. Tree Wind Breaks in Central Asia and Their Effects on Agricultural Water Consumption. *Land* **2019**, *8*, 167. [CrossRef]
3. Thevs, N.; Strenge, E.; Aliev, K.; Eraaliev, M.; Lang, P.; Baibagysov, A.; Xu, J. Tree Shelterbelts as an Element to Improve Water Resource Management in Central Asia. *Water* **2017**, *9*, 842. [CrossRef]
4. Kort, J. Benefits of Windbreaks to Field and Forage Crops. *Agric. Ecosyst. Environ.* **1988**, *22*, 165–190. [CrossRef]
5. Schroeder, R.W.; Kort, J. Shelterbelts in the Soviet Union. *J. Soil Water Conserv.* **1989**, *2*, 130–134.
6. Albenskii, A.V.; Kalashnikov, A.F.; Ozolin, G.P.; Nikitin, P.L.; Surmach, G.P.; Kulik, N.F.; Senkevich, A.A.; Kasyanov, F.M.; Pavlovskii, E.S.; Roslyakov, N.V. *Agroforestry Melioration*; Lesnaya Promyshlennost: Moscow, Russia, 1972.
7. Stepanov, A.M. *Agroforestry Melioration in Irrigated Lands*; Agropromizdat: Moscow, Russia, 1987.
8. Dokuchaev Scientific Research Institute of Agriculture. *Agroforestry Melioration of Vegetable Cultures and Potato*; Voronesh Publishing House: Voronesh, Russia, 1961.
9. Ruppert, D.; Welp, M.; Spies, M.; Thevs, N. Farmers' Perceptions of Tree Shelterbelts on Agricultural Land in Rural Kyrgyzstan. *Sustainability* **2020**, *12*, 1093. [CrossRef]
10. Ozolin, G.P. *Cultivation of Poplars under Conditions of Artificial Irrigation*; Lesnaya Promyslennost: Moscow, Russia, 1966.
11. Usmanov, A.U. Poplar. In *Dendrology of Uzbekistan. Volume III*; Esenina, T.S., Ed.; Fan Uxber SSR: Tashkent, Uzbekistan, 1971.
12. Ceulemans, R.; Deraedt, W. Production Physiology and Growth Potential of Poplars under Short-Rotation Forestry Culture. *For. Ecol. Manag.* **1999**, *121*, 9–23. [CrossRef]
13. Isebrands, J.G.; Richardson, J. *Poplars and Willows—Trees for Society and the Environment*; FAO: Rome, Italy; CABI: Wellingford, UK, 2014.
14. Clifton-Brown, J.; Harfouche, A.; Casler, M.D.; Jones, H.D.; MacAlpine, W.J.; Murphy-Bokern, D.; Smart, L.B.; Adler, A.; Ashman, C.; Awty-Carroll, D.; et al. Breeding progress and preparedness for mass-scale deployment of perennial lignocellulosic biomass crops switchgrass, miscanthus, willow and poplar. *GCB Bioenergy* **2019**, *11*, 118–151. [CrossRef] [PubMed]
15. Annicchiarico, P. *Genotype x Environment Interaction*; FAO Plant Production and Protection Paper 174; FAO: Rome, Italy, 2002.
16. Landgraf, D.; Carl, C.; Neupert, M. Biomass Yield of 37 Different SRC Poplar Varieties Grown on a Typical Site in North Eastern Germany. *Forests* **2020**, *11*, 1048. [CrossRef]
17. Bergante, S.; Facciotto, G.; Marchi, M. Growth dynamics of 'Imola' poplar clone (Populus ×canadensis Mönch) under different cultivation inputs. *Ann. Silvic. Res.* **2020**, *44*, 71.
18. Stanton, B.J.; Bourque, A.; Coleman, M.; Eisenbies, M.; Emerson, R.M.; Espinoza, J.; Gantz, C.; Himes, A.; Rodstrom, A.; Shuren, R.; et al. The practice and economics of hybrid poplar biomass production for biofuels and bioproducts in the Pacific Northwest. *BioEnergy Res.* **2020**, 1–18. [CrossRef]
19. Maier, C.A.; Burley, J.; Cook, R.; Ghezehei, S.B.; Hazel, D.W.; Nichols, E.G. Tree Water Use, Water Use Efficiency, and Carbon Isotope Discrimination in Relation to Growth Potential in Populus deltoides and Hybrids under Field Conditions. *Forests* **2019**, *10*, 993. [CrossRef]
20. Fortier, J.; Gagnon, D.; Truax, B.; Lambert, F. Biomass and volume yield after 6 years in multiclonal hybrid poplar riparian buffer strips. *Biomass Bioenergy* **2010**, *34*, 1028–1040. [CrossRef]
21. Maissupova, I.K.; Sarsekova, D.N.; Weger, J.; Bubeník, J. Comparison of the growth of fast-growing poplar and willow in two sites of Central Kazakhstan. *J. For. Sci.* **2017**, *63*, 239.
22. Albenskii, A.V. *Cultivation of Poplars*; State Forestry Publishing House: Moscow, Russia, 1946.
23. Niemczyk, M.; Kaliszewski, A.; Jewiarz, M.; Wróbel, M.; Mudryk, K. Productivity and biomass characteristics of selected poplar (Populus spp.) cultivars under the climatic conditions of northern Poland. *Biomass Bioenergy* **2018**, *111*, 46–51. [CrossRef]
24. Dillen, S.Y.; Marron, N.; Sabatti, M.; Ceulemans, R.; Bastien, C. Relationships among productivity determinants in two hybrid poplar families grown during three years at two contrasting sites. *Tree Physiol.* **2009**, *29*, 975–987. [CrossRef]
25. Truax, B.; Gagnon, D.; Fortier, J.; Lambert, F. Yield in 8 year-old hybrid poplar plantations on abandoned farmland along climatic and soil fertility gradients. *For. Ecol. Manag.* **2012**, *267*, 228–239. [CrossRef]
26. Truax, B.; Gagnon, D.; Fortier, J.; Lambert, F. Biomass and Volume Yield in Mature Hybrid Poplar Plantations on Temperate Abandoned Farmland. *Forests* **2014**, *5*, 3107–3130. [CrossRef]
27. Barigah, T.; Saugier, B.; Mousseau, M.; Guittet, J.; Ceulemans, R. Photosynthesis, leaf area and productivity of 5 poplar clones during their establishment year. *Ann. For. Sci.* **1994**, *51*, 613–625. [CrossRef]

28. Zalesny, J.R.S.; Stange, C.M.; Birr, B.A. Survival, Height Growth, and Phytoextraction Potential of Hybrid Poplar and Russian Olive (Elaeagnus Angustifolia L.) Established on Soils Varying in Salinity in North Dakota, USA. *Forests* **2019**, *10*, 672. [CrossRef]
29. Shifflett, S.D.; Hazel, D.W.; Nichols, E.G. Sub-Soiling and Genotype Selection Improves Populus Productivity Grown on a North Carolina Sandy Soil. *Forests* **2016**, *7*, 74. [CrossRef]
30. Welham, C.; Van Rees, K.; Seely, B.; Kimmins, H. Projected long-term productivity in Saskatchewan hybrid poplar plantations: Weed competition and fertilizer effects. *Can. J. For. Res.* **2007**, *37*, 356–370. [CrossRef]

*forests*

*Article*

# Economic Modelling of Poplar Short Rotation Coppice Plantations in Hungary

**Endre Schiberna ***, **Attila Borovics and Attila Benke**

Forest Research Institute, University of Sopron, H-9400 Sopron, Hungary; borovics.attila@uni-sopron.hu (A.B.); benke.attila@uni-sopron.hu (A.B.)

* Correspondence: se@erti.hu

**Abstract:** No study has been previously completed on the range of sites, potential yield, and financial characteristics of poplar short rotation coppice plantations (SRC) in Hungary. This paper conducts a literature survey to reveal the biomass production potential of such plantations and presents a model that is used to analyze their financial performance. The results indicate that the break-even-point of production is between 6 and 8 oven-dry tons per hectare per year once a minimum cost level and wood chip price within a 10% range of the 2020 value are considered. The higher the wood chip price, the lower the break-even-point. Since the model excluded the administrative costs that depend on the type and size of the management organization, the break-even-points can be significantly higher in reality, which suggests that short rotation energy plantations can be a financially reasonable land-use option in above average or even superior poplar-growing sites. The rotation period of industrial poplar plantations that produce high quality veneer logs ranges from 12 to 25 years. Though such sites can provide higher returns on investment, short rotation plantations have the advantage of providing a more evenly distributed cash flow. To facilitate the wider application of poplar SRC, the related policies need to apply specific subsidies and allow the rotation cycle to be extended up to 20–25 years, which is currently limited to 15 years.

**Keywords:** *Populus* sp.; biomass; bioenergy; short rotation woody crops; SRC; financial analysis; break-even-point; net present value

**Citation:** Schiberna, E.; Borovics, A.; Benke, A. Economic Modelling of Poplar Short Rotation Coppice Plantations in Hungary. *Forests* **2021**, *12*, 623. https://doi.org/10.3390/f12050623

Academic Editors: Ronald S. Zalesny, Jr. and Andrej Pilipović

Received: 1 April 2021
Accepted: 11 May 2021
Published: 14 May 2021

**Publisher's Note:** MDPI stays neutral with regard to jurisdictional claims in published maps and institutional affiliations.

## 1. Introduction

Short rotation coppice (SRC) is a woody plantation system in which fast-growing tree or shrub species are planted in high density and coppiced in short harvesting cycles (2–8 years) at ground level for a 15–25-year lifecycle. From a management intensity perspective, SRC stands between agricultural crop production and managed forests. The reduction of management intensity originating from converting agricultural land use to SRC cultivation results in additional environmental benefits, especially in soil protection and the enhancement of soil life. These benefits, however, lag the services close-to-nature forests can provide, both in range and magnitude [1,2].

The distinctive characteristic of this type of woody plantation is that under favorable circumstances—including proper site conditions, fast-growing cultivars, and proper technological choices—SRCs can provide high yield in short harvesting cycles and can be maintained for over 20 years depending on their health status [3]. The financial performance of productive plantations is comparable to that of annual crops, which creates opportunity for diversifying farmer income [4]. Furthermore, SRCs have a favorable carbon and energy balance [5], and their products can contribute to available wood supply that might have been reduced by nature conservation measures in close-to-nature forests [6].

However, the above-mentioned ecological and economic benefits can only be realized if SRCs are financially viable. For this reason, financial performance of SRCs should be evaluated under a multifactorial financial environment, which considers the local ecological and financial conditions as determinant factors.

SRC as a land-use option has been thoroughly investigated in recent decades. Numerous international studies have dealt with the economic analysis of SRCs, and the conclusions in these studies highlight the most important aspects and considerations influencing successful SRC application. Biomass yield plays a significant role in generating sufficient revenue. Yield can be increased by selecting superior cultivars [7] and is markedly influenced by plantation site quality [8]. The break-even price of woodchips depends on plantation yield [9] and determines the harvesting cycle [10]. In addition to the natural yield, plantation profitability is particularly influenced by woodchip prices and subsidies [11–15]. Overall, plantation profitability depends both on biomass price and biomass yield [16]. Plantation establishment, maintenance, harvest, and wood transportation costs also affect the financial result. Cultivation costs can contribute 75% to the costs of woodchips at power plants [17,18]. Under less-than-optimal site conditions or less than ideal financial circumstances, alternative methods to increase plantation profitability are available. These include harvesting cycle extension, woodchip drying using waste heat [19], and irrigation and fertilization with wastewater and sewage sludge [20]. The establishment of an SRC system is a long-term investment requiring careful planning based on evaluations of environmental site conditions, financial circumstances (rental costs of land, cash flow, and return), and mathematical models [21,22].

The forest area of Hungary is slightly greater than 2 million hectares [23], which represents a 22.7% forest ratio. Forest stands and plantations consisting of different poplar species (*Populus* sp.) represent a 10.6% share of forested land and an 8.4% share in standing volume [24]. Apart from black locust (*Robinia pseudoacacia* L.), hybrid poplars can be considered as the main species of Hungarian plantation forestry and are cultivated on more than 105 thousand hectares. The number of cultivated hybrid poplar clones exceeds 20, with most belonging to the Euramericana poplar group [*Populus* × *euramericana* (Dode) Guinier]. The current set of cultivated clones in Hungary include many of foreign origin (Italy, Belgium, and the Netherlands), but the Hungarian poplar breeding program has also produced numerous successful clones since its launch following the Second World War. The most frequently used cultivars in afforestations and reforestations in Hungary are I-214, Pannonia, Kopecky, Agathe-F (OP-229), and Koltay.

Hybrid poplar cultivars prefer semi-humid or humid climates combined with loose or medium-hard soils preferably with high humus content, or semi-arid climates with high groundwater levels. Concerning Hungary's ecological conditions, in the absence of available shallow ground water, sites with 600 mm or greater annual precipitation and sandy loam or loamy soils (without any soil failures) can be considered suitable for planting poplar SRCs. Since Hungary is in the xeric limit zone of forests [25], even minor changes in site conditions can result in perceivable wood increment differences; thus, plantations can show these mosaic differences [20].

Hybrid poplar clones are extremely susceptible to weed competition and various damaging agents [26,27], the occurrence and severity of which depend on many factors. Therefore, these are considered risks rather than calculable yield losses or additional costs. Hybrid poplar growth is strongly connected to high precipitation or ground water within reach of the root system. Climate change projections raise concerns of shifting climate zones and the possibility that a large proportion of currently available poplar growing sites will not be suitable for this purpose in the next 50–100 years [28–30]. These prospects are especially worrying since the groundwater level in the Great Plain, which covers approximately two-thirds of the country, is decreasing. Moreover, investigations have not arrived at any clear conclusions concerning the outcome of this change. The interplay between plantations and the changing ground-water-level is also not fully understood [31,32].

Although field experiments of poplar SRCs in Hungary started in the 1980s, their wider utilization began in 2007 following the implementation of an SRC subsidy scheme. This financial support covered establishment costs, which included site preparation, soil fertilization, purchasing/storage of reproductive materials, and planting, as well as initial maintenance, pavement construction, fencing, etc.

Newly bred, fast-growing Italian poplar cultivars (AF-2, Monviso) and well-known forestry cultivars (I-214, Pannonia, Kopecky, Koltay) comprised 55% of the very short rotation plantations. The area of poplar SRC reached a few thousand hectares, but sources recording the actual area contradict each other. This plantation process ceased in 2013 after the subsidy scheme was terminated.

Experiences with poplar SRCs reveal some of the systematic weaknesses of this land-use form. Plantation yield did not meet expectations. Moreover, harvesting services were difficult to reach and turned out to be expensive due to the small scale of the plantations. Expectations on wood chip demand and calculations on transportation costs were overly optimistic. Many poplar SRCs became unprofitable despite the above-mentioned subsidy scheme. These widespread failures triggered the need for deeper financial analysis in this field.

The current study aims to survey the range of natural yields of poplar SRCs in Hungary that are based on sound evidence, to construct realistic and efficient model-technologies that can be applied in various poplar-growing sites, and to evaluate the financial performance of poplar SRCs.

## 2. Materials and Methods

### 2.1. Data Sources

Scientific and other literature from 1996 to 2018 have been used to survey the various combinations of sites, cultivars, and technologies [33–37]. In addition to these data sources, unpublished results of the authors' own field experiments have also been utilized.

Financial data were collected from various sources, as no single source provided all the necessary items. The National Land Centre in Hungary operates a statistical program that collects the contractor fees of various forestry-related operations, as well as the prices of wood products and other forest products such as propagation materials [38]. Items that were not included in this source were collected through oral interviews with producers [39].

### 2.2. Technology

The current paper defines plantation management technology as both the major characteristics of the plantation, such as the poplar cultivar, spacing, rotation cycle, and lifecycle, as well as the series of operations from establishment through maintenance and periodic harvests to liquidation. Good technology choices will result in a near-optimal use of the site potential, which is reflected in the financial results of production. However, such optimization is greatly limited in practice. For instance, weather conditions during the short production period can diverge significantly from predictions, which affects growth rate, health status, weed control, nutrient supply, etc. Available machinery may also determine spacing, maximum dimension of the trees, and harvesting methods.

The model applied in this study is designed to reflect the most typical spacing, lifecycle, and rotation cycle of poplar plantations in Hungary. Moreover, it only contains the minimum number of operations. The model is divided into three yield-categories representing the full spectrum of poplar-sites that have been recorded in the literature.

The model includes three yield categories: high yield, average yield, and low yield with 12 $t_{OD}$/ha/year, 8 $t_{OD}$/ha/year, and 4 $t_{OD}$/ha/year nominal growth rates, respectively. Nominal growth rate refers to the mean growth rate in the first two rotations combined. Growth rates in the various rotation periods are calculated with the nominal growth rate and the modification factor presented in Table 1.

**Table 1.** Modification factors of nominal growth rate over the rotation periods of poplar Short Rotation Coppices (SRC) in this study.

| Rotation Period | 1 | 2 | 3 | 4 | 5 |
|---|---|---|---|---|---|
| Growth rate modification factor | 100% | 100% | 95% | 90% | 85% |

Due to legal restrictions, plantation lifecycles are limited to 15 years in all three models. The high yield and average yield categories have three-year rotation cycles, while in the low yield category has a four-year cycle. This differentiation is justified by the fact that a lower growth rate allows for a longer growth period before the growing space is fully used up and the increasing competition induces significant mortality.

Establishing the plantation includes planning, soil preparation, and planting. Planning is based on site inspection and the plan is submitted to the Forest Authority for approval. Soil preparation depends on the current state of the specific land lot. In the case of poplar, it is essential to provide well-cultivated, loose soil for maximum growth. The model assumes that one ploughing and one disking before planting is sufficient. In total, 5556 poplar cuttings were planted with 3.0 × 0.6 m spacing. The above characteristics of plantation establishment is uniformly applied in all three yield categories.

Harvesting takes place according to the rotation cycle. Harvesting, chipping, storage, and transportation methods may vary in practice, and they are not specified in the model, as we only need to assume that the same methods apply to the three yield categories. The harvest costs are calculated on an oven-dry matter basis, and the unit cost depends on the harvested amount per hectare. The harvest-fee of an oven-dry ton is smaller in the higher yield scenario than it is in the average and low yield categories.

Harvest was followed by plantation maintenance, which includes weed control, plant protection, and nutrient recycling into the soil. Unless applied preventively, plant protection is not a periodic activity; however, since the need for it cannot be predicted, it is incorporated into the model as such.

Under excellent site conditions, the plantation can provide high yield, which implies that precipitation or ground water provide sufficient amounts of water and that there is no nutrient shortage in the soil. To fulfill the latter condition, the model includes nutrient recycling at a rate of 150 kg/ha after each harvest. Soil fertilization is decreased in the average yield category and is not applied in the low yield category because under less favorable conditions, available water poses a stronger limitation than nutrient supply. Therefore, nutrient recycle would not increase the growth rate.

Overhead costs, especially business administration expenses, depend largely on the type and scale of the organization at hand. Since it is not within the focus area of this study, these costs are excluded from the calculations. Only two overhead cost items are considered, both of which are essential regardless of the scale and legal form of the operation. One of these is the cost of a technical advisor, which is constant in all three categories. Another is the land rental fee, which represents site quality. Better site conditions entail higher land rental fees since the growth potential of the site correlates to its humus content. Higher humus content is also an advantage in agriculture.

At the end of the lifecycle, the plantation needs to be liquidated according to legal regulations. This includes the removal or grinding of stumps and roots.

### 2.3. Financial Evaluation

This paper applies four indicators to describe and compare the financial performance of the yield categories in the model. The cash flow of the three yield categories can be calculated based on the land management operations in the model and their unit costs.

The mean annual net income (MANI) is the quotient of the total net income and the lifecycle of the investment. The total net income of the investment is the cash flow sum. Net present value (NPV) is applied to describe the absolute financial value of the plantation as an investment at the time of its establishment. Annuity (A) is used to represent a theoretical series of constant annual income for the life period of the investment that could replace the actual cash flow without changing its NPV. Internal rate of return (IRR) is the interest rate at which the NPV of the investment is zero. The equations of these financial parameters are as follows:

$$\text{MANI} = \frac{\sum_{i=0}^{n} CF_i}{n} \tag{1}$$

$$NPV = \sum_{i=0}^{n} \frac{CF_i}{(1+p)^i} \tag{2}$$

$$A = NPV \frac{p*(1+p)^n}{(1+p)^n - 1} \tag{3}$$

$$NPV = 0 = \sum_{i=0}^{n} \frac{CF_i}{(1+IRR)^i} \tag{4}$$

where $i$ is years, $n$ is lifecycle of investment, $p$ is interest rate, $CF_i$ is total cash flow in year $i$, MANI is Mean Annual Net Income, NPV is Net Present Value, A is annuity of the Net Present Value, and $IRR$ is internal rate of return.

## 3. Results

### 3.1. Growth Rate of Poplar SRCs

To evaluate biomass yield levels that can be considered typical under Hungarian ecological circumstances, biomass yield data were collected from articles and other scientific works. The data sources and the parameters of the experiments are summarized and referenced in Table 2.

The cited experimental plantations represent all major geographical regions (Great Plain, Transdanubia, and Northern Hungary) and cover the range of sites suitable for poplar growing in the country. The most frequently used poplar clones were AF-2, Agathe-F, I-214, and Pannónia. Most of the sites have a sand soil texture, but loamy sites can also be found (Hanságliget and Bajti). Three sites (Dejtár, Karancslapujtő, Tiszakécske) are characterized by seasonal natural additional water supplementation, which in the site classification terminology in Hungary means that groundwater in springs is no deeper than 1.5 m, implying that it remains accessible to tree roots during most of the growing season. No irrigation was reported at any site. Soil fertilization was applied in two cases; in Dejtár the experiment focused on the effects of manure fertilization and combined manure and ash fertilization compared to no fertilization. The planting densities and spacing show great variations: the highest planting density was 13,333 pieces per hectare (1.5 × 0.5 m spacing), while the lowest was 2500 trees per hectare (2.0 × 2.0 m spacing); both were used in Hanságliget where the primary aim of the experiment was to study the effects of spacing on growth. The plantations were harvested in different rotation periods between 2 and 8 years. The Helvécia, Karancslapujtő, and Tiszakécske experiments were dedicated to investigating optimal harvesting cycles.

In addition to data gained from the literature, biomass yields measured in an experimental plantation in Bajti Nursery, located next to Sárvár, in Western Transdanubia, were used for evaluation as well. The experimental short-rotation plantation was planted in 2007 using a randomized block design, three repetitions, and a planting density of 8333 tree/hectare (3.0 m between the rows and 0.4 m in the rows). In addition to black locust and willow cultivars, 59 different poplar clones were planted in single-row parcels, with 100 cuttings per row. The soil type of the garden is loamy forest soil on alluvial subsoil with low humus content. The second and the third repetitions were harvested after the first vegetation period. Following the second vegetation period, the third repetition was harvested again. As a result, the plantation had 1-, 2-, and 3-year-old shoots after the third year. Only the 3-year-old shoots (repetition) were harvested after the third vegetation period. The biomass yield of most of the clones was measured each year after harvesting.

Based on the results of the literature survey, the lowest measured yield in Hungary was 0.8 $t_{OD}$/ha/year, while the highest was 14.1 $t_{OD}$/ha/year. Though other sources [40] from similar bio-geographic conditions suggest that even higher yields could have been achieved, we accept the results above for the case of Hungary.

Table 2. The summary of experimental poplar SRCs used in this study.

| Sources | Cultivars | Soil Texture | Add.Water Suppl. | Nutrition Supplement | Spacing (m × m) | Planting Density (Stems/ha) | Rotation Length (Year) | Yield ($t_{OD}$/ha/Year) | Location |
|---|---|---|---|---|---|---|---|---|---|
| [33] | n.d. | n.d. | n.d. | n.d. | 3.0 × 0.5 | 6667 | 2 | 12.0 * | Szakoly ** |
| [34] | AF-2 | Sand | Seasonal | Manure | 3.0 × 1.0 | 3333 | 6 | 13.1 | Dejtár |
| [35] | AF-2 | Sand | Seasonal | Manure, wood ash | 3.0 × 1.0 | 3333 | 2 | 6.6–8.2 | Dejtár |
| | Pannónia | n.d. | Seasonal | n.d. | 1.5 × 1.0 | 6667 | 3, 7, 8 | 1.7–9.1 | Karancslapujtő |
| | | | | | 2.0 × 1.0 | 5000 | 3, 7, 8 | 0.9–7.4 | |
| | | | | | 2.0 × 1.5 | 3333 | 3, 7 | 0.8–7.9 | |
| | | | | | 2.0 × 3.0 | 1667 | 3, 7, 8 | 1.5–7.0 | |
| | Agathe-F | Sand | No | n.d. | 1.5 × 0.5 | 13,333 | 4, 5, 8 | 2.4–6.3 | Helvécia |
| | | | | | 1.5 × 1.0 | 6667 | 4, 5, 8 | 1.8–5.5 | |
| | | | | | 1.5 × 2.0 | 3333 | 4, 5, 8 | 1.6–6.4 | |
| [36] | BL Constanzo | Sand | No | n.d. | 1.5 × 1.0 | 6667 | 3, 4, 5, 8 | 2.2–5.9 | Helvécia |
| | I-214 | | | | 1.5 × 1.0 | 6667 | 3, 4, 5, 8 | 1.8–3.2 | |
| | Pannónia | | | | 1.5 × 1.0 | 6667 | 3, 4, 5, 8 | 4.1–6.1 | |
| | Agathe-F | | | | 1.5 × 1.0 | 6667 | 3, 4, 5, 8 | 3.2–6.2 | |
| | S-298-8 | | | | 1.5 × 1.0 | 6667 | 3, 4, 5, 8 | 1.7–3.5 | |
| | H-328 | Sand | Seasonal | n.d. | 1.5 × 1.0 | 6667 | 5, 6, 7, 8 | 6.7–7.2 | Tiszakécske |
| | Kornik-21 | | | | 1.5 × 1.0 | 6667 | 5, 6 | 7.3–8.1 | |
| | S-298-8 | | | | 1.5 × 1.0 | 6667 | 5, 6 | 7.4–8.5 | |
| [37] | I-214 | Loam | No | n.d. | 1.5 × 0.5 | 13,333 | 4 | 11.5 | Hánságliget |
| | | | | | 1.0 × 1.0 | 10,000 | 4 | 9.6 | |
| | | | | | 2.0 × 2.0 | 2500 | 4 | 7.5 | |
| | Agathe-F | | | | 1.5 × 0.5 | 13,333 | 4 | 14.1 | |
| | | | | | 1.0 × 1.0 | 10,000 | 4 | 12.9 | |
| | | | | | 2.0 × 2.0 | 2500 | 4 | 12.3 | |
| | I 45/51 | | | | 1.5 × 0.5 | 13,333 | 4 | 11.6 | |
| | | | | | 1.0 × 1.0 | 10,000 | 4 | 8.7 | |
| | | | | | 2.0 × 2.0 | 2500 | 4 | 6.0 | |
| | Blanc du Poitou | | | | 1.5 × 0.5 | 13,333 | 4 | 9.4 | |
| | | | | | 1.0 × 1.0 | 10,000 | 4 | 9.8 | |
| | | | | | 2.0 × 2.0 | 2500 | 4 | 6.9 | |
| **** | Koltay I-214 Pannónia | Loam | No | No | 3.0 × 0.4 | 8333 | 3 | 8.0 *** 5.9 *** 5.2 *** | Bajti (Sárvár) |

* Calculated from fresh biomass weight, using an average water content of 55%. ** The original locations are unpublished. The most probable nearest location is shown. *** Average annual yield measured between 2011 and 2015. **** unpublished.

Three yield categories were created for further modelling:

- Low yield category: below 6 $t_{OD}$/ha/year;
- Average yield category: 6 $t_{OD}$/ha/year and above but less than 10 $t_{OD}$/ha/year;
- High yield category: 10 $t_{OD}$/ha/year and above.

*3.2. Financial Model*

The costs of the three yield categories of poplar SRCs are presented in Table 3. The establishment and liquidation of the plantation have the exact same costs in all categories, as site quality or the yield of the plantation have no influence on these operations. The maintenance costs decrease from high yield to low yield, while the harvesting costs per oven-dry matter increase in the same direction. Overhead costs also decrease toward lower yield categories due to the decreasing land rental fees.

**Table 3.** Financial model of poplar SRCs in three yield categories applied in this study [38,39].

| Costs | Unit | High Yield | Average Yield | Low Yield |
|---|---|---|---|---|
| **Total establishment costs** | | **726.3** | **726.3** | **726.3** |
| Site inspection | | 42.9 | 42.9 | 42.9 |
| Planning | €/ha | 42.9 | 42.9 | 42.9 |
| Soil preparation | | 128.6 | 128.6 | 128.6 |
| Propagation material | | 256.0 | 256.0 | 256.0 |
| Planting | | 256.0 | 256.0 | 256.0 |
| **Total maintenance costs** | | **242.9** | **185.7** | **128.6** |
| Weed control | €/ha | 100.0 | 100.0 | 100.0 |
| Plant protection | | 28.6 | 28.6 | 28.6 |
| Nutrient supply | | 114.3 | 57.1 | 0.0 |
| **Total harvesting costs** | | **11.4** | **14.3** | **17.1** |
| Cutting | €/$t_{OD}$ | 5.7 | 8.6 | 11.4 |
| Transporting | | 5.7 | 5.7 | 5.7 |
| **Liquidation costs** | €/ha | **257.1** | **257.1** | **257.1** |
| **Total overhead costs** | | **200.0** | **171.4** | **142.9** |
| Rental fee | €/ha | 171.4 | 142.9 | 114.3 |
| Technical advisor | | 28.6 | 28.6 | 28.6 |

Bold rows represent cost-type-groups, which are the sum of the cost items below.

Ranging from 43.5% to 51.3%, overhead costs have the largest share in the total costs over the lifecycle of the poplar SRCs. Harvesting and transportation costs are the second largest contributors to total costs, and their share ranges between 24.0% and 26.4%. There is a larger difference in the maintenance cost distribution, as it decreases from 16.6% in the high yield category to 2.6% in the low yield category. Although the establishment and liquidation costs are the same in all yield categories, they play a less important role in the high yield category with only a 13.4% share compared to the 22.1% share in the low yield category. Table 4 summarizes the data above.

**Table 4.** The share of cost types within the total costs of the poplar SRCs yield categories in this study.

| Yield Categories | Establishment and Liquidation | Harvesting and Transportation | Maintenance | Overheads | Total Costs |
|---|---|---|---|---|---|
| High yield | 13.4% | 26.4% | 16.6% | 43.6% | 100% |
| Average yield | 15.6% | 26.1% | 14.7% | 43.5% | 100% |
| Low yield | 22.1% | 24.0% | 2.6% | 51.3% | 100% |

The cash flow balance of each yield category is calculated every year and presented in Table 5. In all cases, the price of the wood chips delivered to the place of use is 60.0 €/$t_{OD}$. The MANI of the low yield category is −77.0 €/ha/year, which means that even if the overhead costs are not entirely incorporated into the model, the cash flow balance over the lifecycle of the plantation is negative. As a result of this, there is no reason to consider the remaining financial parameters for this category. The average yield and high yield categories provide 44.9 €/ha/year and 201.4 €/ha/year MANI, respectively.

**Table 5.** The cash flow and the financial parameters of the poplar SRC yield categories in this study.

| Parameters | | Unit | High Yield | Average Yield | Low Yield |
|---|---|---|---|---|---|
| Reference yield | | $t_{OD}$/ha/year | 12 | 8 | 4 |
| Rotation cycle | | y | 3 | 3 | 4 |
| Interest rate | | % | | 3.0 | |
| Price of wood chips | | €/$t_{OD}$ | | 60.0 | |
| Cash Flow | year 0 | €/ha | −726.3 | −726.3 | −726.3 |
| | year 1 | | −442.9 | −357.1 | −271.4 |
| | year 2 | | −200.0 | −171.4 | −142.9 |
| | year 3 | | 1548.6 | 925.7 | −142.9 |
| | year 4 | | −442.9 | −357.1 | 542.9 |
| | year 5 | | −200.0 | −171.4 | −271.4 |
| | year 6 | | 1548.6 | 925.7 | −142.9 |
| | year 7 | | −442.9 | −357.1 | −142.9 |
| | year 8 | | −200.0 | −171.4 | 542.9 |
| | year 9 | | 1461.1 | 870.9 | −271.4 |
| | year 10 | | −442.9 | −357.1 | −142.9 |
| | year 11 | | −200.0 | −171.4 | −142.9 |
| | year 12 | | 1373.7 | 816.0 | 508.6 |
| | year 13 | | −442.9 | −357.1 | −271.4 |
| | year 14 | | −200.0 | −171.4 | −142.9 |
| | year 15 | | 1029.1 | 504.0 | 62.9 |
| Total Lifecycle Income | | €/ha | 10,152.0 | 6768.0 | 3480.0 |
| Total Lifecycle Cost | | €/ha | 7131.4 | 6094.9 | 4634.9 |
| Lifecycle Cash Flow Balance | | €/ha | 3020.6 | 673.1 | −1154.9 |
| Mean Annual Net Income (MANI) | | €/ha/y | 201.4 | 44.9 | −77.0 |
| Net Present Value (NPV) | | €/ha | 2238.9 | 314.0 | −1114.0 |
| Annuity (A) | | €/ha/y | 187.5 | 26.3 | −93.3 |
| Internal Rate of Return (IRR) | | % | 21.6 | 6.8 | * |

* Not calculable.

For the calculation of NPV, A, and IRR, a 3.0% reference net interest rate was applied, which is understood as an interest rate above inflation. The NPV of the high yield category is 2238.9 €/ha, which equals 187.5 €/ha/year annuity and 21.6% IRR. Average yield category provides 314.0 €/ha NPV, 26.3 €/ha/year annuity, and 6.8% IRR.

Although the cost structure in this calculation is incomplete, the break-even-point (BEP) can be calculated as identifying yield-price combinations that provide zero (or near zero) MANI. Tables 6 and 7 show the result of this calculation with constant cost structure both with and without a 75% plantation establishment subsidy, respectively.

**Table 6.** The MANI of poplar SRCs by yield and the price of the wood chips in this study (break-even-points marked with grey background).

| Price (€/$t_{OD}$) | Yield ($t_{OD}$/ha/Year) | | | | | | | | | | |
|---|---|---|---|---|---|---|---|---|---|---|---|
| | Low Yield | | | | Average Yield | | | High Yield | | | |
| | 3 | 4 | 5 | 6 | 7 | 8 | 9 | 10 | 11 | 12 | 13 |
| 54 | −135.8 | −100.2 | −64.6 | −74.9 | −37.6 | −0.2 | 37.1 | 53.7 | 93.7 | 133.7 | 173.7 |
| 57 | −127.1 | −88.6 | −50.1 | −58.0 | −17.8 | 22.3 | 62.5 | 81.9 | 124.7 | 167.5 | 210.4 |
| 60 | −118.4 | −77.0 | −35.6 | −41.1 | 1.9 | 44.9 | 87.8 | 110.1 | 155.7 | 201.4 | 247.0 |
| 63 | −109.7 | −65.4 | −21.1 | −24.1 | 21.6 | 67.4 | 113.2 | 138.3 | 186.7 | 235.2 | 283.7 |
| 66 | −101.0 | −53.8 | −6.6 | −7.2 | 41.4 | 90.0 | 138.6 | 166.5 | 217.8 | 269.1 | 320.3 |

**Table 7.** The MANI of poplar SRCs by yield and the price of the wood chips in this study (break-even-points marked with grey background) with 75% establishment subsidy.

| Price ($€/t_{OD}$) | Low Yield | | | Yield ($t_{OD}$/ha/year) Average Yield | | | | High Yield | | | |
|---|---|---|---|---|---|---|---|---|---|---|---|
| | 3 | 4 | 5 | 6 | 7 | 8 | 9 | 10 | 11 | 12 | 13 |
| 54 | −99.5 | −63.9 | −28.2 | −38.6 | −1.3 | 36.1 | 73.4 | 90.0 | 130.0 | 170.0 | 210.0 |
| 57 | −90.8 | −52.3 | −13.7 | −21.7 | 18.5 | 58.6 | 98.8 | 118.2 | 161.0 | 203.8 | 246.7 |
| 60 | −82.1 | −40.7 | 0.8 | −4.8 | 38.2 | 81.2 | 124.2 | 146.4 | 192.0 | 237.7 | 283.3 |
| 63 | −73.4 | −29.1 | 15.3 | 12.2 | 58.0 | 103.8 | 149.5 | 174.6 | 223.0 | 271.5 | 320.0 |
| 66 | −64.7 | −17.5 | 29.8 | 29.1 | 77.7 | 126.3 | 174.9 | 202.8 | 254.1 | 305.4 | 356.7 |

## 4. Discussion

SRCs have been on the agenda for decades. Biomass, and especially woody biomass, is a renewable resource that can provide a positive energy balance and a low carbon output. Furthermore, it can be produced and used locally, thereby avoiding the negative effects of long transport distances. It is also available on a scale that is sufficient to make a meaningful contribution to the total energy supply. From a producer perspective, this land utilization form is intended to be flexible in terms of converting arable lands and plantations back and forth according to the producer needs and market demands. Furthermore, it is intended to be suitable for low quality land that would not be utilized otherwise.

Despite the long history of such plantations, Hungary lacks well-recorded field experiments that would cover various combinations of sites and technologies for the whole lifecycle. It must be noted that a short rotation period makes the growth rate highly dependent on weather variables; therefore, experiments on the same site can show significantly varying results over different periods of time.

The profitability of poplar SRCs is highly dependent on subsidies. The availability and conditions of such financial sources change frequently in Hungary; however, there are three basic types: an establishment subsidy paid by area at a fixed rate; the EU Single Area Payments Scheme (SAPS); and the subsidies for energy producers paid by the volume of production. SAPS plays the most important role, as it is comparable to the overhead costs applied in the model, namely the land rental fee and the technical advisor cost. Since this subsidy is available in almost all forms of agricultural land-use, including the no-use option, it is a neutral factor in land utilization decisions. The subsidies on the energy producer side influence the purchase price of the wood chips; thus, there is no need to incorporate these into the model separately.

The financial analysis of a specific SRC investment must consider actual conditions including the site conditions; suitable poplar cultivars; available propagation material; machinery, especially harvesting machines; storage needs; transportation methods and distances; and potential buyer requirements such as delivery frequency, moisture content, chip size, quality, etc. These factors may greatly influence the financial results and the investment decision. The currently applied agro-technology at a specific farm may be suitable for SRC cultivation, and it is also possible that the wood chips produced could be utilized within the farm. In the latter case, the substitution of other energy sources on the farm can bring additional benefits. On the other hand, the technology applied in this calculation may not be feasible in some areas due to lack of harvesting capacity or due to the small scale of a specific SRC. In such cases, the production costs would be significantly higher.

Specific environmental conditions can dramatically influence SRC yield, even under optimal management decisions. For instance, water supply can greatly affect the growth rate. The site analysis and the subsequent poplar cultivar selection is based on long-term climatic data, while the weather during the production period may vary and significantly deviate from long-term observations.

All these individual circumstances influence the investment decision and shed a different light on the potential role of SRC as a land management option at a regional or national level. This financial analysis was conducted to evaluate the financial performance of poplar SRCs under a common framework incorporating best practices.

Site quality has a moderate influence on the cost structure. Establishment and liquidation costs are the same in each yield category; maintenance costs decrease, while harvesting costs slightly increase as they move from the high yield category to the low yield category. The total lifecycle cost in the high yield category is 7131.4 €/ha, which drops to 4634.9 €/ha for the low yield category, signifying a 35% cost decrease. The yield, however, has a great impact on the total lifecycle income, which drops from 10,152.0 €/ha in the high yield category to 3480.0 €/ha in the low yield category, for a decrease of 65.7%. According to the financial model, it can be concluded that the lower yield of low-quality sites and the lower income is not compensated by lower costs.

The results show that poplar SRCs reach the BEP at approximately 7 $t_{OD}$/ha/year under current financial conditions and the presented cost structure. Since the cost structure in the model excludes substantial cost items such as business administration, infrastructure maintenance, etc., the real BEP is assumed to be significantly higher. This means that plantations on lower-average and low-quality sites are not financially viable without subsidies. In the case of a 75% establishment subsidy, BEP is approximately 5 to 6 $t_{OD}$/ha/year.

It is evident that the subsidy system influences the financial performance of short rotation poplar plantations considerably. Alternative land-use forms compete for land. In the European Union, agricultural policy is one of the highest priorities and is empowered by a massive subsidy system. In Hungary, the proportion of direct subsidies in traditional agricultural crop production is above 60% [41] (p. 39). Land-use forms also include wilderness, sanctuaries, and many other areas for nature conservation and recreational purposes, which further increases land-use competition, especially on low-quality lands where poplar SRCs were primarily meant to be grown.

This result also challenges the former claim that poplar SRCs would be suitable for the utilization of low-quality land. It reflects the fact that poplar hybrids are capable of high-rate biomass production, but are susceptible to natural conditions, especially to temperature and water supply, the latter of which is a more critical limiting factor in Hungary. However, higher-average and high-quality poplar growing sites are more suitable for industrial wood production, where the main product is veneer log coupled with pulp wood and wood chips as side products. Such industrial wood plantations have a 12–25-year lifecycle, which is comparable to that of short rotation plantations.

One of the most important comparative advantages of SRC is that its production technology is less sophisticated, and fewer operations are needed; therefore, less experienced farmers can manage them. Since quality requirements in wood chip production do not play a major role, technological failures or damage from natural disasters or damaging agents do not result in significant income loss. The shorter rotation period also decreases the risk of damage and provides a more balanced income-flow.

Due to legal restrictions, the plantation lifecycle in our model is 15 years. Extending the lifecycle with a few more rotations to 20–25 years could increase the net income and, consequently, the return on investment. Nevertheless, this is severely limited by the gradual decrease of yield over time and depends on the health condition of the individual plantation.

This paper concluded that short rotation poplar plantations in Hungary have an approximate BEP of 7 $t_{OD}$/ha/year, which is only possible in above average, good quality poplar sites in which veneer production is also an option. Subsidies and extended rotation periods could lower BEP significantly, but both would also require changes in the current related policies.

**Author Contributions:** Conceptualization, E.S.; methodology, E.S.; validation, A.B. (Attila Borovics), and A.B. (Attila Benke); formal analysis, E.S.; investigation, E.S., A.B. (Attila Borovics), and A.B. (Attila Benke); writing—original draft preparation, E.S. and A.B. (Attila Benke); writing—review and editing, E.S., A.B. (Attila Borovics), and A.B. (Attila Benke). All authors have read and agreed to the published version of the manuscript.

**Funding:** This research received no external funding.

**Data Availability Statement:** All data that are necessary for the reconstruction of this study can be found in this paper and in the referenced sources.

**Acknowledgments:** The authors wish to thank the hard work of those managing poplar plantation experiments in the Forest Research Institute, Hungary.

**Conflicts of Interest:** The authors declare no conflict of interest.

## References

1. Aylott, M.J.; Casella, E.; Tubby, I.; Street, N.R.; Smith, P.; Taylor, G. Yield and spatial supply of bioenergy poplar and willow short-rotation coppice in the UK. *New Phytol.* **2008**, *178*, 897. [CrossRef]
2. Vanbeveren, S.P.; Ceulemans, R. Biodiversity in short-rotation coppice. *Renew. Sustain. Energy Rev.* **2019**, *111*, 34–43. [CrossRef]
3. Štochlová, P.; Novotná, K.; Costa, M.; Rodrigues, A. Biomass production of poplar short rotation coppice over five and six rotations and its aptitude as a fuel. *Biomass Bioenergy* **2019**, *122*, 183–192. [CrossRef]
4. Busch, G. A spatial explicit scenario method to support participative regional land-use decisions regarding economic and ecological options of short rotation coppice (SRC) for renewable energy production on arable land: Case study application for the Göttingen dist. *Energy. Sustain. Soc.* **2017**, *7*, 372. [CrossRef]
5. McKenney, D.W.; Yemshanov, D.; Fox, G.; Ramlal, E. Cost estimates for carbon sequestration from fast growing poplar plantations in Canada. *For. Policy Econ.* **2004**, *6*, 345–358. [CrossRef]
6. Updegraff, K.; Baughman, M.J.; Taff, S.J. Environmental benefits of cropland conversion to hybrid poplar: Economic and policy considerations. *Biomass Bioenergy* **2004**, *27*, 411–428. [CrossRef]
7. Ceulemans, R.; Deraedt, W. Production physiology and growth potential of poplars under short-rotation forestry culture. *For. Ecol. Manag.* **1999**, *121*, 9–23. [CrossRef]
8. Faasch, R.J.; Patenaude, G. The economics of short rotation coppice in Germany. *Biomass Bioenergy* **2012**, *45*, 27–40. [CrossRef]
9. Miller, R.O. Financial modeling of short rotation poplar plantations in michigan, USA. *Conf. Short Rotat. Woody Crop. Oper. Work. Gr.* **2016**, *2016*, 1–12.
10. Christersson, L. Poplar plantations for paper and energy in the south of Sweden. *Biomass Bioenergy* **2008**, *32*, 997–1000. [CrossRef]
11. Manzone, M.; Airoldi, G.; Balsari, P. Energetic and economic evaluation of a poplar cultivation for the biomass production in Italy. *Biomass Bioenergy* **2009**, *33*, 1258–1264. [CrossRef]
12. Manzone, M.; Bergante, S.; Facciotto, G. Energy and economic evaluation of a poplar plantation for woodchips production in Italy. *Biomass Bioenergy* **2014**, *60*, 164–170. [CrossRef]
13. El Kasmioui, O.; Ceulemans, R. Financial analysis of the cultivation of poplar and willow for bioenergy. *Biomass Bioenergy* **2012**, *43*, 52–64. [CrossRef]
14. El Kasmioui, O.; Ceulemans, R. Financial Analysis of the Cultivation of Short Rotation Woody Crops for Bioenergy in Belgium: Barriers and Opportunities. *Bioenergy Res.* **2013**, *6*, 336–350. [CrossRef]
15. Mitchell, C.P.; Stevens, E.A.; Watters, M.P. Short-rotation forestry-Operations, productivity and costs based on experience gained in the UK. *For. Ecol. Manag.* **1999**, *121*, 123–136. [CrossRef]
16. Hauk, S.; Knoke, T.; Wittkopf, S. Economic evaluation of short rotation coppice systems for energy from biomass-A review. *Renew. Sustain. Energy Rev.* **2014**, *29*, 435–448. [CrossRef]
17. Gasol, C.M.; Martínez, S.; Rigola, M.; Rieradevall, J.; Anton, A.; Carrasco, J.; Ciria, P.; Gabarrell, X. Feasibility assessment of poplar bioenergy systems in the Southern Europe. *Renew. Sustain. Energy Rev.* **2009**, *13*, 801–812. [CrossRef]
18. Allen, D.; McKenney, D.W.; Yemshanov, D.; Fraleigh, S. The economic attractiveness of Short Rotation Coppice biomass plantations for bioenergy in Northern Ontario. *For. Chron.* **2013**, *89*, 66–78. [CrossRef]
19. Schweier, J.; Becker, G. Economics of poplar short rotation coppice plantations on marginal land in Germany. *Biomass Bioenergy* **2013**, *59*, 494–502. [CrossRef]
20. Csiha, I.; Rásó, J.; Keserű, Z.; Kamandiné Végh, Á.; Kovács, C. A termőhelyi változások hatása a nemesnyár klónok produktumára. In Proceedings of the Alföldi Erdőkért Egyesület Kutatói Nap, Püspökladány, Hungary, 9 November 2012; Volume 1, pp. 19–27.
21. San Miguel, G.; Corona, B.; Ruiz, D.; Landholm, D.; Laina, R.; Tolosana, E.; Sixto, H.; Cañellas, I. Environmental, energy and economic analysis of a biomass supply chain based on a poplar short rotation coppice in Spain. *J. Clean. Prod.* **2015**, *94*, 93–101. [CrossRef]
22. Wolbert-Haverkamp, M.; Musshoff, O. Is short rotation coppice economically interesting? An application to Germany. *Agrofor. Syst.* **2014**, *88*, 413–426. [CrossRef]
23. UN FAO. *Global Forest Resources Assessment 2020 Report Hungary*; UN FAO: Rome, Italy, 2020.
24. National Food Chain Safety Office. *Forest Resources and Forest Management in Hungary 2015*; National Food Chain Safety Office: Budapest, Hungary, 2016.
25. Mátyás, C. Forecasts needed for retreating forests. *Nature* **2010**, *464*, 1271. [CrossRef] [PubMed]
26. Hirka, A.; Csóka, G.; Koltay, A.; Janik, G. A nyárak és füzek biotikus és abiotikus kárai. *Erdészeti Kut.* **2008**, *93*, 258–280.
27. Koltay, A. Az energetikai faültetvények növényvédelmi vonatkozásai. *Mezőgazdasági Tech.* **2010**, *51*, 66.

28. Führer, E. A klímaértékelés erdészeti vonatkozásai. *Erdészettudományi Közlemények* **2018**, *8*, 27–42. [CrossRef]
29. Führer, E.; László, H.; Anikó, J.; Machon, A.; Ildikó, S. Application of a new aridity index in Hungarian forestry practice. *Időjrás* **2011**, *115*, 205–216.
30. Gálos, B.; Ernő, F. A klíma erdészeti célú előrevetítése. *Erdészettudományi Közlemények* **2018**, *8*, 43–55. [CrossRef]
31. Norbert, M.; Tibor, T.; Kitti, B.; András, S.; Ervin, R.; Zoltán, G. Groundwater uptake of forest and agricultural land covers in regions of recharge and discharge. *iForest* **2016**, *9*, 696–701.
32. András, S.; Zoltán, G.; Tibor, T. Erdők hatása a talaj és altalaj sóforgalmára, valamint a talajvíz szintjére. *Agrokémia és Talajt.* **2012**, *61*, 195–209.
33. Szajkó, G.; Mezősi, A.; Pató, Z.; Orsolya, S.; Sugár, A.; András Istvá, T. *Erdészeti és Ültetvény Eredetű fás Szárú Energetikai Biomassza Magyarországon*; Regionális Energiagazdasági Kutatóközpont: Budapest, Hungary, 2009.
34. Heilig, D.; Heil, B.; Kovács, G. A növőtér-szabályozás hatása fás szárú nemesnyár ültetvény dendromassza-hozamára. *Erdészettudományi Közlemények* **2018**, *8*, 51–59. [CrossRef]
35. Kondor, A. A Földhasználat Átalakításának Lehetősége az Energiafűz (*Salix viminalis* L.) Termesztésbe Vonásával SZABOLCS-Szatmár-Bereg Megyében. Ph.D Thesis, Debreceni Egyetem, Debrecen, Hungary, 2015.
36. Marosvölgyi, B.; Halupa, L.; Wesztergom, I. Poplars as biological energy sources in Hungary. *Biomass Bioenergy* **1999**, *16*, 245–247. [CrossRef]
37. Szendrödi, L.B. Growing space-age related three-dimensional modeling of biomass production of hybrid poplar. *Biomass Bioenergy* **1996**, *10*, 251–259. [CrossRef]
38. Agrárminisztérium. Erdészeti Árstatisztikák. Available online: https://agrarstatisztika.kormany.hu/erdogazdalkodas2 (accessed on 20 October 2020).
39. Vass, J.; (Celldömölk, Hungary); Falvay, S.; (Szigetszentmiklós, Hungary); Hegedűs, A.; (Mogyoróska, Hungary); Barkóczy, Z.; (Sopron, Hungary). Personal communication, 2020.
40. Posavec, S.; Kajba, D.; Beljan, K.; Boric, D. Economic analysis of short rotation coppice investment. *Austrian J. For. Sci.* **2017**, *134*, 163–176.
41. Keszthelyi, S.; Kiss Csatári, E. *A Tesztüzemi Információs Rendszer Eredményei 2018*; Nemzeti Agrárkutatási és Innovációs Központ: Budapest, Hungary, 2020.

*Article*

# Productivity of Black Locust (*Robinia pseudoacacia* L.) Grown on a Varying Habitats in Southeastern Poland

Artur Kraszkiewicz 

Department of Machinery Exploitation and Management of Production Processes, University of Life Sciences in Lublin, Głęboka Street 28, 20-612 Lublin, Poland; artur.kraszkiewicz@up.lublin.pl; Tel.: +48-815-319-731

**Abstract:** This study investigated growth performances of black locust (*Robinia pseudoacacia* L.) tree species in various soil and agro-climatic conditions in Poland. Implementing of research was based on monoculture black locust stands in which it was possible to carry out dendrometric tests allowing us to learn about their volume. These stands were located on marginal soils. In the sample plots selected for the study, the parameters of stands (main and secondary) were determined, such as number and social structure of trees, average tree height, average diameter at breast height (DBH), and volume. The volume was determined with division into trunks and branches and wood thickness classes (0.0–1.0 cm, 1.1–5.0 cm, 5.1–10.0 cm and then every 5 cm). During the research, it was found that sunlight and moisture conditions mainly affect the volume. It has been noticed that the content of nutrients in the soil plays a minor role because black locust grows very well in poorly fertile soils, often subject to erosion processes. Black locust grows well on damp, shaded slopes with northern exposures. In such areas, the stand volume was the highest (353.8 m$^3$ ha$^{-1}$), exceeding the average volume of the remaining 35-year-old stands on sandy soils by 60%. Along with the increase in the age of stands, the share of trunk wood increased with the wood of branches. The share of wood up to 5.0 cm was small in older stands, at most a dozen or so percent. However, in young stands (4- and 8-year-old), the share of the thickness class up to 5 cm was even 65% of the stand volume. In 35-year-old stands, wood fractions of 15.1–20.0 cm were dominant. In the oldest, 64-year-old stand, over 30 cm thick wood constituted 44% of the stand volume. However, statistical analysis showed, with $p = 0.1644$, no differences existed between the thickness of the individual thickness classes.

**Keywords:** tree growth; tree biomass; volume forest stand; thickness classes

**Citation:** Kraszkiewicz, A. Productivity of Black Locust (*Robinia pseudoacacia* L.) Grown on a Varying Habitats in Southeastern Poland. *Forests* **2021**, *12*, 470. https://doi.org/10.3390/f12040470

Academic Editors: Ronald S. Zalesny Jr. and Andrej Pilipović

Received: 25 February 2021
Accepted: 8 April 2021
Published: 12 April 2021

**Publisher's Note:** MDPI stays neutral with regard to jurisdictional claims in published maps and institutional affiliations.

## 1. Introduction

The black locust (*Robinia pseudoacacia* L.) in Europe is an introduced species from North America [1]. It was initially used for ornament purpose in parklands or as melliferous [2,3]. Later the species has been grown for economic uses specially woods (venial wood, utility wood, mine wood, and firewood) [4–8] and environmental conservation (soil erosion control) [9–17]. Recently, there has been an increase in interest in this species for cultivation in short rotation energy crops [18–24]. This approach is to reduce the use of forest wood for energy purposes [25,26]. In line with the concept of sustainable development, the possibility of establishing plantations in degraded, erosion-endangered, and unsuitable areas for agriculture should be increased [27,28]. In depth knowledge on the species growth performances under different soil and agroclimatic conditions would help policy decision makes, tree growers, and farmers to make right decision of growing and managing the tree species.

The results of these studies indicate that the production and properties of wood depend on the habitat conditions and coexistence of trees in the group. Black locust grows best on moisture, fertile, calcareous clay soils [1]. Also optimum for black locust is tufted and brown soils: plump, airy, fresh, rich in calcium, with a pH in the range of 4.6–8.2, but also grows well on others (except marshland), including on stony, non-cariogenic, sand

and dry ones [2–4,11,13]. Additionally, studies by Mantovani et al. [29,30] indicate that this species is not from the group of water-savers, and with its abundance it corresponds to an increase in weight gain—most intensively in August. In addition to the habitat, the productivity and structure of forest stand classes is affected by coexistence in the context of tree competition. These studies indicate that, along with the deterioration of the biosocial position of trees in the stand, in the shade under the canopy of the main trees, the annual growth rate of wood decreases [31–35].

In research on the productivity of black locust stands relating to the habitat and biosocial position of trees, there is often no reference to their impact on the structure of individual wood thickness classes. Much attention in the literature is devoted to the creation of mathematical models to estimate the abundance of stands [35–43]. The most popular method of creating biomass models is combining the biomass data of a tree or its components, such as stems, branches, leaves or roots, with one or more easily measurable dendrometric variables, e.g., breast height diameter, wood thickness. Therefore, under the conditions of the Polish agroclimatic, it was decided to conduct research updating the knowledge on the thickness of the black locust stands. Taking into account also factors such as moisture and soil fertility. It is also important to what extent age cause changes in the thickness of individual classes of wood biomass. Such data should supplement the pool of information forming the basis for the development of mathematical models estimating the amount of biomass produced.

The study aimed to assess the productivity and quantitative structure in individual thickness classes of the black locust wood biomass obtained in 14 stands of different age and habitat conditions of marginal soils.

## 2. Materials and Methods

### 2.1. Study Sites

The research was carried out based on monoculture black locust stands in six sites (including single or groups of stands) in the Małopolska Kraina (according to the nature and forest regionalization of Poland [44])—Figure 1. Monoculture stands guided the selection of sites with an area and number of trees enabling separation of representative sample areas and collection of representative sample trees, and the existence of source materials enabling the reconstruction of the history of stands.

Upland areas dominate in the Małopolska Kraina (VI), but there are also significant areas of valley bottoms and terraces with dunes. The climate is relatively diversified due to the diversification of the topography, with the features of continentalism increasing to the east. The average annual air temperature in the western and southern parts of the region ranges from 8.0 to 8.5 °C, and in the eastern part—Up to 7.5 °C. In most areas, the growing season lasts 200–210 days, and the average annual rainfall amounts to 650 mm [44].

Due to the diversity of the topography, soil formation, and moisture in the Małopolska Kraina, there is a large diversity of stands. The subcontinental oak–hornbeam forests are characteristic for this land—i.e., linden, oak, and hornbeam forests, mainly in the Lesser Poland variety. They are located in half of the area of this region. Mixed forests are more numerous in its western part. Forests are mainly found in upland areas and wet and marshy areas. The forest cover in the region is 24.9%. At the same time, in the mesoregions, it ranges from 2%–4% in Podgórze Rzeszowskie (VI.34) and the Lower San River Valley (VI.30) to 60% in the Świętokrzyski Forest (VI.23) and 65% in Solski Forest (VI.13). The average stand volume is 239 $m^3$ $ha^{-1}$. In mesoregions, it ranges from 166 $m^3$ $ha^{-1}$ in Górny Śląsk to 287 $m^3$ $ha^{-1}$ in the West-Lublin Upland mesoregion. In the Małopolska Kraina, protective forests (mainly water-proof, damaged, and around towns) constitute over 63% of the forest area of the State Forests (SF). Damaged forests are over 80% of the SF areas in mesoregions with well-developed industries, including Górny Śląsk (VI.16) [44].

**Figure 1.** Location of the Małopolska Kraina (VI) on the territory of Poland with marked mesoregions (Arabic numerals) and research stands (red dots). Important markings mesoregion: 4—West-Lublin Upland, 5—East-Lublin Upland, 27—Chmielnicko-Staszowski, 28—Opatowski, 29—Vistula Lowland. Prepared on the basis of [44].

The research sites were located in several mesoregions of the natural and forest districts marked with the numbers VI.4, VI.5, VI.27, VI.28, and VI. 29 (Figure 1), differing in the natural conditions of forest production.

The Dębno site is located in the West-Lublin Upland mesoregion (VI.4). The forest cover is small and amounts to 16%. The forests form small and medium-sized complexes; they occupy about 516 km$^2$, 45% of which is on the Regional Directorate of State Forests (RDSF) board in Lublin. Snopków and Lublin sites are located in the East-Lublin Upland mesoregion (VI.5)—Figure 1. It is distinguished by fertile soils made of loess and "loess-like" dusts and, therefore, the lowest forest cover amounting to 14%. The forests form small complexes; they occupy about 291 km$^2$, of which over 73% is managed by RDSF in Lublin [44].

The Skrzypaczowice and Zawidza sites are located in the Opatowski (VI.28) and Chmielnicko-Staszowski (VI.27) mesoregions. The Opatów area is covered with loess, which is the dominant geological formation. In the southern part of the mesoregion, there are Pleistocene tills, sands, and glacial gravels of the South Polish Glaciation, and occasionally—Cambrian deposits, on small areas mainly sandstones, claystones, conglomerates, and duststones. The mesoregion is dominated almost exclusively by broadleaf vegetation with the participation of luminous oak trees. The forest cover is very low, amounting to 4%. Forests in the form of very small complexes cover about 52 km$^2$, of which RDSF Radom manages 51%. However, the Chmielnik–Staszów area is covered with Pleistocene tills, sands, and glacial gravels of the South Polish Glaciation. Much smaller areas are occupied by the formations of the Neogene period—Organo-detritus and sulfur-bearing limestones, gravel, sandstone, and gypsum. Besides, eolian sands—locally in the dunes and loess—occur sporadically. The forest cover is average and amounts to 30%. The forests form small and medium-sized complexes; the largest are located in the eastern part of the mesoregion. The forests cover approximately 476 km$^2$, of which 67% is managed by the RDSF in Radom.

Piaseczno site is located in the Vistula Lowland mesoregion (VI.29). There are definitely Holocene geological formations—sands, gravels, river bogs, peat, and dust. The dominant vegetation landscape in this area is the ash and riparian elm forests. The mesoregion's forest cover is low and amounts to 6%. About 93 km$^2$ area covered by forests, 39% of which is managed by the State Forests. The mesoregion is narrow and elongated, and within its borders, there is a small area of RDSF in Radom [42]. However, it should be noted that the Piaseczno site is an artificial structure—an external dump of the sulfur mine in Piaseczno, 40 m high, and slope inclination 60%–80%, built of various formations and with a large variety of habitats [11,45].

In each of the sites, tree stands with one or more features influencing wood production's natural and economic conditions were selected: habitat factors, age, origin, the intensity of care, and breeding treatments. Fourteen stands were distinguished, including:

- ➢ five on sands (loose sands)—No. 1–5—All in the Piaseczno site, derived from planting, at the age of 35, on the slopes of the dump with various exposures (N—Surfaces No. 1 and 2 located in the upper and lower parts of the slope, respectively, SE—No. 3 and 4—The upper and lower part of the slope, S—No. 5—The upper part of the slope), without felling (without harvesting);
- ➢ three in the clay—No. 6–8—two in the Piaseczno site, derived from planting, at the age of 35, (located in the upper parts of the slopes, No. 6—N exposure, and No. 7—S exposure) under development conditions such as stands No. 1–5, and one in the Zawidza site (No. 8) in the managed forest (with intermediate cutting—Breeding cuts), at the age of 41, in the plain;
- ➢ six on dust formations (loess and loess-like)—In the sites: Dębno (No. 9, the lower part of the valley slope, with S exposure and 40% slope)—33-year-old stand from self-seeding, systematically cut; Skrzypaczowice (No. 10)—A 64-year-old commercial stand, planted, located in the central part of the eroded slope with an SE exposure and a 15% slope; Lublin (No. 11)—4-year self-seeding trees located on a plain area; Snopków (No. 12–14)—3-row mid-field, 8-year-old plantings, No. 12 and 13 located in a flat area, and No. 14 in the upper part of the valley slope, with S exposure and a 15% slope.

Detailed characteristics of the Piaseczno site and the history of tree stands are provided in the publications of Ziemnicki et al. [45], Węgorek [34], and Kraszkiewicz and Węgorek [46]. Developing tree stands in the Snopków site are described by Orlik et al. [39] and Węgorek and Kraszkiewicz [40]. The data on stands in the Zawidza and Skrzypaczowice sites were given according to the source materials of the Zawidza forestry in the Dębno site—according to information from the farm owner, and in the Lublin site—based on own observations.

*2.2. Biomass Sampling*

In the black locust stands selected for research to assess fertility, the following was determined:

(a). Stand parameters—By the method of sample plots of 500 m$^2$ (20 × 25 m$^2$) in stands No. 1–11 and 400 m$^2$ in stands No. 12–14 (rows of trees 80 m long and 5 m wide):
- ➢ the number and social structure of trees according to the tree biological classification of Kraft, considering the main and secondary stands;
- ➢ average height of trees (in main and secondary stands)—as the arithmetic mean of DBH measured with the SUUNTO altimeter with an accuracy of 0.25 m;

> mean DBH (in main and secondary stands)—as the arithmetic mean of DBH measured with a precision HAGLOF caliper with an accuracy of 0.5 cm;

(b). Volume with division into trunks and branches and wood thickness classes (0.0–1.0 cm, 1.1–5.0 cm, 5.1–10.0 cm and then every 5 cm)—using the sample tree's method. One tree of average height and DBH each and average conformation from each sample plot (main and secondary stands) [47]; a conductor was considered as a trunk from the point of cut (5–10 cm above the ground) to a diameter of 5 cm in the bark (in the upper end); the remaining, thinner part (top) was classified as a branch; the sample trees were cut with a chainsaw at the end of December:

> trunk volume in the bark—sectional method (section length 1 m);
> branch volume in the bark—using the xylometric method.

*2.3. Statistical Analysis*

The obtained results, were subjected to analysis of variance (ANOVA) analysis for factorial systems. The qualitative factors were the age of the stands as well as the individual thickness classes. Statistical analyzes verifying differences in wood volume in individual thickness classes were carried out individually within a single age group (where there was an appropriate amount of data—only stands at the age of 8 and 35), as well as in the following groups: all together; 4 and 8; 35 and 41; 33 and 35; 33, 35, and 41 as well as 33, 35, 41 and 64. Additionally, the probability of differences between the volume of wood in all stands between the following pairs of thickness classes was determined: 0.0–1.0 cm and 1.1–5.0 cm; 1.1–5.0 cm and 5.1–10.0 cm; etc. Prior to these analyses, the consistency of results with the normal distribution was verified using the Shapiro–Wilk method, and the homogeneity of variance was estimated using the Brown–Forsyth test. The observed differences were considered statistically significant at the significance level of $p$-value $< 0.05$. Additionally, for the distribution of wood volume in individual stands with division into thickness classes, trend lines were determined, which were described with a second-degree polynomial and the coefficient of determination $R^2$.

## 3. Results

*3.1. Characteristics of Soils*

Table 1 shows the surface coverage by layers of vegetation (covering plant layers) and the soil profiles' basic features under the stands.

**Table 1.** Covering plant layers and selected soil characteristics.

| Forest Stand/Area Number | Soil | Surface Coverage (%) | | | Litter (cm) | Humus Layer (cm) |
|---|---|---|---|---|---|---|
| | | Trees | Shrubs | Undergrowth | | |
| 1 | | 80 | 60 | 30 | 2 | 4 |
| 2 | | 80 | 60 | 30 | 2 | 2 |
| 3 | sand | 80 | 20 | 70 | 4 | 5 |
| 4 | | 80 | 20 | 70 | 5 | 7 |
| 5 | | 80 | 20 | 100 | 5 | 8 |
| 6 | | 70 | 50 | 30 | 1 | 5 |
| 7 | clay | 80 | 30 | 70 | 2 | 6 |
| 8 | | 50 | 50 | 80 | 3 | 10 |
| 9 | | 80 | 20 | 80 | 3 | 8 |
| 10 | | 50 | 50 | 60 | 5 | 10 |
| 11 | | – | 70 | 90 | 1 | 20 |
| 12 | dust | 90 | 60 | 90 | 1 | 24 |
| 13 | | 90 | 60 | 90 | 1 | 24 |
| 14 | | 90 | 60 | 90 | 1 | 22 |

The layer of trees was generally characterized by good compactness. The most common surface coverage was 80%, and in the stands in the Snopków site (No. 12–14)—90%.

It should be noted, however, that the compactness in stands No. 12–14 was so high only in the rows of trees, and there was full access of sunlight on both sides of the rows. The lowest coverage (50%) of the tree layer was in stands used for pole-cuts (No. 8 and 10). The shrubs (as an under-emergent layer) cover from 20% to 60% of the area, and the ground cover—30%–100%. In the four-year-old stand in the Lublin site (No. 11), the layer of trees has not yet developed—from the viewpoint of the height achieved by black locust trees—and they constituted a shrub layer (Table 1).

The litter was 1–5 cm thick, depending on the site in the relief, species composition, and the degree of surface coverage by plant layers (Table 1). The humus layer of the soil under very young tree stands established on agricultural land exceeded 20 cm, in old stands (No. 8 and 10), it was 10 cm, and in stands in the wasteland (depending on the site)—2–8 cm (Table 1).

Table 2 shows the abundance of soils determined based on the average content of basic nutrients in a layer of 0–50 cm. The nitrogen volume was determined by the Kowalkowski method [48] and the phosphorus and potassium content according to the scale provided by Baule and Fricker [49].

**Table 2.** Content of nutrients and soil abundance.

| Forest Stand/Area Number | Content of Nutrients (g·kg$^{-1}$) and Soil Abundance | | | | C:N |
|---|---|---|---|---|---|
| | N$_{total}$ | P | K | C$_{organic}$ | |
| 1 | 0.36 insufficient | 0.004 insufficient | 0.021 insufficient | 2.09 | 5.81 |
| 2 | 0.38 insufficient | 0.005 insufficient | 0.027 insufficient | 2.59 | 6.82 |
| 3 | 0.24 insufficient | 0.004 insufficient | 0.024 insufficient | 2.21 | 9.21 |
| 4 | 0.57 insufficient | 0.004 insufficient | 0.023 insufficient | 2.71 | 4.75 |
| 5 | 0.41 insufficient | 0.005 insufficient | 0.024 insufficient | 2.90 | 7.07 |
| 6 | 0.77 medium | 0.054 medium | 0.142 good | 6.60 | 8.57 |
| 7 | 0.84 medium | 0.004 insufficient | 0.030 insufficient | 4.32 | 5.14 |
| 8 | 0.78 medium | 0.007 insufficient | 0.025 insufficient | 6.05 | 7.76 |
| 9 | x | x | x | x | x |
| 10 | 1.23 medium | 0.015 medium | 0.028 insufficient | 6.79 | 5.52 |
| 11 | 0.43 insufficient | 0.059 medium | 0.063 medium | 3.62 | 8.42 |
| 12 | 1.95 good | 0.176 good | 0.260 good | 10.37 | 5.32 |
| 13 | 1.34 good | 0.101 good | 0.126 good | 8.74 | 6.52 |
| 14 | 0.57 insufficient | 0.109 good | 0.087 medium | 4.18 | 7.33 |

Abbreviations: x—Was not done.

On the sand slopes of the sulfur mine dump (tree stands No. 1–5), the soil abundance in nutrients was insufficient, and the organic carbon content was very low, ranging from slightly over 2 to almost 3 g kg$^{-1}$. In clay formations, only nitrogen supply was at an average level, and there was twice as much carbon as in sand formations. Forest stands on agricultural land with soils made of dusty formations (No. 12–14) were supplied with nutrients, usually at a good level (others at an average level), with a soil carbon content of about 4 to over 10 g kg$^{-1}$ (Table 2). The ratio of carbon to nitrogen (C:N) was low (below 10—Table 2), which indicates the rapid mineralization of organic matter [50], and the insufficient nitrogen content in the soil indicates this element was incorporated quickly into the stand production process.

*3.2. Characteristics of Forest Stands*

Table 3 presents the characteristics of the tree stands influencing their volume.

**Table 3.** Number of trees per 1 ha and average tree dimensions in stands.

| Forest Stand/Area Number | Age (Years) | Number of Trees (Pcs. ha$^{-1}$) | | Average Height (m) | | Average Diameter at Breast Height (cm) | |
|---|---|---|---|---|---|---|---|
| | | Main | Secondary | Main | Secondary | Main | Secondary |
| 1 | 35 | 919 | 588 | 16.5 | 12.5 | 19.0 | 9.5 |
| 2 | 35 | 1029 | 882 | 18.0 | 13.0 | 19.5 | 10.0 |
| 3 | 35 | 750 | 1525 | 15.5 | 11.0 | 17.5 | 8.5 |
| 4 | 35 | 580 | 1440 | 17.0 | 11.5 | 23.5 | 8.5 |
| 5 | 35 | 943 | 1057 | 15.5 | 10.0 | 16.0 | 8.5 |
| 6 | 35 | 650 | 575 | 19.0 | 12.0 | 21.0 | 9.0 |
| 7 | 35 | 870 | 1111 | 18.0 | 11.5 | 18.0 | 9.0 |
| 8 | 41 | 410 | - | 24.5 | - | 26.5 | - |
| 9 | 33 | 1840 | - | 11.5 | - | 11.0 | - |
| 10 | 64 | 320 | - | 24.0 | - | 38.5 | - |
| 11 | 4 | 1720 | - | 2.0 | - | 4.5 | - |
| 12 | 8 | 1905 | - | 7.5 | - | 12.0 | - |
| 13 | 8 | 1828 | - | 8.0 | - | 11.5 | - |
| 14 | 8 | 1715 | - | 7.5 | - | 12.0 | - |

In the 35-year-old stands No. 1–7 (Piaseczno), where no compacting treatments were performed (no felling was performed), under the canopy of trees forming the main stands (under the conditions of the research, these were trees belonging to biological class II and III), secondary tree stands formed composed of captured, jammed and dead trees (IV and V biological classes). The number of trees in the main stands ranged from 580 to 1029, and in the secondary stands—575–1525 pcs. ha$^{-1}$. In more than half of the stands, the numbers of trees in the secondary stands were greater than in the main stands, and in stands No. 3 and 4, they constituted 67% and 71% of the total number of trees, respectively. The average height of the main stands was 15.5–19.0 m. The trees reached the highest heights on clay soils (stands No. 6 and 7). Average tree heights in the secondary stands were 4–7 m lower than in the main stands. The average DBH in the main stands ranged from 16.0 cm on the sand slope with southern exposure (stand No. 5) to 23.5 cm on the lower part of the sand embankment with the south-eastern exposure. The mean DBH in the secondary stands was even—8.5–10.0 cm (Table 3).

In stands No. 8–14, there were no secondary stands due to intensive thinning (stands No. 8 and 10), the ongoing removal of mastered trees (stand No. 9), or too young age (stands No. 11–14). In these stands, as in stands No. 1–7 in the Piaseczno site, there were no towering trees (biological class I) and dominant trees (biological class II) with a small share of co-dominant trees (biological class III). The stands No. 8 and 10 (41- and 64-year-old) were characterized by a minimal number of trees per 1 ha; they reached similar heights (24.5 and 24.0 m), and the average DBH was 26.5 and 38.5 cm, respectively (Table 3). The tree stand No. 9 (33-year-old) was characterized by very many trees—1840 pcs. ha$^{-1}$ (all in the main stand), and therefore, compared to 35-year-old stands (No. 1–7), it was characterized by a small DBH but also a small height (Table 3). The 4-year-old stand (No. 11) had a small population—There were fewer trees per 1 ha than in the 33-year-old stand. For this reason, the average height of the trees was small (2.0 m), and they were heavily branched. Row stands No. 12–14 (8-year-old) occupied 5 m wide strips of land, and therefore, after calculating the number of trees per 1 ha, they were characterized by a small density—1715–1905 pcs. ha$^{-1}$, comparable to stand No. 9 (33-year-old). Average dimensions of the trees were even and amounted to 11.5–12.0 cm DBH; 7.5–8.0 m high.

### 3.3. Characteristics Volume of Forest Stand

The volume of forest stands (main and secondary stands together), including the volume of trunks and branches, as shown in Figure 2—Full numerical data with the breakdown of the volume of trunks and branches (in m$^3$ ha$^{-1}$) into thickness classes are given in Table A1 (Appendix A).

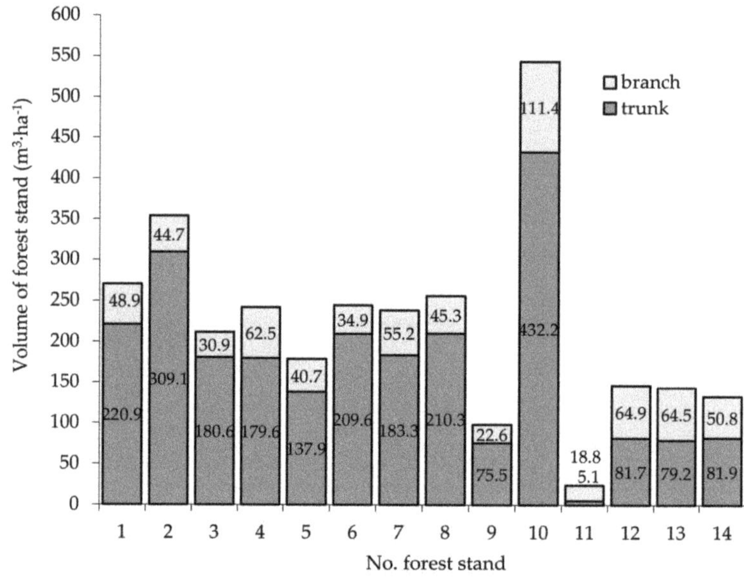


**Figure 2.** Volume of forest stand.

The research was carried out in tree stands of different ages and origins, in very different habitats, and with different intensities of breeding treatments (loosening cuts). For this reason, the volumes of stands are very diverse. In stands where no breeding treatments were performed, and due to their age, thinning should be performed several times—stands No. 1–7. The volume is 'overstated' relating to stands where intensive cuts were made—stands No. 8 and 10.

With such a high variability of the factors determining the volume, this parameter can be compared within stands No. 1–7 and No. 12–14, while within the remaining stands with the probability of making significant errors, despite considering the differences in age and intensity of breeding treatments. The stands No. 1–7 (35-year-old) had a volume of about 180 to 355 (on average almost 250 m$^3$ ha$^{-1}$, including the volume of the branches ranging from about 31 to 62 m$^3$ ha$^{-1}$ (Figure 2). Tree stand No. 2, occupying a lower part of the sand slope with northern exposure, with the highest soil moisture, had the largest volume—354 m$^3$ ha$^{-1}$ (142% of the average volume of stands No. 1–7). The lowest volume was found in stand No. 5, located in the upper part of the sand slope with southern exposure, with the lowest soil moisture—179 m$^3$ ha$^{-1}$ (72% of the average volume).

In the 8-year-old stands (No. 12–14), the volume was relatively even—133–147 m$^3$ ha$^{-1}$, with the average being 141 m$^3$ ha$^{-1}$. It results from excellent habitat conditions, and slight differences in the volume may result from the diverse tree density (Table 3). The branch volume ranged from 51 to 65 m$^3$ ha$^{-1}$.

Stand No. 9 (33-year-old) had a surprisingly low volume—Only 98 m$^3$ ha$^{-1}$ (Figure 2). This stand grew on a slope with soils formed from dust deposits. It was only two years younger than stands in a sulfur mine dump (stands No. 1–7), and despite this, its volume accounted for less than 40% of the average volume of stands No. 1–7. The reason was very poor or lack of nursing procedures in youth and destitute breeding procedures. Despite the currently conducted clearing cuts and removal of mastered trees (therefore the lack of a secondary tree stand), the tree density in the main stand was very high—1840 pcs. ha$^{-1}$ (Table 3). This forest stand was created from a very dense self-seeding, and competition for food caused the increments of trees to be very small—as evidenced by the average height and DBH (Table 3), which directly translated into the stand volume.

In stand No. 8 (41-year-old), the volume—256 m$^3$ ha$^{-1}$—Was almost the same as the average volume of 35-year-old stands (No. 1–7), amounting to 250 m$^3$ ha$^{-1}$. However, it should be noted that the volume mentioned above of 35-year-old stands includes the main and secondary stands, while the volume of the main stands themselves is on average 206 m$^3$ ha$^{-1}$. After considering age differences (between stands No. 1–7 and 8), it can be considered comparable with the volume of a 41-year-old stand.

The volume of the tree stand No. 10 (64-year-old)—544 m$^3$ ha$^{-1}$ (Figure 2)—was more than twice (112%) higher than that of No. 8 (41-year-old). Considering that both stands grew on good soils, had the same density, both exceeded the age of peak growth, and the age of stand No. 10 was only 50% higher than that of stand No. 8, the volume of stand No. 10 should be considered high.

The youngest (four-year-old), stand No. 11, had very low fertility, despite good habitat conditions. It was almost six times smaller than the volume of eight-year-old stands (no. 12–14)—Figure 2. To a large extent, these differences result from a minimal tree density relating to the age of stand No. 11 (Table 3). Due to the loose density, in the four-year-old stand, sections of conductors over 5 cm thick (trunks) in the amount of slightly over 5 m$^3$ ha$^{-1}$ have developed (Figure 2).

The shares of branches and trunks in the volume of stands in the bark, expressed as a percentage, are presented in Figure 3.

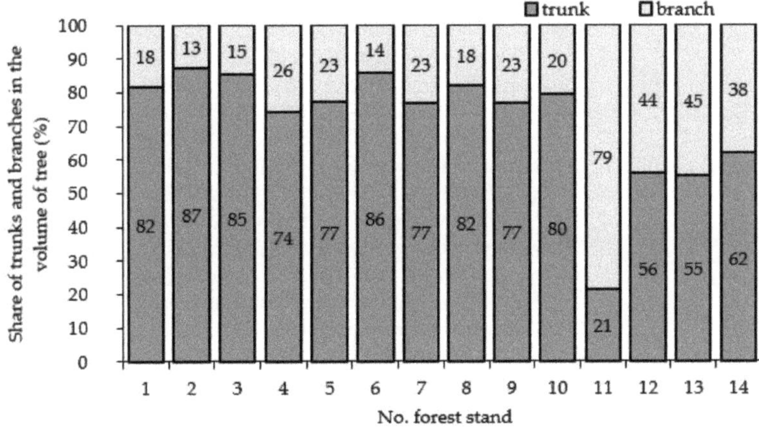


**Figure 3.** Share of trunks and branches in the volume of forest stands.

In the youngest (four- and eight-year-old) tree stands, the share of branches in their volume was very large. In the four-year-old stand (No. 11), the volume of branches reached 80% of that of the entire stand. In the 8-year-old stands (No. 12–14), the share of branches was 38%–45%. In older stands (33–64 years old), the share of branches was much smaller and accounted for 13%–26% of their total volume. The very large volume of branches relating to the volume of trunks in a four-year-old stand (3.7:1.0) results partly from loose compactness and partly from the principle adopted in this paper—related to the purpose of the study—that sections of trunks with a thickness of less than 5.1 cm are included in the branches. There was a significant amount of thinner stem sections in a specific stand (No. 11), with an average DBH of 4.5 cm (Table 3). The relatively large share of branches in the mass of eight-year-old stands (No. 12–14) is the effect of crown expansion due to the lack of lateral cover—these are row plantings. The ratio of the volume of branches to the volume of trunks was 0.6-0.8:1. In older stands, the proportions were 0.15–0.35:1 (Figure 3).

*3.4. Stand Thickness Classes*

Table A1 shows the wood volume in the bark according to thickness classes, and Figure 4 shows the share of wood of individual thickness classes in the volume of stands.

In older stands (33–64-year-old), the share of the 0.0-1.0 cm class was small and amounted to 1%–4% of the stand volume, and in the 1.1–5.0 cm class—7%–20%. However, in the young (four- and eight-year-old) stands, the share of the thickness class 0.0–1.0 cm was higher and amounted to 8%–16%, and in class 1.1–5.0 cm—18%–65% of the stand volume.

Considering the thickness structure of the main stands No. 1–7 (excluding the secondary ones)—Table A1—The share of the class 15.1–20.0 cm was 32%–61%, and the share of the class 20.1–25.0 cm—9%–29%. In the 41-year-old stand (No. 8), almost one-third of the volume was in the classes 15.1–20.0 and 20.1–25.0 cm, and 11% of the volume was 25.1–30.0 cm thick. The significant share of wood with a thickness of 5.1–15.0 cm (17% in total) was because more than half of the volume of these fractions were branches (Table A1). In the oldest, a 64-year-old stand (No. 10), the stems of the trunks exceeded the thickness of 40 cm (the share of the 40.1–45.0 cm class was 4%), and the timber over 30 cm thick accounts for 44% of the stand volume (Figure 4). The branches (limbs) were over 20 cm thick, and their share in the thickness of the fraction 20.1–25.0 cm, was over 40% (Table A1). Stand No. 9 differed from the general tendency to expand the range of wood with the age of the stand, in which, for the reasons described above, the thickness of the trunks in the butt part (thickest) did not exceed 15 cm, although the stand was 33 years old.

When analyzing the volume of the stands, it can be concluded that the productivity of black locust is mainly affected by sunlight and moisture conditions. For this species, the abundance of nutrients plays a less important role, while black locust grows well on poorly abundant soils, often subject to erosion processes. Studies in equate-age stands located on the same soils with slight differences in the nutrient content (stands No. 1–5, Table 2) indicated sunlight and moisture conditions as factors determining the growth of stands. Black locust grows well on damp, shaded slopes with northern exposure, an example of which is the volume of the stand No. 2 growing in the lower part of the slope (good moisture conditions) with northern exposure (Figure 2). The stand volume in this area was the highest—354 m$^3$ ha$^{-1}$, exceeding the average volume of the remaining 35-year-old stands on sandy soils (No. 1 and 3–5) by 60%. In stand No. 1, growing in similar sunlight conditions, but in the upper part of the slope (worse moisture conditions), the volume was lower, amounting to 270 m$^3$ ha$^{-1}$, and higher than the average (by 20%). The same trends in the differentiation of stand volume depending on the location on the slope (at the same exposure) were found in stands No. 3 and 4 (Figure 2).

The performed statistical analysis confirmed differences in the volume between stands. Also, this analysis showed the significance of differences between the volume of individual thickness classes in stands of the same age with the probability of $p = 0.0000$. However, when comparing the wood volume between all of them individual thickness classes, regardless of age, no significant differences were found at the level of $p = 0.1644$. On the other hand, in the group of stands aged 33, 35, 41, and 64 years, the probability was $p = 0.1226$. In groups 33, 35, and 41—$p = 0.3719$. In the group of stands aged 35 and 41, $p = 0.2198$, while in the group of 33 and 35—$p = 0.6289$. The highest probability was in the group of 4 and 8 years—$p = 0.80331$. Probability ($p$) of differences in wood volume in all stands between the following pairs of thickness classes: 0.0–1.0 and 1.1–5.0; 1.1–5.0 and 5.1–10.0; etc. were respectively: 0.8048; 0.5165; 0.1004; 0.2265; 0.2760 and further in the next four pairs below 0.05. The wood volume distributions in individual thickness classes for each of the analyzed stands showed a second-degree polynomial trend with the coefficient of determination from slightly over 30% (stand 6) to 100% (stand 11). The resulting equations of the trend line are presented in Table 4.

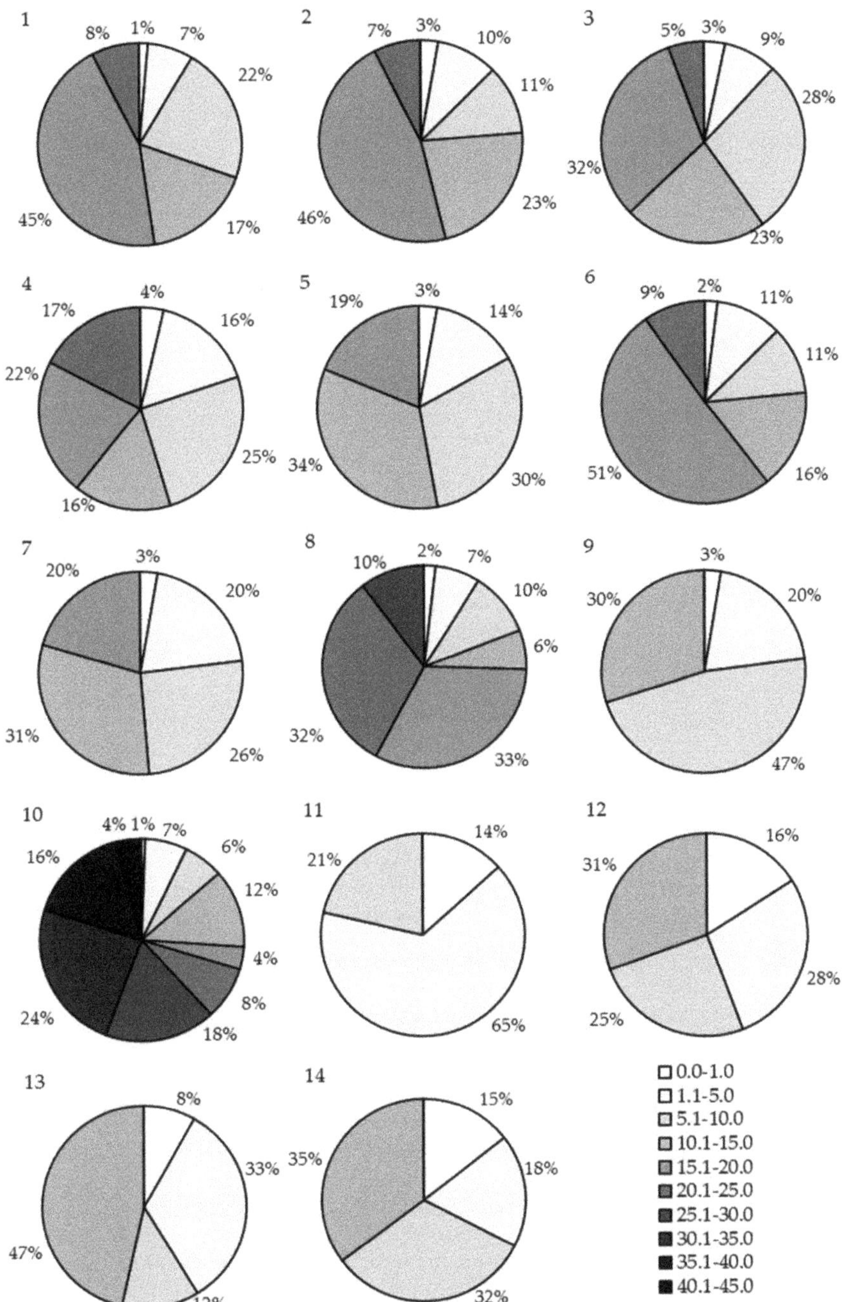



**Figure 4.** Share of thickness classes in the volume of forest stands (numbers from 1-14 correspond to the numbering of forest stands).

**Table 4.** Trendline equations of the volume distribution in individual thickness classes.

| Forest Stand/Area Number | Age (Years) | Soil | Trendline Equation | Coefficient of Determination $R^2$ |
|---|---|---|---|---|
| 1 | 35 | | $y = -6.2696x^2 + 55.296x - 57.73$ | 0.4264 |
| 2 | 35 | | $y = -6.6857x^2 + 62.937x - 66.98$ | 0.3563 |
| 3 | 35 | sand | $y = -3.7768x^2 + 33.055x - 34.01$ | 0.4516 |
| 4 | 35 | | $y = -1.9679x^2 + 21.644x - 14.04$ | 0.8575 |
| 5 | 35 | | $y = -3.7571x^2 + 32.223x - 26.68$ | 0.6532 |
| 6 | 35 | | $y = -3.5107x^2 + 37.035x - 39.66$ | 0.3206 |
| 7 | 35 | clay | $y = -3.3x^2 + 31.66x - 21.04$ | 0.5409 |
| 8 | 41 | | $y = -2.8286x^2 + 31.457x - 32.743$ | 0.4643 |
| 9 | 33 | | $y = -8.575x^2 + 53.505x - 44.925$ | 0.8645 |
| 10 | 64 | | $y = -1.9674x^2 + 28.404x - 26.117$ | 0.3982 |
| 11 | 4 | | $y = -11.45x^2 + 46.75x - 32.1$ | 1 |
| 12 | 8 | dust | $y = -2.5x^2 + 18.4x + 9.4$ | 0.775 |
| 13 | 8 | | $y = 3.425x^2 - 3.635x + 19.325$ | 0.4789 |
| 14 | 8 | | $y = -0.225x^2 + 11.215x + 6.825$ | 0.9203 |

Abbreviations: x—variable wood thickness, y—correlation equation, wood volume depending on its thickness.

## 4. Discussion

The research was conducted in 14 stands of different ages (from 4 to 64 years old) and different origins, in very different soils and with different intensities of nursery treatments (loosening cuts). Under the conditions of the research, the volume of tree stands in the discussed areas varied. The number of trees in the main stands ranged from 580 to 1029 and in the secondary—575–1525 pcs. ha$^{-1}$. The main stands' average height was 15.5-19.0 m, while the average DBH in the main stands ranged from 16.0 cm on the sand slope with southern exposure (stand No. 5) to 23.5 cm on the lower part of the sand embankment with the south-eastern exposure.

During the study, it was found that sunlight and moisture conditions mainly influenced the volume of stands. The black locust grows remarkably in less fertile soils, often subject to erosion processes. The studies in stands of equal age located on the same soils with slight differences in the content of nutrients (stands No. 1–5, Table 2) indicated light and moisture conditions as factors determining the growth of stands. Black locust grows well on damp, shaded slopes with northern exposure, an example of which is the volume of the stand No. 2 growing in the lower part of the slope (good moisture conditions) with northern exposure. The stand volume in this area was the highest—353.8 m$^3$ ha$^{-1}$, exceeding the average volume of the remaining 35-year-old stands on sandy soils (No. 1 and 3–5) by 60%. In stand No. 1, growing in similar light conditions, but in the upper part of the slope (worse moisture conditions), the volume was lower, amounting to 270 m$^3$ ha$^{-1}$, and higher than the average (by 20%). The same trends in the differentiation of the volume of stands depending on the location on the slope (at the same exposure) were found in stands No. 3 and 4. The same dependencies in terms of location on the slope and soil moisture were observed by Ziemnicki et al. [45] and Węgorek [11]. Their research was carried out on the same stands, but they were carried out in the earlier years of growth. Montovani et al. [29,30] write extensively about the demand for water when the black locust grows in their works. However, excessive soil moisture is not advisable. Huntley and Pacyniak indicate poor tolerance of black locust to heavy and wet soils. The fact that this species performs very well in marginal soils was shown in their research by Ziemnicki et al. [45] and Gilewska [10].

The volume of the analyzed stands, due to their different age, was very diversified and ranged from less than 30 (No. 11, 4-year-old stand) to less than 550 m$^3$ ha$^{-1}$ (No. 10, 64-year-old stand). However, the values observed were comparable to those reported in the literature. According to research by Huntley [1], 126 m$^3$ of wood per 1 ha in the USA are obtained from 27-year-old plantations. In the study by Andrašev [3] carried out in 15 stands of black locust in predominant age of 21–43 years, located in Vojvodina (Serbia) on the chernozem subtype (A), the minimum volume was 160 m$^3$ ha$^{-1}$ and the highest 459 m$^3$ ha$^{-1}$. However, on the chernozem subtype (B), the smallest thickness was 145 m$^3$ ha$^{-1}$ and the highest 368 m$^3$ ha$^{-1}$. According to the research by Pacyniak [4], the volume of 50-year-old tree stands (in forest habitats in Poland) amounted to slightly over 292 m$^3$ ha$^{-1}$.

Among woody plants used for energy, willows and poplars are the most widespread in Poland [51–54]. Hence, an attempt was made to relate the research results to the relevant yields of wood obtained from energy plantations of these plants given in the literature. However, these comparisons should be considered approximate, especially in assessing black locust productivity under the conditions of own research, due to the different nature of the habitats, production conditions, and production cycle. In Poland, a significant yield of bush willows was obtained by Szczukowski et al. [51] in the experiments carried out on the wheat complex's brown soil, where the average dry matter yield of shrub willow wood in the four-year production cycle was 79.31 Mg ha$^{-1}$. Zajączkowski et al. [52] and Niemczyk et al. [53] indicate that the average dry matter yield of willow wood in two three-year production cycles on the soil made of lightweight clay was 30.92–42.48 Mg ha$^{-1}$, and the dry weight yield of poplar wood (in three-year cycles) 43.45–50.11 Mg ha$^{-1}$, while for poplars in the five-year cutting cycle, it was from 10 to 40 Mg ha$^{-1}$. These results indicate, however, that the black locust under the research conditions is a species with lower productivity.

Own research shows that as the age of stands increased, the share of trunk wood relating to the wood of branches increased. The share of wood up to 5.0 cm was small in older stands, at most a dozen or so percent. However, in young stands (4 and 8 years old), the share of the thickness class up to 5 cm was even 65% of the stand volume. In 35-year-old stands, wood fractions of 15.1–20.0 cm were dominant. In the oldest, the 64-year-old stand, over 30 cm thick wood constituted 44% of the stand volume. It is not easy to compare these results with the works of other researchers in terms of the wealth identified and assessed in individual classes. The results of the studies by Cui et al. [36] as cited by Dimobe [55] and Júnior et al. [56] indicate that the amount of biomass in terms of thickness, e.g. branches and trunks, is very dependent on the age of the trees. Additionally, Dong et al. [57] showed that the share of trunks and branches is very dependent on the DBH. On the other hand, Riofrío et al. [58] show that different soils and age of stands influence the differentiation of the share of biomass in each element of trees. Own research shows that between the same classes repeated in stands of different ages, the variation in the amount of wood in m$^3$ ha$^{-1}$ was statistically insignificant with $p = 0.1644$. The strongest relationship was between the two smallest classes with $p = 0.8048$. The designated trend lines and the equations describing them indicate a polynomial distribution of the second degree of wood volume in individual thickness classes. The matching coefficients take values over 50%, only in some cases they were values close to 30%.

## 5. Conclusions

The black locust grows well in degraded habitats subject to erosion processes. The biomass volume from such stands was comparable to that of forest habitats. Importantly, these tests made it possible to show the share of thickness in individual classes of wood thickness. The results of the research indicate that between individual stands, different habitats and age, the individual thickness of the individual classes did not differ statistically significantly. The smallest differences in volume were noted for stands aged 4 and 8 years, as well as between thicknesses classes 0.0–1.0 cm and 1.1–5.0 cm. Additionally, the distribution of wood thickness in individual thickness classes for the stands in question was described by the trend line equations.

Considering the habitat conditions, competition for agricultural production space, using marginal soils, and the intensity of breeding energy plantations of willow and poplar faced with the growth conditions of the stands based on which the research was carried out, it seems that black locust can be recommended for establishing tree stands and energy crops on various forms of wasteland to obtain medium-sized timber. This species does better than other similar purpose species in these site conditions.

It would be advisable to extend the research to other types of wastelands and to determine the optimal felling age (length of the production cycle) and the method of renewing black locust plantations (stem suckers, root suckers), which would verify the dendrometric measurements made and increase the accuracy of the developed models for estimating the biomass volume as a function of wood thickness.

**Funding:** This research was co-financed under the research topics of the by the Polish Ministry of Science and Higher Education No. 2 P06S 047 30 and TKR/S/4/2021. Additionally funded from the 'Excellent science' program of the Ministry of Education and Science as a part of the contract no. DNK/SP/465641/2020 "The role of the agricultural engineering and environmental engineering in the sustainable agriculture development".

**Data Availability Statement:**
https://katalog.bg.up.lublin.pl/cgi-bin/koha/opac-detail.pl?biblionumber=44152.

**Acknowledgments:** I want to sincerely thank the employees of the Department of Environmental Engineering and Geodesy of the University of Life Sciences in Lublin for their assistance during data collection.

**Conflicts of Interest:** The authors declare no conflict of interest.

## Appendix A

**Table A1.** Abundance of forest stand bark ($m^3\ ha^{-1}$).

| No. Forest Stand | Thickness Classes (cm) | Main Forest Stand | | | Secondary Forest Stand | | | Total Forest Stand | | |
|---|---|---|---|---|---|---|---|---|---|---|
| | | Trunk | Branch | Σ | Trunk | Branch | Σ | Trunk | Branch | Σ |
| 1 | 0.0-1.0 | - | 3.1 | 3.1 | - | 0.7 | 0.7 | - | 3.8 | 3.8 |
| | 1.1–5.0 | - | 17.7 | 17.7 | - | 2.0 | 2.0 | - | 19.7 | 19.7 |
| | 5.1–10.0 | 13.8 | 25.4 | 39.2 | 19.5 | - | 19.5 | 33.3 | 25.4 | 58.7 |
| | 10.1–15.0 | 43.2 | - | 43.2 | 3.3 | - | 3.3 | 46.5 | - | 46.5 |
| | 15.1–20.0 | 120.8 | - | 120.8 | - | - | - | 120.8 | - | 120.8 |
| | 20.1–25.0 | 20.3 | - | 20.3 | - | - | - | 20.3 | - | 20.3 |
| | Σ | 198.1 | 46.2 | 244.3 | 22.8 | 2.7 | 25.5 | 220.9 | 48.9 | 269.8 |
| 2 | 0.0-1.0 | - | 7.5 | 7.5 | - | 2.4 | 2.4 | - | 9.9 | 9.9 |
| | 1.1–5.0 | - | 24.9 | 24.9 | - | 9.9 | 9.9 | - | 34.8 | 34.8 |
| | 5.1–10.0 | 17.7 | - | 17.7 | 21.6 | - | 21.6 | 39.3 | - | 39.3 |
| | 10.1–15.0 | 70.9 | - | 70.9 | 8.5 | - | 8.5 | 79.4 | - | 79.4 |
| | 15.1–20.0 | 164.1 | - | 164.1 | - | - | - | 164.1 | - | 164.1 |
| | 20.1–25.0 | 26.3 | - | 26.3 | - | - | - | 26.3 | - | 26.3 |
| | Σ | 279.0 | 32.4 | 311.4 | 30.1 | 12.3 | 42.4 | 309.1 | 44.7 | 353.8 |
| 3 | 0.0-1.0 | - | 4.1 | 4.1 | - | 3.1 | 3.1 | - | 7.2 | 7.2 |
| | 1.1–5.0 | - | 10.4 | 10.4 | - | 7.9 | 7.9 | - | 18.3 | 18.3 |
| | 5.1–10.0 | 9.2 | 5.4 | 14.6 | 44.8 | - | 44.8 | 54.0 | 5.4 | 59.4 |
| | 10.1–15.0 | 38.7 | - | 38.7 | 9.3 | - | 9.3 | 48.0 | - | 48.0 |
| | 15.1–20.0 | 66.9 | - | 66.9 | - | - | - | 66.9 | - | 66.9 |
| | 20.1–25.0 | 11.7 | - | 11.7 | - | - | - | 11.7 | - | 11.7 |
| | Σ | 126.5 | 19.9 | 146.4 | 54.1 | 11.0 | 65.1 | 180.6 | 30.9 | 211.5 |
| 4 | 0.0-1.0 | - | 3.3 | 3.3 | - | 5.7 | 5.7 | - | 9.0 | 9.0 |
| | 1.1–5.0 | - | 29.4 | 29.4 | - | 9.9 | 9.9 | - | 39.3 | 39.3 |
| | 5.1–10.0 | 11.9 | 14.2 | 26.1 | 35.3 | - | 35.3 | 47.2 | 14.2 | 61.4 |
| | 10.1–15.0 | 37.5 | - | 37.5 | - | - | - | 37.5 | - | 37.5 |
| | 15.1–20.0 | 52.9 | - | 52.9 | - | - | - | 52.9 | - | 52.9 |
| | 20.1–25.0 | 42.0 | - | 42.0 | - | - | - | 42.0 | - | 42.0 |
| | Σ | 144.3 | 46.9 | 191.2 | 35.3 | 15.6 | 50.9 | 179.6 | 62.5 | 242.1 |
| 5 | 0.0-1.0 | - | 4.1 | 4.1 | - | 1.0 | 1.0 | - | 5.1 | 5.1 |
| | 1.1–5.0 | - | 22.8 | 22.8 | - | 2.5 | 2.5 | - | 25.3 | 25.3 |
| | 5.1–10.0 | 11.8 | 10.3 | 22.1 | 31.8 | - | 31.8 | 43.6 | 10.3 | 53.9 |
| | 10.1–15.0 | 60.8 | - | 60.8 | - | - | - | 60.8 | - | 60.8 |
| | 15.1–20.0 | 33.5 | - | 33.5 | - | - | - | 33.5 | - | 33.5 |
| | Σ | 106.1 | 37.2 | 143.3 | 31.8 | 3.5 | 35.3 | 137.9 | 40.7 | 178.6 |

**Table A1.** *Cont.*

| No. Forest Stand | Thickness Classes (cm) | Main Forest Stand | | | Secondary Forest Stand | | | Total Forest Stand | | |
|---|---|---|---|---|---|---|---|---|---|---|
| | | Trunk | Branch | Σ | Trunk | Branch | Σ | Trunk | Branch | Σ |
| 6 | 0.0–1.0 | - | 4.3 | 4.3 | - | 0.7 | 0.7 | - | 5.0 | 5.0 |
| | 1.1–5.0 | - | 20.3 | 20.3 | - | 5.8 | 5.8 | - | 26.1 | 26.1 |
| | 5.1–10.0 | 6.3 | 3.8 | 10.1 | 16.4 | - | 16.4 | 22.7 | 3.8 | 26.5 |
| | 10.1–15.0 | 37.6 | - | 37.6 | 1.3 | - | 1.3 | 38.9 | - | 38.9 |
| | 15.1–20.0 | 124.5 | - | 124.5 | - | - | - | 124.5 | - | 124.5 |
| | 20.1–25.0 | 23.5 | - | 23.5 | - | - | - | 23.5 | - | 23.5 |
| | Σ | 191.9 | 28.4 | 220.3 | 17.7 | 6.5 | 24.2 | 209.6 | 34.9 | 244.5 |
| 7 | 0.0–1.0 | - | 4.9 | 4.9 | - | 1.7 | 1.7 | - | 6.6 | 6.6 |
| | 1.1–5.0 | - | 42.5 | 42.5 | - | 6.1 | 6.1 | - | 48.6 | 48.6 |
| | 5.1–10.0 | 18.5 | - | 18.5 | 42.5 | - | 42.5 | 61.0 | - | 61.0 |
| | 10.1–15.0 | 73.7 | - | 73.7 | - | - | - | 73.7 | - | 73.7 |
| | 15.1–20.0 | 48.6 | - | 48.6 | - | - | - | 48.6 | - | 48.6 |
| | Σ | 140.8 | 47.4 | 188.2 | 42.5 | 7.8 | 50.3 | 183.3 | 55.2 | 238.5 |
| 8 | 0.0–1.0 | - | 4.7 | 4.7 | - | - | - | - | 4.7 | 4.7 |
| | 1.1–5.0 | - | 18.2 | 18.2 | - | - | - | - | 18.2 | 18.2 |
| | 5.1–10.0 | 11.5 | 14.8 | 26.3 | - | - | - | 11.5 | 14.8 | 26.3 |
| | 10.1–15.0 | 8.4 | 7.6 | 16.0 | - | - | - | 8.4 | 7.6 | 16.0 |
| | 15.1–20.0 | 82.9 | - | 82.9 | - | - | - | 82.9 | - | 82.9 |
| | 20.1–25.0 | 81.4 | - | 81.4 | - | - | - | 81.4 | - | 81.4 |
| | 25.1–30.0 | 26.1 | - | 26.1 | - | - | - | 26.1 | - | 26.1 |
| | Σ | 210.3 | 45.3 | 255.6 | - | - | - | 210.3 | 45.3 | 255.6 |
| 9 | 0.0–1.0 | - | 2.6 | 2.6 | - | - | - | - | 2.6 | 2.6 |
| | 1.1–5.0 | - | 20.0 | 20.0 | - | - | - | - | 20.0 | 20.0 |
| | 5.1–10.0 | 46.2 | - | 46.2 | - | - | - | 46.2 | - | 46.2 |
| | 10.1–15.0 | 29.3 | - | 29.3 | - | - | - | 29.3 | - | 29.3 |
| | Σ | 75.5 | 22.6 | 98.1 | - | - | - | 75.5 | 22.6 | 98.1 |
| 10 | 0.0–1.0 | - | 3.4 | 3.4 | - | - | - | - | 3.4 | 3.4 |
| | 1.1–5.0 | - | 36.7 | 36.7 | - | - | - | - | 36.7 | 36.7 |
| | 5.1–10.0 | 4.2 | 29.9 | 34.1 | - | - | - | 4.2 | 29.9 | 34.1 |
| | 10.1–15.0 | 57.0 | 10.8 | 67.8 | - | - | - | 57.0 | 10.8 | 67.8 |
| | 15.1–20.0 | 8.2 | 11.4 | 19.6 | - | - | - | 8.2 | 11.4 | 19.6 |
| | 20.1–25.0 | 26.6 | 19.2 | 45.8 | - | - | - | 26.6 | 19.2 | 45.8 |
| | 25.1–30.0 | 95.7 | - | 95.7 | - | - | - | 95.7 | - | 95.7 |
| | 30.1–35.0 | 131.2 | - | 131.2 | - | - | - | 131.2 | - | 131.2 |
| | 35.1–40.0 | 87.9 | - | 87.9 | - | - | - | 87.9 | - | 87.9 |
| | 40.1–45.0 | 21.4 | - | 21.4 | - | - | - | 21.4 | - | 21.4 |
| | Σ | 432.2 | 111.4 | 543.6 | - | - | - | 432.2 | 111.4 | 543.6 |
| 11 | 0.0–1.0 | - | 3.2 | 3.2 | - | - | - | - | 3.2 | 3.2 |
| | 1.1–5.0 | - | 15.6 | 15.6 | - | - | - | - | 15.6 | 15.6 |
| | 5.1–10.0 | 5.1 | - | 5.1 | - | - | - | 5.1 | - | 5.1 |
| | Σ | 5.1 | 18.8 | 23.9 | - | - | - | 5.1 | 18.8 | 23.9 |
| 12 | 0.0–1.0 | - | 23.6 | 23.6 | - | - | - | - | 23.6 | 23.6 |
| | 1.1–5.0 | - | 41.3 | 41.3 | - | - | - | - | 41.3 | 41.3 |
| | 5.1–10.0 | 37.0 | - | 37.0 | - | - | - | 37.0 | - | 37.0 |
| | 10.1–15.0 | 44.7 | - | 44.7 | - | - | - | 44.7 | - | 44.7 |
| | Σ | 81.7 | 64.9 | 146.6 | - | - | - | 81.7 | 64.9 | 146.6 |
| 13 | 0.0–1.0 | - | 11.9 | 11.9 | - | - | - | - | 11.9 | 11.9 |
| | 1.1–5.0 | - | 47.4 | 47.4 | - | - | - | - | 47.4 | 47.4 |
| | 5.1–10.0 | 12.4 | 5.2 | 17.6 | - | - | - | 12.4 | 5.2 | 17.6 |
| | 10.1–15.0 | 66.8 | - | 66.8 | - | - | - | 66.8 | - | 66.8 |
| | Σ | 79.2 | 64.5 | 143.7 | - | - | - | 79.2 | 64.5 | 143.7 |
| 14 | 0.0–1.0 | - | 19.3 | 19.3 | - | - | - | - | 19.3 | 19.3 |
| | 1.1–5.0 | - | 23.9 | 23.9 | - | - | - | - | 23.9 | 23.9 |
| | 5.1–10.0 | 35.3 | 7.6 | 42.9 | - | - | - | 35.3 | 7.6 | 42.9 |
| | 10.1–15.0 | 46.6 | - | 46.6 | - | - | - | 46.6 | - | 46.6 |
| | Σ | 81.9 | 50.8 | 132.7 | - | - | - | 81.9 | 50.8 | 132.7 |

# References


1. Huntley, J.C. Robinia pseudoacacia L. black locust. In *Silvics of North. America: Volume 2—Hardwoods*; Agriculture Handbook 654; United States Department of Agriculture: Washington, DC, USA, 1990; pp. 755–761.
2. Rédei, K.; Csiha, I.; Keserû, Z.; Gál, J. Influence of Regeneration Method on the Yield and Stem Quality of Black Locust (*Robinia pseudoacacia* L.) Stands: A Case Study. *Acta Silv. Lign. Hung.* **2012**, *8*, 103–111. [CrossRef]
3. Andrašev, S.; Rončević, S.; Ivanišević, P.; Pekeč, S.; Bobinac, M. Productivity of black locust (*Robinia pseudoacacia* L.) stands on chernozem in Vojvodina. *Bull. Fac. For.* **2014**, *10*, 9–32. [CrossRef]
4. Pacyniak, C. Black locust (*Robinia pseudoacacia* L.) in the conditions of the forest environment in Poland. *Rocz. AR Pozn. Rozpr. Nauk.* **1981**, *111*, 1–85. (In Polish)
5. Strączyńska, S.; Strączyński, S. Characteristics of the habitat conditions of black locust (*Robinia pseudoacacia* L.) planting in a combustion waste landfill. *Zesz. Probl. Post. Nauk Rol.* **2004**, *501*, 417–423. (In Polish)

6.   Kraszkiewicz, A. Chemical composition and selected energy properties of black locust bark (*Robinia pseudoacacia* L.). *Agric. Eng.* **2016**, *20*, 117–124. [CrossRef]

7.   Kraszkiewicz, A. The Combustion of Wood Biomass in Low Power Coal-Fired Boilers. *Combust. Sci. Technol.* **2015**, *188*, 389–396. [CrossRef]

8.   Bilkic, B.; Haykiri-Acma, H.; Yaman, S. Combustion reactivity estimation parameters of biomass compared with lignite based on thermogravimetric analysis. *Energy Sources Part A Recover. Util. Environ. Eff.* **2020**, 1–14. [CrossRef]

9.   Bender, J.; Gilewska, M.; Wójcik, A. Usefulness of black locust for woodlot in post-mining land. *Arch. Ochr. Środow.* **1985**, *3–4*, 113–133. (In Polish)

10.  Gilewska, M. Biological reclamation of ash landfills from lignite. *Rocz. Glebozn.* **2004**, *55*, 103–110. (In Polish)

11.  Węgorek, T. Changes in some properties of the earth material and development of phytocoenoses in the external dump of sulfur mine as a result of target forest reclamation. *Rozpr. Nauk. AR Lublinie* **2003**, *275*, 1–140. (In Polish)

12.  Carl, C.; Biber, P.; Veste, M.; Landgraf, D.; Pretzsch, H. Key drivers of competition and growth partitioning among Robinia pseudoacacia L. trees. *For. Ecol. Manag.* **2018**, *430*, 86–93. [CrossRef]

13.  Seserman, D.M.; Veste, M.; Freese, D. Optimisation of biomass productivity of black locust (Robinia pseudoacacia L.) on marginal lands—A case study in Lower Lusatia, NE Germany. In Proceedings of the 19th EGU General Assembly (EGU2017), Vienna, Austria, 23–28 April 2017.

14.  Nicolescu, V.N.; Buzatu-Goant, C.; Bakti, B.; Keserű, Z.; Antal, B.; Rédei, K. Black locust (Robinia pseudoacacia L.) as a multi-purpose tree species in Hungary and Romania: A review. *J. For. Res.* **2018**, *29*, 1449–1463. [CrossRef]

15.  González, I.; Sixto, H.; Rodríguez-Soalleiro, R.; Oliveira, N. Nutrient Contribution of Litterfall in a Short Rotation Plantation of Pure or Mixed Plots of *Populus alba* L. and *Robinia pseudoacacia* L. *Forests* **2020**, *11*, 1133. [CrossRef]

16.  Orlik, T.; Węgorek, T.; Zubala, T. Efficiency and growth of warty birch, European larch and black locust in mid-field strip plantations. *Folia Univ. Agric. Stetin., Agric.* **2001**, No. 87(217). 167–170. (In Polish)

17.  Węgorek, T.; Kraszkiewicz, A. Growth dynamics of false acacia (*Robinia pseudacacia* L.) in middle-field shelterbelt on loess soils. *Acta Agrophys.* **2005**, *5*, 211–218. (In Polish)

18.  European Commission. Communication from The Commission to The European Parliament, The Council, The European Economic and Social Committee and The Committee of The Regions Energy Roadmap 2050. Available online: https://eur-lex.europa.eu/legal-content/PL/TXT/PDF/?uri=CELEX:52011DC0885&from=EN (accessed on 28 December 2020).

19.  Renewable Energy Development Strategy, 2000: Report of the Ministry of the Environment Republic of Poland, Warsaw. Available online:https://ieo.pl/pl/raporty/82-strategia-rozwoju-energetyki-odnawialnej/file (accessed on 27 January 2021). (In Polish).

20.  Poland's Energy Policy until 2030. Ministry of Economy: Warsaw, Poland, 10 November 2009. Available online: http://nfosigw.gov.pl/download/gfx/nfosigw/pl/nfoopisy/1328/1/4/polityka_energetyczna_polski_do_2030r.pdf (accessed on 11 January 2021). (In Polish)

21.  Abbasi, T.; Abbasi, S.A. Biomass energy and the environmental impacts associated with its production and utilization. *Renew. Sust. Energ. Rev.* **2010**, *14*, 919–937. [CrossRef]

22.  Evans, A.; Strezov, V.; Evans, T.J. Sustainability considerations for electricity generation from biomass. *Renew. Sust. Energ. Rev.* **2010**, *14*, 1419–1427. [CrossRef]

23.  Ferrè, C.; Comolli, R. Comparison of soil $CO_2$ emissions between short-rotation coppice poplar stands and arable lands. *iForest Biogeosci. For.* **2018**, *11*, 199–205. [CrossRef]

24.  Landgraf, D.; Carl, C.; Neupert, M. Biomass Yield of 37 Different SRC Poplar Varieties Grown on a Typical Site in North Eastern Germany. *Forests* **2020**, *11*, 1048. [CrossRef]

25.  Cichy, W. Wood in combined heat and power plants—Innovation or sabotage? *Ekopartner* **2005**, *2*, 8–9. (In Polish)

26.  Szpil, Z. Energy prices, market—Ecology—Diversification. *Aura* **2006**, *3*, 18–19. (In Polish)

27.  Update of the National Program for Increasing Forest Cover 2014. Ministry of the Environment, Sękocin Stary 2014. Available online:https://nfosigw.gov.pl/download/gfx/nfosigw/pl/nfoekspertyzy/858/184/1/2013-772.pdf (accessed on 19 January 2021). (In Polish)

28.  Lewandowski, P. *Renewable Energy in Western Pomerania*; Wyd. Hogben: Szczecin, Poland, 2005; p. 326. (In Polish)

29.  Mantovani, D.; Veste, M.; Freese, D. How Much Water is Used by a Black Locust (*Robinia pseudoacacia* L.) Short-Rotation Plantation on Degraded Soil? Tagungsbeitrag zu: Jahrestagung der DBG Kommission IV Titel der Tagung: Böden verstehen—Böden nutzen—Böden fit machen Veranstalter: DBG, September 2011, Berlin Berichte der DBG. Available online:https://www.researchgate.net/publication/235695650 (accessed on 1 April 2021).

30.  Mantovani, D.; Veste, M.; Freese, D. Black locust (*Robinia pseudoacacia* L.) ecophysiological and morphological adaptations to drought andtheir consequence on biomass production and water-use efficiency. *N. Z. J. For. Sci.* **2014**, *44*, 29. [CrossRef]

31.  Plomion, C.; Leprovost, G.; Stokes, A. Wood Formation in Trees. *Plant. Physiol.* **2001**, *127*, 1513–1523. [CrossRef] [PubMed]

32.  Fabisiak, E. Variability of basic anatomical elements and wood density of selected tree species. *Rocz. AR Poznaniu. Rozpr. Nauk.* **2005**, *369*, 1–176. (In Polish)

33.  Rocha, M.F.V.; Veiga, T.R.L.A.; Soares, B.C.h.D.; Araújo, A.C.C.; Carvalho, A.M.M.; Hein, P.R.G. Do the Growing Conditions of Trees Influence the Wood Properties? *Floresta Ambient.* **2019**, *26*, e20180353. [CrossRef]

34.  Bach, E.M.; Ramirez, K.S.; Fraser, T.D.; Wall, D.H. Soil Biodiversity Integrates Solutions for a Sustainable Future. *Sustainability* **2020**, *12*, 2662. [CrossRef]

35. Bruchwald, A. Natural basics of building growth models. *Sylwan* **1988**, *11*, 1–10. (In Polish)
36. Cui, Y.; Bi, H.; Liu, S.; Hou, G.; Wang, N.; Ma, X.; Zhao, D.; Wang, S.; Yun, H. Developing Additive Systems of Biomass Equations for *Robinia pseudoacacia* L. in the Region of Loess Plateau of Western Shanxi Province, China. *Forests* **2020**, *11*, 1332. [CrossRef]
37. Dong, L.; Zhang, L.; Li, F. Developing Two Additive Biomass Equations for Three Coniferous Plantation Species in Northeast China. *Forests* **2016**, *7*, 136. [CrossRef]
38. Yuen, J.Q.; Fung, T.; Ziegler, A.D. Review of allometric equations for major land covers in SE Asia: Uncertainty and implications for above- and below-ground carbon estimates. *For. Ecol. Manag.* **2016**, *360*, 323–340. [CrossRef]
39. Chave, J.; Réjou-Méchain, M.; Búrquez, A.; Chidumayo, E.; Colgan, M.S.; Delitti, W.B.; Duque, A.; Eid, T.; Fearnside, P.M.; Goodman, R.C.; et al. Improved allometric models to estimate the aboveground biomass of tropical trees. *Glob. Change Biol.* **2014**, *20*, 3177–3190. [CrossRef] [PubMed]
40. Luo, Y.; Wang, X.; Ouyang, Z.; Lu, F.; Feng, L.; Tao, J. A review of biomass equations for China's tree species. *Earth Syst. Sci. Data* **2020**, *12*, 21–40. [CrossRef]
41. Paul, K.I.; Roxburgh, S.H.; Chave, J.; England, J.R.; Zerihun, A.; Specht, A.; Lewis, T.; Bennett, L.T.; Baker, T.G.; Adams, M.A.; et al. Testing the generality of above-ground biomass allometry across plant functional types at the continent scale. *Glob. Chang. Biol.* **2016**, *22*, 2106–2124. [CrossRef] [PubMed]
42. Basuki, T.; Van Laake, P.; Skidmore, A.; Hussin, Y. Allometric equations for estimating the above-ground biomass in tropical lowland Dipterocarp forests. *For. Ecol. Manag.* **2009**, *257*, 1684–1694. [CrossRef]
43. Kuyah, S.; Dietz, J.; Muthuri, C.; Jamnadass, R.; Mwangi, P.; Coe, R.; Neufeldt, H. Allometric equations for estimating biomass in agricultural landscapes: I. Aboveground biomass. *Agric. Ecosyst. Environ.* **2012**, *158*, 216–224. [CrossRef]
44. Zielony, R.; Kliczkowska, A. *Poland's Natural and Forest Regionalization 2010*; CILP: Warsaw, Poland, 2012; pp. 1–359. (In Polish)
45. Ziemnicki, S.; Fijałkowski, D.; Repelewska-Pękalowa, J.; Węgorek, T. *Reclamation of an Opencast Mine Heap (on the Example of Piaseczno)*; PWN: Warsaw, Poland, 1980; pp. 1–100. (In Polish)
46. Kraszkiewicz, A.; Węgorek, T. Structure of robinia stands on sandy slopes threatened with erosion. *Zesz. Nauk. Akad. Rol. Poznaniu. CCCLXXV Rol.* **2006**, *65*, 69–74. (In Polish)
47. Bruchwald, A. *Dendrometry*; Wyd. SGGW: Warsaw, Poland, 1999; pp. 1–261. (In Polish)
48. Kowalkowski, A. *Mineral. Fertilization of Stands*; SGGW-AR: Warsaw, Poland, 1982; pp. 1–86. (In Polish)
49. Baule, H.; Fricker, C. *Fertilization of Forest Trees*; PWRiL: Warsaw, Poland, 1971; pp. 1–314. (In Polish)
50. Pokojska, U. The role of humus in shaping the reaction, buffer properties and ion-exchange capacity of forest soils. *Rocz. Glebozn.* **1986**, *37*, 249–262. (In Polish)
51. Szczukowski, S.; Tworkowski, J.; Stolarski, M.; Grzelczyk, M. Productivity of willow (*Salix* spp.) and characterization of willow biomass as fuel. *Zesz. Probl. Post. Nauk Rol.* **2005**, *507*, 495–503.
52. Zajączkowski, K.; Kwiecień, R.; Zajączkowska, B.; Wajda, T.; Zawadzki, M. *Production Possibilities of Selected Varieties of Poplar and Willow in Plantations with a Shortened Cycle*; IBL: Warsaw, Poland, 2001; pp. 1–66. (In Polish)
53. Niemczyk, M.; Wojda, T.; Kantorowicz, W. The breeding usefulness of selected poplar varieties in energy plantations with a short production cycle. *Sylwan* **2016**, *160*, 292–298. (In Polish)
54. Szostak, A.; Bidzińska, G.; Ratajczak, E.; Herbeć, M. Wood biomass from plantations of fast-growing trees as an alternative source of wood raw material in Poland. *Drewno* **2013**, *56*, 85–112. [CrossRef]
55. Dimobe, K.; Mensah, S.; Goetze, D.; Ouédraogo, A.; Kuyah, S.; Porembski, S.; Thiombiano, A. Aboveground biomass partitioning and additive models for Combretum glutinosum and Terminalia laxiflora inWest Africa. *Biomass Bioenergy* **2018**, *115*, 151–159. [CrossRef]
56. Júnior, L.R.N.; Engel, V.L.; Parrotta, J.A.; De Melo, A.C.G.; Ré, D.S. Allometric equations for estimating tree biomass in restored mixed-species Atlantic Forest stands. *Biota Neotrop.* **2014**, *14*, 1–9.
57. Dong, L.H.; Zhang, L.J.; Li, F. Additive Biomass Equations Based on Di_erent Dendrometric V ariables for Two Dominant Species (*Larix gmelini* Rupr. and *Betula platyphylla* Suk.) in Natural Forests in the Eastern Daxing'an Mountains, Northeast China. *Forests* **2018**, *9*, 261. [CrossRef]
58. Riofrío, J.; Herrero, C.; Grijalva, J.; Bravo, F. Aboveground tree additive biomass models in Ecuadorian highland agroforestry systems. *Biomass Bioenergy* **2015**, *80*, 252–259. [CrossRef]

*Article*

# Productivity and Profitability of Poplars on Fertile and Marginal Sandy Soils under Different Density and Fertilization Treatments

Solomon B. Ghezehei [1,*], Alexander L. Ewald [1], Dennis W. Hazel [1], Ronald S. Zalesny, Jr. [2] and Elizabeth Guthrie Nichols [1]

[1]  Department of Forestry and Environmental Resources, North Carolina State University, Raleigh, NC 27695, USA; alewald@ncsu.edu (A.L.E.); hazeld@ncsu.edu (D.W.H.); egnichol@ncsu.edu (E.G.N.)
[2]  Institute for Applied Ecosystem Studies, Northern Research Station, USDA Forest Service, Rhinelander, WI 54501, USA; ron.zalesny@usda.gov
*  Correspondence: sbghezeh@ncsu.edu; Tel.: +1-919-513-1371

**Citation:** Ghezehei, S.B.; Ewald, A.L.; Hazel, D.W.; Zalesny, R.S., Jr.; Nichols, E.G. Productivity and Profitability of Poplars on Fertile and Marginal Sandy Soils under Different Density and Fertilization Treatments. *Forests* **2021**, *12*, 869. https://doi.org/10.3390/f12070869

Academic Editor: Jason G. Vogel

Received: 28 April 2021
Accepted: 28 June 2021
Published: 30 June 2021

**Publisher's Note:** MDPI stays neutral with regard to jurisdictional claims in published maps and institutional affiliations.

**Abstract:** We evaluated the productivity and profitability of four highly productive poplars including *Populus deltoides* × *P. deltoides* (DD '140' and '356'), *P. deltoides* × *P. maximowiczii* (DM '230'), and *P. trichocarpa* × *P. deltoides* (TD '185') under two densities (2500 and 5000 trees ha$^{-1}$), and three fertilization treatments (0, 113, 225 kg nitrogen ha$^{-1}$) at three sandy coastal sites varying in soil quality. Green stem biomass (GSB) was estimated from the sixth-year stem diameter. Leaf-rust (*Melampsora castagne*) and beetle damage (by *Chrysomela scripta* Fabricius), the leaf area index (LAI) and foliar nitrogen, were measured in year two. At all sites, DD and DM had higher survival (>93%) than TD (62–83%). DD produced greater GSB (92.5–219.1 Mg ha$^{-1}$) than DM (54–60.2 Mg ha$^{-1}$) and TD (16.5–48.9 Mg ha$^{-1}$), and this was greater under the higher density (85.9–148.6 Mg ha$^{-1}$ vs. 55.9–124.9 Mg ha$^{-1}$). Fertilization significantly increased GSB on fertile soil but not marginal soils; a higher rate did not significantly enhance GSB. Leaf rust was higher for fertile soil (82%) than marginal soils (20–22%), and TD '185' (51% vs. others 34%). *C. scripta* damage was higher for the higher density (+42%) than lower density, and TD '185' (50% vs. others >38%). LAI was higher on fertile soil (1.85 m$^2$ m$^{-2}$) than marginal soils (1.35–1.64 m$^2$ m$^{-2}$), and under the lower density (1.67 m$^2$ m$^{-2}$ vs. 1.56 m$^2$ m$^{-2}$). The high GSB producer DD '356' had the lowest LAI (1.39 m$^2$ m$^{-2}$ vs. 1.80 m$^2$ m$^{-2}$). Foliar nitrogen varied among genomic groups (DD '140' 1.95%; TD '185' 1.80%). Our plots were unprofitable at a 27 USD Mg$^{-1}$ delivered price; the biggest profitability barriers were the high costs of higher density establishment and weed control. The best-case treatment combinations of DD ('140', '356') would be cost-effective if the price increased by 50% (USD 37.54 Mg$^{-1}$) or rotations were 12 years (fertile-soil) and longer (marginal soils). The requirement for cost-effectiveness of poplars includes stringent and site-specific weed control which are more important than fertilizer applications.

**Keywords:** cottonwood leaf beetle (*Chrysomela scripta*); stand density; fertilizer application; *Populus*; soil quality; *Melampsora* rust

## 1. Introduction

The Energy Independence and Security Act (EISA) of 2007 mandates an increase in biofuel use, from 34.1 billion liters in 2008 to 136.3 billion liters in 2022 [1], and targets a 9% decrease in greenhouse gas emissions [2]. Currently, more than 137 million tonnes of corn (*Zea mays* L.) are used for ethanol to be blended into transportation fuels in the United States (US), which dramatically increases the demand for field corn to meet the EISA standards [3]. Increased corn-based ethanol production raises environmental concerns for erosion, water and fertilizer uses, pollution related to pesticide use, and nutrient run-off, and the use of agricultural food-crop lands for energy production [3–5]. Concurrently, the southeastern US is the largest exporter of wood pellets to Europe due to abundant feedstock inventory, manufacturing proximity, and accessible shipping ports [6]. Increased wood pellet market growth, and European renewable energy targets [7–9] are expected to catalyze the need for more noncontentious lands and sustainable feedstock supplies from the US.

One opportunity to expand biofuels and bioenergy production in the US, while minimizing environmental concerns, is to develop second generation feedstocks such as tall grasses, *Panicum* L. and *Miscanthus* (Nees) Andersson, and short rotation woody crops (SRWCs) such as *Populus* L. or *Salix* L. [10,11]. When compared to annual and perennial crops, woody species grown in short rotation have higher energy densities, lower transportation costs, and reduced needs for annual inputs such as fertilizer [12,13]. Of the SRWCs produced in the US, poplars (*Populus* sp. and their hybrids) are among the most commonly analyzed bioenergy crops due to their high productivity and decades of genetic improvement [12,14].

Prior clonal studies have shown that poplars, when intensively cultured as SRWCs, can produce substantially greater biomass than other temperate species [7,14–18]. Ecosystem demography models have estimated potential yields of poplar plantations across the temperate regions of the US ranging between 10 and 18 Mg ha$^{-1}$ year$^{-1}$ (dry mass) [19]. However, growth projections for poplars are heavily dependent on climate, soil condition, clonal parentage, fertilization rates, and stand densities [16,19].

Planting densities vary widely for poplars, typically ranging from 1000 to 20,000 stems ha$^{-1}$, based on end uses of the trees [20]. When poplars are planted at high densities, they produce smaller individual trees but a greater cumulative stand biomass [21,22]. Determining the optimal stand density depends on the biological and nutrient limitations of a site, such as climate, precipitation, and soil quality [20]. Because density influences the amount of biomass produced on a "per tree" and "per area" basis, the intended use of the biomass is important for site establishment, rotation lengths, and profitability for landowners.

To avoid SRWC competition with agriculture and conventional forestry for land use, marginal lands are an important resource for SRWC production [1,23]. Abandoned and degraded lands in SRWC production could be utilized to produce 10 to 52% of the current liquid fuel consumption [23,24]. However, marginal lands often have lower quality soils with limited fertility [25], and poplar biomass yields often decline significantly on marginal soils [26–28]. One study reported poplar biomass yields of 22.4 m$^3$ ha$^{-1}$ year$^{-1}$ on fertile soils compared with 1.1 m$^3$ ha$^{-1}$ year$^{-1}$ on marginal soils [27].

Understanding poplar responses to fertilizer application rates on marginal lands is essential to optimizing yields and enhancing economic returns for biofuels and bioenergy production [29,30]. Prior studies have evaluated nitrogen (N) fertilizer application rates, ranging from 60 to 250 kg N ha$^{-1}$ for poplars across a variety of different site-specific conditions, and optimum nitrogen fertilization rates are site- and genotype-specific [31–33]. In the southeastern US, estimated optimal N application rates for optimal growth were 70 to 90 kg N ha$^{-1}$ year$^{-1}$ for two, non-irrigated cottonwood (*Populus deltoides* Bartr. ex Marsh) clones [32].

Fertilization and densely-planted stands increase input costs and impact profitability if biomass yields, and market prices do not yield revenue greater than the establishment and production costs. Poplars and other fast-growing trees can be economically feasible when costs are minimized by effective stand establishment and management, especially the suppression of weed competition, and when revenues are maximized [34]. Advancing woody feedstocks for bioenergy production requires economic profitability for landowners [35], and subsidization can improve the economic viability of SRWC stands [34,36]. In the southeastern US, prices of bioenergy feedstocks are low compared to other markets, and understanding what management practices (e.g., stand density, fertilization, and rotation length) will produce the best economic returns from poplar genotypes is needed to appropriately promote and position SRWCs as a viable bioenergy feedstock [34].

Our objectives were to examine if first-year fertilization and its interactive effects with site/soil quality, stand density, and poplar genetics could lead to greater biomass yields and improved cost-effectiveness of poplar biomass production systems. For this purpose, we selected four poplar clones ('140', '185', '230', '356') that were the most productive genotypes after eight years of clonal trials in North Carolina, USA [7,17,37,38]. We tested poplar productivity and economic returns under two stand densities and three fertilization regimes on fertile and marginal coastal sandy soils. We hypothesized that fertilizer application in year-one enhances poplar productivity and leads to improved profitability under high and low stand densities on fertile and marginal lands. Genotype responses to the early-age incidence of disease and pests, foliar N content, and leaf area index (LAI) were also evaluated across sites and treatments. Economic returns were evaluated based on the estimated stem biomass at six, and twelve years for current and potential market prices.

## 2. Materials and Methods

### 2.1. Site Description

Trial stands were established at coastal sites that varied in soil quality and cropping history (Table 1) using four poplar genotypes belonging to three genomic groups: *P. deltoides* × *P. deltoides* 'DD' clones '140' and '356'; *P. deltoides Populus maximowiczii* A. Henry 'DM' clone '230'; *Populus trichocarpa* Torr. et. Gray × *P. deltoides* 'TD' clone '185'. Williamsdale, NC, USA had highly productive, fertile soils [39] while the two sites in Clinton, NC, USA (Clinton A and Clinton B) had marginal soils [40]. Prior to establishing poplars, the sites were planted with a winter wheat (*Triticum aestivum* L.) cover crop in fall 2014 and routinely mowed.

**Table 1.** Climate and site characteristics of Clinton A, Clinton B and Williamsdale in North Carolina, USA.

| Site Characteristics | Clinton A | Clinton B | Williamsdale |
|---|---|---|---|
| Soil Quality | Marginal | Marginal | Fertile |
| Location | 35° 1′ 20.47″ N; 78° 16′ 24.62″ W | 35° 1′ 20″ N; 78° 16′ 23.90″ W | 34° 45′ 51.06″ N; 78°5′ 59.04″ W |
| Soil Series [c] | Wagram loamy sand | Orangeburg loamy fine sand | Noboco loamy fine sand |
| Mean Annual Rainfall (mm) [a,b] | 1371 | 1371 | 1480 |
| Elevation (msl) [a] | 50.6 | 50.6 | 17 |
| Mean Annual T (C) [a,b] | 16.7 | 16.7 | 16.7 |
| Mean Daily Humidity (%) [a,b] | 71.9 | 71.9 | 73.4 |
| Mean Plant Available Water (cm$^3$ cm$^{-3}$) [a,b] | 0.18 | 0.18 | 0.2 |
| Soil pH [d] | 6.2 | 6.1 | 6 |
| Cation Exchange Capacity (Meq/100g) [d] | 2.3 | 2.1 | 6.9 |
| Soil P (kg ha$^{-1}$) [d] | 504 | 605 | 1011 |
| Soil K (kg ha$^{-1}$) [d] | 131 | 112 | 410 |
| Soil Mg (kg ha$^{-1}$) [d] | 102 | 92 | 308 |
| Soil Ca (kg ha$^{-1}$) [d] | 559 | 517 | 1630 |
| Soil NO-3 (kg ha$^{-1}$) [d] | 3.7 | 2.5 | 4.1 |
| Prior Crop | Sorghum | Sorghum | Corn |

[a] Climate data provided by the State Climate Office of North Carolina; [b] December 2014–December 2016; [c] Data provided by USDA Web Soil Survey (last accessed on 15 June 2020)); [d] collected at a depth of 15 cm.

### 2.2. Experimental Setup

The experimental design was a cluster-randomized design with two densities (2500 and 5000 trees ha$^{-1}$), three fertilization levels (0, 113, and 225 kg N ha$^{-1}$) and four clones. Each cluster contained four trees of a clone; the clusters were replicated three times, and placed randomly within the study sites. At each site, 18 experimental plots (nine plots per density), and a total of 288 experimental trees were used. Border trees were planted around the perimeter of each plot to reduce border effects.

Site preparation was performed in February 2015 when soils were subsoiled (i.e., ripped) then banded with a 30-cm band of 41% glyphosate (4.67 L ha$^{-1}$) and 37.4%

pendimethalin (9.35 L ha$^{-1}$) solution along the rip lines for weed control prior to planting. Poplar cuttings were purchased from ArborGen, LLC (Ridgeville, SC, USA), soaked for 24 h, and planted as 25-cm, non-rooted cuttings in March 2015. Nitrogen fertilizer (YaraLiva Calcinit 15.5-0-0, Yara North America, Tampa, FL, USA) was applied by hand in a 0.6-m radius around the individual trees within the fertilized plots in April 2015. Plots were banded with glyphosate/pendimethalin along the tree lines approximately five times per growing season for weed control. Non-banded areas were maintained by mowing once a month. All sites were treated for cottonwood leaf beetle (*Chrysomela scripta* Fabricius) infestation with Sevin® (Carbaryl, Bayer, Leverkusen, Germany), which was applied by hand sprayer in the early summer of year one and mist blower in early- and mid-summer of year two. Soil samples were collected in the spring of 2015 at a depth of 15 cm for each of the unfertilized plots and combined by site. The soils were analyzed at Waters Agricultural Laboratory (Warsaw, NC, USA) using the Mehlich III method, which is a method of extracting multiple soil micronutrients and macronutrients using a weak acid (Table 1).

### 2.3. Data Collection

The stem diameter at breast height (DBH) of six-year-old trees was measured at a height of 1.3 m using a Lufkin Executive Thin Line DBH tape (Apex Tool Group, Cleveland, OH, USA). The green stem biomass (GSB) per tree was estimated using the equation [41],

$$GSB = 0.1375 \, DBH^{2.3681} \tag{1}$$

where GSB is in kg and DBH is in cm.

During the second year of growth (July 2016), the presence of *Melampsora castagne* rust and *C. scripta* leaf damage were inventoried by scoring the presence or absence of damage for the whole tree. Leaf Area Index (LAI) measurements were also taken using an LAI-2000 LI-COR Plant Canopy Analyzer (LI-COR, Lincoln, Nebraska, NE, USA); the measurements were taken within the tree clusters with the same treatment combinations (density × fertilizer rate × clones). Leaf samples for foliar percent nitrogen (N) were also collected by compositing four leaves taken from each tree, one from each ordinal direction while increasing in height, for a total of 16 leaves per clone, per plot. All foliar samples were analyzed at Waters Agricultural Laboratory (Warsaw, NC, USA).

### 2.4. Data Analysis

Tree stem biomass values within each density × fertilization × clone split blocks (clusters) were summed to obtain total GSB values (Mg ha$^{-1}$), which were analyzed (as completely randomized design) using a generalized linear model (PROC GLM, $p = 0.05$; SAS, Cary, NC, USA) to examine the effects of the levels of stand density, fertilization rate, and clone at the three sites. The following statistical model was used:

$$y_{ijk} = \mu + \alpha_D + \beta_F + \gamma_G + \alpha\beta_{DF} + \alpha\gamma_{DG} + \beta\gamma_{FG} + \alpha\beta\gamma_{DFG} + \varepsilon_{DFG} \tag{2}$$

where: $\mu$ is the overall average of the experiment, $\alpha_D$ is the effect of density treatment (fixed), $\beta_F$ is the effect of fertilization treatment (fixed), $\gamma_G$ is the fixed effect of clones, $\alpha\beta_{DF}$, $\alpha\gamma_{DG}$, $\beta\gamma_{FG}$, and $\alpha\beta\gamma_{DFG}$ are effects of treatment interactions (density × fertilization, density × genetics, fertilization × genetics, density × fertilization × genetics, respectively) and $\varepsilon_{DFG}$ is the random error.

LAI, foliar N, *Melampsora* rust, and *C. scripta* leaf damage data were also evaluated within the levels of stand density, fertilization rate, and clone at the three sites. *Melampsora* rust and *C. scripta* damage were scored using a modified Schreinder index with the percentage of leaves scored on a scale of 0 to 4 (0 = 100% of foliage with observable damage; 1 = 75%; 2 = 50%; 3 = 25%; 4 = no evidence of *Melampsora* rust or *C. scripta* damage) [42]. Field precision was determined by the duplication of field measurements for DBH, LAI, and foliar N. The resulting relative percent differences were 1 ± 3% for DBH (7.2% duplicated), $5.3 \pm 5.6$ % for LAI (21.2% duplicated) and $16 \pm 12$% for foliar nitrogen percent (11.5% duplicated).

### 2.5. Economic Analysis

The economic viability of the stands was examined using net present value (NPV) and break-even price analysis. Stand establishment and maintenance costs included chemical weed suppression, sub-soiling, cover crops, pesticide applications, and associated labor and material costs. Harvest and transport costs were determined assuming the harvested biomass was hauled for an average hauling distance of 84 km using a 40-tonne net log truck with a diesel consumption of 2.13 km $L^{-1}$. A diesel price of 0.62 USD $L^{-1}$ (assumed) was used, and poplar seedlings were assumed to cost 0.375 per seedling, the same as the '1000+' hardwood price package in the North Carolina tree seedling catalog for 2020–2021 [43]. No costs of land rent or property taxes were included due to the unavailability of such cost information for marginal and abandoned lands. Stand revenues after six growing years were determined based on GSB values, and a delivered price of 27.01 USD $Mg^{-1}$, which was the ten-year (2011–2020) average hardwood pulpwood delivered price for eastern North Carolina [44]. To determine revenues from the stands in 12-year rotations, mean annual increments (MAI, in Mg $ha^{-1}$ $year^{-1}$) of GSB at the age of 12 years were estimated using equations of the best-fit (logarithmic) curves for MAI versus age for treatment combinations during the study years (six). The treatment combinations used were selected based on GSB analyses (showing significant differences), and included density $\times$ genomic group at the Clinton sites (Supplementary Figure S1) and fertilizer rate $\times$ genomic group at the Williamsdale site (Supplementary Figure S2).

## 3. Results and Discussion

### 3.1. Clonal Productivity Responses to Stand Density and Fertilizer Application

The four poplar clones selected for this study had demonstrated superior growth and survival compared to other poplar clones as a function of genetics and site quality for sandy coastal soils [17,37]. There were differences in GSB among genomic groups in the fertile and marginal sites; however, there were no significant interaction effects of genotype with density or fertilization ($p > 0.1371$; Tables 2 and 3). DD genotypes produced significantly greater GSB (Clinton A: 109.3 Mg $ha^{-1}$; Clinton B: 92.5 Mg $ha^{-1}$; Williamsdale: 219.1 Mg $ha^{-1}$) than the DM (Clinton A: 54 Mg $ha^{-1}$; Clinton B: 56.3 Mg $ha^{-1}$; Williamsdale: 60.2 Mg $ha^{-1}$) and TD (Clinton A: 48.9 Mg $ha^{-1}$; Clinton B: 38.5 Mg $ha^{-1}$; Williamsdale: 16.5 Mg $ha^{-1}$) clones. Only clone ($p < 0.0001$) and density ($p < 0.0001$) were significant factors at both Clinton sites, while at Williamsdale, fertilization ($p = 0.0003$, Table 2 and Table S1) and density $\times$ fertilization $\times$ clone interaction effects ($p = 0.0093$) were also significant (Figure 1). At the three sites, clones '140' (95.1 to 214.3 Mg $ha^{-1}$) and '356' (89.8 to 223.8 Mg $ha^{-1}$) produced the greatest GSB and clone '185' had lowest GSB (16.5 to 48.9 Mg $ha^{-1}$). GSB after six years was greatest under the higher stand density at all sites (85.9 to 148.6 Mg $ha^{-1}$ versus 55.9 to 124.9 Mg $ha^{-1}$), although biomass productivity of both densities was comparable at early ages after the second growing season. Effects of site between Clinton A and B, and site interactions with stand density, fertilizer rates and clones were insignificant ($p \geq 0.1254$; Table S2).

**Table 2.** Green stem biomass (Mg ha$^{-1}$) of four six-year-old hybrid poplar clones established under three fertilization rates and two stand densities at three sites in the Coastal region of North Carolina, USA. Treatment means within sites with the same letter are not significantly different.

| Treatments | | Mean Stem Biomass (Mg ha$^{-1}$) | | |
|---|---|---|---|---|
| | | **Clinton A** | **Clinton B** | **Williamsdale** |
| **Stand Density** | **5000 trees ha$^{-1}$** | 93.9 A | 85.6 A | 148.6 A |
| | **2500 trees ha$^{-1}$** | 65.3 B | 55.9 B | 124.9 A |
| | **MSD ($\alpha = 0.05$)** | 18.02 | 16.69 | 27.98 |
| **Fertilizer Rate** | **225 kg N ha$^{-1}$** | 73.5 A | 69.6 A | 164.8 A |
| | **113 kg N ha$^{-1}$** | 93.6 A | 78.4 A | 154.8 A |
| | **0 kg N ha$^{-1}$** | 71.8 A | 63.7 A | 93.1 B |
| | **MSD ($\alpha = 0.05$)** | 26.6 | 24.6 | 36.06 |
| **Clone** | **'140'** | 114.8 A | 95.1 A | 214.3 A |
| | **'356'** | 103.2 A | 89.8 A | 223.8 A |
| | **'230'** | 54.0 B | 56.3 B | 60.2 B |
| | **'185'** | 48.9 B | 38.5 B | 16.5 B |
| | **MSD ($\alpha = 0.05$)** | 33.80 | 31.30 | 52.81 |
| **Overall** | | 79.4 | 69.8 | 129.7 |

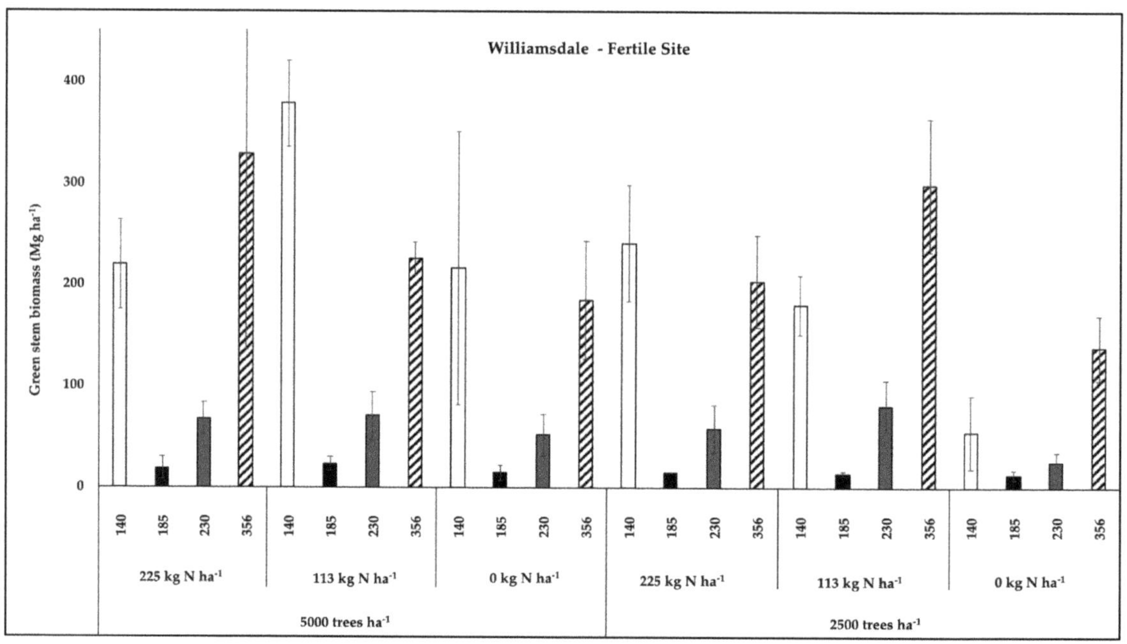


**Figure 1.** Mean green stem biomass (Mg ha$^{-1}$) for *Populus* clones established at a fertile Coastal site (Williamsdale) in North Carolina, USA by stand density and fertilization treatment after six years of growth. Error bars denote one standard deviation of the mean per treatment ($\alpha = 0.05$, ANOVA).

**Table 3.** Estimated green stem biomass (Mg ha$^{-1}$) after six years, and leaf area index (LAI, m$^2$ m$^{-2}$) and foliar nitrogen content (%) during the second growing season of poplar clones established at two stand densities with three fertilization rates at three sites in the Coastal region of North Carolina, USA.

| | | | Stands Age: Six Years | | | Stand Density: Two Years | | | | | |
| | | Clone (Genomic Group) | Mean Green Stem Biomass ± 1SD (Mg ha$^{-1}$) | | | LAI (m$^2$ m$^{-2}$) (SD) | | | Foliar N% (SD) | | |
| Stand Density | Fertilizer Rates | | Clinton A | Clinton B | Williamsdale | Clinton A | Clinton B | Williamsdale | Clinton A | Clinton B | Williamsdale |
|---|---|---|---|---|---|---|---|---|---|---|---|
| 5000 trees ha$^{-1}$ | 225 (kg N ha$^{-1}$) | '140' (DD) | 107.3 ± 48.8 | 114.4 ± 52.3 | 219.1 ± 44.3 | 1.2 (0.5) | 0.9 (0.3) | 1.2 (0.2) | 2 (0.2) | 1.7 (0.2) | 2.1 (0.1) |
| | | '185' (TD) | 62.1 ± 78.4 | 55.0 ± 31.5 | 18.5 ± 11.0 | 1.8 (0.9) | 1.2 (0.3) | 1.4 (0.2) | 2.1 (0.1) | 1.7 (0.3) | 2 (0.2) |
| | | '230' (DM) | 52.3 ± 14.6 | 68.0 ± 28.5 | 66.8 ± 16.1 | 1.9 (0.8) | 1 (0.2) | 1.2 (0.3) | 1.9 (0.2) | 1.8 (0.2) | 2 (0.2) |
| | | '356' (DD) | 103.2 ± 52.8 | 136.3 ± 39.4 | 329 ±193.4 | 1.3 (0.6) | 0.9 (0.3) | 1.3 (0.2) | 1.8 (0.4) | 1.8 (0.2) | 2 (0.1) |
| | 113 (kg N ha$^{-1}$) | '140' (DD) | 146.2 ± 41.6 | 140.5 ± 51.5 | 378.1 ± 42.5 | 1 (0.3) | 1.2 (0.4) | 2.2 (0.1) | 1.7 (0.3) | 1.6 (0.1) | 2.2 (0.1) |
| | | '185' (TD) | 61.5 ± 42.0 | 48.0 ± 12.2 | 22.8 ± 6.8 | 1.6 (0.4) | 1.8 (0.5) | 2.5 (0.2) | 1.8 (0.2) | 1.6 (0.3) | 2.1 (0.2) |
| | | '230' (DM) | 90.2 ± 16.8 | 74.5 ± 19.7 | 70.2 ± 23.5 | 1.4 (0.5) | 1.5 (0.3) | 2.7 (0.1) | 1.7 (0.2) | 1.6 (0.1) | 2 (0.3) |
| | | '356' (DD) | 170.2 ± 60.3 | 80.6 ± 57.0 | 225.2 ± 15.9 | 1.5 (0.5) | 1 (0.1) | 2.5 (0.1) | 1.3 (0.3) | 1.9 (0.1) | 2.1 (0.1) |
| | 0 (kg N ha$^{-1}$) | '140' (DD) | 141.7 ± 61.4 | 112.2 ± 27.9 | 215.7±135.2 | 1.5 (0.4) | 1 (0.1) | 1.9 (0.4) | 2.1 (0.3) | 1.9 (0.1) | 2.2 (0.2) |
| | | '185' (TD) | 44.6 ± 27.2 | 44.9 ± 14.4 | 14.3 ± 7.8 | 2.1 (0.4) | 2.1 (0.1) | 2.2 (0.6) | 1.8 (0.1) | 1.8 (0.1) | 2.1 (0.6) |
| | | '230' (DM) | 43.5 ± 17.8 | 62.7 ± 19.8 | 51.4 ± 20.7 | 2.2 (0.2) | 1.1 (0.4) | 1.8 (0.3) | 1.9 (0) | 1.8 (0.2) | 2.1 (0.2) |
| | | '356' (DD) | 93.2 ± 55.6 | 76.4 ± 71.8 | 184.7 ± 58.3 | 1.4 (0.3) | 1.1 (0.1) | 2.2 (0.6) | 2 (0.2) | 1.8 (0.1) | 2.1 (0.1) |
| 2500 trees ha$^{-1}$ | 225 (kg N ha$^{-1}$) | '140' (DD) | 102.1 ± 14.8 | 57.4 ± 25.7 | 240.6 ± 57.3 | 1.8 (0.3) | 1.3 (0.5) | 1.2 (0.5) | 1.8 (0.1) | 1.7 (0.4) | 2 (0.2) |
| | | '185' (TD) | 30.0 ± 13.5 | 38.4 ± 7.1 | 15.2 | 1.5 (0.5) | 1.1 (0.4) | 1.3 (0.6) | 1.6 (0.4) | 2.2 (0.3) | 1.9 (0) |
| | | '230' (DM) | 44.0 ± 10.9 | 38.8 ± 5.7 | 57.8 ± 23.5 | 1.3 (0.4) | 1.1 (0.5) | 1.7 (0.9) | 1.8 (0.1) | 2.1 (0.3) | 1.9 (0.2) |
| | | '356' (DD) | 70.1 ± 43.2 | 43.2 ± 12.8 | 203.2 ± 45.1 | 1.1 (0.4) | 1.6 (0.6) | 1.5 (0.7) | 1.7 (0.3) | 1.9 (0.2) | 1.9 (0.1) |
| | 113 (kg N ha$^{-1}$) | '140' (DD) | 83.9 ± 39.7 | 80.4 ± 54.6 | 180.1 ± 28.8 | 1.8 (0.3) | 1.8 (0.4) | 1.8 (0.8) | 1.8 (0.1) | 1.7 (0.1) | 2.1 (0.5) |
| | | '185' (TD) | 46.3 ± 4.0 | 24.3 ± 15.4 | 13.1 ± 2.4 | 1.5 (0.5) | 1.9 (0.3) | 1.5 (0.6) | 1.6 (0.4) | 1.6 (0) | 1.8 (0.2) |
| | | '230' (DM) | 45.3 ± 5.4 | 53.1 ± 11.2 | 80.1 ± 24.9 | 1.3 (0.4) | 1.5 (0.2) | 1.9 (1) | 1.8 (0.1) | 2 (0.3) | 2.1 (0.2) |
| | | '356' (DD) | 89.7 ± 33.4 | 107.9 ± 58.8 | 298.0 ± 65.9 | 1.1 (0.4) | 1.4 (0.3) | 1.4 (0.6) | 1.7 (0.3) | 1.9 (0.2) | 2.1 (0.1) |
| | 0 (kg N ha$^{-1}$) | '140' (DD) | 107.7 ± 12.1 | 65.6 ± 46.4 | 54.1 ± 36.0 | 1.5 (0.6) | 1.3 (0.5) | 2.3 (0.5) | 1.9 (0.2) | 1.7 (0.2) | 2.1 (0.2) |
| | | '185' (TD) | 51.0 ± 37.2 | 21.3 ± 16.0 | 12.2 ± 4.4 | 1.8 (0.8) | 1.4 (0.3) | 2.3 (0.8) | 2 (0.3) | 1.9 (0.1) | 2.1 (0.1) |
| | | '230' (DM) | 48.6 ± 21.1 | 40.4 ± 12.7 | 25.5 ± 9.2 | 2.1 (0.2) | 1.9 (0.3) | 2.4 (0.4) | 1.8 (0.2) | 1.6 (0.3) | 1.9 (0.1) |
| | | '356' (DD) | 71.4 | 90.0 ± 22.5 | 137.5 ± 32.1 | 2.1 (0.6) | 2.1 (0.1) | 2.1 (0.6) | 1.6 (0.2) | 1.4 (0.2) | 2.1 (0.1) |

GSB under high and lower fertilization rates was similar at the Williamsdale site (164.8 Mg ha$^{-1}$ and 154.8 Mg ha$^{-1}$, respectively; Table 2) but significantly higher than GSB in unfertilized plots (93.1 Mg ha$^{-1}$). At the marginal sites, GSB did not significantly vary with fertilization rates but GSB was greater under the lower fertilizer treatments (78.4–93.6 Mg ha$^{-1}$) than the high- (69.6 to 73.5 Mg ha$^{-1}$) and no-fertilizer (63.7 to 71.8 Mg ha$^{-1}$) treatment plots. The high fertilizer rate (225 kg N ha$^{-1}$) had the greatest effect on GSB (> 29%) at the fertile Williamsdale site (all but one treatment combination) and greater effects on low productivity clones ('185' and '230') and the higher stand density (5000 trees ha$^{-1}$) at the marginal sites. Compared to the no-fertilizer treatment, the lower fertilizer rate (113 kg N ha$^{-1}$) led to higher GSB in all treatment combinations at Williamsdale (>7%) and for more clones ('185', '230', '356') in the high than low stand density (only clone '356') at Clinton-A. At Clinton B, the lower fertilization rate led to increases in GSB of clone '140' under both densities (>22.5%). Across stand densities in the marginal sites, DD clones produced greater GSB under the lower fertilization rate (clone '356' at Clinton A, clone '140' at Clinton B) compared to unfertilized plots.

The density × fertilization × clone interaction effects on GSB were significant for all density ($p < 0.0001$), and fertilizer ($p < 0.0001$) treatment levels. However, only clones '140' and '356' showed significant interactions ($p < 0.0001$ and $p = 0.0027$, respectively) with the other density and fertilization treatments. Site comparisons between Clinton A and B (where the only difference was for the soil series) showed that soils of the marginal sites ($p = 0.1254$) or site interactions with the study treatment did not have significant effects on GSB ($p = 0.1254$ to 0.8577).

The site-specific responses of poplar productivity were not surprising and have been documented for coastal sandy soils [17,37]. Miller and Bender [45] reported that most of the variability in poplar growth responses was due to site effects, while genetics accounted for 23% of growth variability. Our study observed greater GSB differences among genotypes and due to genotype × site interactions, than biomass differences between the marginal and fertile sites, which were mainly attributed to soil quality differences. Our results supported that GSB yields can be improved by using poplar genotypes uniquely suited to site conditions. Estimated GSB yields of our non-fertilized, fertile site were similar to findings from previous poplar studies on moderately to highly fertile lands [16,21,35,38,41] that reported biomass values of 4.6 to 33.9 green Mg ha$^{-1}$ year$^{-1}$ (assuming 50% moisture content), and indicated that a green biomass greater than 40 Mg ha$^{-1}$ year$^{-1}$ is possible from site-matched genotypes [46]. In our study, GSB of 3.6 to 18 Mg ha$^{-1}$ year$^{-1}$ on marginal soils without fertilization (at 2500 trees ha$^{-1}$ after six years of growth) were comparable to or greater than biomass values of 1.1 to 15.1 Mg ha$^{-1}$ year$^{-1}$ from previous poplar studies on marginal lands [27,47].

Interestingly, fertilization significantly improved GSB yields at the fertile site of our study, but not at the marginal sites. Studies addressing the site-specific response of poplars to fertilization have reported mixed results, with some having improved yields with fertilization [48] and others showing no improvements with fertilization [49]. The variable response of clones to fertilization between fertile and marginal soils emphasizes site-specific and soil quality differences on genetic responses. Marginal soils at our sites had a lower cation exchange capacity (CEC) than fertile soils (Table 1), in addition to lower phosphorus, potassium, and magnesium. The low CEC can result in rapid loss of nutrients due to leaching, particularly when fertilizer is applied in a single application to sandy soils [50].

*3.2. Early Growth/Productivity in Relation to the Incidence of Pests and Disease*

Productivity depends on tree survival and health, particularly tree responses to predation, weed control, and disease [16,21,27,49]. Neither stand density ($p = 0.3664$), fertilization ($p = 0.2341$), nor their interactions ($p > 0.117$) were significant for survival, an outcome in agreement with Ghezehei et al. [7] and in contrast to other studies [18,28,51,52]. However, survival was significantly affected by clones ($p < 0.0001$). Regardless of the stand density and sites, survival in non-fertilized plots was >92% for clones '140', '230' and '356' and ranged from 75% to 92% for clone '185'. Survival in fertilized plots was not consistent among clones, particularly for TD clone '185', that had 58% to 100% of its trees alive, depending on the stand density and site. Regardless of the density or fertilization rate, DD ('140' and '356') and DM ('230') clones had greater survival variability on the marginal soils ('140': 83 to 100%; '356' and '230':67 to 100%) than the fertile site where GSB was the highest. However, lower clonal survival did not always result in significantly lower biomass in fertilized plots, which corroborated a previous poplar study in the southeast USA [41].

Diseases and predation impact poplar productivity [37,53]. *Melampsora* rust, *C. scripta*, and white-tailed deer (*Odocoileus virginianus* Zimmermann) damage were observed at all three sites for all clones. The presence of deer damage was minimal (<1%) and did not impact tree health in this study. *Melampsora* rust was prevalent across all sites but differed significantly among sites ($p < 0.0001$) and clones ($p = 0.0036$) (Figure 2, Table S5). Stands on the fertile soil at Williamsdale had the highest rate of *Melampsora* rust (82%) compared to the stands on marginal soils for Clinton A (20%) and Clinton B (22%). Clone '185' had a significantly higher presence of *Melampsora* rust (51%) across all sites than the other clones (34%). The stand density ($p = 0.7090$) and fertilization rate ($p = 0.2031$) did not significantly influence the presence of *Melampsora* rust (Figure 2). These results support prior findings that site and genotype are important factors for the extent of *Melampsora* rust damage [54].

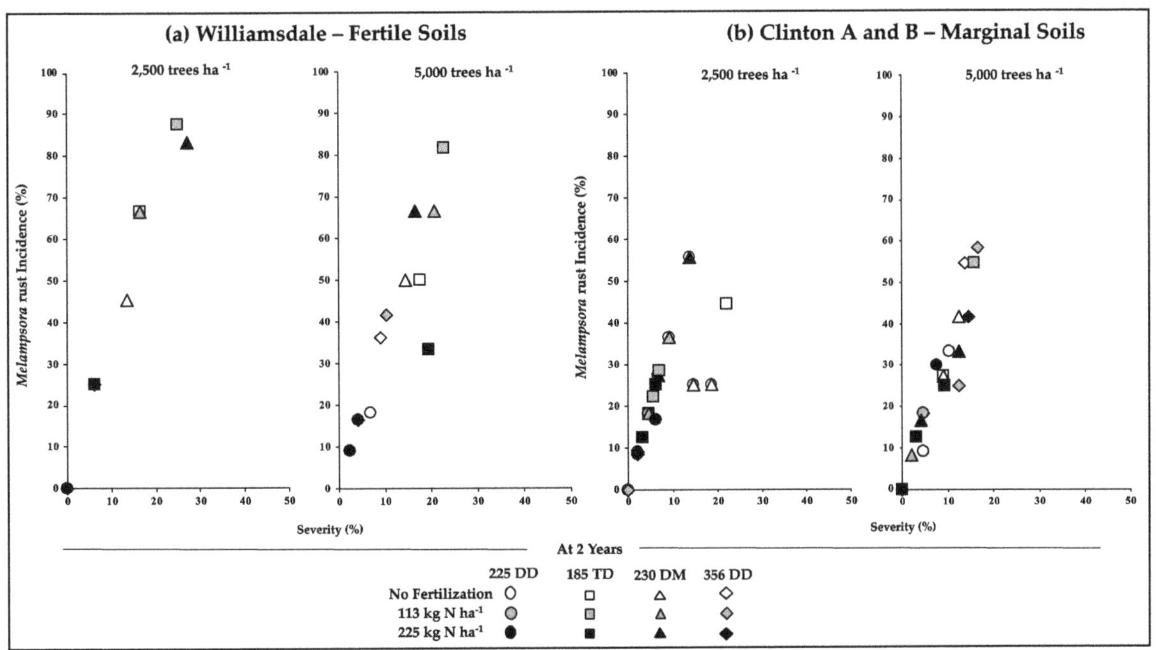


**Figure 2.** *Melampsora* rust incidence (%) and severity (%) for four *Populus* clones established on (**a**) fertile soil at Williamsdale and (**b**) marginal soils in Clinton in the Coastal region in North Carolina, USA, by site, stand density, and fertilization treatment after two years of growth.

Stand density significantly affected the presence of *C. scripta* ($p < 0.0001$; Figure 3 and Table S6). Overall, the stand density of 5000 trees ha$^{-1}$ had 42% higher rates of *C. scripta* beetle damage than the density of 2500 trees ha$^{-1}$. All clones exhibited a mean beetle damage incidence greater than 38%, but the TD clone '185' had a higher damage incidence of 50%. Interestingly, low-fertilizer plots (113 kg N ha$^{-1}$) had more beetle damage (52%) than the no-fertilization plots (33.6%) or the high-fertilizer (225 kg N ha$^{-1}$) plots (44.5%), for all clones across the three sites. All three sites were treated for beetle infestation for both growing seasons, but denser stands may pose physical challenges to pesticide distribution, thus limiting the effectiveness of application [55].

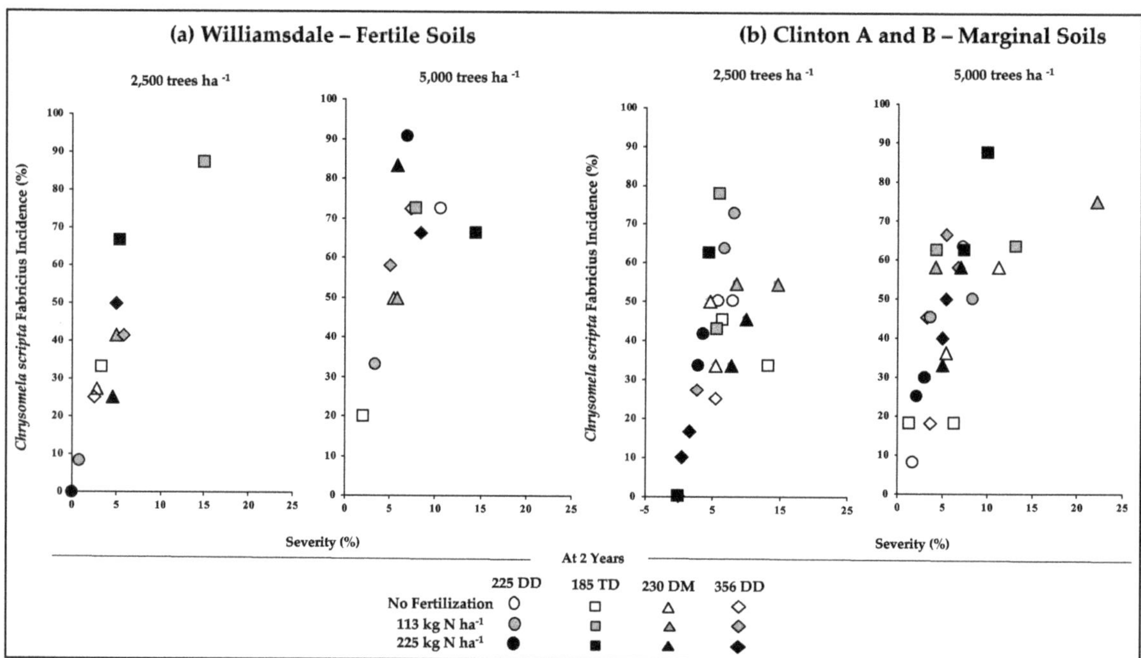



**Figure 3.** *Chrysomela scripta* Fabricius incidence (%) and severity (%) of four *Populus* clones established on (**a**) fertile soil at Williamsdale and (**b**) marginal soils in Clinton in the Coastal region in North Carolina, USA, by stand density, and fertilization treatment after two years of growth.

The severity of the *Melampsora* rust and *C. scripta* damage can decrease photosynthesis efficiency, cause early defoliation, and increase susceptibility to other pests and fungal diseases [54]. Thus, *Melampsora* rust and *C. scripta* damage can reduce annual growth by as much as 50% [54], which impacts stand productivity and profitability for landowners. Trees with observed *Melampsora* rust and *C. scripta* incidence had a mean severity below 25% of the total crown area for any given treatment combinations of clone or stand density (Figures 2 and 3). The differences in the severity of damage of treatment combinations were within 10% of one another, which led to weak correlations between biomass productivity (after two years) and *Melampsora* rust ($R^2 = 0.0886$) or *C. scripta* incidence ($R^2 = 0.0056$).

### 3.3. Productivity in relation to LAI and Foliar N

Biomass production in forest stands is directly related to the amount of solar radiation intercepted by the foliage or leaf area [56–58]. In our study, poplar LAI was significantly affected by stand density ($p = 0.0012$), fertilization ($p < 0.0001$), and clone ($p < 0.0001$), but not their interaction ($p = 0.9461$, Table S3). In contrast to previous studies [21,56,57], the correlation between LAI and biomass productivity at two years after planting was not strong ($R^2 = 0.12$). The lack of correlation most likely reflects the influence of stand density, fertilization, and clone. The LAI (1.85 m² m⁻²) of all trees on fertile soils (Williamsdale) was greater than the LAI for marginal soils (1.64 m² m⁻² for Clinton A and 1.35 m² m⁻² for Clinton B). Despite being one of the two highest stem biomass producers at the sites, DD clone '356' had a significantly lower LAI (1.39 m² m⁻²) than the other three clones

(1.80 m$^2$ m$^{-2}$) (Table 3). In contrast to the findings of Fang et al. [21], the stand density of 5000 trees ha$^{-1}$ had a lower average LAI (1.56 m$^2$ m$^{-2}$) than that of 2500 trees ha$^{-1}$ (1.67 m$^2$ m$^{-2}$).

Foliar N (%) analyses can indicate whether nutrient concentrations are sufficient for metabolic requirements for tree growth [59]. This study found that foliar N of the young stands was significantly affected by fertilization ($p < 0.0001$), clone ($p < 0.0001$), and density $\times$ fertilization $\times$ clone interactions ($p = 0.0227$), but not stand density ($p = 0.4443$). Among unfertilized trees, foliar N was significantly higher at Williamsdale (1.97%) than Clinton A and B (1.84%). There were significant differences in foliar N among genotypes ($p < 0.0001$). DD clone '140' had the highest foliar N (1.95%) while TD clone '185' had the lowest foliar N (1.80%). Significant correlations were present between foliar N and biomass in this study ($p < 0.0001$, Table S7), which agreed with previous poplar results [59,60].

The influence of fertilization on foliar N content has varied among earlier studies. Pope et al. [61] reported that fertilization can be an important factor for foliar N poplar content on a certain soil series. Foliar N concentrations increased by approximately 5% in the fertilized treatments on fertile soils (Williamsdale), regardless of the stand density or clone. However, foliar N concentrations decreased by approximately 3% in the fertilized treatments of marginal soils, regardless of the stand density or clone, which may reflect a low N availability due to poor soil quality. Wilson et al. [60] observed that fertilized trees had higher foliar N%, but that N leachate increased with the fertilizer application rate. Thus, increased fertilization rates may not produce expected increases in poplar growth and foliar N, potentially due to soil processes such as microbial competition, leaching, and adsorption [60].

*3.4. Profitability*

At the current delivered price of 27.01 USD Mg$^{-1}$, none of the study plantations would be economically feasible at the age of six years (Table 4). Economic feasibility of all stands in our study improved at longer rotations (Table 4), and the findings agreed with previous poplar studies [16,41,45,51]. The cost-effectiveness of plantations is also enhanced by selecting the best clones for a region. Based on our study, plantations of the best clones ('140' and '356') would be feasible after at least 12 years of growth on the fertile site, enhanced with a medium fertilizer rate (113 kg N ha$^{-1}$), and with longer than 12-year rotations on the marginal or fertile soils, with or without fertilization. Poplar biomass yields of 7 to 8 dry Mg ha$^{-1}$ year$^{-1}$ would not be profitable, even in markets supportive of short-rotation forestry [62], and our six-year-old stands would not be cost-effective even with a GSB of 63 Mg ha$^{-1}$ (GSB ranged from 2 Mg ha$^{-1}$ for clone '185' to 63 Mg ha$^{-1}$ clone '356'). It should be noted that trial-based productivity estimates, such as the year-12 MAI values in this study, are likely to overestimate the productivity that can be expected from large-scale productions.

Stand density has significant implications on the biomass productivity and cost-effectiveness of poplar plantations [16,45]. In the current study, stand density greatly impacted NPV. After six years of growth, GSB production was generally greater for the density of 5000 trees ha$^{-1}$ than the density of 2500 trees ha$^{-1}$ (Table 4), but the biomass gains were negated due to the higher establishment cost of the former (Figure 4). Consequently, non-fertilized and fertilized plots were generally less economically feasible at the higher density due to the added cost of more cuttings. Bioenergy plantations can be planted at a high stand density to maximize woody biomass production per unit area; it is not unreasonable to expect that higher density plantations can be viable for this purpose by producing more trees but with a smaller diameter, due to high intraspecific competition. In contrast, lower density trees produce fewer but bigger diameter trees; the production of larger diameters could lead to a higher income since a larger diameter can enhance stem grade for many applications.

**Table 4.** Net present values (NPV, USD ha$^{-1}$), break-even prices (USD per green Mg stem biomass), and mean annual increments of stem biomass (MAI, in Mg ha$^{-1}$ year$^{-1}$) of poplar stands established at three sites in the Coastal region of North Carolina, USA, as a function of stand density, fertilization rate, and rotation length (six and twelve years). Prices used in NPV calculations, and break-even prices represent hardwood pulpwood prices at delivery.

| | | | Stand Age: 6 Years | | | | | | Stand Age: 12 Years | | | | | | | | |
| Stand Density | Fertilizer Rates | Clone | NPV ($USD ha$^{-1}$) at (Price = $27 Mg$^{-1}$) | | | Break-Even Price ($USD Mg$^{-1}$ Green Biomass) | | | NPV ($USD ha$^{-1}$) (Price = $27 USD Mg$^{-1}$) | | | Break-Even Price ($USD Mg$^{-1}$ Green Biomass) | | | Stem MAI, (Mg ha$^{-1}$ year$^{-1}$) | | |
| | | | Clinton A | Clinton B | Williamsdale | Clinton A | Clinton B | Williamsdale | Clinton A | Clinton B | Williamsdale | Clinton A | Clinton B | Williamsdale | Clinton A | Clinton B | Williamsdale |
|---|---|---|---|---|---|---|---|---|---|---|---|---|---|---|---|---|---|
| 2500 Trees ha$^{-1}$ | 0 (kg N ha$^{-1}$) | '140' | ($1300) | ($1509) | ($1838) | 38.0 | 48.3 | 58.9 | $33 | ($1195) | ($1939) | - | 32.9 | 40.5 | 25.3 | 15.4 | 11.3 |
| | | '185' | ($1686) | ($1810) | ($2123) | 57.8 | 106.7 | 191.8 | ($1630) | ($2045) | ($2651) | 37.6 | 59.2 | 110.2 | 12.0 | 5.0 | 2.5 |
| | | '230' | ($1702) | ($1681) | ($2032) | 59.7 | 65.9 | 102.3 | ($1677) | ($1679) | ($2425) | 38.4 | 40.8 | 63.3 | 11.4 | 9.5 | 5.3 |
| | | '356' | ($1547) | ($1343) | ($1270) | 47.1 | 40.7 | 35.4 | ($1240) | ($728) | ($520) | 32.6 | 29.4 | 28.2 | 16.8 | 21.2 | 28.7 |
| | 113 (kg N ha$^{-1}$) | '140' | ($1485) | ($1431) | ($1004) | 43.4 | 43.4 | 32.0 | ($1023) | ($935) | $181 | 30.9 | 30.6 | - | 19.7 | 18.9 | 37.5 |
| | | '185' | ($1741) | ($1813) | ($2139) | 62.2 | 97.0 | 181.7 | ($1744) | ($2010) | ($2659) | 39.5 | 54.7 | 104.7 | 10.9 | 5.7 | 2.7 |
| | | '230' | ($1748) | ($1617) | ($1684) | 63.1 | 55.3 | 46.6 | ($1763) | ($1458) | ($1520) | 39.9 | 36.0 | 34.0 | 10.6 | 12.5 | 16.7 |
| | | '356' | ($1446) | ($1244) | ($202) | 41.9 | 37.5 | 27.3 | ($912) | ($408) | $2185 | 30.2 | 28.0 | 23.9 | 21.1 | 25.3 | 62.1 |
| | 225 (kg N ha$^{-1}$) | '140' | ($1372) | ($1598) | ($606) | 39.3 | 52.9 | 29.1 | ($684) | ($1386) | $1,195 | 29.0 | 34.9 | - | 24.0 | 13.5 | 50.1 |
| | | '185' | ($1862) | ($1727) | ($2139) | 85.3 | 69.1 | 160.3 | ($2067) | ($1750) | ($2,636) | 50.1 | 42.1 | 93.4 | 7.1 | 9.0 | 3.2 |
| | | '230' | ($1767) | ($1724) | ($1849) | 64.6 | 68.5 | 57.1 | ($1799) | ($1742) | ($1,913) | 40.6 | 41.9 | 39.5 | 10.3 | 9.1 | 12.0 |
| | | '356' | ($1589) | ($1694) | ($860) | 48.1 | 63.6 | 30.7 | ($1298) | ($1658) | $559 | 33.0 | 39.7 | 25.7 | 16.5 | 10.2 | 42.3 |
| 5000 Trees ha$^{-1}$ | 0 (kg N ha$^{-1}$) | '140' | ($1963) | ($2085) | ($1629) | 39.3 | 43.6 | 33.5 | ($772) | ($1182) | ($69) | 28.4 | 30.0 | - | 33.3 | 26.4 | 44.9 |
| | | '185' | ($2623) | ($2543) | ($2998) | 80.8 | 78.6 | 220.9 | ($2634) | ($2473) | ($3492) | 46.2 | 44.8 | 118.4 | 10.5 | 10.6 | 3.0 |
| | | '230' | ($2630) | ($2422) | ($2746) | 82.3 | 62.1 | 76.0 | ($2655) | ($2132) | ($2861) | 46.8 | 37.8 | 47.5 | 10.2 | 14.7 | 10.7 |
| | | '356' | ($2292) | ($2329) | ($1839) | 49.2 | 54.6 | 35.8 | ($1702) | ($1869) | ($595) | 32.6 | 34.6 | 27.8 | 21.9 | 18.0 | 38.5 |
| | 113 (kg N ha$^{-1}$) | '140' | ($1969) | ($1930) | ($561) | 38.9 | 39.1 | 27.9 | ($723) | ($676) | $2656 | 28.2 | 28.1 | - | 34.3 | 33.0 | 78.8 |
| | | '185' | ($2545) | ($2559) | ($2977) | 64.8 | 75.7 | 147.9 | ($2348) | ($2451) | ($3385) | 39.3 | 43.5 | 82.5 | 14.4 | 11.3 | 4.7 |
| | | '230' | ($2349) | ($2379) | ($2655) | 50.6 | 55.9 | 61.6 | ($1796) | ($1942) | ($2579) | 33.2 | 35.2 | 40.4 | 21.2 | 17.5 | 14.6 |
| | | '356' | ($1806) | ($2337) | ($1601) | 36.3 | 53.2 | 33.1 | ($263) | ($1825) | $57 | 27.1 | 34.0 | - | 40.0 | 18.9 | 46.9 |
| | 225 (kg N ha$^{-1}$) | '140' | ($2257) | ($2131) | ($1670) | 46.0 | 43.7 | 33.6 | ($1493) | ($1201) | ($75) | 31.2 | 30.0 | - | 25.2 | 26.9 | 45.6 |
| | | '185' | ($2565) | ($2535) | ($3033) | 64.7 | 69.1 | 178.8 | ($2360) | ($2341) | ($3484) | 39.2 | 40.7 | 97.5 | 14.6 | 12.9 | 3.9 |
| | | '230' | ($2631) | ($2447) | ($2705) | 73.0 | 59.7 | 64.2 | ($2547) | ($2090) | ($2664) | 42.8 | 36.7 | 41.6 | 12.3 | 16.0 | 13.9 |
| | | '356' | ($2285) | ($1982) | ($922) | 47.0 | 39.9 | 29.1 | ($1571) | ($781) | $1795 | 31.6 | 28.5 | - | 24.3 | 32.0 | 68.6 |

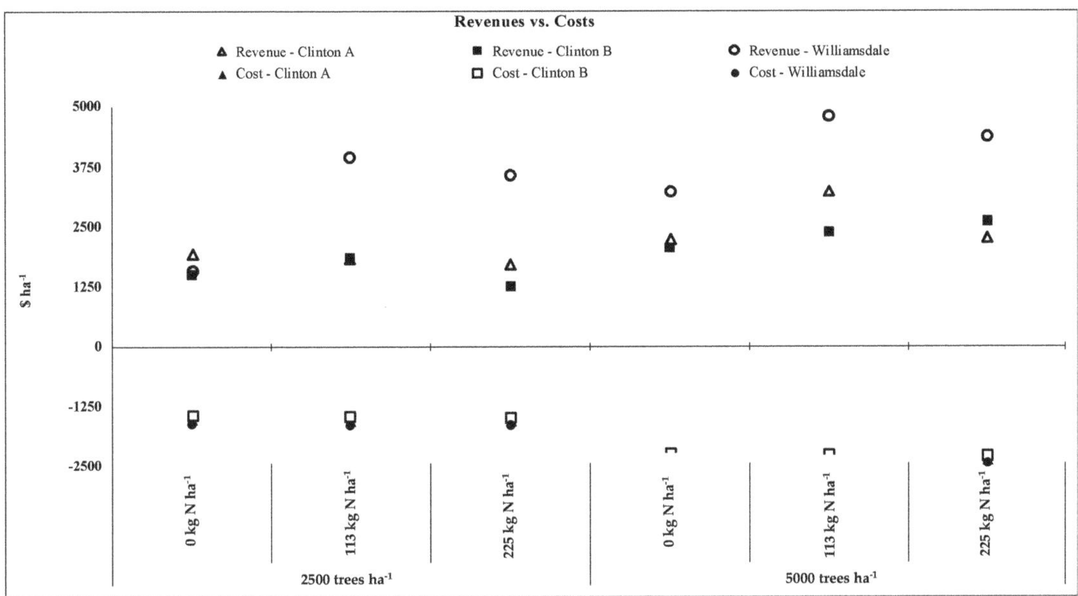


**Figure 4.** Comparisons of revenues and costs (establishment and management) of six-year-old hybrid poplar plantations established in the Coastal region in North Carolina, USA, by site, stand density, and fertilization treatment. Mean clonal stem biomass values were used to calculate revenues.

Limited nutrient availability can reduce growth in SRWC plantations, and prior studies have shown increased biomass yields with nitrogen fertilization [7,16,35,48,61]. In our study, fertilization consistently improved growth, especially on fertile soils; the benefits, however, were marginalized due to the high cost of the fertilizer (USD 427.25 t$^{-1}$) and subsequent additional weed management control. For fertile soils, where fertilization resulted in significant GSB increases ('140' and '356'), break-even prices, which are the market prices needed to simply recoup costs, decreased substantially from non-fertilized soils (Table 4). Medium fertilizer application rates seemed to have the best economic potential at both densities since it led to a higher GSB than the no-fertilization treatment, but was less costly than the higher-rate fertilization treatment (225 kg N ha$^{-1}$).

Site fertility was a major factor for cost-effectiveness of the stands. For the same density, fertilizer rate, and growth (e.g., 13.4 Mg ha$^{-1}$ year$^{-1}$ from clone '140' at Clinton B and clone '230' at Williamsdale), marginal soils at Clinton had lower break-even prices than the fertile soil at Williamsdale because of the lower weed control costs on marginal soils. At the current price, profitability appears possible in rotations longer than 12 years on non-fertilized marginal and fertile soils in the region. A 50% increase in the current price (to USD 40.51 Mg$^{-1}$) would make the best cases of density × fertilization × clone in the three sites we studied economically feasible; a 25% price increase (to USD 33.76 Mg$^{-1}$) would lead to the cost-effectiveness of plantations at both the marginal and the fertile sites in a 12-year rotation (Table 4).

The most substantial challenges to profitability in our study were greater establishment costs for higher density plantings at all sites and weed control costs of the fertile site. Management greatly affected the cost-effectiveness of stands, a result that agreed with a previous study by Ghezehei et al. [34]. Average production costs of the study ranged from USD 1450 ha$^{-1}$ for low density, lower fertilization marginal stands, to USD 2467 ha$^{-1}$ for the high density, lower fertilization fertile stand (Williamsdale). Based on our current results, more stringent weed control was more effective than fertilization, whereas, a high fertilization rate is not as important for marginal or fertile soils to maximize productivity or economic feasibility. We also observed that fertilization and genotype can have significant effects on biomass productivity. These findings highlight the importance of selecting the correct poplar genomic groups or clones for site-specific conditions, as well as stand management and density, in order to optimize productivity with costs to achieve economic feasibility. This is especially true on marginal lands where biomass yields are hindered. These findings agree with prior economic analyses for poplars in coastal sandy soils [7].

## 4. Conclusions

Appropriate stand density, fertilization, and genotype selection can provide substantial biomass yields for poplars in coastal sandy soils. This study demonstrated that fertilization of poplars on marginal lands can improve biomass to be similar to stem biomass observed on non-fertilized, fertile soils; however, stem biomass was significantly greater on fertilized stands on fertile soil. The influence of genotype was specific to the fertilization, disease, pest, and foliar N response on fertile and marginal soils. Higher levels of fertilization did not necessarily correlate to higher biomass or profitability. At current market prices, profitability was not feasible for the study stands, even for those stands that produced a high stem biomass on fertile soil increased with fertilizer in the planting year. Higher stand density generally increased stem biomass on fertile and marginal lands, yet the most likely path to profitability would be at least 12-year rotations of both higher and lower density stands of DD clones treated with fertilizer. To be profitable, the marginal sites would require 12-year rotations, a medium fertilization rate, and proper clonal selection (i.e., '356'). The economic viability of poplars depends on site-specific management practices and genotype selection between fertile and marginal soils, as well as selection of the optimal rotation and significant increases of market prices.

**Supplementary Materials:** The following are available online at https://www.mdpi.com/article/ 10.3390/f12070869/s1; Figure S1: Mean annual increment (MAI, Mg ha$^{-1}$) of green stem biomass versus stand age at Clinton A and Clinton B (in the Coastal region of North Carolina, USA) during the first six years of growth; Figure S2: Mean annual increment (MAI, Mg ha$^{-1}$) of green stem biomass versus stand age at Williamsdale site in North Carolina, during the first six years of growth; Table S1: Results of analyses of variance for stem biomass (green) of poplar clones established at two stand densities with three fertilization rates at Clinton A, Clinton B, and Williamsdale (in the Coastal region of North Carolina, USA) after six years of growth ( = 0.05); Table S2: Results of analyses of variance for site effects and interaction on stem biomass (green) of poplar clones established at two stand densities with three fertilization rates at Clinton A, Clinton B, and Williamsdale (in the Coastal region of North Carolina, USA) after six years of growth ( = 0.05); Table S3: Results of analyses of variance for LAI of poplar clones during the second year of growth. The stands were established at two stand densities with three fertilization rates at three sites in the Coastal region of North Carolina, USA (Clinton A, Clinton B, and Williamsdale; = 0.05); Table S4: Results of analyses of variance for foliar nitrogen (%) of poplar clones during the second year of growth. The stands were established at two stand densities with three fertilization rates at three sites in the Coastal region of North Carolina, USA (Clinton A, Clinton B, and Williamsdale; = 0.05); Table S5: Results of analyses of variance for Melampsora (rust) presence on poplar clones during the second year of growth. The stands were established at two stand densities with three fertilization rates at three sites in the Coastal region of North Carolina, USA ( = 0.05); Table S6: Results of analyses of variance for Chrysomela scripta Fabricius damage presence on poplar clones during the second year of growth. The stands were established at two stand densities with three fertilization rates at three sites in the Coastal region of North Carolina, USA ( = 0.05); Table S7: Analysis results of stem biomass correlations with LAI, foliar N, Melampsora spp. rust, and Chrysomela scripta damage for poplars ( = 0.05). The poplar were two years old and established at two stand densities with three fertilization rates at three sites in the Coastal region of North Carolina, USA.

**Author Contributions:** Conceptualization and methodology, A.L.E., E.G.N. and D.W.H.; investigation, S.B.G., A.L.E., D.W.H. and E.G.N.; data curation and formal analysis, S.B.G. and A.L.E.;

writing—original draft preparation, S.B.G. and A.L.E.; writing—review and editing, S.B.G., A.L.E., D.W.H., R.S.Z.J. and E.G.N.; project administration, E.G.N. and S.B.G.; funding acquisition, E.G.N. and D.W.H. All authors have read and agreed to the published version of the manuscript.

**Funding:** This research was funded by the North Carolina Department of Agriculture and Consumer Services, Bioenergy Research Initiatives (Grant numbers: G40100285014RSD and 17-072-4001).

**Data Availability Statement:** Data are available on request from the corresponding author. The data are not publicly available due to data management and storage agreement with the project sponsor, the North Carolina Department of Agriculture & Consumer Services, Bioenergy Research Initiatives.

**Acknowledgments:** We would like to thank the Williamsdale Biofuels Field Lab and the Horticultural Crops Research Station at Clinton for accommodating the study sites, and John Garner, Stewart Biegler and Charles Burrow for their cooperation and assistance in managing the research stands. We also thank Erica Simmons (NC State University student) for data analyses for the current study.

**Conflicts of Interest:** The authors declare no conflict of interest.

# References

1.  U.S. Department of Energy. 2016 Billion-Ton Report: Advancing Domestic Resources for a Thriving Bioeconomy. In *Economic Availability of Feedstocks*; Langholtz, M.H., Stokes, B.J., Eaton, L.M., Eds.; ORNL/TM-2016/160; Oak Ridge National Laboratory: Oak Ridge, TN, USA, 2016; Volume 1, 448p. Available online: http://energy.gov/eere/bioenergy/2016-billion-ton-report (accessed on 23 February 2021). [CrossRef]
2.  Alternative Fuels Data Center. Energy Independence and Security Act of 2007. US Department of Energy, Energy Efficiency and Renewable Energy Office. Available online: https://afdc.energy.gov/laws/eisa.html (accessed on 29 March 2021).
3.  Hay, J.F. Corn for Biofuel Production. Farm Energy. 3 April 2019. Available online: https://farm-energy.extension.org/corn-for-biofuel-production/ (accessed on 18 March 2021).
4.  Larson, J.A.; English, B.C.; De La Torre Ugarte, D.G.; Menard, R.J.; Hellwinckel, C.M.; West, T.O. Economic and environmental impacts of the corn grain ethanol industry on the United States agricultural sector. *J. Soil Water Conserv.* **2010**, *65*, 267–279. [CrossRef]
5.  Chen, L.; Debnath, D.; Zhong, J.; Ferin, K.; Van Loock, A.; Khanna, M. The economic, and environmental costs and benefits of the renewable fuel standard. *Environ. Res. Lett.* **2021**, *16*, 034021. [CrossRef]
6.  Ghezehei, S.B.; Nichols, E.G.; Hazel, D.W. Early Clonal survival and growth of poplars grown on North Carolina Piedmont and Mountain marginal lands. *BioEnergy Res.* **2016**, *9*, 548–558. [CrossRef]
7.  Official Journal of the European Union. Commission Implementing Regulation (EU) 2020/1294 of 15 September 2020 on the Union Renewable Energy Financing Mechanism. Official Journal of the European Union, L303/1-L303/16. 17 September 2020. Available online: https://eur-lex.europa.eu/legal-content/EN/TXT/PDF/?uri=CELEX:32020R1294&from=EN (accessed on 27 February 2021).
8.  Matzenberger, J.; Kranzl, L.; Tromborg, E.; Junginger, M.; Daioglou, V.; Sheng Goh, C.; Keramidas, K. Future perspectives of international bioenergy trade. *Renew. Sustain. Energy Rev.* **2015**, *43*, 926–941. [CrossRef]
9.  FORISK Consulting. Wood bioenergy US. *Forisk Res. Q.* **2021**. Quarter 1. Available online: https://forisk.com/wordpress//wp-content/assets/2021_Q1_WBUS_Free-Summary_20210203.pdf (accessed on 28 April 2021).
10. Townsend, P.A.; Haider, N.; Boby, L.; Heavey, J.; Miller, T.A.; Volk, T.A. *A Roadmap for Poplar and Willow to Provide Environmental Services and to Build the Bioeconomy*; WSU Extension-EM115E; Washington State University (WSU): Washington, WA, USA, 2019; Peer-Reviewed.
11. Cser, H.; Boby, L. Biomass feedstocks characteristics. *South Reg. Ext. For.* **2015**. SREF-BE-005. Available online: https://sref.info/resources/publications/biomass-feedstocks-characteristics-1 (accessed on 18 April 2021).
12. Balatinecz, J.J.; Kretschmann, D.E. Properties and utilization of poplar wood. In *Poplar Culture in North America*; Dickmann, D.I., Isebrands, J.G., Eckenwalder, J.E., Richardson, J., Eds.; National Research Council of Canada: Ottawa, ON, Canada, 2001; pp. 277–291.
13. Tyndall, J.; Schulte, L.; Hall, R. Expanding the US Cornbelt Biomass Portfolio: Forester perceptions of the potential for woody biomass. *Small Scale For.* **2011**, *10*, 287–303. [CrossRef]
14. Zalesny, R.S., Jr.; Cunningham, M.W.; Hall, R.B.; Mirck, J.; Rockwood, D.L.; Stanturf, J.A.; Volk, T.A. Woody biomass from short rotation energy crops. In *Sustainable Production of Fuels, Chemicals and Fibers from Forest Biomass*; Zhu, J.Y., Zhang, X., Pan, X., Eds.; American Chemical Society: Washington, DC, USA, 2011; pp. 27–63. [CrossRef]
15. Wright, J.; Gallagher, T. Sustainable Production Practices of Short Rotation Woody Crops in the Southeastern United States: A Guidebook for Cottonwood and Hybrid Poplar. IBSS Southeastern Partnership for Integrated Biomass Supply Systems. Available online: http://www.se-ibss.org/publications-and-patents/extension-and-outreach-publications/sustainable-production-practices-of-short-rotation-woody-crops-in-the-southeastern-united-states-a-guidebook-for-cottonwood-and-hybrid-poplar (accessed on 15 February 2021).

16. Miller, R.O.; Bender, B.A. *Planting Density Effects on Biomass Growth of Hybrid Poplar Varieties in Michigan*; Forest Biomass Innovation Center Research Report; Michigan State University: East Lansing, MI, USA, 2016; Available online: https://www.canr.msu.edu/uploads/396/36452/2016c.pdf (accessed on 3 March 2021).
17. Shifflett, S.; Hazel, D.; Frederick, D.; Nichols, E.G. Species Trials of short rotation woody crops on two wastewater application sites in North Carolina, USA. *BioEnergy Res.* **2014**, *7*, 157–173. [CrossRef]
18. Coyle, D.R.; Coleman, M.D.; Durant, J.A.; Newman, L.A. Survival and growth of 31 *Populus* clones in South Carolina. *Biomass Bioenergy* **2006**, *30*, 750–758. [CrossRef]
19. Wang, D.; LeBauer, D.; Dietze, M. Predicting yields of short-rotation hybrid poplar (*Populous* spp.) for the United States through model-data synthesis. *Ecol. Soc. Am.* **2013**. [CrossRef]
20. Hinchee, M.; Rottmann, W.; Mullinax, L.; Zhang, C.; Shujun, C.; Cunningham, M.; Pearson, L.; Nehra, N. Short-rotation woody crops for bioenergy and biofuels applications. In *Biofuels*; Tomes, D., Lakshmanan, P., Songstad, D., Eds.; Springer: New York, NY, USA, 2011; pp. 139–156. [CrossRef]
21. Fang, S.; Xu, X.; Lu, S.; Tang, L. Growth dynamics and biomass production in short-rotation poplar plantations: 6-year results for three clones at four spacings. *Biomass Bioenergy* **2000**, *41*, 415–425. [CrossRef]
22. Amichev, B.Y.; Johnston, M.; Van Rees, K.C.J. Hybrid poplar growth in bioenergy production systems: Biomass prediction with a simple process-based model (3PG). *Biomass Bioenergy* **2010**, *34*, 687–702. [CrossRef]
23. Mehmood, M.A.; Ibrahim, M.; Rashid, U.; Nawaz, M.; Ali, A.; Hussain, A.; Gull, M. Biomass production for bioenergy using marginal lands. *Sustain. Prod. Consum.* **2017**, *9*, 3–21. [CrossRef]
24. Campbell, J.E.; Lobell, D.B.; Genova, R.C.; Field, C.B. The global potential of bioenergy on abandoned agriculture lands. *J. Environ. Sci. Technol.* **2008**, *42*, 5791–5794. [CrossRef]
25. Finzi, A.C.; Norby, R.J.; Calfapietra, C.; Gallet-Budynek, A.; Gielen, B.; Holmes, W.E.; Hoosbeek, M.R.; Iversen, C.M.; Jackson, R.B.; Kubiske, M.E.; et al. Increases in Nitrogen Uptake Rather than Nitrogen-Use Efficiency Support Higher Rates of Temperate Forest Productivity under Elevated $CO_2$. *Proc. Natl. Acad. Sci. USA* **2007**, *104*, 14014–14019. [CrossRef]
26. Liu, Z.; Dickmann, D.I. Responses of two hybrid *Populus* clones to flooding, drought, and nitrogen availability. I. Morphology and growth. *Can. J. Bot.* **1992**, *70*, 2265–2270. [CrossRef]
27. Truax, B.; Gagnon, D.; Fortier, J.; Lambert, F. Yield in 8 year-old hybrid poplar plantations on abandoned farmland along climatic and soil fertility gradients. *For. Ecol. Manag.* **2012**, *267*, 228–239. [CrossRef]
28. Dipesh, K.; Blazier, M.; Pelkki, M.; Liechty, H. Genotype influences survival and growth of eastern cottonwood (*Populus deltoides* L.) managed as a bioenergy feedstock on retired agricultural sites of the Lower Mississippi Alluvial Valley. *New For.* **2017**, *48*, 95–114. [CrossRef]
29. Crouse, D.A. Soils and Plant Nutrients. North Carolina State Extension Publications. Available online: https://content.ces.ncsu.edu/extension-gardener-handbook/1-soils-and-plant-nutrients (accessed on 21 March 2021).
30. Flynn, R. *Calculating Fertilizer Costs*; Cooperative Extension Service; College of Agricultural, Consumer and Environmental Sciences, New Mexico State University: Las Cruces, NM, USA, 2014.
31. Brown, K.; Van den Driessche, R. Effects of nitrogen and phosphorus fertilization on the growth and nutrition of hybrid poplars on Vancouver Island. *New For.* **2005**, *29*, 89–104. [CrossRef]
32. Coyle, D.R.; Aubrey, D.P.; Siry, J.P.; Volfovicz-Leon, R.R.; Coleman, M.D. Optimal nitrogen application rates for three intensively-managed hardwood tree species in the southeastern USA. *For. Ecol. Manag.* **2013**, *303*, 131–142. [CrossRef]
33. Lasa, B.; Jauregui, I.; Aranjuelo, I.; Sakalauskiene, S.; Aparicio-Tejo, P.M. Influence of stage of development in the efficiency of nitrogen fertilization on poplar. *J. Plant Nutr.* **2016**, *39*, 87–98. [CrossRef]
34. Ghezehei, S.B.; Nichols, E.G.; Hazel, D.W. Productivity and cost-effectiveness of short-rotation hardwoods on various land types in the Southeastern USA. *Int. J. Phytorem.* **2019**. [CrossRef]
35. Long, H.; Li, X.; Wang, H.; Jia, J. Biomass resources and their bioenergy potential estimation: A review. *Renew. Sustain. Energy Rev.* **2013**, *26*, 344–352. [CrossRef]
36. El Kasmioui, O.; Ceulemans, R. Financial analysis of the cultivation of short rotation woody crops for bioenergy in Belgium: Barriers and opportunities. *BioEnergy Res.* **2013**, *6*, 336–350. [CrossRef]
37. Shifflett, S.D.; Hazel, D.W.; Nichols, E.G. Sub-soiling and genotype selection improves *Populus* productivity grown on a North Carolina sandy soil. *Forests* **2016**, *794*, 74. [CrossRef]
38. Ghezehei, S.B.; Nichols, E.G.; Maier, C.A.; Hazel, D.W. Adaptability of *Populus* to physiography and growing conditions in the Southeastern USA. *Forests* **2019**, *10*, 118. [CrossRef]
39. U.S. Department of Agriculture, Agricultural Statistics Service. 2012 Census of Agriculture, Volume 1, Chapter 1: National Level Data, Land: 2012 and 2007 (Table 8). Available online: https://www.nass.usda.gov/Publications/AgCensus/2012/Full_Report/Volume_1,_Chapter_1_US/st99_1_008_008.pdf (accessed on 16 April 2021).
40. Web Soil Survey. 2016. Available online: https://websoilsurvey.sc.egov.usda.gov/App/HomePage.htm (accessed on 1 March 2021).
41. Ghezehei, S.B.; Wright, J.; Zalesny, R.S., Jr.; Nichols, E.G.; Hazel, D.W. Matching site-suitable poplars to rotation length for optimized productivity. *For. Ecol. Manag.* **2020**, *45*, 117670. [CrossRef]
42. Schreiner, E.J. *Rating Poplars for Melampsora Leaf Rust Infection*; Forest Research Note NE-90; U.S. Department of Agriculture, Forest Service, Northeastern Forest Experiment Station: Upper Darby, PA, USA, 1959; pp. 1–3.

43. North Carolina Tree Seedling Catalog 2020–2021. Available online: https://www.ncforestservice.gov/nursery/pdf/NCFS_Tree_Seedling_Catalog.pdf (accessed on 15 February 2021).

44. Historic North Carolina Delivered Timber Prices, 1988–2020. NC State Extension. Available online: https://content.ces.ncsu.edu/historic-north-carolina-delivered-timber-prices-1988-2014 (accessed on 21 April 2021).

45. Miller, R.O.; Bender, B.A. Sources of variation in hybrid poplar biomass production throughout Michigan, USA. Michigan State University, Forest Biomass Innovation Center Research Report, 2016 (d). In Proceedings of the A Presentation in the 25th International Poplar Symposium, Berlin, Germany, 13–16 September 2016.

46. Zalesny, R.S., Jr.; Stanturf, J.A.; Gardiner, E.S.; Perdue, J.H.; Young, T.M.; Coyle, D.R.; Headlee, W.L.; Bañuelos, G.S.; Hass, A. Ecosystem services of woody crop production systems. *BioEnergy Res.* **2016**, *9*, 465–491. [CrossRef]

47. Dillen, S.Y.; Djomo, S.N.; Al Afas, N.; Vanbeveren, S.; Ceulemans, R. Biomass yield and energy balance of a short-rotation poplar coppice with multiple clones on degraded land during 16 years. *Biomass Bioenergy* **2013**, *56*, 157–165. [CrossRef]

48. Christersson, L. Biomass production of intensively grown poplars in the southernmost part of Sweden: Observations of characters, traits and growth potential. *Biomass Bioenergy* **2006**, *30*, 497–508. [CrossRef]

49. Vande Walle, I.; Van Camp, N.; Van de Casteele, L.; Verheyen, K.; Lemeur, R. Short-rotation forestry of birch, maple, poplar and willow in Flanders (Belgium) I-Biomass production after 4 years of tree growth. *Biomass Bioenergy* **2007**, *31*, 267–275. [CrossRef]

50. Sonon, L.S.; Kissel, D.E.; Saha, U. Cation Exchange Capacity and Base Saturation. University of Georgia Extension. 2017. Available online: https://secure.caes.uga.edu/extension/publications/files/pdf/C%201040_2.PDF (accessed on 27 March 2021).

51. Ghezehei, S.B.; Shifflett, S.D.; Hazel, D.W.; Nichols, E.G. SRWC bioenergy productivity and economic feasibility on marginal lands. *J. Environ. Manage.* **2015**, *160*, 57–66. [CrossRef]

52. Navarro, A.; Stellacci, A.; Campi, P.; Vitti, C.; Modugno, F.; Mastrorilli, M. Feasibility of SRC species for growing in Mediterranean conditions. *BioEnergy Res.* **2016**, *9*, 208–223. [CrossRef]

53. Verlinden, M.S.; Broeckx, L.S.; Van den Bulcke, J.; Van Acker, J.; Ceulemans, R. Comparative study of biomass determinants of 12 poplar (*Populus*) genotypes in a high-density short-rotation culture. *For. Ecol. Manag.* **2013**, *307*, 101–111. [CrossRef]

54. Covarelli, L.; Beccari, G.; Tosi, L.; Fabre, B.; Frey, P. Three-year investigations on leaf rust of poplar cultivated for biomass production in Umbria, Central Italy. *Biomass Bioenergy* **2013**, *49*, 315–322. [CrossRef]

55. Coyle, D.R.; Hart, E.R.; McMillin, J.D.; Rule, L.C.; Hall, R.B. Effects of repeated cottonwood leaf beetle defoliation on *Populus* growth and economic value over an 8-year harvest rotation. *For. Ecol. Manag.* **2008**, *255*, 3365–3373. [CrossRef]

56. Ceulemans, R.; Pontailler, J.; Mau, F.; Guittet, J. Leaf allometry in young poplar stands: Reliability of leaf area index estimation, site and clone effects. *Biomass Bioenergy* **1993**, *4*, 315–321. [CrossRef]

57. Ridge, C.R.; Hinckley, T.M.; Stettler, R.F.; Van Volkenburgh, E. Leaf growth characteristics of fast-growing poplar hybrids *Populus trichocarpa* × *P. deltoides*. *Tree Physiol.* **1986**, *1*, 209–216. [CrossRef] [PubMed]

58. Waring, R.; Landsberg, J.; Linder, S. Tamm Review: Insights gained from light use and leaf growth efficiency indices. *For. Ecol. Manag.* **2016**, *379*, 232–342. [CrossRef]

59. dLteif, A.; Whalen, J.K.; Bradley, R.L.; Camire, C. Diagnostic tools to evaluate the foliar nutrition and growth of hybrid poplars. *Can. J. For. Res.* **2008**, *38*, 2138–2147.

60. Wilson, A.; Nzokou, P.; Güney, D.; Kulaç, Ş. Growth response and nitrogen use physiology of Fraser fir (*Abies fraseri*), red pine (*Pinus resinosa*), and hybrid poplar under amino acid nutrition. *New For.* **2013**, *44*, 281–295. [CrossRef]

61. Pope, P.E.; McLaughlin, R.A.; Hansen, E.A. Biomass and nitrogen dynamics of hybrid poplar on two different soils: Implications for fertilization strategy. *Can. J. For. Res.* **1988**, *18*, 223–230.

62. Schweier, J.; Becker, G. Economics of poplar short rotation coppice plantations on marginal land in Germany. *Biomass Bioenergy* **2013**, *59*, 494.

*Article*

# Reed Canary Grass for Energy in Sweden: Yields, Land-Use Patterns, and Climatic Profile

Blas Mola-Yudego [1,2,*], Xiaoqian Xu [1,*], Oskar Englund [3] and Ioannis Dimitriou [2]

1   School of Forest Sciences, University of Eastern Finland (UEF), P.O. Box 111, FI-80101 Joensuu, Finland
2   Department of Crop Production Ecology, Swedish University of Agricultural Sciences (SLU),
    75007 Uppsala, Sweden; Ioannis.Dimitriou@slu.se
3   Department of Ecotechnology and Sustainable Building Engineering, Mid Sweden University,
    83125 Östersund, Sweden; oskar.englund@miun.se
*   Correspondence: blas.mola@uef.fi (B.M.-Y.); xiaoqian.xu@uef.fi (X.X.)

**Abstract:** Research Highlights: (1) Reed canary grass (RCG) is analysed in Sweden compared to willow and poplar for 2001–2020. (2) Each crop presents a different land-use and climatic profile. (3) Average yield records of RCG are similar to willow and poplar. (4) There are divergences between trial-based and commercial yields. (5) Existing land-use change patterns suggest meadow > RCG and RCG > cereal. (6) RCG land area is very sensitive to policy incentives. Background and objectives: RCG is an alternative crop for biomass-to-energy due to high yield and frost tolerance. We assess the cultivation in Sweden by using an extensive compilation of data, with emphasis on the extent of the cultivation, climatic profile, land-use patterns, and productivity. Material and methods: RCG plantations are analysed for 2001–2020. A geostatistical analysis is performed to characterize where it is cultivated and the land uses associated. Climatic, productivity, and yield profiles are compared to willow and poplar plantations from experiments and from commercial plantations. Results: The results show that the cultivation of RCG expanded after 2005, with a maximum of 800 ha in 2009, to then decrease to the current levels of about 550 ha. It is mainly grown in colder climatic areas, with lower agricultural productivity than willow and poplar. Mean yields from trials are 6 oven dry tonnes (odt) ha$^{-1}$ year$^{-1}$; commercial yields are 3.5 odt ha$^{-1}$ year$^{-1}$. RCG replaces meadow land and then is replaced by cereals when abandoned. Conclusions: RCG is an interesting alternative with similar yields (commercial and trials) as other energy crops, but its success is more sensitive to policy incentives.

**Keywords:** energy crops; land use; biomass; bioenergy; reed canary grass (RCG); *Phalaris arundinacea* L.

**Citation:** Mola-Yudego, B.; Xu, X.; Englund, O.; Dimitriou, I. Reed Canary Grass for Energy in Sweden: Yields, Land-Use Patterns, and Climatic Profile. *Forests* **2021**, *12*, 897. https://doi.org/10.3390/f12070897

Academic Editor: Timothy A. Martin

Received: 26 April 2021
Accepted: 30 June 2021
Published: 9 July 2021

**Publisher's Note:** MDPI stays neutral with regard to jurisdictional claims in published maps and institutional affiliations.

## 1. Introduction

Perennial grasses have been considered as promising energy crops due to several characteristics that make them attractive for intensive biomass production compared to annual crops, i.e., high yield potential, high lignin and cellulose contents in their biomass, high heating value, low water content, lower management inputs such as soil tillage, and others [1]. They also offer advantages compared to perennial trees for energy since they do not need special equipment for management practices and can use common existing equipment for annual crops. In addition, they can enhance conditions for biodiversity and provide several ecosystem services, e.g., phytoremediation, erosion control, enhanced soil organic carbon, mediation of water flows, and retention of nutrients and other agrochemicals [2–4].

Among them, reed canary grass (RCG) (*Phalaris arundinacea* L.) has shown a great energy potential in Europe for direct combustion; as a feedstock for pellets and other solid biofuels [5]; in biochemical technologies like bioethanol and biogas [6,7]; in other thermochemical applications, like pyrolysis [8]; and additional added value applications being currently considered [9]. In Northern Europe, RCG has been used at a commercial level in Finland [10–12] and Sweden [13,14], presenting advantages due to its frost resistance and adaptability potential to hard climatic conditions. In Sweden, its use has been documented since 1749 for forage in Scania, and studies in the 1800s highlighted its high yield potential, particularly in the northernmost areas [15]. In the 1980s, research was performed aiming to grow RCG as an alternative biomass source for the pulp industry [16] and later for large-scale industrial production for energy [17]. In fact, it was considered as one of the most interesting energy crops in the country [18], and in the early 2000s, new varieties started to be dedicated exclusively for this purpose [19]. Since then, RCG has been established along the whole country [18] in large stands with a height of about 1.5–2 m. The main commercial varieties have been *Palaton* and *Venture* and, more recently, *Bamse* [17].

Crop management activities are regarded as easier compared to other lignocellulosic energy crops, such as willow and poplar, with lower establishment costs due to the use of existing conventional equipment and the use of seeds in the establishment phase [20]. Under Swedish conditions, the soil is prepared by ploughing before sowing. Perennial weeds, such as couch grass, are controlled the previous year. Sowing occurs in early spring with a row spacing of 10–15 cm. Growth is rather slow in the beginning until the root system is established, and weeding could be necessary during the first year. The first harvest occurs the second year after sowing, and a well-managed field can be productive for 10–15 years before re-establishment is required [17]. RCG grows well on most kinds of soils but particularly in poorly drained soils, as it tolerates waterlogging better than many other grasses [21]. Due to its deep root system once well established, it is also more drought-resistant than other grasses [1].

In the 1990s, the fertilisation recommendations were 150, 100, and 30 kg per hectare nitrogen (N), phosphorus (P), and potassium (K), respectively, in the first year, and 80, 30, and 10 kg per hectare during the rest of the production period [19]. In recent decades, this has been changed to 40, 15, and 50 kg in the year of sowing; 100, 15, and 80 kg the following year; and 50, 5, and 20 kg in spring [22]. To reduce fertilisation costs, the application of mixtures of sewage sludge, wood, and grass ash have also been practiced [19,23]. Harvest takes place in the second year after sowing, preferably in early spring because the grass presents the lowest moisture content (ca. 10%–15%) and can be used in power plants without drying. The first harvest can be 20% lower than subsequent harvests [17,19]. Moreover, sodium (Na), K, and chlorine (Cl) concentrations are the lowest in early spring, which makes it a better fuel with decreased corrosion risks for the boiler [17]. Harvesting is a critical operation, as increasing the harvest height from 5 cm to 10 cm can result in harvest losses exceeding 25% of biomass [24]. In general, ordinary hay harvesting equipment is used, and transportation from the field usually occurs in bales. The final removal of the crop is often performed by conventional soil tillage operations [19].

As a perennial grass, RCG can complement the options for energy crops in the country, today largely based on woody plantations. In this sense, research has been focused on trials and management regimes (e.g., [25,26]), the biology of the crop (e.g., [27]), and even biomass properties for energy use (e.g., [28]). However, despite its current commercial use, there have been few attempts to provide a comprehensive assessment of the cultivation, especially with regards to the other lignocellulosic energy crops grown for similar purposes, and the land use changes in the areas cultivated, linked to the policy framework. This paper analyses the present situation of RCG cultivation in Sweden using an extensive compilation of records, with emphasis on the current extent of the crop, land-use change patterns, and overall productivity. The main goal is to assess its performance and development compared with other biomass production systems, such as willow, poplar, or hybrid aspen, in order to better define its role in the mix of energy crops.

## 2. Material and Methods

### 2.1. Data Sources

Several databases for agricultural production were combined for the analysis. The location of the commercial plantations for the period 2001–2017 was retrieved from the land register using the IACS (Integrated Agricultural Control System) database maintained by the Swedish Board of Agriculture. This database permits to extract the land use of each polygon (*blocks* in the databases), defined as a uniform land area that remains quite constant from one year to the next [29] although the use of the land may be altered. Land-use data from 2001 to 2016 were included in the analysis. The method to deal with the land uses was based on Xu and Mola-Yudego [30]. The total area cultivated for 2017–2020 was retrieved from Statistics Sweden but, in this case, was aggregated.

A database of existing trials was constructed, including records from RCG, willow, and poplar/hybrid aspen (given the limited area planted, in this paper, will be referred together). For RCG, trial records were retrieved from Landström et al. [18], Lindvall et al. [23], Nilsson et al. [14], Lindvall et al. [25], and Lindvall [21] during the period 1991–2015 (N = 201). For willow, a trial database was used based on Mola-Yudego et al. [31] (N = 290) and for poplar, from Dimitriou and Mola-Yudego [32] (N = 58). In the case of RCG, the observations were annual harvests, whereas for willow and poplar, they were the annual yield of the biomass produced during the cutting season or rotation (4–20 years). Concerning commercial records, the annual yield for RCG for Sweden was extracted from the Eurostat database [33] as well as from records supplied by Statistics Sweden and, for Finland, from Luke [34].

To analyze the agricultural profile of the cultivations, the data was based on the standard yield estimates by agricultural districts [35]. For the climatic profile, data were retrieved from the WorldClim database for Sweden using the last standard period 1960–1990 at a resolution of $1 \times 1$ km for the monthly minimum and maximum temperatures and precipitation [36].

*2.2. Data Analysis*

All plantations of RCG existing in Sweden were identified and geo-located for the period 2001–2017. The series was completed till 2020 with the aggregated area available. The total cultivated area and the average size of the plantations were quantified for each year during that period and compared to similar records of willow and poplar. The geographical distribution of the plantations was further analysed by using spatial kernel methods [37,38]. Kernel density estimation is a non-parametric method that allows to define *core areas* (areas with the highest density of the crop) and *home areas* (area entailing most of the cultivated area). The method was applied following Mola-Yudego and González-Olabarria [39]. The *core area* was the smallest area to include 50% of all existing plantations for a given period, and the *home area* defined the area including nearly all plantations (90%). The same analysis was performed to willow and poplar plantations.

For each plantation, monthly estimates of temperature (maximum and minimum) and precipitation were calculated in order to provide a climatic profile. The monthly mean values for all plantations were then averaged for the whole country in order to provide a climatic profile for each biomass production system. Similarly, the estimated standard yield of cereals was used as an indicator of land productivity. Among the options, barley is the most common crop in most areas where plantations are established. To reduce the effect of climatic variations on specific years, the average was estimated for several years (2003–2017) using the same approach as in Xu and Mola-Yudego [30]. This yield was assigned to each plantation, and the country's average was calculated on a yearly basis for the three main plantation systems in the same way as for the climatic variables.

The productivity of RCG were assessed using yield records from trials and commercial yields and compared to the performance of the other plantation systems. The estimates from experimental plots were investigated by observing the ranges and geographical distribution of the trials compared to the equivalent levels of willow and poplar from similar trials and experimental plots. These values were also contrasted to the official averages resulting from commercial plantations both in Sweden and in nearby areas in Finland.

Finally, changes in land uses were also investigated; prior land uses in each plantation were identified and grouped in three main categories (cereal production, fallow land, and meadows), and the changes in area were estimated annually.

## 3. Results

Prior to the data available in the land registry, there were records of ca. 4000–5000 ha sown with RCG in 1991, which were mainly dedicated to forage and animal use [17]. These plantations nearly disappeared by the end of the 1990s, as the area under cultivation was around 675 ha in the first year of detailed records. After 2005, the area increased significantly to 800 ha in 2009 (Figure 1), mainly for energy purposes. Prior to this year, RCG could get support only when there was a contract for industrial use (mainly energy) when grown on land without support rights for set-aside land [40]. This level lasted until 2013; after that, there was a steady decline in the area to the most recent figure (ca. 550 ha). The distribution of the size of the fields followed a logarithmic distribution with the prevalence of small plantations over large ones. Fields larger than 5 ha were uncommon and over 10 ha were very rare. The average plantation size increased over time from ca. 2.18 ha in 2005 to 2.4 ha until recent years.

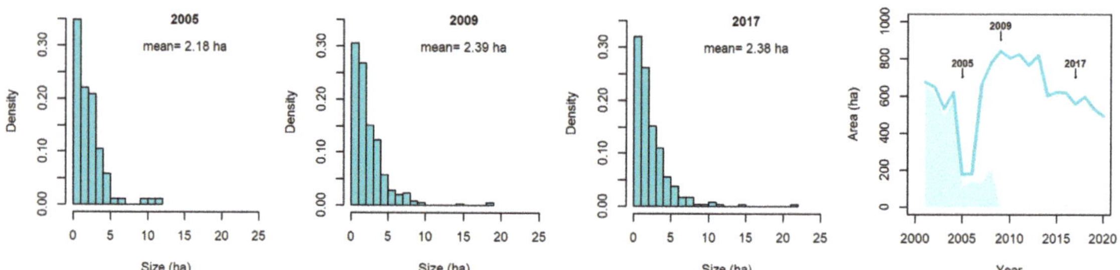

**Figure 1.** Evolution of area cultivated with reed canary grass land for the period 2001–2020 and distribution of the plantations according to their size. Shaded area refers to non-contractual plantations (largely before 2009).

Although RCG plantations are distributed along the whole country, the largest concentration is at the northeast, around the regions of West and North Bothnia and, to a lesser extent, in the central and southern parts of the country, where the share of agricultural land is larger, and other lignocellulosic energy crops, such as willow and poplar, are typically planted (Figure 2).

The geographical location of the plantations is reflected in the climatic profiles. The mean annual precipitation of a plantation of RCG in Sweden was 582 mm compared with 605.5 mm and 654.7 mm for willow and poplar aspen, respectively. The mean annual temperatures were between −0.44 °C (minimum) and 7.56 °C (maximum), which represents a colder average than the 2.78 °C (minimum) and 9.99 °C (maximum) for willow and 3.16 °C (minimum) and 10.05 °C (maximum) for poplar (Figure 3).

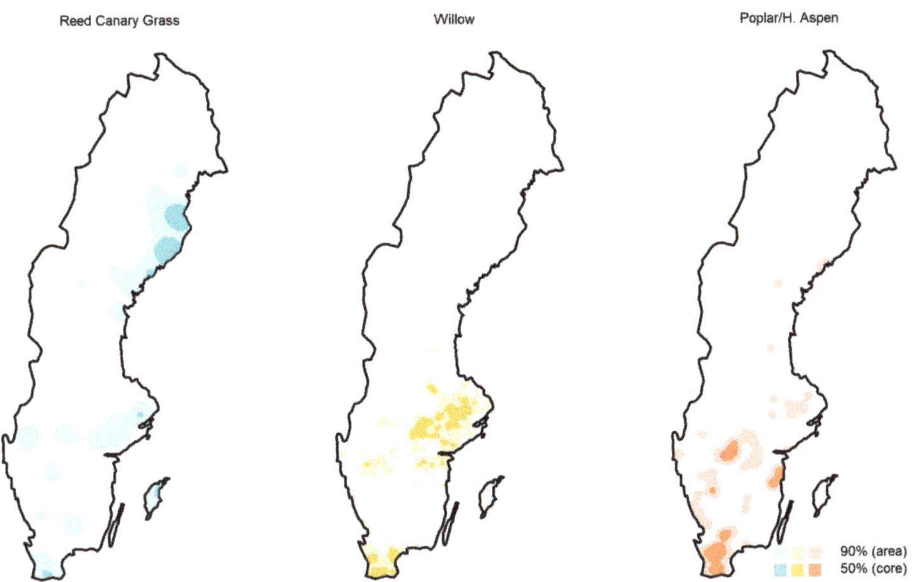

**Figure 2.** Location of reed canary grass plantations in Sweden, including all fields and extension of the cultivation area and *core areas* with the highest concentration of plantations compared to willow and poplar/hybrid aspen for the same year.

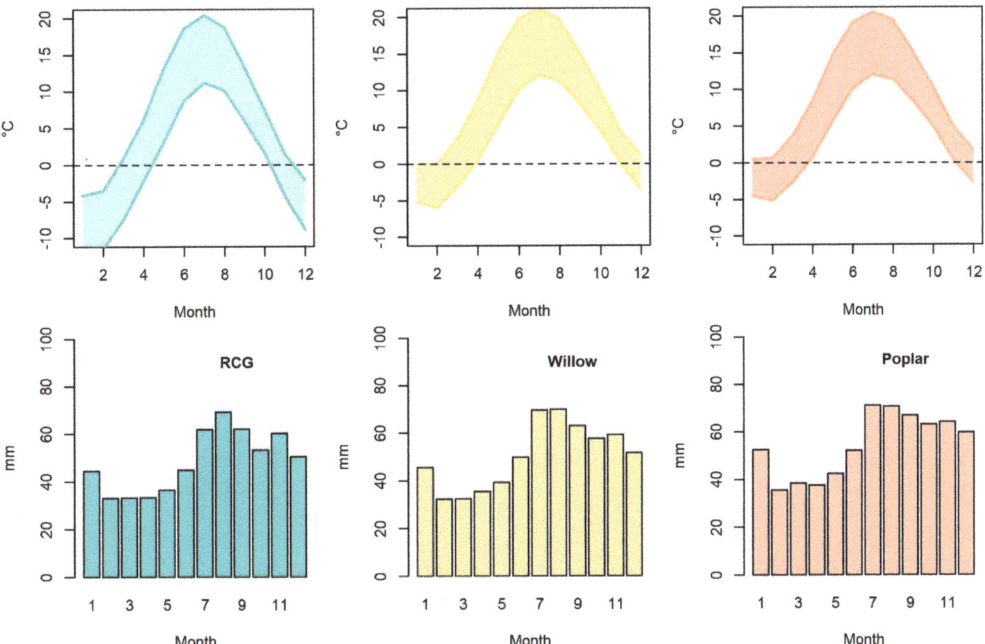

**Figure 3.** Agro-climatic profile of reed canary grass (RCG) compared to other fast-growing plantations in Sweden. The profiles represent the mean monthly maxima and minima temperatures and precipitation records of all plantations of RCG (N = 350), willow (N = 3305), and poplar/hybrid aspen (N = 253) for a reference year (2009).

The analysed trials represent the geographical distribution of the commercial plantations (Figure 4), and the results show large ranges. The maximum record from the trials is close to 15 oven dry tonnes (odt) ha$^{-1}$ year$^{-1}$ (comparable to poplar). Yields over 10 odt ha$^{-1}$ year$^{-1}$ are more common in the case of willow plantations. In general, the mean yields from trials for RCG, willow, and poplar are similar despite the fact that RCG is often located in less favourable climatic areas, with lower average agricultural productivity (Figure 4).

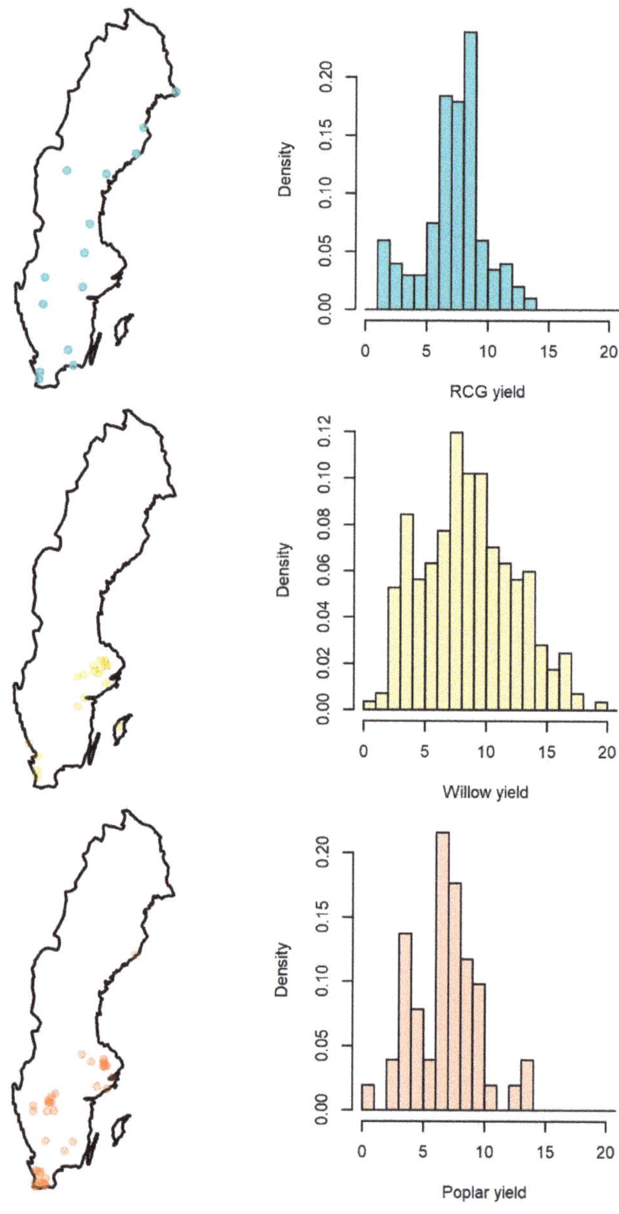

**Figure 4.** Estimated distribution of yields (odt ha$^{-1}$ year$^{-1}$) for reed canary grass compared to willow and poplar in different trials along the country (see maps) for the period of 1991–2010.

The yields from trials, however, are largely overestimating the commercial yield of RCG, estimated around 3.37 odt ha$^{-1}$ year$^{-1}$ in Sweden (Figure 5). Nearby areas in Finland present similar values for commercial plantations, for example, around 3.6 in Lapland and 4.2 in South Ostrobothnia. The Finnish average is even lower (3.1 odt ha$^{-1}$ year$^{-1}$). The corresponding values for commercial willow plantations in Sweden are 2.6 odt ha$^{-1}$ year$^{-1}$ and 4.2 odt ha$^{-1}$ year$^{-1}$ for the first and second rotation, respectively.

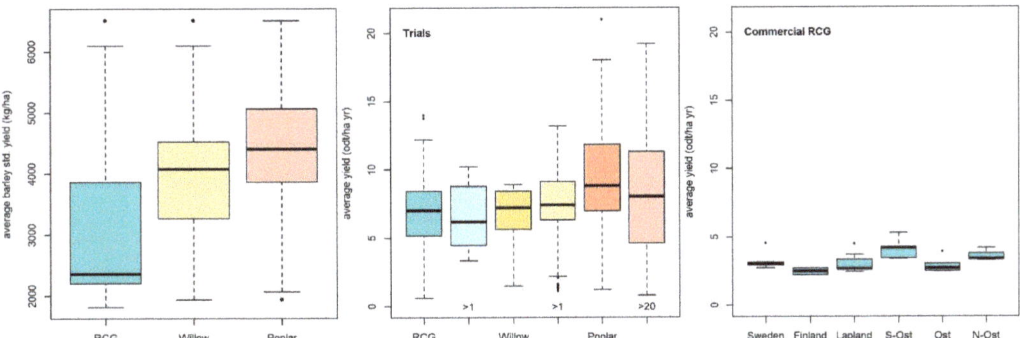

**Figure 5.** Land productivity yields of the trials for reed canary grass (RCG), willow, and poplar in Sweden. **Left**: Averaged agricultural productivity of all established plantations using the standard yields of barley by agricultural district. **Centre**: Average yield (oven dry tonnes, odt) from trials in Sweden for reed canary grass (1991–2010), willow (1986–2004), and poplar (1980–2015), where >1 refers to second harvest and subsequent and >20 refers to plantations over 20 years. **Right**: Average yield for RCG (odt) from commercial plantations for the period 2011–2017 in Sweden and Finland (counties: Lapland, South Ostrobothnia, Ostrobothnia, and North-Ostrobothnia).

Finally, changes in land use show that RCG is mainly replacing meadow and cereal land, in that order, starting in 2005. However, after 2009, RCG plantations were replaced by cereals to a larger extent than meadows (Figure 6).

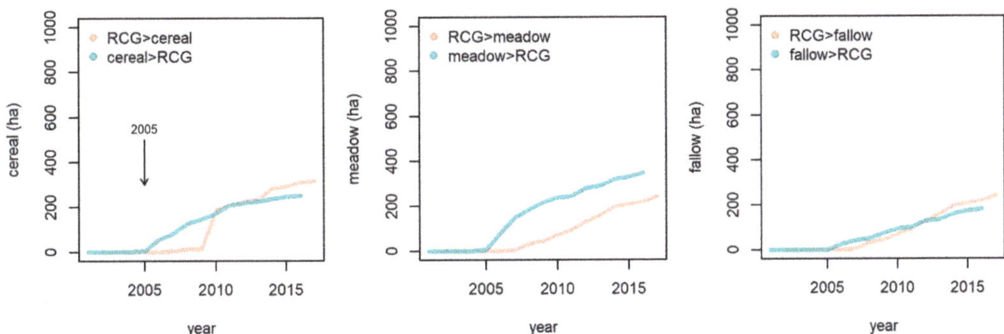

**Figure 6.** Changes in land use related to reed canary grass (RCG) used for energy in Sweden. The line *cereal > RGC* refers to the area with cereal replaced by RCG, and *RCG > cereal* refers to plantations being abandoned and replaced by cereal fields and likewise for meadow and fallow land.

## 4. Discussion

RCG has had a long history in Sweden. This study aims at providing a detailed overview of the crop based on multiple sources of different character. The crop is also compared with the other two existing lignocellulosic biomass production systems, willow and poplar (including hybrid aspen), grown in the country.

In 1991, the area with RCG peaked at around 4000–5000 ha, mainly used for forage [17]. The expansion was driven by earlier support schemes for converting land from food into non-food crops [19]. These early incentives were established in order to deregulate the agricultural sector, reduce the overplanted food crop areas, and restructure agricultural land use. Subsidies for energy crops were introduced as a compensation tool in the period 1991–1996 [41], stimulating the establishment of energy crops such as willow and opening up the use of RCG for energy rather than fodder [13,42]. As bioenergy markets had not matured before the incentives were removed in the mid-1990s, combined with increased food crop subsidies, the area of both crops decreased [41], leading to the stagnation of new willow plantation areas and the removal of most of the RCG established by that time.

According to the results, by the beginning of the 2000s, there were less than 800 ha of RCG in Sweden compared to over 10,000 ha of willow and 1000 ha poplar in the same period [30]. There was a steady decline that almost supposed the removal of all cultivated area in 2005, followed by a new rapid expansion, reaching nearly 1000 ha by 2009. The same increment was observed in Finland, parallel to subsidies and policy incentives for its cultivation, which resulted in about 20,000 ha by 2009 [34]. In 2009, the set-aside requirement, i.e., the EU requirement for farmers producing high quantities of cereals to leave a percentage of their land (ca. 10%) out of production or grow it with industrial crops, was decided to be removed [43], which is one explanation for the progressive decline after that period.

Concerning productivity, RCG compares well with the other lignocellulosic energy crops in the country [32]. Yields from experiments are at similar ranges for all three crops compared and are at similar levels as indicated in other studies in Finland, Estonia, and Lithuania, with spring harvest yields of ca. 6 odt ha$^{-1}$ year$^{-1}$ [44–46]. These levels must be taken as an upper threshold in optimal management conditions, as yield observations resulting from experimental plots tend to largely overestimate commercial levels [31]. In addition, there are important harvest losses, which are significant in this case; the effective harvest yield can be only 45%–56% of the biological yields [12,47]. In fact, the results confirm this divergence, as the official commercial averages for Sweden and for Finland (in similar climatic regions) are ca. 50% lower (3.3–3.5 odt ha$^{-1}$ year$^{-1}$). The resulting yields match the commercial averages for willow [48], indicating a similar performance of both energy crops in terms of biomass.

This can be seen as a competitive advantage: RCG has a lower establishment cost compared with willow plantations [20], and it is growing in less favourable areas for agriculture than other energy crops (the results reflect shorter growing seasons and less cereal productivity than the areas where poplar or willow are planted) thus replacing less fertile land. However, at the same time, the abrupt changes in planting areas suggest the crop is more sensitive than willow to the regulations derived from the policy framework. This is also confirmed by previous studies highlighting that farmers' willingness to grow RCG is highly sensitive to subsidy levels [49]. This could be related to the lower establishment costs, limiting potential losses when abandoned earlier, and to the shorter lifespan, around 10 years, while willow cultivation extends over 25 years [50]. A more dramatic example took place in Finland in the same period of study, as the plantation area decreased from ca. 19,000 ha (2007) to merely 6600 ha (2013) and further down to 3000 ha (2020) [34,51].

It is noticeable, however, that although the total area cultivated in Finland decreased abruptly, the total number of farms cultivating RCG remained stable, from 367 (2013) to 317 (2019), a reduction by a factor of 2 in area but only by 15% in the number of farms [34]. This suggests that farmers already cultivating RCG decided to continue growing it after the subsidies were removed but chose to reduce the land dedicated to the crop. Overall, the dependency on subsidies can be explained by the narrow economic margins of the crop. Larsson [13] estimated that the minimum farm gate price of RCG required for being profitable would be 56 Swedish krona (SEK) MWh$^{-1}$ at that time in Sweden. Whereas transportation cost could be reduced by baling [52], losses still occurred due to outdoor storage, especially in long rainy and snowy seasons [19]. In addition, management costs can be relatively high when RCG is established on marginal land due to, e.g., deep soil preparation [53].

There were ambitious plans for the expansion of RCG cultivation. For example, the area nearby Skellefteå planned to expand to 3000 ha of RCG in the late 2000s [16,54], and in Finland, it was expected to reach 100,000 ha by 2015 [55,56], which in both cases did not occur since most of the policy incentives were removed before these goals were reached. Additional challenges that precluded RCG from reaching these levels include larger-than-expected harvest losses and lower fuel quality in terms of lower heat values [57], higher ash content, and higher alkaline concentrations, increasing corrosion risks for the boiler [17]. In addition, RCG has a negative perception among some farmers [49], and it is often regarded as an invasive species [3].

Notwithstanding the negative perception, RCG can have multiple positive environmental effects. Besides the production of sustainable biomass, it has demonstrated its role in carbon sequestration [58], enabling a net carbon sink on organic soils [59]. The effect on soil carbon is, however, dependent on the land where it is established (e.g., [60]). For example, positive effects on soil carbon can, in general, be expected when established on former cropland, while such effects should be less significant when established on previous pastures. The results show that RCG was originally established on previous meadows although, after a few years, many plantations were being replaced by cereals. This pattern is likely linked to the changes related to set-aside land as well as the increase in cereal prices after 2007, as observed in Xu and Mola-Yudego [30]; the larger implications of this land-use change pattern should be subject to further study.

Despite some studies suggesting its landscape dominance to have a negative effect on biodiversity for their invasive character [61], the typical cultivation in small areas separated by other land uses (with due planning and control) may indeed favour a mosaic-based land-use pattern, creating diverse niches for fauna and flora [62]. In addition, the strategic introduction of perennial crops into agricultural landscapes can, in general, result in multiple positive effects by supporting ecosystem services that can mitigate existing environmental impacts related to, e.g., soil and water, while supposing minor effects on food production [4,63,64]. For instance, the potential of RCG to mitigate GHG emissions from abandoned peat extraction areas has been shown [65].

The success of RCG will depend on a predictable and sustainable economic profit for the farmers. In general, further cost reduction in management practices and higher revenues in terms of energy prices will be required in order to avoid the excessive dependence on direct policy incentives. Technical and management aspects related to the reduction of harvesting losses can also contribute to a better economic output. Finally, financial compensation for environmental benefits associated with its cultivation would enable the crop to compete with fallow land when grown on marginal land, which is usually the case in Sweden [14].

## 5. Conclusions

Compared to willow and poplar cultivation in Sweden, RCG presents good commercial yields despite harvesting losses and being established on less productive locations. The different profiles of the three main biomass production systems for energy show regional complementary features. However, there are important limitations that prevent the expansion of the crop, such as insufficient markets for the biomass and lack of compensation for environmental benefits; the establishment of new plantations is, therefore, currently highly sensitive to direct support schemes. Finally, there are ongoing land-use trends towards cereal cultivation, possibly due to more favourable prices. The analysis provided in the study concerning cultivation areas, land use, yield performance, and climatic profiles can serve as a basis for future analysis of the status of energy crops in Northern Europe and elsewhere.

**Author Contributions:** Conceptualization, B.M.-Y. and X.X.; methodology, B.M.-Y. and X.X.; software, B.M.-Y. and X.X.; validation, B.M.-Y. and X.X.; formal analysis, B.M.-Y. and X.X.; investigation, B.M.-Y. and X.X.; resources, B.M.-Y. and X.X.; data curation, B.M.-Y. and X.X.; writing—original draft preparation, B.M.-Y., X.X., I.D. and O.E.; writing—review and editing, B.M.-Y., X.X., I.D. and O.E.; visualization, B.M.-Y. and X.X.; supervision, B.M.-Y., I.D.; project administration, B.M.-Y.; funding acquisition, B.M.-Y., X.X. and I.D. All authors have read and agreed to the published version of the manuscript. B.M.-Y. and X.X. shared first authorship.

**Funding:** This research was funded by IEA Task 43 Bioenergy and Niemi foundation, grant number 20200009.

**Data Availability Statement:** The original data can be found in the cited publications; the maps and land uses data is available at https://jordbruksverket.se/.

**Acknowledgments:** We would like to thank Gerda Landell at Statistics Sweden for her kind contribution to the data used in the paper and to Seikh Sharif for his help compiling trial data. Finally, we thank the help and support by the IEA Bioenergy Task 43 and the Niemi foundation.

**Conflicts of Interest:** The authors declare no conflict of interest.

# References

1. Lewandowski, I.; Scurlock, J.M.; Lindvall, E.; Christou, M. The development and current status of perennial rhizomatous grasses as energy crops in the US and Europe. *Biomass Bioenergy* **2003**, *25*, 335–361. [CrossRef]
2. Antonkiewicz, J.; Kołodziej, B.; Bielińska, E.J. The use of reed canary grass and giant miscanthus in the phytoremediation of municipal sewage sludge. *Environ. Sci. Pollut. Res.* **2016**, *23*, 9505–9517. [CrossRef]
3. Lavergne, S.; Molofsky, J. Reed canary grass (*Phalaris arundinacea*) as a biological model in the study of plant invasions. *Crit. Rev. Plant Sci.* **2004**, *23*, 415–429. [CrossRef]
4. Englund, O.; Börjesson, P.; Berndes, G.; Scarlat, N.; Dallemand, J.F.; Grizzetti, B.; Dimitriou, I.; Mola-Yudego, B.; Fahl, F. Beneficial land use change: Strategic expansion of new biomass plantations can reduce environmental impacts from EU agriculture. *Glob. Environ. Chang.* **2020**, *60*, 101990. [CrossRef]
5. Jasinskas, A.; Streikus, D.; Šarauskis, E.; Palšauskas, M.; Venslauskas, K. Energy evaluation and greenhouse gas emissions of reed plant pelletizing and utilization as solid biofuel. *Energies* **2020**, *13*, 1516. [CrossRef]
6. Roj-Rojewski, S.; Wysocka-Czubaszek, A.; Czubaszek, R.; Kamocki, A.; Banaszuk, P. Anaerobic digestion of wetland biomass from conservation management for biogas production. *Biomass Bioenergy* **2019**, *122*, 126–132. [CrossRef]
7. Tilvikiene, V.; Kadziuliene, Z.; Dabkevicius, Z.; Venslauskas, K.; Navickas, K. Feasibility of tall fescue, cocksfoot and reed canary grass for anaerobic digestion: Analysis of productivity and energy potential. *Ind. Crop. Prod.* **2016**, *84*, 87–96. [CrossRef]
8. Alhumade, H.; da Silva, J.C.G.; Ahmad, M.S.; Çakman, G.; Yıldız, A.; Ceylan, S.; Elkamel, A. Investigation of pyrolysis kinetics and thermal behavior of Invasive Reed Canary (*Phalaris arundinacea*) for bioenergy potential. *J. Anal. Appl. Pyrolysis* **2019**, *140*, 385–392. [CrossRef]
9. Scordia, D.; Cosentino, S.L. Perennial energy grasses: Resilient crops in a changing European agriculture. *Agriculture* **2019**, *9*, 169. [CrossRef]
10. Lankoski, J.; Ollikainen, M. Bioenergy crop production and climate policies: A von Thunen model and the case of reed canary grass in Finland. *Eur. Rev. Agric. Econ.* **2008**, *35*, 519–546. [CrossRef]
11. Mikkola, H.J.; Ahokas, J. Renewable energy from agro biomass. *Agron. Res.* **2011**, *9*, 159–164.
12. Pahkala, K.; Aalto, M.; Isolahti, M.; Poikola, J.; Jauhiainen, L. Large-scale energy grass farming for power plants—A case study from Ostrobothnia, Finland. *Biomass Bioenergy* **2008**, *32*, 1009–1015. [CrossRef]
13. Larsson, S. Supply curves of reed canary grass (*Phalaris arundinacea* L.) in Västerbotten County, northern Sweden, under different EU subsidy schemes. *Biomass Bioenergy* **2006**, *30*, 28–37. [CrossRef]
14. Nilsson, D.; Rosenqvist, H.; Bernesson, S. Profitability of the production of energy grasses on marginal agricultural land in Sweden. *Biomass Bioenergy* **2015**, *83*, 159–168. [CrossRef]
15. Alway, F.J. Early trials and use of reed-canary grass as a forage plant. *J. Am. Soc. Agron.* **1931**, *23*, 64–66. [CrossRef]
16. Finell, M. The Use of Reed Canary-Grass (*Phalaris arundinacea*) as a Short Fibre Raw Material for the Pulp and Paper Industry. Ph.D. Thesis, Swedish University of Agricultural Sciences, Umeå, Sweden, 2003. Available online: http://urn.kb.se/resolve?urn=urn: nbn:se:slu:epsilon-95 (accessed on 16 January 2021).
17. Landfors, K.; Hollsten, R. *Energigräs–en Kunskapssammanställning*; Jordbruksverket: Jönköping, Sweden, 2011; 20p. (In Swedish)
18. Landström, S.; Lomakka, L.; Andersson, S. Harvest in spring improves yield and quality of reed canary grass as a bioenergy crop. *Biomass Bioenergy* **1996**, *11*, 333–341. [CrossRef]
19. Venendaal, R.; Jørgensen, U.; Foster, C.A. European energy crops: A synthesis. *Biomass Bioenergy* **1997**, *13*, 147–185. [CrossRef]
20. Shield, I.F.; Barraclough, T.J.P.; Riche, A.B.; Yates, N.E. The yield response of the energy crops switchgrass and reed canary grass to fertiliser applications when grown on a low productivity sandy soil. *Biomass Bioenergy* **2012**, *42*, 86–96. [CrossRef]
21. Lindvall, E. Nutrient Supply to Reed Canary Grass as a Bioenergy Crop. *Acta Univ. Agric. Suec.* **2014**, *54*, 53.

22. Lewandowski, I. Biomass Production from Lignocellulosic Energy Crops. In *Encyclopedia of Applied Plant Sciences*; Thomas, B., Murray, B.G., Murphy, D.J., Eds.; Elsevier: Amsterdam, Netherlands, 2017; pp. 159–163, ISBN 978-0-12-394808-3.

23. Lindvall, E.; Gustavsson, A.M.; Samuelsson, R.; Magnusson, T.; Palmborg, C. Ash as a phosphorus fertilizer to reed canary grass: Effects of nutrient and heavy metal composition on plant and soil. *GCB Bioenergy* **2015**, *7*, 553–564. [CrossRef]

24. Pahkala, K.; Partala, A.; Suokannas, A.; Klemola, E.; Kalliomäki, T.; Kirkkari, A.M.; Flyktman, M. *Odling och Skörd av Rörflen för Energiproduktion*; MTT: Jokioinen, Finland, 2003; 27p, ISBN 951-729-744-0. (In Swedish)

25. Lindvall, E.; Gustavsson, A.M.; Palmborg, C. Establishment of reed canary grass with perennial legumes or barley and different fertilization treatments: Effects on yield, botanical composition and nitrogen fixation. *GCB Bioenergy* **2012**, *4*, 661–670. [CrossRef]

26. Kandel, T.P.; Elsgaard, L.; Karki, S.; Lærke, P.E. Biomass yield and greenhouse gas emissions from a drained fen peatland cultivated with reed canary grass under different harvest and fertilizer regimes. *BioEnergy Res.* **2013**, *6*, 883–895. [CrossRef]

27. Xiong, S.; Zhang, Q.G.; Zhang, D.Y.; Olsson, R. Influence of harvest time on fuel characteristics of five potential energy crops in northern China. *Bioresour. Technol.* **2008**, *99*, 479–485. [CrossRef] [PubMed]

28. Larsson, S.H.; Thyrel, M.; Geladi, P.; Lestander, T.A. High quality biofuel pellet production from pre-compacted low density raw materials. *Bioresour. Technol.* **2008**, *99*, 7176–7182. [CrossRef] [PubMed]

29. JBB, Swedish Board of Agriculture (Jordbruksverket). Kartor och Geografiska Informations System. Available online: https://nya.jordbruksverket.se/ (accessed on 20 November 2019). (In Swedish)

30. Xu, X.; Mola-Yudego, B. Where and when are plantations established? Land-use replacement patterns of fast-growing plantations on agricultural land. *Biomass Bioenergy* **2021**, *144*, 105921. [CrossRef]

31. Mola-Yudego, B.; Díaz-Yáñez, O.; Dimitriou, I. How much yield should we expect from fast-growing plantations for energy? Divergences between experiments and commercial willow plantations. *BioEnergy Res.* **2015**, *8*, 1769–1777. [CrossRef]

32. Dimitriou, I.; Mola-Yudego, B. Poplar and willow plantations on agricultural land in Sweden: Area, yield, groundwater quality and soil organic carbon. *For. Ecol. Manag.* **2017**, *383*, 99–107. [CrossRef]

33. Eurostat 2021, Crop Production in EU Standard Humidity, Code: Apro_cpsh1. Available online: https://ec.europa.eu/eurostat (accessed on 6 April 2021).

34. LUKE 2021, Natural Resources Institute Finland, Statistical Databases, Crop Production Statistics, Luke_Maa_Sato_0. Available online: www.statdb.luke.fi (accessed on 6 April 2021).

35. Swedish Official Statistics, Standard Yields for Yield Survey Districts, Counties and the Whole Country, Several years). Available online: https://www.scb.se (accessed on 23 October 2019). (In Swedish)

36. Fick, S.E.; Hijmans, R.J. WorldClim 2: New 1–km spatial resolution climate surfaces for global land areas. *Int. J. Climatol.* **2017**, *37*, 4302–4315. [CrossRef]

37. Worton, B.J. Kernel methods for estimating the utilization distribution in home-range studies. *Ecology* **1989**, *70*, 164–168. [CrossRef]

38. Silverman, B.W. *Density Estimation for Statistics and DATA Analysi*; CRC Press: Boca Raton, FL, USA, 1986; Volume 26.

39. Mola-Yudego, B.; González-Olabarria, J.R. Mapping the expansion and distribution of willow plantations for bioenergy in Sweden: Lessons to be learned about the spread of energy crops. *Biomass Bioenergy* **2010**, *34*, 442–448. [CrossRef]

40. Jordbruksverket. *Stöd för Odling av Grödor för Industri- och Energiändamål*; Jönköping, Sweden, 2008; 24p. Available online: http://www2.jordbruksverket.se/webdav/files/SJV/trycksaker/Jordbruksstod/JS13.pdf (accessed on 26 January 2021). (In Swedish)

41. Hadders, G.; Olssen, R. European Energy Crops Overview, Country Report for Sweden. 1996. Available online: https://www.diva-portal.org/smash/get/diva2:959639/FULLTEXT01.pdf (accessed on 26 January 2021).

42. Mola-Yudego, B.; Pelkonen, P. The effects of policy incentives in the adoption of willow short rotation coppice for bioenergy in Sweden. *Energy Policy* **2008**, *36*, 3062–3068. [CrossRef]

43. Official Journal of the European Union. Communication from the Commission to All Farmers Concerning Set Aside from 2009 (2008/C 324/07). 2008. Available online: https://eur-lex.europa.eu/LexUriServ/LexUriServ.do?uri=OJ:C:2008:324:0011:0011:EN:PDF (accessed on 26 January 2021).

44. Pahkala, K.; Pihala, M. Different plant parts as raw material for fuel and pulp production. *Ind. Crop. Prod.* **2000**, *11*, 119–128. [CrossRef]

45. Heinsoo, K.; Hein, K.; Melts, I.; Holm, B.; Ivask, M. Reed canary grass yield and fuel quality in Estonian farmers' fields. *Biomass Bioenergy* **2001**, *35*, 617–625. [CrossRef]

46. Jasinskas, A.; Zaltauskas, A.; Kryzeviciene, A. The investigation of growing and using of tall perennial grasses as energy crops. *Biomass Bioenergy* **2008**, *32*, 981–987. [CrossRef]

47. Lötjönen, T.; Paappanen, T. Bale density of reed canary grass spring harvest. *Biomass Bioenergy* **2013**, *51*, 53–59. [CrossRef]

48. Mola-Yudego, B.; Aronsson, P. Yield models for commercial willow biomass plantations in Sweden. *Biomass Bioenergy* **2008**, *32*, 829–837. [CrossRef]

49. Paulrud, S.; Laitila, T. Farmers' attitudes about growing energy crops: A choice experiment approach. *Biomass Bioenergy* **2010**, *34*, 1770–1779. [CrossRef]

50. Börjesson, P. Environmental effects of energy crop cultivation in Sweden—I: Identification and quantification. *Biomass Bioenergy* **1999**, *16*, 137–154. [CrossRef]

51. TIKE. *Yearbook of Farm Statistics, Utilized Agricultural Area, 2001–2007*; ISSN: Helsinki, Finland, 2007.

52. Lindh, T.; Paappanen, T.; Rinne, S.; Sivonen, K.; Wihersaari, M. Reed canary grass transportation costs—Reducing costs and increasing feasible transportation distances. *Biomass Bioenergy* **2009**, *33*, 209–212. [CrossRef]

53. Larsson, S.; Nilsson, C. A remote sensing methodology to assess the costs of preparing abandoned farmland for energy crop cultivation in northern Sweden. *Biomass Bioenergy* **2005**, *28*, 1–6. [CrossRef]

54. Olsson, J.; Salomon, E.; Baky, E.; Palm, O. *Energigrödor–en Möjlighet för Jordbruksanvändning av Slam*; Svenskt Vatten Utveckling, Svenskt Vatten AB: Bromma, Sweden, 2008; Volume 12, p. 53. (In Swedish)

55. MMM, (Ministry of Agriculture and Forestry). Press Release Interim Report. 2006. Available online: https://mmm.fi/sv/-/ruokohelven-viljelyala-noussut-nopeasti-suomessa (accessed on 26 January 2006).

56. Hyvönen, N.P.; Huttunen, J.T.; Shurpali, N.J.; Tavi, N.M.; Repo, M.E.; Martikainen, P.J. Fluxes of nitrous oxide and methane on an abandoned peat extraction site: Effect of reed canary grass cultivation. *Bioresour. Technol.* **2009**, *100*, 4723–4730. [CrossRef] [PubMed]

57. Burvall, J. Influence of harvest time and soil type on fuel quality in reed canary grass (*Phalaris arundinacea* L.). *Biomass Bioenergy* **1997**, *12*, 149–154. [CrossRef]

58. Xiong, S.; Kätterer, T. Carbon-allocation dynamics in reed canary grass as affected by soil type and fertilization rates in northern Sweden. *Acta Agric. Scand. Sect. B Soil Plant Sci.* **2010**, *60*, 24–32. [CrossRef]

59. Shurpali, N.J.; Hyvönen, N.P.; Huttunen, J.T.; Clement, R.J.; Reichstein, M.; Nykänen, H.; Nykänen, H.; Biasi, C.; Martikainen, P.J. Cultivation of a perennial grass for bioenergy on a boreal organic soil–carbon sink or source? *GCB Bioenergy* **2009**, *1*, 35–50. [CrossRef]

60. Krol, D.J.; Jones, M.B.; Williams, M.; Choncubhair, O.N.; Lanigan, G.J. The effect of land use change from grassland to bioenergy crops Miscanthus and reed canary grass on nitrous oxide emissions. *Biomass Bioenergy* **2019**, *120*, 396–403. [CrossRef]

61. Spyreas, G.; Wilm, B.W.; Plocher, A.E.; Ketzner, D.M.; Matthews, J.W.; Ellis, J.L.; Heske, E.J. Biological consequences of invasion by reed canary grass (*Phalaris arundinacea*). *Biol. Invasions* **2010**, *12*, 1253–1267. [CrossRef]

62. Hersperger, A.M. Spatial adjacencies and interactions: Neighborhood mosaics for landscape ecological planning. *Landsc. Urban Plan.* **2006**, *77*, 227–239. [CrossRef]

63. Englund, O.; Börjesson, P.; Mola-Yudego, B.; Berndes, G.; Dimitriou, I.; Cederberg, C.; Scarlat, N. Beneficial Land-Use Change in Europe: Deployment Scenarios for Multifunctional Riparian Buffers and Windbreaks. 2021. Available online: https://www.researchsquare.com/article/rs-128604/v1 (accessed on 10 May 2021). [CrossRef]

64. Englund, O.; Dimitriou, I.; Dale, V.H.; Kline, K.L.; Mola-Yudego, B.; Murphy, F.; English, B.; McGrath, J.; Busch, G.; Negri, M.C.; et al. Multifunctional perennial production systems for bioenergy: Performance and progress. *Wiley Interdiscip. Rev. Energy Environ.* **2020**, *9*, e375. [CrossRef]

65. Mander, Ü.; Järveoja, J.; Maddison, M.; Soosaar, K.; Aavola, R.; Ostonen, I.; Salm, J.O. Reed canary grass cultivation mitigates greenhouse gas emissions from abandoned peat extraction areas. *GCB Bioenergy* **2012**, *4*, 462–474. [CrossRef]

*Article*

# Utilization of Fish Farm Effluent for Irrigation Short Rotation Willow (*Salix alba* L.) under Lysimeter Conditions

Ildikó Kolozsvári [1,*], Ágnes Kun [1], Mihály Jancsó [1], Beatrix Bakti [2], Csaba Bozán [1] and Csaba Gyuricza [3]

[1] Research Center of Irrigation and Water Management, Institute of Environmental Sciences, Hungarian Agricultural and Life Science University, Anna Liget u. 35., H-5540 Szarvas, Hungary; Kun.Agnes@uni-mate.hu (Á.K.); Jancso.Mihaly@uni-mate.hu (M.J.); Bozan.Csaba@uni-mate.hu (C.B.)

[2] Department of Plantation Forestry, Forest Research Institute, University of Sopron, Farkasszigeti u. 3., H-4150 Püspökladány, Hungary; Bakti.Beatrix@uni-sopron.hu

[3] Institute of Agronomy, Hungarian Agricultural and Life Science University, Páter Károly utca 1., H-2100 Gödöllő, Hungary; Gyuricza.Csaba@uni-mate.hu

* Correspondence: Kolozsvari.Ildiko@uni-mate.hu; Tel.: +36-70-440-3187

**Abstract:** Efficient utilization, treatment, and disposal of agricultural wastewater and sewage sludge are important environmental risks. In our research, effluent water from intensive aquaculture was evaluated for the irrigation of short rotation energy willow in a lysimeter experiment. Two different water types and their combinations were applied with weekly doses of 15, 30, and 60 mm, respectively. Our results revealed that implementing effluent water instead of fresh water could potentially increase the yield of the willow due to its higher nitrogen content (29 N mg/L). The biomass of irrigated short rotation coppice (SRC) willow plants were between 493–864 g/plant, 226–482 g/plant, and 268–553 g/plant dry weight during experiment period (2015–2017), respectively. However, due to the chemical properties (Na concentration, SAR value) of effluent water, the increase of the soil exchangeable sodium percentage (ESP) was significant and it can lead to soil degradation in the long term. The current study also investigated the relationship between chemical composition of the plant tissue and the irrigation water. In the case of K-levels of willow clones, an increasing trend was observed year-by-year. In terms of N and Na content was localized in leaf parts, especially in samples irrigated with effluent. Less N and Na values were detected in the stem and in the samples irrigated with surface water. In SRC willow plants, phosphorus was mostly localized in the stem, to a lower extent in the leaf part. The difference is mostly observed in the case of the amount of irrigation water, where the P content of the examined plant parts decreased with the increase of the amount of irrigation water. In the case of phenological observations, higher values of plant height were measured during diluted and effluent irrigation. Moreover, the SPAD of the plants irrigated with effluent water exceeded the irrigated ones with surface water.

**Keywords:** effluent water treatment; short rotation coppice willow; irrigation; growth response; biomass crops; mineral content

**Citation:** Kolozsvári, I.; Kun, Á.; Jancsó, M.; Bakti, B.; Bozán, C.; Gyuricza, C. Utilization of Fish Farm Effluent for Irrigation Short Rotation Willow (*Salix alba* L.) under Lysimeter Conditions. *Forests* **2021**, *12*, 457. https://doi.org/10.3390/f12040457

Academic Editors: Ronald S. Zalesny Jr. and Andrej Pilipovic

Received: 26 February 2021
Accepted: 6 April 2021
Published: 9 April 2021

**Publisher's Note:** MDPI stays neutral with regard to jurisdictional claims in published maps and institutional affiliations.

## 1. Introduction

Efficient utilization, treatment, and disposal of increasing amounts of wastewater and sewage sludge are major environmental risks in these times. Approximately 4.5 million hectares of arable lands are used for agricultural production in Hungary. Of which, the size of arable lands where the cultivation of traditional crops (wheat, corn, sunflower) is unprofitable can be estimated at 100,000 hectares [1]. These areas are generally waterlogged, prone to inland excess water formation—which is a temporary water inundation on the agricultural lands due to the heavy rainy activities, sudden snow melting, and heavy soil textures with limited water permeability [2]. The cultivation of fast-growing and short rotation woody crops is possible on all types of soils used for agricultural cultivation [3].

Short rotation coppice (SRC) willow species have significant growth potential and biomass product among them [4].

Most of the approximately 400 species of fast-growing and short rotation woody crops live in the northern half of the Earth, the *Salix* is a characteristic woody genus of the Holarctic realm. Most of them grow on the alluvium soils of streams and riverbanks. White willow (*S. alba* L.) is a tree of fluvisol soils along the slow-flowing waters of the greater plains. It reaches its climatic optimum in the steppe, wooded steppe, and semi-desert belt. It is a heat-demanding, hygrophytic species that prefers high soil moisture. It also endures prolonged drought when roots reach groundwater. For rapid growth, it requires flooding in early summer. It also tolerates permanent summer flooding. In this case, it develops respiratory roots in the submerged section of its trunk [5]. It grows properly on loose alluvial soil, but it is characterized by more delayed initial development on bound clay soil [6].

Nowadays, the use of biomass energy is appreciating again, thus, it is expected to maintain its share of almost 10% of world energy consumption in the future. The simplest version of bioenergy utilization is the energy use of biomass in its original state or close to its original state. With this in mind, the use of forest and agricultural crops and by-products suitable for direct combustion, as well as woody and herbaceous energy crops from the various biomasses, for the purpose of the most favorable heat and electricity production and bioethanol production [7,8]. *Salix* spp. can be used for manufacturing, energy production, and medicinal purposes [9]. In case of willow cultivation, yield of up to 25–30 t/ha/year (10–15 t/dry weight/year) can be achieved. As willow tolerates poor air/water conditions of soils well, it would be considered to be the most suitable for regularly flooded areas (floodplains, inland excess water hazarded areas). SRC willow clones can be harvested annually or every 2–3 years in the same crop area, where their cultivation can remain profitable for up to 15–20 years [10]. In this case, it is necessary to ensure nutrient replenishment in order to establish good soil condition and achieve a satisfactory biomass yield [10,11].

In order to reduce the eutrophication of surface water, irrigation utilization of wastewater can provide an alternative solution. SRC plants, as biofilters, can save water, reduce the high organic matter content, micro and macro elements of effluents, especially the concentrations of N and P. Wastewater irrigation provides an opportunity to apply a lower dose of fertilizers or even to eliminate conventional nutrient replenishment as well [12].

The aim of our research was to determine the N and Na content of the soil, determination of phenological parameters and the amount of macroelements (N, K, P) and Na found in the plant parts. Additionally our goal to determine the biomass of energy willow plants in addition to the applied treatments.

## 2. Materials and Methods

### 2.1. Study Site and Climatic Conditions

The experiment was carried out at the Lysimeter Research Station (46°51′49″ N 20°31′39″ E Szarvas, Hungary) of the Hungarian Agricultural and Life Science University (HUALS), Institute of Environmental Sciences (IES), Research Center of Irrigation and Water Management (ÖVKI). Sixty-four non-weighing lysimeters (1 m$^3$) were used to determine the effect of effluent water irrigation on the development of willow clones. The lysimeters were 1 m deep and 1 m$^2$ in surface. The soil of the lysimeter is not stratified disturbed soil, where the soil properties in lysimeters where clay texture, 0.08% total salinity, 0.41% calcium carbonate, and 1.172% carbon content. At the bottom of all lysimeters, a 10 cm layer of fine gravel was placed for the collection of leachate water.

The climate of Hungary is influenced by continental and oceanic effects, the specific area of the experimental site is described as warm dry climate region. Based on long term local data (1981–2010), the mean annual air temperature is 10.8 °C (Table 1), while the mean temperatures in July and January are 21.9 °C and −1.0 °C, respectively. The average annual precipitation is 515.3 mm. The meteorological data during the three years experiment

(2015–2017) were collected by an automatic weather station maintained by the HUALS ÖVKI in Szarvas. Its distance to the Lysimeter Station is 600 m. In 2015, lower values of temperature were recorded only in October. While in 2016 and in 2017, May, August, October, December, and January were colder, respectively. The year of 2015 was dry, the total precipitation was only 400.6 mm while in the year of 2016, 633 mm was measured. However, the distribution was heterogeneous which a dry spring and a very wet early summer characterized. In 2017, the amount of precipitation was close to the average.

**Table 1.** Meteorological data of 2015–2017 during the irrigation period.

| | Average Temperature (°C) | | | | Precipitation Amount (mm) | | | |
|---|---|---|---|---|---|---|---|---|
| | **1981–2010** | **2015** | **2016** | **2017** | **1981–2010** | **2015** | **2016** | **2017** |
| January | −1.0 | 2.2 | −0.9 | −6.7 | 29.1 | 58.8 | 61.6 | 28.3 |
| February | 0.5 | 2.4 | 6.0 | 2.6 | 29.9 | 17.3 | 88.5 | 30.2 |
| March | 5.6 | 7.4 | 7.3 | 9.4 | 27.8 | 25.5 | 20.0 | 13.4 |
| April | 11.5 | 11.5 | 13.4 | 11.0 | 42.0 | 8.2 | 12.3 | 49.7 |
| May | 16.8 | 17.1 | 16.6 | 17.2 | 50.6 | 53.7 | 18.8 | 40.9 |
| June | 19.8 | 21.2 | 21.3 | 22.1 | 61.3 | 21.0 | 124.4 | 69.3 |
| July | 21.9 | 24.4 | 22.5 | 22.8 | 57.5 | 31.4 | 123.6 | 31.8 |
| August | 21.4 | 24.2 | 21.1 | 23.7 | 50.7 | 40.9 | 50.5 | 33.3 |
| September | 16.6 | 18.7 | 18.3 | 16.6 | 47.8 | 64.0 | 9.8 | 74.2 |
| October | 11.2 | 10.4 | 9.8 | 11.6 | 32.4 | 105.2 | 72.7 | 33.7 |
| November | 5.0 | 6.3 | 5.0 | 6.0 | 41.3 | 3.2 | 49.6 | 39.6 |
| December | 0.3 | 2.6 | −1.2 | 2.7 | 44.8 | 4.5 | 1.0 | 89.2 |
| Average/Summa | **10.8** | **12.4** | **11.6** | **11.6** | **515.3** | **433.7** | **633.6** | **533.6** |

### 2.2. The Plant Material

The SRC willow (*Salix alba* L.) 'Naperti' candidate variety of the National Agricultural Research and Innovation Centre, Forest Research Institute (NARIC FRI), Department of Plantation Forestry was planted into the lysimeters in 2014. Eight lysimeter containers were used for one treatment. Two plants were planted into each lysimeter with 50 cm of plant spacing and 100 cm of row spacing. In order to reduce the edge effects, additional willow clones were planted around the containers with the same plant and row spacing. The first cutting took place in December 2015, the second in January 2017, and the third harvest in January 2018.

### 2.3. Experimental Design for Effluent Water Irrigation

Two different water types and their combinations were applied for the irrigation experiment of the energy willow clones. Untreated effluent water from a local intensive African catfish farm was used directly collected from the outflow of fish rearing tanks with the weekly doses of 15 mm (E15), 30 mm (E30), and 60 mm (E60) during the vegetation season in eight replications, respectively (Table 2). The flow-through system of the fish tanks is supported by a geothermal well from a confined aquifer. This system has the main role of temperature and water quality maintenance since the African catfish are fed high-protein diet and need warm water (above 16 °C) to grow. The daily amount of effluent water from the farm exceeds 1000 m$^3$. That effluent water contains large amount of metabolites as fish feces, organic materials and rarely chemicals or antibiotics depending on the fish rearing technology [13]. Because of the geothermal origin, the effluent water also carrying high content of total salinity including high percent sodium (Table 3). The type of water meets the classification of sodium hydrogen carbonate.

As an irrigated control treatment, freshwater was applied from the local oxbow lake of the River Körös (46°51′38.6″ N 20°31′28.0″ E, Szarvas, Hungary). Irrigation schedule was planned as weekly doses of 15 mm (K15), 30 mm (K30), and 60 mm (K60) in eight replications, respectively (Table 2). Additionally, a non-irrigated control (C) treatment was also applied with eight replications.

**Table 2.** The amount of irrigation water applied per year and distribution of precipitation during studied vegetation period.

| | Irrigation Water Doses | Possibility of Irrigation during the Investigated Period | Amount of Water Applied by Irrigation (mm) | Precipitation during the Investigated Period (mm) | Amount of Available Water Quantity during the Investigated Period (mm) |
|---|---|---|---|---|---|
| 2015 | 15 mm | 15 * | 310 | 105 | 415 |
| | 30 mm | | 520 | | 625 |
| | 60 mm | | 940 | | 1045 |
| 2016 | 15 mm | 6 | 90 | 308 | 398 |
| | 30 mm | | 180 | | 488 |
| | 60 mm | | 360 | | 668 |
| 2017 | 15 mm | 9 | 135 | 184 | 319 |
| | 30 mm | | 270 | | 454 |
| | 60 mm | | 540 | | 724 |

* During the first irrigation, each treatment received uniformly 100 mm of River Körös irrigation water.

**Table 3.** Average major quality parameters of irrigation water used under experiment.

| | EC ($\mu$S/cm) | NH4-N (mg/L) | N (mg/L) | P (mg/L) | K (mg/L) | Na (mg/L) | SAR |
|---|---|---|---|---|---|---|---|
| Effluent water | 1306.7 | 21.9 | 29 | 3.9 | 7.2 | 273.5 | 11.9 |
| Körös River water | 388.3 | 0.4 | 1.2 | 0.2 | 4.3 | 31.3 | 1.2 |
| Diluted water | 1073.0 | 10.3 | 13.3 | 1.7 | 5.4 | 132.3 | 3.5 |

To reduce the negative effects of this high salinity, effluent water was pretreated and it was used with the weekly dose of 60 mm (diluted treatment D) after a pretreatment. Before irrigation, effluent water was diluted four times (1:3) by adding river water to meet the recommended upper limit of total salinity in irrigation water (500 mg/L). Moreover, gypsum (calcium sulfate) was also added (312 mg/L) to reduce the percent sodium for diluted treatment. A micro-sprinkler irrigation system was used for all irrigation treatment.

Soil samples were collected before the first irrigation on 3 July 2015 and after the last irrigation period on 5 October 2017 from all treatments from three soil depths (0–20 cm, 20–40 cm, 40–60 cm) with three replications. Soil analyses were made according to Hungarian standards for five parameters: plant available nitrogen, calcium, potassium, magnesium and exchangeable sodium. The available nitrogen content of the soil was characterized by the sum of the nitrite and nitrate contents of the soil ($KCL\text{-}NO_2^- + NO_3^-\text{-}N$). Nitrite and nitrate were extracted with potassium chloride and the concentration was measured using FIA spectrophotometer (according to Hungarian standard MSZ 20135:1999). Exchangeable cations (K, Na, Ca, Mg) were extracted with barium-chloride and triethanolamine and their concentrations were measured using flame photometer (according to Hungarian standard MSZ-08-0214-2:1978).

From the results of the analyses of soil exchangeable basis, the exchangeable sodium percentage (ESP) and its changes during the experiment was calculated according to the equation

$$\text{ESP (exchangeable sodium percentage, \%)} = (Na/(Na + K + Ca + Mg)) \times 100$$

where, $Na^+$, $K^+$, $Ca^{2+}$ and $Mg^{2+}$—concentrations are expressed in milliequivalents per 100 g of soil [14].

$$\Delta ESP_{2015\text{-}2017} \text{ (exchangeable sodium percentage, \%)} = ESP_{2017} - ESP_{2015}$$

*2.4. Determination of Phenological Paramteres and Mineral Content*

From the plant phenology measurements, plant height (measuring rod) and SPAD values (Konica Minolta SPAD-502) were measured on a weekly basis during the growing seasons. For height measurement, three plants were selected per treatment and the current value was determined from their average. During the determination of the SPAD value, it was also generated from the average of 3-3 measurements. In this case, we analyzed the chlorophyll content of the leaves of the lower, middle, and upper branches. The mineral content of the plant parts was analyzed at the end of the growing season. In all cases assayed by the Hungarian and ISO standard methods. For the determination of the sodium, phosphorus, and potassium were extracted with nitric acid+hydrogen peroxide and their concentrations were measured using Inductively coupled plasma-optical emission spectrometry ICP-OES (according to Hungarian standard MSZ 08 1783 28-30:1985 Use of high capacity equipment in plant analyses—quantitative determination of sodium, phosphorus, and potassium content in plant materials by the ICP methods) and at nitrogen applied (ISO 5983 2:2009 Determination of nitrogen content and calculation of crude protein content. Part 2: Block digestion and steam distillation method) methods. In the analytical studies, we worked with six replicates.

### 2.5. Statistical Analyses

Statistical analyses were implemented by IBM SPSS Statistics 25.0 software. Applying one-way analysis of variance (ANOVA), we examined the effect of irrigation water quality and quantity on the phenological and important content properties of willow clones per treatment and plant part. The differences were determined significant, where the Tukey's or Games-Howell tests were considered significant at $p \leq 0.05$ or $p \leq 0.01$. In soil chemical studies during the statistical evaluation, independent $t$-test was used for the 15- and 30-mm irrigated samples, and ANOVA (2) test was used for the 60 mm samples (treatment with 60 mm 3 irrigation water quality were applied and compared). Pearson correlation is used in correlation analysis.

## 3. Results

### 3.1. Results of Chemical Analyses—Changing of Sodium and Nitrogen Content of Soil

The effect of irrigation water quality on the exchangeable sodium content of the soil can be proved in each soil depth and irrigation water amount, (despite 15 mm in 40–60 cm soil layer which may be due to excessive variance) (Table 4). The increasing sodium content in the soil due to the high sodium concentration of the reused water was demonstrably dependent on the amount of irrigation water, the highest $\Delta ESP_{2015-2017}$ (+6.85%) was measured in E60 treatment in the surface layer. This statement is also true for Körös River water, but in the case of K60 the change is already negative, which means that the exchangeable Na content of the soil decreased as a result of irrigation. Examining the results measured in different depths of soil layers, we found that sodium accumulated to a lesser extent in the deeper soil layers compared to the surface layers in treatments irrigated with reused water (Table 4), however it can be proved only in case of $\Delta ESP_{2015-2017}$ value of E30 treatment between 0–20 cm and 40–60 cm, ($n = 3$, $p = 0.041$, independent $t$-test). As a result of the improvement of irrigation water, it was possible to reduce the increase of the Na content of the soil in all soil depths (Table 4).

In accordance with the nitrogen content of the effluent water, the available nitrogen content of the soil was higher in all treatment irrigated with reused water than in treatment irrigated with Körös River water (Table 5). At treatment with 30 and 60 mm irrigation water amount differences between the available N values according to the irrigation water quality were statistically proved (Table 5). Comparing the improved water quality (7.52 mg/kg) to Körös River Water (2.96 mg/kg) higher available N values were detected and differences were proved.

**Table 4.** Changes of the exchangeable sodium adsorption ratio between 2015 (before experiment) and 2017 (after experiment).

| Depth of Soil Layer | Irrigation Water | Exchangeable Sodium Percentage ΔESP (2015–2017) | | | | | | |
| --- | --- | --- | --- | --- | --- | --- | --- | --- |
| | | Irrigation Water Amount | | | | | | |
| | | 15 mm | $p$-Value [1] | 30 mm | $p$-Value [1] | 60 mm | $p$-Value [2] | Non-Irrigated |
| | | Mean ± Std. Deviation | | | | | | |
| 0–20 cm | Effluent water | 4.66 ± 0.6 | *** | 5.9 ± 0.77 | *** | 6.85 ± 0.10 c | *** | Non-irrigated 0–20 cm: 0.36 ± 0.2 |
| | Körös River water | 0.05 ± 0.1 | | 0.5 ± 0.35 | | −0.62 ± 0.16 a | | |
| | Diluted water | - | - | - | - | 2.19 ± 0.30 b | | |
| 20–40 cm | Effluent water | 2.85 ± 1.1 | * | 3.5 ± 1.10 | ** | 5.82 ± 0.64 c | *** | Non-irrigated 20–40 cm: 0.33 ± 0.1 |
| | Körös River water | 0.14 ± 0.1 | | 0.4 ± 0.36 | | −0.68 ± 0.08 a | | |
| | Diluted water | - | - | - | - | 1.85 ± 0.45 b | | |
| 40–60 cm | Effluent water | 1.02 ± 0.8 | n.s. | 1.8 ± 0.05 | *** | 4.38 ± 0.74 c | *** | Non-irrigated 20–40 cm: 0.32 ± 0.4 |
| | Körös River water | 0.02 ± 0.2 | | 0.4 ± 0.09 | | −0.53 ± 0.23 a | | |
| | Diluted water | - | - | - | - | 1.19 ± 0.13 b | | |

Comment: * $p < 0.05$; ** $p < 0.01$; *** $p < 0.001$. The negative values means the decrease during three experimental year. For each treatment, soil sampling was collected from three (0–20, 20–40, 40–60 cm) depth levels. During the statistical evaluation, an independent $t$-test ([1]) was used for the 15 mm and 30 mm irrigated samples, and an ANOVA ([2]) test was used for the 60 mm samples. Results are means ± SD, $n = 3$. Different letters introduce significant difference confirming to the Tukey's post hoc test at $p \leq 0.05$.

**Table 5.** Available nitrogen (KCl-$NO_2^-$+$NO_3^-$-N) content of soil in 2017.

| Irrigation water | Available $N_{2017}$ (mg/kg) | | |
| --- | --- | --- | --- |
| | Irrigation water amount | | |
| | 15 mm | 30 mm | 60 mm |
| | Mean ± Std. Deviation | | |
| Effluent water | 13.43 ± 7.71 a | 16.65 ± 4.04 b | 15.46 ± 3.29 c |
| Diluted water | - | - | 7.52 ± 3.85 b |
| Körös River water | 7.02 ± 3.85 a | 3.65 ± 0.78 a * | 2.96 ± 0.28 a * |
| Non-irrigated control | | 11.70 ± 4.53 | |

For each treatment, soil sampling was collected from three (0–20, 20–40, 40–60 cm) depth levels. During the statistical evaluation, an independent $t$-test (1) was used for the 15 and 30 mm irrigated samples, and an ANOVA (2) test was used for the 60 mm samples. Results are means ± SD, $n = 9$ (the values of the samples from different depths were considered as repetitions of each other). Different letters introduce significant difference among different treatment, confirming to the Tukey's test at $p \leq 0.05$ at 60 mm. The stars are indicated the significant difference from the non-irrigated control ($p < 0.01$).

*3.2. Phenological Results*

### 3.2.1. Changes of Relative Chlorophyll Content

The Figure 1a shows that the values of the energy willow SPAD means ranged from 42.3 to 47.5 during the first growing season. The highest value was measured for E60 treatment. The smallest SPAD values were recorded in the K15 treatment. With the exception of treatment K15, all treatments exceeded the SPAD values of treatment C. In the following growing year (Figure 1b) it can be seen that the chlorophyll values of the irrigated treatments did not reach the control. The highest SPAD mean was 51.5 for the control, while the lowest for K15 treatment was 45.3. In addition, except for the first year, compared to Körös River irrigated treatments, higher values were found in the samples irrigated with effluent water. This difference is due to the excess nutrient content of the effluent.

This trend is no longer observed in 2017 (Figure 1c). Compared to control SPAD values, samples irrigated with surface water show a lower rate. In this case, the K30 plants had the lowest chlorophyll means (43.5), while the E30 treatment had the highest value (49.0).

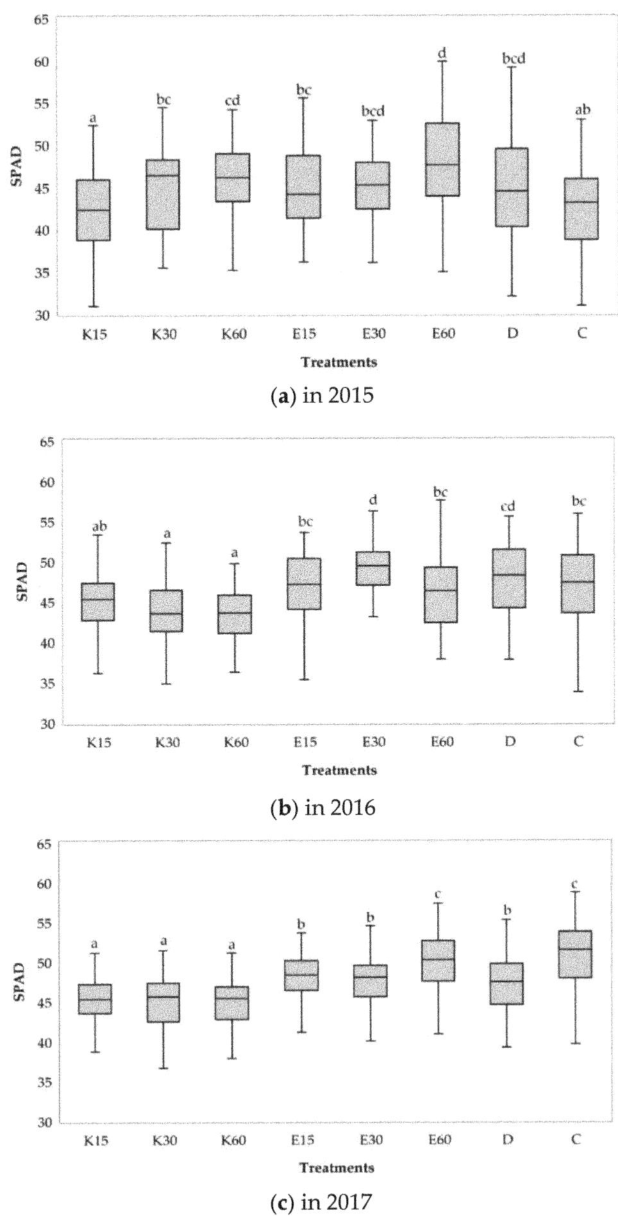

(**a**) in 2015

(**b**) in 2016

(**c**) in 2017

**Figure 1.** Chlorophyll values from 2015 to 2017 growing years: (**a**) SPAD values of *S. alba* energy willow coppice in a year of 2015; (**b**) SPAD values of *S. alba* energy willow coppice in a year of 2016; (**c**) SPAD values of *S. alba.* energy willow coppice in a year of 2017. Average chlorophyll content data are presented from eight treatments. Results are means $\pm$ SD, $n = 6$. Different letters introduce significant differences among irrigation water qualities for the three vegetation season, in year of 2015 and 2016 confirming to the Tukey's test at $p \leq 0.05$, and in year of 2017 corroborating to the Games-Howell test at $p \leq 0.05$.

During the one-way analysis of variance in the 2015 production year, the SPAD values of the willow K15 treatment differed significantly from the chlorophyll values of the other irrigated treatments (K30 $p = 0.021$, K60 $p = 0.001$, E15 $p = 0.030$, E30 $p = 0.003$, E60 = 0.000, D $p = 0.019$). Significantly higher SPAD values were recorded for treatments E60 ($p = 0.000$), D ($p = 0.000$), and K60 ($p = 0.022$) compared to treatment C. In year of 2016, the SPAD values of the treatments showed a significant difference. There was a strong significant value ($p < 0.001$) between the control and the irrigated treatments, where, with the exception of E60 treatment, plants contained significantly less chlorophyll (Figure 1b). In 2017, the SPAD values of plants irrigated with K60 treatment showed a significant difference compared to the other treatments, except for the K15 and K30 samples. In the case of E60 treatment the $p = 0.003$, they had a significantly higher ($p < 0.001$) chlorophyll value with respect to the ones listed above. Comparing the irrigated treatments to the control values, it can be observed that the E30 ($p = 0.042$) treatment had a significantly higher SPAD value, while the K30 ($p = 0.001$) and K60 ($p = 0.000$) treatments had significantly lower chlorophyll values.

### 3.2.2. Growth of Test Plants during the Seasons

Data from the 2015 cultivation year show that the values of irrigated treatments exceeded the height of the control plants (Figure 2a). From the data measured on 17 September, it can be concluded that at the last measurement the tallest energy willows grew to height of 428 cm with D treatment, while the control plants reached 282 cm. During the examined period, compared to the first measurement at the last the control plants grew by 34 cm while in E60 treatment by 178 cm. Throughout the last height determination, in the one-way analysis of variance of the data, all treatments proved to be significantly higher ($p = 0.00$) compared to the control plants ($n = 6$, Tukey's test).

In Figure 2b, plant height values for 2016 developed similarly as in 2015. Throughout the time of last measurement, control proved to be the lowest (289 cm), while treatment D showed the highest (413 cm) values. Following the analysis period, it can be observed that the control willows increased by 75 cm and the plants irrigated with 60 mm effluent water by 176 cm. In a year of 2016 during the final measurement compared to control plants in height, irrigated treatments grew significantly higher ($p < 0.001$, $n = 6$, Tukey's test).

In the 2017 growing year, a slowdown in productivity growth was detected (Figure 2c). The growth rate of control willows is lower than that of treated plants, where by the end of the analysis only 259 cm had been reached. The highest plants in this case were also observed in E60 treatment (370 cm). In the analysis of variance in 11 September, there was a significant difference between the control and the irrigated treatments ($p = 0.000$, $n = 6$, Tukey's test).

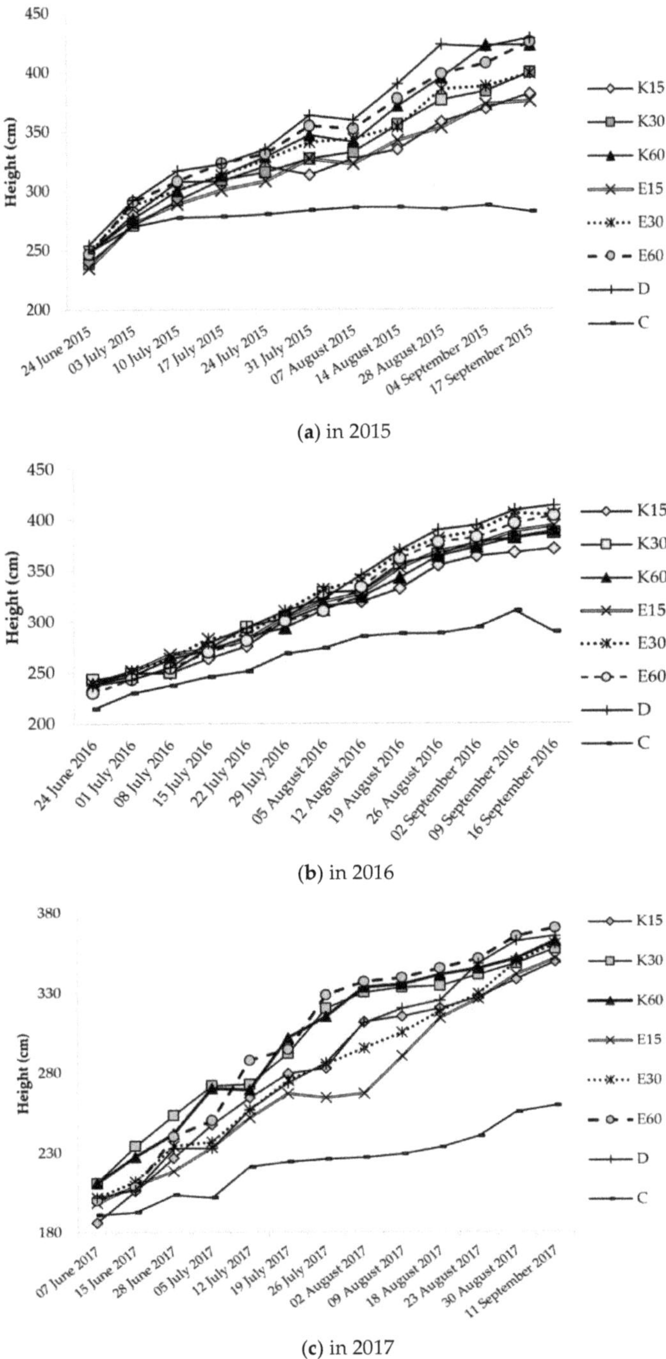



**Figure 2.** Plant height values from 2015 to 2017 growing years: (**a**) height values of *S. alba* energy willow coppice in a year of 2015; (**b**) height values of *S. alba* energy willow coppice in a year of 2016; (**c**) height values of *S. alba* energy willow coppice in a year of 2017.

### 3.3. Results of Mineral Content

3.3.1. Changing of Nitrogen Content in Plant Parts

When comparing the treatments in the 2015 growing year (Figure 3a), it can be seen that the N content measured in the leaf part was significantly higher in the E60 treatment than in the E15, D, K30, and K60 treatments. The highest N (3.5 m/m%) content was found in willows E30 treatment while the lowest values were measured in the K60 (1.6 m/m%) treatment. Leaves contained significantly more N in E60 ($p$ = 0.015), E30 ($p$ = 0.000), K15 (0.005), and C ($p$ = 0.001) treatments than in K60 treatment. In addition, in the case of the stem, the K60 treatment had the lowest N content, where the samples irrigated with effluent, K15 and C contained significantly more N.

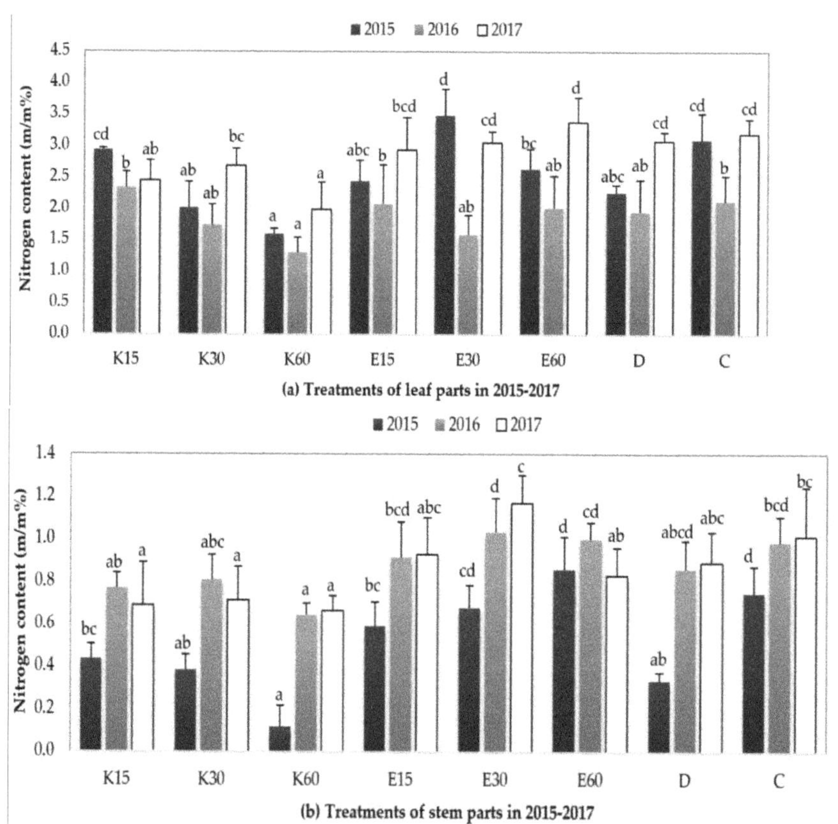

**Figure 3.** Nitrogen content values of leaf and stem parts from 2015 to 2017: (**a**) nitrogen content values of leaf parts in a year of 2015–2017; (**b**) nitrogen content values of stem parts in a year of 2015–17. Average nitrogen content data are presented from eight treatments. Results are means ± SD, $n$ = 6. Treatments were compared annually. Different letters introduce significant differences among irrigation water qualities for the three vegetation season, confirming to the Tukey's test at $p \leq 0.05$.

In 2016, the N content measured in the leaves showed a decrease. The E15 ($p$ = 0.018), K15 ($p$ = 0.002), and C ($p$ = 0.029) leaf samples had significantly more N content than the cases of willows irrigated with K60 treatment. In the case of the stem, it can be seen (Figure 3b) that the values moved in almost the same range in 2016 and 2017. The measurement results of both years exceeded the N content measured in 2015. In the third experimental year—in the case of samples D, C, K30 and irrigated with effluent water—it can be stated that significantly more N was stored in the leaves of willows than in the case of those irrigated with K60 treatment.

### 3.3.2. Changing of Potassium Content in Plant Parts

In the macronutrient analysis of plant parts of SRC willow clones, most K was concentrated in the leaves (Figure 4a). The comparison of the annual data shows that we measured the lowest K content in 2015 and the highest in 2017. In the case of leaves part at the first year, the K value ranged from 11,880 to 15,465 mg/kg d.m., while in the second year, the measured element content was 11,445–18,492 mg/kg d.m., and finally in 2017, 18,187 and 21,627 mg/kg d.m. were detected.

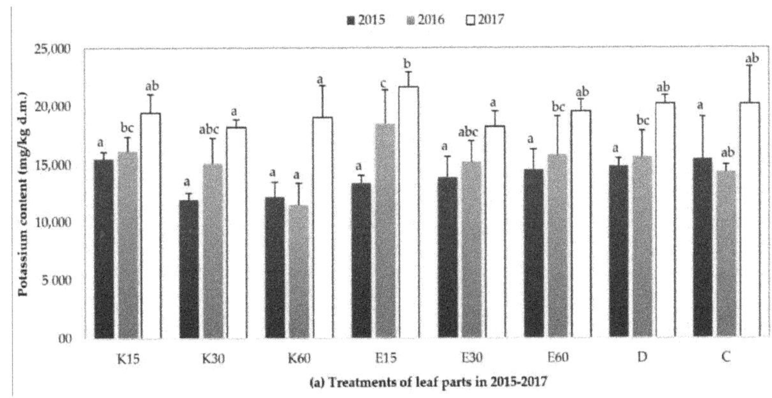


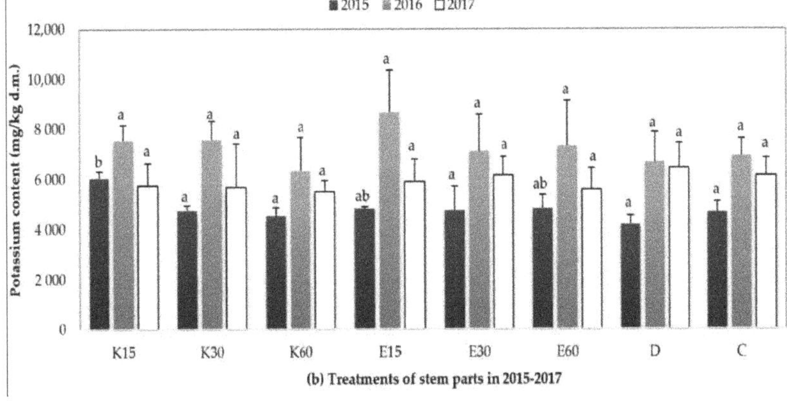


**Figure 4.** Potassium content values of leaf and stem parts from 2015 to 2017: (**a**) potassium content values of leaf parts in a year of 2015–2017; (**b**) potassium content values of stem parts in a year of 2015–2017. Average potassium content data are presented from eight treatments. Results are means ± SD, $n$ = 6. Different letters introduce significant differences among irrigation water qualities for the three vegetation season, confirming to the Tukey's test at $p \leq 0.05$.

It should be noted that in the leaf samples irrigated with E15, significant increase in K level was observed in the last two years of the experiment. During the annual Tukey's multiple comparisons, there were no significant differences between the treatments in 2015; however, in the second year of the study, compared to the K60 treatment values, E15 ($p$ = 0.000), E60 ($p$ = 0.023), D ($p$ = 0.034) and K15 ($p$ = 0.010) leaf samples had significantly higher K levels. Furthermore, in 2017, E15 ($p$ = 0.013) had significantly higher K levels compared to data from K30 samples. In the case of stems part, the same trend is observed as for the leaves. However, the K level of the stem parts was very high in 2016, where the E15 treatment reached 8640 mg/kg d.m. value. In the first and last irrigation years, the K content of the stem parts of SRC willow clones ranged from about 4100 to 6400 mg/kg d.m (Figure 4b). During the one-way analysis of variance, there was a significant difference between the 2015 measurement data. Compared to D analysis, treatments of K15 ($p = 0.001$) had significantly higher K content. There was no detectable significant difference in the other two years.

### 3.3.3. Changing of Sodium Content in Plant Parts

In the first experimental year, the Na content measured in the leaf parts of the test plant ranged from 49 to 79 mg/kg d.m. (Figure 5a). The lowest value was measured for treatment D, while the highest value was detected for sample E30 (Figure 5b). In 2016 and 2017 growing years, the Na level in the leaf parts was similar, where the lower values were recorded by the Körös River water-irrigated samples and the higher values by the effluent irrigation. Statistical analysis in a second year at leaf parts showed significant difference between K15 and K60 treatments ($p = 0.025$).

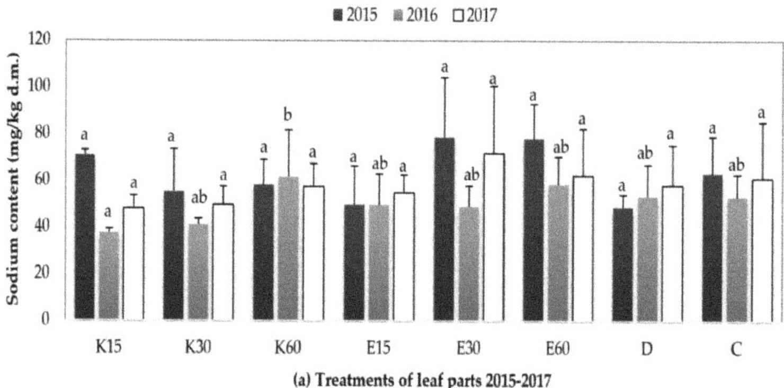


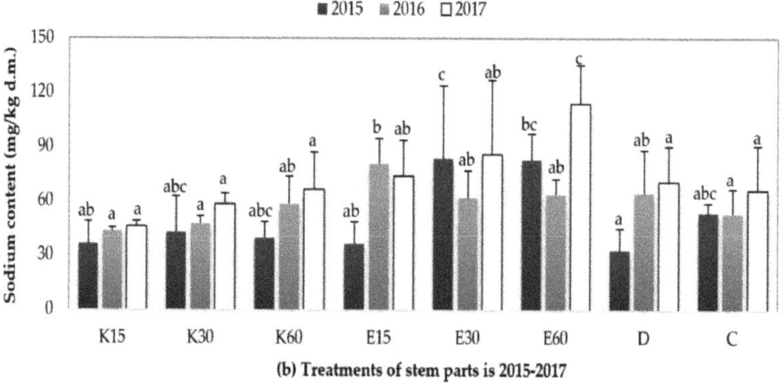


**Figure 5.** Sodium content values of leaf and stem parts from 2015 to 2017 growing years: (**a**) sodium content values of leaf parts in a year of 2015–2017; (**b**) sodium content values of stem parts in a year of 2015–2017. The year of 2015 average sodium content data are presented from eight treatments. Results are means ± SD, $n = 3$. The letters introduce significant differences among irrigation water qualities for the three vegetation season, confirming to the Tukey's test at $p \leq 0.1$. The year of 2016 and 2017 average sodium content data are presented from six samples per treatment. Results are means ± SD, $n = 6$. Different letters introduce significant differences among irrigation water qualities for the three vegetation season, confirming to the Games-Howell's test at $p \leq 0.05$.

In case of the stems, it can be observed that, except for 2015, the lowest Na level was measured in the D treatment, while the highest value was analyzed in the E 30 samples (Figure 5b). At the same time, it can be discovered that the Na content of the stem parts shows an increasing trend from year to year, especially of the samples irrigated with effluent water. The values measured in 2017 are remarkable, where the Na content of the E60 samples reached 114 mg/kg d.m., which is 137% higher than in 2015. In the first study year the one-way analysis of variance E15, D, and K15 treatments contained significantly less ($p < 0.1$) Na compared to the values measured in E30 treatment. During the one-way analysis of variance, significantly lower Na levels were detected in the second vegetation period for stem samples with K15 and K30 treatments. This trend can also be observed in 2017, where the stems of the clones also contained significantly less Na than the samples irrigated with oxbow lake water (15, 30, 60 mm doses).

### 3.3.4. Changing of Phosphorus Content in Plant Parts

The P content of the effluent and diluted water irrigated leaves of the willows in 2015 and 2017 growing years moved between in range 1990 and 3023 mg/kg d.m. In 2015, control had significantly more P content than treatments E60 ($p = 0.004$) and D ($p = 0.001$) (Table 6). In 2016, most of the P content was detected in the samples irrigated D treatment, where significantly lower P levels were observed in the leaves of E30 ($p = 0.046$) and C ($p = 0.043$) samples. In the last experimental year, significantly less P content was detected for the control treatment ($p = 0.033$) compared to the D irrigation. Concerning the P element content measured in the stem part of the energy willows in the three vegetation years, it is identifiable that the measured level was between 813 and 2457 mg/kg dm. In the first year, compared to the control value, the P content was significantly lower in the D ($p = 0.000$) and E30 ($p = 0.005$) treatments.

**Table 6.** Phosphorus content measured in the plant parts of SRC willow clones irrigated with effluent water from an intensive catfish farm. Average phosphorus content data are presented from five treatments. Results are means $\pm$ SD, $n = 6$. Different letters introduce significant differences among irrigation water qualities for the three vegetation seasons, confirming to the Tukey's test at $p \leq 0.05$.

| | | E15 | E30 | E60 | D | C |
|---|---|---|---|---|---|---|
| 2015 | | $3023 \pm 241$ b | $2737 \pm 95$ ab | $2180 \pm 370$ a | $1990 \pm 144$ a | $3340 \pm 419$ b |
| 2016 | *leaf* | $2643 \pm 57$ ab | $1865 \pm 210$ a | $2007 \pm 519$ ab | $2850 \pm 365$ b | $1855 \pm 219$ a |
| 2017 | | $2428 \pm 19$ ab | $2272 \pm 127$ ab | $2532 \pm 196$ ab | $2723 \pm 118$ b | $2123 \pm 66$ a |
| 2015 | | $1537 \pm 35$ c | $1050 \pm 221$ ab | $1330 \pm 193$ bc | $813 \pm 79$ a | $1647 \pm 146$ c |
| 2016 | *stem* | $2192 \pm 201$ bc | $2010 \pm 172$ ab | $1788 \pm 244$ a | $2457 \pm 201$ c | $1740 \pm 146$ a |
| 2017 | | $1616 \pm 209$ a | $1596 \pm 169$ a | $1422 \pm 266$ a | $2003 \pm 239$ b | $1693 \pm 90$ ab |

In the second year, significantly less P content was measured in the stem part of SRC willows compared to D irrigation in the E30 ($p = 0.004$), E60 ($p = 0.000$), and control ($p = 0.000$) treatments. In 2017, with the exception of treatment D, the stem samples of the clones contained significantly fewer P elements.

### 3.4. Biomass Changing over the Three Years of the Experiment

The willows had the highest biomass in 2015, where it reached 864 g/plant dry weight in case of K60 (Figure 6). It can be observed that in all three experimental years the biomass product of the control plants became the lowest. Furthermore, the decreasing trend that occurred in the crop year by year is clearly visible.

In terms of control by experimental years, this resulted in a 56% yield reduction. This decrease was due to the physical limitations of the lysimeters. Namely, the volume of 1 m$^3$ restrained the root growth of the two- and three-year-old willows (Figure 6). At the same time, the trend that irrigation had a positive effect on biomass compared to control values is appeared. It can be observed that each year the plants treated with effluent water had an average higher g/plant dry weight value. In 2015 it was 554–734 g/plant, in 2016 it was

298–482 g/plant), and the last year the data show 313–447 g/plant dry weight. While in the case of those irrigated with Körös River water the harvested dry weight.

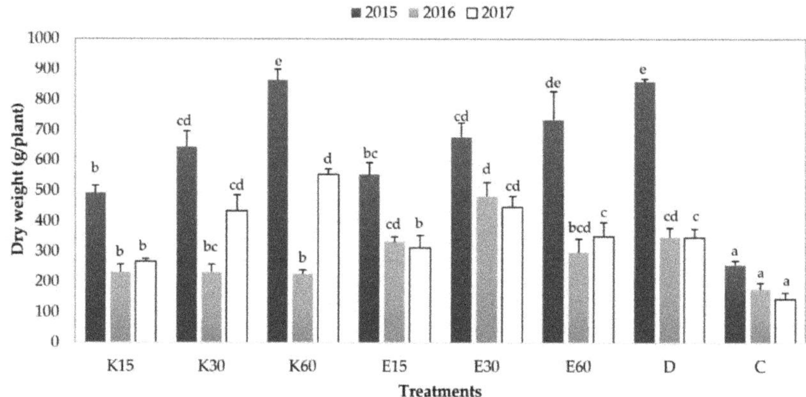


**Figure 6.** Biomass dry weight of short rotation willow coppice from 2015 to 2017. Average dry weight data are presented from eight treatments after harvesting. Results are means $\pm$ SD, $n = 8$. Different letters introduce significant differences among irrigation water qualities for the three vegetation seasons, confirming to the Games-Howell's test at $p \leq 0.05$.

During statistical study the examining the differences between the treatments in each experimental year, it can be stated that compared to the biomass production of the control plants, all irrigated treatments had significantly higher ($p = 0.000$) product.

## 4. Discussion

We investigated the irrigation utilization of effluent water from an intensive African catfish farm in short-rotation energy willow plants in 2015–2017. Within agricultural water expection can provide an ideal solution for the conservation of water resources, as the irrigation utilization of nutrient-rich effluent from freshwater aquaculture systems can be used in many plantations [15,16]. At the same time, the organic matter load of natural recipients and the doses of fertilizer applied during cultivation can also be reduced [17]. Dhawan and Sehdev [18] described in their research that irrigation cultivation experiments with effluent from fish farms show higher yields.

As expected, the sodium content of the soil irrigated with reused water was increased in all treatments and depths (Table 4). Similarly, Jahany and Rezapour [19] stated that the high values of exchangeable Na and ESP in the treated effluent water irrigated soil could be associated with the chemistry of the effluent water used. According to Jahany and Rezapour [19], conditions were probably favorable to Na accumulation on the exchange complexes because of the high amounts of $Na^+$ and $HCO_{3-}$ supplied by effluent water, the combined effects of the increase in bicarbonate from irrigation and evapotranspiration process are likely to help the depletion of $Ca^{+2}$ and $Mg^{+2}$ ions as insoluble carbonates (such as calcite and magnesite) while the more soluble Na remains in the solution and subsequently, results in the over-accumulation of exchangeable Na as well as an increase in the ESP values. In order to reduce sodium increase in soil, in the reused water gypsum was applied as amendments to improve its quality after diluted it with Körös River water in our experiment. According to the results, the sodium accumulation was reduced compared to water from catfish farm due to improved water quality (Table 4). These results in accordance with previous works recording that dilution of irrigation water [20–22] or amendments (mainly materials with calcium) [23–25] for irrigated soil could be appropriate solution to reuse effluent water for irrigation.

The beneficial effect of the reused water on the available N content of the soil can be demonstrated. In addition to the nitrogen concentration of the water from fish farm, the effect of irrigation on nitrogen mineralization may also cause of the increase of the mineral [26]. All this advocates its agricultural reuse thus, irrigation supports water retention and water conservation and helps to protect surface waters from nutrient loading. One of the reasons for the high N content observed in the control soil (compared to the treatments irrigated with Körös River) (Table 5) may be caused by reduced nutrient uptake due to the lower biomass, which was limited by water and/or tree roots habitat [27]. The dilution of the reused water also resulted in a significant increase in nitrogen in the soil (Table 5).

Chlorophyll content is one of the indicators that can provide information on the health status of a plantation. Examining the chlorophyll content of the foliage of the test plants gives a more accurate picture of the changes caused by natural and anthropogenic stressors, as these affect the amount of chlorophyll. The change in the nitrogen content of the plant is also reflected in the chlorophyll content of the leaves. For this reason, a linear relationship is observed between the chlorophyll content and the nitrogen content of the leaves [28,29]. The change in our plant nitrogen content is reflected (r = 0.351, Pearson correlation) in the chlorophyll content of the leaves. Consequently, a linear relationship is observed between the SPAD value of the leaves and the nitrogen content [30]. The SPAD of the plants irrigated with effluent water exceeded the irrigated ones with Körös river water. However, it can be perceived that the quality of irrigation water also influenced this value. In the three years of irrigation, it can be observed that the leaves of willows irrigated with 60 mm of effluent water had the highest chlorophyll content.

The data show that the height of the willows has been decreasing year by year. At the same time, irrigation had a positive effect on plant growth, as we measured higher values in the latest measurement. In 2015, the plants reached 428 cm for treatment D, in 2016 the willows for treatment D were also the highest at 414 cm, and in the last year we measured the highest height data in E60 at 370 cm. Comparing the average highest and lowest plant height data, the difference between plant stands was 141 cm in 2015, 124 cm in 2016, and 120 cm in 2017, which also appears in the biomass product.

The trend is also observed in the N-level of willow leaves. Nitrogen stress is always reflected in the chlorophyll content of the leaves, because in general the chlorophyll content of the leaves is linearly related to the N content of the leaves [31,32]. It also serves as a reliable result for woody plants [33,34]. Furthermore, in the case of effluent irrigation, higher N concentrations are observed in the plant parts.

Potassium is the most abundant cation in plants. Plants actively accumulate large amounts of potassium, and are able to absorb significant amounts even from small concentrations of solution. It is found in greater amounts in meristems in organs with a vigorous metabolism. The K content of older organs decreases. The K-demand and K-content of plants thus change during the vegetation period [35]. In the case of K-levels measured in the plant parts of willow clones, an increasing trend can be observed every year. At the same time, higher element content is more characteristic in the leaf parts, the reason for which can be explained by the $Na^+/K^+$ ratio [36].

Sodium does not specifically activate many enzymes, in which K elements can be substituted; however, the effect of K is specific [37]. The $C_4$ plants require microelement amounts of Na; however, it does not cause deficiency symptoms in $C_3$ plants (SRC) [38]. Sodium is not essential even for extreme halophytes, only required in microelement amounts by $C_4$ and CAM-type plants. Sodium becomes toxic to glycophytons when translocated into the sprout in significant amounts [39]. The development of the Na concentration of the plants was closely monitored, as significant amount Na is released into the area through the effluent water of the intensive African catfish farm. In the case of willow clones irrigated with effluent water, the Na content was most localized in the stem parts, during which an increase from year to year can be observed. The Na content of 114 mg/kg d.m. measured in the E60 treatment in 2017 is remarkable, which was 50% higher than the values measured in the control. However, this amount did not prove to be toxic for SRC plants.

The effluent with a higher P content had correlated (r = −0.579, Pearson correlation) negative effect on the P content of the plants. In SRC willow plants, phosphorus was mostly localized in the stem, to a lower extent in the leaf part. The difference is mostly observed in the case of the amount of irrigation water, where the P content of the examined plant parts decreased with the increase of the amount of irrigation water.

Our hypothesis that irrigation has a positive effect on biomass product has been confirmed. For both irrigation water qualities, the biomass product of non-irrigated control SRC willow clones exceeded [40,41]. Under lysimeter conditions, the biomass and irrigation water quality do not correlate with each other. However, some decrease in biomass production is observed from year to year. This reduction was due to the limited living space, as the 1 m$^3$ vessel size of the lysimeters proved to be small over the years [42]. Although, the limited water supply caused a significant decrease in the biomass product. In the first year the biomass product of irrigated SRC willow plants was between 493–864 g/plant dry weight, in the second year 226–482 g/plant dry weight, in the third year 268–553 g/plant dry weight. Which is 170–250% higher than the average yield of the non-irrigated control.

## 5. Conclusions

The main objective of our study is the yield-enhancing effect of the irrigation utilization of effluent water from an intensive African catfish breeding farm in a short-cut energy willow plantation.

The experiment shows that N content of the effluent provably increased the nitrogen supply of the soil. These provides confidence that application of alternative water sources as irrigation water may reduce the nutrient load on surface waters and at could increase soil quality at the place of use. At the same time, it must be pointed out that the increase of the soil ESP due to the effluent water chemical properties (Na concentration, SAR value) was significant and in the long term it can lead to soil degradation (anthropogenic salinization). Nevertheless, it is suggested to improve the effluent water quality instead of ignore it, because an interesting finding of the present work was that improving the quality of the effluent water (by dilution and adding gypsum) is an effective method to reduce soil sodium accumulation.

Assessing to SPAD values and plant heights of the willow based on irrigation water quality it can be concluded that the sodium content of the effluent water do not cause any harmful effect on plants. In the leaf part of the willow, more N and Na were measured in the tissues than in stems and plant irrigated with effluent water had higher N and Na concentration than those irrigated with surface water. Further research may be required to examine the long term effect of the irrigation water salinity on the elements of the plant to explore the cause of the from year to year decline in biomass independently from the restricted habitat due to lysimeters volume.

The application of irrigation water had a positive effect on the biomass of the plants, significantly higher biomass was produced compared to the non-irrigated control willows.

In summary, the agronomic consequences are that alternative waters can provide an excellent opportunity for water-scarce regions, but paying attention to water quality parameters that limit use (salinization of the soil, nitrate leaching from the soil, salt stress of the cultivated plants etc.) is important.

**Author Contributions:** Conceptualization, I.K., Á.K., B.B., C.B., and M.J.; Data curation, I.K., Á.K., Formal analysis, I.K.; Funding acquisition, C.B.; Supervision, C.G., Methodology, I.K., Á.K., C.B. and M.J.; Writing—original draft preparation, I.K.; Á.K.; Writing—review and editing, I.K., Á.K. and M.J. All authors contributed critically to the drafts and gave final approval for publication. All authors have read and agreed to the published version of the manuscript.

**Funding:** This work was supported by the Hungarian Ministry of Agriculture under Project no. OD001 and by the National Research, Development and Innovation Office (GINOP-2.3.3-15-2016-00042 pro-ject), and by the Ministry of Innovation and Technology within the framework of the Thematic Excellence Programme 2020, Institutional Excellence Subprogramme (TKP2020-IKA-12). We gratefully acknowledge the staff of ÖVKI Lysimeter Station.

**Institutional Review Board Statement:** Not applicable.

**Informed Consent Statement:** Not applicable.

**Data Availability Statement:** The data presented in this study are available on request from the corresponding author. The data are not public, as this study is part of a forthcoming PhD dissertation.

**Conflicts of Interest:** The authors declare no conflict of interest.

## References


1. Barczi, A.; Joó, K.; Pető, Á.; Bucsi, T. Survey of the buried paleosoil under the Lyukas-mound in Hungary. *Eurasian Soil Sci.* **2006**, *39*, 133–140. [CrossRef]
2. Bosco, C.; de Rigo, D.; Dewitte, O.; Poesen, J.; Panagos, P. Corrigendum to modelling soil erosion at European scale: Towards harmonization and reproducibility. *Nat. Hazards Earth Syst. Sci.* **2015**, *15*, 225–245. [CrossRef]
3. Mikó, P.; Kovács, G.P.; Alexa, L.; Balla, I.; Póti, P.; Gyuricza, C. Biomass production of energy willow under unfavourable field conditions. *Appl. Ecol. Environ. Res.* **2014**, *12*, 1–11. [CrossRef]
4. Hanley, S.J.; Karp, A. Genetic strategies for dissecting complex traits in biomass willows (*Salix* spp.). *Tree Physiol.* **2014**, *34*, 1167–1180. [CrossRef]
5. Gencsi, L.; Vancsura, R. *Dendrology*; Mezőgazda Publisher: Budapest, Hungary, 1992. (In Hungarian)
6. Koop, H. Vegetative reproduction of trees in some European natural forests. *Vegetatio* **1987**, *72*, 103–110.
7. Blaskó, L.; Balogh, I.; Ábrahám, É.B. Possibilities of sweet sorghum production for ethanol on the Hungarian plain. *Cereal Res. Commun.* **2008**, *36*, 1251–1254.
8. Szalay, D.; Papp, V.; Hodúr, C.; Czupy, I. Examination of bark content for different species of short rotation coppice. In Proceedings of the IOP Conference Series: Earth and Environmental Science, 5th International Conference on Environment and Renewable Energy, Ho Chi Minh City, Vietnam, 25–28 February 2019; Volume 307. [CrossRef]
9. Shara, M.; Stohs, J.S. Efficacy and safety of white willow bark (Salix alba) extracts. *Phytother. Res.* **2015**, *29*, 1112–1116. [CrossRef]
10. Gyuricza, C.; Nagy, L.; Ujj, A.; Miko, P.; Alexa, L. The impact of composts on the heavy metal content of the soil and plants in energy willow plantations (*Salix* sp.). *Cereal Res. Commun.* **2008**, *36*, 279–282. Available online: https://www.jstor.org/stable/90002695 (accessed on 12 March 2021).
11. Smart, B.L.; Cameron, K.D. Shrub willow. In *Handbook of Bioenergy Crop Plants*; Kole, C., Joshi, C.P., Shonnard, D.R., Eds.; CRC Press: Boca Raton, FL, USA; London, UK; New York, NY, USA, 2012; pp. 687–708.
12. Aronsson, P.; Perttu, K. Willow vegetation filters for wastewater treatment and soil remediation combined with biomass production. *For. Chonicle* **2001**, *77*, 293–299. [CrossRef]
13. Tóth, F.; Zsuga, K.; Kerepeczki, É.; Nagy-Berzi, L.; Körmöczi, L.; Lövei, L.G. Seasonal differences in taxonomic diversity of rotifer communities in a Hungarian lowland oxbow lake esposed to aquaculture effluent. *Water* **2020**, *12*, 1300. [CrossRef]
14. Richards, L.A. Diagnosis and improvement of saline and alkali soils. *Agric. Handb.* **1954**, *60*. [CrossRef]
15. Castro, R.S.; Borges Azevedo, C.M.S.; Bezerra-Neto, F. Increasing cherry tomato yield using fish effluent as irrigation water in Northeast Brazil. *Sci. Hortic.* **2006**, *110*, 44–50. [CrossRef]
16. Miranda, F.R.; Lima, R.N.; Crisóstomo, L.A.; Santana, M.G.S. Reuse of inland low-salinity shrimp farm effluent for melon irrigation. *Aquac. Eng.* **2008**, *39*, 1–5. [CrossRef]
17. Al-Jaloud, A.A.; Hussain, G.; Alsadon, A.A.; Siddiqui, A.Q.; Al-Najada, A. Use of aquaculture effluent as a supplemental source of nitrogen fertilizer to wheat crop. *Arid Soil Res. Rehabil.* **1993**, *7*, 233–241. [CrossRef]
18. Dhawan, A.; Sehdev, R.S. Present status and scope of integrated fish farming in the north-west plains of India. In *Integrated Fish Farming*; Mathias, J.A., Charles, A.T., Baotong, H., Eds.; CRC Press: Boca Raton, FL, USA; New York, NY, USA, 1994; pp. 295–306.
19. Jahany, M.; Rezapour, S. Assessment of the quality indices of soils irrigated with treated wastewater in a calcareous semi-arid environment. *Ecol. Indic.* **2020**, *109*, 105800. [CrossRef]
20. Malash, N.; Flowers, T.J.; Ragab, R. Effect of irrigation systems and water management practices using saline and non-saline water on tomato production. *Agric. Water Manag.* **2005**, *78*, 25–38. [CrossRef]
21. Yu, Y.; Wen, B.; Yang, Y.; Lu, Z.H. The effects of treated wastewater irrigation on soil health. *Adv. Mater. Res.* **2011**, *393–395*, 1545–1549. [CrossRef]
22. Shilpi, S.; Seshadri, B.; Sarkar, B.; Bolan, B.; Lamb, D.; Naidu, R. Comparative values of various wastewater streams as a soil nutrient source. *Chemosphere* **2018**, *192*, 272–281. [CrossRef]
23. Purves, D. Waste materials deliberately added to the soil. In *Trace-Element Contamination of the Environment*; Elsevier: Amsterdam, The Netherlands, 1977; Volume 93.

24. Hopkins, B.G.; Horneck, D.A.; Stevens, R.G.; Ellsworth, J.W.; Sullivan, D.M. *Managing Irrigation Water Quality for Crop Production in the Pacific Northwest*; Technical Report; Oregon State University Extension Service: Corvallis, OR, USA, 2007. Available online: https://catalog.extension.oregonstate.edu/sites/catalog/files/project/pdf/pnw597.pdf (accessed on 8 April 2021).
25. Sheoran, P.; Basak, N.; Kumar, A.; Yadav, R.K.; Singh, R.; Sharma, R.; Kumar, S.; Singh, R.K.; Sharma, P.C. Ameliorants and salt tolerant varieties improve rice-wheat production in soils undergoing sodification with alkali water irrigation in Indo–Gangetic Plains of India. *Agric. Water Manag.* **2021**, *243*, 106492. [CrossRef]
26. Truua, M.; Truua, J.; Heinsoo, K. Changes in soil microbial community under willow coppice: The effect of irrigation with secondary-treated municipal wastewater. *Ecol. Eng.* **2009**, *35*, 1011–1020. [CrossRef]
27. Kun, Á.; Bozán, C.; Oncsik, B.M.; Barta, K. Calculation nitrogen and sodium budget from lysimeter-grown short-rotation willow coppice experiment. *Columella* **2018**, *5*, 43–51. [CrossRef]
28. Carter, G.A. Ratios of leaf reflectances in narrow wavebands as indicators of plant stress. *Int. J. Remote Sens.* **1994**, *15*, 697–703. [CrossRef]
29. Yoder, B.J.; Pettigrew-Crosby, R.E. Predicting nitrogen and chlorophyll content and concentrations from reflectance spectra (400–2500 nm) at leaf and canopy scales. *Remote Sens. Environ.* **1995**, *53*, 199–211. [CrossRef]
30. Peng, Y.; Gitelson, A.A. Application of chlorophyll-related vegetation indices for remote estimation of maize productivity. *Agric. For. Meteorol.* **2011**, *151*, 1267–1276. [CrossRef]
31. Niinemets, U.; Tenhunen, J.D. A model separating leaf structural and physiological effects on carbon gain along light gradients for the shade-tolerant species Acer saccharum. *Plant Cell Environ.* **1997**, *20*, 845–866. [CrossRef]
32. Evans, J.R. Photosynthesis and nitrogen relationships in leaves of C3 plants. *Oecologia* **1989**, *78*, 9–19. [CrossRef]
33. Pinkard, E.A.; Patel, V.; Mohammed, C. Chlorophyll and nitrogen determination for plantation-grown Eucalyptus nitens and E. globulus using a non-destructive meter. *For. Ecol. Manag.* **2006**, *223*, 211–217. [CrossRef]
34. Chang, S.X.; Robinson, D.J. Nondestructive and rapid estimation of hardwood foliar nitrogen status using the SPAD-502 chlorophyll meter. *For. Ecol. Manag.* **2003**, *181*, 331–338. [CrossRef]
35. Geirth, M.; Mäser, P. Potassium transporters in plants—Involvement in K+ acquisition, redistribution and homeostasis. *FEBS Lett.* **2007**, *581*, 2348–2356. [CrossRef]
36. Freitas, S.W.; Oliveira, B.A.; Mesquita, O.R.; Carvalho, H.H.; Prisco, T.J.; Gomes-Filho, E. Sulfur-induced salinity tolerance in lettuce is due to a better P and K uptake, lower Na/K ratio and an efficient antioxidative defense system. *Sci. Hortic.* **2019**, *257*, 108764. [CrossRef]
37. Nieves-Cordones, M.; Al Shiblawi, F.R.; Sentenac, H. Roles and transport of sodium and potassium in plants. In *The Alkali Metal Ions: Their Role for Life*; Sigel, A., Sigel, H., Sigel, R., Eds.; Springer: Cham, Switzerland, 2016; Volume 16, pp. 291–324. [CrossRef]
38. Maathuis, F.J.M. Sodium in plants: Perception, signalling, and regulation of sodium fluxes. *J. Exp. Bot.* **2014**, *65*, 849–858. [CrossRef] [PubMed]
39. Kronzucker, H.J.; Coskun, D.; Schulze, M.L.; Wong, J.R.; Britto, T.D. Sodium as nutrient and toxicant. *Plant Soil* **2013**, *369*, 1–23. [CrossRef]
40. Jerbi, A.; Brereton, N.J.B.; Sas, E.; Amiot, S.; Lacsapelle-T, X.; Comeau, Y.; Pitre, F.E.; Labrecque, M. High biomass yield increases in a primary effluent wastewater phytofiltration are associated to altered leaf morphology and stomal size in Salix miyabeana. *Sci. Total Environ.* **2020**, *738*, 139728. [CrossRef] [PubMed]
41. Aasamaa, K.; Heinsoo, K.; Holm, B. Biomass production, water use and photosynthesis of Salix clones grown in a wastewater purification system. *Biomass Bioenergy* **2010**, *34*, 897–905. [CrossRef]
42. Oddiraju, V.G.; Beyl, C.A.; Barker, P.A.; Stutte, G.W. Container size alters root growth of western black cherry as measured via image analysis. *Hort. Sci.* **1994**, *29*, 910–913. [CrossRef]

*Article*

# Short-Rotation Willows as a Wastewater Treatment Plant: Biomass Production and the Fate of Macronutrients and Metals

Darja Istenič [1,2,*] and Gregor Božič [3]

[1] Faculty of Health Sciences, University of Ljubljana, Zdravstvena pot 5, 1000 Ljubljana, Slovenia
[2] Faculty for Civil and Geodetic Engineering, University of Ljubljana, Jamova cesta 2, 1000 Ljubljana, Slovenia
[3] Slovenian Forestry Institute, Večna pot 2, 1000 Ljubljana, Slovenia; gregor.bozic@gozdis.si
* Correspondence: darja.istenic@zf.uni-lj.si

**Abstract:** Evapotranspirative willow systems (EWS) are zero-discharge wastewater treatment plants that produce woody biomass and have no discharge to surface or groundwater bodies. The influence of wastewater on the growth of three clones of *Salix alba* ('V 093', 'V 051' and 'V 160') and the distribution of macronutrients and metals in a pilot EWS receiving primary treated municipal wastewater was studied under a sub-Mediterranean climate. The influent wastewater, shoot number, stem height, and biomass production at coppicing were monitored in two consecutive two-year rotations. Soil properties and the concentrations of macronutrients and metals in soil and woody biomass were analyzed after the first rotation. *S. alba* clones in EWS produced significantly more woody biomass compared to controls. 'V 052' produced the highest biomass yield in both rotations (38–59 t DM ha$^{-1}$) and had the highest nitrogen and phosphorus uptake (48% and 45%) from wastewater. Nitrogen and phosphorus uptake into the harvestable woody biomass was significantly higher in all clones studied compared to other plant-based wastewater treatment plants, indicating the nutrient recovery potential of EWS. The indigenous white willow clone 'V 160' had the lowest biomass yield but absorbed more nutrients from wastewater compared to 'V 093'. Wastewater composition and load were consistent with the nutrient requirements of the willows; however, an increase in salinity was observed after only two years of operation, which could affect EWS efficiency and nutrient recovery in the long term.

**Keywords:** evapotranspirative willow system; resource recovery; sustainable wastewater treatment; short rotation coppice

**Citation:** Istenič, D.; Božič, G. Short-Rotation Willows as a Wastewater Treatment Plant: Biomass Production and the Fate of Macronutrients and Metals. *Forests* **2021**, *12*, 554. https://doi.org/10.3390/f12050554

Academic Editor: Ronald S. Zalesny Jr. and Andrej Pilipović

Received: 7 March 2021
Accepted: 26 April 2021
Published: 29 April 2021

**Publisher's Note:** MDPI stays neutral with regard to jurisdictional claims in published maps and institutional affiliations.

## 1. Introduction

The coupling of domestic wastewater treatment with short-rotation willow coppice (SRWC) biomass production originated in Scandinavia in the 1980s, when agricultural willow plantations producing woody biomass for energy purposes were recognized as a potential treatment system for domestic wastewater [1,2]. Untreated domestic wastewater and nutrient-rich effluents from central wastewater treatment plants (WWTPs) with secondary treatment resulted in pollution of surface and groundwater, while on the other hand the profitability of SRWC was reduced due to the expensive mineral fertilizer and water requirements. So-called vegetation filter systems consisting of SRWC irrigated with domestic wastewater [3–6], industrial wastewater [4,7] or landfill leachate [8,9] have been explored and applied.

However, the irrigation of SRWC with wastewater cannot completely eliminate wastewater pollution, as some of the applied wastewater runs off or seeps into the environment; therefore, special attention must be paid to irrigation rates and timing [5,10,11]. Moreover, such applications are not suitable for areas with sensitive ground and surface waters. In response, zero-discharge evapotranspirative willow systems (EWS) were developed in Denmark in the 1990s [12,13], and their field of application shifted from agricultural biomass production to the wastewater treatment sector.

EWS are a special type of WWTP used to treat domestic wastewater from small settlements or individual households. The system consists of a 1.5-m deep watertight basin filled with soil and planted with willow clones (*Salix* sp.). The primary treated domestic wastewater is applied under pressure through an inlet pipe located 0.6 m below the ground surface. When properly designed, all influent wastewater and precipitation are evapotranspired on an annual basis, i.e., all influent wastewater is used for willow growth and evaporation [12,13].

EWS are now used in rural areas in all Scandinavian countries, the Baltic States, Poland, England [13], Ireland [14], and China [15] and there are pilot studies in Slovenia [16], France and Greece [4]. EWS provide efficient wastewater treatment and do not require large amounts of energy or skilled personnel for operation and maintenance. As such, they are a suitable technology for decentralized sustainable wastewater management. As EWS are zero-discharge systems that generate woody biomass from wastewater, they allow the direct recovery and reuse of resources and support the goals of the closed-loop concept. In addition, zero-discharge wastewater treatment systems also contribute significantly to the reduction of surface and groundwater pollution.

The potential of EWS for nutrient recovery and the effects of wastewater on willow growth and biomass production require further research attention. Most scientific work focuses on wastewater-irrigated SRWC; however, there is much less scientific research on EWS. In addition, research on EWS pays much attention to evapotranspiration rates and the hydraulic loadings of the system [17–19], but nutrient recovery and the fate of heavy metals are rarely addressed [12]. However, numerous authors have shown that the composition of the wastewater corresponds to the nutrient requirements of willows [12,20,21].

The efficiency in the uptake of pollutants from wastewater into woody biomass, and their accumulation in system media, are critical for planning, management, system performance evaluation and resource recovery. There is a need to estimate the mass balance of nutrients in EWS, i.e., their percentage of harvestable woody biomass and accumulation in the soil compared to their amount in the influent wastewater. The level of nutrients entering the system with domestic wastewater is similar to the level of nutrients in the willow biomass: the proportional requirement of willow for N, P and K (100:14:72) is similar to the proportion of these nutrients usually found in municipal wastewater (100:18:65) [20]. The exception is P, which was reported to be 30% higher in influent wastewater than in biomass; however, the P balance also depends on the use of P-containing detergents in the household(s) producing the wastewater [12]. Consequently, a significant increase in available P in the soil was reported [22], suggesting that soil may become saturated with P after a period of time. This may lead to a problem in SRWC vegetation filter treatment performance and reduces P recovery via woody biomass. Therefore, Lachapelle-T et al. [22] suggest that fertigation should be adjusted according to seasonal transpiration rates and plant nutrient requirements; however, in the case of EWS, wastewater is constantly applied according to the production in the household(s) and stored in the EWS as an elevated water level over the winter. Therefore, in the case of EWS, P accumulation in the system can be expected, resulting in a P-rich substrate that can be reused as fertilizer.

Fertigation with wastewater significantly increases willow yield compared to commercial rainfed- or potable water-irrigated SRWC [21,22]. The differences in yield increase depend on the characterization of the wastewater and the loading rate. Similarly, Curneen and Gill [17] reported the highest biomass and evapotranspiration for willow cultivars receiving septic tank effluent, compared to systems fed with secondary treated effluent and rainwater. The higher biomass production of willows fertigated with wastewater is reflected in larger stem diameter and plant height compared to non-fertigated plants [23]. However, when wastewater is applied to SRWC, not only do the nutrients increase biomass yield, but also the constant water availability [21]. In addition, water use by willows has been shown to be positively affected by N and P application [24]; however, permanent flooding had negative effects on most growth parameters in willows, except for the number of shoots per plant and root biomass [25].

Total woody biomass production is comparable in SRWC irrigated with wastewater and EWS, varying from 10 [2,12] to 22–26 [22] t dry matter (DM) ha$^{-1}$ yr$^{-1}$ in Denmark, Sweden and Canada, while under Mediterranean climatic conditions, aboveground biomass production can reach up to 64 t DM ha$^{-1}$ in a two-year rotation [23], suggesting that climate may have a significant influence on system performance. In addition to climate, planting density, irrigation regimes, willow age and clone choice can also influence woody biomass production [5], while some authors find no differences between clones or irrigation regimes [26].

Plants used in EWS must have similar characteristics to those used in other phytotechnologies. They must be adapted to high nutrient and salinity levels that could increase in the system over time, and should have high transpiration rates, rapid growth and biomass production, high plantation densities, and coppicing ability. Willows and poplars have all these characteristics [27], supported also by their numerous and deep roots that provide a large root surface area [28]; however, willows show better tolerance to permanent flooding and anaerobic conditions than poplars [29,30], enabled by their important adaptations such as hypertrophied lenticels, aerenchyma, and adventitious roots [31]. Additionally, willows have higher water requirements than almost all agricultural crops [32], which is another important requirement for plants used in EWS.

Although willows are the first choice for this type of phytoremediation, different willow species and clones show different efficiencies in terms of growth, nitrogen and water use efficiency [33]. To date, most studies on EWS and SRWC for the treatment of municipal wastewater have been conducted in Europe and North America. In Denmark, the UK, Sweden, and Ireland, *Salix viminalis* has been mainly used, namely, the clones 'Jorr' (*S. viminalis*), 'Tora' (*S. viminalis* × *S. schwerinii* E. Wolf), and 'Bjørn' (*S. schwerinii* E. Wolf × *S. viminalis* L.) [11,12,18], while Rastas Amofah et al. [34], Sweden, used the frost-tolerant *S. viminalis* crossbreed 'Karin' ((*S. schwerinii* × *S. viminalis*) × *S. viminalis*) and 'Gudrun' (*S. burjatica* Nasarow × *S. dasyclados* Wimm). In Canada, research on an SRWC vegetation filter treating municipal wastewater was conducted on *S. miyabeana* 'SX67' [6,22]. *S. purpurea* [9,35], *S. amygdalina* [36] and other willow species were used to treat landfill leachate. *S. alba* var. 'Chermesina' was tested for phytoremediation in Poland [37], while *S. alba* clones used in this study were evaluated for their biomass production potential by Kajba and Andrić [38].

Concentrations of heavy metals in plant tissues and media can be correlated with heavy metal concentrations in the environment; however, most heavy metals are stored in roots, and transport to aboveground tissues may be limited [39,40]. Furthermore, typical domestic wastewater contains low levels of heavy metals [41]; therefore, elevated concentrations are not expected in the woody biomass of EWS. On the other hand, heavy metals entering the EWS may accumulate in the roots and soil media over the long term, which may affect reuse options after the decommissioning of the facility. Therefore, in this study, we analyzed the concentrations of heavy metals in both woody biomass and soil media to assess the risks of reusing the system components.

The objective of this study was to investigate the potential of EWS to recover nutrients from primary treated municipal wastewater through the production of woody biomass and accumulation in the soil. In addition, the growth dynamics, biomass production and response of selected *S. alba* clones to wastewater irrigation in the sub-Mediterranean climate were investigated, to provide data for the proper design and operation of EWS as wastewater treatment plants in rural areas of the sub-Mediterranean region. In addition, *S. alba* was not tested in EWS before.

## 2. Materials and Methods

### 2.1. Pilot Evapotranspirative Willow System

The study was conducted on a 27 m$^2$ pilot EWS built next to a municipal WWTP in Ajdovščina, Slovenia (45°52′32″ N 13°54′20″ E). A detailed description and illustration of the pilot plant can be found in Istenič et al. [16]. Briefly, the pilot EWS consisted of nine watertight treatment beds (each 3 m long, 1 m wide), filled with local soil (1.5 m deep) and planted with three clones of *S. alba* at a density of 1 tree per m$^2$. Each clone was tested in three parallel beds distributed in a Latin square to minimize environmental differences caused by positioning (north/south orientation, prevailing wind direction, etc.). In addition, control trees were planted around the EWS to avoid the edge effect and to be monitored as control plants.

Three *S. alba* clones from a selection of Croatian arborescent willows were tested, namely, two hybrids 'V 052' (*S. alba* L. var. *calva* G.F.W. Mey × *S. alba* L.) and 'V 093' (*S. alba* L. × *S. alba* var. *vitellina* (L.) Stokes) × *S. alba* L.) and one clone of the indigenous white willow 'V 160' (*S. alba* L.). The willows were provided as 1-year-old seedlings, planted and immediately cut back to 10 cm above ground level. The EWS was fed with primary treated municipal wastewater. The amount of water supplied was adjusted according to the water requirements of the willows—the water level in the treatment beds was maintained at approximately 1 m (0.5 m below the surface) during the monitoring period and was allowed to evapotranspire completely before the start of a new season. The pilot system was commissioned in March 2016 and monitored for four consecutive growing seasons until the end of the growing season in 2019. The monitoring period lasted 169, 115, 106 and 134 days, respectively, for the 4 consecutive years.

### 2.2. Wastewater and Soil Analyses

Grab samples of influent wastewater were taken weekly and analyzed for the typical parameters of municipal wastewater, namely, biochemical oxygen demand (BOD$_5$), chemical oxygen demand (COD), total phosphorus (TP), orthophosphate phosphorus (PO$_4$-P), total nitrogen (TN), ammonium nitrogen (NH$_4$-N), nitrite nitrogen (NO$_2$-N), nitrate nitrogen (NO$_3$-N), total suspended solids (TSS), settleable solids (SS), dissolved oxygen and oxygen saturation, temperature (T), pH and electrical conductivity (EC). Analyses followed the Standard methods for the examination of water and wastewater [42].

The mass loading of contaminants for each treatment bed in the EWS was calculated by multiplying the average contaminant concentrations in the wastewater by the total volume of water applied to each bed for each year, divided by the bed area. The mass loading is expressed in grams of added contaminants per m$^2$ of EWS.

A composite sample of the original fill soil was collected during the construction of the EWS. The grain size distribution was determined by sieving and sedimentation, according to SIST ISO 11277:2011, following USDA Textural Soil Classification [43] to define the soil texture class. After the first and second growing seasons, the soil was sampled according to the willow clones. Three samples were collected from each treatment bed using a soil probe and combined into one composite sample for each clone. Samples from the treatment beds and the sample from the original fill soil were analyzed for pH in 0.01 M CaCl$_2$ solution (SIST ISO 10390:2006), soil organic matter and organic carbon (SIST ISO 10694:1996), plant-available phosphorus (P$_2$O$_5$) and potassium (K$_2$O) (ÖNORM L 1087—modification with amonlactate extraction), total nitrogen (SIST ISO 13878:1999) and cation exchange capacity (CEC) (SIST ISO 13536:1996, modification by using KCl instead of BaCl$_2$). Ca, Mg, K and Na, as well as exchangeable acidity, the sum of base cations and base saturation were measured according to the Methods of Soil Analysis, ASA [44]. Heavy metal content was determined by extraction in aqua regia and analysis by ICP-MS.

### 2.3. Estimation of Willow Growth and Biomass Production

The number of shoots per stump was counted on each tree in the EWS and on control trees. Shoots were counted every month during the growing season; however, the number of shoots at the end of the growing season was used as the outcome. Stem height was measured every other week: the highest stem of each tree was measured with a wooden ruler. Mean and standard deviation were calculated for all control and test clones for each season.

The biomass production was measured at the 1st and 2nd harvest. All shoots from each test and control tree were harvested and weighed. Then, the shoots from each treatment bed were pooled and cut into woodchips. A sample of the woodchips was collected from each treatment bed for laboratory analysis of the moisture content. Based on the moisture content determined and the planting density (1 tree per $m^2$), the dry matter (DM) produced per hectare was calculated for each treatment bed. The mean and standard deviation of the beds with the same clone were calculated. Control trees were not grown in separate beds; therefore, all control trees of the same clone were combined into one sample.

The conversion of wastewater to biomass was calculated by dividing the DM produced in each treatment bed by the total amount of wastewater supplied to the bed during the 1st and 2nd rotation periods. The mean and standard deviation of all beds with the same clone were calculated.

### 2.4. Nutrient and Metal Content in the Woody Biomass

Samples of woodchips after the 1st rotation were further analyzed for carbon, nutrient and metal contents, namely, the total C content was analyzed according to SIST ISO 10694:1996, and total N content according to SIST ISO 13878:1999. P and the other 35 elements were analyzed by aqua regia digestion and ICP-MS.

The partitioning of P and N from wastewater (in $g\ m^{-2}$) between woody biomass and other compartments (accumulation in the root and leaf biomass, accumulation in the soil, and denitrification in the case of N) was studied by comparing the mass loads of TN and TP from wastewater, and the N and P content in the woody biomass produced (calculated by multiplying the N and P concentration in the dry matter by the total amount of biomass produced), while the N and P content in the other compartments together was calculated as the difference between the TN and TP, applied with the wastewater, and the content in the woody biomass. A similar procedure was also used by Lachapelle-T et al. [22].

### 2.5. Statistical Analyses

Microsoft Office Excel 2016 was used for statistical analysis of the data. The results are presented as mean and standard deviation of the mean. Significant differences were tested using one-way analyses of variance (ANOVA), with a significance level of 0.05 ($\alpha = 0.05$) between the mean values of shoot numbers in clones, test and control trees, and between the mean values of macronutrient and metal concentrations in control and test trees. When the results showed statistical significance, Student's *t*-test was used to further interpret the results and to show significant differences between the mean macronutrient and metal concentrations of different clones of the test trees.

## 3. Results

### 3.1. Wastewater Characteristics

The influent wastewater at the central WWTP in Ajdovščina is typical municipal wastewater with occasional elevated organic load ($BOD_5$ and COD), mainly originating from the food industry in the catchment area (Table 1). During the experiment, the influent wastewater also showed variable $NH_4$-N, $NO_2$-N and $NO_3$-N concentrations, which can be attributed to occasional nitrification in the primary clarifier from which the wastewater for the EWS was taken.

**Table 1.** Characteristics of municipal wastewater fed to the evapotranspirative willow system during four consecutive seasons.

| | $BOD_5$ | COD | TP | $PO_4$-P | TN | $NH_4$-N | $NO_3$-N | $NO_2$-N | TSS | SS | $O_2$ | $O_2$ | T | pH | EC |
|---|---|---|---|---|---|---|---|---|---|---|---|---|---|---|---|
| Unit | | | | | mg/L | | | | | | mL/L | mg/L | % | °C | | µS/cm |
| Average | 452 | 739 | 5.75 | 2.84 | 50.6 | 23.3 | 0.298 | 0.092 | 207 | 7.4 | 2.87 | 33.0 | 23.9 | 6.68 | 441 |
| SD | 182 | 294 | 2.45 | 2.13 | 14.4 | 10.5 | 0.792 | 0.307 | 161 | 10.3 | 2.47 | 27.0 | 4.8 | 0.58 | 482 |
| Nr | 40 | 40 | 40 | 40 | 36 | 36 | 36 | 36 | 20 | 14 | 19 | 19 | 22 | 39 | 39 |

The three clones of *S. alba* received different amounts of wastewater depending on water use by evapotranspiration; consequently, mass loading rates varied among rotation periods and clones, and according to the influent composition (Table 2). Mass loading rates for all parameters were lowest in the 1st year of the 1st rotation, because water use by the willows was lowest (willows at age 1/2; 1-year-old stem with 2-year-old root system). In the 2nd year, the root systems and trees were more developed, resulting in higher water use. Consequently, more wastewater was supplied, and mass loading rates increased. Mass loading rates also increased slightly after the first coppicing (aged 1/4) and remained in the same range in the 2nd year of the 2nd rotation (aged 2/5). In the 2nd rotation, 'V 160' received lower mass loading rates due to lower water demand.

**Table 2.** Annual mass loading rates of evapotranspirative willow system beds in g m$^{-2}$, according to the three willow clones tested ('V 093', V 052' and 'V 160') for four consecutive years. Mean and standard deviation are given ($N = 3$).

| | $BOD_5$ | COD | TP | $PO_4$-P | TN | $NH_4$-N | $NO_3$-N | $NO_2$-N |
|---|---|---|---|---|---|---|---|---|
| | | | | 1st rotation, (aged 1/2) | | | | |
| 'V 093' | 181 ± 9 | 278 ± 14 | 2.84 ± 0.15 | 1.50 ± 0.08 | 17.2 ± 0.9 | 11.1 + 0.1 | 0.023 ± 0.001 | 0.015 + 0.001 |
| 'V 052' | 183 ± 64 | 281 ± 98 | 2.87 ± 1.0 | 1.51 ± 0.53 | 17.4 ± 6.1 | 11.2 ± 3.9 | 0.023 ± 0.008 | 0.015 ± 0.005 |
| 'V 160' | 170 ± 21 | 262 ± 32 | 2.67 ± 0.32 | 1.41 ± 0.17 | 16.2 ± 2.0 | 10.4 ± 1.3 | 0.022 ± 0.003 | 0.014 ± 0.002 |
| | | | | 1st rotation, (aged 2/3) | | | | |
| 'V 093' | 392 ± 61 | 600 ± 93 | 8.62 ± 1.33 | 6.32 ± 0.98 | 53.6 ± 8.3 | 34.7 ± 5.4 | 0.278 ± 0.043 | 0.072 ± 0.111 |
| 'V 052' | 323 ± 114 | 496 ± 174 | 7.12 ± 2.51 | 5.21 ± 1.84 | 44.3 ± 15.6 | 28.6 ± 10.1 | 0.230 ± 0.081 | 0.060 ± 0.021 |
| 'V 160' | 326 ± 55 | 499 ± 84 | 7.17 ± 1.21 | 5.25 ± 0.89 | 44.6 ± 7.5 | 28.8 ± 4.9 | 0.231 ± 0.039 | 0.060 ± 0.010 |
| | | | | 2nd rotation, (aged 1/4) | | | | |
| 'V 093' | 485 ± 7 | 793 ± 11 | 6.17 ± 0.08 | 3.05 ± 0.04 | 54.3 ± 0.7 | 25.0 ± 0.3 | 0.319 ± 0.004 | 0.099 ± 0.001 |
| 'V 052' | 469 ± 54 | 766 ± 88 | 5.97 ± 0.69 | 2.95 ± 0.34 | 52.4 ± 6.1 | 24.2 ± 2.8 | 0.309 ± 0.036 | 0.095 ± 0.011 |
| 'V 160' | 406 ± 103 | 665 ± 169 | 5.17 ± 1.31 | 2.56 ± 0.65 | 45.5 ± 11.6 | 21.0 ± 5.3 | 0.268 ± 0.068 | 0.083 ± 0.021 |
| | | | | 2nd rotation, (aged 2/5) | | | | |
| 'V 093' | 592 ± 31 | 1019 ± 54 | 4.89 ± 0.26 | 1.30 ± 0.07 | 68.6 ± 3.6 | 18.8 ± 1.0 | 0.599 ± 0.032 | 0.182 ± 0.010 |
| 'V 052' | 619 ± 130 | 1064 ± 224 | 5.11 ± 1.08 | 1.35 ± 0.29 | 71.6 ± 15.1 | 19.7 ± 4.1 | 0.626 ± 0.132 | 0.190 ± 0.040 |
| 'V 160' | 471 ± 148 | 810 ± 254 | 3.89 ± 1.22 | 1.03 ± 0.32 | 54.5 ± 17.1 | 15.0 ± 4.7 | 0.476 ± 0.149 | 0.145 ± 0.045 |

*3.2. Willow Growth and Biomass Production*

As expected, the number of shoots per stump increased significantly in the 2nd rotation period, confirming that coppicing stimulates the formation of multiple shoots (Figure 1). In the 1st rotation period, the test willows of 'V 052' and 'V 160' had significantly more shoots compared to the control trees, while in the case of 'V 093' the situation was reversed, and the control trees produced more shoots. Moreover, the test trees of 'V 093' had significantly fewer shoots (5.8 ± 0.3) compared to 'V 052' and 'V 160' (7.6 ± 0.7 and 7.5 ± 0.7, respectively). In the 2nd rotation, the differences between the clones became more evident: 'V 052' produced 27 ± 7 shoots per stump, which was significantly more than 'V 093' and 'V 160', which produced 17 ± 5 and 17 ± 6 shoots per stump, respectively. All the clones developed significantly more shoots in the EWS compared to control trees. However, in the 2nd year of the 2nd rotation, the number of shoots reduced in all test clones. The greatest reduction was observed in clones 'V 160' and 'V 052' (27% and 25%, respectively), while the reduction in 'V 093' was only 12%. This reduction was not observed in control trees, which already had a lower number of shoots. In 2019, only one control tree of 'V 160' survived.

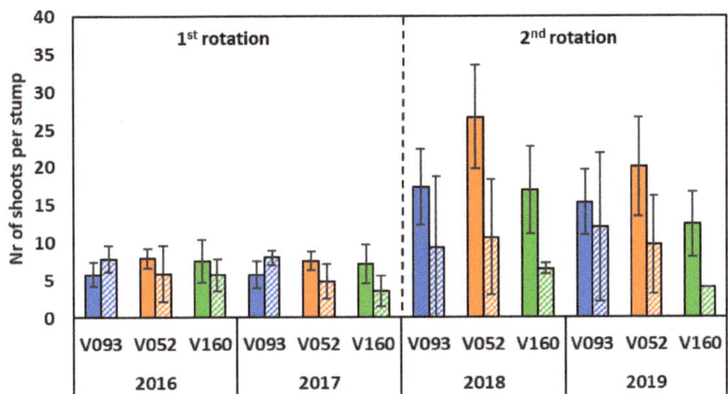


**Figure 1.** Mean number of shoots per stump at the end of each growing season for three clones of *S. alba* ('V 093', 'V 052' and 'V 160') in the evapotranspirative willow system (full columns) and controls (dashed columns). Mean and standard deviation are given ($N = 9$ for test trees and 1–6 for control trees).

In the 1st year of the 1st rotation, the test willows reached about 2.5 m and were 20–40 cm taller than the control trees (Figure 2). In the 2nd growing season, the willows in the EWS grew another 2 m, with 'V 052' being taller ($479 \pm 50$ cm) than 'V 093' ($463 \pm 37$ cm) and 'V 160' ($431 \pm 65$ cm). The difference from the control trees increased: test trees for 'V 052', 'V 093' and 'V 160' were 1.2, 1.6 and 2.1 m taller than the control trees, respectively. After the first coppicing, the willows grew back successfully and rapidly, reaching 3.4 ('V 160') and 3.8 m ('V 093' and 'V 052'). In the 2nd growing season of the 2nd rotation, the difference between the clones increased and showed the same trend as in the 2nd growing season of the 1st rotation, with 'V 052' being the tallest ($501 \pm 52$ cm), followed by 'V 093' ($498 \pm 18$ cm) and 'V 160' ($423 \pm 27$ cm), which was again the shortest.

The clones of *S. alba* in EWS produced on average between 34–38 and 33–59 t DM ha$^{-1}$ for the 1st and 2nd rotation, respectively (Figure 3). Biomass production was much higher compared to the controls, reaching a maximum of 14 t DM ha$^{-1}$. The standard deviations of the mean biomass production between the treatment beds of the same clone were high because the beds in the pilot EWS had different positions to sun and wind, resulting in different growing conditions. In the 2nd rotation, biomass production of 'V 093' and 'V 052' increased by 21% and 55%, respectively, compared to the 1st rotation, while it remained in the same range for 'V 160'. Additionally, the increase in conversion of wastewater to woody biomass was even more obvious: the conversion increased significantly in the 2nd rotation, when the clones produced $5.8 \pm 1.1$, $6.8 \pm 0.5$ and $8.5 \pm 0.6$ kg DM per m$^3$ of wastewater, for 'V 160', 'V 093' and 'V 052', respectively. Moreover, 'V 052' showed significantly higher conversion compared to the other two clones in the 2nd rotation (Figure 4).

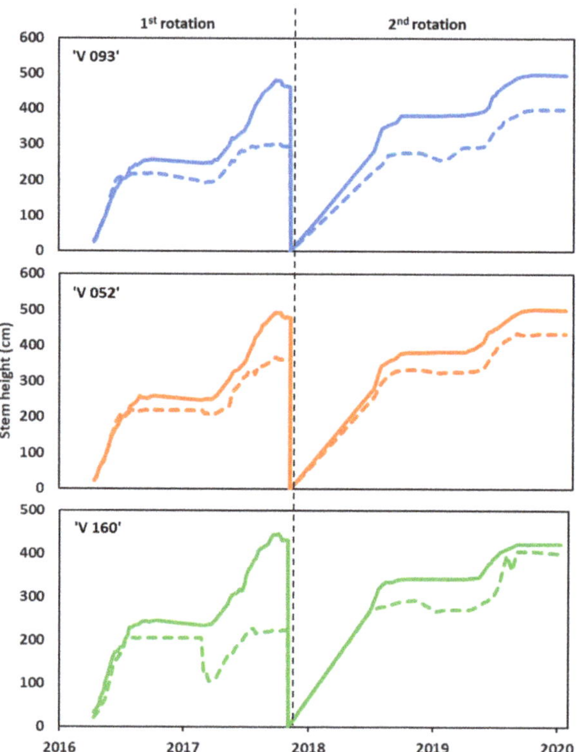

**Figure 2.** Mean stem height in two consecutive two-year rotations for three clones of *S. alba* ('V 093', 'V 052' and 'V 160') in evapotranspirative willow system (solid lines) and controls (dashed lines) ($N = 9$ for test trees and 1–6 for control trees).

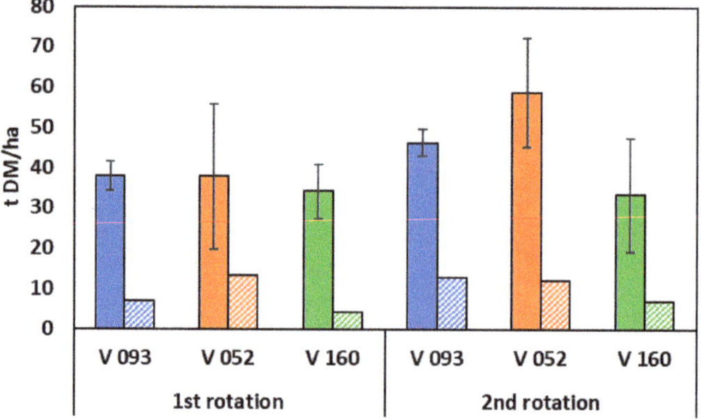

**Figure 3.** Mean biomass production in the 1st and 2nd rotation for three clones of *S. alba* ('V 093', 'V 052' and 'V 160') in evapotranspirative willow system (solid-filled columns) and controls (pattern-filled columns) ($N = 3$ for evapotranspirative willow system and 1 for control).

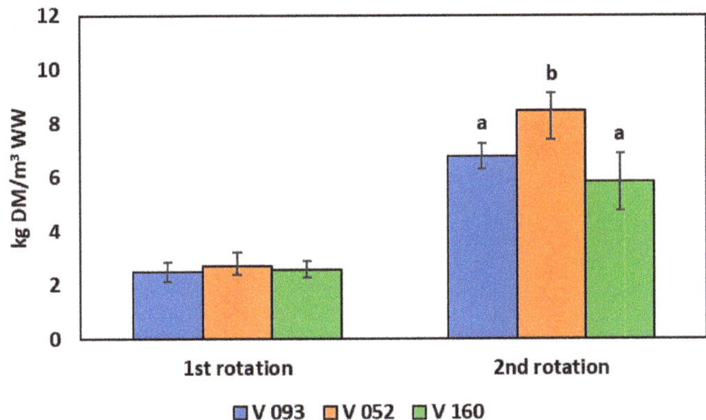



**Figure 4.** Mean biomass dry matter (DM) produced per m$^3$ of added wastewater for three clones of *S. alba* ('V 093', 'V 052', 'V 160') in evapotranspirative willow system for the 1st and 2nd rotation. Letters a, b, show statistically significant differences between clones in the 2nd rotation ($N = 3$).

### 3.3. Fate of Macronutrients

Macronutrients (C, N, P, K, Ca, Mg, and S) are major components of plant biomass. For willows in the EWS, they are derived from the atmosphere (C), soil, and wastewater (N, P, K, Ca, Mg, and S). The EWS was filled with locally available clay soil (Table 3). The properties of a soil composite sample before the addition of wastewater indicate a relatively fertile soil, with high P and organic matter content (Table 4). Irrigation with wastewater for two consecutive seasons of the 1st rotation resulted in an increase in some soil parameters. There was no increase in soil organic matter and organic carbon, indicating the efficient decomposition of organic matter from wastewater. Total N increased after the 1st growing season, indicating excessive N input to the young willows; however, in the 2nd growing season, despite the much greater N load from wastewater, there was no further N accumulation, probably due to the intensive growth of willows and possible denitrification. Due to N uptake, the C/N ratio also increased in the 2nd growing season. Similarly, the P supplied by the wastewater also seemed to meet the P demand of the willows, since there was neither an accumulation of $P_2O_5$ nor a decrease in the soil. On the other hand, $K_2O$ content, which was already relatively low in the original soil, increased with the addition of wastewater, suggesting that the K requirement of the willows was met and excess $K_2O$ accumulated in the soil, or that K uptake was displaced by other minerals. This is consistent with the results of cation analysis, namely, that soil concentrations of Ca and K cations increased in the 2nd year; however, changes in Mg concentrations showed no such trend. In contrast to the other cations, Na concentrations increased over the years, indicating salt accumulation in the system.

**Table 3.** Texture of the soil used to fill up the evapotranspirative willow system.

| Parameter | Percentage/Classification |
| --- | --- |
| Sand | 26.1 |
| Silt—coarse | 18.1 |
| Silt—fine | 30.4 |
| Silt—total | 48.5 |
| Clay | 25.4 |
| Classification * | clay |

* according to USDA textural soil classification.

**Table 4.** Soil properties in the evapotranspirative willow system before wastewater addition (Start), after the 1st and 2nd year of the 1st rotation. Mean and standard deviation are given for the first and second year ($N = 3$).

|  | Unit | Start * | 1st Year | 2nd Year |
|---|---|---|---|---|
| pH in CaCl$_2$ |  | 7.1 | 7.1 ± 0.06 | 7.3 ± 0.06 |
| P$_2$O$_5$ | mg/100 g | 144 | 162 ± 46.6 | 177 ± 3.13 |
| K$_2$O | mg/100 g | 13.0 | 13.7 ± 0.66 | 14.4 ± 0.32 |
| Organic matter | % | 7.0 | 7.5 ± 0.3 | 7.2 ± 0.5 |
| Organic carbon | % | 4.1 | 4.3 ± 0.15 | 4.2 ± 0.3 |
| TN | % | 0.43 | 0.48 ± 0.01 | 0.40 ± 0.04 |
| C/N ratio |  | 9.5 | 9.0 ± 0.32 | 10.5 ± 0.40 |
| Ca | mmol/100 g | 35.2 | 30.0 ± 0.68 | 36.8 ± 1.57 |
| Mg | mmol/100 g | 2.74 | 3.11 ± 0.12 | 2.87 ± 0.09 |
| K | mmol/100 g | 0.30 | 0.20 ± 0.02 | 0.30 ± 0.02 |
| Na | mmol/100 g | 0.05 | 0.17 ± 0.02 | 0.43 ± 0.10 |
| Exchangable acidity | mmol/100 g | 3.55 | NA | 4.80 ± 0.22 |
| Sum of base cations | mmol/100 g | 38.3 | 33.4 ± 0.78 | 40.4 ± 1.74 |
| CEC | mmol/100 g | 41.8 | NA | 45.2 ± 1.54 |
| Base saturation | % | 91.6 | NA | 89.4 ± 0.84 |

* before application of wastewater. .

In the woody biomass, the amount of C did not differ significantly between the control and test trees (Table 5); however, 'V 093' showed a trend of storing more C compared to the other two clones in both control and test trees. There was no statistically significant difference in N and P contents between the test and control trees. When comparing the test clones, 'V 052' appeared to accumulate more N and P compared to the other two clones; however, the difference was not always statistically significant. The test trees showed significantly lower K and Ca contents and significantly higher S concentrations compared to the control trees. There was no significant difference in Mg content between the test and control trees; however, 'V 160' accumulated significantly more Mg in the test trees compared to the other two clones.

**Table 5.** Macronutrient concentrations (in g kg$^{-1}$ DM) in the woody biomass of three clones of *S. alba* ('V 093', 'V 052' and 'V 160') in the evapotranspirative willow system and controls after the 1st rotation. Mean and standard deviation are given ($N = 2$ for control and 3 for test trees). *P*-value indicates statistical significance between control and test trees ($N = 6$ for control and 9 for test trees), and superscripts a and b indicate statistically significant differences between clones of the test trees.

|  | Control Trees | | | Test Trees | | | |
|---|---|---|---|---|---|---|---|
|  | 'V 093' | 'V 052' | 'V 160' | 'V 093' | 'V 052' | 'V 160' | *P* |
| C | 477 ± 1.8 | 473 ± 0.28 | 475 ± 0.0 | 477 ± 0.40 [a] | 474 ± 0.31 [b] | 474 ± 1.8 [ab] | 0.845 |
| N | 7.6 ± 0.07 | 10 ± 0.2 | 7.6 ± 0.21 | 6.5 ± 1.0 | 7.9 ± 0.76 | 7.1 ± 0.73 | 0.081 |
| P | 0.99 ± 0.16 | 1.1 ± 0.40 | 0.91 ± 0.07 | 0.92 ± 0.23 [ab] | 1.2 ± 0.04 [a] | 1.0 ± 0.08 [b] | 0.775 |
| K | 2.4 ± 0.07 | 2.9 ± 0.28 | 2.2 ± 0.14 | 1.8 ± 0.25 | 2.2 ± 0.21 | 2.0 ± 0.25 | 0.029 |
| Ca | 8.1 ± 2.8 | 8.2 ± 2.1 | 9.1 ± 0.07 | 5.0 ± 0.25 | 5.7 ± 0.78 | 5.8 ± 1.2 | 0.005 |
| Mg | 0.93 ± 0.18 | 1.1 ± 0.1 | 0.84 ± 0.16 | 0.75 ± 0.06 [b] | 0.75 ± 0.04 [b] | 0.87 ± 0.04 [a] | 0.060 |
| S | 1.6 ± 0.42 | 1.7 ± 0.07 | 1.1 ± 0.50 | 1.9 ± 0.27 | 1.8 ± 0.23 | 1.9 ± 0.15 | 0.044 |

The results on the distribution of nutrients from the wastewater into the woody biomass of *S. alba* and the other compartments of the EWS show that 52–65% and 55–69% of N and P, respectively, were degraded and accumulated in the soil, root system and leaves, while 35–48% and 31–45% were stored in the woody biomass (Figure 5). 'V 052' showed the highest accumulation of nutrients (48% and 45% for N and P, respectively) and 'V 093' the lowest (35% and 31% for N and P, respectively). The distribution of K could not be presented, as it was not measured in the influent wastewater.

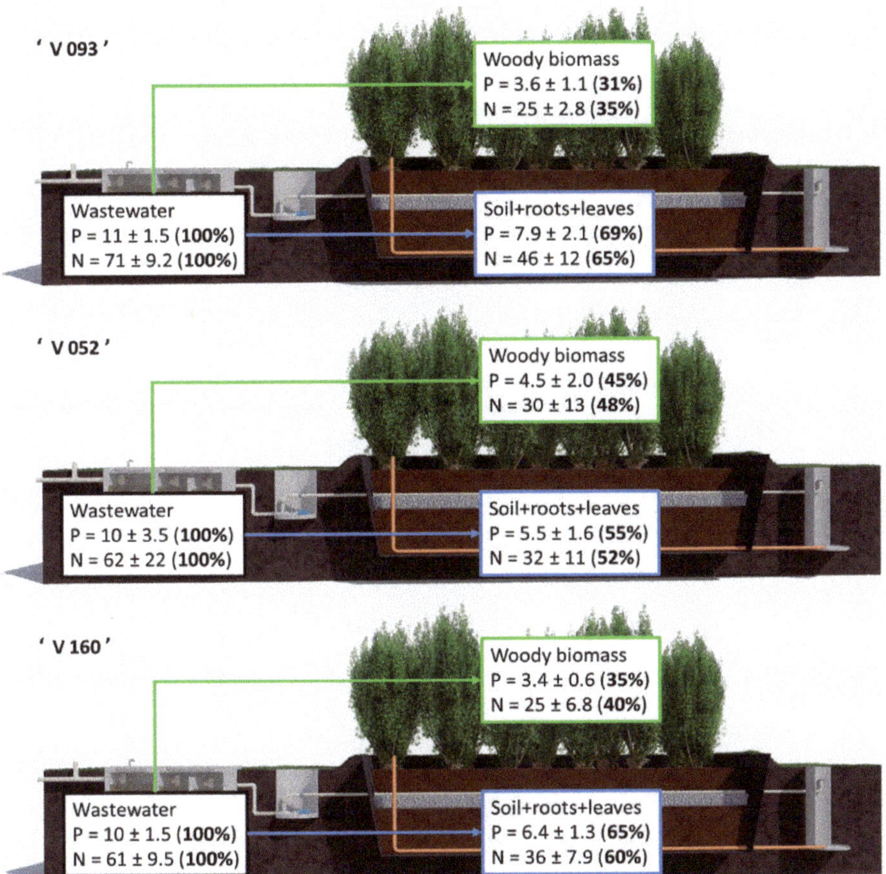


**Figure 5.** Distribution of total nitrogen (N) and phosphorus (P) from wastewater in g m$^{-2}$ between woody biomass of *S. alba* and other compartments (accumulation in soil, roots and leaves and denitrification in the case of N) in the evapotranspirative willow system after the 1st rotation for three investigated clones ('V 093', 'V 052' and 'V 160').

### 3.4. Fate of Metals

The concentrations of metals in the woody biomass after the 1st rotation showed some significant differences between the control and test trees and between the clones of the test trees (Table 6). The concentrations were compared with the heavy metal concentrations in the soil (Table 7). After two growing seasons of irrigation with municipal wastewater, most of the heavy metals measured showed a slight increase in soil concentrations. There was no increase for Cd and As, while Cr and Pb were increased only in the soil of 'V 160'. The heavy metals studied are below the critical levels given in the decree on limit values, alert thresholds and critical levels of dangerous substances into the soil (OG RS, 68/96, 41/04).

The control trees accumulated significantly more Fe, Mo, Sr, Ba, Ti and B compared to the test trees, although the test trees had higher available Fe and Mo concentrations in the soil (Sr, Ba, Ti and B were not measured in the soil). On the other hand, Na and Ag occurred at significantly higher concentrations in the test trees, which may be related to their presence in wastewater. The difference in heavy metal accumulation between clones was significant only for Cu and Mn: 'V 052' accumulated significantly higher concentrations of Cu and 'V 160' accumulated significantly higher concentrations of Mn, but the concentrations are much lower compared to soil, so there is no obvious transport of metals from soil to aboveground tissues.

**Table 6.** Metal concentrations in mg kg$^{-1}$ DM in woody biomass for three clones of *S. alba* ('V 093', 'V 052' and 'V 160') in the evapotranspirative willow system and controls after the 1st rotation. Mean and standard deviation are given (*N* = 2 for control and 3 for test trees). *P*-value indicates statistical significance between control and test trees (*N* = 6 for control and 9 for test trees) and superscripts a, b, c, indicate statistically significant differences between clones of the test trees.

| | Control Trees | | | Test Trees | | | |
| --- | --- | --- | --- | --- | --- | --- | --- |
| | 'V 093' | 'V 052' | 'V 160' | 'V 093' | 'V 052' | 'V 160' | *P* |
| Fe | 105 ± 7 | 95 ± 7 | 75 ± 7 | 57 ± 6 | 63 ± 6 | 67 ± 6 | 0.003 |
| Al | <LOD | <LOD | <LOD | <LOD | <LOD | <LOD | |
| Na | 20 ± 0 | 15 ± 7 | 25 ± 7 | 33 ± 6 | 33 ± 15 | 47 ± 12 | 0.003 |
| Mo | 0.04 ± 0.03 | 0.06 ± 0.00 | 0.05 ± 0.01 | 0.01 ± 0.01 | 0.02 ± 0.00 | 0.03 ± 0.02 | 0.008 |
| Cu | 7.8 ± 1.7 | 11.1 ± 1.0 | 8.8 ± 2.2 | 6.2 ± 0.6 [c] | 10.0 ± 0.5 [a] | 7.9 ± 0.4 [b] | 0.271 |
| Pb | 1.22 ± 1.11 | 0.19 ± 0.04 | 0.16 ± 0.00 | 0.12 ± 0.01 | 1.55 ± 2.48 | 0.14 ± 0.03 | 0.885 |
| Zn | 46 ± 5 | 58 ± 24 | 40 ± 7 | 42 ± 2 | 42 ± 4 | 45 ± 5 | 0.436 |
| Ni | 0.70 ± 0.28 | 0.45 ± 0.07 | 0.30 ± 0.14 | 0.23 ± 0.12 | 0.30 ± 0.17 | 0.53 ± 0.40 | 0.343 |
| Co | 0.10 ± 0.01 | 0.11 ± 0.01 | 0.06 ± 0.00 | 0.35 ± 0.33 | 0.06 ± 0.01 | 0.08 ± 0.02 | 0.322 |
| Mn | 9.0 ± 2.8 | 10.0 ± 4.2 | 9.0 ± 1.4 | 9.3 ± 1.2 [b] | 8.7 ± 0.6 [b] | 15.3 ± 2.5 [a] | 0.265 |
| As | 0.25 ± 0.07 | 0.40 ± 0.00 | <LOD | 0.29 ± 0.36 | 0.20 ± 0.10 | 0.11 ± 0.07 | 0.678 |
| U | <LOD | <LOD | <LOD | <LOD | <LOD | <LOD | 0.254 |
| Th | <LOD | <LOD | <LOD | <LOD | <LOD | <LOD | 1.000 |
| Sr | 10.8 ± 4.6 | 10.9 ± 0.2 | 10.8 ± 1.8 | 6.6 ± 0.8 | 7.0 ± 1.0 | 7.6 ± 0.7 | 0.007 |
| Cd | 0.88 ± 0.56 | 0.72 ± 0.01 | 0.57 ± 0.04 | 0.43 ± 0.07 | 1.00 ± 0.96 | 0.49 ± 0.01 | 0.713 |
| Sb | 0.03 ± 0.01 | 0.02 ± 0.01 | <LOD | <LOD | 0.02 ± 0.00 | <LOD | 0.139 |
| Bi | <LOD | <LOD | <LOD | <LOD | <LOD | <LOD | 0.254 |
| V | <LOD | <LOD | <LOD | <LOD | <LOD | <LOD | |
| La | 0.02 ± 0.02 | <LOD | <LOD | <LOD | <LOD | <LOD | 0.363 |
| Cr | 2.50 ± 0.85 | 2.05 ± 0.07 | 1.95 ± 0.21 | 2.17 ± 0.25 | 2.20 ± 0.20 | 2.50 ± 0.61 | 0.608 |
| Ba | 4.65 ± 2.05 | 4.85 ± 0.78 | 3.65 ± 0.07 | 2.20 ± 0.00 | 3.70 ± 3.03 | 2.40 ± 0.61 | 0.046 |
| Ti | 13.5 ± 0.7 | 27.0 ± 2.8 | 24.5 ± 9.2 | 4.0 ± 2.6 | 11.7 ± 7.8 | 18.7 ± 9.6 | 0.037 |
| B | 15.5 ± 4.9 | 14.5 ± 0.7 | 13.0 ± 0.0 | 9.3 ± 0.6 | 9.7 ± 1.5 | 9.7 ± 1.5 | 0.004 |
| W | <LOD | 0.20 ± 0.00 | 0.14 ± 0.09 | 2.29 ± 2.27 | 0.15 ± 0.13 | 0.15 ± 0.13 | 0.201 |
| Sc | 0.30 ± 0.14 | 0.25 ± 0.07 | 0.20 ± 0.00 | 0.23 ± 0.06 | 0.30 ± 0.00 | 0.27 ± 0.06 | 0.674 |
| Tl | <LOD | <LOD | <LOD | <LOD | <LOD | <LOD | 0.254 |
| Se | 0.30 ± 0.14 | 0.20 ± 0.00 | 0.20 ± 0.14 | 0.20 ± 0.00 | 0.23 ± 0.06 | 0.17 ± 0.06 | 0.488 |
| Te | <LOD | <LOD | <LOD | <LOD | <LOD | <LOD | 0.254 |
| Ga | <LOD | <LOD | <LOD | <LOD | <LOD | <LOD | 1.000 |
| Ag | 3.21 ± 2.54 | 3.50 ± 0.71 | 3.00 ± 0.00 | 4.67 ± 1.53 | 6.67 ± 1.15 | 9.67 ± 1.53 | 0.002 |
| Au | <LOD | <LOD | <LOD | <LOD | 0.11 ± 0.17 | <LOD | 0.347 |
| Hg | 4.00 ± 1.41 | 2.50 ± 0.71 | 3.00 ± 1.41 | 1.90 ± 1.15 | 2.33 ± 1.53 | 3.33 ± 0.58 | 0.321 |

**Table 7.** Heavy metal concentrations in mg kg$^{-1}$ of soil in the evapotranspirative willow system before wastewater addition (Start) and after the 1st rotation for the three clones of *S. alba* ('V 093', 'V 052' and 'V 160').

|  | Start | After 1st Rotation | | |
|---|---|---|---|---|
|  |  | 'V 093' | 'V 052' | 'V 160' |
| Fe | 28,700 | 29,200 | 29,300 | 31,000 |
| Cd | 0.7 | 0.6 | 0.7 | 0.6 |
| Cu | 66.2 | 67.2 | 66.5 | 72.8 |
| Ni | 80.9 | 78.1 | 81.3 | 86.6 |
| Pb | 36.4 | 34.1 | 35.5 | 42.2 |
| Zn | 153 | 161 | 172 | 166 |
| Cr | 63 | 60 | 59 | 67 |
| Hg | 0.56 | 0.79 | 0.76 | 0.65 |
| Co | 16.1 | 16.3 | 16.5 | 18.3 |
| Mo | 1.2 | 1.3 | 1.3 | 1.5 |
| As | 8.7 | 8 | 8.2 | 8.1 |
| Mn | 1012 | 1146 | 1110 | 1069 |

## 4. Discussion

### 4.1. Wastewater

Mass loading rates to the EWS were dependent on the composition of the influent wastewater, willow water use, and root system development. The most significant increase in mass loading rate was from the 1st to the 2nd year of the 1st rotation, when the most vigorous root development is assumed. Since the loading rate increased only slightly or remained in the same range in the following years, it is assumed that the root system developed almost to full capacity in the 2nd growing season. This is also in agreement with Rytter [45], who reported the highest increase in total belowground production (coarse and fine roots) in the 2nd growing season after planting.

The organic and nutrient loading rates in this study, i.e., 1.6–7.9 g COD, 0.10–0.53 g TN, and 0.02–0.08 g TP, applied per m$^2$ per day during the growing season, were in the same range as those reported by Amiot et al. [46], who applied 3.4–5.1 g COD, 0.48–0.72 g TN, and 0.05–0.07 g TP per m$^2$ per day to an SRWC vegetation filter, and reported efficient removal of pollutants from domestic wastewater during the growing season.

### 4.2. Willow Growth and Biomass Production

Cutting back is known to stimulate the formation of multiple shoots; however, different clones, planting densities, irrigation regimes, available nutrients, and plant age may result in significantly different numbers and heights of shoots [5,47]. The *S. alba* clones in this study developed 12–20 shoots per stump at the end of the 2nd rotation, while Holm and Heinsoo [5] reported 1.5–7.1 shoots per stump for clones of *S. viminalis*, *S. burjatica*, *S. dasyclados*, and *S. schwerinii* irrigated with municipal wastewater in Estonia. In addition, the willows in the EWS had higher and significantly more shoots compared to the controls, confirming the positive effects of adequate water and nutrient supply from wastewater. Although the controls were planted at the edges of the system and had more space available, they still developed fewer shoots compared to the densely vegetated willows in the EWS. The reduction in shoot number in the 2nd year of the 2nd rotation might indicate that in some clones a certain number of shoots becomes dominant and overtakes the growth while weaker shoots die off. In our study, this was particularly observed for clones 'V 160' and 'V 052'.

The annual biomass production (17–19 and 17–30 t DM ha$^{-1}$ y$^{-1}$, for willows aged 2/3 and 2/5, respectively) was in the same range or higher than that reported by Kajba and Andrić [38] (10–19 and 19–24 t DM ha$^{-1}$ y$^{-1}$, for willows aged 2/3 and 2/5, respectively), who tested biomass production of several clones, including those tested in this study, under controlled conditions in a nursery in Croatia. Similar biomass production was also reported by Lachapelle-T. et al. [22], namely, 22–26 t DM ha$^{-1}$ y$^{-1}$ for *S. miyabeana* aged

1/8, irrigated with wastewater under Canadian humid continental climatic conditions. However, both studies were conducted at almost the same geographical latitude (45°51′29″ N and 45°52′32″ N for Canadian and this study, respectively) that confirms the concordance between the studies, and shows that the latitude and the corresponding total amount of incoming solar energy might have more impact on biomass production than climate.

Willows in short-rotation plantations usually achieve the highest annual increment three to four years after planting, and it has been reported that an increase in the 2nd rotation is 18–62% compared to the 1st rotation [48]. Similarly, in our study, biomass production in 'V 093' and 'V 052' increased by 21% and 55% respectively, while there was no increase in 'V 160'. In subsequent rotations, the well-established root system supports the growth of aboveground shoots by providing the stored fixed carbon [27], thus maintaining the high annual increment.

The well-established root system together with the increased number of shoots also seems to increase the conversion of wastewater to biomass, as significantly more DM per $m^3$ of wastewater was produced in the 2nd rotation in all three clones, including 'V 160', which had the same total biomass production (34 and 33 t DM $ha^{-1}$), but produced 2.6 and 5.8 kg DM per $m^3$ of wastewater in the 1st and 2nd rotation, respectively.

The stimulatory effect of wastewater on willow growth can be demonstrated by comparing the biomass production of willows irrigated with wastewater and control willows irrigated with fresh water. In this study, control willows received only rainwater (about 1330 mm per year) and produced 5.5 t DM $ha^{-1}$ $y^{-1}$, while biomass yield increased by 318% in EWS, which is much higher compared to other studies: Börjesson and Berndes [21] reported a 30–100% increase in wastewater-fertigated willow plantations compared to rainfed systems in Sweden, and Lachapelle-T. et al. [22] reported an 83–117% increase compared to *S. miyabeana* irrigated with potable water under Canadian climatic conditions. Similarly, Fabio and Smart [49] reported a 61% yield increase in willows irrigated with municipal waste, sludge, or wastewater, which was even higher than the yield increase from synthetic fertilizers (48%). This confirms that water is also a limiting factor for willow growth and that wastewater has a dual stimulating effect—it provides nutrients and increases water availability. The significantly higher yield increase in our study demonstrates the increased importance of water availability in the sub-Mediterranean climate. Despite the high annual precipitation (1330 mm), uneven distribution of rainfall, high average annual temperatures (13.5 °C), high solar radiation and strong winds caused high evapotranspiration (reference ET0 975 mm) and lower water availability for control willows, resulting in lower biomass production.

### 4.3. Fate of Macronutrients

The original soil in EWS was already nutrient-rich before the addition of wastewater. The $P_2O_5$ content of 144 mg in 100 g was far above the target values of fertile soil in agricultural production (13–25 mg in 100 g), and the organic content of 7% was higher than the normal organic content (3%) in the surrounding agricultural lands. To our knowledge, there are no upper limits for available P content in the soil. On the other hand, the values of $K_2O$ were lower (13 mg in 100 g) compared to the target values in agriculture in Slovenia (20–30 mg in 100 g) [50]. The cation exchange capacity (CEC) was relatively high, as expected, due to the high content of clay and organic matter. Irrigation with wastewater caused changes in some soil chemical properties, as is well documented for agricultural wastewater reuse [51].

Many authors reported that wastewater meets the nutrient requirements of willows and that P may be present in excessive concentrations [12,20,22]. Thus, N, P and K are expected to be utilized by willows, while excess amounts are immobilized in the soil and N can be denitrified and released to the atmosphere in the form of $N_2$. Production of $NO_X$ and $N_2O$ in the denitrification process is not expected, due to high concentrations of dissolved organic carbon applied by wastewater and anoxic conditions in the system.

### 4.3.1. Carbon and Organic Matter

The primary treated wastewater applied to the EWS contained mainly organic matter and fewer inorganic nutrients. Despite significant amounts of organic matter applied to the EWS (2620–10,640 kg COD ha$^{-1}$), there was no significant increase in soil organic matter or organic carbon content, as is consistent with Lachapelle-T et al. [22], indicating the efficient decomposition of organic matter in the EWS. The C stored in the willow biomass was derived from the atmosphere. In this study, the C content in dry biomass was equal among clones and between test and control trees, varying between 473 and 477 mg C kg$^{-1}$ DM, which is lower compared to the 502 mg C kg$^{-1}$ DM reported by Stolarski et al. [52] in *S. alba* and the 491–518 mg C kg$^{-1}$ DM reported by Matthews and Lamlom and Savidge [53,54] for different willow species. This variability is consistent with the findings of Stolarski et al. [52] that C content in willows differs according to location and genotype.

### 4.3.2. Nitrogen

The fertilization of commercial SRWC is usually based on N requirements [49]. Additionally, N uptake by willows is an important design parameter for SRWC vegetation filters [22]. The influence of N uptake by willows was also observed in this study, as the intensive growth of willows in the 2nd season reduced the percentage of total N accumulated in the soil during the 1st season. Total N applied via wastewater during the 2nd growing season in this study was 440–540 kg N ha$^{-1}$, which is in the same range as that reported by Lachapelle-T et al. [22] (370–580 kg N ha$^{-1}$). The uptake into woody biomass corresponds to 35–48% of the total N applied via wastewater, and the rest of the applied N (52–65%) was stored in roots and leaves or denitrified. The distribution is similar to that of Lachapelle-T et al. [22], i.e., 18–59% was stored in woody biomass, and 35–70% was accumulated in soil, roots, and leaves or denitrified. In both studies, the amount of N applied was much higher compared to the recommended N fertilization in commercial SRWC, which ranges from 40–180 kg N ha$^{-1}$ [32,49]; however, N uptake was in the same range: 250–300, 88–260 and up to 311 kg N ha$^{-1}$ for this study (*S. alba* aged 2/3), wastewater-irrigated SRWC (*S. viminalis* aged 3/3 and *S. miyabeana* aged 1/8) [17,22], and commercial SRWC (different species and ages) [49], respectively. Additionally, in this study, N concentrations in the woody biomass of control (rainfed) and test (wastewater-irrigated) trees of different clones did not differ significantly. This suggests a universal N uptake capacity of willows regardless of climate, N application and/or willow species/clones, and that excess applied N is subject to denitrification, accumulation in the soil, or leaching in the case of SRWC.

### 4.3.3. Phosphorus

Despite high $P_2O_5$ concentrations in the initial soil, $P_2O_5$ concentrations did not increase during the first two growing seasons in this study. The total P applied via wastewater in the 2nd growing season was higher in this study (71–86 kg P ha$^{-1}$) compared to that in Lachapelle-T et al. [22] (37–58 kg P ha$^{-1}$), while the percentage of total applied P incorporated into the woody biomass was similar (31–45% and 18–59% for this and the Canadian study, respectively) resulting in different P concentrations in the woody biomass: 34–45 and 9–26 kg P ha$^{-1}$ for this study on *S. alba* (aged 2/3) and Lachapelle-T et al. [22] on *S. miyabeana* (age 1/8), respectively. Again, a different P uptake was reported by Curneen and Gill [17], namely, 28–35 kg P ha$^{-1}$ for *S. viminalis* (aged 3/3). Additionally, in our study, there was no significant difference in P accumulation between the control (rainfed) and test (wastewater-irrigated) trees, while 'V 052' accumulated more P compared to the other two clones in EWS. This indicates differences in P uptake between different willow species/clones. The rest of the applied P was stored in leaves and roots and immobilized in the soil in a form unavailable to plants. The recommended P application is 24 kg P ha$^{-1}$ [32], which is much lower compared to our and similar studies, suggesting that EWS might accumulate P in the soil during long-term operation. When the soil is saturated, an increase in P that is available in the soil is expected, as also observed by Lachapelle-T et al. [22].

### 4.3.4. Potassium

Potassium is not a basic wastewater parameter and is therefore rarely monitored in municipal wastewater. Its concentrations in wastewater range from 10–30 mg/L [55], which in the case of our wastewater load would result in 100–353 kg K ha$^{-1}$ applied in the 2nd growing season, which is in the same range or higher than the recommendation for commercial SRWC, i.e., 120–155 kg K ha$^{-1}$ [32]. In this study, K$_2$O levels were slightly elevated in the soil of the EWS, suggesting that the applied wastewater may have met the K requirement of the willows and excess K was accumulated in the soil. Accumulation of K in the woody biomass was 70–87 kg K ha$^{-1}$. According to K concentrations in wastewater as reported in the literature [55], this would result in 20–80% uptake of applied K into woody biomass, which is too wide a range to draw any conclusions regarding the distribution of applied K in the system. However, the accumulation of K in the woody biomass was much higher than the 21–29 kg K ha$^{-1}$ reported by Curneen and Gill [17], and lower than the 85–123 kg K ha$^{-1}$ reported by Gregersen and Brix [12], suggesting possible differences between willow species and clones and/or a response to the elevated environmental K concentrations (loads), as also observed for P but not for N. Similarly, Adegbidi et al. [56] also reported a significant increase in K and P, but also N in the case of fertilizer application. Although the applied wastewater appeared to have sufficiently high K concentrations for the needs of the willows and excess K was accumulated in the soil, the control willows in this study had significantly higher K content in the woody biomass, questioning the availability of K in the EWS.

### 4.3.5. Sulfur

The S content in control and test trees (1.1 to 1.9 mg S kg$^{-1}$ DM) was much higher than the 0.57 mg S kg$^{-1}$ DM determined by Stolarski [57] in shoots of *S. alba*. The difference may be due to the different S uptake of different clones of *S. alba* and the location [52], suggesting that the studied clones 'V 093', 'V 052' and 'V 160', when grown in a sub-Mediterranean climate, may emit more SO$_2$ during combustion when used as an energy crop. On the other hand, they allow a higher uptake of S from wastewater and have better S recycling potential.

### 4.3.6. Calcium and Magnesium

Different willow species can accumulate different concentrations of Ca and Mg [56,58]. In this study, the test trees of 'V 160' accumulated more Mg compared to the other two clones. Since fertilization has been shown not to increase Ca and Mg uptake in willows [56], the excess of these elements supplied via wastewater can be expected to accumulate in EWS in the long term. In our study, there was a slight increase in soil Ca concentration and no apparent changes in Mg concentration, which may indicate an adequate amount of available Ca and Mg for willow growth. An increase in soil Ca was also observed by Lachapelle-T et al. [22] when SRWC was irrigated with higher wastewater loads. Despite the slight increase in soil Ca in the EWS, the test trees had significantly lower Ca concentrations compared to the control trees. This is consistent with significantly lower K concentrations in the test trees and a significant increase in Na concentration in the soil, indicating increasing salinity in a zero-discharge EWS. Salinity can lead to nutrient deficiencies or disproportions, due to the competition of Na$^+$ with K$^+$ and Ca$^{2+}$ in the soil–root system [59]. Increasing salinity in EWS was recognized as a potential problem already when systems were designed [12]; however, the systems in Denmark have been operated for more than 20 years with no apparent detrimental effects of salinity on willow growth or evapotranspiration [13]. According to Gregersen and Brix [12], the electrical conductivity in EWS beds did not increase during the first two years of operation and did not show an increase in salinity; however, the results of the current study on soil Na concentration and Ca and K uptake indicate that an increase in salinity and nutrient imbalance can already be observed during the first years of operation. For an efficient EWS operation, suitable willow species and clones tolerant to increasing salinity need to be selected, and more attention should be paid to the response of willows (evapotranspiration, nutrient and metal uptake) to the combination of long-term flooding and salinity, as the combination of these two plant stressors is known to be stronger than either stress alone [31].

### 4.3.7. Removal of Nutrients by Harvesting

Biomass harvesting removes some nutrients from the system, representing nutrient recovery potential. In the current study, one-third to one-half of the applied N and P was accumulated in the harvestable woody biomass and the remainder was stored in the soil, denitrified, or taken up by roots and leaves, as is consistent with Lachapelle-T et al. [22]. This is a significant amount compared to constructed wetlands as the most common plant-based wastewater treatment systems. For example, harvesting *Phragmites australis* in a constructed wetland can only remove about 4% and 5% of applied N and P, respectively [60]. In SRWC systems, harvesting is usually done after the leaves have fallen, so the nutrients stored in the leaves remain on-site and are recycled through decomposition in the topsoil. This internal nutrient cycling reduces the need for fertilizers. However, to recover more nutrients from the wastewater in EWS, the biomass could be harvested before leaf abscission. In addition, the harvesting cycle can significantly affect nutrient removal and nutrient use efficiency, which also needs to be considered in the operation and management of EWS. In fact, Adegbidi et al. [56] found that the annual harvest cycle had the lowest nutrient efficiency and the highest nutrient removal, and suggested this type of cycle for nutrient phytoremediation in vegetation strips. Similarly, an annual harvest cycle should be considered for higher nutrient recovery in EWS. Willow woodchips produced in this way can be used as a soil amendment in agriculture [61] to return nutrients to the food chain.

Despite the high potential for nutrient removal through harvesting, EWS soil may become saturated with nutrients after a certain period of operation. The nutrient-rich media can be used as a fertilizer or soil amendment [12], which can also return nutrients to the food chain; however, the presence of heavy metals and persistent organic pollutants must be investigated prior to application. By reusing the soil and utilizing the willow biomass, the material cycle of EWS can be closed.

### 4.4. Fate of Metals

Concentrations of heavy metals in the woody biomass and soil media can affect the potential for their reuse. Typical concentrations of heavy metals in municipal wastewater are expectedly low, ranging from a few to a few hundred $\mu g\ L^{-1}$ [41]; however, heavy metals showed a slight increase in soil concentrations, due to high wastewater loads and zero-discharge operation of EWS, which was also observed by Gregersen and Brix [12]. Studies of SRWC irrigated with wastewater or landfill leachate generally do not report elevated heavy metal concentrations in soil, due to leaching to surrounding water bodies or to the subsurface [8,35]; however, in the case of EWS, all influent heavy metals are retained in the system and are available for uptake by willows or accumulate in the soil. As salinity increases over the long term, the availability and uptake of heavy metals may change and the accumulation of heavy metals in the soil between years may not be linear.

The response of willows to increased salinity from irrigation with wastewater and landfill leachate has been reported in many studies [62–64]; however, the effect of salinity on heavy metal uptake by willows has not been directly investigated. It is known from plant nutrition research that salinity can affect plant micronutrient concentrations differently, depending on the plant species and salinity level [59]. In this study, test trees under higher salinity conditions accumulated less Fe, Mo, Sr, Ba, Ti, and B compared to control trees, even when these elements were present in the soil at higher concentrations (e.g., Fe and Mo), indicating a possible competition of these cations with salt ions, particularly Na.

In addition to increasing salinity, the dynamics of heavy metals in EWS are also affected by waterlogging—during the winter period (low evapotranspiration), influent wastewater accumulates in the system, leading to soil saturation and anaerobic conditions that can convert metals such as Fe and Mn to reduced and more soluble forms [31].

The uptake of heavy metals by willows depends on both heavy metal concentrations in the soil and willow species and clones [38,65]. In agreement with this, our study found some significant differences in heavy metal accumulation among clones. In addition, studies on phytoremediation of heavy metal contaminated soils generally report higher heavy metal concentrations in the woody biomass of willows in response to higher concentrations in the adjacent soil. Reports of heavy metal accumulation and uptake in EWS are rare, and report low concentrations of the heavy metals studied in plant tissues [12].

## 5. Conclusions

Evaluation of nutrient recovery from primary treated wastewater by a zero-discharge willow system under a sub-Mediterranean climate was conducted on a pilot EWS, using three *Salix alba* clones ('V 093', 'V 052' and 'V160'). The growth dynamics and biomass production of selected clones were investigated. The study showed that *S. alba* clones were suitable for use in EWS and produced significantly more biomass when irrigated with wastewater. 'V 052' was the highest, produced the highest number of shoots and had the highest biomass yield (38–59 t DM ha$^{-1}$) in both rotations. In addition, 'V 052' had the highest N and P uptake (48 and 45%) from wastewater, and the highest conversion of wastewater to biomass (8.5 kg DM per m$^3$ of wastewater). The indigenous white willow clone 'V 160' was the shortest and had the lowest biomass yield and wastewater to biomass conversion. Nevertheless, 'V 160' took up more nutrients from wastewater compared to 'V 093'. The uptake of N and P from wastewater into harvestable wood biomass was significant compared to other plant-based wastewater treatment systems, indicating a good nutrient recovery potential of EWS. Wastewater composition and loading were consistent with willow nutrient requirements; however, the uptake of macronutrients and metals may be hindered or altered by increasing salinity caused by EWS zero-discharge operation. Increased salinity has been noted as increased Na concentration in the soil and woody biomass, and decreased Ca and K uptake after only two years of operation; however, the plants in Denmark have been in full operation for 20 years, and to our knowledge there have been no reports of decreased willow growth, evapotranspiration, or other deleterious effects that may be caused by increased salinity, indicating the potential adaptability of *Salix* spp. that should be further investigated.

**Author Contributions:** Conceptualization, D.I.; methodology, D.I. and G.B.; validation, D.I. and G.B.; formal analysis, D.I.; investigation, D.I.; resources, D.I. and G.B.; data curation, D.I.; writing—original draft preparation, D.I.; writing—review and editing, D.I. and G.B.; visualization, D.I.; project administration, D.I.; funding acquisition, D.I. and G.B. Both authors have read and agreed to the published version of the manuscript.

**Funding:** The authors acknowledge the projects "Development and efficiency of evapotranspirative zero-emission system for closing wastewater material flows Z2-6751" and "Closing material flows by wastewater treatment with green technologies J2-8162" were financially supported by the Slovenian Research Agency. Additionally, the authors acknowledge the financial support from the Slovenian Research Agency (research core funding No. P3-0388 and P4-0107) and from the Public Forestry Service Program financed by the Ministry for Agriculture, Forestry and Food of the Republic of Slovenia.

**Data Availability Statement:** The data presented in this study are openly available in DiRROS (Digital repository of Slovenian research organizations) at DOI: https://doi.org/10.20315/Data.0002, accessed on 3 May 2021.

**Acknowledgments:** The authors would like to thank Hans Brix and Carlos A. Arias, Department of Biology, Aarhus University, Denmark, for their advice and sharing their knowledge on evapotranspirative willow systems. The authors also thank Davorin Kajba, Faculty of Forestry, University of Zagreb, Croatia, to suggest and provide test clones of *S. alba*.

**Conflicts of Interest:** The authors declare no conflict of interest.

## References


1. Aronsson, P.; Perttu, K. A complete system for wastewater treatment using vegetation filters. In Proceedings of the Willow Vegetation Filters for Municipal Wastewaters and Sludges; Aronsson, P., Perttu, K., Eds.; Swedish University of Agricultural Sciences: Uppsala, Sweden, 1994; pp. 211–213.
2. Dimitriou, I.; Aronsson, P. Willows for energy and phytoremediation in Sweden. *Unasylva* **2005**, *56*, 47–50.
3. Aronsson, P.; Perttu, K. (Eds.) Willow Vegetation Filters for Municipal Wastewaters and Sludges: A Biological Purification System. Swedish University of Agriclutural Sciences: Uppsala, Sweden, 1994; ISBN 91-576-4916-2.
4. Larsson, S.; Arronson, P.; Backlund, A.; Carlander, A. Short-rotation Willow Biomass Plantations Irrigated and Fertilised with Wastewaters. *Dan. Environ. Prot. Agency* **2003**, *37*, 1–58. [CrossRef]
5. Holm, B.; Heinsoo, K. Municipal wastewater application to Short Rotation Coppice of willows—Treatment efficiency and clone response in Estonian case study. *Biomass Bioenergy* **2013**, *57*, 126–135. [CrossRef]
6. Guidi Nissim, W.; Jerbi, A.; Lafleur, B.; Fluet, R.; Labrecque, M. Willows for the treatment of municipal wastewater: Performance under different irrigation rates. *Ecol. Eng.* **2015**, *81*, 395–404. [CrossRef]
7. Jurse, A.; Vrhovsek, D.; Justin, M.Z.; Bozic, G.; Gaberscik, A. Effect of wastewater salinity on treatment performance of vertical constructed wetland and growth of poplars in sand filters. *Sylwan* **2013**, *158*, 1–15.
8. Aronsson, P.; Dahlin, T.; Dimitriou, I. Treatment of landfill leachate by irrigation of willow coppice—Plant response and treatment efficiency. *Environ. Pollut.* **2010**, *158*, 795–804. [CrossRef]
9. Zalesny, R.S.; Bauer, E.O. Evaluation of *Populus* and *Salix* continuously irrigated with landfill leachate II. Soils and early tree development. *Int. J. Phytoremediation* **2007**, *9*, 307–323. [CrossRef]
10. Dimitriou, I.; Mola-Yudego, B. Poplar and willow plantations on agricultural land in Sweden: Area, yield, groundwater quality and soil organic carbon. *For. Ecol. Manag.* **2017**, *383*, 99–107. [CrossRef]
11. Dimitriou, I.; Aronsson, P. Wastewater and sewage sludge application to willows and poplars grown in lysimeters–Plant response and treatment efficiency. *Biomass Bioenergy* **2011**, *35*, 161–170. [CrossRef]
12. Gregersen, P.; Brix, H. Zero-discharge of nutrients and water in a willow dominated constructed wetland. *Water Sci. Technol.* **2001**, *44*, 407–412. [CrossRef]
13. Brix, H.; Arias, C.A. Use of willows in evapotranspirative systems for onsite wastewater management—Theory and experiences from Denmark. In *Proceedings of the "STREPOW" International Workshop*; Orlović, S., Ed.; Institute of Lowland Forestry and Environment: Novi Sad, Serbia, 2011; pp. 15–29.
14. O'Hogain, S.; McCarton, L.; Reid, A.; Turner, J.; Fox, S. A review of zero discharge wastewater treatment systems using reed willow bed combinations in Ireland. *Water Pract. Technol.* **2011**, *6*. [CrossRef]
15. Wu, Y.; Xia, L.; Hu, Z.; Liu, S.; Liu, H.; Nath, B.; Zhang, N.; Yang, L. The application of zero-water discharge system in treating diffuse village wastewater and its benefits in community afforestation. *Environ. Pollut.* **2011**, *159*, 2968–2973. [CrossRef]
16. Istenič, D.; Božič, G.; Arias, C.A.; Bulc, T.G. Growth dynamic of three different white willow clones used in a zero-discharge wastewater treatment system in the sub-Mediterranean region—An early evaluation. *Desalin. Water Treat.* **2017**, *91*, 260–267. [CrossRef]
17. Curneen, S.J.; Gill, L.W. A comparison of the suitability of different willow varieties to treat on-site wastewater effluent in an Irish climate. *J. Environ. Manag.* **2014**, *133*, 153–161. [CrossRef] [PubMed]
18. Curneen, S.; Gill, L.W. Willow-based evapotranspiration systems for on-site wastewater effluent in areas of low permeability subsoils. *Ecol. Eng.* **2016**, *92*, 199–209. [CrossRef]
19. Frédette, C.; Grebenshchykova, Z.; Comeau, Y.; Brisson, J. Evapotranspiration of a willow cultivar (*Salix miyabeana* SX67) grown in a full-scale treatment wetland. *Ecol. Eng.* **2019**, *127*, 254–262. [CrossRef]
20. Perttu, K.L. Biomass Production and Nutrient Removal from Municipal Wastes Using Willow Vegetation Filters. *J. Sustain. For.* **1994**, *1*, 57–70. [CrossRef]
21. Börjesson, P.; Berndes, G. The prospects for willow plantations for wastewater treatment in Sweden. *Biomass Bioenergy* **2006**, *30*, 428–438. [CrossRef]
22. Lachapelle-T, X.; Labrecque, M.; Comeau, Y. Treatment and valorization of a primary municipal wastewater by a short rotation willow coppice vegetation fi lter. *Ecol. Eng.* **2019**, *130*, 32–44. [CrossRef]
23. Guidi, W.; Piccioni, E.; Bonari, E. Evapotranspiration and crop coefficient of poplar and willow short-rotation coppice used as vegetation filter. *Bioresour. Technol.* **2008**, *99*, 4832–4840. [CrossRef]
24. Guidi, W.; Bonari, E.; Superiore, S.; Anna, S.; Bertolacci, M.; Nazionale, L. Water Consumption of Poplar and Willow Short Rotation Forestry Used as Vegetation Filter: Preliminary Results. In Proceedings of the ICID 21st European Regional Conference, Furt (Oder), Germany, 15–19 May 2005; pp. 19–20.
25. Guidi, W.; Labrecque, M. Effects of high water supply on growth, water use, and nutrient allocation in willow and poplar grown in a 1-year pot trial. *Water Air Soil Pollut.* **2009**, *207*, 85–101. [CrossRef]
26. Quaye, A.K.; Volk, T.A. Biomass production and soil nutrients in organic and inorganic fertilized willow biomass production systems. *Biomass Bioenergy* **2013**, *57*, 113–125. [CrossRef]
27. Kuzovkina, Y.A.; Volk, T.A. The characterization of willow (*Salix* L.) varieties for use in ecological engineering applications: Co-ordination of structure, function and autecology. *Ecol. Eng.* **2009**, *35*, 1178–1189. [CrossRef]

28. Licht, L.A.; Isebrands, J.G. Linking phytoremediated pollutant removal to biomass economic opportunities. In *Proceedings of the Biomass d Bioenergy*; Pergamon: Oxford, UK, 2005; Volume 28, pp. 203–218.
29. Fillion, M.; Brisson, J.; Teodorescu, T.I.; Sauvé, S.; Labrecque, M. Performance of *Salix viminalis* and *Populus nigra×Populus maximowiczii* in short rotation intensive culture under high irrigation. *Biomass Bioenergy* **2009**, *33*, 1271–1277. [CrossRef]
30. Francis, R.A.; Gurnell, A.M.; Petts, G.E.; Edwards, P.J. Survival and growth responses of *Populus nigra*, *Salix elaeagnos* and *Alnus incana* cuttings to varying levels of hydric stress. *For. Ecol. Manag.* **2005**, *210*, 291–301. [CrossRef]
31. Kozlowski, T.T. *Responses of Woody Plants to Flooding and Salinity*; Heron Publishing: Victoria, BC, Canada, 1997.
32. Caslin, B.; Finnan, J.; Johnston, C.; McCracken, A.; Walsh, L. *Short Rotation Coppice Willow Best Practice Guide*; Teagasc Agriculture and Food Development Authority: Carlow, Ireland; AFBI Agri-Food and Bioscience Institute: Belfast, Northern Ireland, UK, 2015; ISBN 1841705683.
33. Weih, M.; Nordh, N. Characterising willows for biomass and phytoremediation: Growth, nitrogen and water use of 14 willow clones under different irrigation and fertilisation regimes. *Biomass Bioenergy* **2002**, *23*, 397–413. [CrossRef]
34. Rastas Amofah, L.; Mattsson, J.; Hedström, A. Willow bed fertigated with domestic wastewater to recover nutrients in subarctic climates. *Ecol. Eng.* **2012**, *47*, 174–181. [CrossRef]
35. Justin, M.Z.; Zupančič, M. Combined purification and reuse of landfill leachate by constructed wetland and irrigation of grass and willows. *Desalination* **2009**, *246*, 157–168. [CrossRef]
36. Białowiec, A.; Wojnowska-Baryła, I.; Agopsowicz, M. The efficiency of evapotranspiration of landfill leachate in the soil–plant system with willow *Salix amygdalina* L. *Ecol. Eng.* **2007**, *30*, 356–361. [CrossRef]
37. Mleczek, M.; Rutkowski, P.; Rissmann, I.; Kaczmarek, Z.; Golinski, P.; Szentner, K.; Strażyńska, K.; Stachowiak, A. Biomass productivity and phytoremediation potential of *Salix alba* and *Salix viminalis*. *Biomass Bioenergy* **2010**, *34*, 1410–1418. [CrossRef]
38. Kajba, D.; Andrić, I. Selection of Willows (*Salix* sp.) for biomass production. *S. East Eur. For.* **2014**, *5*, 145–151. [CrossRef]
39. Landberg, T.; Greger, M. Differences in uptake and tolerance to heavy metals in *Salix* from unpolluted and polluted areas. *Appl. Geochem.* **1996**, *11*, 175–180. [CrossRef]
40. Istenič, D.; Arias, C.A.; Vollertsen, J.; Nielsen, A.H.; Wium-Andersen, T.; Hvitved-Jacobsen, T.; Brix, H. Improved urban stormwater treatment and pollutant removal pathways in amended wet detention ponds. *J. Environ. Sci. Health A Toxic Hazard. Subst. Environ. Eng.* **2012**, *47*, 1466–1477. [CrossRef]
41. Henze, M.; Comeau, Y. Wastewater Characterization. In *Biological Wastewater Treatment: Principles Modelling and Design*; Henze, M., van Loosdrecht, M.C.M., Ekama, G.A., Brdjanovic, D., Eds.; IWA Publishing: London, UK, 2008; pp. 33–52. ISBN 9781843391883.
42. Federation, W.E.; APH Association. *Standard Methods for the Examination of Water and Wastewater*, 22nd ed.; American Public Health Association: Washington, DC, USA, 2012.
43. Burt, R. *Soil Survey Laboratory Methods Manual (Soil Survey Investigations Report No. 4, Version 5.02)*; Burt, R., Ed.; U.S. Department of Agriculture, Natural Resources Conservation Service: Washington, DC, USA, 2014.
44. Black, C.A. *Methods of Soil Analysis: Part 1 Physical and Mineralogical Properties, Including Statistics of Measurement and Sampling*; American Society of Agronomy: Madison, WI, USA, 1965.
45. Rytter, R.M. Biomass production and allocation, including fine-root turnover, and annual N uptake in lysimeter-grown basket willows. *For. Ecol. Manag.* **2001**, *140*, 177–192. [CrossRef]
46. Amiot, S.; Jerbi, A.; Lachapelle-T, X.; Frédette, C.; Labrecque, M.; Comeau, Y. Optimization of the wastewater treatment capacity of a short rotation willow coppice vegetation filter. *Ecol. Eng.* **2020**, *158*, 106013. [CrossRef]
47. Pistocchi, C.; Guidi, W.; Piccioni, E.; Bonari, E. Water requirements of poplar and willow vegetation filters grown in lysimeter under Mediterranean conditions: Results of the second rotation. *Desalination* **2009**, *246*, 137–146. [CrossRef]
48. Keoleian, G.A.; Volk, T.A. Renewable energy from willow biomass crops: Life cycle energy, environmental and economic performance. *Crc. Crit. Rev. Plant Sci.* **2005**, *24*, 385–406. [CrossRef]
49. Fabio, E.S.; Smart, L.B. Effects of nitrogen fertilization in shrub willow short rotation coppice production—A quantitative review. *GCB Bioenergy* **2018**, *10*, 548–564. [CrossRef]
50. Leskošek, M.; Mihelič, R.; Grčman, H.; Pavlič, E. Oskrbljenost kmetijskih tal s fosforjem in kalijem v Sloveniji. In *Novi izzivi v Poljedelstvu*; Tajnšek, A., Šantavec, I., Eds.; Slovensko agronomsko društvo: Ljubljana, Slovenia, 1998; pp. 37–41.
51. Leonel, L.P.; Tonetti, A.L. Wastewater reuse for crop irrigation: Crop yield, soil and human health implications based on giardiasis epidemiology. *Sci. Total Environ.* **2021**, *775*, 145833. [CrossRef]
52. Stolarski, M.J.; Krzyżaniak, M.; Warminski, K.; Zaluski, D.; Olba-Ziety, E. Willows biomass as enery feedstock: The effect of habitat, genotype and harvest rotation on thermophysical properties and elemental composition. *Energies* **2020**, *13*, 4130. [CrossRef]
53. Matthews, G. *The Carbon CONTENT of Trees*; Forestry Commission: Edinburgh, UK, 1993; Volume 4.
54. Lamlom, S.H.; Savidge, R.A. A reassessment of carbon content in wood: Variation within and between 41 North American species. *Biomass Bioenergy* **2003**, *25*, 381–388. [CrossRef]
55. Arienzo, M.; Christen, E.W.; Quayle, W.; Kumar, A. A review of the fate of potassium in the soil-plant system after land application of wastewaters. *J. Hazard. Mater.* **2009**, *164*, 415–422. [CrossRef]
56. Adegbidi, H.G.; Volk, T.A.; White, E.H.; Abrahamson, L.P.; Briggs, R.D.; Bickelhaupt, D.H. Biomass and nutrient removal by willow clones in experimental bioenergy plantations in New York State. *Biomass Bioenergy* **2001**, *20*, 399–411. [CrossRef]
57. Stolarski, M. Content of carbon, hydrogen and sulphur in biomass of some shrub willow species. *J. Elem.* **2008**, *13*, 655–663.

58. Hytönen, J.; Saarsalmi, A. Long-term biomass production and nutrient uptake of birch, alder and willow plantations on cut-away peatland. *Biomass Bioenergy* **2009**, *33*, 1197–1211. [CrossRef]
59. Hu, Y.; Schmidhalter, U. Drought and salinity: A comparison of their effects on mineral nutrition of plants. *J. Plant Nutr. Soil Sci.* **2005**, *168*, 541–549. [CrossRef]
60. Kadlec, R.H.; Wallace, S.D. *Treatment Wetlands*; CRC Press: Boca Raton, FL, USA, 2009; ISBN 9781566705264.
61. Pokharel, P.; Chang, S.X. Manure pellet, woodchip and their biochars differently affect wheat yield and carbon dioxide emission from bulk and rhizosphere soils. *Sci. Total Environ.* **2018**, *659*, 463–472. [CrossRef]
62. Zalesny, J.A.; Zalesny, R.S.; Wiese, A.H.; Sexton, B.; Hall, R.B. Sodium and chloride accumulation in leaf, woody, and root tissue of *Populus* after irrigation with landfill leachate. *Environ. Pollut.* **2008**, *155*, 72–80. [CrossRef]
63. Loncnar, M.; Zupancic, M.; Bukovec, P.; Justin, M.Z. Fate of saline ions in a planted landfill site with leachate recirculation. *Waste Manag.* **2010**, *30*, 110–118. [CrossRef]
64. Nissim, W.G.; Palm, E.; Pandolfi, C.; Mancuso, S.; Azzarello, E. Willow and poplar for the phyto-treatment of landfill leachate in Mediterranean climate. *J. Environ. Manag.* **2021**, *277*, 111454. [CrossRef]
65. Meers, E.; Vandecasteele, B.; Ruttens, A.; Vangronsveld, J.; Tack, F.M.G. Potential of five willow species (*Salix* spp.) for phytoextraction of heavy metals. *Environ. Exp. Bot.* **2007**, *60*, 57–68. [CrossRef]

MDPI

St. Alban-Anlage 66

4052 Basel

Switzerland

Tel. +41 61 683 77 34

Fax +41 61 302 89 18

www.mdpi.com

*Forests* Editorial Office

E-mail: forests@mdpi.com

www.mdpi.com/journal/forests